Readings in the
Philosophy of Technology

Readings in the Philosophy of Technology

Second Edition

David M. Kaplan

ROWMAN & LITTLEFIELD PUBLISHERS, INC.
Lanham • Boulder • New York • Toronto • Plymouth, UK

ROWMAN & LITTLEFIELD PUBLISHERS, INC.

Published in the United States of America
by Rowman & Littlefield Publishers, Inc.
A wholly owned subsidary of The Rowman & Littlefield Publishing Group, Inc.
4501 Forbes Boulevard, Suite 200, Lanham, Maryland 20706
www.rowmanlittlefield.com

Estover Road
Plymouth PL6 7PY
United Kingdom

British Library Cataloguing in Publication Information Available

Library of Congress Cataloging-in-Publication Data:

Readings in the philosophy of technology / [edited by] David M. Kaplan.—2nd ed.
   p. cm.
Includes bibliographical references.
ISBN 978-0-7425-6400-8 (cloth : alk. paper)—
ISBN 978-0-7425-6401-5 (pbk. : alk. paper)—
ISBN 978-0-7425-6536-4 (electronic)
1. Technology—Philosophy. I. Kaplan, David M.
T14.R39 2009
601—dc22   2008055410

Printed in the United States of America

♾ ™ The paper used in this publication meets the minimum requirements of American National Standard for
Information Sciences—Permanence of Paper for Printed Library Materials, ANSI/NISO Z39.48-1992.

Milo

Contents

Part VI: Technology and Science

Acknowledgments

I wrote to a number of people asking for their advice on how to improve the second edition of *Readings in the Philosophy of Technology*. The responses were incredibly generous. I received lengthy replies, detailed criticism, new articles recommended to be included, old articles advised to be removed, alternative organizational schemas, and other very helpful suggestions.

My biggest debt of gratitude is to Evan Selinger. He looked at several versions of my table of contents and offered nothing but good advice. Some of it even found its way into this edition.

Others who responded at length include Joe Pitt (who generously granted permission to reprint articles from *Techné*), Peter Kroes, Pieter Veermas, Peter-Paul Verbeek, Katinka Waelbers, Keekok Lee, Carl Mitcham, Paul Thompson, Jan Kyrre Berg-Olsen, Larry Hickman, Inmaculada de Melo-Martin, Andrew Pickering, Julian Savulescu, Nick Bostrom, Andy Clark, and the ACLU.

My editor at Rowman & Littlefield, Ross Miller, has been a delight to work with. So have Evan Wiig, Doug English, and Krista Sprecher. I think we did a great job on this second edition.

Introduction

Our lives are filled with technologies. They are everywhere. We live in them. We prepare food with them. We wear them as clothes. We read and write with them. We work and play with them. We manufacture and purchase them. And we constantly cope with them in one way or another whether we realize it or not. Our world is largely a constructed environment; our technologies and technological systems form the background, context, and medium for our lives. It is hard to imagine a life that doesn't involve at least some tools, devices, or machines. It is even harder to imagine what our lives would be like today without complex technological systems of energy, transportation, communication, and production. The things we make and use shape our culture and environment, alter patterns of human activity, and influence who we are and how we live. In short, we make and use a lot of stuff—and stuff matters.

The philosophy of technology examines the nature of technology as well as the effects and transformation of technologies upon human knowledge, activities, societies, and environments. The aim of philosophy of technology is to understand, evaluate, and criticize the ways in which technologies reflect as well as change human life, individually, socially, and politically. It also examines the transformations effected by technologies on the natural world and broader ecospheres. The assumption underlying the philosophy of technology is that devices and artifacts transform our experience in ways that are philosophically relevant. That is to say, technology not only extends our capacities and effects changes but it does so in ways that are interesting with respect to fundamental areas of philosophical inquiry. Technology poses unique problems of epistemology, metaphysics, moral philosophy, political philosophy, philosophy of science, and environmental philosophy, to name just some of the topics in philosophy affected by technology. The task for a philosophy of technology is to analyze the nature of technology, its significance, and the ways that it mediates and transforms our experience.

Here are examples of the kinds of topics technology raises with respect to some subject areas of philosophy.

> *Epistemology:* The nature of technical knowledge; the nature of a technical explanation; the relationship between technology and science; the role of experimentation in scientific discovery; the effects of technology on perception; the difference between human and artificial intelligence.
>
> *Metaphysics:* The difference between something's being "natural" and "artificial"; how technology transforms human nature; how technology transforms plant and animal life;

the role technology plays in determining what is real; the ontological status of the realities created by technology.

Moral Philosophy: The relationship between ethics and technology; whether things are value-neutral or value-laden; what is appropriate for humans to make and do with technology; the limits (if any) of technology; whether technologies are good or bad in themselves apart from human uses; whether technology challenges our traditional notions of moral conduct.

Political Philosophy: How technology affects our political rights and liberties; how the U.S. Constitution regulates technology; the relationship between democracy and technology; the relationship between capitalism and technology; how decisions about technical design and governance should be made; how the benefits and burdens of technology should be distributed.

Environmental Philosophy: How technology intervenes in the natural world; the appropriate relationship between technology and nature; whether natural environments are superior to artificial environments; on the permissibility of bioengineering plants and animals; on the desirability of genetically modified foods.

The philosophy *of* technology addresses these issues in a different manner than an analysis of philosophy *and* technology. The latter treats philosophical problems as *external* to technology and technological practices. Artifacts themselves do not have philosophical dimensions; only the choices and effects on humans are philosophically interesting. This approach examines the consequences, risks, and impacts of making and using things, but the technology and related practices themselves remain unexamined. They are not subject to a philosophical analysis. By contrast, the philosophy *of* technology treats philosophical problems as *internal* to technology and its practices. It finds traditional areas of philosophy within artifacts and technology-human relations. The world of technology is itself philosophically interesting—and not merely because it has important consequences. The philosophy of technology takes artifacts seriously and subjects them to the same kind of philosophical scrutiny reserved for the topics typically analyzed by philosophers, such as language, logic, and knowledge.

Take, for example, a typical philosophy *and* technology concern, such as the spread of industrial pollution, stem cell research, and the environmental risks of genetically modified food. Typically, the debates focus on the pros and cons of making and using such things. We frame the issue in terms of things like costs and benefits, acceptable and unacceptable risks, desirable and undesirable consequences, or sometimes as a clash between technological innovation and traditional moral or religious convictions. We then analyze the technology in terms of ready-made philosophical concepts, usually moral and political concepts such as "freedom," "general welfare," and "human nature." This approach questions the limits of technology and asks whether or not societies should pursue this or that artifact or technical choice. It contrasts our human capability to create new things with our moral obligation to be prudent given the limits of our wisdom and foresight. We inquire, for example, whether a particular technology is something we as a society want to implement; what the long-term consequences of it might be; who should decide which technologies are desirable or undesirable; and who would benefit and who would suffer. This is the approach of philosophy *and* technology. We take ready-made practical and philosophical concepts and apply them to new situations created by technology.

There is absolutely nothing wrong with this approach. It makes sense to analyze human creations in terms of their risks and consequences. But, from the point of view of the philosophy *of* technology, we can probe the matter more deeply and take into consideration the very nature of a technology and not merely its external contingencies. We can question technology. We can examine it rather than take it for granted. We do not have to treat it as a "black box," whose nature is inaccessible to anyone but an engineer or technician. We can investigate the meaning, nature, and moral character of technology and its practices just as we would any other object or phenomenon. We can examine, for example, why something was designed in a particular way; what technical and non-technical factors were at work; how it functions in relation to other artifacts/users/environments; how it transforms its users; and which ideas and values are embedded in it. Matters like these are internal to artifacts and technological practices, not external to them. These types of issues refuse to treat technology as something merely given and which we can examine only after the fact—as if things magically appear out of thin air. Instead, we can analyze technology like anything else that humans make or do. Philosophers and citizens have the same capacity to question things as experts. The key is to know what kinds of questions to ask and how to apply philosophical concepts to artifacts and devices. We have to question technology. It is our responsibility.

Yet, defining what precisely counts as technology is not easy. There are so many different kinds of technologies, each designed for a different purpose, made from different materials, requiring different skills, and used in different contexts that it is unlikely that a common set of defining properties could possibly apply to all of them. The range of objects included in the class of technologies is enormous. It includes everything from low-tech handheld tools (hammers and nails), medium-tech motorized machines (cars and dishwashers), large-scale constructions (buildings and bridges), high-tech digital devices (laptops and MRI machines), to vastly complex technological systems (satellites and oil refineries). If every humanly made object is a technology (with the arguable exception of art) it is hard to see what such wildly different things have in common. Other than the fact that each is a humanly made artifact, there doesn't seem to be much in common among the diversity of stuff we make.

Imagine trying to teach someone the meaning of the word "technology." Imagine trying to teach a space alien who comes to Earth to learn more about us. How would you teach such a broad and vague term to a being with minimal knowledge of humans? Would you point out different man-made objects? What would be the distinctly technological character of each item you point to? What about manufactured objects designed to be consumed, like food and drugs? What about naturally occurring objects that are used as technology, such as rocks and sticks for smashing or probing? What about the knowledge and skills necessary to make and use things? Would you also have to explain the meaning of the words "use" and "techniques" and other terms related to technological practice? And what about animal technologies, like beehives, spiderwebs, and beaver dams? Aren't they technologies, too!? At this point, our friendly space alien would surely be scratching its green head in confusion. It would conclude that humans have no idea what they're talking about when they use the word "technology."

Obviously people know what technology is, in spite of the ambiguities and difficulties in defining the term. We can usually tell the difference between naturally occurring things and man-made artifacts without any problem. We know how to use the word appropriately in a sentence. And, although we might not apply the word consistently (otherwise food and art would be thought of as technologies), we generally understand the meaning of terms like "technology," "technical," and "technique" even if we cannot define them precisely.

Maybe we should heed the advice of Ludwig Wittgenstein and avoid the problem of definitions altogether. As he says in the *Philosophical Investigations* (1953), just because a concept is inexact does not mean it is meaningless. A concept can be useful and usable without being precisely determined. For example, you can tell someone to "stand roughly here" to have his or her picture taken, and even though the word "here" is imprecise, it works perfectly well as an instruction. Inexact doesn't mean unusable. Wittgenstein suggests that philosophers are too concerned with clarifying the exact meanings of terms; they forget that most of time we understand each other perfectly well even though we rarely take the time to spell things out with precision. He says that a better way to understand the meaning of terms is to examine the different contexts in which they function. It is more important to understand the role that a word plays in conversation than to search for the essence hidden behind a word's meaning.

With technology, perhaps we should look at the various ways that artifacts and technical concepts relate to the world. We might analyze the ways that concepts and things relate to designers, users, markets, governments, and environments rather than attempt to understand them in themselves. We should examine how things are made, how they are used, and how they function in relation to broader cultural practice. Above all, we should question technology philosophically, challenging conventional ideas of it to uncover its full meaning and significance for us and for nature. This approach moves us beyond slippery equivocations over definitions into the rough ground of artifacts, technological practices, and the actual uses of technological concepts.

There are four classic theories in the philosophy of technology. Each takes a broad perspective on the nature of technology in relation to society. The classic theories are *neutrality*, *determinism*, *autonomy*, and *social construction*. That is to say, philosophers have interpreted technology either as, a) a neutral tool that can be used for either good or bad purposes; b) the driving force of social change; c) a Frankenstein's monster that is increasingly beyond our control; or d) man-made things that shape societies, which, in turn, shape technological development. Most of us believe in one or some of these theories. Let's consider each in order.

The theory of technological *neutrality* defines technology in terms of its technical properties. Technology, on this model, is applied science. As such, it is seen to embody a kind of pure, abstract, universal rationality—that is, a rationality governed only by natural laws and technical considerations that are independent of social forces. What matters most in a technology is that it works, so this line of argument goes, and what works can be determined objectively according to universally valid, scientifically established principles. The technology itself is simply a tool. It can be used for a variety of human ends, and good or bad purposes. It is neutral with respect to values and purposes. There is no such thing as morally good or bad technology, only good or bad users. The technology itself obeys only value-free, context-free principles. It is precisely this indifference to ends that makes technology so practical: when it works, it works everywhere and when it breaks down it can be fixed the same way, by anyone with the right technical know-how. The same standards, the same rules, the same techniques, and the same concept of efficiency govern the creation and use of technologies.

The theory of technological neutrality states that artifacts are independent of values. This commonsense understanding of technology is also known as "instrumentalism." Things are mere instruments for human activities. Technology is value-neutral; human users are not. For example, the shooter, not the gun, is responsible for the shooting death. The driver, not the vehicle, is responsible for the traffic jam. Perhaps even the smoothie-maker, not the blender, is responsible for the blended fruit juice and ice. After all, someone has to push the button in order to initiate the blending—and a human, not the blender, decides what the smoothie will be made of. According to the theory of neutrality, devices are subservient to human choices. They serve us, not the other way around. Things are mere means to our human ends, purposes, and values. The neutrality of technology assumes a complete separation of (technical) means and (human) ends. Technical objects and human values have nothing to do with one another.

Next is the theory of technological *determinism*, the idea that technology drives the course of history. Devices and machines rather than people are the primary engines of change. Technological developments precipitate social developments, not the other way around. When we speak, for example, of the Iron Age, the Industrial Revolution, and the Information Age we evoke a hidden theory of technological determinism: different historical eras are caused by technological developments. This theory of technological determinism has a strong and weak version. The strong version states that there is a fixed sequence of technological development and, therefore, a necessary path of social change. Technology *imposes* on a society specific social-political consequences. Technical innovations spark social transformation; society responds more to technology than technology to society. For example, we say that the atomic bomb ended World War II, automation caused a loss of jobs, and the internet has changed the way we do business. We are all familiar with the well-worn theories about the printing press and the Reformation, and the cotton gin and the Civil War. In these cases, the technology is the primary agent of change, not humans.

The weak version states that technology *influences* social relations. It helps to shape and pattern history but this imposition is not so strong as to determine the course of technological progress and social evolution. Technology mediates and steers a society but it does not quite drive it. Examples include communication technologies (like e-mail), which affect the way people communicate; reproductive technologies (like the Pill), which affect sexual practices; and cooking technologies (like microwave ovens), which affect how food is prepared. None of these technologies caused social change but each has had a profound influence on the way we communicate, prevent reproduction, and cook, respectively. These technologies "softly" determine. They exert a great influence on society without strictly causing specific effects.

Next is the theory of *autonomous* technology, the idea that humans no longer control technology; instead, it controls us. It is an independent force that follows its own rules and imperatives. Humans merely respond, adapt, and conform. Technology imposes a way of life on a society: everything is *technicized*. Technical efficiency is the only end in a technological society. In fact, that end is built into things so that users have no choice but to adopt a technicized lifestyle. Technology constitutes a new cultural system that restructures the entire social world as an object of control. It is an independent cultural force that overrides all traditional or competing values. Autonomous technology is similar to technological determinism in that things not humans are the primary force in society, but in the latter things are neutral. Technical means and human ends are completely separate. In the theory of technological autonomy, means and ends are linked. Everything about life in such a world is technological.

This view might sound extreme but it is rather commonplace in science fiction films, where future worlds are depicted as completely mechanized, rigidly ordered, and dehumanized. Everything in these future worlds is either technology or treated like technology; everything is uniform, homogeneous, and part of a machine-like whole—for example, the Death Star and the Imperial Empire in *Star Wars* films, the ruthless machine-world of the *Terminator* films, the computer-controlled simulated world of the *Matrix*, and the specter of a robot-dominated world of *I, Robot*. These films and countless others depict a world of autonomous technology. Machines either control the world or threaten to; societies are rigidly organized with little or no room for individuality or dissent; humans and social values are subservient to technology, not the other way around; and society is like a machine that regiments its members into complete conformity with the established social order. The underlying philosophy of science fiction *dystopias* is the theory of autonomous technology. In fact, this theory of technology is so common it can be found across cinematic genres. We find this pattern in any film where the good guys are down-to-earth, humble, and humane individuals and the bad guys are cool, aloof (possibly wealthy), and uniform. *We* are humans; *they* are machines. Although few films address technology or a technological society explicitly, many have antagonists who embody the characteristics of autonomous technology and protagonists who embody the freedom and individuality that is its antithesis. The theory of autonomous technology is far from being extreme and esoteric. It is the most common and well-known theory of technology.

Finally, the *social construction* of technology is the idea that society simultaneously shapes technology as technology shapes society. Humanity and technology are situated in a circular relationship, each influencing the other. Social constructionists maintain that when we actually consider the diversity of things we find *technologies* (plural), no singular essence that applies in every instance. Humans make, use, and assign meaning to things in a variety of different ways, in relation to a variety of different social contexts. Far from being applied science, technology, on this model, is more like *embodied humanity*. Technologies are part human, part material, and always social. Technological systems are even more complex. They link small devices to massive machinery to social practices to economic and legal institutions. For example, it is hard to make sense of technologies like airplanes, air conditioners, and CAT-scan machines without considering the systems of transportation, energy, and healthcare in which they function. The advantage of viewing technology in this way is that it calls attention to the way that humanity, technology, and the environment are bound up together in a relationship of mutual constitution. Humans, things, and contexts all fit together like pieces of a puzzle.

The social constructionist view was developed by sociologists and philosophers as a response to the dominant approaches to technology, which they viewed as overly abstract and theoretical. From their perspective, the theories of technological neutrality, determinism, and autonomy all suffer from the same weaknesses. They all treat technology as if it is something radically different from humanity; they take the technical qualities of things to be their most important characteristic; and they overlook the obvious fact that if a technology is made and used by human beings then it cannot help but reflect human ends, values, and ideas. Constructionists insist that technology cannot be value-neutral because people are not value-neutral; technology cannot determine history because it never is so independent from society to be in a position to cause it; and technology cannot be autonomous force because technology is a human affair, not a mere technical matter. They conclude that it is more helpful to think of artifacts as a *socially constructed reality* rather than as the application of universally valid scientific principles. Technology cannot be one single thing because society is not one single thing. Technology cannot be entirely good or bad because society is not entirely good or bad. The social construction approach tries to show how technologies are inextricably bound to human interests, social practices, physical laws, and a very long list of other constitutive factors.

Almost all of the contributors to the second edition of *Readings in the Philosophy of Technology* take a constructionist approach. The trend over the last thirty years is to treat technical concepts and things in relation to, not apart from, the social world. Almost all of the chapters in this volume examine technology as a social phenomenon rather than as the application of scientific reasoning. We will only very briefly consider some of the classic theories of technology. They are interesting primarily for their influential role in shaping the popular imagination, but not as viable options today. We hope to move the philosophy of technology beyond gloomy legacies of technological neutrality, determinism, and autonomy. Now that we are in the twenty-first century, it is time to rethink and renew the starting point for a philosophical investigation of technology. The challenge is to avoid the extreme views: it is neither the most important determinant in social life, nor is it merely an innocent tool. There is reason neither to be dopily optimistic about a future technological utopia, nor to be warily pessimistic about a bleak technological dystopia. The aim of this collection is to help us think rationally about the ways in which technologies reflect as well as change life on an individual, social, and ecological level. The goal is to learn how to analyze and criticize the human creations that influence the fate of humanity and nature. We make and use a lot of stuff—and stuff matters.

Part I

PHILOSOPHICAL PERSPECTIVES

The founders of philosophy of technology were existentialists and Marxist theorists like Martin Heidegger, Karl Jaspers, Herbert Marcuse, and Jacques Ellul. For these philosophers, the essence of technology is "technological rationality," or a kind of technological mindset that takes a detached, objective, and overly scientific view of the world. Technological thinking is about problem solving: everything can be analyzed logically and any situation can be managed efficiently. According to this approach, the problem with technology is that it is part of a worldview that treats everything as nothing more than mere objects to be controlled. While that kind of thinking may be useful for dealing with legitimate technical matters, these philosophers worry about the fate of human beings when we are managed and handled as mere things—when we're seen as mere technical problems with technical solutions. The founders of philosophy of technology criticized this kind of thinking because it disconnects us from the world and from each other. The more we rely on technology the more dehumanizing our societies become. This generation of philosophers contrasted the detached objectivity of technological rationality with more humane forms of experience that are connected to, not severed from, the natural and social worlds. Early philosophy of technology took a *transcendental* perspective on technology, creating theories of technological rationality that account for the very conditions of making and using instruments.

Recently, the philosophy of technology has taken an "empirical turn" away from the transcendental orientation toward a more practical, contextual interpretation of artifacts and machines. As opposed to the early pessimistic assessments of a singular technological rationality, philosophers since the 1980s tend to view technology empirically and historically, in terms of its actual uses in social contexts. This approach treats technology as a social construction that interacts with other social forces rather than as an autonomous entity with its own unique rationality. Technology is now seen as *inter*dependent in relation to society rather than independent of it. Technology and society form an inseparable pair; neither is intelligible without reference to the other. Today, philosophers of technology examine the various ways that our technologies (plural) form the background, context, and medium for our lives, shaping our culture and the environment, altering patterns of human activity, and influencing who we are and how we live.

The readings in this section include two of the founders of the philosophy of technology, Heidegger and Marcuse, followed by several examples of empirical, contextual philosophy of technology. The chapters in this section provide basic frameworks for understanding what technology is, how we relate to it, how it embodies moral and political dimensions, how it interacts with other aspects of society, and how we ought to design it.

1

The single most influential philosopher of technology is Martin Heidegger. His 1954 essay "The Question Concerning Technology" defined the field for many years: people either followed or repudiated Heidegger, but everyone read him and took his theory of technology seriously. In his notoriously difficult essay, Heidegger endeavors to uncover the essence of technology so that we may have a "free relationship" with it. Once we understand the real essence of technology we will learn how to cope with it and take it for what it is.

According to Heidegger, what we have until now failed to understand is that the essence of technology is not a tool or device but rather a way of understanding things. We mistakenly assume that a technology is nothing more than a value-free instrument—that is, technology is a means to an end. While somewhat accurate, this definition represents only the causal, instrumental meaning of a "means." Heidegger reminds us that the ancient Greeks had a broader conception of causality. A cause answers the questions "why?" A cause is like an explanation of what brings something about, or that which is responsible for something. Aristotle, for example, said that everything has four causes. A cause can explain what something is made out of (material cause), what it is to be something (formal cause), what produces something (efficient cause), and what it is for (final cause). Together the four causes explain what is responsible for bringing something, as Heidegger says, "into appearance" (*poiēsis*). The four causes make things present; they bring forth "out of concealment into unconcealment." The essence of technology is no different from any other thing. It too can be explained in terms of the way things are brought into appearance. Technology, therefore, cannot be defined as a mere tool. That only accounts for its efficient cause. The broader definition and full essence of technology is a way of looking at things, a way of revealing things (*alētheia*).

Yet, according to Heidegger, the way that technology reveals things (*technē*) is problematic. It is what he calls a "challenging revealing," one that organizes nature into a "standing-reserve" of energy and resources. It places an unreasonable demand that nature supply us endlessly and efficiently. Even humans, the supposed masters of technology, are challenged and ordered into standing-reserve as "human resources." Heidegger calls this way of revealing the world "enframing" (*Ge-stell*). It is a way of ordering people to see the world (and each other) as a mere stockpile of resources to be manipulated. Enframing happens both in us and in the world; it is the revelation of being (human beings and nature) as standing-reserve. Particular technologies (in the ordinary sense of tools and machines) only respond to the enframing. They are the consequence not the cause of it; they merely help reveal things as standing-reserve. The same is true of modern science. It, too, is derivative of enframing.

The danger of the technological understanding of being is twofold. First, Heidegger says that we ourselves become mere standing-reserve. Second, in our role as human standing-reserve, we tend to think we are the masters of everything. But in truth, we cannot see ourselves or understand the world clearly. *Ge-stell* keeps the essence of things concealed. The danger of technology (as *technē*, not the ordinary sense of it!) is that it is a partial and incomplete understanding of being; one that seeks more and more efficiency for its own sake. We interpret everything, including ourselves, as resources to be dealt with as effectively and efficiently as possible. The technological understanding of being obscures other ways of seeing things. Above all, it obscures *poiēsis* as an alternate way of revealing. *Poēisis* is a broader form of revealing than *technē*. The essence of technology is ultimately *poiēsis*.

Heidegger argues that we need to see through the limits of instrumentality and efficiency, recognize the true essence of technology as only one way among many to see the world, and overcome our pretension to complete mastery and control over things. The "saving power" of technology is that its essence is ambiguous. The very instrumentality (*technē*) that threatens us also has the potential to save us (as *poiēsis*). Heidegger exhorts us to become more open to understanding things differently and to embrace non-efficient living. The key is to neither reject nor accept techno-logical thinking without criticism but to take a new, artful attitude toward it that affirms a broader, more inclusive understanding of reality.

The next chapter, "Heidegger on Gaining a Free Relationship to Technology," is by Herbert Dreyfus, an eminent Heidegger scholar and one of the leading philosophers of technology. Dreyfus very helpfully clarifies what Heidegger's notion of the essence of technology is, what is wrong with it, and what we can do about it. He explains that the essence of technology for Heidegger is a limiting and restricting understanding of reality. The problem with it is that we interpret everything as mere material. The solution is to have a more accurate understanding of the limits of technology and take it for what it is: just one way of seeing things. Once we realize that we have been living with this technological understanding of things we have already stepped out of that framework. We are then free to appreciate that there is more to life than efficiency. This new attitude toward tech-nology is what Heidegger calls "releasement." Dreyfus reads Heidegger's claim (at the end of "Question Concerning Technology") that only a God that will save us as a call for a new cultural paradigm that celebrates non-efficient practices like friendship, love, community, music, art, and living closer to nature. The new God might be interpreted as a call to help foster a new understand-ing of the world and new ways of living together.

In "New Forms of Social Control," Herbert Marcuse echoes Heidegger's belief that the essence of technology is a technological rationality geared toward the control and domination of people and nature. It dissolves traditional rationality—including science, ethics, and politics—into a rationality that employs efficiency as the single standard of judgment. But unlike Heidegger, Marcuse claims that technological rationality is a political rationality. Economic and political power has co-opted the full sense of rationality, transformed it into technological rationality, and used it in the function of social control.

In this reading, taken from his most influential book, *One-Dimensional Man* (1964), Marcuse argues that advanced industrialized societies employ science and technology to serve existing sys-tems of production and consumption. The scientific and technical aspects of a society are used both to increase productivity and to dominate humans and nature. The result is a carefully managed society that creates a one-dimensional person who willingly conforms to a society that limits free-dom, imposes false needs, stifles creativity, and co-opts all resistance. We have come to accept mass production, standardization, and bureaucracy as the (seeming) embodiments of rationality and efficiency. But we accept this only because technological rationality manipulates human needs through advertising, marketing, and mass media. We are caught up in a systematic deception; our needs have become the needs of the techno-political apparatus. Marcuse argues that appeals to enlightened self-interest, freedom, and autonomy have come to appear quaint and irrational. We are stripped of our individuality by a technological rationality that makes conformity seem reasonable and resistance seem unreasonable. The one-dimensional society has eroded the capacity for individ-uality and critical thinking.

At the end of *One-Dimensional Man* (not excerpted here) Marcuse holds out the possibility that technological rationality could be harnessed by a society to fully realize (rather than repress) human capacities. He is pessimistic about the prospects for such a transformation because the tech-nological apparatus has incorporated and subsumed all critical and oppositional thought, but it is, nevertheless, possible. It would take a new beginning—a "Great Refusal" of the old systems of domination—to transform science and technology into agencies that would satisfy human needs and help us realize our human potential. Technological rationality itself requires transformation; it cannot remain value-neutral but rather must embody the values that permit human flourishing.

Next is a reading by Larry Hickman on John Dewey's philosophy of technology. Dewey's remarks on technology are scattered throughout his massive corpus. Hickman collects, condenses, and constructs a Deweyan philosophy of technology in a more coherent and satisfying way than Dewey himself ever does. In this reading, Hickman focuses on Dewey's idea that philosophical knowledge is a form of technology, and that, properly understood, technology is a kind of intelligence. Dewey replaces traditional theories of knowledge with the notion of "inquiry," a practical, problem-solving activity that involves both concrete and conceptual tools whose product, knowing, exhibits the traits of other technological artifacts in general. Inquiry is instrumental for producing knowledge; that is to say, inquiry leads to the production of new outcomes. It is a technological activity that produces artifacts useful for further implementation. In turn, technology involves not only tools and machines, but the knowledge and social contexts necessary to make things possible. Technology and instrumentalism are synonymous on this model.

To avoid interpreting Dewey as saying that all activity and all knowing is technological (hence rendering the very concept vacuous), Hickman explains that we can distinguish between activities that involve tools and artifacts, and activities that do not. Activity that involves tools and artifacts can be further divided into two types. What is *technological* involves cognitive and deliberate activity; what is merely *technical* is non-cognitive and habitual. Activity that *does not* involve tools and artifacts can also be divided into two types. Some activity is *non-instrumental but cognitive* (e.g., use of limbs to perform basic tasks); other activity is *non-instrumental and non-cognitive* (e.g., immediate perceptions, pleasures, and pains). The boundaries between the technological and the technical, the cognitive and the non-cognitive are, of course, blurry. But for Dewey, technology means only activity that involves the cognitive use and production of new artifacts. It involves all of the intelligent techniques, tools, and social practices that humans have evolved to use to accommodate themselves to their environments.

One of the advantages of Dewey's theory of technology, according to Hickman, is the enormous "ecological power" gained by "naturalizing" technological activity. Humans are technological animals who use a variety of mental and physical tools to adjust to their natural and social environments. We come to see technological activity as continuous with the other activities of natural organisms geared toward adapting to their environments. Another advantage is that Dewey undercuts a number of traditional philosophical problems of the alleged gulf between subjectivity and objectivity, individual and community, the natural and the social. Dewey's original theory of technology gives us some powerful tools for understanding and improving our technological culture.

Albert Borgmann's "Focal Things and Practices," excerpted from his *Technology and the Character of Contemporary Life* (1984), develops Heidegger's analysis by specifying in greater detail what it would mean to launch a "reform of technology." Borgmann's key distinction is between "focal things" and "devices." A focal thing involves people, technologies, and places. It engages users with things, with each other, and with their environments. Focal things "gather us," as Heidegger would say. By contrast, a device is merely an instrument for producing whatever the device is for. It produces a commodity (e.g., furnaces make heat, dishwashers make clean dishes). The device functions inconspicuously by disburdening us and making a commodity readily available. The "promise of modern technology," Borgmann explains, is that the use of devices will free us from the misery and toil imposed by nature thereby enriching our lives. But, as Borgmann points out, technology has failed to live up to its promise because it is silent as to the ends, purposes, and goods that make up an enriched, fulfilled life. To reform technology we need to revive focal things and practices, not simply make better devices. Focal things and practices contain within them a vision of the good life missing from the device paradigm. We need to become more not less engaged; our technologies need to become more not less conspicuous.

Borgmann's examples of focal practices include performing music (rather than listening to it on a device), jogging outside (rather than working out at a gym), and enjoying a traditional home-cooked meal with friends and family (rather than consuming fast food alone). These activities combine humans and technologies, means and ends in a more engaging and satisfying way than the mere use of devices. Like Heidegger, Borgmann invites us to question technology and become free from our traditional notions of it. But he moves beyond Heidegger by suggesting more specific and practical ways to reform technology.

Don Ihde, in "A Phenomenology of Technics," excerpted from *Technology and the Lifeworld: From Garden to Earth* (1990), makes no sweeping claims about technology as such. Instead he provides a framework to analyze patterns of our experience of technology. The method he uses is phenomenology, a descriptive approach premised on the idea that experience is always relational. The core principle of phenomenology is the doctrine of "intentionality." Every experience, every act of consciousness (seeing, hearing, smelling, remembering, etc.) is intentional: it is always an "experience of" or "consciousness of" something. All awareness is directed toward objects in the world. The "intentionality analysis," of which Ihde writes, aims to identify the essential or invariant features of experienced phenomena. According to this analysis, we never experience technological objects in themselves, but rather objects in relation to us. Furthermore, our experience of technological objects is ambiguous; their meaning is never fixed but flexible and relative to use and context.

Idhe describes several unique sets of human-technology relations, each positioning us in a slightly different relation to things. "Embodiment relations" describe our experience of devices we use to encounter and manipulate things. (These include devices like reading glasses, hearing aids, writing implements, and handheld tools.) "Hermeneutic relations" describe our experience of instruments that we read rather than use as tools. (These include devices like clocks, thermometers, spectrographic devices, and other technologies with visual displays that must be interpreted to be understood.) "Alterity relations" describe how technologies appear as "other" to us, possessing a kind of independence. (These include things like robots, ATM machines, and video games that we interact with as if they were autonomous beings.) Finally, "background relations" describe our experience of technologies that form the context of our lives in a way that is seldom consciously perceived. (This set of devices includes things like the lighting, air conditioning, clothing, and automatic machines that operate in the background, subtly affecting our experience.) The virtue of Ihde's phenomenological approach is its emphasis on an active sense of the human-technology relationship. In place of a single essence of technology we find different experiences of technology and subtle ways in which technologies mediate our lives.

In "Philosophy of Technology Meets Social Constructivism," Philip Brey surveys several theories of "social constructivism" to analyze how this sociological approach contributes to the philosophy of technology. Social constructivism (also known as "social constructionism") is an empirical investigation into the way that technologies are socially fabricated realities. From the perspective of sociology or science-technology studies (STS), technologies are more fruitfully understood as social things rather than technical things. Their properties and meanings are best explained by reference to specific actors and relevant social groups rather than as technical things that are explained by laws of nature. Technology, on this model, has no objective or fixed properties but rather is flexible and open to interpretation.

Brey identifies three main versions of social constructivism: strong, weak, and actor-network theory. Strong social constructivism maintains that technology is entirely socially constructed. Technology itself has no properties, effects, or powers; only people do. Weak social constructivism maintains that non-social factors play a role in technological change. Social biases and politics may be built into and embodied in material things. Actor-network theory studies networks of human actors and natural and technical objects as the primary agents of technological change. Both technical devices and natural forces are actors (or *actants*) in resolving controversies surrounding technology.

Brey then addresses some of the perceived shortcomings of these sociological approaches, which have been accused of failing to analyze the consequences of technological change, ignoring how social groups are impacted by technology, and omitting of moral or political criticism. He finds that, for the most part, analyses by sociologists and STS can address these concerns. Brey argues that philosophy of technology should pay more attention to empirical studies of technology. Social constructivist approaches provide analyses that are more concrete, detailed, and realistic than those typically advanced by the philosophy of technology.

In "Women and the Assessment of Technology," Corlann Gee Bush employs a version of constructivism to argue that technology is about equality: who benefits and who suffers, whose opportunities increase and whose decrease, and who creates and who accommodates. In societies characterized by a sex-role division of labor, technologies have different effects on men than on women. They reinforce rather than challenge long-standing patterns of gender inequality. The main culprit, according to Bush, is the "tech-fix," the belief that technology can solve all of our problems, including social problems. She contends that the tech-fix has not worked well for women—if anything, it has made things worse by being used to rationalize inequality.

Bush encourages us to "unthink" common myths about technology so that we might "rethink" the values and hidden biases that it often conceals. Among the myths she debunks are the beliefs that technology represents the triumph of human intelligence, that technology represents a threat to our humanity, and that technology is a neutral tool that can fix whatever problems we need solved. These myths oversimplify the complex nature of technology. They ignore the "valence" or tendencies built into things, which inclines them to interact in similar situations in identifiable and predictable ways.

The next step after unthinking is to rethink technology by paying close attention to the contexts in which technical decisions are made and technical things adopted. Bush finds four contexts in which technology operates: a design context, a user context, an environmental context, and a cultural context. Taken together these four contexts form a holistic approach that can uncover how a technology advantages or disadvantages women. Such an approach is what she calls an "equity analysis," designed to interpret the effects of technological change on women's lives. A feminist assessment of technology analyzes the way that society and technology interact with the aim of making both more equitable.

Peter Kroes, in "Design Methodology and the Nature of Technical Artifacts," looks at technology from the perspective of design theory, a normative (i.e., moral) field that aims to analyze and improve design practices. Yet he finds the literature on design methodology to be surprisingly undeveloped. Basic questions are unanswered such as, what is a design? What is a good or successful design? What are the criteria? Are there design methods? Kroes believes that because design methodology aims at improvement it has focused mainly on design processes while ignoring the outcome of design, that is to say, it ignores technical artifacts. He takes a closer look into design methodology and finds that the design process and the thing designed are so intimately related that it is impossible to make full sense of one without the other. We understand the nature of design in terms of artifacts designed, and we understand the nature of artifacts in terms of the design process. Furthermore, the very normative stance taken by design methodology toward design processes implies an improved design product, and so it has to account for the criteria for success in design.

Kroes argues that design theory needs to have a better understanding of the nature of the technical artifacts it produces. He defends the view that artifacts have a dual nature, part physical and part "intentional," corresponding to two common conceptualizations of the world: we can see the world as consisting of physical objects that interact causally with other physical things in the natural world, or we can see the world as consisting of human beings (agents) who act intentionally in the world and whose behavior is explained in terms of reasons and motives and not mere causes. Technology is natural and social, physical and intentional. Kroes defines technology as an artifact with "a technical function and with a physical structure consciously designed, produced and used by humans to realize its function." On this model, we can only make complete sense of a technical artifact when we consider both its physical structure and the context of intentional human action in which it is used. Yet, it will take a good deal of conceptual and empirical work to answer questions like: How should designers take into account both physical and intentional properties of things? What does the context of design have to do with the context of use? What is the relationship between quality in design process and quality in design outcome? The model of the dual nature of technical artifacts creates a whole research agenda for design methodology and helps form a bridge between the world of design and the philosophy of technology.

In "Democratic Rationalization," Andrew Feenberg provides an alternative to a theory of technological rationalization (the spread of uniform technical reasoning to every aspect of social life) with a version of (weak) social constructivism. Feenberg maintains that technology is partially composed by technical properties and functions, partly by social influences. The "double aspect theory" that he proposes explains that social meaning and technological/functional rationality are intertwined as two aspects of same object (not unlike Kroes's dual nature theory). Feenberg's critical theory of technology is based on a strategy of identifying the ways that advanced industrialized societies systematically decontextualize the technical aspects of things in order to secure their illegitimate power and authority. The danger of the apparent neutrality of technical rationality is that it is often enlisted in support of hegemony (a form of social and political control). The technical aspects of a device or system are written into its "technical code," which embodies social values and interests and takes the form of technical rules and procedures. These rules typically secure power and advantage for hegemony over the interests of the public. Hegemony can *play the technical card* in order to give the illusion that the technological regimes it relies on are universally valid and necessary—and far too complex for ordinary people to understand.

A critical understanding of technology endeavors to uncover the social horizon in which a device is produced, remove any illusion of its necessity, and expose the relativity of technical choices. Feenberg maintains that we have to recognize the indeterminate, contextual character of technical things as well as the social and political stakes of technical design so that we can change the values designed into our technologies. Then we can begin to criticize our society's technological-political practices and imagine alternatives that would foster a more democratic and livable environment. He concludes with an appeal for a "democratic rationalization" that would incorporate democratic values into industrial design.

Bruno Latour, in "A Collective of Humans and NonHumans," seeks to overcome the dualistic paradigm that defines modernity: the separation of subjectivity from objectivity, facts from values, and humans from technology. Writing in response to the "science wars" that occurred between sociologists (who claim that scientific facts are mere social constructions) and philosophers of science (who claim that scientific facts at the very least aspire to objectivity), Latour argues that both sides miss the point. There has never been such a thing as humanity without technology nor technology without humanity. Consequently, the very idea of the *social* construction of objectivity in science and technology is incoherent. Society does not exist apart from science and technology. It never has. Instead Latour believes that humans and nonhumans, (social) actors and (objective) networks are "symmetrical." Neither is more important than the other; both are always bound up together. Humans and technology are active agents (or rather, *actants*). Obviously, humans have goals and intentions, but so do technologies. They have what Latour calls a "program of action," or series of goals, steps, and intentions that unfold like a story when used. Each artifact has its "script" that forces (or at least inclines) a user to play a role in its story. For example, I become a motorist when I drive a car; a gardener when use a rake. As a result, it is more helpful to understand our lives as social-technical; our lives are composed of actants.

Latour describes human-technology relations in terms of various figures of mediation. There are four ways that agents and technologies act jointly together. The first form of mediation is "translation," or the creation of new actants. For example, a person and a gun become a-person-with-a-gun-in-his-hand, and the gun becomes a-weapon-wielded. Both the person and the device share responsibility for acting. Next is "composition," or the combination and association of actants. For example, driving is a property not just of cars and drivers but auto manufacturers and dealers, oil refineries and highways, and so on. Next is "reversible black-boxing," or the process that renders invisible the networks of relations that contribute to the artifact. The final form of mediation is "delegation," or the shifting/displacement of agency to things. An engineer inscribes a program of action, for example, in a mound of concrete that delegates the role of a traffic sign (getting cars to slow down) to a speed bump. These forms of mediation describe patterns of human-technology relations. Latour suggests we abandon the unhelpful noun "technology" and replace it with the helpful adjective "technical" so that we might have a better understanding of the various ways humans and nonhuman things are bound up together.

1

The Question Concerning Technology

Martin Heidegger

In what follows we shall be *questioning* concerning technology. Questioning builds a way. We would be advised, therefore, above all to pay heed to the way, and not to fix our attention on isolated sentences and topics. The way is a way of thinking. All ways of thinking, more or less perceptibly, lead through language in a manner that is extraordinary. We shall be questioning concerning *technology,* and in so doing we should like to prepare a free relationship to it. The relationship will be free if it opens our human existence to the essence of technology. When we can respond to this essence, we shall be able to experience the technological within its own bounds.

Technology is not equivalent to the essence of technology. When we are seeking the essence of "tree," we have to become aware that that which pervades every tree, as tree, is not itself a tree that can be encountered among all the other trees.

Likewise, the essence of technology is by no means anything technological. Thus we shall never experience our relationship to the essence of technology so long as we merely conceive and push forward the technological, put up with it, or evade it. Everywhere we remain unfree and chained to technology, whether we passionately affirm or deny it. But we are delivered over to it in the worst possible way when we regard it as something neutral; for this conception of it, to which today we particularly like to do homage, makes us utterly blind to the essence of technology.

According to ancient doctrine, the essence of a thing is considered to be *what* the thing is. We ask the question concerning technology when we ask what it is. Everyone knows the two statements that answer our question. One says: Technology is a means to an end. The other says: Technology is a human activity. The two definitions of technology belong together. For to posit ends and procure and utilize the means to them is a human activity. The manufacture and utilization of equipment, tools, and machines, the manufactured and used things themselves, and the needs and ends that they serve, all belong to what technology is. The whole complex of these contrivances is technology. Technology itself is a contrivance, or, in Latin, an *instrumentum.*

The current conception of technology, according to which it is a means and a human activity, can therefore be called the instrumental and anthropological definition of technology.

Who would ever deny that it is correct? It is in obvious conformity with what we are envisioning when we talk about technology. The instrumental definition of technology is indeed so uncannily correct that it even holds for modern technology, of which, in other respects, we maintain with some justification that it is, in contrast to the older handwork technology, something completely different and therefore new. Even the power plant with its turbines and generators is a man-made means to an end established by man. Even the jet aircraft and the high-frequency apparatus are means to ends. A radar station is of course less simple than a weather vane. To be sure, the construction of a high-frequency apparatus requires the interlocking of various processes of technical-industrial production. And certainly a sawmill in a secluded valley of the Black Forest is a primitive means compared with the hydroelectric plant in the Rhine River.

But this much remains correct: modern technology too is a means to an end. That is why the instrumental conception of technology conditions every attempt to bring man into the right relation to technology. Everything depends on our manipulating technology in the proper manner as a means. We will, as we say, "get" technology "spiritually in hand." We will master it. The will to mastery becomes all the more urgent the more technology threatens to slip from human control.

But suppose now that technology were no mere means, how would it stand with the will to master it? Yet we said, did we not, that the instrumental definition of technology is correct? To be sure. The correct always fixes upon something pertinent in whatever is under consideration. However, in order to be correct, this fixing by no means needs to uncover the thing in question in its essence. Only at the point where such an uncovering happens does the true come to pass. For that reason the merely correct is not yet the true. Only the true brings us into a free relationship with that which concerns us from out of its essence. Accordingly, the correct instrumental definition of technology still does not show us technology's essence. In order that we may arrive at this, or at least come close to it, we must seek the true by way of the correct. We must ask: What is the instrumental itself? Within what do such things as means and end belong? A means is that whereby something is effected and thus attained. Whatever has an effect as its consequence is called a cause. But not only that by means of which something else is effected is a cause. The end in keeping with which the kind of means to be used is determined is also considered a cause. Wherever ends are pursued and means are employed, wherever instrumentality reigns, there reigns causality.

For centuries philosophy has taught that there are four causes: (1) the *causa materialis,* the material, the matter out of which, for example, a silver chalice is made; (2) the *causa formalis,* the form, the shape into which the material enters; (3) the *causa finalis,* the end, for example, the sacrificial rite in relation to which the chalice required is determined as to its form and matter; and (4) the *causa efficiens,* which brings about the effect that is the finished, actual chalice, in this instance, the silversmith. What technology is, when represented as a means, discloses itself when we trace instrumentality back to fourfold causality.

But suppose that causality, for its part, is veiled in darkness with respect to what it is? Certainly for centuries we have acted as though the doctrine of the four causes had fallen from heaven as a truth as clear as daylight. But it might be that the time has come to ask, Why are there just four causes? In relation to the aforementioned four, what does "cause" really mean? From whence does it come that the causal *character* of the four causes is so unifiedly determined that they belong together?

So long as we do not allow ourselves to go into these questions, causality, and with it instrumentality, and with the latter the accepted definition of technology, remain obscure and groundless.

For a long time we have been accustomed to representing cause as that which brings something about. In this connection, to bring about means to obtain results, effects. The *causa efficiens,* but one among the four causes, sets the standard for all causality. This goes so far that we no longer even count the *causa finalis,* telic finality, as causality. *Causa, casus,* belongs to the verb *cadere,* "to fall," and means that which brings it about that something falls out as a result in such and such a way. The doctrine of the four causes goes back to Aristotle. But everything that later ages seek in Greek thought under the conception and rubric "causality," in the realm of Greek thought and for Greek thought per se has simply nothing at all to do with bringing about and effecting. What we call cause [*Ursache*] and the Romans call *causa* is called *aition* by the Greeks, that to which something else is indebted [*das, was ein anderes verschuldet*]. The four causes are the ways, all belonging at once to each other, of being responsible for something else. An example can clarify this.

Silver is that out of which the silver chalice is made. As this matter (*hyle*), it is co-responsible for the chalice. The chalice is indebted to, that is, owes thanks to, the silver for that out of which it consists. But the sacrificial vessel is indebted not only to the silver. As a chalice, that which is indebted to the silver appears in the aspect of a chalice and not in that of a brooch or a ring. Thus the sacrificial vessel is at the same time indebted to the aspect (*eidos*) of chaliceness. Both the silver into which the aspect is admitted as chalice and the aspect in which the silver appears are in their respective ways co-responsible for the sacrificial vessel.

But there remains yet a third that is above all responsible for the sacrificial vessel. It is that which in advance confines the chalice within the realm of consecration and bestowal. Through this the chalice is circumscribed as sacrificial vessel. Circumscribing gives bounds to the thing. With the bounds the thing does not stop; rather from out of them it begins to be what, after production, it will be. That which gives bounds, that which completes, in this sense is called in Greek *telos,* which is all too often translated as "aim" or "purpose," and so misinterpreted. The *telos* is responsible for what as matter and for what as aspect are together co-responsible for the sacrificial vessel.

Finally there is a fourth participant in the responsibility for the finished sacrificial vessel's lying before us ready for use, that is, the silversmith—but not at all because he, in working, brings about the finished sacrificial chalice as if it were the effect of a making; the silversmith is not a *causa efficiens.*

The Aristotelian doctrine neither knows the cause that is named by this term nor uses a Greek word that would correspond to it.

The silversmith considers carefully and gathers together the three aforementioned ways of being responsible and indebted. To consider carefully [*überlegen*] is in Greek *legein, logos. Legein* is rooted in *apophainesthai,* to bring forward into appearance. The silversmith is co-responsible as that from whence the sacrificial vessel's bringing forth and resting-in-self take and retain their first departure. The three previously mentioned ways of being responsible owe thanks to the pondering of the silversmith for the "that" and the "how" of their coming into appearance and into play for the production of the sacrificial vessel.

Thus four ways of being responsible hold sway in the sacrificial vessel that lies ready before us. They differ from one another, yet they belong together. What unites them from the beginning? In what does this playing in unison of the four ways of being responsible play? What is the source of the unity of the four causes? What, after all, does this owing and being responsible mean, thought as the Greeks thought it?

Today we are too easily inclined either to understand being responsible and being indebted moralistically as a lapse, or else to construe them in terms of effecting. In either case we bar to ourselves the way to the primal meaning of that which is later called causality. So long as this way is not opened up to us we shall also fail to see what instrumentality, which is based on causality, actually is.

In order to guard against such misinterpretations of being responsible and being indebted, let us clarify the four ways of being responsible in terms of that for which they are responsible. According to our example, they are responsible for the silver chalice's lying ready before us as a sacrificial vessel. Lying before and lying ready (*hypokeisthai*) characterize the presencing of something that presences. The four ways of being responsible bring something into appearance. They let it come forth into presencing [*An-wesen*]. They set it free to that place and so start it on its way, namely, into its complete arrival. The principal characteristic of being responsible is this starting something on its way into arrival. It is in the sense of such a starting something on its way into arrival that being responsible is an occasioning or an inducing to go forward [*ver-an-lassen*]. On the basis of a look at what the Greeks experienced in being responsible, in *aitia,* we now give this verb "to occasion" a more inclusive meaning, so that it now is the name for the essence of causality thought as the Greeks thought it. The common and narrower meaning of "occasion" in contrast is nothing more than striking against and releasing, and means a kind of secondary cause within the whole of causality.

But in what, then, does the playing in unison of the four ways of occasioning play? They let what is not yet present arrive into presencing. Accordingly, they are unifiedly ruled over by a bringing that brings what presences into appearance. Plato tells us what this bringing is in a sentence from the *Symposium* (205b): *hē gar toi ek tou mē onton eis to on ionti hotōioun aitia pasa esti poiēsis.* "Every occasion for whatever passes over and goes forward into presencing from that which is not presencing is *poiēsis,* is bringing-forth [*her-vor-bringen*]."

It is of utmost importance that we think bringing-forth in its full scope and at the same time in the sense in which the Greeks thought it. Not only handcraft manufacture, not only artistic and poetical bringing into appearance and concrete imagery, is a bringing-forth, *poiēsis. Physis* also, the arising of something from out of itself, is a bringing-forth, *poiēsis. Physis* is indeed *poiēsis* in the highest sense. For what presences by means of *physis* has the bursting open belonging to bringing-forth, for example, the bursting of a blossom into bloom, in itself (*en heautōi*). In contrast, what is brought forth by the artisan or the artist, for example, the silver chalice, has the bursting open belonging to bringing-forth not in itself, but in another (*en allōi*), in the craftsman or artist.

The modes of occasioning, the four causes, are at play, then, within bringing-forth. Through bringing-forth, the growing things of nature as well as whatever is completed through the crafts and the arts come at any given time to their appearance.

But how does bringing-forth happen, be it in nature or in handwork and art? What is the bringing-forth in which the fourfold way of occasioning plays? Occasioning has to do with the presencing [*Anwesen*] of that which at any given time comes to appearance in bringing-forth. Bringing-forth brings hither out of concealment forth into unconcealment. Bringing-forth comes to pass only insofar as something concealed comes into unconcealment. This coming rests and moves freely within what we call revealing [*das Entbergen*]. The Greeks have the word *alētheia* for revealing. The Romans translate this with *veritas.* We say "truth" and usually understand it as the correctness of an idea.

But where have we strayed to? We are questioning concerning technology, and we have arrived now at *alētheia,* at revealing. What has the essence of technology to do with revealing? The answer: everything. For every bringing-forth is grounded in revealing. Bringing-forth, indeed, gathers within itself the four modes of occasioning—causality—and rules them throughout. Within its domain belong end and means, belongs instrumentality. Instrumentality is considered to be the fundamental characteristic of technology. If we inquire, step by step, into what technology, represented as means, actually is, then we shall arrive at revealing. The possibility of all productive manufacturing lies in revealing.

Technology is therefore no mere means. Technology is a way of revealing. If we give heed to this, then another whole realm for the essence of technology will open itself up to us. It is the realm of revealing, that is, of truth.

This prospect strikes us as strange. Indeed, it should do so, should do so as persistently as possible and with so much urgency that we will finally take seriously the simple question of what the name "technology" means. The word stems from the Greek. *Technikon* means that which belongs to *technē*. We must observe two things with respect to the meaning of this word. One is that *technē* is the name not only for the activities and skills of the craftsman, but also for the arts of the mind and the fine arts. *Technē* belongs to bringing-forth, to *poiēsis;* it is something poietic.

The other point that we should observe with regard to *technē* is even more important. From earliest times until Plato the word *technē* is linked with the word *epistēmē*. Both words are names for knowing in the widest sense. They mean to be entirely at home in something, to understand and be expert in it. Such knowing provides an opening up. As an opening up it is a revealing. Aristotle, in a discussion of special importance (*Nicomachean Ethics,* Bk. VI, chaps. 3 and 4), distinguishes between *epistēmē* and *technē* and indeed with respect to what and how they reveal. *Technē* is a mode of *alētheuein.* It reveals whatever does not bring itself forth and does not yet lie here before us, whatever can look and turn out now one way and now another. Whoever builds a house or a ship or forges a sacrificial chalice reveals what is to be brought forth, according to the perspectives of the four modes of occasioning. This revealing gathers together in advance the aspect and the matter of ship or house, with a view to the finished thing envisioned as completed, and from this gathering determines the manner of its construction. Thus what is decisive in *technē* does not lie at all in making and manipulating nor in the using of means, but rather in the aforementioned revealing. It is as revealing, and not as manufacturing, that *technē* is a bringing-forth.

Thus the clue to what the word *technē* means and to how the Greeks defined it leads us into the same context that opened itself to us when we pursued the question of what instrumentality as such in truth might be.

Technology is a mode of revealing. Technology comes to presence in the realm where revealing and unconcealment take place, where *alētheia,* truth, happens.

In opposition to this definition of the essential domain of technology, one can object that it indeed holds for Greek thought and that at best it might apply to the techniques of the handcraftsman, but that it simply does not fit modern machine-powered technology. And it is precisely the latter and it alone that is the disturbing thing, that moves us to ask the question concerning technology per se. It is said that modern technology is something incomparably different from all earlier technologies because it is based on modern physics as an exact science. Meanwhile we have come to understand more clearly that the reverse holds true as well: Modern physics, as experimental, is dependent upon technical apparatus and upon progress in the building of apparatus. The establishing of this mutual relationship between technology and physics is correct. But it remains a merely historiographical establishing of facts and says nothing about that in which this mutual relationship is grounded. The decisive question still remains: Of what essence is modern technology that it happens to think of putting exact science to use?

What is modern technology? It too is a revealing. Only when we allow our attention to rest on this fundamental characteristic does that which is new in modern technology show itself to us.

And yet the revealing that holds sway throughout modern technology does not unfold into a bringing-forth in the sense of *poiēsis.* The revealing that rules in modern technology is challenging [*herausfordern*], which puts to nature the unreasonable demand that it supply energy that can be extracted and stored as such. But does this not hold true for the old windmill as well? No. Its sails do indeed turn in the wind; they are left entirely to the wind's blowing. But the windmill does not unlock energy from the air currents in order to store it.

In contrast, a tract of land is challenged into the putting out of coal and ore. The earth now reveals itself as a coal mining district, the soil as a mineral deposit. The field that the peasant formerly cultivated and set in order appears differently than it did when to set in order still meant to take care of and to maintain. The work of the peasant does not challenge the soil of the field. In the sowing of the grain it places the seed in the keeping of the forces of growth and watches over its increase. But meanwhile even the cultivation of the field has come under the grip of another kind of setting-in-order, which *sets* upon nature. It sets upon it in the sense of challenging it. Agriculture is now the mechanized food industry. Air is now set upon to yield nitrogen, the earth to yield ore, ore to yield uranium, for example; uranium is set upon to yield atomic energy, which can be released either for destruction or for peaceful use.

This setting-upon that challenges forth the energies of nature is an expediting, and in two ways. It expedites in that it unlocks and exposes. Yet that expediting is always itself directed from the beginning toward furthering something else, i.e., toward driving on to the maximum yield at the minimum expense. The coal that has been hauled out in some mining district has not been supplied in order that it may simply be present somewhere or other. It is stockpiled; that is, it is on call, ready to deliver the sun's warmth that is stored in it. The sun's warmth is challenged forth for heat, which in turn is ordered to deliver steam whose pressure turns the wheels that keep a factory running.

The hydroelectric plant is set into the current of the Rhine. It sets the Rhine to supplying its hydraulic pressure, which then sets the turbines turning. This turning sets those machines in motion whose thrust sets going the electric current for which the long-distance power station and its network of cables are set up to dispatch electricity. In the context of the interlocking processes pertaining to the orderly disposition of electrical energy, even the Rhine itself appears as something at our command. The hydroelectric plant is not built into the Rhine River as was the old wooden bridge that joined bank with bank for hundreds of years. Rather the river is dammed up into the power plant. What the river is now, namely, a water power supplier, derives from out of the essence of the power station. In order that we may even remotely consider the monstrousness that reigns here, let us ponder for a moment the contrast that speaks out of the two titles, "The Rhine" as dammed up into the *power* works, and "The Rhine" as uttered out of the *art* work, in Hölderlin's hymn by that name. But, it will be replied, the Rhine is still a river in the landscape, is it not? Perhaps. But how? In no other way than as an object on call for inspection by a tour group ordered there by the vacation industry.

The revealing that rules throughout modern technology has the character of a setting-upon, in the sense of a challenging-forth. That challenging happens in that the energy concealed in nature is unlocked, what is unlocked is transformed, what is transformed is stored up, what is stored up is, in turn, distributed, and what is distributed is switched about ever anew. Unlocking, transforming, storing, distributing, and switching about are ways of revealing. But the revealing never simply comes to an end. Neither does it run off into the indeterminate. The revealing reveals to itself its own manifoldly interlocking paths, through regulating their course. This regulating itself is, for its part, everywhere secured. Regulating and securing even become the chief characteristics of the challenging revealing.

What kind of unconcealment is it, then, that is peculiar to that which comes to stand forth through this setting-upon that challenges? Everywhere everything is ordered to stand by, to be immediately at hand, indeed to stand there just so that it may be on call for a further ordering. Whatever is ordered about in this way has its own standing. We call it the standing-reserve [*Bestand*]. The word expresses here something more, and something more essential, than mere "stock." The name "standing-reserve" assumes the rank of an inclusive rubric. It designates nothing less than the way in which everything presences that is wrought upon by the challenging revealing. Whatever stands by in the sense of standing-reserve no longer stands over against us as object.

Yet an airliner that stands on the runway is surely an object. Certainly. We can represent the machine so. But then it conceals itself as to what and how it is. Revealed, it stands on the taxi strip only as standing-reserve, inasmuch as it is ordered to ensure the possibility of transportation. For this it must be in its whole structure and in every one of its constituent parts, on call for duty, that is, ready for takeoff. (Here it would be appropriate to discuss Hegel's definition of the machine as an autonomous tool. When applied to the tools of the craftsman, his characterization is correct. Characterized in this way, however, the machine is not thought at all from out of the essence of technology within which it belongs. Seen in terms of the standing-reserve, the machine is completely unautonomous, for it has its standing only from the ordering of the orderable.)

The fact that now, wherever we try to point to modern technology as the challenging revealing, the words "setting-upon," "ordering," "standing-reserve," obtrude and accumulate in a dry, monotonous, and therefore oppressive way, has its basis in what is now coming to utterance.

Who accomplishes the challenging setting-upon through which what we call the real is revealed as standing-reserve? Obviously, man. To what extent is man capable of such a revealing? Man can indeed conceive, fashion, and carry through this or that in one way or another. But man does not have control over unconcealment itself, in which at any given time the real shows itself or withdraws. The fact that the real has been showing itself in the light of Ideas ever since the time of Plato, Plato did not bring about. The thinker only responded to what addressed itself to him.

Only to the extent that man for his part is already challenged to exploit the energies of nature can this ordering revealing happen. If man is challenged, ordered, to do this, then does not man himself belong even more originally than nature within the standing-reserve? The current talk about human resources, about the supply of patients for a clinic, gives evidence of this. The forester who, in the wood, measures the felled timber and to all appearances walks the same forest path in the same way as did his grandfather is today commanded by profit-making in the lumber industry, whether he knows it or not. He is made subordinate to the orderability of cellulose, which for its part is challenged forth by the need for paper, which is then delivered to newspapers and illustrated magazines. The latter, in their turn, set public opinion to swallowing what is printed, so that a set configuration of opinion becomes available on demand. Yet precisely because man is challenged more originally than are the energies of nature, that is, into the process of ordering, he never is transformed into mere standing-reserve. Since man drives technology forward, he takes part in ordering as a way of revealing. But the unconcealment itself, within which ordering unfolds, is never a human handiwork, any more than is the realm through which man is already passing every time he as a subject relates to an object.

Where and how does this revealing happen if it is no mere handiwork of man? We need not look far. We need only apprehend in an unbiased way that which has already claimed man and has done so, so decisively that he can only be man at any given time as the one so claimed. Wherever man opens his eyes and ears, unlocks his heart, and gives himself over to meditating and striving, shaping and working, entreating and thanking, he finds himself everywhere already brought into the unconcealed. The unconcealment of the unconcealed has already come to pass whenever it calls man forth into the modes of revealing allotted to him. When man, in his way, from within unconcealment reveals that which presences, he merely responds to the call of unconcealment even when he contradicts it. Thus when man, investigating, observing, ensnares nature as an area of his own conceiving, he has already been claimed by a way of revealing that challenges him to approach nature as an object of research, until even the object disappears into the objectlessness of standing-reserve.

Modern technology as an ordering revealing is, then, no merely human doing. Therefore we must take that challenging that sets upon man to order the real as standing-reserve in accordance with the way in which it shows itself. That challenging gathers man into ordering. This gathering concentrates man upon ordering the real as standing-reserve.

That which primordially unfolds the mountains into mountain ranges and courses through them in their folded togetherness is the gathering that we call *Gebirg* [mountain chain].

That original gathering from which unfold the ways in which we have feelings of one kind or another we name *Gemüt* [disposition].

We now name that challenging claim which gathers man thither to order the self-revealing as standing-reserve: *Ge-stell* [Enframing].

We dare to use this word in a sense that has been thoroughly unfamiliar up to now.

According to ordinary usage, the word *Gestell* [frame] means some kind of apparatus, for example, a bookrack. *Gestell* is also the name for a skeleton. And the employment of the word *Ge-stell* [Enframing] that is now required of us seems equally eerie, not to speak of the arbitrariness with which words of a mature language are thus misused. Can anything be more strange? Surely not. Yet this strangeness is an old usage of thinking. And indeed thinkers accord with this usage precisely at the point where it is a matter of thinking that which is highest. We, late born, are no longer in a position to appreciate the significance of Plato's daring to use the word *eidos* for that which in everything and in each particular thing endures as present. For *eidos*, in the common speech, meant the outward aspect [*Ansicht*] that a visible thing offers to the physical eye. Plato exacts of this word, however, something utterly extraordinary: that it name what precisely is not and never will be perceivable with physical eyes. But even this is by no means the full extent of what is extraordinary here. For *idea* names not only the nonsensuous aspect of what is physically visible. Aspect (*idea*) names and is, also, that which constitutes the essence in the audible, the tasteable, the tactile, in everything that is in any way accessible. Compared with the demands that Plato makes on language and thought in this and other instances, the use of the word *Gestell* as the name for the essence of modern technology, which we now venture here, is almost harmless. Even so, the usage now required remains something exacting and is open to misinterpretation.

Enframing means the gathering together of that setting-upon which sets upon man, i.e., challenges him forth, to reveal the real, in the mode of ordering, as standing-reserve. Enframing means that way of revealing which holds sway in the essence of modern technology and which is itself nothing technological. On the other hand, all those things that are so familiar to us and are standard parts of an assembly, such as rods, pistons, and chassis, belong to the technological. The assembly itself, however, together with the aforementioned stockparts, falls within the sphere of technological activity; and this activity always merely responds to the challenge of Enframing, but it never comprises Enframing itself or brings it about.

The word *stellen* [to set upon] in the name *Ge-stell* [Enframing] not only means challenging. At the same time it should preserve the suggestion of another *Stellen* from which it stems, namely, that producing and presenting [*her- und dar-stellen*] which, in the sense of *poiēsis*, lets what presences come forth into unconcealment. This producing that brings forth—for example, the erecting of a statue in the temple precinct—and the challenging ordering now under consideration are indeed fundamentally different, and yet they remain related in their essence. Both are ways of revealing, of *alētheia*. In Enframing, that unconcealment comes to pass in conformity with which the work of modern technology reveals the real as standing-reserve. This work is therefore neither only a human activity nor a mere means within such activity. The merely instrumental, merely anthropological definition of technology is therefore in principle untenable. And it cannot be rounded out by being referred back to some metaphysical or religious explanation that undergirds it.

It remains true, nonetheless, that man in the technological age is, in a particularly striking way, challenged forth into revealing. That revealing concerns nature, above all, as the chief storehouse of the standing energy reserve. Accordingly, man's ordering attitude and behavior display themselves first in the rise of modern physics as an exact science. Modern science's way of representing pursues and entraps nature as a calculable coherence of forces. Modern physics is not experimental physics because it applies apparatus to the questioning of nature. Rather the reverse is true. Because physics, indeed already as pure theory, sets nature up to exhibit itself as a coherence of forces calculable in advance, it therefore orders its experiments precisely for the purpose of asking whether and how nature reports itself when set up in this way.

But after all, mathematical physics arose almost two centuries before technology. How, then, could it have already been set upon by modern technology and placed in its service? The facts testify to the contrary. Surely technology got under way only when it could be supported by exact physical science. Reckoned chronologically, this is correct. Thought historically, it does not hit upon the truth.

The modern physical theory of nature prepares the way first not simply for technology but for the essence of modern technology. For already in physics the challenging gathering-together into ordering revealing holds sway. But in it that gathering does not yet come expressly to appearance. Modern physics is the herald of Enframing, a herald whose origin is still unknown. The essence of modern technology has for a long time been concealing itself, even where power machinery has been invented, where electrical technology is in full swing, and where atomic technology is well under way.

All coming to presence, not only modern technology, keeps itself everywhere concealed to the last. Nevertheless, it remains, with respect to its holding sway, that which precedes all: the earliest. The Greek thinkers already knew of this when they said: That which is earlier with regard to the arising that holds sway becomes manifest to us men only later. That which is primally early shows itself only ultimately to men. Therefore, in the realm of thinking, a painstaking effort to think through still more primally what was primally thought is not the absurd wish to revive what is past, but rather the sober readiness to be astounded before the coming of what is early.

Chronologically speaking, modern physical science begins in the seventeenth century. In contrast, machine-power technology develops only in the second half of the eighteenth century. But modern technology, which for chronological reckoning is the later, is, from the point of view of the essence holding sway within it, the historically earlier.

If modern physics must resign itself ever increasingly to the fact that its realm of representation remains inscrutable and incapable of being visualized, this resignation is not dictated by any committee of researchers. It is challenged forth by the rule of Enframing, which demands that nature be orderable as standing-reserve. Hence physics, in all its retreating from the representation turned only toward objects that has alone been standard till recently, will never be able to renounce this one thing: that nature reports itself in some way or other that is identifiable through calculation and that it remains orderable as a system of information. This system is determined, then, out of a causality that has changed once again. Causality now displays neither the character of the occasioning that brings forth nor the nature of the *causa efficiens,* let alone that of the *causa formalis.* It seems as though causality is shrinking into a reporting—a reporting challenged forth—of standing-reserves that must be guaranteed either simultaneously or in sequence. To this shrinking would correspond the process of growing resignation that Heisenberg's lecture depicts in so impressive a manner.*

Because the essence of modern technology lies in Enframing, modern technology must employ exact physical science. Through its so doing, the deceptive illusion arises that modern technology is applied physical science. This illusion can maintain itself only so long as neither the essential origin of modern science nor indeed the essence of modern technology is adequately found out through questioning.

We are questioning concerning technology in order to bring to light our relationship to its essence. The essence of modern technology shows itself in what we call Enframing. But simply to point to

---

*W. Heisenberg, "Das Naturbild in der heutigen Physik," in *Die Künste im technischen Zeitalter* (Munich: Bayerische Akademie der schönen Künste, 1954), pp. 43 ff.

this is still in no way to answer the question concerning technology, if to answer means to respond, in the sense of correspond, to the essence of what is being asked about.

Where do we find ourselves brought to, if now we think one step further regarding what Enframing itself actually is? It is nothing technological, nothing on the order of a machine. It is the way in which the real reveals itself as standing-reserve. Again we ask: Does this revealing happen somewhere beyond all human doing? No. But neither does it happen exclusively *in* man, or decisively *through* man.

Enframing is the gathering together that belongs to that setting-upon which sets upon man and puts him in position to reveal the real, in the mode of ordering, as standing-reserve. As the one who is challenged forth in this way, man stands within the essential realm of Enframing. He can never take up a relationship to it only subsequently. Thus the question as to how we are to arrive at a relationship to the essence of technology, asked in this way, always comes too late. But never too late comes the question as to whether we actually experience ourselves as the ones whose activities everywhere, public and private, are challenged forth by Enframing. Above all, never too late comes the question as to whether and how we actually admit ourselves into that wherein Enframing itself comes to presence.

The essence of modern technology starts man upon the way of that revealing through which the real everywhere, more or less distinctly, becomes standing-reserve. "To start upon a way" means "to send" in our ordinary language. We shall call that sending-that-gathers [*versammelde Schicken*] which first starts man upon a way of revealing, *destining* [*Geschick*]. It is from out of this destining that the essence of all history [*Geschichte*] is determined. History is neither simply the object of written chronicle nor simply the fulfillment of human activity. That activity first becomes history as something destined.* And it is only the destining into objectifying representation that makes the historical accessible as an object for historiography, that is, for a science, and on this basis makes possible the current equating of the historical with that which is chronicled.

Enframing, as a challenging-forth into ordering, sends into a way of revealing. Enframing is an ordaining of destining, as is every way of revealing. Bringing-forth, *poiēsis,* is also a destining in this sense.

Always the unconcealment of that which is goes upon a way of revealing. Always the destining of revealing holds complete sway over man. But that destining is never a fate that compels. For man becomes truly free only insofar as he belongs to the realm of destining and so becomes one who listens and hears, and not one who is simply constrained to obey.

The essence of freedom is *originally* not connected with the will or even with the causality of human willing.

Freedom governs the open in the sense of the cleared and lighted up, i.e., of the revealed. It is to the happening of revealing, that is, of truth, that freedom stands in the closest and most intimate kinship. All revealing belongs within a harboring and a concealing. But that which frees—the mystery—is concealed and always concealing itself. All revealing comes out of the open, goes into the open, and brings into the open. The freedom of the open consists neither in unfettered arbitrariness nor in the constraint of mere laws. Freedom is that which conceals in a way that opens to light, in whose clearing there shimmers that veil that covers what comes to presence of all truth and lets the veil appear as what veils. Freedom is the realm of the destining that at any given time starts a revealing upon its way.

The essence of modern technology lies in Enframing. Enframing belongs within the destining of revealing. These sentences express something different from the talk that we hear more fre-

---

*See *Vom Wesen der Wahrheit,* 1930; 1st ed., 1943, pp. 16 ff. [English translation, "On the Essence of Truth," in *Existence and Being,* ed. Werner Brock (Chicago: Regnery, 1949), pp. 308 ff.]

quently, to the effect that technology is the fate of our age, where "fate" means the inevitableness of an unalterable course.

But when we consider the essence of technology, then we experience Enframing as a destining of revealing. In this way we are already sojourning within the open space of destining, a destining that in no way confines us to a stultified compulsion to push on blindly with technology or, what comes to the same thing, to rebel helplessly against it and curse it as the work of the devil. Quite to the contrary, when we once open ourselves expressly to the *essence* of technology, we find ourselves unexpectedly taken into a freeing claim.

The essence of technology lies in Enframing. Its holding sway belongs within destining. Since destining at any given time starts man on a way of revealing, man, thus under way, is continually approaching the brink of the possibility of pursuing and pushing forward nothing but what is revealed in ordering, and of deriving all his standards on this basis. Through this the other possibility is blocked, that man might be admitted more and sooner and ever more primally to the essence of that which is unconcealed and to its unconcealment, in order that he might experience as his essence his needed belonging to revealing.

Placed between these possibilities, man is endangered from out of destining. The destining of revealing is as such, in every one of its modes, and therefore necessarily, *danger.*

In whatever way the destining of revealing may hold sway, the unconcealment in which everything that is shows itself at any given time harbors the danger that man may misconstrue the unconcealed and misinterpret it. Thus where everything that presences exhibits itself in the light of a cause-effect coherence, even God can, for representational thinking, lose all that is exalted and holy, the mysteriousness of his distance. In the light of causality, God can sink to the level of a cause, of *causa efficiens.* He then becomes, even in theology, the god of the philosophers, namely, of those who define the unconcealed and the concealed in terms of the causality of making, without ever considering the essential origin of this causality.

In a similar way the unconcealment in accordance with which nature presents itself as a calculable complex of the effects of forces can indeed permit correct determinations; but precisely through these successes the danger can remain that in the midst of all that is correct the true will withdraw.

The destining of revealing is in itself not just any danger, but *the* danger.

Yet when destining reigns in the mode of Enframing, it is the supreme danger. This danger attests itself to us in two ways. As soon as what is unconcealed no longer concerns man even as object, but does so, rather, exclusively as standing-reserve, and man in the midst of objectlessness is nothing but the orderer of the standing-reserve, then he comes to the very brink of a precipitous fall; that is, he comes to the point where he himself will have to be taken as standing-reserve. Meanwhile man, precisely as the one so threatened, exalts himself to the posture of lord of the earth. In this way the impression comes to prevail that everything man encounters exists only insofar as it is his construct. This illusion gives rise in turn to one final delusion: It seems as though man everywhere and always encounters only himself. Heisenberg has with complete correctness pointed out that the real must present itself to contemporary man in this way.* *In truth, however, precisely nowhere does man today any longer encounter himself, that is, his essence.* Man stands so decisively in attendance on the challenging-forth of Enframing that he does not apprehend Enframing as a claim, that he fails to see himself as the one spoken to, and hence also fails in every way to hear in what respect he ek-sists, from out of his essence, in the realm of an exhortation or address, and thus *can never* encounter only himself.

But Enframing does not simply endanger man in his relationship to himself and to everything

---

*"Das Naturbild," pp. 60 ff.

that is. As a destining, it banishes man into that kind of revealing which is an ordering. Where this ordering holds sway, it drives out every other possibility of revealing. Above all, Enframing conceals that revealing which, in the sense of *poiēsis,* lets what presences come forth into appearance. As compared with that other revealing, the setting-upon that challenges forth thrusts man into a relation to that which is, that is at once antithetical and rigorously ordered. Where Enframing holds sway, regulating and securing of the standing-reserve mark all revealing. They no longer even let their own fundamental characteristic appear, namely, this revealing as such.

Thus the challenging Enframing not only conceals a former way of revealing, bringing-forth, but it conceals revealing itself and with it That wherein unconcealment, that is, truth, comes to pass.

Enframing blocks the shining-forth and holding-sway of truth. The destining that sends into ordering is consequently the extreme danger. What is dangerous is not technology. There is no demonry of technology, but rather there is the mystery of its essence. The essence of technology, as a destining of revealing, is the danger. The transformed meaning of the word "Enframing" will perhaps become somewhat more familiar to us now if we think Enframing in the sense of destining and danger.

The threat to man does not come in the first instance from the potentially lethal machines and apparatus of technology. The actual threat has already affected man in his essence. The rule of Enframing threatens man with the possibility that it could be denied to him to enter into a more original revealing and hence to experience the call of a more primal truth.

Thus, where Enframing reigns, there is *danger* in the highest sense.

> But where danger is, grows
> The saving power also.

Let us think carefully about these words of Hölderlin. What does it mean "to save"? Usually we think that it means only to seize hold of a thing threatened by ruin, in order to secure it in its former continuance. But the verb "to save" says more. "To save" is to fetch something home into its essence, in order to bring the essence for the first time into its genuine appearing. If the essence of technology, Enframing, is the extreme danger, and if there is truth in Hölderlin's words, then the rule of Enframing cannot exhaust itself solely in blocking all lighting-up of every revealing, all appearing of truth. Rather, precisely the essence of technology must harbor in itself the growth of the saving power. But in that case, might not an adequate look into what Enframing is as a destining of revealing bring into appearance the saving power in its arising?

In what respect does the saving power grow there also where the danger is? Where something grows, there it takes root, from thence it thrives. Both happen concealedly and quietly and in their own time. But according to the words of the poet we have no right whatsoever to expect that there where the danger is we should be able to lay hold of the saving power immediately and without preparation. Therefore we must consider now, in advance, in what respect the saving power does most profoundly take root and thence thrive even in that wherein the extreme danger lies, in the holding sway of Enframing. In order to consider this, it is necessary, as a last step upon our way, to look with yet clearer eyes into the danger. Accordingly, we must once more question concerning technology. For we have said that in technology's essence roots and thrives the saving power.

But how shall we behold the saving power in the essence of technology so long as we do not consider in what sense of "essence" it is that Enframing is actually the essence of technology?

Thus far we have understood "essence" in its current meaning. In the academic language of philosophy, "essence" means *what* something is; in Latin, *quid. Quidditas,* whatness, provides the answer to the question concerning essence. For example, what pertains to all kinds of trees—oaks, beeches, birches, firs—is the same "treeness." Under this inclusive genus—the "universal"—fall all real and possible trees. Is then the essence of technology, Enframing, the common genus for everything technological? If that were the case then the steam turbine, the radio transmitter, and the cyclotron would each be an Enframing. But the word "Enframing" does not mean here a tool or any kind of apparatus. Still less does it mean the general concept of such resources. The machines and apparatus are no more cases and kinds of Enframing than are the man at the switch-board and the engineer in the drafting room. Each of these in its own way indeed belongs as stock-part, available resource, or executer, within Enframing; but Enframing is never the essence of technology in the sense of a genus. Enframing is a way of revealing having the character of destin-ing, namely, the way that challenges forth. The revealing that brings forth (*poiēsis*) is also a way that has the character of destining. But these ways are not kinds that, arrayed beside one another, fall under the concept of revealing. Revealing is that destining which, ever suddenly and inexplica-bly to all thinking, apportions itself into the revealing that brings forth and that also challenges, and which allots itself to man. The challenging revealing has its origin as a destining in bringing-forth. But at the same time Enframing, in a way characteristic of a destining, blocks *poiēsis*.

Thus Enframing, as a destining of revealing, is indeed the essence of technology, but never in the sense of genus and *essentia*. If we pay heed to this, something astounding strikes us: It is technology itself that makes the demand on us to think in another way what is usually understood by "essence." But in what way?

If we speak of the "essence of a house" and the "essence of a state," we do not mean a generic type; rather we mean the ways in which house and state hold sway, administer themselves, develop and decay—the way in which they "essence" [*Wesen*]. Johann Peter Hebel in a poem, "Ghost on Kanderer Street," for which Goethe had a special fondness, uses the old word *die Wes-erei*. It means the city hall inasmuch as there the life of the community gathers and village existence is constantly in play, that is, comes to presence. It is from the verb *wesen* that the noun is derived. *Wesen* understood as a verb is the same as *währen* [to last or endure], not only in terms of meaning, but also in terms of the phonetic formation of the word. Socrates and Plato already think the essence of something as what essences, what comes to presence, in the sense of what endures. But they think what endures as what remains permanently (*aei on*). And they find what endures permanently in what, as that which remains, tenaciously persists throughout all that happens. That which remains they discover, in turn, in the aspect (*eidos, idea*), for example, the Idea "house."

The Idea "house" displays what anything is that is fashioned as a house. Particular, real, and possible houses, in contrast, are changing and transitory derivatives of the Idea and thus belong to what does not endure.

But it can never in any way be established that enduring is based solely on what Plato thinks as *idea* and Aristotle thinks as *to ti ēn einai* (that which any particular thing has always been), or what metaphysics in its most varied interpretations thinks as *essentia*.

All essencing endures. But is enduring only permanent enduring? Does the essence of tech-nology endure in the sense of the permanent enduring of an Idea that hovers over everything tech-nological, thus making it seem that by technology we mean some mythological abstraction? The way in which technology essences lets itself be seen only from out of that permanent enduring in which Enframing comes to pass as a destining of revealing. Goethe once uses the mysterious word *fortgewähren* [to grant permanently] in place of *fortwähren* [to endure permanently].* He hears *währen* [to endure] and *gewähren* [to grant] here in one unarticulated accord. And if we now pon-der more carefully than we did before what it is that actually endures and perhaps alone endures,

---

*"Die Wahlverwandtschaften" [Congeniality], pt. II, chap. 10, in the novelette *Die wunderlichen Nachbars-kinder* [The strange neighbor's children].

we may venture to say: *Only what is granted endures. That which endures primally out of the earliest beginning is what grants.*

As the essencing of technology, Enframing is that which endures. Does Enframing hold sway at all in the sense of granting? No doubt the question seems a horrendous blunder. For according to everything that has been said, Enframing is, rather, a destining that gathers together into the revealing that challenges forth. Challenging is anything but a granting. So it seems, so long as we do not notice that the challenging-forth into the ordering of the real as standing-reserve still remains a destining that starts man upon a way of revealing. As this destining, the coming to presence of technology gives man entry into That which, of himself, he can neither invent nor in any way make. For there is no such thing as a man who, solely of himself, is only man.

But if this destining, Enframing, is the extreme danger, not only for man's coming to presence, but for all revealing as such, should this destining still be called a granting? Yes, most emphatically, if in this destining the saving power is said to grow. Every destining of revealing comes to pass from out of a granting and as such a granting. For it is granting that first conveys to man that share in revealing which the coming-to-pass of revealing needs. As the one so needed and used, man is given to belong to the coming-to-pass of truth. The granting that sends in one way or another into revealing is as such the saving power. For the saving power lets man see and enter into the highest dignity of his essence. This dignity lies in keeping watch over the unconcealment—and with it, from the first, the concealment—of all coming to presence on this earth. It is precisely in Enframing, which threatens to sweep man away into ordering as the supposed single way of revealing, and so thrusts man into the danger of the surrender of his free essence—it is precisely in this extreme danger that the innermost indestructible belongingness of man within granting may come to light, provided that we, for our part, begin to pay heed to the coming to presence of technology.

Thus the coming to presence of technology harbors in itself what we least suspect, the possible arising of the saving power.

Everything, then, depends upon this: that we ponder this arising and that, recollecting, we watch over it. How can this happen? Above all through our catching sight of what comes to presence in technology, instead of merely staring at the technological. So long as we represent technology as an instrument, we remain held fast in the will to master it. We press on past the essence of technology.

When, however, we ask how the instrumental comes to presence as a kind of causality, then we experience this coming to presence as the destining of a revealing.

When we consider, finally, that the coming to presence of the essence of technology comes to pass in the granting that needs and uses man so that he may share in revealing, then the following becomes clear:

The essence of technology is in a lofty sense ambiguous. Such ambiguity points to the mystery of all revealing, that is, of truth.

On the one hand, Enframing challenges forth into the frenziedness of ordering that blocks every view into the coming-to-pass of revealing and so radically endangers the relation to the essence of truth.

On the other hand, Enframing comes to pass for its part in the granting that lets man endure—as yet unexperienced, but perhaps more experienced in the future—that he may be the one who is needed and used for the safekeeping of the coming to presence of truth. Thus does the arising of the saving power appear.

The irresistibility of ordering and the restraint of the saving power draw past each other like the paths of two stars in the course of the heavens. But precisely this, their passing by, is the hidden side of their nearness.

When we look into the ambiguous essence of technology, we behold the constellation, the stellar course of the mystery.

The question concerning technology is the question concerning the constellation in which revealing and concealing, in which the coming to presence of truth, comes to pass.

But what help is it to us to look into the constellation of truth? We look into the danger and see the growth of the saving power.

Through this we are not yet saved. But we are thereupon summoned to hope in the growing light of the saving power. How can this happen? Here and now and in little things, that we may foster the saving power in its increase. This includes holding always before our eyes the extreme danger.

The coming to presence of technology threatens revealing, threatens it with the possibility that all revealing will be consumed in ordering and that everything will present itself only in the unconcealedness of standing-reserve. Human activity can never directly counter this danger. Human achievement alone can never banish it. But human reflection can ponder the fact that all saving power must be of a higher essence than what is endangered, though at the same time kindred to it.

But might there not perhaps be a more primally granted revealing that could bring the saving power into its first shining forth in the midst of the danger, a revealing that in the technological age rather conceals than shows itself?

There was a time when it was not technology alone that bore the name *technē*. Once that revealing that brings forth truth into the splendor of radiant appearing also was called *technē*.

Once there was a time when the bringing-forth of the true into the beautiful was called *technē*. And the *poiēsis* of the fine arts also was called *technē*.

In Greece, at the outset of the destining of the West, the arts soared to the supreme height of the revealing granted them. They brought the presence [*Gegenwart*] of the gods, brought the dialogue of divine and human destinings, to radiance. And art was simply called *technē*. It was a single, manifold revealing. It was pious, *promos,* i.e., yielding to the holding-sway and the safekeeping of truth.

The arts were not derived from the artistic. Art works were not enjoyed aesthetically. Art was not a sector of cultural activity.

What, then, was art—perhaps only for that brief but magnificent time? Why did art bear the modest name *technē?* Because it was a revealing that brought forth and hither, and therefore belonged within *poiēsis*. It was finally that revealing which holds complete sway in all the fine arts, in poetry, and in everything poetical that obtained *poiēsis* as its proper name.

The same poet from whom we heard the words

> *But where danger is, grows*
> *The saving power also.*

says to us:

> *. . . poetically dwells man upon this earth.*

The poetical brings the true into the splendor of what Plato in the *Phaedrus* calls to *ekphanestaton,* that which shines forth most purely. The poetical thoroughly pervades every art, every revealing of coming to presence into the beautiful.

Could it be that the fine arts are called to poetic revealing? Could it be that revealing lays claim to the arts most primally, so that they for their part may expressly foster the growth of the saving power, may awaken and found anew our look into that which grants and our trust in it?

Whether art may be granted this highest possibility of its essence in the midst of the extreme danger, no one can tell. Yet we can be astounded. Before what? Before this other possibility: that the frenziedness of technology may entrench itself everywhere to such an extent that someday, throughout everything technological, the essence of technology may come to presence in the coming-to-pass of truth.

Because the essence of technology is nothing technological, essential reflection upon technology and decisive confrontation with it must happen in a realm that is, on the one hand, akin to the essence of technology and, on the other, fundamentally different from it.

Such a realm is art. But certainly only if reflection on art, for its part, does not shut its eyes to the constellation of truth after which we are *questioning*.

Thus questioning, we bear witness to the crisis that in our sheer preoccupation with technology we do not yet experience the coming to presence of technology, that in our sheer aesthetic-mindedness we no longer guard and preserve the coming to presence of art. Yet the more questioningly we ponder the essence of technology, the more mysterious the essence of art becomes.

The closer we come to the danger, the more brightly do the ways into the saving power begin to shine and the more questioning we become. For questioning is the piety of thought.

2

Heidegger on Gaining a Free Relation to Technology

Hubert Dreyfus

INTRODUCTION: WHAT HEIDEGGER IS NOT SAYING

In *The Question Concerning Technology* Heidegger describes his aim:

"We shall be questioning concerning technology, and in so doing we should like to prepare a free relationship to it."

He wants to reveal the essence of technology in such a way that "in no way confines us to a stultified compulsion to push on blindly with technology or, what comes to the same thing, to rebel helplessly against it."[1] Indeed, he claims that "When we once open ourselves expressly to the *essence* of technology, we find ourselves unexpectedly taken into a freeing claim."[2]

We will need to explain essence, opening, and freeing before we can understand Heidegger here. But already Heidegger's project should alert us to the fact that he is not announcing one more reactionary rebellion against technology, although many respectable philosophers, including Jürgen Habermas, take him to be doing just that; nor is he doing what progressive thinkers such as Habermas want him to do, proposing a way to get technology under control so that it can serve our rationally chosen ends.

The difficulty in locating just where Heidegger stands on technology is no accident. Heidegger has not always been clear about what distinguishes his approach from a romantic reaction to the domination of nature, and when he does finally arrive at a clear formulation of his own original view, it is so radical that everyone is tempted to translate it into conventional platitudes about the evils of technology. Thus Heidegger's ontological concerns are mistakenly assimilated to humanistic worries about the devastation of nature.

Those who want to make Heidegger intelligible in terms of current anti-technological banalities can find support in his texts. During the war he attacks consumerism:

> The circularity of consumption for the sake of consumption is the sole procedure which distinctively characterizes the history of a world which has become an unworld.[3]

And as late as 1955 he holds that:

> The world now appears as an object open to the attacks of calculative thought. . . . Nature becomes a gigantic gasoline station, an energy source for modern technology and industry.[4]

In this address to the Schwartzwald peasants he also laments the appearance of television antennae on their dwellings.

> Hourly and daily they are chained to radio and television. . . . All that with which modern techniques of communication stimulate, assail, and drive man—all that is already much closer to man today than his fields around his farmstead, closer than the sky over the earth, closer than the change from night to day, closer than the conventions and customs of his village, than the tradition of his native world.[5]

Such statements suggest that Heidegger is a Luddite who would like to return from the exploitation of the earth, consumerism, and mass media to the world of the pre-Socratic Greeks or the good old Schwartzwald peasants.

HEIDEGGER'S ONTOLOGICAL APPROACH TO TECHNOLOGY

As his thinking develops, however, Heidegger does not deny these are serious problems, but he comes to the surprising and provocative conclusion that focusing on loss and destruction is still technological.

> All attempts to reckon existing reality . . . in terms of decline and loss, in terms of fate, catastrophe, and destruction, are merely technological behavior.[6]

Seeing our situation as posing a problem that must be solved by appropriate action turns out to be technological too:

> [T]he instrumental conception of technology conditions every attempt to bring man into the right relation to technology. . . . The will to mastery becomes all the more urgent the more technology threatens to slip from human control.[7]

Heidegger is clear this approach cannot work:

> No single man, no group of men, no commission of prominent statesmen, scientists, and technicians, no conference of leaders of commerce and industry, can brake or direct the progress of history in the atomic age.[8]

His view is both darker and more hopeful. He thinks there is a more dangerous situation facing modern man than the technological destruction of nature and civilization, yet a situation about which something *can* be done—at least indirectly. The threat is not a *problem* for which there can be a *solution* but an ontological *condition* from which we can be *saved*.

Heidegger's concern is the human distress caused by the *technological understanding of being,* rather than the destruction caused by specific technologies. Consequently, Heidegger distinguishes the current problems caused by technology—ecological destruction, nuclear danger, consumerism, et cetera—from the devastation that would result if technology solved all our problems.

> What threatens man in his very nature is the . . . view that man, by the peaceful release, transformation, storage, and channeling of the energies of physical nature, could render the human condition . . . tolerable for everybody and happy in all respects.[9]

The "greatest danger" is that

> the approaching tide of technological revolution in the atomic age could so captivate, bewitch, dazzle, and beguile man that calculative thinking may someday come to be accepted and practiced as *the only* way of thinking.[10]

The danger, then, is not the destruction of nature or culture but a restriction in our way of thinking—a leveling of our understanding of being.

To evaluate this claim we must give content to what Heidegger means by an understanding of being. Let us take an example. Normally we deal with things, and even sometimes people, as resources to be used until no longer needed and then put aside. A styrofoam cup is a perfect example. When we want a hot or cold drink it does its job, and when we are through with it we throw it away. How different this understanding of an object is from what we can suppose to be the everyday Japanese understanding of a delicate teacup. The teacup does not preserve temperature as well as its plastic replacement, and it has to be washed and protected, but it is preserved from generation to generation for its beauty and its social meaning. It is hard to picture a tea ceremony around a styrofoam cup.

Note that the traditional Japanese understanding of what it is to be human (passive, contented, gentle, social, etc.) fits with their understanding of what it is to be a thing (delicate, beautiful, traditional, etc.). It would make no sense for us, who are active, independent, and aggressive—constantly striving to cultivate and satisfy our desires—to relate to things the way the Japanese do; or for the Japanese (before their understanding of being was interfered with by ours) to invent and prefer styrofoam teacups. In the same vein *we* tend to think of politics as the negotiation of individual desires while the Japanese seek consensus. In sum the social practices containing an understanding of what it is to be a human self, those containing an interpretation of what it is to be a thing, and those defining society fit together. They add up to an understanding of being.

The shared practices into which we are socialized, then, provide a background understanding of what counts as things, what counts as human beings, and ultimately what counts as real, on the basis of which we can direct our actions toward particular things and people. Thus the understanding of being creates what Heidegger calls a *clearing* in which things and people can show up for us. We do not produce the clearing. It produces us as the kind of human beings that we are. Heidegger describes the clearing as follows:

> [B]eyond what is, not away from it but before it, there is still something else that happens. In the midst of beings as a whole an open place occurs. There is a clearing, a lighting. . . . This open center is . . . not surrounded by what is; rather, the lighting center itself encircles all that is. . . . Only this clearing grants and guarantees to human beings a passage to those entities that we ourselves are not, and access to the being that we ourselves are.[11]

What, then, is the essence of technology, that is, the technological understanding of being, that is, the technological clearing, and how does opening ourselves to it give us a free relation to technological devices? To begin with, when we ask about the essence of technology we are able to see that Heidegger's question cannot be answered by defining technology. Technology is as old as civilization. Heidegger notes that it can be correctly defined as "a means and a human activity." He calls this "the instrumental and anthropological definition of technology."[12] But if we ask about the *essence* of technology (the technological understanding of being) we find that modern technology is "something completely different and . . . new."[13] Even different from using styrofoam cups to serve our desires. The essence of modern technology, Heidegger tells us, is to seek more and more flexibility and efficiency *simply for its own sake.* "[E]xpediting is always itself directed from the beginning . . . towards driving on to the maximum yield at the minimum expense."[14] That is, our only goal is optimization:

> Everywhere everything is ordered to stand by, to be immediately at hand, indeed to stand there just so that it may be on call for a further ordering. Whatever is ordered about in this way has its own standing. We call it standing-reserve. . . .[15]

No longer are we subjects turning nature into an object of exploitation:

> The subject-object relation thus reaches, for the first time, its pure "relational," i.e., ordering, character in which both the subject and the object are sucked up as standing-reserves.[16]

A modern airliner is not an object at all, but just a flexible and efficient cog in the transportation system.[17] (And passengers are presumably not subjects but merely resources to fill the planes.) Heidegger concludes: "Whatever stands by in the sense of standing-reserve no longer stands over against us as object."[18]

All ideas of serving God, society, our fellow men, or even our own calling disappear. Human beings, on this view, become a resource to be used, but more important to be enhanced—like any other.

> Man, who no longer conceals his character of being the most important raw material, is also drawn into this process.[19]

In the film *2001,* the robot HAL, when asked if he is happy on the mission, answers: "I'm using all my capacities to the maximum. What more could a rational entity desire?" This is a brilliant expression of what anyone would say who is in touch with our current understanding of being. We pursue the growth or development of our potential simply for its own sake—it is our only goal. The human potential movement perfectly expresses this technological understanding of being, as does the attempt to better organize the future use of our natural resources. We thus become part of a system which no one directs but which moves toward the total mobilization of all beings, even us. This is why Heidegger thinks the perfectly ordered society dedicated to the welfare of all is not the solution of our problems but the distressing culmination of the technological understanding of being.

WHAT THEN CAN WE DO?

But, of course, Heidegger uses and depends upon modern technological devices. He is no Luddite and he does not advocate a return to the pre-technological world.

> It would be foolish to attack technology blindly. It would be shortsighted to condemn it as the work of the devil. We depend on technical devices; they even challenge us to ever greater advances.[20]

Instead, Heidegger suggests that there is a way we can keep our technological devices and yet remain true to ourselves:

> We can affirm the unavoidable use of technical devices, and also deny them the right to dominate us, and so to warp, confuse, and lay waste our nature.[21]

To understand how this might be possible we need an illustration of Heidegger's important distinction between technology and the technological understanding of being. Again we can turn to Japan.

In contemporary Japan a traditional, non-technological understanding of being still exists alongside the most advanced high-tech production and consumption. The TV set and the household gods share the same shelf—the styrofoam cup co-exists with the porcelain one. We can thus see that one can have technology without the technological understanding of being, so it becomes clear that the technological understanding of being can be dissociated from technological devices.

To make this dissociation, Heidegger holds, one must rethink the history of being in the West. Then one will see that although a technological understanding of being is our destiny, it is not our fate. That is, although our understanding of things and ourselves as resources to be ordered, enhanced, and used efficiently has been building up since Plato and dominates our practices, we are not stuck with it. It is not the way things have to be, but nothing more or less than our current cultural clearing.

Only those who think of Heidegger as opposing technology will be surprised at his next point. Once we see that technology is our latest understanding of being, we will be grateful for it. We did not make this clearing nor do we control it, but if it were not given to us to encounter things and ourselves as resources, nothing would show up *as* anything at all and no possibilities for action would make sense. And once we realize—in our practices, of course, not just in our heads—that we *receive* our technological understanding of being, we have stepped out of the technological understanding of being, for we then see that what is most important in our lives is not subject to efficient enhancement. This transformation in our sense of reality—this overcoming of calculative thinking—is precisely what Heideggerian thinking seeks to bring about. Heidegger seeks to show how we can recognize and thereby overcome our restricted, willful modern clearing precisely by recognizing our essential receptivity to it.

> [M]odern man must first and above all find his way back into the full breadth of the space proper to his essence. That essential space of man's essential being receives the dimension that unites it to something beyond itself . . . that is the way in which the safekeeping of being itself is given to belong to the essence of man as the one who is needed and used by being.[22]

But precisely how can we experience the technological understanding of being as a gift to which we are receptive? What is the phenomenon Heidegger is getting at? We can break out of the technological understanding of being whenever we find ourselves gathered by things rather than controlling them. When a thing like a celebratory meal, to take Heidegger's example, pulls our practices together and draws us in, we experience a focusing and a nearness that resists technological ordering. Even a technological object like a highway bridge, when experienced as a gathering and focusing of our practices, can help us resist the very technological ordering it furthers. Heidegger describes the bridge so as to bring out both its technological ordering function and its continuity with pre-technological things.

> The old stone bridge's humble brook crossing gives to the harvest wagon its passage from the fields into the village and carries the lumber cart from the field path to the road. The highway bridge is tied into the network of long-distance traffic, paced as calculated for maximum yield. Always and ever differently the bridge escorts the lingering and hastening ways of men to and fro. . . . The bridge *gathers* to itself in *its own way* earth and sky, divinities and mortals.[23]

Getting in sync with the highway bridge in its technological functioning can make us sensitive to the technological understanding of being as the way our current clearing works, so that we experience our role as receivers, and the importance of receptivity, thereby freeing us from our compulsion to force all things into one efficient order.

This transformation in our understanding of being, unlike the slow process of cleaning up the environment which is, of course, also necessary, would take place in a sudden Gestalt switch.

> The turning of the danger comes to pass suddenly. In this turning, the clearing belonging to the essence of being suddenly clears itself and lights up.[24]

The danger, when grasped as the danger, becomes that which saves us. "The self-same danger is, when it is *as* the danger, the saving power."[25]

This remarkable claim gives rise to two opposed ways of understanding Heidegger's response to technology. Both interpretations agree that once one recognizes the technological understanding of being for what it is—a historical understanding—one gains a free relation to it. We neither push forward technological efficiency as our only goal nor always resist it. If we are free of the techno-logical imperative we can, in each case, discuss the pros and cons. As Heidegger puts it:

> We let technical devices enter our daily life, and at the same time leave them outside . . . as things which are nothing absolute but remain dependent upon something higher [the clearing]. I would call this comportment toward technology which expresses "yes" and at the same time "no," by an old word, *releasement towards things.*[26]

One way of understanding this proposal—represented here by Richard Rorty—holds that once we get in the right relation to technology, viz. recognize it as a clearing, it is revealed as just as good as any other clearing. Efficiency—getting the most out of ourselves and everything else—is fine, so long as we do not think that efficiency for its own sake is the *only* end for man, dictated by reality itself, to which all others must be subordinated. Heidegger seems to support this acceptance of the technological understanding of being when he says:

> That which shows itself and at the same time withdraws [i.e., the clearing] is the essential trait of what we call the mystery. I call the comportment which enables us to keep open to the mean-ing hidden in technology, *openness to the mystery.* Releasement toward things and openness to the mystery belong together. They grant us the possibility of dwelling in the world in a totally different way. They promise us a new ground and foundation upon which we can stand and endure in the world of technology without being imperiled by it.[27]

But acceptance of the mystery of the gift of understandings of being cannot be Heidegger's whole story, for he immediately adds:

> Releasement toward things and openness to the mystery give us a vision of a new rootedness which *someday* might even be fit to recapture the old and now rapidly disappearing rootedness in a changed form.[28]

We then look back at the preceding remark and realize *releasement* gives only a "possibility" and a "promise" of "dwelling in the world in a totally different way."

Mere openness to technology, it seems, leaves out much that Heidegger finds essential to human being: embeddedness in nature, nearness or localness, shared meaningful differences such as noble and ignoble, justice and injustice, salvation and damnation, mature and immature—to name those that have played important roles in our history. *Releasement,* while giving us a free relation to technology and protecting our nature from being distorted and distressed, cannot give us any of these.

For Heidegger, there are, then, two issues. One issue is clear:

The issue is the saving of man's essential nature. Therefore, the issue is keeping meditative thinking alive.[29]

But that is not enough:

> If releasement toward things and openness to the mystery awaken within us, then we should
> arrive at a path that will lead to a new ground and foundation.[30]

Releasement, it turns out, is only a stage, a kind of holding pattern, awaiting a new understanding
of being, which would give some content to our openness—what Heidegger calls a new rootedness.
That is why each time Heidegger talks of *releasement* and the saving power of understanding technology as a gift he then goes on to talk of the divine.

> Only when man, in the disclosing coming-to-pass of the insight by which he himself is beheld
> . . . renounces human self-will . . . does he correspond in his essence to the claim of that insight.
> In thus corresponding man is gathered into his own, that he . . . may, as the mortal, look out
> toward the divine.[31]

The need for a new centeredness is reflected in Heidegger's famous remark in his last interview:
"Only a god can save us now."[32] But what does this mean?

THE NEED FOR A GOD

Just preserving pre-technical practices, even if we could do it, would not give us what we need. The
pre-technological practices no longer add up to a shared sense of reality and one cannot legislate a
new understanding of being. For such practices to give meaning to our lives, and unite us in a
community, they would have to be focused and held up to the practitioners. This function, which
later Heidegger calls "truth setting itself to work," can be performed by what he calls a work of
art. Heidegger takes the Greek temple as his illustration of an artwork working. The temple held
up to the Greeks what was important, and so let there be heroes and slaves, victory and disgrace,
disaster and blessing, and so on. People whose practices were manifested and focused by the temple
had guidelines for leading good lives and avoiding bad ones. In the same way, the medieval cathedral made it possible to be a saint or a sinner by showing people the dimensions of salvation and
damnation. In either case, one knew where one stood and what one had to do. Heidegger holds that
"there must always be some being in the open [the clearing], something that is, in which the openness takes its stand and attains its constancy."[33]

We could call such special objects cultural paradigms. A cultural paradigm focuses and collects the scattered practices of a culture, unifies them into coherent possibilities for action, and
holds them up to the people who can then act and relate to each other in terms of the shared exemplar.

When we see that for later Heidegger only those practices focused in a paradigm can establish
what things can show up as and what it makes sense to do, we can see why he was pessimistic
about salvaging aspects of the Enlightenment or reviving practices focused in the past. Heidegger
would say that we should, indeed, try to preserve such practices, but they can save us only if they
are radically transformed and integrated into a new understanding of reality. In addition we must
learn to appreciate marginal practices—what Heidegger calls the saving power of insignificant
things—practices such as friendship, back-packing into the wilderness, and drinking the local wine
with friends. All these practices are marginal precisely because they are not efficient. They can, of

course, be engaged in for the sake of health and greater efficiency. This expanding of technological efficiency is the greatest danger. But these saving practices could come together in a new cultural paradigm that held up to us a new way of doing things, thereby focusing a world in which formerly marginal practices were central and efficiency marginal. Such a new object or event that grounded a new understanding of reality Heidegger would call a new god. This is why he holds that "only another god can save us."[34]

Once one sees what is needed, one also sees that there is not much we can do to bring it about. A new sense of reality is not something that can be made the goal of a crash program like the moon flight—a paradigm of modern technological power. A hint of what such a new god might look like is offered by the music of the sixties. The Beatles, Bob Dylan, and other rock groups became for many the articulation of new understanding of what really mattered. This new understanding almost coalesced into a cultural paradigm in the Woodstock Music Festival, where people actually lived for a few days in an understanding of being in which mainline contemporary concern with rationality, sobriety, willful activity, and flexible, efficient control were made marginal and subservient to Greek virtues such as openness, enjoyment of nature, dancing, and Dionysian ecstasy along with a neglected Christian concern with peace, tolerance, and love of one's neighbor without desire and exclusivity. Technology was not smashed or denigrated but all the power of the electronic media was put at the service of the music which focused all the above concerns.

If enough people had found in Woodstock what they most cared about, and recognized that all the others shared this recognition, a new understanding of being might have coalesced and been stabilized. Of course, in retrospect we see that the concerns of the Woodstock generation were not broad and deep enough to resist technology and to sustain a culture. Still we are left with a hint of how a new cultural paradigm would work, and the realization that we must foster human receptivity and preserve the endangered species of pre-technological practices that remain in our culture, in the hope that one day they will be pulled together into a new paradigm, rich enough and resistant enough to give new meaningful directions to our lives.

To many, however, the idea of *a* god which will give us a unified but open community—one set of concerns which everyone shares if only as a focus of disagreement—sounds either unrealistic or dangerous. Heidegger would probably agree that its open democratic version looks increasingly unobtainable and that we have certainly seen that its closed totalitarian form can be disastrous. But Heidegger holds that given our historical essence—the kind of beings we have become during the history of our culture—such a community is necessary to us. This raises the question of whether our need for one community is, indeed, dictated by our historical essence, or whether the claim that we can't live without a centered and rooted culture is simply romantic nostalgia.

It is hard to know how one could decide such a question, but Heidegger has a message even for those who hold that we, in this pluralized modern world, should not expect and do not need one all-embracing community. Those who, from Dostoievsky, to the hippies, to Richard Rorty, think of communities as local enclaves in an otherwise impersonal society still owe us an account of what holds these local communities together. If Dostoievsky and Heidegger are right, each local community still needs its local god—its particular incarnation of what the community is up to. In that case we are again led to the view that releasement is not enough, and to the modified Heideggerian slogan that only some new *gods* can save us.

NOTES

1. Martin Heidegger, "The Question Concerning Technology," *The Question Concerning Technology* (New York: Harper Colophon, 1977), pp. 25–26.

2. Ibid.

3. Heidegger, "Overcoming Metaphysics," *The End of Philosophy* (New York: Harper and Row, 1973), p. 107.

4. Heidegger, *Discourse on Thinking* (New York: Harper and Row, 1966), p. 50.

5. Ibid., p. 48.

6. Heidegger, "The Turning," *The Question Concerning Technology*, p. 48.

7. Heidegger, "The Question Concerning Technology," *The Question Concerning Technology*, p. 5.

8. Heidegger, *Discourse on Thinking*, p. 52.

9. Martin Heidegger, "What Are Poets For?" *Poetry, Language, Thought* (New York: Harper and Row, 1971), p. 116.

10. Heidegger, *Discourse on Thinking*, p. 56.

11. Heidegger, "The Origin of the Work of Art," *Poetry, Language, Thought*, p. 53.

12. Heidegger, "The Question Concerning Technology," p. 5.

13. Ibid.

14. Ibid., p. 15.

15. Ibid., p. 17.

16. Heidegger, "Science and Reflection," *The Question Concerning Technology*, p. 173.

17. Heidegger, "The Question Concerning Technology," p. 17.

18. Ibid.

19. Heidegger, "Overcoming Metaphysics," *The End of Philosophy*, p. 104.

20. Heidegger, *Discourse on Thinking*, p. 53.

21. Ibid., p. 54.

22. Heidegger, "The Turning," *The Question Concerning Technology*, p. 39.

23. Heidegger, *Poetry, Language, Thought*, pp. 152–53.

24. Ibid., p. 44.

25. Heidegger, "The Turning," *The Question Concerning Technology*, p. 39.

26. Heidegger, *Discourse on Thinking*, p. 54.

27. Ibid., p. 55.

28. Ibid. (My italics.)

29. Ibid., p. 56.

30. Ibid.

31. Heidegger, "The Turning," *The Question Concerning Technology*, p. 47.

32. "Nur noch ein Gott kann uns retten," *Der Spiegel*, May 31, 1976.

33. Heidegger, "The Origin of the Work of Art," *Poetry, Language, Thought*, p. 61.

34. This is an equally possible translation of the famous phrase from *Der Spiegel*.

3

The New Forms of Control

Herbert Marcuse

$\mathbf{A}$ comfortable, smooth, reasonable, democratic un-freedom prevails in advanced industrial civilization, a token of technical progress. Indeed, what could be more rational than the suppression of individuality in the mechanization of socially necessary but painful performances; the concentration of individual enterprises in more effective, more productive corporations; the regulation of free competition among unequally equipped economic subjects; the curtailment of prerogatives and national sovereignties which impede the international organization of resources. That this technological order also involves a political and intellectual coordination may be a regrettable and yet promising development.

The rights and liberties which were such vital factors in the origins and earlier stages of industrial society yield to a higher stage of this society: they are losing their traditional rationale and content. Freedom of thought, speech, and conscience were—just as free enterprise, which they serves to promote and protect—essentially *critical* ideas, designed to replace an obsolescent material and intellectual culture by a more productive and rational one. Once institutionalized, these rights and liberties shared the fate of the society of which they had become an integral part. The achievement cancels the premises.

To the degree to which freedom from want, the concrete substance of all freedom, is becoming a real possibility, the liberties which pertain to a state of lower productivity are losing their former content. Independence of thought, autonomy, and the right to political opposition are being deprived of their basic critical function in a society which seems increasingly capable of satisfying the needs of the individuals through the way in which it is organized. Such a society may justly demand acceptance of its principles and institutions, and reduce the opposition to the discussion and promotion of alternative policies *within* the status quo. In this respect, it seems to make little difference whether the increasing satisfaction of needs is accomplished by an authoritarian or a non-authoritarian system. Under the conditions of the rising standard of living, non-conformity with the system itself appears to be socially useless, and the more so when it entails tangible economic and political disadvantages and threatens the smooth operation of the whole. Indeed, at least insofar as the necessities of life are involved, there seems to be no reason why the production and distribution of goods and services should proceed through the competitive concurrence of individual liberties.

Freedom of enterprise was from the beginning not altogether a blessing. As the liberty to work or to starve, it spelled toil, insecurity, and fear for the vast majority of the population. If the individual were no longer compelled to prove himself on the market, as a free economic subject, the disappearance of this kind of freedom would be one of the greatest achievements of civilization. The technological processes of mechanization and standardization might release individual energy into a yet uncharted realm of freedom beyond necessity. The very structure of human existence would be altered; the individual would be liberated from the work world's imposing upon him alien needs and alien possibilities. The individual would be free to exert autonomy over a life that would be his own. If the productive apparatus could be organized and directed toward the satisfaction of the vital needs, its control might well be centralized; such control would not prevent individual autonomy, but render it possible.

This is a goal within the capabilities of advanced industrial civilization, the "end" of technological rationality. In actual fact, however, the contrary trend operates: the apparatus imposes its economic and political requirements for defense and expansion on labor time and free time, on the material and intellectual culture. By virtue of the way it has organized its technological base, contemporary industrial society tends to be totalitarian. For "totalitarian" is not only a terroristic political coordination of society, but also a non-terroristic economic-technical coordination which operates through the manipulation of needs by vested interests. It thus precludes the emergence of an effective opposition against the whole. Not only a specific form of government or party rule makes for totalitarianism, but also a specific system of production and distribution which may well be compatible with a "pluralism" of parties, newspapers, "countervailing powers," et cetera.

Today political power asserts itself through its power over the machine process and over the technical organization of the apparatus. The government of advanced and advancing industrial societies can maintain and secure itself only when it succeeds in mobilizing, organizing, and exploiting the technical, scientific, and mechanical productivity available to industrial civilization. And this productivity mobilizes society as a whole, above and beyond any particular individual or group interests. The brute fact that the machine's physical (only physical?) power surpasses that of the individual, and of any particular group of individuals, makes the machine the most effective political instrument in any society whose basic organization is that of the machine process. But the political trend may be reversed; essentially the power of the machine is only the stored-up and projected power of man. To the extent to which the work world is conceived of as a machine and mechanized accordingly, it becomes the *potential* basis of a new freedom for man.

Contemporary industrial civilization demonstrates that it has reached the stage at which "the free society" can no longer be adequately defined in the traditional terms of economic, political, and intellectual liberties, not because these liberties have become insignificant, but because they are too significant to be confined within the traditional forms. New modes of realization are needed, corresponding to the new capabilities of society.

Such new modes can be indicated only in negative terms because they would amount to the negation of the prevailing modes. Thus economic freedom would mean freedom *from* the economy—from being controlled by economic forces and relationships; freedom from the daily struggle for existence, from earning a living. Political freedom would mean liberation of the individuals *from* politics over which they have no effective control. Similarly, intellectual freedom would mean the restoration of individual thought now absorbed by mass communication and indoctrination, abolition of "public opinion" together with its makers. The unrealistic sound of these propositions is indicative, not of their utopian character, but of the strength of the forces which prevent their realization. The most effective and enduring form of warfare against liberation is the implanting of material and intellectual needs that perpetuate obsolete forms of the struggle for existence.

The intensity, the satisfaction and even the character of human needs, beyond the biological level, have always been preconditioned. Whether or not the possibility of doing or leaving, enjoying or destroying, possessing or rejecting something is seized as a *need* depends on whether or not it can be seen as desirable and necessary for the prevailing societal institutions and interests. In this sense, human needs are historical needs and, to the extent to which the society demands the repressive development of the individual, his needs themselves and their claim for satisfaction are subject to overriding critical standards.

We may distinguish both true and false needs. "False" are those which are superimposed upon the individual by particular social interests in his repression: the needs which perpetuate toil, aggressiveness, misery, and injustice. Their satisfaction might be most gratifying to the individual, but this happiness is not a condition which has to be maintained and protected if it serves to arrest the development of the ability (his own and others) to recognize the disease of the whole and grasp the chances of curing the disease. The result then is euphoria in unhappiness. Most of the prevailing needs to relax, to have fun, to behave and consume in accordance with the advertisements, to love and hate what others love and hate, belong to this category of false needs.

Such needs have a societal content and function which are determined by external powers over which the individual has no control; the development and satisfaction of these needs is heteronomous. No matter how much such needs may have become the individual's own, reproduced and fortified by the conditions of his existence; no matter how much he identifies himself with them and finds himself in their satisfaction, they continue to be what they were from the beginning—products of a society whose dominant interests demands repression.

The prevalence of repressive needs is an accomplished fact, accepted in ignorance and defeat, but a fact that must be undone in the interest of any happy individual as well as all those whose misery is the price of his satisfaction. The only needs that have an unqualified claim for satisfaction are the vital ones—nourishment, clothing, lodging at the attainable level of culture. The satisfaction of these needs is the prerequisite for the realization of *all* needs, of the unsublimated as well as the sublimated ones.

For any consciousness and conscience, for any experience which does not accept the prevailing societal interest as the supreme law of thought and behavior, the established universe of needs and satisfactions is a fact to be questioned—questioned in terms of truth and falsehood. These terms are historical. The judgment of needs and their satisfaction, under the given conditions, involves standards of *priority*—standards which refer to the optimal development of the individual, of all individuals, under the optimal utilization of the material and intellectual resources available to man. The resources are calculable. "Truth" and "falsehood" of needs designate objective conditions to the extent to which the universal satisfaction if vital needs and, beyond it, the progressive alleviation of toil and poverty, are universally valid standards. But as historical standards, they do not only vary according to area and stage of development, they also can be defined only in (greater or lesser) *contradiction* to the prevailing ones. What tribunal can possibly claim the authority of decision?

In the last analysis, the question of what are true and false needs must be answered by the individuals themselves, but only in the last analysis; that is, if and when they are free to give their own answer. As long as they are kept incapable of being autonomous, as long as they are indoctrinated and manipulated (down to their very instincts), their answer to this question cannot be taken as their own. By the same token, however, no tribunal can justly arrogate to itself the right to decide which needs should be developed and satisfied. Any such tribunal is reprehensible, although our revulsion does not do away with the question: how can the people who have been the object of effective and productive domination by themselves create the conditions of freedom?

The more rational, productive, technical, and total the repressive administration of society becomes, the more unimaginable the means and ways by which the administered individuals might break their servitude and seize their own liberation. To be sure, to impose Reason upon an entire society is a paradoxical and scandalous idea—although one might dispute the righteousness of a society which ridicules this idea while making its own population into objects of total administration. All liberation depends on the consciousness of servitude, and the emergence of this consciousness is always hampered by the predominance of needs and satisfactions which, to a great extent, have become the individual's own. The process always replaces one system of preconditioning by another; the optimal goal is the replacement of false needs by true ones, the abandonment of repressive satisfaction.

The distinguishing feature of advanced industrial society is its effective suffocation of those needs which demand liberation—liberation also from that which is tolerable and rewarding and comfortable—while it sustains and absolves the destructive power and repressive function of the affluent society. Here, the social controls exact the overwhelming need for the production and consumption of waste; the need for stupefying work where it is no longer a real necessity; the need for modes of relaxation which soothe and prolong this stupefication; the need for maintaining such deceptive liberties as free competition at administered prices, a free press which censors itself, free choice between brands and gadgets.

Under the rule of a repressive whole, liberty can be made into a powerful instrument of domination. The range of choice open to the individual is not the decisive factor in determining the degree of human freedom, but *what* can be chosen and what *is* chosen by the individual. The criterion for free choice can never be an absolute one, but neither is it entirely relative. Free election of masters does not abolish the masters or the slaves. Free choice among a wide variety of goods and services does not signify freedom if these goods and services sustain social controls over a life of toil and fear—that is, if they sustain alienation. And the spontaneous reproduction of superimposed needs by the individual does not establish autonomy; it only testifies to the efficacy of the controls.

Our insistence on the depth and efficacy of these controls is open to the objection that we overrate greatly the indoctrinating power of the "media," and that by themselves the people would feel and satisfy the needs which are now imposed upon them. The objection misses the point. The preconditioning does not start with the mass production of radio and television and with the centralization of their control. The people enter this stage as preconditioned receptacles of long-standing; the decisive differences is in the flattening out of the contrast (or conflict) between the given and the possible, between the satisfied and the unsatisfied needs. Here, the so-called equalization of class distinctions reveals its ideological function. If the worker and his boss enjoy the same television program and visit the same resort places, if the typist is as attractively made up as the daughter of her employer, if the Negro owns a Cadillac, if they all read the same newspaper, then this assimilation indicates not the disappearance of classes, but the extent to which the needs and satisfactions that serve the preservation of the Establishment are shared by the underlying population.

Indeed, in the most highly developed areas of contemporary society, the transplantation of social into individual needs is so effective that the difference between them seems to be purely theoretical. Can one really distinguish between mass media as instruments of information and entertainment, and as agents of manipulation and indoctrination? Between the automobile as nuisance and as convenience? Between the horrors and the comforts of functional architecture? Between the work for national defense and the work for corporate gain? Between the private pleasure and the commercial and political utility involved in increasing the birth rate?

We are again confronted with one of the most vexing aspects of advanced industrial civilization: the rational character of its irrationality. Its productivity and efficiency, its capacity to increase and spread comforts, to turn waste into need, and destruction into construction, the extent to which this civilization transforms the object world into are extension of man's mind and body makes the very notion of alienation questionable. The people recognize themselves in their commodities; they find their soul in their automobile, hi-fi set, split level home, kitchen equipment. The very mechanism which ties the individual to his society has changed, and social control is anchored in the new needs which it has produced.

The prevailing forms of social control are technological in a new sense. To be sure, the technical structure and efficacy of the productive and destructive apparatus has been a major instrumentality for subjecting the population to the established social division of labor throughout the modern period. Moreover, such integration has always been accompanied by more obvious forms of compulsion: loss of livelihood, the administration of justice, the police, the armed forces. It still is. But in the contemporary period, the technological controls appear to be the very embodiment of Reason for the benefit of all social groups and interests—to such an extent that all contradiction seems irrational and all counteraction impossible.

No wonder then that, in the most advanced areas of this civilization, the social controls have been introjected to the point where even individual protest is affected at its roots. The intellectual and emotional refusal "to go along" appears neurotic and impotent. This is the socio-psychological aspect of the political event that marks the contemporary period: the passing of the historical forces which, at the preceding stage of industrial society, seemed to represent the possibility of new forms of existence.

But the term "introjection" perhaps no longer describes the way in which the individual by himself reproduces and perpetuates the external controls exercised by his society. Introjection suggests a variety of relatively spontaneous processes by which a Self (Ego) transposes the "outer" into the "inner." Thus introjection implies the existence of an inner dimension distinguished from and even antagonistic to the external exigencies—an individual consciousness and an individual unconscious *apart from* public opinion and behavior.[1] The idea of "inner freedom" here has its reality: it designates the private space in which man may become and remain "himself."

Today this private space has been invaded and whittled down by technological reality. Mass production and mass distribution claim the *entire* individual, and industrial psychology has long since ceased to be confined to the factory. The manifold processes of introjection seem to be ossified in almost mechanical reactions. The result is, not adjustment but *mimesis:* an immediate identification of the individual with *his* society and, through it, with the society as a whole.

This immediate, automatic identification (which may have been characteristic of primitive forms of association) reappears in high industrial civilization; its new "immediacy," however, is the product of a sophisticated, scientific management and organization. In this process, the "inner" dimension of the mind in which opposition to the status quo can take root is whittled down. The loss of this dimension, in which the power of negative thinking—the critical power of Reason—is at home, is the ideological counterpart to the very material process in which advanced industrial society silences and reconciles the opposition. The impact of progress turns Reason into submission to the facts of life, and to the dynamic capability of producing more and bigger facts of the same sort of life. The efficiency of the system blunts the individuals' recognition that it contains no facts which do not communicate the repressive power of the whole. If the individuals find themselves in the things which shape their life, they do so, not by giving, but by accepting the law of things—not the law of physics but the law of their society.

I have just suggested that the concept of alienation seems to become questionable when the individuals identify themselves with the existence which is imposed upon them and have in it their own development and satisfaction. This identification is not illusion but reality. However, the reality constitutes a more progressive stage of alienation. The latter has become entirely objective; the subject which is alienated is swallowed up by its alienated existence. There is only one dimension, and it is everywhere and in all forms. The achievements of progress defy ideological indictment as well as justification; before their tribunal, the "false consciousness" of their rationality becomes the true consciousness.

This absorption of ideology into reality does not, however, signify the "end of ideology." On the contrary, in a specific sense advanced industrial culture is *more* ideological than its predecessor, inasmuch as today the ideology is in the process of production itself.[2] In a provocative form, this proposition reveals the political aspects of the prevailing technological rationality. The productive apparatus and the goods and services which it produces "sell" or impose the social system as a whole. The means of mass transportation and communication, the commodities of lodging, food, and clothing, the irresistible output of the entertainment and information industry carry with them prescribed attitudes and habits, certain intellectual and emotional reactions which bind the consumers more or less pleasantly to the producers and, through the latter, to the whole. The products indoctrinate and manipulate; they promote a false consciousness which is immune against its false-hood. And as these beneficial products become available to more individuals in more social classes, the indoctrination they carry ceases to be publicity; it becomes a way of life. It is a good way of life—much better than before—and as a good way of life, it militates against qualitative change. Thus emerges a pattern of *one-dimensional thought and behavior* in which ideas, aspirations, and objectives that, by their content, transcend the established universe of discourse and action are either repelled or reduced to terms of this universe. They are redefined by the rationality of the given system and of its quantitative extension.

The trend may be related to a development in scientific method: operationalism in the physical, behaviorism in the social sciences. The common feature is a total empiricism in the treatment of concepts; their meaning is restricted to the representation of particular operations and behavior. The operational point of view is well illustrated by P. W. Bridgman's analysis of the concept of length:[4]

> We evidently know what we mean by length if we can tell what the length of any and every object is, and for the physicist nothing more is required. To find the length of an object, we have to perform certain physical operations. The concept of length is therefore fixed when the operations by which length is measured are fixed: that is, the concept of length involves as much and nothing more than the set of operations by which length is determined. In general, we mean by any concept nothing more than a set of operations; *the concept is synonymous with the corresponding set of operations.*

Bridgman has seen the wide implications of this mode of thought for the society at large:

> To adopt the operational point of view involves much more than a mere restriction of these senses in which we understand "concept," but means a far-reaching change in all our habits of thought, in that we shall no longer permit ourselves to use as tools in our thinking concepts of which we cannot give an adequate account in terms of operations.

Bridgman's prediction has come true. The new mode of thought is today the predominant tendency in philosophy, psychology, sociology, and other fields. Many of the most seriously troublesome concepts are being "eliminated" by showing that no adequate account of them in terms of operations or behavior can be given. The radical empiricist onslaught . . . thus provides the methodological justification for the debunking of the mind by the intellectuals—a positivism which, in its denial of the transcending elements of Reason, forms the academic counterpart of the socially required behavior.

Outside the academic establishment, the "far-reaching change in all our habits of thought" is more serious. It serves to coordinate ideas and goals with those exacted by the prevailing system, to enclose them in the system, and to repel those which are irreconcilable with the system. The reign of such a one-dimensional reality does not mean that materialism rules, and that the spiritual, metaphysical, and bohemian occupations are petering out. On the contrary, there is a great deal of "Worship together this week," "Why not try God," Zen, existentialism, and beat ways of life, etc. But such modes of protest and transcendence are no longer negative. They are rather the ceremonial part of practical behaviorism, its harmless negation, and are quickly digested by the status quo as part of its healthy diet.

One-dimensional thought is systematically promoted by the makers of politics and their pur-veyors of mass information. Their universe of discourse is populated by self-validating hypotheses which, incessantly and monopolistically repeated, become hypnotic definitions or dictations. For example, "free" are the institutions which operate (and are operated on) in the countries of the Free World; other transcending modes of freedom are, by definition, either anarchism, communism, or propaganda. "Socialistic" are all encroachments on private enterprises not undertaken by private enterprise itself (or by government contracts), such as universal and comprehensive health insur-ance, or the protection of nature from all too sweeping commercialization, or the establishment of public services which may hurt private profit. This totalitarian logic of accomplished facts has its Eastern counterpart. There, freedom is the way of life instituted by a communist regime, and all other transcending modes of freedom are either capitalistic, or revisionist, or leftist sectarianism. In both camps, non-operational ideas are non-behavioral and subversive. The movement of thought is stopped at barriers which appear as the limits of Reason itself.

Such limitation of thought is certainly not new. Ascending modern rationalism, in its specula-tive as well as empirical form, shows a striking contrast between extreme critical radicalism in scientific and philosophic method on the one hand, and an uncritical quietism in the attitude toward established and functioning social institutions. Thus Descartes' *ego cogitans* was to leave the "great public bodies" untouched, and Hobbes held that "the present ought always to be preferred, main-tained, and accounted best." Kant agreed with Locke in justifying revolution *if and when* it has succeeded in organizing the whole and in preventing subversion.

However, these accommodating concepts of Reason were always contradicted by the evident misery and injustice of the "great public bodies" and the effective, more or less conscious rebellion against them. Societal conditions existed which provoked and permitted real dissociation from the established state of affairs; a private as well as political dimension was present in which dissocia-tion could develop into effective opposition, testing its strength and the validity of its objectives.

With the gradual closing of this dimension by the society, the self-limitation of thought assumes a larger significance. The interrelation between scientific-philosophical and societal proc-esses, between theoretical and practical Reason, asserts itself "behind the back" of the scientists and philosophers. The society bars a whole type of oppositional operations and behavior; conse-quently, the concepts pertaining to them are rendered illusory or meaningless. Historical transcen-dence appears as metaphysical transcendence, not acceptable to science and scientific thought. The operational and behavioral point of view, practiced as a "habit of thought" at large, becomes the view of the established universe of discourse and action, needs and aspirations. The "cunning of Reason" works, as it so often did, in the interest of the powers that be. The insistence on operational and behavioral concepts turns against the efforts to free thought and behavior *from* the given reality and *for* the suppressed alternatives. Theoretical and practical Reason, academic and social behaviorism meet on common ground: that of an advanced society which makes scientific and technical progress into an instrument of domination.

"Progress" is not a neutral term; it moves toward specific ends, and these ends are defined by the possibilities of ameliorating the human condition. Advanced industrial society is approaching the stage where continued progress would demand the radical subversion of the prevailing direction and organization of progress. This stage would be reached when material production (including the necessary services) becomes automated to the extent that all vital needs can be satisfied while necessary labor time is reduced to marginal time. From this point on, technical progress would transcend the realm of domination and exploitation which thereby limited its rationality; technology would become subject to the free play of faculties in the struggle for the pacification of nature and of society.

Such a state is envisioned in Marx's notion of the "abolition of labor." The term "pacification of existence" seems better suited to designate the historical alternative of a world which—through an international conflict which transforms and suspends the contradictions within the established societies—advances on the brink of a global war. "Pacification of existence" means the development of man's struggle with man and with nature, under conditions where the competing needs, desires, and aspirations are no longer organized by vested interests in domination and scarcity—an organization which perpetuates the destructive forms of this struggle.

Today's fight against this historical alternative finds a firm mass basis in the underlying population, and finds its ideology in the rigid orientation of thought and behavior to the given universe of facts. Validated by the accomplishments of science and technology, justified by its growing productivity, the status quo defies all transcendence. Faced with the possibility of pacification on the grounds of its technical and intellectual achievements, the mature industrial society closes itself against this alternative. Operationalism, in theory and practice, becomes the theory and practice of *containment*. Underneath its obvious dynamics, this society is a thoroughly static system of life: self-propelling in its oppressive productivity and in its beneficial coordination. Containment of technical progress goes hand in hand with its growth in the established direction. In spite of the political fetters imposed by the status quo, the more technology appears capable of creating the conditions for pacification, the more are the minds and bodies of man organized against this alternative.

The most advanced areas of industrial society exhibited throughout these two features: a trend toward consummation of technological rationality, and intensive efforts to contain the trend within the established institutions. Here is the internal contradiction of this civilization: the irrational element in its rationality. It is the token of its achievements. The industrial society which makes technology and science its own is organized for ever-more-effective domination of man and nature, for the ever-more-effective utilization of its resources. It becomes irrational when the success of these efforts opens new dimensions of human realization. Organization for peace is different from organization for war; the institutions which served the struggle for existence cannot serve the pacification of existence. Life as an end is qualitatively different from life as a means.

Such a qualitatively new mode of existence can never be envisaged as the mere by-product of economic and political changes, as the more or less spontaneous effect of the new institutions which constitute the necessary prerequisite. Qualitative change also involves a change in the *technical* basis on which this society rests—one which sustains the economic and political institutions through which the "second nature" of man as an aggressive object of administration is stabilized. The techniques of industrialization are political techniques; as such, they prejudge the possibilities of Reason and Freedom.

To be sure, labor must precede the reduction of labor, and industrialization must precede the development of human needs and satisfactions. But as all freedom depends on the conquest of alien necessity, the realization of freedom depends on the *techniques* of this conquest. The highest productivity of labor can be used for the perpetuation of labor, and the most efficient industrialization can serve the restriction and manipulation of needs.

When this point is reached, domination—in the guise of affluence and liberty—extends to all spheres of private and public existence, integrates all authentic opposition, absorbs all alternatives. Technological rationality reveals its political character as it becomes the great vehicle of better domination, creating a truly totalitarian universe in which society and nature, mind and body are kept in a state of permanent mobilization for the defense of this universe.

NOTES

1. The change in the function of the family here plays a decisive role: its "socializing" functions are increasingly taken over by outside groups and media. See my *Eros and Civilization* (Boston: Beacon Press, 1955), p. 96 ff.

2. Theodor W. Adorno, *Prismen. Kulturkritif und Gesellschaft* (Frankfort: Suhrkamp, 1955), p. 24 f.

3. P. W. Bridgman, *The Logic of Modern Physics* (New York: Macmillan, 1928), p. 5. The operational doctrine has since been refined and qualified. Bridgman himself has extended the concept of "operation" to include the "paper-and-pencil" operations of the theorist (in Philipp J. Frank, *The Validation of Scientific Theories* [Boston: Beacon Press, 1954], chap. II). The main impetus remains the same: it is "desirable" that the paper-and-pencil operations "be capable of eventual contact, although perhaps indirectly, with instrumental operations."

4. P. W. Bridgman, *The Logic of Modern Physics,* loc. cit., p. 31.

4

John Dewey as a Philosopher of Technology

Larry Hickman

The reigning historian of the philosophy of technology, Carl Mitcham, has written that the first publication in the field was Friedrich Dessauer's *Philosophie der Technik*, published in 1927.[1] That year also marked the appearance of Martin Heidegger's *Sein und Zeit* (*Being and Time*), which is widely accepted as the first major contribution to the field. Works on the subject by Ernst Jünger in 1932 and by José Ortega y Gasset in 1939 quickly followed.

Until recently, however, no one seemed to notice that American philosophy, or more specifically classical American pragmatism, had also made a solid contribution to the field.[2] I have argued that John Dewey's treatments of education, aesthetics, social and political philosophy, logic, and the philosophy of nature should also be read as contributions to a cultural critique of technology.[3] Some twenty years prior to the publication of the works of Dessauer and Heidegger, Dewey was already writing about a whole range of topics that today are considered central concerns within the philosophy of technology. Later, Dewey's books *Essays in Experimental Logic* (1916), *Experience and Nature* (1925), and *Art as Experience* (1934) all contained incisive critiques of technological culture.

To put this matter in perspective, it may help to recall that Dewey was born in 1859, the year of America's first successful oil well in Titusville, Pennsylvania, and the publication of Darwin's *Origin of Species*. He died in 1952, the year of the first hydrogen bomb test and the first mass marketing of the birth control pill. Dewey's ninety-two years thus spanned two major technological revolutions in America. At the time of his birth, America's economy was based to a great extent on wind, water, and wood. As he grew to maturity, he observed the shift to an economy of steel, coal, and steam. At the time of his death, America had entered the age of synthetics, electronics, and nuclear energy. The post-industrial society in which we now live was already present in rudimentary form.

I draw attention to these details because Dewey's work as philosopher of technology is of more than just historical interest. His analysis of human experience as transactional with, and within, its various overlapping contexts holds the promise of stimulating new ways of thinking about many of the concerns—especially the ones that involve our environment—that have only recently received the attention of professional philosophers.

The key to understanding Dewey's work as a contribution to the philosophy of technology is, I suggest, an appreciation of his contention that all inquiry or deliberation that involves tools and artifacts, whether those tools and artifacts be abstract or concrete, tangible or intangible, should be viewed as instrumental: in other words, as a form of technology. In short, he understood that technology involves more than just tangible tools, machines, and factories. It also involves the abstract thought and cultural practices that provide the contexts for such things and make them possible. His view of this matter was based upon his broad characterization of technology, which served as the basis for the functional taxonomy of types of activity that I developed earlier and that may also be formulated as the *invention, development, and cognitive deployment of tools and other artifacts, brought to bear on raw materials and intermediate stock parts, with a view to the resolution of perceived problems.*[4]

This is my gloss on thousands of words that Dewey devoted to his characterization of technology. It is also quite close to his statement, provided as the epigraph to this chapter, that "technology" signifies all the intelligent techniques by which the energies of nature and man are directed and used in satisfaction of human needs; it cannot be limited to a few outer and comparatively mechanical forms. In the face of its possibilities, the traditional conception of experience is obsolete" (LW.5.270).

It might be objected that this characterization begs the question by identifying technology with "intelligent techniques." But what Dewey in fact accomplished by putting matters as he did was the very distinction between technology and technique that I attempted to work out earlier. He was also distinguishing between cases in which it appears that technology is being done but in which in fact something else, such as economic self-interest, has intervened. On this radical view, when such interventions occur, it is intelligence itself that suffers. I shall later discuss the factors that led to the resurgence of rubella as a public health problem in the 1980s as an example of just such a failure of intelligence.

Dewey's view of these matters constitutes a radical departure from the epistemology of the modern period of philosophy. At least since Descartes it had been generally accepted that the central problem of epistemology was the problem of skepticism: how is it that we can have certain or reliable knowledge of the world? Although the story of modern epistemology is long and complex, certain of its features stand out in high profile. As Descartes and other modern philosophers attempted to move out from under the influence of medieval scholastic thought, they faced the difficulty of constructing a foundation for science that offered the same level of certitude that scholasticism had claimed. Since their move was toward naturalism, however, they were obligated to locate certitude within nature, as opposed to the supernatural.

The best recourse seemed to Descartes and others to treat certainty as knowledge possessed by an *individual thinking mind.* Modern theories of knowledge and belief were thus designed to find ways of depicting states of affairs in a world that was assumed to exist separately from a thinking mind, and this in a way that would ensure that such depictions were reliable.

Like the late-nineteenth-century photographers who attempted to get ever better emulsions for ever more accurate photographs of a world outside and independent of their cameras, these epistemologists were attempting to get ever more accurate mental representations of a world that they thought was outside and independent of their minds. They characterized that world not just as independent of mind, but also as whatever it was without respect to whether or not it would ever be known by an individual mind. Now, some 350 years later, some epistemologists and philosophers of science are still doing this.

Dewey thought that this "picture theory" or "spectator theory" or knowledge was deeply flawed. He reasoned that knowing is not just the capturing of a picture or impression, but an active and experimental involvement of an entire organism (not just a "thinking substance" or even a brain) with the raw materials of its experience in such a manner that tools—including habits and concepts, for example—are brought to bear on those materials and new products are formed. And he thought that the point of making these new products was not to take a more accurate picture representation of what was or had been the case (an external "state of affairs"), but rather to deal with felt problems and difficulties in ways that effected their resolution. He thought that inquiry is always launched for the sake of resolving some specific felt difficulty. When inquiry is successful, he argued, it produces a new product—a new outcome.

For Dewey there is no such thing as knowledge in general, but the production of new knowledge in specific cases, ranging from the most quotidian to the most abstract, involves technology just as surely as cases of problem-solving in chemical engineering. This is because we live forward in time in a world that is perilous at best and in continual need of being "tuned up." We have to keep turning out new knowledge-products, including new tools and methods, if we are to convert conditions that range all the way from what is merely irritating to what is life-threatening into situations that are stable, harmonious, and more nearly what we wish them to be.

For Dewey, therefore, one of the most important concerns of philosophy was not so much epistemology, or the attempt to deal with the problem of skepticism, but logic, or the theory of inquiry. Inquiry, he once wrote, is not so much a matter of "grasping antecedently given sureties" as it is a matter of experimentation, or "making sure" (LW.1.123).

Unlike modernist epistemology, Dewey's notion of inquiry emphasizes the use of raw materials and the tools that have been designed for the refinement of those materials. It also involves other tools whose purpose it is to refine and reconstruct tools that already exist, but that are simpler and more primitive. Inquiry also requires the production and stockpiling of intermediate parts, among which are relatively secure concepts and objects. The end or goal of inquiry is products that can be said to be finished in a relative sense of that term, that is, satisfactory until they are challenged by further experience and demonstrated to be in need of reworking or reconstruction.

It was by means of this view of the instrumental or productive role and function of inquiry in human experience that Dewey avoided the problems that had vitiated the work of many of his predecessors. His view avoids the problems of the empiricism advanced by John Locke, for example, since the central place that his instrumentalism gives to production allows it to undercut both the sensory atomism and the associationism on which such empiricism depends. The problem with putative sensory atoms, Dewey argued, is that they are not primitive at all. They are the products of reflection. And the problem with associationism is that its associations tend to be arbitrary if they are based on nothing more than an arrangement of sensory atoms.

His view avoids the difficulties of Cartesian rationalism, moreover, by treating productive inquiry as a public, observable enterprise that takes place within a community, and not as something that takes place within private, non-extended, albeit reified mind. Dewey called inquiry "an outdoor fact," and thought it no less natural and observable than activities such as chewing or walking.

It also avoids the pitfalls generated by the Kantian treatment of knowledge, especially the view that perceptual and conceptual contents have different origins, by treating perceptual and conceptual materials as functional aspects of ongoing inquiry, even as different portions or aspects of judgments. In Dewey's view, the perceptual is concerned with marking out and locating a problem in inquiry, whereas the conceptual is concerned with setting out possible methods of solution. That both types of materials function correlatively within organized inquiry is apparent from the structure of judgments, whose subjects, Dewey pointed out, tend to be perceptual and whose predicates tend to be conceptual.

Dewey worked out his extended technological metaphor for inquiry at great length in the introduction to his 1916 *Essays in Experimental Logic*. That essay is pervaded by technological figures. Here is a typical example:

Hence, while all meanings are derived from things which antedate suggestion or thinking or "consciousness"—not all qualities are equally fitted to be meanings of a wide efficiency, and it is a work of art to select the proper qualities for doing the work. This corresponds to the working over of raw material into an effective tool. A spade or a watch-spring is made out of antecedent material, but does not pre-exist as a ready-made tool; and the more delicate and complicated the work which it has to do, the more art intervenes. (MW.10.354)

In the same essay Dewey asserted that "there is no problem of why and how the plow fits, or applies to, the garden, or the watch-spring to time-keeping. They were made for those respective purposes; the question is how well they do their work, and how they can be reshaped to do it better" (MW.10.354–55).

This passage contains several points that are important to the issue at hand, namely the relevance of philosophy as a tool for tuning up technological culture.

First, Dewey wanted to demystify those entities traditionally called "logical objects," "essences," and "ideals," by taking them out of the psychical or meta-physical realms they had occupied in the works of Plato and Frege, for example, and by treating them as so many tools in a toolbox. These tools include logical connectives and numbers, abstract terms such as "democracy," and essences such as "the family."[5] When it is understood that these entities are tools and the products of tools, then it will also be understood that they are open to reconstruction and reconfiguration. They will not be honored as essences that are deemed to be fixed and finished for all time.

Since Dewey's program is radical, its application would involve certain casualties. Among the big losers, to name just a few examples, would be Platonism in mathematics and the doctrine of original intent in constitutional law. This is because each of these positions, as it is usually articulated, depends upon the premise that its respective essence or ideal is absolute and fixed, and not instrumental and consequently in need of continuing reconstruction as circumstances dictate.

So Dewey argued that essences and ideals should be treated not as absolute and fixed, but instead as just more artifacts, constructed not so much *by* inquiry as arising *from* inquiry. They are not found within a chain of inference, but are instead the by-products of inference. In this way they are like agricultural implements that are developed and improved not as a direct consequence of farming but incidentally, as the by-products of tilling, planting, and harvesting.

In all this Dewey was developing a metaphor that would allow him to bring the various types of inquiry we term "successful" under one general formula. He worked out what was already implicit in the work of his fellow pragmatists Charles Sanders Peirce and William James. For those philosophers, all successful inquiry is productive of new outcomes that are more secure than the situations that occasioned the inquiry that produced them. This is true in the sciences, in the arts, in engineering, in agriculture, and in quotidian or everyday enterprises as well.

As Dewey argued in his 1938 *Logic*, the subject matter and the specific tactical methods of inquiry may be, and most likely are, different from one of these enterprises to the next; but each enterprise nevertheless participates within a more general strategic form of inquiry that he called the "general method of intelligence." Because his root metaphor was technological, however, Dewey was able to do explicitly what Peirce and James had done only implicitly. He was able, for example, to reconstruct the important categories of human activity traditionally termed "theory," "practice," and "production."

He did this by reconstructing the Aristotelian hierarchy of types of knowledge. Aristotle had lived in a world in which science was still only empirical and not yet experimental. In other words, Aristotle's science was observational, and not yet instrumental. Instrumentation was not yet viewed as an essential ingredient in science, nor as a source of insights into the pattern of successful inquiry. Aristotle therefore held theory, or contemplation, to be the highest form of knowledge and as such he regarded it as superior to practice, which he in turn regarded superior to production.

But because Dewey's emphasis was on the production of successful outcomes as the end of inquiry, he treated theory and practice as component parts within inquiry and as instruments for further production. He did not completely invert the Aristotelian schema, however, since he regarded theory and practice as phases of inquiry, whose outcome is the production of something new. In Dewey's view, theory and practice must cooperate if there is to be success in the production of new knowledge.

THREE OBJECTIONS

In talking to people about Dewey's program for tuning up our technological culture as I have sought to articulate it, several objections have been raised. I believe that they are based on misunderstandings not just of Dewey's critique of technology, but also of the problems and possibilities of our technological culture.

1) Some have claimed that it is an exaggeration to say that philosophical inquiry *is* a form of technology—an instrumentality—for the transformation of our technological culture. This objection seems to reflect the traditional view that philosophy has its own areas of interest, that technology has its own concrete areas of interest, and that despite some occasional areas of overlap, the two activities are fundamentally separate. What has philosophy got to do with the space program or the construction of bridges? The former has to do with human values, and the latter has to do with instrumental rationality.

A version of this view has been advanced by Jürgen Habermas, for example, who has tended to drive a wedge between what he has called the "knowledge constitutive interests" of science and technology on the one hand and the "communicative" and "emancipatory" interests of the human sciences on the other. Put more simply, this is the old "fact-value" split that was lamented by C. P. Snow in *The Two Cultures*.[6]

There are three things I want to say in response to this. First, one of Dewey's great insights was that philosophy has a special kind of productive function, since philosophy is a kind of general "liaison officer," as he put it, "making reciprocally intelligible voices speaking provincial tongues, and thereby enlarging as well as rectifying the meanings with which they are charged" (LW.1.306). In other words, philosophy can serve as a kind of translator that helps the various arts, sciences, engineering, and agriculture continue their discussions with one another. Just as philosophers of science help scientists within different disciplines talk to one another and learn from one another's methods, philosophers as critics of technological culture are in a position to perform this function on a more inclusive scale.

There are several very good reasons why it is up to philosophy to perform this task. As I have already indicated, philosophy contains as one of its parts logic, or the theory of the most general patterns of inquiry. And whereas inquiry within computer aesthetics and inquiry within materials science have different subject matters and different tactical methods, each contributes to and in turn receives the contributions of more general strategic methods of inquiry. Logic, as the theory of this general method of inquiry, serves as a facilitator.

Second, philosophy also involves metaphysics, which Dewey reconstructed as "a statement of the generic traits manifested by existence of all kinds without regard to their differentiation into physical and mental" (LW.1.308). "Any theory," he wrote, "that detects and defines these traits is therefore but a ground-map of the province of criticism, establishing base lines to be employed in more intricate triangulations (LW.1.309). For Dewey, metaphysics is anything but arcane: it has a connection to the objective world. The importance of the generic traits, he wrote, "lies in their application in the conduct of life: that is, in their *moral* bearing provided *moral* be taken in its basic broad human sense" (LW.16.389). In short, philosophers at their best are not only involved in a criticism of culture, but, because the process is self-correcting, they are also involved in a criticism of criticisms of culture as well.

The material just quoted comes from well-known passages from Dewey's great book *Experience and Nature*. I therefore find it remarkable that several generations of philosophers could have read them without grasping their implications for technological culture.

Third, as I have already indicated, philosophy as a critique of technology does not honor the traditional dualisms of body and mind, tangible and intangible, concrete and abstract, except as they are required as tools of inquiry. The general pattern of inquiry, as laid out by Dewey in his numerous books and essays on logic, is a technological enterprise precisely because it utilizes raw materials upon which tools are brought to bear in a cognitive fashion in order to produce novel artifacts, namely situations that are determined to be more desirable than the ones with which it started.

But this general pattern applies to inquiry of all types, whether the primary focus is that part of our experience we call tangible or that other portion we call the intangible. In other words, this general pattern of inquiry fits cases that involve what we would call hardware, and it also fits cases that are patently conceptual. It applies to descriptions of how manufacturers proceed from iron ore and coal to intermediate and finished steel products, and it applies to descriptions of how writers move from the raw materials of their experiences and research interests to working drafts and thence to finished works of fiction and nonfiction. It applies to the construction of logical and mathematical proofs, and it applies in social and political inquiry.

This concern with the means and ends involved in the production of novel artifacts seems to me to be one of the most important of Dewey's insights about technology. Whenever and wherever techniques of production and construction are utilized, no matter whether the sphere is conceptual or material, there is, in Dewey's view, productive work being done. This is why Dewey regarded the public, or better yet, the many publics that make up what we normally call "the public," as products. They are created as responses to issues of common interest, and their members seek to secure the ends-in-view that they hold in common. It is hardly a secret that billions of dollars are spent each year, from Madison Avenue to Pennsylvania Avenue, to create, manage, and reconstruct such publics precisely as artifacts.

2) A second objection comes from people who are interested in the arts. A colleague once objected that it is a mistake to say that a writer at work on a novel is doing anything "technological." There is in back of this second objection, I think, just the same confusion of terms that plagued Dewey during his long career. When I call writing a novel a problem-solving or technological activity I mean only that there is inquiry going forward and that it is technological because just as in other types of inquiry there are raw materials, there are tolls that are deliberately or cognitively deployed and further refined for tasks at hand, there are artifacts produced, and those artifacts are the responses to perceived goals as those goals are themselves developed and refined during the course of inquiry.

Applied to the work of the novelist the pattern is clear. The raw materials are the experiences of the novelist and the experiences of others that she has at second hand. But the novelist doesn't utilize *all* her experiences, and so there is involved a process of abstraction, selection, and reconfiguration. Dewey thought that this happens in all types of inquiry. As a goal or procedure is set up to solve some problem, in this case the writing of a novel, some things are taken as the facts of the case. Then they are weighed, tested, tried, and refined, all with respect to the task at hand. During this process, the task itself is usually modified. This calls for a reevaluation of what have been taken as the facts of the case. Some formerly pertinent data are discarded; other data are seen for the first time to be relevant.

In the case of writing a novel, characters emerge and are developed, plots thicken and then thin again, and there is the production of a new artifact: a novel. (Beyond that, the novel takes its place as an artifact that is used in the construction of further products or artifacts: various publics that will be motivated to purchase the novel, as well as the lives that will be altered as a consequence of reading it.)

Although there is a confusion of terms present in this objection, I believe that there is something else as well. The objection betrays a concern that the "fine" arts be held in higher esteem—or at least a different kind of esteem—than those that are "merely technological." But to treat the fine arts in this manner is to cut short their full reach as instrumental to an enhanced appreciation of the materials with which they are concerned.

Another variety of this type of objection might take the following form: if writing a novel in fact falls under the definition of technology as it has been advanced (namely, *the invention, development, and cognitive development of tools and artifacts brought to bear on raw materials and intermediate stock parts, with a view to the resolution of perceived problems*), then why shouldn't the editors of a journal of automotive engineering accept for publication an essay on literary criticism? Writing novels and designing automobiles are, after all, both forms of technology.

This objection misses the point on two counts. First, even if we were to employ the popular and uncritical notion of technology as having exclusively to do with material culture, we still would not expect the editors of the journal of automotive engineering to publish essays on hydrology or coal research. Although both disciplines fit the common definition of technology, their practitioners have different interests and ends-in-view. Second, it might in fact be appropriate under certain circumstances for the editors of the automotive engineering journal to publish a literary-essay that explores some aspect of automobiled life in a way that would inform and expand the horizon of automotive engineers. To deny this would be to honor the "fact-value" split about which Dewey continually complained, and which has retarded the resolution of many of our most pressing social problems.

Dewey took a significant risk when he reconstructed the term "technology" in the way that I have described. He took the risk that he would be labeled an uncritical follower of what some have termed "Enlightenment rationality." He also took the risk that he would be thought to have attempted a reduction of all human cognitive activity to one grey, amorphous discipline. But he seems to have thought the risk worth taking since the perceived benefits were so great. Repairing the old fact-value, technology-culture split was one such benefit. And naturalizing technology was another.

3) A third and related objection is that if we treat technology as inclusive of conceptual tools and artifacts as well as those that are tangible and material, then we have just taken technology so broadly that *everything* is included. Drawing the net of this objection somewhat more tightly than Carl Mitcham's articulation of it, however, the intuition is that we must reserve the term technology for operations with hardware, or perhaps also for the kind of software that can be held in the hand, or put on a bookshelf, or loaded in a computer, so that we can differentiate what happens in those regions from what happens in religion or poetry, for example. The idea behind this objection is that religion and poetry are "spiritual," whereas technology is not.

As I hope to have demonstrated, what is strictly technological—what involves inquiry into technique, tools, and artifacts—constitutes but a small part of the experience of most people. That portion or phase of experience that I called "technical" is a much larger part, to be sure; but the most prevalent feature of experience is what is immediate, that is *non-cognitive and non-instrumental* organic. This is a far cry from "just turning everything into technology."[7] But because the misunderstanding has been so profound, perhaps more needs to be said.

First, I believe that this objection rests on an explicit ontological dualism that is itself untenable. If what is "spiritual" is of value, then it would seem worthwhile to find ways of allowing it to penetrate all of our experiences. And if "technology" fails to be "spiritual," then its development has somewhat been cut short. Dewey rejected dualities of this type because he thought that they "formulated recognition of an impasse in life; an impotence in interaction, inability to make effective transition, limitation of power to regulate and thereby to understand" (LW.1.186).

Second, we cannot identify the technological with the cognitive *as such*, since there is cognitive work that does not involve tools except in a highly attenuated and analogous sense of the term. In retrospect, anthropologists may wish to speak metaphorically of the opposed thumb as a tool that the higher primates used to make the transition from savanna to forest. But the notion of an organic structure as tool is parasitic on the notion of extra-organic structure as tool. To reverse the relation would be anachronistic.

Nevertheless, once we begin to reflect on the ways in which tools are invented, developed, and utilized, it is possible to read the script forward in such a way that mathematical and logical objects, for example, are accepted as legitimate cases of tools. When this occurs, then the last nail goes into the coffin of Platonism. These are more or less the conclusions that Dewey reached during his decade at the University of Chicago, 1894–1904, and that formed the core of his productive pragmatism.[8]

Third, whether or not we use the term "spiritual" to designate religious practice, the undeniable fact is that religions, too, utilize tools, instruments, and artifacts of various types to effect their chosen ends. The leaders of the Roman Catholic Church long ago understood the importance of relics, the bread and wine of the Eucharist, incense, gilded altars, and other material artifacts, together with certain techniques such as the confession, as tools that could be used for the maintenance and enlargement of a believing public. Moreover, the cases in which the Church has retarded or rejected the advances of science in the name of what is "spiritual" have represented some of its greatest embarrassments. The case of Galileo, who was finally pardoned in 1992, some 359 years after being condemned as a heretic, is but one example of this phenomenon.

FOUR ADVANTAGES

I believe that there are several advantages of thinking about philosophy in the sense in which Dewey understood it, and as I have tried to expand upon that understanding, that is, as a tool for tuning up technology. I shall discuss four of these advantages. The first is what I shall call the *felicities of genetic analysis*; the second is the enormous *ecological power* gained by treating human technological activity as continuous with other natural activities; the third is that we get *off the foundationalist hook*; and the fourth is that we are able to generate *stable platforms for social action*.

1) First, this broad view of philosophy as criticism of technology opens up a whole new area of inquiry, namely the genetic analysis of conceptual tools. Just as there is a vestige in the modern plow of the bent stick, there is a vestige in the square root of minus one of the marks made on the wall of an ancient shepherd's fold in order to compare the number of outgoing sheep in the morning to the number of incoming sheep in the evening. And it is hardly surprising that organisms with ten fingers, counting thumbs, would operate in much of the world with number systems of base ten.[9]

This genetic approach to technology rejects the claims of scientific realism, namely that there is a prefigured reality "out there" waiting to be discovered, just as it is, in and of itself, apart from any contribution on the part of inquiry. It argues instead that the conceptual tools of science, including those we call scientific laws, are constructed, but not that they are constructed out of nothing.

When they are sophisticated and complex, they are constructed out of tools and intermediate stock parts that are already on hand. In some cases, such as in mathematics, they are primarily relations of relations, or abstractions of abstractions. And the most primitive of such tools are constructed out of the rawest of raw empirical materials, namely, felt needs and desires and flashes of insight or accident.

Why is this felicitous? Because it helps get philosophy out of the box it has often found itself in during its long career and out into the world of human affairs where it can do the work of criticism and reconstruction. It helps philosophy to link up with disciplines such as sociology, anthropology, archeology, and paleontology and thereby to focus its considerable energies upon real problems. It is also felicitous because it helps us get out from under the positivist-scientistic burden, the one that claims that the methods of the physical sciences provide "master narratives" that are somehow independent of such histories.

2) A second advantage of the view I am advancing is that it leads us to look for continuities between the adjustive activities of human beings and the adjustive activities of other natural organisms. This has profound consequences for environmental philosophy. Technology "naturalized" as I have described it, as inquiry into the techniques that human beings utilize to accommodate themselves to their environments and to alter those environments to their needs, functions as a kind of linkage or bridge to similar activities undertaken by higher primates, and even by "lower" non-human animals. It is not something above or apart from nature, but rather the cutting edge of evolutionary development.

I wish I could report that this last point is a minor one and that it has little import for the future of technoscientific education. James Moore, one of the team that worked with Martin Marty and R. Scott Appleby on the "Fundamentalisms" project, reported that by 1984 the Institute for Creation Research had a mailing list of some 75,000, an annual budget of $1.2 million, and a publication list of some fifty-five books that together had sold over one million copies.[10] As later as 1993, one of the largest technical universities in the United States, on whose faculty I was employed for two decades, still had engineering faculty who publicly defended "creation science," thus denying the type of continuity thesis that I have just put forward. In its place they argued for a strong version of supernaturalism that cuts technology off from its roots in the evolution of non-human nature. It is difficult to determine how successful these engineers were in moving their students to accept their arguments, but when the campus newspaper polled students regarding which one book they would choose to have with them in the event of a major disaster that destroyed their civilization, the majority of those polled chose the Bible over other presumably more practical tomes such as *The Foxfire Book*.[11]

This point directly addresses a different sort of objection, namely that if we treat philosophy as a tool for tuning up technological culture, as Dewey recommended that we do, then we have thereby become too preoccupied with one kind of philosophical activity, namely the type that is designed to alter the physical environment, at the expense of another kind of philosophical activity, namely the one by means of which we accommodate ourselves to our environments by means of certain "spiritual" exercises. This is similar to a charge that was brought against Dewey by first-generation critical theorists and others during his lifetime, and it is a charge that is still advanced against him during our own time. Put succinctly, it is that Dewey was a latter-day proponent of "Enlightenment rationality" who urged the domination of nature, and who ignored "spiritual" values or thought them nothing more than impediments to greater levels of efficiency.

It is correct to say that an awareness of this split between what have been called "technologies of environmental domination" and what some have called "technologies of the self" is important for understanding the history of technology, as well as the history of the philosophy of technology. But this is also a point on which Dewey's critics have profoundly misunderstood his work.

The fact is that we can identify two poles or dimensions *within* human experience. One is concerned with the alteration of circumstances that are relatively external to us, organically speaking. Another is the pole that is primarily concerned with the accommodation of ourselves as organisms to such circumstances. Although the first of these poles has sometimes been characterized as the domination of nature, it has also been characterized in some technophobic circles as "technology" *simpliciter*. Because Dewey lived in the wake of Darwin, however, and because he was interested in constructing a new form of naturalism that would take into account continuities within nature, he looked for a way to define technology with sufficient breadth that it could include this second pole of experience. This second pole has been the concern of thinkers such as Max Scheler and Michel Foucault, and it has been advanced in some strains of Buddhism. It also had an important place in Dewey's thinking.

In the first few pages of his 1934 book, *A Common Faith*, Dewey made this point clear. It is significant that such a clear statement of the matter appears in Dewey's only book on the philosophy of religious experience. Here is Dewey's remark:

> While the words "accommodation," "adaptation," and "adjustment" are frequently employed as synonyms, attitudes exist that are so different that for the sake of clear thought they should be discriminated. There are conditions we meet that cannot be changed. If they are particular and limited, we modify our own particular attitudes in accordance with them. Thus we accommodate ourselves to changes in weather, to alterations in income when we have no other recourse. When the external conditions are lasting we become inured, habituated. . . . The two main traits of this attitude, which I should like to call accommodation, are that it affects *particular* modes of conduct, not the entire self, and that the process is mainly *passive*. It may, however, become general and then it becomes fatalistic resignation or submission. There are other attitudes toward the environment that are also particular but that are more active. . . . Instead of accommodating ourselves to conditions, we modify conditions so that they will be accommodated to our wants and purposes. This process may be called adaptation.
>
> Now both of these processes are often called by the more general name of adjustment. But there are also changes in ourselves in relation to the world in which we live that are much more inclusive and deep seated. They relate not to this and that want in relation to this and that condition of our surroundings, but pertain to our being in its entirety. Because of their scope, this modification of ourselves is enduring. . . . It is a change *of* will conceived as the organic plenitude of our being, rather than any special change *in* will. (LW.9.12–13)

In this passage Dewey deftly undercuts the traditional philosophical problem of the inner and the outer, the mental and the physical, by locating it in the context of his critique of technology. Viewed as a part of a larger picture, habits are tools of adjustment. A habit is something that has a certain generality of application. It is something that has been tried out and found to be capable of serving certain purposes. Viewed from this perspective, as habits of a sort, hammers and saws become continuous with the other habits developed over millennia by higher order primates, for example, in their attempts to adjust to changing environmental conditions. Viewed in this perspective, to say that human beings are uniquely technological animals is not to place them outside and above nature, but within nature and a part of it. Our activities differ from those of our non-human relatives and ancestors not in kind, but only in level of complexity.

Habits are found throughout nature, but only human beings have reached the level of complexity that allows such a high level of self-control with respect to their deliberate formation, development, retention, and modification. It is for this reason—our ability to engage in the self-controlled manipulation of habits—that we human beings are able to reach very high levels of efficiency. We not only accommodate ourselves to environing conditions, but we also adapt environing conditions to our needs. These two activities taken together Dewey calls *adjustment* or *growth*, and he identifies the inquiry that is involved with such adjustment with technology in his broad sense of the term.

3) Here is a third advantage of Dewey's view of philosophy as a tool for tuning up technology. If knowing is a technological activity, then we are off the foundationalist hook. "Certainty" becomes an honorific term that is restricted to narrow non-existential doctrine. The laws of mathematical addition and subtraction are "certain" in this honorific sense not because they correspond to "the furniture of the world," to use Bertrand Russell's infelicitous phase, but because a great deal of work has been focused on a very narrow area of inquiry, that is, one that is so narrow as to exclude actual existence. As for the remaining domains of inquiry, which constitute the vast majority of the locations where technoscientific work is done, reconstruction continues to be done on the assumption that further improvements can be made in existential affairs and in the laws that are developed and employed to characterize them. "Fallibilism" and "probability" replace "certainty" as key operational terms.

4) Fourth, this view has the advantage of providing secure and steady platforms for the improvement of situations that are not as we wish them to be. It is not that we "look for" solutions in the sense of keeping our eyes open, or even that we wait for them to appear, as Heidegger told us that a "*Holzweg*" or clearing in a forest might just appear. If we are to flourish, we must construct hypotheses in a deliberate and intelligent fashion. Knowing is not so much a matter of "finding out" as it is a matter of "making sure." On this view, the kind of inquiry that leads to greater control of problematic social and political situations is also a type of technological undertaking, since it involves an active construction of desirable outcomes through the use of the tools and artifacts that are proper to that domain of knowledge-getting. Not only science itself, but the philosophy, sociology, and politics of science become important technological undertakings.

It is instructive to note the ways in which Dewey's view on this matter contrasts with that of Heidegger. Heidegger writes of a waiting readiness for a clearing to appear in the forest. Dewey writes of sharpening our tools in order to engage conditions that are not what we wish them to be. In one case we get a kind of watchfulness before the incomprehensibility of Being. In the other we get active management of problematic situations.

Critics of technology, such as Heidegger and his followers, have often said that it is technology that constitutes the major human problem. But what they have usually meant is that there are too many techniques, tools, and artifacts and that those things prevent our involvement in more proper occupations such as those that are religious, or "spiritual" in a broad sense, that is, that are concerned with what Heidegger termed "the shepherding of Being." I believe that Dewey would have agreed that technology constitutes the major human problem, but for reasons that are radically different from the ones just given. He thought of technology as inquiry into techniques, tools, and artifacts. And he thought that techniques are among the habits that are necessary to the continuance and growth of human life. He therefore thought that the major human problem was improving intelligence, which he identified with technology. And this means no more or less than developing better and more productive methods of inquiry into our techniques, our tools, and our artifacts.

Following Dewey's lead, I have characterized technology as *the invention, development, and cognitive deployment of tolls and other artifacts, brought to bear on raw materials and intermediate stock parts, to resolve perceived problems*. I have also argued that philosophy is one of the most effective tools we have for tuning up technology.

In addition, I have argued that what are commonly called the "theoretical sciences" such as chemistry and biology are no less cases of this type of activity than what are commonly called "material technologies" such as mechanical engineering and crop science. Theoretical knowing, such as that involved in mathematics, is no less a case of technological activity than is the type of knowing that is involved with concrete, practical outcomes such as building bridges. Because the theoretical is also artifactual, even what is sometimes called "pure research" is a type of technology.

So whereas the narrow characterizations of technology often tend to draw a line between material artifacts and everything else, which is commonly called science or even culture, and whereas some phenomenological accounts often tend to draw a line between what is practical and what is theoretical, I want to draw a line between what is involved in and a conscious result of intelligent, reconstructive activity, on the one side, and what is merely passive, rote, and uncritically accepted on the other. It seems to me that by dividing things up as I have, we achieve a kind of continuity within the domain of human enterprises that increases our power to effect meaningful adaptive change, that we are able to develop a wider appreciation for the ways that human beings function in and as a part of nature, and that we are able to see the relevance and make more sense out of genetic or historical studies.

If the program that I have outlined is a viable one, then philosophy is indeed an important and effective instrument for tuning up our technological culture. In the chapters that follow, this program will be examined in more detail.

NOTES

1. Mitcham points out that two earlier works had "philosophy of technology" in their titles, but their aims were really quite restricted. These were Ernst Kapp's book *Grundlinien einer Philosophie der Technik* (1877) and Eberhard Zschimmer's *Philosophie der Technik* (1913). See *Philosophy and Technology*, ed. Carl Mitcham and Robert Mackey (New York: Free Press, 1972), 22.

2. Some very recent monographs on the philosophy of technology still ignore Dewey's contribution to the field. As I was writing this chapter, for example, I received a copy of Joseph C. Pitt's *Thinking about Technology: Foundations of the Philosophy of Technology* (New York: Seven Bridges Press, 2000), which contains no mention of Dewey.

3. Hickman, *Dewey's Pragmatic Technology*.

4. This characterization has certain advantages over some of its alternatives. In *Thinking about Technology*, for example, Joseph C. Pitt defines technology as "humanity at work." (p. xi) Pitt's definition does, of course, have the advantage of generality. Further, as he indicates, it also obviates the problems that Jacques Ellul generated when he treated technology as a thing with an essence. On the downside, however, Pitt's definition does not appear on its face to preserve the distinction that I established in the first section of the chapter, namely, the distinction between technology and technique. In other words, it does not preserve the distinction between cognitive and non-cognitive deployment of tools and other artifacts.

On pages 10 and 11, Pitt criticizes the definition advanced by Emmanuel Mesthene, whose work I will discuss in chapter 7, "Populism and the Cult of the Expert." Curiously, Pitt objects to Mesthene's notion that technology is "the organization of knowledge for the achievement of practical purposes" (Mesthene *Technological*, 25) on the grounds that the phrase "organized knowledge" is redundant. Given the fact that our culture is currently suffering the splintering effects of increased specialization, this is a remarkable claim. One of the great needs of our milieu is precisely that what currently counts as knowledge be not only expanded, but better organized as well. Pitt then repeats with emphasis his definition, "*technology is humanity at work*" (p. 11). The idea, he writes, is that technology must involve the activity of humans, as opposed to organisms such as beavers or aliens, and that it must also involve "their deliberate and purposeful use of tools, taken in the general sense" (p. 11).

Two things about Pitt's gloss on his own definition are striking. First, his gloss seems to amplify what is in the definition to the point of significant revision. It adds the terms "deliberate" and "purposeful," for example. Nevertheless, the amplified definition still fails to capture the distinction I made between what is technological (cognitive) and what is technical (habitual), since it is quite possible to work mechanically in ways that are both deliberate and purposeful. Assembly line workers and farm laborers must do this daily. Of course play can be purposeful and deliberate as well.

Second, there is no acknowledgment that technology is involved not just in the use of tools but also in their invention and development; that tools must be applied in certain ways and not others; and that the problem that initiates inquiry is a function of a situation that involves inquiry with a particular perspective.

David Rothenberg, in an interview published in *A Parliament of Minds: Philosophy for a New Millennium*, ed. Michael Tobias, J. Patrick Fitzgerald, and David Rothenberg (Albany: SUNY Press, 2000), does not provide much help in this regard. "What is technology? It's really the whole history of tools that human beings have used to live in the world" (p. 169). This definition, if it is intended to be one, provides scant guidance concerning how to sort out the underlying differences, for example, between Greek warships and contemporary spacecraft.

5. See Larry A. Hickman, "Making the Family Functional: The Case for Legalized Same-Sex Domestic Partnerships," *Philosophy of the Social Sciences* 29, no. 2 (June 1999): 231–47; a revised and enlarged version of "Making the Family Functional: The Case for Same-Sex Marriage," in *Same-Sex Marriage: The Moral and Legal Debate*, ed. Robert M. Baird and Stuart E. Rosenbaum (Amherst, N.Y.: Prometheus Books, 1997), 192–202.

6. C. P. Snow, *The Two Cultures* (Cambridge: Cambridge University Press, 1959).

7. Carl Mitcham has noted that one response to my interpretation of Dewey's critique of technology in *John Dewey's Pragmatic Technology* involved the claim of reductionism. He thinks that "[my] reply to one possible formulation of the charge of reductionism does not consider the possibility that if all life is technological then the concept of technology becomes vacuous." See Mitcham, *Thinking through Technology*, 75. I hope to have put that objection to rest with the fourfold taxonomy I have developed in this chapter.

8. Mathematical objects have been developed within the sphere of the philosophy of mathematics. See Philip J. Davis and Reuben Hersh, *The Mathematical Experience* (New York: Houghton Mifflin, 1981).

9. See John D. Barrow, *Pi in the Sky: Counting, Thinking, and Being* (Oxford: Clarendon Press, 1992). See especially chapter 2, "The Counter Culture." This is an excellent introduction to the history of counting.

10. James Moore, "The Creationist Cosmos of Protestant Fundamentalism," in *Fundamentalisms and Society: Reclaiming the Sciences, the Family, and Education*, ed. Martin E. Marty and R. Scott Appleby (Chicago: University of Chicago Press, 1993), 12–49.

11. Eliot Wigginton, ed., *The Foxfire Book* (New York: Doubleday, 1972).

5

Focal Things and Practices

Albert Borgmann

THE DEVICE PARADIGM

We must now provide an explicit account of the pattern or paradigm of technology. I begin with two clear cases and analyze them in an intuitive way to bring out the major features of the paradigm. And I attempt to raise those features into sharper relief against the sketch of a pretechnological setting and through the consideration of objections that may be advanced against the distinctiveness of the pattern.

Technology, as we have seen, promises to bring the forces of nature and culture under control, to liberate us from misery and toil, and to enrich our lives. To speak of technology making promises suggests a substantive view of technology and is misleading. But the parlance is convenient and can always be reconstructed to mean that implied in the technological mode of taking up with the world there is a promise that this approach to reality will, by way of the domination of nature, yield liberation and enrichment. Who issues the promise to whom is a question of political responsibility; and who the beneficiaries of the promise are is a question of social justice. These questions are taken up in later chapters. What we must answer first is the question of how the promise of liberty and prosperity was specified and given a definite pattern of implementation.

As a first let us note that the notions of liberation and enrichment are joined in that of availability. Goods that are available to us enrich our lives and, if they are technologically available, they do so without imposing burdens on us. Something is available in this sense if it has been rendered instantaneous, ubiquitous, safe, and easy.[1] Warmth, for example, is now available. We get a first glimpse of the distinctiveness of availability when we remind ourselves that warmth was not available, for example, in Montana a hundred years ago. It was not instantaneous because in the morning a fire first had to be built in the stove or the fireplace. And before it could be built, trees had to be felled, logs had to be sawed and split, the wood had to be hauled and stacked. Warmth was not ubiquitous because some rooms remained unheated, and none was heated evenly. The coaches and sleighs were not heated, nor were the boardwalks or all of the shops and stores. It was

Albert Borgmann, *Technology and the Character of Contemporary Life: A Philosophical Inquiry*, Chicago: University of Chicago Press, 1984, pp. 40–44, 196–210, 221–226. Copyright © 1984 by the University of Chicago Press. Reprinted by permission of the University of Chicago Press.

not entirely safe because one could get burned or set the house on fire. It was not easy because work, some skills, and attention were constantly required to build and sustain a fire.

Such observations, however, are not sufficient to establish the distinctiveness of availability. In the common view, technological progress is seen as a more or less gradual and straightforward succession of lesser by better implements.[2] The wood-burning stove yields to the coal-fired central plant with heat distribution by convection, which in turn gives way to a plant fueled by natural gas and heating through forced air, and so on.[3] To bring the distinctiveness of availability into relief we must turn to the distinction between things and devices. A thing, in the sense in which I want to use the word here, is inseparable from its context, namely, its world, and from our commerce with the thing and its world, namely, engagement. The experience of a thing is always and also a bodily and social engagement with the thing's world. In calling forth a manifold engagement, a thing necessarily provides more than one commodity. Thus a stove used to furnish more than mere warmth. It was a *focus,* a hearth, a place that gathered the work and leisure of a family and gave the house a center. Its coldness marked the morning, and the spreading of its warmth the beginning of the day. It assigned to the different family members tasks that defined their place in the household. The mother built the fire, the children kept the firebox filled, and the father cut the firewood. It provided for the entire family a regular and bodily engagement with the rhythm of the seasons that was woven together of the threat of cold and the solace of warmth, the smell of wood smoke, the exertion of sawing and of carrying, the teaching of skills, and the fidelity to daily tasks. These features of physical engagement and of family relations are only first indications of the full dimensions of a thing's world. Physical engagement is not simply physical contact but the experience of the world through the manifold sensibility of the body. That sensibility is sharpened and strengthened in skill. Skill is intensive and refined world engagement. Skill, in turn, is bound up with social engagement. It molds the person and gives the person character.[4] Limitations of skill confine any one person's primary engagement with the world to a small area. With the other areas one is immediately engaged through one's acquaintance with the characteristic demeanor and habits of the practitioners of the other skills. That acquaintance is importantly enriched through one's use of their products and the observation of their working. Work again is only one example of the social context that sustains and comes to be focused in a thing. If we broaden our focus to include other practices, we can see similar social contexts in entertainment, in meals, in the celebration of the great events of birth, marriage, and death. And in these wider horizons of social engagement we can see how the cultural and natural dimensions of the world open up.

We have now sketched a background against which we can outline a specific notion of the device. We have seen that a thing such as a fireplace provides warmth, but it inevitably provides those many other elements that compose the world of the fireplace. We are inclined to think of these additional elements as burdensome, and they were undoubtedly often so experienced. A device such as a central heating plant procures mere warmth and disburdens us of all other elements. These are taken over by the machinery of the device. The machinery makes no demands on our skill, strength, or attention, and it is less demanding the less it makes its presence felt. In the progress of technology, the machinery of a device has therefore a tendency to become concealed or to shrink. Of all the physical properties of a device, those alone are crucial and prominent which constitute the commodity that the device procures. Informally speaking, the commodity of a device is "what a device is there for." In the case of a central heating plant it is warmth, with a telephone it is communication, a car provides transportation, frozen food makes up a meal, a stereo set furnishes music. "Commodity" for the time being is to be taken flexibly. The emphasis lies on the commodious way in which devices make goods and services available. There are at first unavoidable ambiguities in the notion of the device and the commodity; they can gradually be resolved through substantive analyses and methodological reflections.[5] Tentatively, then, those aspects or properties of a device that provide the answer to "What is the device for?" constitute its commodity, and they remain relatively fixed. The other properties are changeable and are changed, normally on the basis of scientific insight and engineering ingenuity, to make the commodity still more available. Hence every device has functional equivalents, and equivalent devices may be physically and structurally very dissimilar from one another.

The development of television provides an illustration of these points. The bulky machinery of the first sets was obtrusive in relation to the commodity it procured, namely, the moving two-dimensional picture which appeared in fuzzy black and white on a screen with the size and shape of a bull's-eye. Gradually the screens became larger, more rectangular; the picture became sharper and eventually colored. The sets became relatively smaller and less conspicuous in their machinery. And this development continues and has its limit in match-box-sized sets which provide arbitrarily large and most finely grained moving and colored pictures. The example also shows how radical changes in the machinery amounted to continuous improvements of the function as tubes gave way to transistors and these yielded to silicon chips. Cables and satellites were introduced as communication links. Pictures could be had in recorded rather than transmitted form, and recordings can be had on tapes or discs. These considerations in turn show how the technical development of a device increases availability. Increasingly, video programs can be seen nearly everywhere—in bars, cars, in every room of a home. Every conceivable film can be had. A program broadcast at an inconvenient time can be recorded and played later. The constraints of time and place are more and more dissolved. It is an instructive exercise to see how in the implements that surround us daily the machinery becomes less conspicuous, the function more prominent, how radical technical changes in the machinery are but degrees of advancement in the commodity, and how the availability of the commodities increases all the while.

The distinction in the device between its machinery and its function is a specific instance of the means-ends distinction. In agreement with the general distinction, the machinery or the means is subservient to and validated by the function or the end. The technological distinction of means and ends differs from the general notion in two respects. In the general case, it is very questionable how clearly and radically means and ends can be distinguished without doing violence to the phenomena.[6] In the case of the technological device, however, the machinery can be changed radically without threat to the identity and familiarity of the function of the device. No one is confused when one is invited to replace one's watch, powered by a spring, regulated by a balance wheel, displaying time with a dial and pointers, with a watch that is powered electrically, is regulated by a quartz crystal, and displays time digitally. This concomitance of radical variability of means and relative stability of ends is the first distinguishing feature. The second, closely tied to the first, is the concealment and unfamiliarity of the means and the simultaneous prominence and availability of the ends.[7]

The concealment of the machinery and the disburdening character of the device go hand in hand. If the machinery were forcefully present, it would eo ipso make claims on our faculties. If claims are felt to be onerous and are therefore removed, then so is the machinery. A commodity is truly available when it can be enjoyed as a mere end, unencumbered by means. It must be noted that the disburdenment resting on a feudal household is ever incomplete. The lord and the lady must always reckon with the moods, the insubordination, and the frailty of the servants.[8] The device provides social disburdenment, i.e., anonymity. The absence of the master-servant relation is of course only one instance of social anonymity. The starkness of social anonymity in the technological universe can be gauged only against a picture of the social relations in a world of things. Such a picture will also show that social anonymity necessarily shades off into one of nature, culture, and history.

FOCAL THINGS AND PRACTICES

To see that the force of nature can be encountered analogously in many other places, we must develop the general notions of focal things and practices. This is the first point of this chapter. The Latin word *focus,* its meaning and etymology, are our best guides to this task. But once we have learned tentatively to recognize the instances of focal things and practices in our midst, we must acknowledge their scattered and inconspicuous character too. Their hidden splendor comes to light when we consider Heidegger's reflections on simple and eminent things. But an inappropriate nostalgia clings to Heidegger's account. It can be dispelled, so I will argue, when we remember and realize more fully that the technological environment heightens rather than denies the radiance of genuine focal things and when we learn to understand that focal things require a practice to prosper within. These points I will try to give substance in the subsequent parts of this chapter by calling attention to the focal concerns of running and of the culture of the table.

The Latin word *focus* means hearth. We came upon it earlier where the device paradigm was first delineated and where the hearth or fireplace, a thing, was seen as the counterpart to the central heating plant, a device. It was pointed out that in a pretechnological house the fireplace constituted a center of warmth, of light, and of daily practices. For the Romans the *focus* was holy, the place where the housegods resided. In ancient Greece, a baby was truly joined to the family and household when it was carried about the hearth and placed before it. The union of a Roman marriage was sanctified at the hearth. And at least in the early periods the dead were buried by the hearth. The family ate by the hearth and made sacrifices to the housegods before and after the meal. The hearth sustained, ordered, and centered house and family.[9] Reflections of the hearth's significance can yet be seen in the fireplace of many American homes. The fireplace often has a central location in the house. Its fire is now symbolical since it rarely furnishes sufficient warmth. But the radiance, the sounds, and the fragrance of living fire consuming logs that are split, stacked, and felt in their grain have retained their force. There are no longer images of the ancestral gods placed by the fire; but there often are pictures of loved ones on or above the mantel, precious things of the family's history, or a clock, measuring time.[10]

The symbolical center of the house, the living room with the fireplace, often seems forbidding in comparison with the real center, the kitchen with its inviting smells and sounds. Accordingly, the architect Jeremiah Eck has rearranged homes to give them back a hearth, "a place of warmth and activity" that encompasses cooking, eating, and living and so is central to the house whether it literally has a fireplace or not.[11] Thus we can satisfy, he says, "the need for a place of focus in our family lives."[12]

"Focus," in English, is now a technical term of geometry and optics. Johannes Kepler was the first to use it, and he probably drew on the then already current sense of focus as the "burning point of lens or mirror."[13] Correspondingly, an optic or geometric focus is a point where lines or rays converge or from which they diverge in a regular or lawful way. Hence "focus" is used as a verb in optics to denote moving an object in relation to a lens or modifying a combination of lenses in relation to an object so that a clear and well-defined image is produced.

These technical senses of "focus" have happily converged with the original one in ordinary language. Figuratively they suggest that a focus gathers the relations of its context and radiates into its surroundings and informs them. To focus on something or to bring it into focus is to make it central, clear, and articulate. It is in the context of these historical and living senses of "focus" that I want to speak of focal things and practices. Wilderness on this continent, it now appears, is a focal thing. It provides a center of orientation; when we bring the surrounding technology into it, our relations to technology become clarified and well-defined. But just how strong its gathering and radiating force is requires further reflection. And surely there will be other focal things and practices: music, gardening, the culture of the table, or running.

We might in a tentative way be able to see these things as focal; what we see more clearly and readily is how inconspicuous, homely, and dispersed they are. This is in stark contrast to the focal things of pretechnological times, the Greek temple or the medieval cathedral that we have mentioned before. Martin Heidegger was deeply impressed by the orienting force of the Greek temple. For him, the temple not only gave a center of meaning to its world but had orienting power in the strong sense of first originating or establishing the world, of disclosing the world's essential dimensions and criteria.[14] Whether the thesis so extremely put is defensible or not, the Greek temple was certainly more than a self-sufficient architectural sculpture, more than a jewel of well-articulated and harmoniously balanced elements, more, even, than a shrine for the image of the goddess or the god. As Vincent Scully has shown, a temple or a temple precinct gathered and disclosed the land in which it was situated. The divinity of land and sea was focused in the temple.[15]

To see the work of art as the focus and origin of the world's meaning was a pivotal discovery for Heidegger. He had begun in the modern tradition of Western philosophy where, the sense of reality is to be grasped by determining the antecedent and controlling conditions of all there is (the *Bedingungen der Möglichkeit* as Immanuel Kant has it). Heidegger wanted to outdo this tradition in the radicality of his search for the fundamental conditions of being. Perhaps it was the relentlessness of his pursuit that disclosed the ultimate futility of it. At any rate, when the universal conditions are explicated in a suitably general and encompassing way, what truly matters still hangs in the balance because everything depends on how the conditions come to be actualized and instantiated.[16] The preoccupation with antecedent conditions not only leaves this question unanswered; it may even make it inaccessible by leaving the impression that, once the general and fundamental matters are determined, nothing of consequence remains to be considered. Heidegger's early work, however, already contained the seeds of its overcoming. In his determination to grasp reality in its concreteness, Heidegger had found and stressed the inexorable and unsurpassable givenness of human existence, and he had provided analyses of its pretechnological wholeness and its technological distraction though the significance of these descriptions for technology had remained concealed to him.[17] And then he discovered that the unique event of significance in the singular work of art, in the prophet's proclamation, and in the political deed was crucial. This insight was worked out in detail with regard to the artwork. But in an epilogue to the essay that develops this point, Heidegger recognized that the insight comes too late. To be sure, our time has brought forth admirable works of art. "But," Heidegger insists, "the question remains: is art still an essential and necessary way in which that truth happens which is decisive for historical existence, or is art no longer of this character?"[18]

Heidegger began to see technology (in his more or less substantive sense) as the force that has eclipsed the focusing powers of pretechnological times. Technology becomes for him the final phase of a long metaphysical development. The philosophical concern with the conditions of the possibility of whatever is now itself seen as a move into the oblivion of what finally matters. But how are we to recover orientation in the oblivious and distracted era of technology when the great embodiments of meaning, the works of art, have lost their focusing power? Amidst the complication of conditions, of the *Bedingungen,* we must uncover the simplicity of things, of the *Dinge.*[19] A jug, an earthen vessel from which we pour wine, is such a thing. It teaches us what it is to hold, to offer, to pour, and to give. In its clay, it gathers for us the earth as it does in containing the wine that has grown from the soil. It gathers the sky whose rain and sun are present in the wine. It refreshes and animates us in our mortality. And in the libation it acknowledges and calls on the divinities. In these ways the thing (in agreement with its etymologically original meaning) gathers and discloses what Heidegger calls the fourfold, the interplay of the crucial dimensions of earth and sky, mortals and divinities.[20] A thing, in Heidegger's eminent sense, is a focus; to speak of focal things is to emphasize the central point twice.

Still, Heidegger's account is but a suggestion fraught with difficulties. When Heidegger described the focusing power of the jug, he might have been thinking of a rural setting where wine jugs embody in their material, form, and craft a long and local tradition; where at noon one goes down to the cellar to draw a jug of table wine whose vintage one knows well; where at the noon meal the wine is thoughtfully poured and gratefully received.[21] Under such circumstances, there might be a gathering and disclosure of the fourfold, one that is for the most part understood and in the background and may come to the fore on festive occasions. But all of this seems as remote to most of us and as muted in its focusing power as the Parthenon or the Cathedral of Chartres. How can so simple a thing as a jug provide that turning point in our relation to technology to which Heidegger is looking forward? Heidegger's proposal for a reform of technology is even more programmatic and terse than his analysis of technology.[22] Both, however, are capable of fruitful development.[23] Two points in Heidegger's consideration of the turn of technology must particularly be noted. The first serves to remind us of arguments already developed which must be kept in mind if we are to make room for focal things and practices. Heidegger says, broadly paraphrased, that the orienting force of simple things will come to the fore only as the rule of technology is raised from its anonymity, is disclosed as the orthodoxy that heretofore has been taken for granted and allowed to remain invisible.[24] As long as we overlook the tightly patterned character of technology and believe that we live in a world of endlessly open and rich opportunities, as long as we ignore the definite ways in which we, acting technologically, have worked out the promise of technology and remain vaguely enthralled by that promise, so long simple things and practices will seem burdensome, confining, and drab. But if we recognize the central vacuity of advanced technology, that emptiness can become the opening for focal things. It works both ways, of course. When we see a focal concern of ours threatened by technology, our sight for the liabilities of mature technology is sharpened.

A second point of Heidegger's is one that we must develop now. The things that gather the fourfold, Heidegger says, are inconspicuous and humble. And when we look at his litany of things, we also see that they are scattered and of yesterday: jug and bench, footbridge and plow, tree and pond, brook and hill, heron and deer, horse and bull, mirror and clasp, book and picture, crown and cross.[25] That focal things and practices are inconspicuous is certainly true; they flourish at the margins of public attention. And they have suffered a diaspora; this too must be accepted, at least for now. That is not to say that a hidden center of these dispersed focuses may not emerge some day to unite them and bring them home. But it would clearly be a forced growth to proclaim such a unity now. A reform of technology that issues from focal concerns will be radical not in imposing a new and unified master plan on the technological universe but in discovering those sources of strength that will nourish principled and confident beginnings, measures, i.e., which will neither rival nor deny technology.

But there are two ways in which we must go beyond Heidegger. One step in the first direction has already been taken. It led us to see that the simple things of yesterday attain a new splendor in today's technological context. The suggestion in Heidegger's reflections that we have to seek out pretechnological enclaves to encounter focal things is misleading and dispiriting. Rather we must see any such enclave itself as a focal thing heightened by its technological context. The turn to things cannot be a setting aside and even less an escape from technology but a kind of affirmation of it. The second move beyond Heidegger is in the direction of practice, into the social and, later, the political situation of focal things.[26] Though Heidegger assigns humans their place in the fourfold when he depicts the jug in which the fourfold is focused, we scarcely see the hand that holds the jug, and far less do we see of the social setting in which the pouring of the wine comes to pass.

In his consideration of another thing, a bridge, Heidegger notes the human ways and works that are gathered and directed by the bridge.[27] But these remarks too present practices from the viewpoint of the focal thing. What must be shown is that focal things can prosper in human practices only. Before we can build a bridge, Heidegger suggests, we must be able to dwell.[28] But what does that mean concretely?

The consideration of the wilderness has disclosed a center that stands in a fruitful counterposition to technology. The wilderness is beyond the procurement of technology, and our response to it takes us past consumption. But it also teaches us to accept and to appropriate technology. We must now try to discover if such centers of orientation can be found in greater proximity and intimacy to the technological everyday life. And I believe they can be found if we follow up the hints that we have gathered from and against Heidegger, the suggestions that focal things seem humble and scattered but attain splendor in technology if we grasp technology properly, and that focal things require a practice for their welfare. Running and the culture of the table are such focal things and practices. We have all been touched by them in one way or another. If we have not participated in a vigorous or competitive run, we have certainly taken walks; we have felt with surprise, perhaps, the pleasure of touching the earth, of feeling the wind, smelling the rain, of having the blood course through our bodies more steadily. In the preparation of a meal we have enjoyed the simple tasks of washing leaves and cutting bread; we have felt the force and generosity of being served a good wine and homemade bread. Such experiences have been particularly vivid when we came upon them after much sitting and watching indoors, after a surfeit of readily available snacks and drinks. To encounter a few simple things was liberating and invigorating. The normal clutter and distraction fall away when, as the poet says,

> there, in limpid brightness shine,
> on the table, bread and wine.[29]

If such experiences are deeply touching, they are fleeting as well. There seems to be no thought or discourse that would shelter and nurture such events; not in politics certainly, nor in philosophy where the prevailing idiom sanctions and applies equally to lounging and walking, to Twinkies, and to bread, the staff of life. But the reflective care of the good life has not withered away. It has left the profession of philosophy and sprung up among practical people. In fact, there is a tradition in this country of persons who are engaged by life in its concreteness and simplicity and who are so filled with this engagement that they have reached for the pen to become witnesses and teachers, speakers of deictic discourse. Melville and Thoreau are among the great prophets of this tradition. Its present health and extent are evident from the fact that it now has no overpowering heroes but many and various more or less eminent practitioners. Their work embraces a spectrum between down-to-earth instruction and soaring speculation. The span and center of their concerns vary greatly. But they all have their mooring in the attention to tangible and bodily things and practices, and they speak with an enthusiasm that is nourished by these focal concerns. Pirsig's book is an impressive and troubling monument in this tradition, impressive in the freshness of its observations and its pedagogical skill, troubling in its ambitious and failing efforts to deal with the large philosophical issues. Norman Maclean's *A River Runs through It* can be taken as a fly-fishing manual, a virtue that pleases its author.[30] But it is a literary work of art most of all and a reflection on technology inasmuch as it presents the engaging life, both dark and bright, from which we have so recently emerged. Colin Fletcher's treatise of *The Complete Walker* is most narrowly a book of instruction about hiking and backpacking.[31] The focal significance of these things is found in the interstices of equipment and technique; and when the author explicitly engages in deictic discourse he has "an unholy awful time" with it.[32] Roger B. Swain's contemplation of gardening in *Earthly*

Pleasures enlightens us in cool and graceful prose about the scientific basis and background of what we witness and undertake in our gardens.[33] Philosophical significance enters unbidden and easily in the reflections on time, purposiveness, and the familiar. Looking at these books, I see a stretch of water that extends beyond my vision, disappearing in the distance. But I can see that it is a strong and steady stream, and it may well have parts that are more magnificent than the ones I know.[34]

To discover more clearly the currents and features of this, the other and more concealed, American mainstream, I take as witnesses two books where enthusiasm suffuses instruction vigorously, Robert Farrar Capon's *The Supper of the Lamb* and George Sheehan's *Running and Being*.[35] Both are centered on focal events, the great run and the great meal. The great run, where one exults in the strength of one's body, in the ease and the length of the stride, where nature speaks powerfully in the hills, the wind, the heat, where one takes endurance to the breaking point, and where one is finally engulfed by the goodwill of the spectators and the fellow runners.[36] The great meal, the long session as Capon calls it, where the guests are thoughtfully invited, the table has been carefully set, where the food is the culmination of tradition, patience, and skill and the presence of the earth's most delectable textures and tastes, where there is an invocation of divinity at the beginning and memorable conversation throughout.[37]

Such focal events are compact, and if seen only in their immediate temporal and spatial extent they are easily mistaken. They are more mistakable still when they are thought of as experiences in the subjective sense, events that have their real meaning in transporting a person into a certain mental or emotional state. Focal events, so conceived, fall under the rule of technology. For when a subjective state becomes decisive, the search for a machinery that is functionally equivalent to the traditional enactment of that state begins, and it is spurred by endeavors to find machineries that will procure the state more instantaneously, ubiquitously, more assuredly and easily. If, on the other hand, we guard focal things in their depth and integrity, then, to see them fully and truly, we must see them in context. Things that are deprived of their context become ambiguous.[38] The letter "a" by itself means nothing in particular. In the context of "table" it conveys or helps to convey a more definite meaning. But "table" in turn can mean many things. It means something more powerful in the text of Capon's book where he speaks of "The Vesting of the Table."[39] But that text must finally be seen in the context and texture of the world. To say that something becomes ambiguous is to say that it is made to say less, little, or nothing. Thus to elaborate the context of focal events is to grant them their proper eloquence.

"The distance runner," Sheehan says, "is the least of all athletes. His sport the least of all sports."[40] Running is simply to move through time and space, step-by-step. But there is splendor in that simplicity. In a car we move of course much faster, farther, and more comfortably. But we are not moving on our own power and in our own right. We cash in prior labor for present motion. Being beneficiaries of science and engineering and having worked to be able to pay for a car, gasoline, and roads, we now release what has been earned and stored and use it for transportation. But when these past efforts are consumed and consummated in my driving, I can at best take credit for what I have done. What I am doing now, driving, requires no effort, and little or no skill or discipline. I am a divided person; my achievement lies in the past, my enjoyment in the present. But in the runner, effort and joy are one; the split between means and ends, labor and leisure is healed.[41] To be sure, if I have trained conscientiously, my past efforts will bear fruit in a race. But they are not just cashed in. My strength must be risked and enacted in the race which is itself a supreme effort and an occasion to expand my skill.

This unity of achievement and enjoyment, of competence and consummation, is just one aspect of a central wholeness to which running restores us. Good running engages mind and body. Here the mind is more than an intelligence that happens to be housed in a body. Rather the mind is the sensitivity and the endurance of the body.[42] Hence running in its fullness, as Sheehan stresses over and over again, is in principle different from exercise designed to procure physical health. The difference between running and physical exercise is strikingly exhibited in one and the same issue of the *New York Times Magazine.* It contains an account by Peter Wood of how, running the New York City Marathon, he took in the city with body and mind, and it has an account by Alexandra Penney of corporate fitness programs where executives, concerned about their Coronary Risk Factor Profile, run nowhere on treadmills or ride stationary bicycles.[43] In another issue, the *Magazine* shows executives exercising their bodies while busying their dissociated minds with reading.[44] To be sure, unless a runner concentrates on bodily performance, often in an effort to run the best possible race, the mind wanders as the body runs. But as in free association we range about the future and the past, the actual and the possible, our mind, like our breathing, rhythmically gathers itself to the here and now, having spread itself to distant times and faraway places.

It is clear from these reflections that the runner is mindful of the body because the body is intimate with the world. The mind becomes relatively disembodied when the body is severed from the depth of the world, i.e., when the world is split into commodious surfaces and inaccessible machineries. Thus the unity of ends and means, of mind and body, and of body and world is one and the same. It makes itself felt in the vividness with which the runner experiences reality. "Somehow you feel more in touch," Wood says, "with the realities of a massive inner-city housing problem when you are running through it slowly enough to take in the grim details, and, surprisingly, cheered on by the remaining occupants."[45] As this last remark suggests, the wholeness that running establishes embraces the human family too. The experience of that simple event releases an equally simple and profound sympathy. It is a natural goodwill, not in need of drugs nor dependent on a common enemy. It wells up from depths that have been forgotten, and it overwhelms the runners ever and again.[46] As Wood recounts his running through streets normally besieged by crime and violence, he remarks: "But we can only be amazed today at the warmth that emanates from streets usually better known for violent crime." And his response to the spectators' enthusiasm is this: "I feel a great proximity to the crowd, rushing past at all of nine miles per hour; a great affection for them individually; a commitment to run as well as I possibly can, to acknowledge their support."[47] For George Sheehan, finally, running discloses the divine. When he runs, he wrestles with God.[48] Serious running takes us to the limits of our being. We run into threatening and seemingly unbearable pain. Sometimes, of course, the plunge into that experience gets arrested in ambition and vanity. But it can take us further to the point where in suffering our limits we experience our greatness too. This, surely, is a hopeful place to escape technology, metaphysics, and the God of the philosophers and reach out to the God of Abraham, Isaac, and Jacob.[49]

If running allows us to center our lives by taking in the world through vigor and simplicity, the culture of the table does so by joining simplicity with cosmic wealth. Humans are such complex and capable beings that they can fairly comprehend the world and, containing it, constitute a cosmos in their own right. Because we are standing so eminently over against the world, to come in touch with the world becomes for us a challenge and a momentous event. In one sense, of course, we are always already in the world, breathing the air, touching the ground, feeling the sun. But as we can in another sense withdraw from the actual and present world, contemplating what is past and to come, what is possible and remote, we celebrate correspondingly our intimacy with the world. This we do most fundamentally when in eating we take in the world in its palpable, colorful, nourishing immediacy. Truly human eating is the union of the primal and the cosmic. In the simplicity of bread and wine, of meat and vegetable, the world is gathered.

The great meal of the day, be it at noon or in the evening, is a focal event par excellence. It gathers the scattered family around the table. And on the table it gathers the most delectable things nature has brought forth. But it also recollects and presents a tradition, the immemorial experiences of the race in identifying and cultivating edible plants, in domesticating and butchering animals; it brings into focus closer relations of national or regional customs, and more intimate traditions still of family recipes and dishes. This living texture is being rent through the procurement of food as a commodity and the replacement of the culture of the table by the food industry. Once food has become freely available, it is only consistent that the gathering of the meal is shattered and disintegrates into snacks, T.V. dinners, bites that are grabbed to be eaten; and eating itself is scattered around television shows, late and early meetings, activities, overtime work, and other business. This is increasingly the normal condition of technological eating. But it is within our power to clear a central space amid the clutter and distraction. We can begin with the simplicity of a meal that has a beginning, a middle, and an end and that breaks through the superficiality of convenience food in the simple steps of beginning with raw ingredients, preparing and transforming them, and bringing them to the table. In this way we can again become freeholders of our culture. We are disfranchised from world citizenship when the foods we eat are mere commodities. Being essentially opaque surfaces, they repel all efforts at extending our sensibility and competence into the deeper reaches of the world. A Big Mac and a Coke can overwhelm our tastebuds and accommodate our hunger. Technology is not, after all, a children's crusade but a principled and skillful enterprise of defining and satisfying human needs. Through the diversion and busyness of consumption we may have unlearned to feel constrained by the shallowness of commodities. But having gotten along for a time and quite well, it seemed, on institutional or convenience food, scales fall from our eyes when we step up to a festively set family table. The foods stand out more clearly, the fragrances are stronger, eating has once more become an occasion that engages and accepts us fully.

To understand the radiance and wealth of a festive meal we must be alive to the interplay of things and humans, of ends and means. At first a meal, once it is on the table, appears to have commodity character since it is now available before us, ready to be consumed without effort or merit. But though there is of course in any eating a moment of mere consuming, in a festive meal eating is one with an order and discipline that challenges and ennobles the participants. The great meal has its structure. It begins with a moment of reflection in which we place ourselves in the presence of the first and last things. It has a sequence of courses; it requires and sponsors memorable conversation; and all this is enacted in the discipline called table manners. They are warranted when they constitute the respectful and skilled response to the great things that are coming to pass in the meal. We can see how order and discipline have collapsed when we eat a Big Mac. In consumption there is the pointlike and inconsequential conflation of a sharply delimited human need with an equally contextless and closely fitting commodity. In a Big Mac the sequence of courses has been compacted into one object and the discipline of table manners has been reduced to grabbing and eating. The social context reaches no further than the pleasant faces and quick hands of the people who run the fast-food outlet. In a festive meal, however, the food is served, one of the most generous gestures human beings are capable of. The serving is of a piece with garnishing; garnishing is the final phase of cooking, and cooking is one with preparing the food. And if we are blessed with rural circumstances, the preparation of food draws near the harvesting and the raising of the vegetables in the garden close by. This context of activities is embodied in persons. The dish and the cook, the vegetable and the gardener tell of one another. Especially when we are guests, much of the meal's deeper context is socially and conversationally mediated. But that mediation has translucence and intelligibility because it extends into the farther and deeper recesses without break and with a bodily immediacy that we too have enacted or at least witnessed firsthand. And what seems to be a mere receiving and consuming of food is in fact the enactment of generosity and gratitude, the affirmation of mutual and perhaps religious obligations. Thus eating in a focal setting differs sharply from the social and cultural anonymity of a fast-food outlet.

The pretechnological world was engaging through and through, and not always positively.

There also was ignorance, to be sure, of the final workings of God and king; but even the unknown engaged one through mystery and awe. In this web of engagement, meals already had focal character, certainly as soon as there was anything like a culture of the table.[50] Today, however, the great meal does not gather and order a web of thoroughgoing relations of engagement; within the technological setting it stands out as a place of profound calm, one in which we can leave behind the narrow concentration and one-sided strain of labor and the tiring and elusive diversity of consumption. In the technological setting, the culture of the table not only focuses our life; it is also distinguished as a place of healing, one that restores us to the depth of the world and to the wholeness of our being.

As said before, we all have had occasion to experience the profound pleasure of an invigorating walk or a festive meal. And on such occasions we may have regretted the scarcity of such events; we might have been ready to allow such events a more regular and central place in our lives. But for the most part these events remain occasional, and indeed the ones that still grace us may be slipping from our grasp. We have seen various aspects of this malaise, especially its connection with television. But why are we acting against our better insights and aspirations? This at first seems all the more puzzling as the engagement in a focal activity is for most citizens of the technological society an instantaneous and ubiquitous possibility. On any day I can decide to run or to prepare a meal after work. Everyone has some sort of suitable equipment. At worst one has to stop on the way home to pick up this or that. It is of course technology that has opened up these very possibilities. But why are they lying fallow for the most part? There is a convergence of several factors. Labor is exhausting, especially when it is divided. When we come home, we often feel drained and crippled. Diversion and pleasurable consumption appear to be consonant with this sort of disability. They promise to untie the knots and to soothe the aches. And so they do at a shallow level of our existence. At any rate, the call for exertion and engagement seems like a cruel and unjust demand. We have sat in the easy chair, beer at hand and television before us; when we felt stirrings of ambition, we found it easy to ignore our superego.[51] But we also may have had our alibi refuted on occasion when someone to whom we could not say no prevailed on us to put on our coat and to step out into cold and windy weather to take a walk. At first our indignation grew. The discomfort was worse than we had thought. But gradually a transformation set in. Our gait became steady, our blood began to flow vigorously and wash away our tension, we smelled the rain, began thoughtfully to speak with our companion, and finally returned home settled, alert, and with a fatigue that was capable of restful sleep.

But why did such occurrences remain episodes also? The reason lies in the mistaken assumption that the shaping of our lives can be left to a series of individual decisions. Whatever goal in life we entrust to this kind of implementation we in fact surrender to erosion. Such a policy ignores both the frailty and strength of human nature. On the spur of the moment, we normally act out what has been nurtured in our daily practices as they have been shaped by the norms of our time. When we sit in our easy chair and contemplate what to do, we are firmly enmeshed in the framework of technology with our labor behind us and the blessings of our labor about us, the diversions and enrichments of consumption. This arrangement has had our lifelong allegiance, and we know it to have the approval and support of our fellows. It would take superhuman strength to stand up to this order ever and again. If we are to challenge *the rule of technology,* we can do so only through *the practice of engagement.*

The human ability to establish and commit oneself to a practice reflects our capacity to comprehend the world, to harbor it in its expanse as a context that is oriented by its focal points. To found a practice is to guard a focal concern, to shelter it against the vicissitudes of fate and our frailty. John Rawls has pointed out that there is decisive difference between the justification of a practice and of a particular action falling under it.[52] Analogously, it is one thing to decide for a focal practice and quite another to decide for a particular action that appears to have focal character.[53] Putting the matter more clearly, we must say that without a practice an engaging action or event can momentarily light up our life, but it cannot order and orient it focally. Competence, excellence, or virtue, as Aristotle first saw, come into being as an *éthos,* a settled disposition and a way of life.[54] Through a practice, Alasdaire MacIntyre says accordingly, "human powers to achieve excellence, and human conceptions of the ends and goods involved, are systematically extended."[55] Through a practice we are able to accomplish what remains unattainable when aimed at in a series of individual decisions and acts.

How can a practice be established today? Here, as in the case of focal things, it is helpful to consider the foundation of pretechnological practices. In mythic times the latter were often established through the founding and consecrating act of a divine power or mythic ancestor. Such an act set up a sacred precinct and center that gave order to a violent and hostile world. A sacred practice, then, consisted in the regular reenactment of the founding act, and so it renewed and sustained the order of the world. Christianity came into being this way; the eucharistic meal, the Supper of the Lamb, is its central event, established with the instruction that it be reenacted. Clearly a focal practice today should have centering and orienting force as well. But it differs in important regards from its grand precursors. A mythic focal practice derived much force from the power of its opposition. The alternative to the preservation of the cosmos was chaos, social and physical disorder and collapse. It is a reduction to see mythic practices merely as coping behavior of high survival value. A myth does not just aid survival; it defines what truly human life is. Still, as in the case of pretechnological morality, economic and social factors were interwoven with mythic practices. Thus the force of brute necessity supported, though it did not define, mythic focal practices. Since a mythic focal practice united in itself the social, the economic, and the cosmic, it was naturally a prominent and public affair. It rested securely in collective memory and in the mutual expectations of the people.

This sketch, of course, fails to consider many other kinds of pretechnological practices. But it does present one important aspect of them and more particularly one that serves well as a backdrop for focal practices in a technological setting. It is evident that technology is itself a sort of practice, and it procures its own kind of order and security. Its history contains great moments of innovation, but it did not arise out of a founding event that would have focal character; nor has it produced focal things. Thus it is not a *focal* practice, and it has indeed, so I have urged, a debilitating tendency to scatter our attention and to clutter our surroundings. A focal practice today, then, meets no tangible or overtly hostile opposition from its context and is so deprived of the wholesome vigor that derives from such opposition. But there is of course an opposition at a more profound and more subtle level. To feel the support of that opposing force one must have experienced the subtly debilitating character of technology, and above all one must understand, explicitly or implicitly, that the peril of technology lies not in this or that of its manifestations but in *the pervasiveness and consistency of its pattern.* There are always occasions where a Big Mac, an exercycle, or a television program are unobjectionable and truly helpful answers to human needs. This makes a case-by-case appraisal of technology so inconclusive. It is when we attempt to take the measure of technological life in its normal totality that we are distressed by its shallowness. And I believe that the more strongly we sense and the more clearly we understand the coherence and the character of technology, the more evident it becomes to us that technology must be countered by an equally patterned and social commitment, that is, by a practice.

At this level the opposition of technology does become fruitful to focal practices. They can now be seen as restoring a depth and integrity to our lives that are in principle excluded within the paradigm of technology. MacIntyre, though his foil is the Enlightenment more than technology, captures this point by including in his definition of practice the notion of "goods internal to a practice."[56] These are one with the practice and can only be obtained through that practice. The split between means and ends is healed. In contrast "there are those goods externally and contingently attached" to a practice; and in that case there "are always alternative ways for achieving such goods, and their achievement is never to be had *only* by engaging in some particular kind of practice."[57] Thus practices (in a looser sense) that serve external goods are subvertible by technology. But MacIntyre's point needs to be clarified and extended to include or emphasize not only the essential unity of human being and a particular sort of doing but also the tangible things in which the world comes to be focused. The importance of this point has been suggested by the consideration of running and the culture of the table. There are objections to this suggestion. Here I want to advance the thesis by considering Rawls's contention that a practice is defined by rules. We can take a rule as an instruction for a particular domain of life to act in a certain way under specified circumstances. How important is the particular character of the tangible setting of the rules? Though Rawls does not address this question directly he suggests in using baseball for illustration that "a peculiarly shaped piece of wood" and a kind of bag become a bat and base only within the confines defined by the rules of baseball.[58] Rules and the practice they define, we might argue in analogy to what Rawls says about their relation to particular cases, are logically prior to their tangible setting. But the opposite contention seems stronger to me. Clearly the possibilities and challenges of baseball are crucially determined by the layout and the surface of the field, the weight and resilience of the ball, the shape and size of the bat, etc. One might of course reply that there are rules that define the physical circumstances of the game. But this is to take "rule" in broader sense. Moreover it would be more accurate to say that the rules of this latter sort reflect and protect the identity of the original tangible circumstances in which the game grew up. The rules, too, that circumscribe the actions of the players can be taken as ways of securing and ordering the playful challenges that arise in the human interplay with reality. To be sure there are developments and innovations in sporting equipment. But either they quite change the nature of the sport as in pole vaulting, or they are restrained to preserve the identity of the game as in baseball.

It is certainly the purpose of a focal practice to guard in its undiminished depth and identity the thing that is central to the practice, to shield it against the technological diremption into means and end. Like values, rules and practices are recollections, anticipations, and, we can now say, guardians of the concrete things and events that finally matter. Practices protect focal things not only from technological subversion but also against human frailty. It was emphasized that the ultimately significant things to which we respond in deictic discourse cannot be possessed or controlled. Hence when we reach out for them, we miss them occasionally and sometimes for quite some time. Running becomes unrelieved pain and cooking a thankless chore. If in the technological mode we insisted on assured results or if more generally we estimated the value of future efforts on the basis of recent experience, focal things would vanish from our lives. A practice keeps faith with focal things and saves for them an opening in our lives. To be sure, eventually the practice needs to be empowered again by the reemergence of the great thing in its splendor. A practice that is not so revived degenerates into an empty and perhaps deadening ritual.

We can now summarize the significance of a focal practice and say that such a practice is required to counter technology in its patterned pervasiveness and to guard focal things in their depth and integrity. Countering technology through a practice is to take account of our susceptibility to technological distraction, and it is also to engage the peculiarly human strength of comprehension, i.e., the power to take in the world in its extent and significance and to respond through an enduring commitment. Practically a focal practice comes into being through resoluteness, either an explicit resolution where one vows regularly to engage in a focal activity from this day on or in a more implicit resolve that is nurtured by a focal thing in favorable circumstances and matures into a settled custom.

In considering these practical circumstances we must acknowledge final difference between focal practices today and their eminent pretechnological predecessors. The latter, being public and prominent, commanded elaborate social and physical settings: hierarchies, offices, ceremonies, and choirs; edifices, altars, implements, and vestments. In comparison our focal practices are humble and scattered. Sometimes they can hardly be called practices, being private and limited. Often they begin as a personal regimen and mature into a routine without ever attaining the social richness that distinguishes a practice. Given the often precarious and inchoate nature of focal practices, evidently focal things and practices, for all the splendor of their simplicity and their fruitful opposition to technology, must be further clarified in their relation to our everyday world if they are to be seen as a foundation for the reform of technology.

WEALTH AND THE GOOD LIFE

Strong claims have been made for focal things and practices. Focal concerns supposedly allow us to center our lives and to launch a reform of technology and so to usher in the good life that has eluded technology. We have seen that focal practices today tend to be isolated and rudimentary. But these are marginal deficiencies, due to unfavorable circumstances. Surely there are central problems as well that pertain to focal practices no matter how well developed. Before we can proceed to suggestions about how technology may be reformed to make room for the good life, the most important objections regarding focal practices, the pivots of that reform, must be considered and, if possible, refuted. These disputations are not intended to furnish the impregnable defense of focal concerns which is neither possible nor to be wished for. The deliberations of this chapter are rather efforts to connect the notion of a focal practice more closely with the prevailing conceptual and social situation and so to advance the standing of focal concerns in our midst. To make the technological universe hospitable to focal things turns out to be the heart of the reform of technology.

Let me now draw out the concrete consequences of this kind of reform. I begin with particular illustrations and proceed to broader observations. Sheehan's focal concern is running, but he does not run everywhere he wants to go. To get to work he drives a car. He depends on that technological device and its entire associated machinery of production, service, resources, and roads. Clearly, one in Sheehan's position would want the car to be as perfect a technological device as possible: safe, reliable, easy to operate, free of maintenance. Since runners deeply enjoy the air, the trees, and the open spaces that grace their running, and since human vigor and health are essential to their enterprise, it would be consistent of them to want an environmentally benign car, one that is free of pollution and requires a minimum of resources for its production and operation. Since runners express themselves through running, they would not need to do so through the glitter, size, and newness of their vehicles.[59]

At the threshold of their focal concern, runners leave technology behind, technology, i.e., as a way of taking up with the world. The products of technology remain ubiquitous, of course: clothing, shoes, watches, and the roads. But technology can produce instruments as well as devices, objects that call forth engagement and allow for a more skilled and intimate contact with the world.[60] Runners appreciate shoes that are light, firm, and shock absorbing. They allow one to move faster, farther, and more fluidly. But runners would not want to have such movement procured by a motorcycle, nor would they, on the other side, want to obtain merely the physiological benefit of such bodily movement from a treadmill.

A focal practice engenders an intelligent and selective attitude toward technology. It leads to a simplification and perfection of technology in the background of one's focal concern and to a discerning use of technological products at the center of one's practice. I am not, of course, describing an evident development or state of affairs. It does appear from what little we know statistically of the runners in this country, for instance, that they lead a more engaged, discriminating, and a socially more profound life.[61] I am rather concerned to draw out the consequences that naturally follow for technology from a focal commitment and from a recognition of the device pattern. There is much diffidence, I suspect, among people whose life is centered, even in their work, around a great concern. Music is surely one of these. But at times, it seems to me, musicians confine the radiance, the rhythm, and the order of music and the ennobling competence that it requires to the hours and places of performance. The entrenchment of technology may make it seem quixotic to want to lead a fully musical life or to change the larger technological setting so that it would be more hospitable and attentive to music. Moreover, as social creatures we seek the approval of our fellows according to the prevailing standards. One may be a runner first and most of all; but one wants to prove too that one has been successful in the received sense. Proof requires at least the display, if not the consumption, of expensive commodities. Such inconsistency is regrettable, not because we just have to have reform of technology but because it is a partial disavowal of one's central concern. To have a focal thing radiate transformatively into its environment is not to exact some kind of service from it but to grant it its proper eloquence.

There is of course intuitive evidence for the thesis that a focal commitment leads to an intelligent limitation of technology. There are people who, struck by a focal concern, remove much technological clutter from their lives. In happy situations, the personal and private reforms take three directions. The first is of course to clear a central space for the focal thing, to establish an inviolate time for running, or to establish a hearth in one's home for the culture of the table. And this central clearing goes hand in hand, as just suggested, with a newly discriminating use of technology.[62] The second direction of reform is the simplification of the context that surrounds and supports the focal area. And then there is a third endeavor, that of extending the sphere of engagement as far as possible. Having experienced the depth of things and the pleasure of full-bodied competence at the center, one seeks to extend such excellence to the margins of life. "Do it yourself" is the maxim of this tendency and "self-sufficiency" its goal. But the tendencies for which these titles stand also exhibit the dangers of this third direction of reform. Engagement, however skilled and disciplined, becomes disoriented when it exhausts itself in the building, rebuilding, refinement, and maintenance of stages on which nothing is ever enacted. People finish their basements, fertilize their lawns, fix their cars. What for? The peripheral engagement suffocates the center, and festivity, joy, and humor disappear. Similarly, the striving for self-sufficiency may open up a world of close and intimate relations with things and people. But the demands of the goal draw a narrow and impermeable boundary about that world. There is no time to be a citizen of the cultural and political world at large and no possibility of assuming one's responsibility in it. The antidote to such disorientation and constriction is the appropriate acceptance of technology. In one or another area of one's life one should gratefully accept the disburdenment from daily and time-consuming chores and allow celebration and world citizenship to prosper in the time that has been gained.

What emerges here is a distinct notion of the good life or more precisely the private or personal side of one. Clearly, it will remain crippled if it cannot unfold into the world of labor and the public realm. To begin on the side of leisure and privacy is to acknowledge the presently dispersed and limited standing of focal powers. It is also to avail oneself of the immediate and undeniably large discretion one has in shaping one's free time and private sphere.[63] Even within these boundaries the good life that is centered on focal concerns is distinctive enough. Evidently, it is a favored and prosperous life. It possesses the time and the implements that are needed to devote oneself to a great calling. Technology provides us with the leisure, the space, the books, the instruments, the equipment, and the instruction that allow us to become equal to some great thing that has beckoned us from afar or that has come to us through a tradition. The citizen of the technological society has been spared the abysmal bitterness of knowing himself or herself to be capable of some excellence or achievement and of being at the same time worn-out by poor and endless work, with no time to spare and no possibility of acquiring the implements of one's desire. That bitterness is aggravated when one has a gifted child that is similarly deprived, and is exacerbated further through class distinctions where one sees richer but less gifted and dedicated persons showered with opportunities of excellence. There is prosperity also in knowing that one is able to engage in a focal practice with a great certainty of physical health and economic security. One can be relatively sure that the joy that one receives from a focal thing will not be overshadowed by the sudden loss of a loved one with whom that joy is shared. And one prospers not only in being engaged in a profound and living center but also in having a view of the world at large in its essential political, cultural, and scientific dimensions. Such a life is centrally prosperous, of course, in opening up a familiar world where things stand out clearly and steadily, where life has a rhythm and depth, where we encounter our fellow human beings in the fullness of their capacities, and where we know ourselves to be equal to that world in depth and strength.

This kind of prosperity is made possible by technology, and it is centered in a focal concern. Let us call it wealth to distinguish it from the prosperity that is confined to technology and that I want to call affluence. Affluence consists in the possession and consumption of the most numerous, refined, and varied commodities. This superlative formulation betrays its relative character. "Really" to be affluent is to live now and to rank close to the top of the hierarchy of inequality. All of the citizens of a typical technological society are more affluent than anyone in the Middle Ages. But this affluence, astounding when seen over time, is dimmed or even insensible at any one time for all but those who have a disproportionately large share of it. Affluence, strictly defined, has an undeniable glamour. It is the embodiment of the free, rich, and imperial life that technology has promised. So at least it appears from below whence it is seen by most people. Wealth in comparison is homely, homely in the sense of being plain and simple but homely also in allowing us to be at home in our world, intimate with its great things, and familiar with our fellow human beings. This simplicity, as said before, has its own splendor that is more sustaining than the glamour of affluence which leaves its beneficiaries, so we hear, sad and bored.[64] Wealth is a romantic notion also in that it continues and develops a tradition of concerns and of excellence that is rooted on the other side of the modern divide, i.e., of the Enlightenment. A life of wealth is certainly not romantic in the sense of constituting an uncomprehending rejection of the modern era and a utopian reform proposal.[65]

I will conclude by considering the narrower sphere of wealth and by connecting it with the traditional notions of excellence and of the family. The virtues of world citizenship, of gallantry, musicianship, and charity still command an uneasy sort of allegiance and it is natural, therefore, to measure the technological culture by these standards. Perhaps people are ready to accept the distressing results of such measurement with a rueful sort of agreement. But obviously the acceptance of the standards, if there is one, is not strong enough to engender the reforms that the pursuit of traditional excellence would demand. This, I believe, is due to the fact that the traditional virtues have for too long been uprooted from the soil that used to nourish them. Values, standards, and rules, I have urged repeatedly, are recollections and anticipations of great things and events. They provide bonds of continuity with past greatness and allow us to ready ourselves and our children for the great things we look forward to. Rules and values inform and are acted out in practices. A virtue is the practiced and accomplished faculty that makes one equal to a great event. From such considerations it is evident that the real circumstances and forces to which the traditional values, virtues, and rules used to answer are all but beyond recollection, and there is little in the technological universe that they can anticipate and ready us for. The peculiar character of technological reality has escaped the attention of the modern students of ethics.

To sketch a notion of excellence that is appropriate to technology is, in one sense, simply to present another version of the reform of technology that has been developed so far. But it is also to uncover and to strengthen ties to a tradition that the modern era has neglected to its peril. As regards world citizenship today, the problem is not confinement but the proliferation of channels of communication and of information. From the mass of available information we select by the criteria of utility and entertainment. We pay attention to information that is useful to the maintenance and advancement of technology, and we consume those news items that divert us. In the latter case the world is shredded into colorful bits of entertainment, and the distracted kind of knowledge that corresponds to that sort of information is the very opposite of the principled appropriation of the world that is meant by world citizenship.[66] The realm of technically useful information does not provide access to world citizenship either. Technical information is taken up primarily in one's work. Since most work in technology is unskilled, the demands on technical knowledge are low, and most people know little of science, engineering, economics, and politics. The people at the leading edge of technology have difficulty in absorbing and integrating the information that pertains to their field.[67] But even if the flood of technical information is appropriately channeled, as I think it can be, its mastery still constitutes knowledge of the social machinery, of the means rather than the ends of life. What is needed if we are to make the world truly and finally ours again is the recovery of a center and a standpoint from which one can tell what matters in the world and what merely clutters it up. A focal concern is that center of orientation. What is at issue here comes to the fore when we compare the simple and authentic world appropriation of someone like Mother Teresa with the shallow and vagrant omniscience of a technocrat.

Gallantry in a life of wealth is the fitness of the human body for the greatness and the playfulness of the world. Thus it has a grounding and a dignity that are lost in traditional gallantry, a loss that leaves the latter open to the technological concept of the perfect body where the body is narcissistically stylized into a glamorous something by whatever scientific means and according to the prevailing fashion. In the case of musicianship the tradition of excellence is unbroken and has expanded into jazz and popular music. What the notion of wealth can contribute to the central splendor and competence of music is to make us sensible to the confinement and the procurement of music. Confinement and procurement are aspects of the same phenomenon. The discipline and the rhythmic grace and order that characterize music are often confined, as said above, to the performance proper and are not allowed to inform the broader environment. This is because the unreformed structure of the technological universe leaves no room for such forces. Accordingly, music is allowed to conform to technology and is procured as a commodity that is widely and inconsequentially consumed. A focal concern for musicianship, then, will curtail the consumption of music and secure a more influential position for the authentic devotion to music.

Finally, one may hope that focal practices will lead to a deepening of charity and compassion. Focal practices provide a profounder commerce with reality and bring us closer to that intensity of experience where the world engages one painfully in hunger, disease, and confinement. A focal practice also discloses fellow human beings more fully and may make us more sensitive to the plight of those persons whose integrity is violated or suppressed. In short, a life of engagement may dispel the astounding callousness that insulates the citizens of the technological societies from the well-known misery in much of the world. The crucial point has been well made by Duane Elgin:

> When people deliberately choose to live closer to the level of material sufficiency, they are
> brought closer to the reality of material existence for a majority of persons on this planet. There

is not the day-to-day insulation from material poverty that accompanies the hypnosis of a culture of affluence.[68]

The plight of the family, finally, consists in the absorption of its tasks and substance by technology. The reduction of the household to the family and the growing emptiness of family life leave the parents bewildered and the children without guidance. Since less and less of vital significance remains entrusted to the family, the parents have ceased to embody rightful authority and a tradition of competence, and correspondingly there is less and less legitimate reason to hold children to any kind of discipline. Parental love is deprived of tangible and serious circumstances in which to realize itself. Focal practices naturally reside in the family, and the parents are the ones who should initiate and train their children in them. Surely parental love is one of the deepest forms of sympathy. But sympathy needs enthusiasm to have substance. Families, I have found, that we are willing to call healthy, close, or warm turn out, on closer inspection, to be centered on a focal concern. And even in families that exhibit the typical looseness of structure, the diffidence of parents, and the impertinence of children, we can often discover a bond of respect and deep affection between parent and youngster, one that is secured in a common concern such as a sport and keeps the family from being scattered to the winds.

NOTES

1. Earlier versions of this notion of technology can be found in "Technology and Reality," *Man and World* 4 (1971): 59–69; "Orientation in Technology," *Philosophy Today* 16 (1972): 135–47; "The Explanation of Technology," *Research in Philosophy and Technology* 1 (1978): 99–118. Daniel J. Boorstin similarly describes the character of everyday America in terms of availability and its constituents. See his *Democracy and Its Discontents* (New York: Random House, 1974).

2. See Emmanuel G. Mesthene, *Technological Change* (Cambridge, Mass: Harvard University Press, 1970), p. 28.

3. See Melvin M. Rotsch, "The Home Environment," in *Technology in Western Civilization,* ed. Melvin Kranzberg and Carroll W. Pursell, Jr. (New York: Oxford University Press, 1967), 2: 226–28. For the development of the kitchen stove (the other branch into which the original fireplace or stove developed), see Siegfried Giedion, *Mechanization Takes Command* (New York: Norton, 1969 [first published in 1948]), pp. 527–47.

4. See George Sturt's description of the sawyers in *The Wheelwright's Shop* (London: Cambridge University Press, 1974 [first published in 1923]), pp. 32–40.

5. In economics, "commodity" is a technical term for a tradable (and usually movable) economic good. In social science, it has become a technical term as a translation of Marx's *Ware* (merchandise). Marx's use and the use here suggested and to be developed agree inasmuch as both are intended to capture a novel and ultimately detrimental transformation of a traditional (pretechnological) phenomenon. For Marx, a commodity of the negative sort is the result of the reification of social relations, in particular of the reification of the workers' labor power, into something tradable and exchangeable which is then wrongfully appropriated by the capitalists and used against the workers. This constitutes the exploitation of the workers and their alienation from their work. It finally leads to their pauperization. I disagree that this transformation is at the center of gravity of the modern social order. The crucial change is rather the splitting of the pretechnological fabric of life into machinery and commodity according to the device paradigm. Though I concede and stress the tradable and exchangeable character of commodities, as I use the term, their primary character, here intended, is their commodious and consumable availability with the technological machinery as their basis and with disengagement and distraction as their recent consequences. On Marx's notion of commodity and commodity fetishism, see Paul M. Sweezy, *The Theory of Capitalist Development* (New York: Monthly Review Press, 1968), pp. 34–40.

6. See Morton Kaplan, "Means/Ends Rationality," *Ethics* 87 (1976): 61–65.

7. Martin Heidegger gives a careful account of the interpenetration of means and ends in the pretechnological disclosure of reality. But when he turns to the technological disclosure of being (*das Gestell*) and to the

device in particular (*das Gerät*), he never points out the peculiar technological diremption of means and ends though he does mention the instability of the machine within technology. Heidegger's emphasis is perhaps due to his concern to show that technology as a whole is not a means or an instrument. See his "The Question Concerning Technology," in *The Question Concerning Technology and Other Essays*, trans. William Lovitt (New York: Harper & Row, 1977).

8. It also turns out that a generally rising standard of living makes personal services disproportionately expensive. See Staffan B. Linder, *The Harried Leisure Class* (New York: Columbia University, 1970), pp. 34–37.

9. See *Paulys Realencyclopädie der classischen Altertumswissenschaft* (Stuttgart: Druckenmuller, 1893–1963), 15: 615–17; See also Fustel de Coulanges, "The Sacred Fire," in *The Ancient City*, trans. Willard Small (Garden City, N.Y., n.d. [first published in 1864]), pp. 25–33.

10. See Kent C. Bloomer and Charles W. Moore, *Body, Memory, and Architecture* (New Haven: Yale University Press, 1977), pp. 2–3 and 50–51.

11. See Jeremiah Eck, "Home Is Where the Hearth Is," *Quest* 3 (April 1979): 12.

12. Ibid., p. 11.

13. See *The Oxford English Dictionary*.

14. See Martin Heidegger, "The Origin of the Work of Art," in *Poetry, Language, Thought*, trans. Albert Hofstadter (New York: Harper & Row, 1971), pp. 15–87.

15. See Vincent Scully, *The Earth, the Temple, and the Gods* (New Haven: Yale University Press, 1962).

16. See my *The Philosophy of Language* (The Hague: Nijoff, 1974), pp. 126–31.

17. See Heidegger, *Being and Time*, trans. John Macquarrie and Edward Robinson (New York: Harper, 1962), pp. 95–107, 163–68, 210–24.

18. See Heidegger, "The Origin of the Work of Art," p. 80.

19. See Heidegger, "The Thing," in *Poetry, Language, Thought*, pp. 163–82. Heidegger alludes to the turn from the *Bedingungen to the Dinge* on p. 179 of the original, "Das Ding," in *Vorträge und Aufsätze* (Pfullingen, 1959). He alludes to the turn from technology to (focal) things in "The Question Concerning Technology."

20. See Heidegger, "The Thing."

21. See M. F. K. Fisher, *The Cooking of Provincial France* (New York: Time-Life Books, 1968), p. 50.

22. Though there are seeds for a reform of technology to be found in Heidegger as I want to show, Heidegger insists that "philosophy will not be able to effect an immediate transformation of the present condition of the world. Only a god can save us." See "Only a God Can Save Us: Der Spiegel's Interview with Martin Heidegger," trans. Maria P. Alter and John D. Caputo, *Philosophy Today* 20 (1976): 277.

23. I am not concerned to establish or defend the claim that my account of Heidegger or my development of his views is authoritative. It is merely a matter here of acknowledging a debt.

24. See Heidegger, "The Question Concerning Technology"; Langdon Winner makes a similar point in "The Political Philosophy of Alternative Technology," in *Technology and Man's Future*, ed. Albert H. Teich, 3d ed. (New York: St. Martin's Press, 1981), pp. 369–73.

25. See Heidegger, "The Thing," pp. 180–82.

26. The need of complementing Heidegger's notion of the thing with the notion of practice was brought home to me by Hubert L. Dreyfus's essay, "Holism and Hermeneutics," *Review of Metaphysics* 34 (1980): 22–23.

27. See Heidegger, "Building Dwelling Thinking," in *Poetry, Language, Thought*, pp. 152–53.

28. Ibid., pp. 148–49.

29. Georg Trakl, quoted by Heidegger in "Language," in *Poetry, Language, Thought*, pp. 194–95 (I have taken some liberty with Hofstadter's translation).

30. See Normal Maclean, *A River Runs through It and Other Stories* (Chicago: University of Chicago Press, 1976). Only the first of the three stories instructs the reader about fly fishing.

31. See Colin Fletcher, *The Complete Walker* (New York: Knopf, 1971).

32. Ibid., p. 9.

33. See Roger B. Swain, *Earthly Pleasures: Tales from a Biologist's Garden* (New York: Scribner, 1981).

34. Here are a few more: Wendell Berry, *Farming: A Handbook* (New York, 1970); Stephen Kiesling, *The Shell Game: Reflections on Rowing and the Pursuit of Excellence* (New York: Morrow, 1982); John Richard Young, *Schooling for Young Riders* (Norman: University of Oklahoma Pres, 1970); W. Timothy Gallwey, *The*

Inner Game of Tennis (New York: Random House, 1974); Ruedi Bear, *Pianta Su: Ski Like the Best* (Boston: Little, Brown, 1976). Such books must be sharply distinguished from those that promise to teach accomplishments without effort and in no time. The latter kind of book is technolgical in intent and fraudulent in fact.

35. See Robert Farrar Capon, *The Supper of the Lamb: A Culinary Reflection* (Garden City, N.Y.: Doubleday, 1969); and George Sheehan, *Running and Being: The Total Experience* (New York: Simon and Schuster, 1978).

36. See Sheehan, pp. 211–20 and elsewhere.

37. See Capon, pp. 167–181.

38. See my "Mind, Body, and World," *Philosophical Forum* 8 (1976): 76–79.

39. See Capon, pp. 176–77.

40. See Sheehan, p. 127.

41. On the unity of achievement and enjoyment, see Alasdair MacIntyre, *After Virtue* (Notre Dame, Ind.: University of Notre Dame Press, 1981), p. 184.

42. See my "Mind, Body, and World," pp. 68–86.

43. See Peter Wood, "Seeing New York on the Run," *New York Times Magazine,* 7 October 1979; Alexandra Penney, "Health and Grooming: Shaping Up the Corporate Image," ibid.

44. See *New York Times Magazine,* 3 August 1980, pp. 20–21.

45. See Wood, p. 112.

46. See Sheehan, pp. 211–17.

47. See Wood, p. 116.

48. See Sheehan, pp. 221–31 and passim.

49. There is substantial anthropological evidence to show that running has been a profound focal practice in certain pretechnological cultures. I am unable to discuss it here. Nor have I discussed the problem, here and elsewhere touched upon, of technology and religion. The present study, I believe, has important implications for that issue, but to draw them out would require more space and circumspection than are available now. I have made attempts to provide an explication in "Christianity and the Cultural Center of Gravity," *Listening* 18 (1983): 93–102; and in "Prospects for the Theology of Technology," *Theology and Technology,* ed. Carl Mitcham and Jim Grote (Lanham, Md.: University Press of America, 1984), pp. 305–22.

50. See M. F. K. Fisher, pp. 9–31.

51. Some therapists advise lying down till these stirrings go away.

52. See John Rawls, "Two Concepts of Rules," *Philosophical Review* 64 (1955): 3–32.

53. Conversely, it is one thing to break a practice and quite another to omit a particular action. For we define ourselves and our lives in our practices; hence to break a practice is to jeopardize one's identity while omitting a particular action is relatively inconsequential.

54. See Aristotle's *Nicomachean Ethics,* the beginning of Book Two in particular.

55. See MacIntyre, p. 175.

56. Ibid., pp. 175–77.

57. Ibid., p. 176.

58. See Rawls, p. 25.

59. On the general rise and decline of the car as a symbol of success, see Daniel Yankelovich, *New Rules: Searching for Self-Fulfillment in a World Turned Upside Down* (Toronto: Bantam Books, 1982), pp. 36–39.

60. Although these technological instruments are translucent relative to the world and so permit engagement with the world, they still possess an opaque machinery that mediates engagement but is not itself experienced either directly or through social mediation. See also the remarks in n. 12 above.

61. See "Who Is the American Runner?" *Runner's World* 15 (December 1980): 36–42.

62. Capon's book is the most impressive document of such discriminating use of technology.

63. A point that is emphatically made by E. F. Schumacher in *Small Is Beautiful* (New York: Harper & Row, 1973) and in *Good Work* (New York, 1979); by Duane Elgin in *Voluntary Simplicity* (New York: Morrow, 1981); and by Yankelovich in *New Rules.*

64. See Roger Rosenblatt, "The Sad Truth about Big Spenders," *Time,* 8 December 1980, pp. 84 and 89.

65. On the confusions that beset romanticism in its opposition to technology, see Lewis Mumford, *Technics and Civilization* (New York: Harcourt, Brace, and World, 1963), pp. 285–303.

66. See Daniel J. Boorstin, *Democracy and Its Discontents* (New York: Random House, 1975), pp. 12–25.

67. See Elgin, pp. 251–71. In believing that the mass of complex technical information poses a mortal threat to bureaucracies, Elgin, it seems to me, indulges in the unwarranted pessimism of the optimists.

68. Ibid., p. 71.

6

A Phenomenology of Technics

Don Ihde

The task of a phenomenology of human-technology relations is to discover the various structural features of those ambiguous relations. In taking up this task, I shall begin with a focus upon experientially recognizable features that are centered upon the ways we are bodily engaged with technologies. The beginning will be within the various ways in which I-as-body interact with my environment by means of technologies.

A. TECHNICS EMBODIED

If much of early modern science gained its new vision of the world through optical technologies, the process of embodiment itself is both much older and more pervasive. To embody one's praxis *through* technologies is ultimately an *existential* relation with the world. It is something humans have always—since they left the naked perceptions of the Garden—done.

I have previously and in a more suggestive fashion already noted some features of the visual embodiment of optical technologies. Vision is technologically transformed through such optics. But while the fact *that* optics transform vision may be clear, the variants and invariants of such a transformation are not yet precise. That becomes the task for a more rigorous and structural phenomenology of embodiment. I shall begin by drawing from some of the previous features mentioned in the preliminary phenomenology of visual technics.

Within the framework of phenomenological relativity, visual technics first may be located within the intentionality of seeing.

I see—through the optical artifact—the world

This seeing is, in however small a degree, at least minimally distinct from a direct or naked seeing.

I see—the world

Don Ihde, *Technology and the Lifeworld,* Bloomington: Indiana University Press, 1990, pp. 72–100, 105–12.

I call this first set of existential technological relations with the world *embodiment relations,* because in this use context I take the technologies *into* my experiencing in a particular way by way of perceiving *through* such technologies and through the reflexive transformation of my perceptual and body sense.

In Galileo's use of the telescope, he embodies his seeing through the telescope thusly:

Galileo—telescope—Moon

Equivalently, the wearer of eyeglasses embodies eyeglass technology:

I—glasses—world

The technology is actually *between* the seer and the seen, in a *position of mediation.* But the referent of the seeing, that towards which sight is directed, is "on the other side" of the optics. One sees *through* the optics. This, however, is not enough to specify this relation as an embodiment one. This is because one first has to determine *where* and *how,* along what will be described as a continuum of relations, the technology is experienced.

There is an initial sense in which this positioning is doubly ambiguous. First, the technology must be *technically* capable of being seen through; it must be transparent. I shall use the term *technical* to refer to the physical characteristics of the technology. Such characteristics may be designed or they may be discovered. Here the disciplines that deal with such characteristics are informative, although indirectly so for the philosophical analysis per se. If the glass is not transparent enough, seeing-through is not possible. If it is transparent enough, approximating whatever "pure" transparency could be empirically attainable, then it becomes possible to embody the technology. This is a material condition for embodiment.

Embodying as an activity, too, has an initial ambiguity. It must be learned or, in phenomenological terms, constituted. If the technology is good, this is usually easy. The very first time I put on my glasses, I see the now-corrected world. The adjustments I have to make are not usually focal irritations but fringe ones (such as the adjustment to backglare and the slight changes in spatial motility). But once learned, the embodiment relation can be more precisely described as one in which the technology becomes maximally "transparent." It is, as it were, taken into my own perceptual-bodily self experience thus:

(I-glasses)-world

My glasses become part of the way I ordinarily experience my surroundings; they "withdraw" and are barely noticed, if at all. I have then actively embodied the technics of vision. Technics is the symbiosis of artifact and user within a human action.

Embodiment relations, however, are not at all restricted to visual relations. They may occur for any sensory or microperceptual dimension. A hearing aid does this for hearing, and the blind man's cane for tactile motility. Note that in these corrective technologies *the same structural features of embodiment* obtain as with the visual example. Once learned, cane and hearing aid "withdraw" (if the technology is good—and here we have an experiential clue for the perfecting of technologies). I hear the world through the hearing aid and feel (and hear) it through the cane. The juncture (I-artifact)-world is through the technology and brought close by it.

Such relations *through* technologies are not limited to either simple or complex technologies. Glasses, insofar as they are engineered systems, are much simpler than hearing aids. More complex than either of these monosensory devices are those that entail whole-body motility. One such common technology is automobile driving. Although driving an automobile encompasses more than embodiment relations, its pleasurability is frequently that associated with embodiment relations.

One experiences the road and surroundings *through* driving the car, and motion is the focal activity. In a finely engineered sports car, for example, one has a more precise feeling of the road and of the traction upon it than in the older, softer-riding, large cars of the fifties. One embodies the car, too, in such activities as parallel parking: when well embodied, one feels rather than sees the distance between car and curb—one's bodily sense is "extended" to the parameters of the driver-car "body." And although these embodiment relations entail larger, more complex artifacts and entail a somewhat longer, more complex learning process, the bodily tacit knowledge that is acquired is perceptual-bodily.

Here is a first clue to the polymorphous sense of bodily extension. The experience of one's "body image" is not fixed but malleably extendable and/or reducible in terms of the material or technological mediations that may be embodied. I shall restrict the term embodiment, however, to those types of mediation that can be so experienced. The same dynamic polymorphousness can also be located in non-mediational or direct experience. Persons trained in the martial arts, such as karate, learn to feel the vectors and trajectories of the opponent's moves within the space of the combat. The near space around one's material body is charged.

Embodiment relations are a particular kind of use-context. They are technologically relative in a double sense. First, the technology must "fit" the use. Indeed, within the realm of embodiment relations one can develop a quite specific set of qualities for design relating to attaining the requisite technological "withdrawal." For example, in handling highly radioactive materials at a distance, the mechanical arms and hands which are designed to pick up and pour glass tubes inside the shielded enclosure have to "feed back" a delicate sense of touch to the operator. The closer to invisibility, transparency, and the extension of one's own bodily sense this technology allows, the better. Note that the design perfection is not one related to the machine alone but to the combination of machine and human. The machine is perfected along a bodily vector, molded to the perceptions and actions of humans.

And when such developments are most successful, there may arise a certain romanticizing of technology. In much anti-technological literature there are nostalgic calls for returns to simple tool technologies. In part, this may be because long-developed tools are excellent examples of bodily expressivity. They are both direct in actional terms and immediately experienced; but what is missed is that such embodiment relations may take any number of directions. Both the sports car driver within the constraints of the racing route and the bulldozer driver destroying a rainforest may have the satisfactions of powerful embodiment relations.

There is also a deeper desire which can arise from the experience of embodiment relations. It is the doubled desire that, on one side, is a wish for *total transparency*, total embodiment, for technology to truly "become me." Were this possible, it would be equivalent to there being no technology, for total transparency would *be* my body and senses; I desire the face-to-face that I would experience without the technology. But that is only one side of the desire. The other side is the desire to have the power, the transformation that the technology makes available. Only by using the technology is my bodily power enhanced and magnified by speed, through distance, or by any of the other ways in which technologies change my capacities. These capacities are always *different* from my naked capacities. The desire is, at best, contradictory. I want the transformation that the technology allows, but I want it in such a way that I am basically unaware of its presence. I want it in such a way that it becomes me. Such a desire both secretly *rejects* what technologies are and overlooks the transformational effects which are necessarily tied to human-technology relations. This illusory desire belongs equally to pro- and anti-technology interpretations of technology.

The desire is the source of both utopian and dystopian dreams. The actual, or material, technology always carries with it only a partial or quasi-transparency, which is the price for the extension of magnification that technologies give. In extending bodily capacities, the technology also transforms them. In that sense, all technologies in use are non-neutral. They change the basic situation, however subtly, however minimally; but this is the other side of the desire. The desire is simultaneously a desire for a change in situation—to inhabit the earth, or even to go beyond the earth—while sometimes inconsistently and secretly wishing that this movement could be without the mediation of the technology.

The direction of desire opened by embodied technologies also has its positive and negative thrusts. Instrumentation in the knowledge activities, notably science, is the gradual extension of perception into new realms. The desire is to see, but seeing is seeing through instrumentation. Negatively, the desire for pure transparency is the wish to escape the limitations of the material technology. It is a platonism returned in a new form, the desire to escape the newly extended body of technological engagement. In the wish there remains the contradiction: the user both wants and does not want the technology. The user wants what the technology gives but does not want the limits, the transformations that a technologically extended body implies. There is a fundamental ambivalence toward the very human creation of our own earthly tools.

The ambivalence that can arise concerning technics is a reflection of one kind upon the *essential ambiguity* that belongs to technologies in use. But this ambiguity, I shall argue, has its own distinctive shape. Embodiment relations display an essential magnification/reduction structure which has been suggested in the instrumentation examples. Embodiment relations simultaneously magnify or amplify and reduce or place aside what is experienced through them.

The sight of the mountains of the moon, through all the transformational power of the telescope, removes the moon from its setting in the expanse of the heavens. But if our technologies were only to replicate our immediate and bodily experience, they would be of little use and ultimately of little interest. A few absurd examples might show this:

In a humorous story, a professor bursts into his club with the announcement that he has just invented a reading machine. The machine scans the pages, reads them, and perfectly reproduces them. (The story apparently was written before the invention of photocopying. Such machines might be said to be "perfect reading machines" in actuality.) The problem, as the innocent could see, was that this machine leaves us with precisely the problem we had prior to its invention. To have reproduced through mechanical "reading" all the books in the world leaves us merely in the library.

A variant upon the emperor's invisible clothing might work as well. Imagine the invention of perfectly transparent clothing through which we might technologically experience the world. We could see through it, breathe through it, smell and hear through it, touch through it. Indeed, it effects no changes of any kind, since it is *perfectly* invisible. Who would bother to pick up such clothing (even if the presumptive wearer could find it)? Only by losing some invisibility—say, with translucent coloring—would the garment begin to be usable and interesting. For here, at least, fashion would have been invented—but at the price of losing total transparency—by becoming that through which we relate to an environment.

Such stories belong to the extrapolated imagination of fiction, which stands in contrast to even the most minimal actual embodiment relations, which in their material dimensions simultaneously extend and reduce, reveal and conceal.

In actual human-technology relations of the embodiment sort, the transformational structures may also be exemplified by variations: In optical technologies, I have already pointed out how spatial significations change in observations through lenses. The entire gestalt changes. When the apparent size of the moon changes, along with it the apparent position of the observer changes. Relativistically, the moon is brought "close"; and equivalently, this optical near-distance applies to both the moon's appearance and my bodily sense of position. More subtly, every dimension of spatial signification also changes. For example, with higher and higher magnification, the well-known phenomenon of depth, instrumentally mediated as a "focal plane," also changes. Depth diminishes in optical near-distance.

A related phenomenon in the use of an optical instrument is that it transforms the spatial significations of vision in an instrumentally focal way. But my seeing without instrumentation is a full bodily seeing—I see not just with my eyes but with my whole body in a unified sensory experience of things. In part, this is why there is a noticeable irreality to the apparent position of the observer, which only diminishes with the habits acquired through practice with the instrument. But the optical instrument cannot so easily transform the entire sensory gestalt. The focal sense that is magnified through the instrument is monodimensioned.

Here may be the occasion (although I am not claiming a cause) for a certain interpretation of the senses. Historians of perception have noted that, in medieval times, not only was vision not the supreme sense but sound and smell may have had greatly enhanced roles so far as the interpretation of the senses went. Yet in the Renaissance and even more exaggeratedly in the Enlightenment, there occurred the reduction to sight as the favored sense, and within sight, a certain reduction *of* sight. This favoritism, however, also carried implications for the other senses.

One of these implications was that each of the senses was interpreted to be clear and distinct from the others, with only certain features recognizable through a given sense. Such an interpretation impeded early studies in echo location.

In 1799 Lazzaro Spallanzani was experimenting with bats. He noticed not only that they could locate food targets in the dark but also that they could do so blindfolded. Spallanzani wondered if bats could guide themselves by their ears rather than by their eyes. Further experimentation, in which the bats' ears were filled with wax, showed that indeed they could not guide themselves without their ears. Spallanzani surmised that either bats locate objects through hearing or they had some sense of which humans knew nothing. Given the doctrine of separate senses and the identification of shapes and objects through vision alone, George Montagu and Georges Cuvier virtually laughed Spallanzani out of the profession.

This is not to suggest that such an interpretation of sensory distinction was due simply to familiarity with optical technologies, but the common experience of enhanced vision through such technologies was at least the standard practice of the time. Auditory technologies were to come later. When auditory technologies did become common, it was possible to detect the same amplification/reduction structure of the human-technology experience.

The telephone in use falls into an auditory embodiment relation. If the technology is good, I hear *you* through the telephone and the apparatus "withdraws" into the enabling background:

(I-telephone)-you

But as a monosensory instrument, your phenomenal presence is that of a voice. The ordinary multidimensioned presence of a face-to-face encounter does not occur, and I must at best imagine those dimensions through your vocal gestures. Also, as with the telescope, the spatial significations are changed. There is here an auditory version of visual near-distance. It makes little difference whether you are geographically near or far, none at all whether you are north or south, and none with respect to anything but your bodily relation to the instrument. Your voice retains its partly irreal near-distance, reduced from the full dimensionality of direct perceptual situations. This telephonic distance is different both from immediate face-to-face encounters and from visual or geographical distance as normally taken. Its distance is a mediated distance with its own identifiable significations.

While my primary set of variations is to locate and demonstrate the invariance of a magnification/reduction structure to any embodiment relation, there are also secondary and important effects noted in the histories of technology. In the very first use of the telephone, the users were fascinated and intrigued by its auditory transparency. Watson heard and recognized Bell's *voice,* even though the instrument had a high ratio of noise to message. In short, the fascination attaches to magnification, amplification, enhancement. But, contrarily, there can be a kind of forgetfulness that equally attaches to the reduction. What is *revealed* is what excites; what is concealed may be forgotten. Here lies one secret for technological trajectories with respect to development. There are *latent telics* that occur through inventions.

Such telics are clear enough in the history of optics. Magnification provided the fascination. Although there were stretches of time with little technical progress, this fascination emerged from time to time to have led to compound lenses by Galileo's day. If some magnification shows the new, opens to what was poorly or not at all previously detected, what can greater magnification do? In our own time, the explosion of such variants upon magnification is dramatic. Electron enhancement, computer image enhancement, CAT and NMR internal scanning, "big-eye" telescopes—the list of contemporary magnificational and visual instruments is very long.

I am here restricting myself to what may be called a *horizontal* trajectory, that is, optical technologies that bring various micro- or macro-phenomena to vision through embodiment relations. By restricting examples to such phenomena, one structural aspect of embodiment relations may be pointed to concerning the relation to microperception and its Adamic context. While *what* can be seen has changed dramatically—Galileo's New World has now been enhanced by astronomical phenomena never suspected and by micro-phenomena still being discovered—there remains a strong phenomenological constant in *how* things are seen. All lenses and optical technologies of the sort being described bring what is to be seen into a normal bodily space and distance. Both the macroscopic and the microscopic appear within the same near-distance. The "image size" of galaxy or amoeba is the *same.* Such is the existential condition for visibility, the counterpart to the technical condition, that the instrument makes things visually present.

The mediated presence, however, must fit, be made close to my actual bodily position and sight. Thus there is a reference within the instrumental context to my face-to-face capacities. These remain primitive and central within the new mediational context. Phenomenological theory claims that for every change in what is seen (the object correlate), there is a noticeable change in how (the experiential correlate) the thing is seen.

In embodiment relations, such changes retain both an equivalence and a difference from nonmediated situations. What remains constant is the bodily focus, the reflexive reference back to my bodily capacities. What is seen must be seen from or within my visual field, from the apparent distance in which discrimination can occur regarding depth, etc., just as in face-to-face relations. But the range of what can be brought into this proximity is transformed by means of the instrument.

Let us imagine for a moment what was never in fact a problem for the history of instrumentation: If the "image size" of both a galaxy and an amoeba is the "same" for the observer using the instrument, how can we tell that one is macrocosmic and the other microcosmic? The "distance" between us and these two magnitudes, Pascal noted, was the same in that humans were interpreted to be between the infinitely large and the infinitely small.

What occurs through the mediation is not a problem *because our construction of the observation presupposes ordinary praxical spatiality.* We handle the paramecium, placing it on the slide and then under the microscope. We aim the telescope at the indicated place in the sky and, before looking through it, note that the distance is at least that of the heavenly dome. But in our imagination experiment, what if our human were *totally immersed* in a technologically mediated world? What if, from birth, all vision occurred only through lens systems? Here the problem would become more difficult. But in our distance from Adam, it is precisely the presumed difference that makes it possible for us to see both nakedly *and* mediately—and thus to be able to locate the difference—that places us even more distantly from any Garden. It is because we retain this ordinary spatiality that we have a reflexive point of reference from which to make our judgments.

The noetic or bodily reflexivity implied in all vision also may be noticed in a magnified way in the learning period of embodiment. Galileo's telescope had a small field, which, combined with early hand-held positioning, made it very difficult to locate any particular phenomenon. What must have been noted, however, even if not commented upon, was the exaggerated sense of bodily motion experienced through trying to fix upon a heavenly body—and more, one quickly learns something about the earth's very motion in the attempt to use such primitive telescopes. Despite the apparent fixity of the stars, the hand-held telescope shows the earth-sky motion dramatically. This magnification effect is within the experience of one's own bodily viewing.

This bodily and actional point of reference retains a certain privilege. All experience refers to it in a taken-for-granted and recoverable way. The bodily condition of the possibility for seeing is now twice indicated by the very situation in which mediated experience occurs. Embodiment relations continue to locate that privilege of my being here. The partial symbiosis that occurs in well-designed embodied technologies retains that motility which can be called expressive. Embodiment relations constitute one existential form of the full range of the human-technology field.

B. HERMENEUTIC TECHNICS

Heidegger's hammer in use displays an embodiment relation. Bodily action through it occurs within the environment. But broken, missing, or malfunctioning, it ceases to be the means of praxis and becomes an obtruding *object* defeating the work project. Unfortunately, that negative derivation of objectness by Heidegger carries with it a block against understanding a second existential human-technology relation, the type of relation I shall term *hermeneutic.*

The term hermeneutic has a long history. In its broadest and simplest sense it means "interpretation," but in a more specialized sense it refers to *textual* interpretation and thus entails *reading.* I shall retain both these senses and take hermeneutic to mean a special interpretive action within the technological context. That kind of activity calls for special modes of action and perception, modes analogous to the reading process.

Reading is, of course, a reading of _____; and in its ordinary context, what fills the intentional blank is a text, something *written.* But all writing entails technologies. Writing has a product. Historically, and more ancient than the revolution brought about by such crucial technologies as the clock or the compass, the invention and development of writing was surely even more revolutionary than clock or compass with respect to human experience. Writing transformed the very perception and understanding we have of language. Writing is a technologically embedded form of language.

There is a currently fashionable debate about the relationship between speech and writing, particularly within current Continental philosophy. The one side argues that speech is primary, both historically and ontologically, and the other—the French School—inverts this relation and argues for the primacy of writing. I need not enter this debate here in order to note the *technological difference* that obtains between oral speech and the materially connected process of writing, at least in its ancient forms.

Writing is inscription and calls for both a process of writing itself, employing a wide range of technologies (from stylus for cuneiform to word processors for the contemporary academic), and other material entities upon which the writing is recorded (from clay tablet to computer printout). Writing is technologically mediated language. From it, several features of hermeneutic technics may be highlighted. I shall take what may at first appear as a detour into a distinctive set of human-technology relations by way of a phenomenology of reading and writing.

Reading is a specialized perceptual activity and praxis. It implicates my body, but in certain distinctive ways. In an ordinary act of reading, particularly of the extended sort, what is read is placed before or somewhat under one's eyes. We read in the immediate context from some miniaturized bird's-eye perspective. What is read occupies an expanse within the focal center of vision, and I am ordinarily in a somewhat rested position. If the object-correlate, the "text" in the broadest sense, is a chart, as in the navigational examples, what is represented retains a representational isomorphism with the natural features of the landscape. The chart represents the land- (or sea)scape and insofar as the features are isomorphic, there is a kind of representational "transparency." The chart in a peculiar way "refers" beyond itself to what it represents.

Now, with respect to the embodiment relations previously traced, such an isomorphic representation is both similar and dissimilar to what would be seen on a larger scale from some observation position (at bird's-eye level). It is similar in that the shapes on the chart are reduced representations of distinctive features that can be directly or technologically mediated in face-to-face or embodied perceptions. The reader can compare these similarities. But chart reading is also different in that, during the act of reading, the perceptual focus is the chart itself, a substitute for the landscape.

I have deliberately used the chart-reading example for several purposes. First, the "textual" isomorphism of a representation allows this first example of hermeneutic technics to remain close to yet differentiated from the perceptual isomorphism that occurs in the optical examples. The difference is at least perceptual in that one sees *through* the optical technology, but now one *sees* the chart as the visual terminus, the "textual" artifact itself.

Something much more dramatic occurs, however, when the representational isomorphism disappears in a printed text. There is no isomorphism between the printed word and what it "represents," although there is some kind of *referential* "transparency" that belongs to this new technologically embodied form of language. It is apparent from the chart example that the chart itself becomes the *object of perception* while simultaneously referring beyond itself to what is not immediately seen. In the case of the printed text, however, the referential transparency is distinctively different from technologically embodied perceptions. *Textual transparency is hermeneutic transparency, not perceptual transparency.*

Once attained, like any other acquisition of the lifeworld, writing could be read and understood in terms of its unique linguistic transparency. Writing becomes an embodied hermeneutic technics. Now the descriptions may take a different shape. What is referred to is referred by the text and is referred to *through* the text. What now presents itself is the "world" of the text.

This is not to deny that all language has its unique kind of transparency. Reference beyond itself, the capacity to let something become present through language, belongs to speech as well. But here the phenomenon being centered upon is the new embodiment of language in writing. Even more thematically, the concern is for the ways in which writing as a "technology" transforms experiential structures.

Linguistic transparency is what makes present the *world* of the text. Thus, when I read Plato, Plato's "world" is made present. But this presence is a *hermeneutic* presence. Not only does it occur *through* reading, but it takes its shape in the interpretative context of my language abilities. His world is linguistically mediated, and while the words may elicit all sorts of imaginative and perceptual phenomena, it is through language that such phenomena occur. And while such phenomena may be strikingly rich, they do not appear as word-like.

We take this phenomenon of reading for granted. It is a sedimented acquisition of the literate lifeworld and thus goes unnoticed until critical reflection isolates its salient features. It is the same with the wide variety of hermeneutic technics we employ.

The movement from embodiment relations to hermeneutic ones can be very gradual, as in the history of writing, with little-noticed differentiations along the human-technology continuum. A series of wide-ranging variants upon readable technologies will establish the point. First, a fairly explicit example of a readable technology: Imagine sitting inside on a cold day. You look out the window and notice that the snow is blowing, but you are toasty warm in front of the fire. You can clearly "see" the cold in Merleau-Ponty's pregnant sense of perception—but you do not actually *feel* it. Of course, you could, were you to go outside. You would then have a full face-to-face verification of what you had seen.

But you might also see the thermometer nailed to the grape arbor post and *read* that it is 28°F. You would now "know" how cold it was, but you still would not feel it. To retain the full sense of an embodiment relation, there must also be retained some isomorphism with the felt sense of the cold—in this case, tactile—that one would get through face-to-face experience. One could invent such a technology; for example, some conductive material could be placed through the wall so that the negative "heat," which is cold, could be felt by hand. But this is not what the thermometer does.

Instead, you read the thermometer, and in the immediacy of your reading you *hermeneutically* know that it is cold. There is an instantaneity to such reading, as it is an already constituted intuition (in phenomenological terms). But you should not fail to note that *perceptually* what you have seen is the dial and the numbers, the thermometer "text." And that text has hermeneutically delivered its "world" reference, the cold.[1]

Such constituted immediacy is not always available. For instance, although I have often enough lived in countries where Centigrade replaces Fahrenheit, I still must translate from my intuitive familiar language to the less familiar one in a deliberate and self-conscious hermeneutic act. Immediacy, however, is not the test for whether the relation is hermeneutic. A hermeneutic relation mimics sensory perception insofar as it is also a kind of seeing as _____ ; but it is a referential seeing, which has as its immediate perceptual focus seeing the thermometer.

Now let us make the case more complex. In the example cited, the experiencer had both embodiment (seeing the cold) and hermeneutic access to the phenomenon (reading the thermometer). Suppose the house were hermetically sealed, with no windows, and the only access to the weather were through the thermometer (and any other instruments we might include). The hermeneutic character of the relation becomes more obvious. I now clearly have to know how to read the instrumentation and from this reading knowledge get hold of the "world" being referred to.

This example has taken actual shape in nuclear power plants. In the Three Mile Island incident, the nuclear power system was observed only through instrumentation. Part of the delay that caused a near meltdown was *misreadings* of the instruments. There was no face-to-face, independent access to the pile or to much of the machinery involved, nor could there be.

An intentionality analysis of this situation retains the mediational position of the technology:

I-technology-world
(engineer-instruments-pile)

The operator has instruments between him or her and the nuclear pile. But—and here, an essential difference emerges between embodiment and hermeneutic relations—what is immediately perceived is the instrument panel itself. It becomes the object of my microperception, although in the special sense of a hermeneutic transparency, I *read* the pile through it. This situation calls for a different formalization:

I-(technology-world)

The parenthesis now indicates that the immediate *perceptual* focus of my experience *is* the control panel. I read through it, but this reading is now dependent upon the semi-opaque connection between the instruments and the referent object (the pile). This *connection* may now become enigmatic.

In embodiment relations, what allows the partial symbiosis of myself and the technology is the capacity of the technology to become perceptually transparent. In the optical examples, the glass-maker's and lens-grinder's arts must have accomplished this end if the embodied use is to become possible. Enigmas which may occur regarding embodiment-use transparency thus may occur within the parenthesis of the embodiment relation:

$$(\text{I-technology}) \rightarrow \text{World}$$
$$\underset{\text{enigma position}}{\uparrow\underline{\qquad\qquad}}$$

(This is not to deny that once the transparency is established, thus making microperception clear, the observer may still fail, particularly at the macroperceptual level. For the moment, however, I shall postpone this type of interpretive problem.) It would be an oversimplification of the history of lens-making were not problems of this sort recognized. Galileo's instrument not only was hard to look through but was good only for certain "middle range" sightings in astronomical terms (it did deliver the planets and even some of their satellites). As telescopes became more powerful, levels, problems with chromatic effects, diffraction effects, etc., occurred. As Ian Hacking has noted,

> Magnification is worthless if it magnifies two distinct dots into one big blur. One needs to resolve the dots into two distinct images. . . . It is a matter of diffraction. The most familiar example of diffraction is the fact that shadows of objects with sharp boundaries are fuzzy. This is a consequence of the wave character of light.[2]

Many such examples may be found in the history of optics, technical problems that had to be solved before there could be any extended reach within embodiment relations. Indeed, many of the barriers in the development of experimental science can be located in just such limitations in instrumental capacity.

Here, however, the task is to locate a parallel difficulty in the emerging new human-technology relation, hermeneutic relations. The location of the technical problem in hermeneutic relations lies in the *connector* between the instrument and the referent. Perceptually, the user's visual (or other) terminus is *upon* the instrumentation itself. To read an instrument is an analogue to reading a text. But if the text does not correctly refer, its reference object or its world cannot be present. Here is a new location for an enigma:

$$\text{I} \rightarrow (\text{technology-world})$$
$$\underset{\text{enigma position}}{\downarrow}$$

While breakdown may occur at any part of the relation, in order to bring out the graded distinction emerging between embodiment and hermeneutic relations, a short pathology of connectors might be noted.

If there is nothing that impedes my direct perceptual situation with respect to the instrumentation (in the Three Mile Island example, the lights remain on, etc.), interpretive problems in reading a strangely behaving "text" at least occur in the open; but the technical enigma may also occur within the text-referent relation. How could the operator tell if the instrument was malfunctioning or that to which the instrument refers? Some form of *opacity* can occur within the technology-referent pole of the relation. If there is some independent way of verifying which aspect is malfunctioning (a return to unmediated face-to-face relations), such a breakdown can be easily detected. Both such occurrences are reasons for instrumental redundancy. But in examples where such independent verification is not possible or untimely, the opacity would remain.

Let us take a simple mechanical connection as a borderline case. In shifting gears on my boat, there is a lever in the cockpit that, when pushed forward, engages the forward gear; upward, neutral; and backwards, reverse. Through it, I can ordinarily feel the gear change in the transmission (embodiment) and recognize the simple hermeneutic signification (forward for forward) as immediately intuitive. Once, however, on coming in to the dock at the end of the season, I disengaged the forward gear—and the propeller continued to drive the boat forward. I quickly reversed—and again the boat continued. The hermeneutic significance had failed; and while I also felt a difference in the way the gear lever felt, I did not discover until later that the clasp that retained the lever itself had corroded, thus preventing any actual shifting at all. But even at this level there can be opacity within the technology-object relation.

The purpose of this somewhat premature pathology of human-technology relations is not to cast a negative light upon hermeneutic relations in contrast to embodiment ones but rather to indicate that there are different locations where perceptual and human-technology relations interact. Normally, when the technologies work, the technology-world relation would retain its unique hermeneutic transparency. But if the I-(technology-world) relation is far enough along the continuum to identify the relation as a hermeneutic one, the intersection of perceptual-bodily relations with the technology changes.

Readable technologies call for the extension of my hermeneutic and "linguistic" capacities *through* the instruments, while the reading itself retains its bodily perceptual location as a relation *with* or *towards* the technology. What is emerging here is the first suggestion of an emergence of the technology as "object" but without its negative Heideggerian connotation. Indeed, the type of special capacity as a "text" is a condition for hermeneutic transparency.

The transformation made possible by the hermeneutic relation is a transformation that occurs precisely through *differences* between the text and what is referred to. What is needed is a particular set of textually clear perceptions that "reduce" to that which is immediately readable. To return to the Three Mile Island example, one problem uncovered was that the instrument panel design was itself faulty. It did not incorporate its dials and gauges in an easily readable way. For example, in airplane instrument panel design, much thought has been given to pattern recognition, which occurs as a perceptual gestalt. Thus, in a four-engined aircraft, the four dials indicating r.p.m. will be coordinated so that a single glance will indicate which, if any, engine is out of synchronization. Such technical design accounts for perceptual structures.

There is a second caution concerning the focus upon connectors and pathology. In all the examples I have used to this point, the hermeneutic technics have involved material connections. (The thermometer employs a physical property of a bimetallic spring or mercury in a column; the instrument panel at TMI employs mechanical, electrical, or other material connections; the shift lever, a simple mechanical connection.) If reading does not employ any such material connections, it might seem that its referentiality is essentially different, yet not even all technological connections are strictly material. Photography retains representational isomorphism with the object, yet does not "materially" connect with its object; it is a minimal beginning of action at a distance.

I have been using contemporary or post-scientific examples, but non-material hermeneutic relations do not obtain only for contemporary humans. As existential relations, they are as "old" as post-Garden humanity. Anthropology and the history of religions have long been familiar with a wide variety of shamanistic praxes which fall into the pattern of hermeneutic technics. In what may at first seem a somewhat outrageous set of examples, note the various "reading" techniques employed in shamanism. The reading of animal entrails, of thrown bones, of bodily marks—all are hermeneutic techniques. The patterns of the entrails, bones, or whatever are taken to *refer* to some state of affairs, instrumentally or textually.

Not only are we here close to a familiar association between magic and the origins of technology suggested by many writers, but we are, in fact, closer to a wider hermeneutic praxis in an intercultural setting. For that reason, the very strangeness of the practice must be critically examined. If the throwing of bones is taken as a "primitive" form of medical diagnosis—which does play a role in shamanism—we might conclude that it is indeed a poor form of hermeneutic relations. What we might miss, however, is that the entire gestalt of what is being diagnosed may differ radically from the other culture and ours.

It may well be that as a focused form of diagnosis upon some particular bodily ailment (appendicitis, for example), the diagnosis will fail. But since one important element in shamanism is a wider diagnosis, used particularly as the occasion of locating certain communal or social problems, it may work better. The sometimes socially contextless emphasis of Western medicine upon a presumably "mechanical" body may overlook precisely the context which the shaman so clearly recognizes. The entire gestalt is different and differently focused, but in both cases there are examples of hermeneutic relations.

In our case, the very success of Western medicine in certain diseases is due to the introduction of technologies into the hermeneutic relation (fever/thermometer; blood pressure/manometer, etc.) The point is that hermeneutic relations are as commonplace in traditional and ancient social groups as in ours, even if they are differently arranged and practiced.

By continuing the intentionality analysis I have been following, one can now see that hermeneutic relations vary the continuum of human-technology-world relations. Hermeneutic relations maintain the general mediation position of technologies within the context of human praxis towards a world, but they also change the variables within the human-technology-world relation. A comparative formalism may be suggestive:

General intentionality relations
Human-technology-world

Variant A: embodiment relations
(I-technology) → world

Variant B: hermeneutic relations
I → (technology-world)

While each component of the relation changes within the correlation, the overall shapes of the variants are distinguishable. Nor are these matters of simply how technologies are experienced.

Another set of examples from the set of optical instruments may illustrate yet another way in which instrumental intentionalities can follow new trajectories. Strictly embodiment relations can be said to work best when there is both a transparency and an isomorphism between perceptual and bodily action within the relation. I have suggested that a trajectory for development in such cases may often be a horizontal one. Such a trajectory not only follows greater and greater degrees of magnification but also entails all the difficulties of a technical nature that go into allowing what is to be seen as though by direct vision. But not all optical technologies follow this strategy. The introduction of hermeneutic possibilities opens the trajectory into what I shall call *vertical* directions, possibilities that rely upon quite deliberate hermeneutic transformations.

It might be said that the telescope and microscope, by extending vision while transforming it, remained *analogue* technologies. The enhancement and magnification made possible by such technologies remain visual and transparent to ordinary vision. The moon remains recognizably the moon, and the microbe—even if its existence was not previously suspected—remains under the microscope a beastie recognized as belonging to the animate continuum. Here, just as the capacity to magnify becomes the foreground phenomenon to the background phenomenon of the reduction necessarily accompanying the magnification, so the similitude of what is seen with ordinary vision remains central to embodiment relations.

Not all optical technologies mediate such perceptions. In gradually moving towards the visual "alphabet" of a hermeneutic relation, deliberate variations may occur which enhance previously undiscernible *differences*:

1) Imagine using spectacles to correct vision, as previously noted. What is wanted is to *return* vision as closely as possible to ordinary perception, not to distort or modify it in any extreme micro- or macroperceptual direction. But now, for snowscapes or sun on the water or desert, we modify the lenses by coloring or polarizing them to cut glare. Such a variation transforms *what* is seen in some degree. Whether we say the polarized lens removes glare or "darkens" the landscape, what is seen is now clearly different from what may be seen through untinted glasses. This difference is a clue which may open a new *telic direction* for development.

2) Now say that somewhere, sometime, someone notes that certain kinds of tinting reveal unexpected results. Such is a much more complex technique now used in infrared satellite photos. (For the moment, I shall ignore the fact that part of this process is a combined embodiment and hermeneutic relation.) If the photo is of the peninsula of Baja California, it will remain recognizable in shape. Geography, whatever depth and height representations, etc., remain but vary in a direction different from any ordinary vision. The infrared photo enhances the difference between vegetation and non-vegetation beyond the limits of any isomorphic color photography. This difference corresponds, in the analogue example, to something like a pictograph. It simultaneously leaves certain analogical structures there and begins to modify the representation into a different, non-perceived "representation."

3) Very sophisticated versions of still representative but non-ordinary forms of visual recognition occur in the new heat-sensitive and light-enhanced technologies employed by the military and police. Night scopes which enhance a person's heat radiation still look like a person but with entirely different regions of what stands out and what recedes. In high-altitude observations, "heat shadows" on the ground can indicate an airplane that has recently had its engines running compared to others which have not. Here visual technologies bring into visibility what was not visible, but in a distinctly now perceivable way.

4) If now one takes a much larger step to spectrographic astronomy, one can see the acceleration of this development. The spectrographic picture of a star no longer "resembles" the star at all. There is no point of light, no disk size, no spatial isomorphism at all—merely a band of differently colored rainbow stripes. The naive reader would not know that this was a picture of a star at all—the reader would have to know the language, the alphabet, that has coded the star. The astronomer-hermeneut does know the language and "reads" the visual "ABCs" in such a way that he knows the chemical composition of the star, its internal makeup, rather than its shape or external configuration. We are here in the presence of a more fully hermeneutic relation, the star mediated not only instrumentally but in a transformation such that we must now thematically *read* the result. And only the informed reader can do the reading.

There remains, of course, the *reference* to the star. The spectrograph is *of* Rigel or *of* Polaris, but the individuality of the star is now made present hermeneutically. Here we have a beginning of a special transformation of perception, a transformation which deliberately enhances differences rather than similarities in order to get at what was previously unperceived.

5) Yet even the spectrograph is but a more radical transformation of perception. It, too, can be transformed by a yet more radical *hermeneutic* analogue to the *digital* transformation which lies embedded in the preferred quantitative praxis of science. The "alphabet" of science is, of course, mathematics, a mathematics that separates itself by yet another hermeneutic step from perception embodied.

There are many ways in which this transformation can and does occur, most of them interestingly involving a particular act of *translation* that often goes unnoticed. To keep the example as simple as possible, let us assume *mechanical* or *electronic* "translation." Suppose our spectrograph is read by a machine that yields not a rainbow spectrum but a set of numbers. Here we would arrive at the final hermeneutic accomplishment, the transformation of even the analogue to a digit. But in the process of hermeneuticization, the "transparency" to the object referred to becomes itself enigmatic. Here more explicit and thematic interpretation must occur.

C. ALTERITY RELATIONS

Beyond hermeneutic relations there lie *alterity relations*. The first suggestions of such relations, which I shall characterize as relations *to* or *with* a technology, have already been suggested in different ways from within the embodiment and hermeneutic contexts. Within embodiment relations, were the technology to intrude upon rather than facilitate one's perceptual and bodily extension into the world, the technology's objectness would necessarily have appeared negatively. Within hermeneutic relations, however, there emerged a certain positivity to the objectness of instrumental technologies. The bodily-perceptual focus *upon* the instrumental text is a condition of its own peculiar hermeneutic transparency. But what of a positive or presentential sense of relations *with* technologies? In what phenomenological senses can a technology be *other*?

The analysis here may seem strange to anyone limited to the habits of objectivist accounts, for in such accounts technologies as objects usually come first rather than last. The problem for a phenomenological account is that objectivist ones are non-relativistic and thus miss or submerge what is distinctive about human-technology relations.

A naive objectivist account would likely begin with some attempt to circumscribe or define technologies by object characteristics. Then, what I have called the technical properties of technologies would become focal. Some combination of physical and material properties would be taken to be definitional. (This is an inherent tendency of the standard nomological positions such as those of Bunge and Hacking). The definition will often serve a secondary purpose by being stipulative: only those technologies that are obviously dependent upon or strongly related to contemporary scientific and industrial productive practices will count.

This is not to deny that objectivist accounts have their own distinctive strengths. For example, many such accounts recognize that technological or "artificial" products are different from the simply found object or the natural object. But the submergence of the human-technology relation remains hidden, since either object may enter into praxis and both will have their material, and thus limited, range of technical usability within the relation. Nor is this to deny that the objectivist accounts of types of technologies, types of organization, or types of designed purposes should be considered. But the focus in this first program remains the phenomenological derivation of the set of human-technology relations.

There is a tactic behind my placing alterity relations last in the order of focal human-technology relations. The tactic is designed, on the one side, to circumvent the tendency succumbed to by Heidegger and his more orthodox followers to see the otherness of technology only in negative terms or through negative derivations. The hammer example, which remains paradigmatic for this approach, is one that derives objectness from breakdown. The broken or missing or malfunctioning technology could be *discarded.* From being an obtrusion it could become *junk.* Its objectness would be clear—but only partly so. Junk is not a focal object of use relations (except in certain limited situations). It is more ordinarily a background phenomenon, that which has been put out of use.

Nor, on the other side, do I wish to fall into a naively objectivist account that would simply concentrate upon the material properties of the technology as an object of knowledge. Such an account would submerge the relativity of the intentionality analysis, which I wish to preserve here. What is needed is an analysis of the positive or presentential senses in which humans relate to technologies as relations *to* or with technologies, to technology-as-other. It is this sense which is included in the term "alterity."

Philosophically, the term "alterity" is borrowed from Emmanuel Levinas. Although Levinas stands within the traditions of phenomenology and hermeneutics, his distinctive work, *Totality and Infinity,* was "anti-Heideggerian." In that work, the term "alterity" came to mean the radical difference posed to any human by another human, an *other* (and by the ultimately other, God). Extrapolating radically from within the tradition's emphasis upon the non-reducibility of the human to either objectness (in epistemology) or as a means (in ethics), Levinas poses the otherness of humans as a kind of *infinite* difference that is concretely expressed in an ethical, face-to-face encounter.

I shall retain but modify this radical Levinasian sense of human otherness in returning to an analysis of human-technology relations. How and to what extent do technologies become other or, at least, *quasi-other?* At the heart of this question lie a whole series of well-recognized but problematic interpretations of technologies. On the one side lies the familiar problem of anthropomorphism, the personalization of artifacts. This range of anthropomorphism can reach from serious artifact-human analogues to trivial and harmless affections for artifacts.

An instance of the former lies embedded in much AI research. To characterize computer "intelligence" as human-like is to fall into a peculiarly contemporary species of anthropomorphism, however sophisticated. An instance of the latter is to find oneself "fond" of some particular technofact as, for instance, a long-cared-for automobile which one wishes to keep going and which may be characterized by quite deliberate anthropomorphic terms. Similarly, in ancient or non-Western cultures, the role of sacredness attributed to artifacts exemplifies another form of this phenomenon.

The religious object (idol) does not simply "represent" some absent power but is endowed with the sacred. Its aura of sacredness is spatially and temporally present within the range of its efficacy. The tribal devotee will defend, sacrifice to, and care for the sacred artifact. Each of these illustrations contains the seeds of an alterity relation.

A less direct approach to what is distinctive in human-technology alterity relations may perhaps better open the way to a phenomenologically relativistic analysis. My first example comes from a comparison to a technology and to an animal "used" in some practical (although possibly sporting) context: the spirited horse and the spirited sports car.

To ride a spirited horse is to encounter a lively animal *other.* In its pre- or nonhuman context, the horse has a life of its own within the environment that allowed this form of life. Once domesticated, the horse can be "used" as an "instrument" of human praxis—but only to a degree and in a way different from counterpart technologies; in this case, the "spirited" sports car.

There are, of course, analogues which may at first stand out. Both horse and car give the rider/driver a magnified sense of power. The speed and the experience of speed attained in riding/

driving are dramatic extensions of my own capacities. Some prominent features of embodiment relations can be found analogously in riding/driving. I experience the trail/road through horse/car and guide/steer the mediating entity under way. But there are equally prominent differences. No matter how well trained, no horse displays the same "obedience" as the car. Take malfunction: in the car, a malfunction "resists" my command—I push the accelerator, and because of a clogged gas line, there is not the response I expected. But the animate resistance of a spirited horse is more than such a mechanical lack of response—the response is more than malfunction, it is *dis*obedience. (Most experienced riders, in fact, prefer spirited horses over the more passive ones, which might more nearly approximate a mechanical obedience.) This life of the other in a horse may be carried much further—it may live without me in the proper environment; it does not need the *deistic* intervention of turning the starter to be "animated." The car will not shy at the rabbit springing up in the path any more than most horses will obey the "command" of the driver to hit the stone wall when he is too drunk to notice. The horse, while approximating some features of a mediated embodiment situation, never fully enters such a relation in the way a technology does. Nor does the car ever attain the sense of animation to be found in horseback riding. Yet the analogy is so deeply embedded in our contemporary consciousness (and perhaps the lack of sufficient experience with horses helps) that we might be tempted to emphasize the similarities rather than the differences.

Anthropomorphism regarding the technology on the one side and the contrast with horseback riding on the other point to a first approximation to the unique type of otherness that relations to technologies hold. Technological otherness is a *quasi-otherness,* stronger than mere objectness but weaker than the otherness found within the animal kingdom or the human one; but the phenomenological derivation must center upon the positive experiential aspects outlining this relation.

In yet another familiar phenomenon, we experience technologies as *toys* from childhood. A widely cross-cultural example is the spinning top. Prior to being put into use, the top may appear as a top-heavy object with a certain symmetry of design (even early tops approximate the more purely functional designs of streamlining, etc.), but once "deistically" animated through either stick motion or a string spring, the now spinning top appears to take on a life of its own. On its tip (or "foot") the top appears to defy its top-heaviness and gravity itself. It traces unpredictable patterns along its pathway. It is an object of *fascination.*

Note that once the top has been set to spinning, what was imparted through an embodiment relation now exceeds it. What makes it fascinating is this property of quasi-animation, the life of its own. Also, of course, once "automatic" in its motion, the top's movements may be entered into a whole series of possible contexts. I might enter a game of warring tops in which mine (suitably marked) represents me. If I-as-top am successful in knocking down the other tops, then this game of hermeneutics has the top winning for me. Similarly, if I take its quasi-autonomous motion to be a hermeneutic predictor, I may enter a divination context in which the path traced or the eventual point of stoppage indicates some fortune. Or, entering the region of scientific instrumentation, I may transform the top into a gyroscope, using its constancy of direction within its now-controlled confines as a better-than-magnetic compass. But in each of these cases, the top may become the focal center of attention as a quasi-other to which I may relate. Nor need the object of fascination carry either an embodiment or hermeneutic referential transparency.

To the ancient and contemporary top, compare briefly the fascination that occurs around video games. In the actual use of video games, of course, the embodiment and hermeneutic relational dimensions are present. The joystick that embodies hand and eye coordination skills extends the player into the displayed field. The field itself displays some hermeneutic context (usually either some "invader" mini-world or some sports analogue), but this context does not refer beyond itself into a worldly reference.

In addition to these dimensions, however, there is the sense of *interacting with* something other than me, the technological *competitor.* In competition there is a kind of dialogue or exchange. It is the quasi-animation, the quasi-otherness of the technology that fascinates and challenges. I must beat the machine or it will beat me.

Although the progression of the analysis here moves from embodiment and hermeneutic relations to alterity ones, the interjection of film or cinema examples is of suggestive interest. Such technologies are transitional between hermeneutic and alterity phenomena. When I first introduced the notion of hermeneutic relations, I employed what could be called a "static" technology: writing. The long and now ancient technologies of writing result in fixed texts (books, manuscripts, etc., all of which, barring decay or destruction, remain stable in themselves). With film, the "text" remains fixed only in the sense that one can repeat, as with a written text, the seeing and hearing of the cinema text. But the mode of presentation is dramatically different. The "characters" are now animate and theatrical, unlike the fixed alphabetical characters of the written text. The dynamic "world" of the cinema-text, while retaining many of the functional features of writing, also now captures the semblance of real-time, action, etc. It remains to be "read" (viewed and heard), but the object-correlate necessarily appears more "life-like" than its analogue—written text. This factor, naively experienced by the current generations of television addicts, is doubtless one aspect in the problems that emerge between television watching habits and the state of reading skills. James Burke has pointed out that "the majority of the people in the advanced industrialized nations spend more time watching television than doing anything else beside work."[3] The same balance of time use also has shown up in surveys regarding students. The hours spent watching television among college and university students, nationally, are equal to or exceed those spent in doing homework or out-of-class preparation.

Film, cinema, or television can, in its hermeneutic dimension, refer in its unique way to a "world." The strong negative response to the Vietnam War was clearly due in part to the virtually unavoidable "presence" of the war in virtually everyone's living room. But films, like readable technologies, are also *presentations,* the focal terminus of a perceptual situation. In that emergent sense, they are more dramatic forms of perceptual immediacy in which the presented display has its own characteristics conveying quasi-alterity. Yet the engagement with the film normally remains short of an engagement with an *other.* Even in the anger that comes through in outrage about civilian atrocities or the pathos experienced in seeing starvation epidemics in Africa, the emotions are not directed to the screen but, indirectly, through it, in more appropriate forms of political or charitable action. To this extent there is retained a hermeneutic reference elsewhere than at the technological instrument. Its quasi-alterity, which is also present, is not fully focal in the case of such media technologies.

A high-technology example of breakdown, however, provides yet another hint at the emergence of alterity phenomena. Word processors have become familiar technologies, often strongly liked by their users (including many philosophers who fondly defend their choices, profess knowledge about the relative abilities of their machines and programs, etc.). Yet in breakdown, this quasi-love relationship reveals its quasi-hate underside as well. Whatever form of "crash" may occur, particularly if some fairly large section of text is involved, it occasions frustration and even rage. Then, too, the programs have their idiosyncrasies, which allow or do not allow certain movements; and another form of human-technology competition may emerge. (Mastery in the highest sense most likely comes from learning to program and thus overwhelm the machine's previous brain-power. "Hacking" becomes the game-like competition in which an entire system is the alterity correlate.) Alterity relations may be noted to emerge in a wide range of computer technologies that, while failing quite strongly to mimic bodily incarnations, nevertheless display a quasi-otherness within the limits of linguistics and, more particularly, of logical behaviors. Ultimately, of course, whatever contest emerges, its sources lie opaquely with other humans as well but also with the transformed technofact, which itself now plays a more obvious role within the overall relational net.

I have suggested that the computer is one of the stronger examples of a technology which may be positioned within alterity relations. But its otherness remains a quasi-otherness, and its genuine usefulness still belongs to the borders of its hermeneutic capacities. Yet in spite of this, the tendency to fantasize its quasi-otherness into an authentic otherness is pervasive. Romanticizations such as the portrayal of the emotive, speaking "Hal" of the movie *2001: A Space Odyssey,* early fears that the "brain power" of computers would soon replace human thinking, fears that political or military decisions will not only be informed by but also made by computers—all are symptoms revolving around the positing of otherness to the technology.

These romanticizations are the alterity counterparts to the previously noted dreams that wish for total embodiment. Were the technofact to be genuinely an other, it would both be and not be a *technology.* But even as quasi-other, the technology falls short of such totalization. It retains its unique role in the human-technology continuum of relations as the medium of transformation, but as a recognizable medium.

The wish-fulfillment desire occasioned by embodiment relations—the desire for a fully transparent technology that would *be* me while at the same time giving me the powers that the use of the technology makes available—here has its counterpart fantasy, and this new fantasy has the same internal contradiction: It both reduces or, here, extrapolates the technology into that which is not a technology (in the first case, the magical transformation is *into me;* in this case, *into the other*), and at the same time, it desires what is not identical with me or the other. The fantasy is for the transformational effects. Both fantasies, in effect, deny technologies playing the roles they do in the human-technology continuum of relations; yet it is only on the condition that there be some detectable differentiation within the relativity that the unique ways in which technologies transform human experience can emerge.

In spite of the temptation to accept the fantasy, what the quasi-otherness of alterity relations does show is that humans may relate positively or presententially *to* technologies. In that respect and to that degree, technologies emerge as focal entities that may receive the multiple attentions humans give the different forms of the other. For this reason, a third formalization may be employed to distinguish this set of relations:

$$I \rightarrow technology\text{-}(\text{-world})$$

I have placed the parentheses thusly to indicate that in alterity relations there may be, but need not be, a relation through the technology to the world (although it might well be expected that the *usefulness* of any technology will necessarily entail just such a referentiality). The world, in this case, may remain context and background, and the technology may emerge as the foreground and focal quasi-other with which I momentarily engage.

This disengagement of the technology from its ordinary-use context is also what allows the technology to fall into the various disengaged engagements which constitute such activities as play, art, or sport.

A first phenomenological itinerary through direct and focal human-technology relations may now be considered complete. I have argued that the three sets of distinguishable relations occupy a continuum. At the one extreme lie those relations that approximate technologies to a quasi-me (embodiment relations). Those technologies that I can so take into my experience that through their semi-transparency they allow the world to be made immediate thus enter into the existential relation which constitutes my self. At the other extreme of the continuum lie alterity relations in which the technology becomes quasi-other, or technology "as" other *to* which I relate. Between lies the relation with technologies that both mediate and yet also fulfill my perceptual and bodily relation with technologies, hermeneutic relations. The variants may be formalized thus:

Human-technology-World Relations
Variant 1, Embodiment Relations
(Human-technology) → World
Variant 2, Hermeneutic Relations
Human → (technology-World)
Variant 3, Alterity Relations
Human → technology-(-World)

Although I have characterized the three types of human-technology relations as belonging to a continuum, there is also a sense in which the elements within each type of relation are differently distributed. There is a *ratio* between the objectness of the technology and its transparency in use. At the extreme height of embodiment, a background presence of the technology may still be detected. Similarly but with a different ratio, once the technology has emerged as a quasi-other, its alterity remains within the domain of human invention through which the world is reached. Within all the types of relations, technology remains artifactual, but it is also its very artifactual formation which allows the transformations affecting the earth and ourselves.

All the relations examined heretofore have also been focal ones. That is, each of the forms of action that occur through these relations have been marked by an implicated self-awareness. The engagements through, with, and to technologies stand within the very core of praxis. Such an emphasis, while necessary, does not exhaust the role of technologies nor the experiences of them. If focal activities are central and foreground, there are also fringe and background phenomena that are no more neutral than those of the foreground. It is for that reason that one final foray in this phenomenology of technics must be undertaken. That foray must be an examination of technologies in the background and at the horizons of human-technology relations.

D. BACKGROUND RELATIONS

With background relations, this phenomenological survey turns from attending to technologies in a foreground to those which remain in the background or become a kind of near-technological environment itself. Of course, there are discarded or no-longer-used technologies, which in an extreme sense occupy a background position in human experience—junk. Of these, some may be recuperated into non-use but focal contexts such as in technology museums or in the transformation into junk art. But the analysis here points to specifically functioning technologies which ordinarily occupy background or field positions.

First, let us attend to certain individual technologies designed to function in the background—automatic and semiautomatic machines, which are so pervasive today—as good candidates for this analysis. In the mundane context of the home, lighting, heating, and cooling systems, and the plethora of semiautomatic appliances are good examples. In each case, there is some necessity for an instant of deistic intrusion to program or set the machinery into motion or to its task. I set the thermostat; then, if the machinery is high-tech, the heating/cooling system will operate independently of ongoing action. It may employ time-temperature changes, external sensors to adjust to changing weather, and other cybernetic operations. (While this may function well in the home situation, I remain amused at the still-primitive state of the art in the academic complex I occupy. It takes about two days for the system to adjust to the sudden fall and spring weather changes, thus making offices which actually have opening windows—a rarity—highly desirable.)

Once operating, the technology functions as a barely detectable background presence; for example, in the form of background noise, as when the heating kicks in. But in operation, the technology does not call for focal attention.

Note two things about this human-technology relation: First, the machine activity in the role of background presence is not displaying either what I have termed a transparency or an opacity. The "withdrawal" of this technological function is phenomenologically distinct as a kind of "absence." The technology is, as it were, "to the side." Yet as a present absence, it nevertheless becomes part of the experienced field of the inhabitant, a piece of the immediate environment.

Somewhat higher on the scale of semiautomatic technologies are task-oriented appliances that call for explicit and repeated deistic interventions. The washing machine, dryer, microwave, toaster, etc., all call for repeated programming and then for dealing with the processed product (wash, food, etc.). Yet like the more automated systems, the semiautomatic machine remains in the background while functioning.

In both systems and appliances, however, one also may detect clues to the ways in which background relations texture the immediate environment. In the electric home, there is virtually a constant hum of one sort or the other, which is part of the technological texture. Ordinarily, this "white noise" may go unnoticed, although I am always reassured that it remains part of fringe awareness, as when guests visit my mountain home in Vermont. The inevitable comment is about the silence of the woods. At once, the absence of background hum becomes noticeable.

Technological texturing is, of course, much deeper than the layer of background noise which signals its absent presence. Before turning to further implications, one temptation which could occur through the too-narrow selection of contemporary examples must be avoided. It might be thought that only, or predominantly, the high-technology contemporary world uses and experiences technologies as backgrounds. That is not the case, even with respect to automated or semiautomatic technologies.

The scarecrow is an ancient "automated" device. Its mimicry of a human, with clothes flapping in the breeze, is a specifically designed automatic crow scarer, made to operate in the absence of humans. Similarly, in ancient Japan there were automated deer scarers, made of bamboo tubes, pivoted on a pin and placed so that a waterfall or running stream would slowly fill the tube. When it is full enough, the device would trip and its other end strike a sounding board or drum, the noise of which would frighten away any marauding deer. We have already noted the role automation plays in religious rituals (prayer wheels and worship representations thought to function continuously).

Interpreted technologically, there are even some humorous examples of "automation" to be found in ancient religious praxes. The Hindu prayer windmill "automatically" sends its prayers when the wind blows; and in the ancient Sumerian temples there were idols with large eyes at the altars (the gods), and in front of them were smaller, large-eyed human statues representing worshipers. Here was an ancient version of an "automated" worship. (Its contemporary counterpart would be the joke in which the professor leaves his or her lecture on a tape recorder for the class—which students could also "automatically" hear, by leaving their own cassettes to tape the master recording.)

While we do not often conceptualize such ancient devices in this way, part of the purpose of an existential analysis is precisely to take account of the identity of function and of the "ancientness" of all such existential relations. This is in no way to deny the differences of context or the degree of complexity pertaining to the contemporary, as compared to the ancient, versions of automation.

Another form of background relation is associated with various modalities of the technologies that serve to insulate humans from an external environment. Clothing is a borderline case. Clothing clearly insulates our bodies from temperature, wind, and other external weather phenomena that could become dangerous to life; but clothing experienced is borderline with embodiment relations, for we do feel the external environment through clothing, albeit in a particularly damped-down mode. Clothing is not designed, in most cases, to be "transparent" in the way the previous instrument examples were but rather to have a certain opacity without restricting movement. Yet clothing is part of a fringe awareness in most of our daily activities (I am obviously not addressing fashion aspects of clothing here).

A better example of a background relation is a shelter technology. Although shelters may be found (caves) and thus enter untransformed into human praxis, most are constructed, as are most technological artifacts; but once constructed and however designed to insulate or account for external weather, they become a more field-like background phenomenon. Here again, human cultures display an amazing continuum from minimalist to maximalist strategies with respect to this version of a near-background.

Many traditional cultures, particularly in Southern Hemisphere areas, practice an essentially open shelter technology, perhaps with primarily a roof to keep off rain and sun. Such peoples frequently find distasteful such items as windows and, particularly, glassed windows. They do not wish to be too isolated or insulated from the elements. At the other extreme is the maximalist strategy, which most extremely wishes to totalize shelter technology into a virtual life-support system, autonomous and enclosed. I shall call this a technological cocoon.

A contemporary example of a near-cocoon is the nuclear submarine. Its crew lives inside, and the vessel is designed to remain at sea for prolonged periods, even underwater for long stretches of time. There are sophisticated recycling systems for waste, water, and air. Contact with the outside, obviously important in this case, is primarily through monitoring equally sophisticated hermeneutic devices (sonar, low-frequency radio, etc.). All ordinary duties take place in the cocoon-like interior. A multibillion-dollar projection to a greater degree of cocoonhood is the long-term space station now under debate.

Part of the very purpose of the space station is to experiment with creating a mini-environment, or artificial "earth," which would be totally technologically mediated. Yet contemporary high-tech suburban homes show similar features. Fully automated for temperature and humidity, tight air structures, some with glass that adjusts to glare, all such homes lie on the same trajectory of self-containment. But while these illustrations are uniquely high-technology textured, there remain, as before, differently contexted but similar examples from the past.

Totally enclosed spaces have frequently been associated with ritual and religious praxis. The Kiva of past southwestern native American cultures was dug deep into the ground, windowless and virtually sealed. It was the site for important initiatory and secret societies, which gathered into such ancient cocoons for their own purposes. The enclosure bespeaks different kinds of totalization.

What is common to the entire range of examples pointed to here is the position occupied by such technology, background position, the position of an absent presence as a part of or a total field of immediate technology.

In each of the examples, the background role is a field one, not usually occupying focal attention but nevertheless conditioning the context in which the inhabitant lives. There are, of course, great differences to be detailed in terms of the types of contexts which such background technologies play. Breakdown, again, can play a significant indexical role in pointing out such differences.

The involvement implications of contemporary, high-technology society are very complex and often so interlocked as to fall into major disruption when background technology fails. In 1985 Long Island was swept by Hurricane Gloria with massive destruction of power lines. Most areas went without electricity for at least a week, and in some cases, two. Lighting had to be replaced by older technologies (lanterns, candles, kerosene lamps), supplies for which became short immediately. My own suspicion is that a look at birth statistics at the proper time after this radical change in evening habits will reveal the same glitch which actually did occur during the blackouts of earlier years in New York.

Similarly, with the failure of refrigeration, eating habits had to change temporarily. The example could be expanded quite indefinitely; a mass purchase of large generators by university buyers kept a Minnesota company in full production for several months after, to be prepared the "next time." In contrast, while the same effects on a shorter-term basis were experienced in the grid-wide blackouts of 1965, I was in Vermont at my summer home, which is lighted by kerosene lamps and even refrigerated with a kerosene refrigerator. I was simply unaware of the massive disruption until the Sunday *Times* arrived. Here is a difference between an older, loose-knit and a contemporary, tight-knit system.

Despite their position as field or background relations, technologies here display many of the same transformational characteristics found in the previous explicit focal relations. Different technologies texture environments differently. They exhibit unique forms of non-neutrality through the different ways in which they are interlinked with the human lifeworld. Background technologies, no less than focal ones, transform the gestalts of human experience and, precisely because they are absent presences, may exert more subtle indirect effects upon the way a world is experienced. There are also involvements both with wider circles of connection and amplification/reduction selectivities that may be discovered in the roles of background relations; and finally, the variety of minimalist to maximalist strategies remains as open to this dimension of human-technology relations as each of the others.

NOTES

1. This illustration is my version of a similar one developed by Patrick Heelan in his more totally hermeneuticized notion of perception in *Space Perception and the Philosophy of Science* (Berkeley: University of California Press, 1983), p. 193.

2. Ian Hacking, *Representing and Intervening* (Cambridge: Cambridge University Press, 1983), p. 195. Hacking develops a very excellent and suggestive history of the use of microscopes. His focus, however, is upon the technical properties that were resolved before microscopes could be useful in the sciences. He and Heelan, however, along with Robert Ackermann, have been among the pioneers dealing with perception and instrumentation in instruments. Cf. also my *Technics and Praxis* (Dordrecht: Reidel Publishers, 1979).

3. James Burke, *Connections* (Boston: Little, Brown and Co., 1978), p. 5.

7

Philosophy of Technology Meets Social Constructivism: A Shopper's Guide

Philip Brey

1. PHILOSOPHY OF TECHNOLOGY MEETS SOCIAL CONSTRUCTIVISM

Social constructivist approached in technology studies have recently gained the attention of philosophers of technology, as is shown by a number of publications (e.g., Mitcham, 1995; Feenberg and Hannay, 1995; Winner, 1991, 1994; Feenberg, 1992, 1995). Whereas the aim of some of these studies is to provide a philosophical critique of social constructivism (e.g., Winner, 1991), others aim to incorporate notions and ideas of social constructivism into the philosophy of technology (e.g., Feenberg, 1992, 1995). The aim of this essay is not to (merely) critique social constructivism, nor is it to incorporate social constructivist notions into a philosophical analysis of technology. Its aim is, rather, to ask and (provisionally) answer two questions concerning the potential implications of social constructivism for philosophy of technology: (1) Could the philosophy of technology benefit from social constructivist approaches in technology studies through an incorporation of some of their analyses, concepts, and theories? (2) If so, how would the philosophy of technology be transformed as a result?[2] These two questions cannot be answered properly without an evaluation of the weak and strong points of both current philosophy of technology and current social constructivist studies. A large part of this essay will be devoted to such an assessment.

In asking how the philosophy of technology may benefit from social constructivism, I am assuming that an agenda of relevant issues and research questions in the philosophy of technology already exists. The philosophy of technology was and is concerned with philosophical questions concerning the nature of technology, and the impact of technology on things of value: the human psyche, society, culture, and the environment. The expected role of social constructivist studies would therefore be to better help the philosophy of technology answer such questions. The possibility should not be excluded, however, that a consideration of these studies shows that certain traditional questions in the philosophy of technology are misconceived because they are based on false

From *Techné: Journal for the Society of Philosophy and Technology* 2, nos. 3–4 (Spring–Summer 1997). Reprinted by permission of *Techné: Research in Philosophy and Technology.*

empirical presuppositions and hence need to be discarded, that other questions need to be rephrased, and that novel philosophical questions present themselves.

In the next section, the case will be made that the philosophy of technology ought to pay more serious attention to empirical studies of technology, and that, among such studies, social constructivist studies have special appeal for the philosophy of technology. In section 3, social constructivist approaches in technology studies will be characterized briefly, and three varieties of social constructivism, broadly defined, will be distinguished: strong and mild social constructivism, and actor-network theory. Section 4 contains a critique of current social constructivist technology studies, taking as its point of departure an influential earlier critique of social constructivism by Langdon Winner (1991). Section 5 provides a critical discussion of both mild social constructivism and actor-network theory, their divergences from mainstream philosophy of technology, and their potential implications for the philosophy of technology. Section 6 does the same for strong constructivism. The balance is drawn in section 7.

2. THE POTENTIAL RELEVANCE OF SOCIAL CONSTRUCTIVIST STUDIES FOR THE PHILOSOPHY OF TECHNOLOGY

One criticism sometimes leveled at the philosophy of technology is that its theories tend to be abstract, and say a lot about "technology," "society," and "humanity," but little about particular technologies and their impacts, and particular social controversies in which technology plays a role. A second criticism that is sometimes voiced is that theories in the philosophy of technology often make or presuppose empirically testable claims, but that these claims are often not based on, or supported by, empirical evidence. Worse, some of its recurring empirical claims have been claimed to be false. In particular, technological determinist conceptions of technological change presupposed in many philosophical studies of technology (e.g., Ellul, 1954; Winner, 1977; Gehlen, 1980) have been claimed to be empirically inadequate (e.g., MacKenzie and Wajcman, 1985a; Pinch and Bijker, 1987; Nobel, 1984). As Pinch and Bijker (1987) have claimed, the philosophy of technology is in need of "more realistic models of both science and technology" (p. 19).

Empirical studies of technologies and their impacts may be useful to the philosophy of technology, I claim, by aiding the philosophy of technology in arriving at analyses that are more concrete and detailed, and that are empirically more realistic. They can help the philosophy of technology to arrive at empirically more realistic theories by supporting or rejecting empirical claims made or presupposed by theories in the philosophy of technology, such as claims about technological change and technological innovation, the way technology impacts society, and the characteristics of different types of technology, and by suggesting alternative empirical claims. These two functions of empirical studies of technology may be summed up by claiming that such studies are able to provide philosophical theories with *micro-elaborations* of their claims and concepts: insofar as philosophical claims and concepts have an empirical component, this empirical component may be corroborated, amended, or replaced by the empirical concepts and claims of empirical studies of technology. Micro-elaborations are particularly important for studies in social and political philosophy of technology and technology ethics, because such studies typically presuppose some empirical model of technology dynamics. They can also prove relevant for other areas in the philosophy of technology.[3]

Philosophical studies of technology that presuppose some conception of technological change would consequently be improved, I claim, by incorporating empirically informed models of technological change. Because the currently most influential models of technological change in technology studies are arguable social constructivist models, these models are a prime candidate for incorporation into the philosophy of technology. Moreover, the potential relevance of social constructivist models of technological change for the philosophy of technology does not remain limited to their analysis of technological innovation. These models also contain (often implicit) accounts of the way in which technology impacts society. They show that technological innovation does not take a linear path from theory to application to introduction of the technology into society, but is instead influenced by social choices at every point. Consequently, technologies bear the imprint of the social processes that have brought them forth.

Because it is during its development stage that many of the social and cultural effects of a new technology are determined, through various processes of social negotiation and interpretation, it becomes important for philosophical studies of the impact of technology on society and culture to take a closer look at this developmental stage. Only if technology evolved according to some internal logic, and had its social and cultural effects conditioned by this logic, or if technologies were strictly neutral, would it be justified to ignore this development stage, because it would then suffice to study this logic, or to study the choices that societies make after a technology has been developed. If their models of technological change are correct, however, social constructivist studies could be helpful in revealing how the social and cultural impacts of a technology correspond to decisions made during its development stage. In this way, they could help the philosophy of technology to better understand these impacts.

3. A BRIEF GUIDE TO SOCIAL CONSTRUCTIVIST TECHNOLOGY STUDIES

Social constructivist approaches are currently influential in both science studies and technology studies. The label "social constructivism" is used to refer to a variety of related, predominantly sociological approaches in science and technology studies. The roots of many, though not all, of these approaches lie in the sociology of knowledge (Bloor, 1976), and many social constructivists who now study technology have their roots in science studies, only to have turned to technology later on (see Woolgar, 1991). The starting point of social constructivist technology studies can be placed in the mid-eighties (see Bijker, Hughes, and Pinch, 1987). Since then, this paradigm has yielded dozens of books and hundreds of articles, most of them socio-historical case studies of technological innovation and technological change.

The term "social constructivism" is sometimes used in a narrow sense, to refer to the influential Social Construction of Technology (SCOT) approach that was outlined originally in Pinch and Bijker (1987) and Bijker (1987), and a number of related approaches, such as those of Collins (1985) and Woolgar (1991). In a broader sense, which will be used throughout this essay, the term also includes what are called "social shaping" approaches (e.g., MacKenzie and Wajcman, 1985a, 1985b; MacKenzie, 1990), and the actor-network approach of Bruno Latour, Michel Callon, and John Law, and their followers (e.g., Callon, 1987; Latour, 1987).

Social constructivist approaches typically employ a principle of *methodological symmetry,* or *methodological relativism* (Pinch and Bijker, 1987; see Pels, 1996). This principle, in its most common form, implies that the analyst remains impartial as to the "real" properties of her object of analysis, viz. technology. This implies, among other things, that the analyst does not evaluate any of the knowledge claims made by different social groups about the "real" properties of the technology under study. This principle was originally formulated in the sociology of knowledge (Bloor, 1976), where it was motivated by the idea that in a sociological explanation of claims to (scientific) knowledge, it is both possible and desirable to remain agnostic about any role of "the world" in settling scientific controversies. Instead, the analyst should analyze putatively true and false claims symmetrically, explaining them by reference to similar (sociological) factors. Such agnosticism is held to be *desirable* because the analyst is claimed to be *possible* because it is conjectured that the world plays a small or even nonexistent role in settling controversies between different knowledge claimants, and that social factors are much more important.

As a consequence of this principle, when applied to technology, the analyst will generally avoid making claims about the true nature of technology, including claims about the (in)operativity of artifacts, technological (in)efficiency, success or failure in technical change, the (ir)rationality of technological choices and procedures, technological progress, the real function of purpose of an artifact, and intrinsic effects of technology. Because the analyst avoids reference to real properties of a technology, moreover, such properties cannot be invoked to explain technological change. For example, no reference should be made to the actual properties of an artifact in explaining its commercial success, or its selection of our pool of several other designs (see Staudenmaier, 1995).

The outcome of the process of controversy and strategy mapping that surrounds technical change is the *stabilization* of a technology, together with concomitant ("co-produced") social relations. Stabilization of a technology implies that its contents are "black-boxed," and are no longer a site for controversy. Its stabilized properties come to determine the way that the technology functions in society. Most social constructivists, including SCOT scholars, attribute the stabilization of an artifact to an agreement or settlement between different social groups, which arrive at a similar interpretation of a technology, as the result of a series of controversies and negotiations. Technology is claimed by these social constructivists to have *interpretive flexibility:* it has no objective, fixed properties, but allows for different interpretations, not only of its functional and social-cultural properties but also of its technical content, that is, the way it works. Facts about a technology are, hence, not objectively given by the technology itself, but are determined by the interpretations of relevant social groups. The rhetorical process of agreement on the true nature of a technology as the outcome of negotiation and social action is called *closure.* Technology is hence socially shaped or socially constructed: its properties are largely if not exclusively determined by the interpretive frameworks and negotiations of relevant social groups.

The above broad characterization of social constructivist technology studies obscures the fact that a variety of approaches exist, between which there are important differences. There have been various attempts at classifying different approaches within social constructivism (e.g., Bijker and Law, 1992a; Sismondo, 1993; Collins and Yearly, 1992; Woolgar, 1991; Grint and Woolgar, 1995). The following taxonomy of three (broad) approaches is loosely based on these attempts.[4]

The most characteristic variety of social constructivism in technology studies may be called *strong social constructivism.* This approach is the one aligned most closely with the sociology of scientific knowledge, and includes the SCOT-approach, as well as the work of such scholars as H. M. Collins and Steve Woolgar. It vigorously upholds the principle of symmetry, and hence avoids all reference to the actual character of technology in its analyses. Technological change is to be explained by reference to social practices, particularly by reference to processes of interpretation, negotiation, and closure by different actors and social groups. Technology is a genuine social construction, that is, a stabilized technology can only be explained by reference to the social elements (including other socially constructed entities) that have produced its stabilization. No "properties," "powers," or "effects" can be attributed to technologies themselves.

Mild social constructivism is the label that will be used to characterize more moderate approaches, that sometimes go under the name of "social shaping" approaches (e.g., MacKenzie and Wajcman, 1985a, 1985b; MacKenzie, 1990).[5] Social shaping approaches retain conventional distinctions, between the social and the natural, and between the social and the technical, and study the way in which social factors shape technology. They do not reject a role for nonsocial factors in technological change, and are also willing to attribute properties and effects to technology, although these properties and effects are usually claimed to be defined relative to a particular social context. Because technologies are socially shaped, these properties and effects are in large part social properties and social effects, which can be attributed to social biases or politics "built into" or "embodied by" these technologies.

Actor-network theory, sometimes simply called "constructivism" (without the "social"), is a third influential approach. It studies stabilization processes of technical and scientific objects as these result from the building of *actor networks,* which are networks of human actors and natural and technical phenomena. Actor-network theorists employ a principle of *generalized* symmetry, according to which any element (social, natural, or technical) in a heterogeneous network of entities that participate in the stabilization of a technology has a similar explanatory role (Callon, 1987; Latour, 1987; Callon and Latour, 1992). Strong social constructivism is criticized for giving special preference to social elements, such as social groups and interpretation processes, on which its explanations are based, whereas natural or technical elements, such as natural forces and technical devices are prohibited from being explanatory elements in explanations. Actor-network theory also allows for technical devices and natural forces to be actors (or "actants") in networks through which technical or scientific objects are stabilized. By an analysis of actor networks, any entity can be shown to be a post hoc construction, but entities are not normally *socially* constructed, because stabilization is not the result only of social factors.

4. UPON OPENING THE BLACK BOX SIX YEARS LATER AND FINDING IT FILLING UP: CONTEMPORARY SOCIAL CONSTRUCTIVISM AND THE PHILOSOPHY OF TECHNOLOGY

Social constructivist technology studies have been under attack from different quarters for being an inadequate approach to technology studies. Before applying such studies to their own work, therefore, philosophers of technology should carefully consider arguments about their flaws and limitations. Some of the main criticisms against social constructivist technology studies were voiced six years ago, in an influential article by Langdon Winner called, "Upon Opening the Black Box and Finding It Empty: Social Constructivism and the Philosophy of Technology" (Winner, 1991). Winner's criticisms, in summary, are as follows:

1) By focusing on processes of technological innovation, social constructivist studies tend to disregard the social consequences of technical choice.
2) Social constructivism tends to recognize only social groups that have a role in "constructing" technology, and not social groups that are impacted by technology but have been suppressed or even excluded during its construction; it hence ignored deep-seated political biases in technological choice, and power struggles by which the initial agenda of technological development was set.
3) Social constructivism disregards that technological change involves dynamics beyond those revealed by studying the characteristics and actions of relevant social groups, such as deeper cultural, intellectual, or social origins of social choices about technology, and autonomous properties of technology.
4) Social constructivism does not take evaluative stances or invoke moral or political principles; indeed it apparently disdains evaluative stances.

Winner's criticism took as its main target what I have identified as strong social constructivism, particularly the SCOT approach. Most of these criticisms, however, also apply to actor-network and mild constructivist approaches.

At the time that Winner's article was published, these four statements about (strong) social constructivism were for the most part accurate. Since then, however, there have been some significant changes in social constructivist practice. Before discussing these changes, I will first consider the extent to which Winner's criticisms do indeed reveal fundamental flaws in the social constructivist approach to technology studies. Most of Winner's criticisms, I claim, do not point to internal methodological flaws in social constructivism, but criticize the narrowness of its scope and the consequently limited social and political relevance of its studies. Apparently, social constructivists have chosen to draw the scope of their field so as to exclude analyses of consequences, analyses of impacted social groups and initial settling of the agenda, and evaluative and normative claims. They would argue that their principal aim, to explain technical change, turns out to be possible without such analyses. These delimitation, then, may not point to inherent flaws in their methodology, but only to a narrowness in their methodology and in their aims.

Only Winner's third criticism questions the adequacy of the explanatory framework of constructivist technology studies on strictly methodological grounds. This criticism is important, however, as it questions the power of social constructivist micro-level sociological analyses to explain the dynamics of technological choice, and suggests that these need to be supplemented with macro-level analyses or analyses that involve reference to nonsocial factors. Whether this criticism by Winner is justified will be discussed in sections 5 and 6.

Although the narrowness of the scope of social constructivism is for the most part defensible on methodological grounds, its result is that the use of social constructivist studies for addressing issues in the philosophy of technology is more limited than it could conceivably have been. The greatest worth of social constructivist technology studies for the philosophy of technology lies in their detailed empirical analyses of the way in which technological development is a contingent, heterogeneous process involving interpretation and social negotiation, and the way in which the resulting technology is socially shaped. As indicated in section 3, such studies are relevant for studies in philosophy of technology in as far as these depend on some conception of technological development and the character of technology. They imply a corrective to technological determinist theories of development, and indicate how during the development stage of technology, many of its social and cultural effects are already built in. In this way, they also indicate the possibilities for alternative technologies (which has been a theme in the philosophy of technology), as well as possible points of intervention in the process of technological development.

Because of the narrowness in scope of constructivist technology studies, however, philosophers of technology will often have to look elsewhere if they are looking for empirical studies of impacts of technology and of initial settings of the agenda and the exclusion of social groups in technological innovation, or for "deeper" social and cultural factors that play a role in technological development. Their own (macro-level) evaluative or political analyses, moreover, will not be able to take a leaf from any (micro-level) evaluative analyses performed by social constructivists.

It should be noted, though, that in the past six years, some significant changes have taken place in the content of social constructivism. It looks as if some social constructivists have taken Winner's critique seriously. First, there are now more studies in which the social consequences of technical choice are considered. Social (and environmental) impacts are still not a main concern of social constructivist studies, however, and their analyses of social consequences tend to be unconventional. Some studies, mainly occurring within a social shaping or actor-network approach, analyze the way in which social consequences are "built into" technologies (e.g., Akrich, 1992; Latour, 1992; MacKenzie and Wajcman, 1985a). Others study the way in which "truths" about the consequences of a technology are socially negotiated and constructed (e.g., Bruheze, 1992; Bijker, 1992, 1995). What both approaches have in common is that they reject a conventional, technological-determinist conception of technological impacts according to which technologies "impinge on" societies and bring about changes. Instead, they adopt a conception of consequences as resulting wholly or in part from social interpretation and negotiation, rather than (just) from intrinsic features of the technology in question.

Second, more attention has been paid by social constructivists to excluded social groups in technical choice and initial settings of technological agendas. As argued by Aibar (1995, 1996) in a defense of the SCOT approach, excluded social groups can be accounted for by analyzing the statistics and dynamics of their *technological frame.* A technological frame is the repository of knowledge, cultural values, goals, practices, and exemplary artifacts shared by a social group, which structures their attributions of meaning to objects and processes in technical innovation, and their subsequent actions.[6] In analyzing a particular process of technological innovation, the analyst can choose to include not only the technological frame of social groups that have been influential in determining the outcome of this process, but also the technological frame, and changes therein, of groups that have failed to have their voice heard.

A related way in which social groups and choices excluded in the setting of the technological agenda can be included in social constructivist analysis is by the notion of the *script* of technical artifacts, a notion that has been introduced into the actor-network approach by Madeleine Akrich (1992) and Bruno Latour (1992). In technological design, design constituencies inscribe a vision of the world into their designs. Designs consequently embody a script: they harbor expectations about the characteristics of users, social relations, the use environment, and so forth, and stimulate or even demand conformity to this vision. Studying the process of inscription and the resulting script of an artifact enables the analyst to reveal how designs exclude certain social groups, or work against their interests in other ways.[7]

Third, influential proponents of social constructivist technology studies recognized early on that its micro-level analysis, in which the technical content of design is explained by reference to the characteristics and actions of relevant social groups, need to be placed in a broader, macro-level context, in which technical content and the characteristics and actions of social groups are related to the wider social, political, and cultural milieu in which they are found (e.g., Pinch and Bijker, 1987, p. 46). Initially, little attention was given to this item on the agenda of constructivist technology studies. Recently, though, such studies have started to appear (e.g., Rosen, 1993; Carlson, 1992; Bijker, 1992, 1995, chaps. 4 and 5; Pfaffenberger, 1992). Typically, these studies translate macro-level variables, such as power relations or characteristics of the culture, into cultural values and goals in the technical frames of relevant social groups. What still remains difficult for most social constructivist approaches is to account for any "autonomous" features of technology: ways in which the use of technology can have consequences that are neither intended nor anticipated by any social group. This issue will be discussed at length in the coming sections.

Fourth, there has been an increasing interest by social constructivists in normative and political issues concerning the role of technology in society (e.g., Bijker, 1993, 1995; Aibar, 1995, 1996), and in the actualities and potentialities of normative and political analysis within a social constructivist framework (e.g., Radder, 1992; Grint and Woolgar, 1995, 1996; Bijker, 1993, 1993; Aibar, 1996; Jasanoff, 1996). It must be observed, however, that many of these studies reject conventional normative and political analyses of the sort often found in mainstream philosophy of technology. The principle of symmetry obeyed in most of these studies prevents the analyst from taking a stand and prescribes methodological neutrality. Many social constructivists have argued that in spite of the principle of symmetry, or perhaps of it, their analyses have political consequences, and argue for the possibility of a social constructivist politics. The very possibility of such a politics will be discussed in section 6.

All in all, it appears that the scope of constructivist technology studies is widening in a way that is interesting to philosophers of technology wishing to apply constructivist studies to their own research. However, there appear to be important tensions between some of the key assumptions of social constructivist technology studies and those of mainstream philosophy of technology. These differences must be overcome before such applications are rendered unproblematic. Three such tensions have been indicated so far. First, social constructivist studies appear to have an unconventional conception of technological effects. Second, social constructivist studies often seem to deny the possibility of unintended and unanticipated consequences of technical choice, whereas philosophers of technology tend to affirm the existence of such consequences. Third, social constructivists tend to have a different conception of political and normative analysis than do philosophers of technology, and often seem to reject conventional normative analyses as incompatible with social constructivist approach.

Philosophers of technology may reject social constructivist models because of these deviances, but they may also see such models as posing a challenge to conventional assumptions in the philosophy of technology. In the remainder of this essay, I choose to meet this challenge by analyzing social constructivist conceptions of technological effects, unintended consequences of technical choice, and politics. I will also try to point out how the corresponding social constructivist models can be employed by philosophers of technology to arrive at novel analyses of the (intended and unintended) impacts of technology, and at novel kinds of evaluative and political studies of technology. This analysis needs to take place for each of the three varieties of social constructivism that have been distinguished in section 3. Mild social constructivist and actor-network approaches will be discussed in section 5, and strong social constructivist approaches will be discussed in section 6.

5. MILD SOCIAL CONSTRUCTIVISM, ACTOR-NETWORK THEORY, AND THE PHILOSOPHY OF TECHNOLOGY

Mild social constructivism acknowledges that technologies are capable of having effects, although such effects are strongly dependent on the social context in which the technology is used. It tends to avoid reference to effects, though, because its focus tends to be on technological innovation, and not on the impacts of technology. It is often concerned with deconstructing the way in which new technologies are stabilized as the result of the heterogeneous action of different actors. Mild social constructivism also appears compatible with there being unintended consequences of technical choice. Because it does not adopt the view that technologies are wholly socially constructed, it is not committed to the view that any effect must be explained by reference to (conscious) social choices.

Moreover, mild social constructivist analyses can take a normative or political slant by analyzing the way in which particular technologies, designed for use within a particular use environment, come to embody a particular politics or particular social effects. The political significance of such studies can be exploited by breaking with the symmetry principle, and by using the studies to make explicitly political and normative statements, such as statements about the "success" of certain social groups in promoting their interests through a particular technology, or the "suppression" of other social groups through a technological innovation. Winner's famous article on political artifacts (1980), for example, is presented in MacKenzie and Wajcman (1985a) as an example of a "social shaping" approach in technology studies. However, Winner's analysis breaks with the principle of methodological symmetry upheld in many studies in the social shaping approach by privileging some of the many effects of technologies over others because of their claimed political relevance, relating these effects to a definite cause that is found in the design history of the technology, and making evaluative statements about the political significance of these effects.

In his recent philosophical work, Andrew Feenberg (1992, 1995) also adopts a social constructivism that is probably best classified as mild.[8] In Feenberg (1992), for example, he uses social constructivist doctrines to update the Frankfurt School approach, and to argue that technology is subject to conscious social control. He argues that modern technology embodies political values that promote hierarchy and domination, whereas social constructivist studies show that a radically different, democratized technology is possible. Such an alternative technology is possible if more social groups participate in technical choice, and if technological development is consequently brought under democratic control. Feenberg concludes by arguing that there is a need for such democratic control, and for challenges to prevailing conceptions of technological rationality. Like Winner, Feenberg transcends the methodological symmetry in social constructivist studies by making evaluative claims. It hence appears that, by selectively breaking with the symmetry principle, philosophers can use mild social constructivist studies as a starting point for evaluative and political analyses.

In spite of its pretenses to being radical (e.g., Callon and Latour, 1992), actor-network theory often treats technology in a way that superficially resembles the analyses of social shaping approaches, by liberally assigning properties, powers and effects to technologies. Artifacts can have effects because they can act, just like human beings. Consequently, they can also have unintended effects, just like an individual can perform actions that were neither intended nor anticipated by others.

Although studies in actor-network theory do not normally contain political or evaluative claims about technologies and their impacts, these studies can provide an empirical basis for such claims by philosophers of technology in much the same way as do studies within a social shaping approach. For example, the notion of the "script" of an artifact (described above, in section 4), appears to be an actor-network idiom referring to the politics of artifacts. As Latour (1992) claims, artifacts harbor a large part of the morality of a society in their scripts or "programs" (Latour, 1995). They issue prescriptions for the behavior of their users, and help to impose a moral structure on society. If Latour's analysis is correct, then actor-network studies of scripts provide a starting point for normative and political analyses of the scripts of artifacts and their inscription into artifacts by design constituencies.

6. STRONG SOCIAL CONSTRUCTIVISM
AND THE PHILOSPHY OF TECHNOLOGY

Reference to technologies having effects or politics or indeed to them as having any fixed property is difficult within strong social constructivism, because of its strict adherence to the symmetry principle. References to unintended effects are even more problematic, because to the extent that technologies can even be claimed to have effects, these effects are claimed to result from social choices, and are therefore, it would seem, not wholly unintended; strong social constructivism seems to imply that every event is socially determined, and therefore within social control. Reference to properties or effects of technology is not only problematic in strong social constructivism, it is often also not seen as part of the task of the analyst. The task of the analyst is to deconstruct technologies by analyzing the processes by which technologies are stabilized and by which "closure" is reached on their properties. Technology is sometimes metaphorically described as a "text" that is "read" by different actors in different ways, and the task of the analyst is to analyze how the text of technology is "written" by different actors, and how particular "readings" of it come to prevail (Woolgar, 1991; Grint and Woolgar, 1995); the task of the analyst is not to select a particular "reading" of a technology, i.e., her own reading, and present it as a "correct" reading.

Still, as Bijker (1995) has argued, a strong social constructivism does not require a complete abandonment of the notion of technologies having effects. In the SCOT approach, a "social impact" of a technology is defined as a modification of the technological frame of a social group (see section 2), as a consequence of the "stabilization" of a particular technology within this technological frame. As an example, Bijker analyzes the development of fluorescent lighting. Different interpretations of fluorescent lighting existed, but the particular social construction that was settled on was a "high-intensity daylight fluorescent lamp." This social construction required changes in the technological frames of various relevant social groups, such as the adoption of new scientific theories, goals, and practices. These changes, then, are social impacts of the introduction of fluorescent lighting. Notice, however, that according to Bijker's analysis these impacts were not generated by any intrinsic properties of fluorescent lamps but instead derive from the particular way in which fluorescent lighting was socially constructed.

The notion of "artifacts having politics," which is rejecting by strong social constructivists, can also be seen to survive in a different form. Pfaffenberger (1992) adopts the "technology-as-text" metaphor, and argues against Winner that artifacts do not have political, not even relative to a particular social context. Instead, Pfaffenberger argues, if an artifact is to have political effects if "must be *discursively regulated* by surrounding it with symbolic media that mystify and therefore constitute the political aims" (p. 294). In other words, a technology, as a mere text, does not force a particular reading. Readings are determined, instead, by dominant discourses surrounding a technology, which prescribe how the technology should be read. The political impact of a new technology therefore cannot be attributed to this technology itself, but must be attributed instead to the "symbolic discourses" that compel a particular interpretation and usage of it. An attempt to change the politics of a technology therefore does not require its substitution by a different technology (a "rewriting" of the text) but can be achieved by challenging the symbolic discourses surrounding the technology and by introducing alternative readings.[9]

The notion of "unintended effects" finally, may also be argued to survive in some form in a strong social constructivist approach. As Bijker (1995) has argued, actors are not always fully in control of their technological frame, and cannot change it at will. Consequently, "stabilized" technologies may transform technological frames in ways that no actor fully controls:

An artifact in the role of exemplar (that is, after closure, when it is part of a technological frame) has become obdurate. The relevant social groups have, in building up the technological frame, invested so much in the artifact that its meaning has become quite fixed—it cannot be changed easily, and it forms part of a hardened network of practices, theories, and social institutions. From this time on it may indeed happen that, naively spoken, an artifact "determines" social development (p. 282).

Notice, however, that what is having an impact on society is here not an independently existing artifact, but instead a socially constructed artifact that affects other social constructions in the technological frames of social groups, in a way not fully controlled by these social groups.[10]

Although the asymmetry principle seems to rule out evaluative and political analyses, many of those who adopt a strong constructivism nevertheless argue that its analyses embody, or are able to result in a kind of politics. Bijker (1995), for example, argues that social constructivist studies are able to support a social constructivist "politics of technology" even whey they obey the principle of symmetry and merely deconstruct particular social constructions.[11] Such a politics does not require that the analyst make evaluative statements or prescribe courses of action. Rather, the political agenda of social constructivist studies should be to show "the malleability of technology, the possibility for choice, the basic insight that *things could have been otherwise*" (p. 280), and also to point to the obduracy of stabilized technologies and other stabilized objects, and the limitations that these impose on attempts to change technology and its social impacts. Strong social constructivist studies are hence political by revealing the contingency or politics contained in technological choice. This information can subsequently be used by actors with a political agenda to influence technical change, including "social impacts" of technologies.

Bijker admits that there is no guarantee that social constructivist studies will have political impacts that are desirable. Ideally, social constructivist studies would aid less privileged social groups by showing them how stronger parties impose a particular political hegemony, and they could resist this hegemony and exert more influence over technology. However, as Bijker points out, these studies may also work against less privileged groups by undermining their attempts at stabilizing certain social technology. Bijker's hope, however, is that social constructivist studies will have a political bias towards stimulating democratic control of technology, by showing to citizens that influence on technology's course is possible, even in the "diffusion" stage of a technology (see Bijker, 1993, 131).

There is no convincing evidence, however, that social constructivist analyses systematically favor less privileged groups. Instead, Jasanoff (1996) presents examples of the use of social constructivist studies by powerful actors to promote their interests; these confirm Bijker's worry that social constructivist analyses do not necessarily favor less privileged groups. Moreover, Martin (1993) has argued that there is little evidence that social constructivist studies work to aid less privileged groups. Perhaps social constructivist politics is a good idea in theory, but there is no convincing evidence that it worked to stimulate positive change in practice.

I tentatively conclude that social constructivist politics, in its current form, is unsatisfactory. If political analysis is desired, it seems more attractive for authors of social constructivist studies to study powerful and less privileged groups asymmetrically, siding with the less privileged group in their analyses (Martin, 1993; Scott, Richards, and Martin, 1990). More specifically, the analyst may attempt to adopt the technological frame of less privileged groups, and present analyses from this perspective that are claimed to represent the "actual character" of a technology and its "real impacts" even though the analyst may be aware that her own analysis is also a "mere" social construction. Such a realist analysis may suggest specific courses of action to these groups, and be more directly helpful in this way than analyses that are merely deconstructive (Soper, 1995; Kling, 1992; Gill, 1996).[12]

7. CONCLUSION

Social constructivist studies pose interesting challenges to the philosophy of technology, presenting nondeterministic models of technological change, and arguing that the choices made in technical innovation in large part determine the social impacts of technologies. This emphasis on the development stage of technology has been shown to result in interesting analyses of technologies, as being "socially shaped" or having a "script," that provide a potentially fruitful basis for normative and evaluative philosophical analysis of technology and its impacts. Even the strong variety of social constructivism allows for evaluative and political analysis, by studying the ways in which technologies are socially constructed by different parties and exploring the possibilities for alternative social constructions by "reading" technologies differently, thus subverting dominant technological frames. (In these cases, normative and evaluate philosophical analysis requires that the symmetry principle upheld by social constructivists be disobeyed.) Although a full investigation of the methodological and empirical adequacy of social constructivist approached is beyond the scope of this essay, these approaches, if valid, do suggest new directions for the philosophy of technology.

NOTES

1. I would like to thank Annemieke Nelis, Martijntje Smits, Tsjalling Swierstra, and Peter Paul Verbeek for their comments on an earlier draft.

2. The converse question, of how social constructivist analyses could benefit from work in the philosophy of technology, will not be addressed.

3. Empirical technology studies are not the only empirical discipline helpful to the philosophy of technology. For philosophical analyses of the social and cultural implications of technology, a good knowledge of general sociology, anthropology, and cultural studies could also prove beneficial.

4. To be precise, my analysis adopts Sismondo's distinction between mild and strong social constructivism, but hold, unlike Sismondo's typology, that the SCOT approach (Pinch and Bijker, 1987) is best understood as a form of strong social constructivism. It also adopts the common distinction between a social constructivism as exemplified by the SCOT approach, and actor-network theory. Moreover, strong social constructivism, in my analysis, corresponds with what Grint and Woolgar (1995) call "post-essentialism" and "the constitutive variant of anti-essentialism." Mild social constructivism corresponds with the remaining forms of auto-essentialism. My typology should be understood as describing ideal types, and approaches exist that fall in between these ideal types.

5. Included in social shaping approaches may be the systems approach that has been developed by Hughes (1987), as well as most work in feminist technology studies (e.g., Wajcman, 1991).

6. The notion of technological frame has been introduced by Bijker (1987, 1993, 1995).

7. A third, promising notion within (strong) social constructivist technology studies was recently presented by Pfaffenberger (1992). Pfaffenberger argues that the introduction of new technologies is normally accompanied by a series of moves and countermoves by relevant social groups, a series of events that he calls a *technological drama*. The analysis of technological dramas does not just reveal the role of social groups that are successful in shaping the technology according to their interests, but also the role of impacted groups that are unsuccessful in doing so.

8. Although Feenberg (1992) mostly draws from the SCOT approach, which is here classified as a type of strong social constructivism, Feenberg only adopts the assumptions made in SCOT that there is room for social choice in technological innovation, and that technologies have interpretive flexibility. Both these claims do not go beyond those of mild social constructivism (see Sismondo, 1993). Some of the analysis in Feenberg (1995), however, seem to come close to embracing the strong social constructivism discussed in section 6.

9. A similar analysis is presented by Bijker (1995, 262–264), who argues for a semiotic conception of power, according to which power is "the apparent order of taken-for-granted categories of existence, as they are fixed and represented in technological frames" (263). Artifacts hence are not political in themselves, but derive their political power from the semiotic structure in which they are embedded.

10. The acknowledgement that technological change and the social impacts of technology are not just the outcome of "explicitly planned, rationally decided, conscious action" has a price, as Bijker is aware. Social constructivist studies cannot fully account for technical change by reference to the actions of different social groups. They also need to recognize structural constraints on technical change. According to Bijker, these structural constraints are found in the semiotic structures, or categories, that stabilize elements in technological frames. Notice, however, that this reference to semiotic structure seems to open the door to functionalist and structuralist accounts of technical change (see Elster, 1983), which is probably not what Bijker has in mind.

11. See Grint and Woolgar (1995, 1996) for a related proposal for a social constructivist politics, which they call "post-essentialist politics." See also Mol and Mesman (1996).

12. Some social constructivists have criticized politically motivated realist analyses because they argue that realist analyses are politically dangerous. Thus, Elam (1994), in a critique of Winner (1991), argues that adherence to "the truth" even in the name of a "politically correct" analysis, goes against the very foundations of a liberal politics. According to Elam, liberal politics requires one to refrain from enforcing one's views on others, and hence from presenting any of one's views as true. Similarly, Woolgar (1993) claims: "Definitive versions of the 'actual political' character of, say, Moses' bridges must be resisted, because there is a very real danger of accepting any political interpretation in the guise of its being true" (p. 527). Woolgar holds that the very denial that any view qualifies as "true" or as superior to other views is in the interest of protecting fundamental liberties. Neither party, however, acknowledges a difference between merely claiming that a

statement is true and dogmatic adherence to the truth of a statement, and neither party presents an argument that presenting statements as true works to undermine fundamental liberties.

REFERENCES

Aibar, E. 1995. "Technological Frames in a Town Planning Controversy: Why We Do Not Have to Drop Constructivism to Avoid Political Abstinence." In Mitcham.

Aibar, E. 1996. "The Evaluative Relevance of Social Studies of Technology." *Society for Philosophy and Technology,* no. 1–2.

Akrich, M. 1992. "The De-Scription of Technical Objects." In Bijker and Law.

Bijker, W. 1987. "The Social Construction of Bakelite: Toward a Theory of Invention." In Bijker, Pinch, and Hughes.

Bijker, W. 1992. "The Social Construction of Flourescent Lighting, or How an Artifact was Invented in its Diffusion Stage." In Bijker and Law.

Bijker, W. 1993. "Do Not Despair: There Is Life after Constructivism." *Science, Technology, & Human Values,* 18:113–138.

Bijker, W. 1995. *Bikes, Bakelite, and Bulbs: Steps Toward a Theory of Socio-Technical Change.* Cambridge: MIT Press.

Bijker, W., and J. Law. 1992a. "General Introduction." In Bijker and Law.

Bijker, W., and J. Law, eds. 1992b. *Shaping Technology/Building Society: Studies in Sociotechnical Change.* Cambridge: MIT Press.

Bijker, W., T. Hughes, and T. Pinch, eds. 1987. *The Social Construction of Technological Systems: New Directions in the Sociology and History of Technology.* Cambridge: MIT Press.

Bloor, D. 1976. *Knowledge and Social Imagery.* London: Routledge and Kegan Paul.

Bruhèze, A. de la. 1992. "Closing the Ranks: Definition and Stabilization of Radioactive Wastes in the U.S. Atomic Energy Commission, 1945–1960." In Bijker and Law.

Callon, M. 1987. "Society in the Making: The Study of Technology as a Tool for Sociological Analysis." In Bijker, Pinch, and Hughes, eds., Cambridge: MIT.

Callon, M., and Latour, B. 1992. "Don't Throw the Baby Out with the Bath School! A Reply to Collins and Yearly." In A. Pickering, ed., *Science as Practice and Culture.* Chicago: University of Chicago Press.

Carlson, W. B. 1992. "Artifacts and Frames of Meaning: Thomas A. Edison, His Managers, and the Cultural Construction of Motion Pictures." In Bijker and Law.

Collins, H. M. 1985. *Changing Order: Replication and Induction in Scientific Practice.* Beverly Hills, Calif.: Sage.

Collins, H., and Yearly, S. 1992. "Epistemological Chicken." In A. Pickering, ed., *Science as Practice and Culture.* Chicago: University of Chicago Press.

Elam, M. 1994. "Anti Anticonstructivism or Laying the Fears of a Langdon Winner to Rest." *Science, Technology, & Human Values,* 19:101–106.

Elster, J. 1993. *Explaining Technical Change.* Cambridge: Cambridge University Press.

Ellul, J. 1954. *La Technique ou l'enjeu du siècle.* Paris: Armand Collin.

Feenberg, A. 1992. "Subversive Rationalization: Technology, Power, and Democracy." *Inquiry,* 35. Reprinted in Feenberg and Hannay, 1995.

Feenberg, A. 1995. *Alternative Modernity: The Technical Turn in Philosophy and Social Theory.* Berkeley: University of California Press.

Feenberg, A., and A. Hannay, eds. 1995. *Technology and the Politics of Knowledge.* Bloomington: Indiana University Press.

Gehlen, A. 1980. *Man in the Age of Technology.* Trans. Patricia Lipscomb. New York: Columbia University Press.

Gill, R. 1996. "Power, Social Transformation, and the New Determinism: A Comment on Grint and Woolgar." *Science, Technology, & Human Values,* 21:347–353.

Grint, K., and Woolgar, S. 1995. "On Some Failures of Nerve in Constructivist and Feminist Analyses of Technology." *Science, Technology, & Human Values* 20:286–310.

Grint, L., and Woolgar, S. 1996. "A Further Decisive Refutation of the Assumption That Political Action

Depends on the 'Truth' and Suggestion That We Need to Go beyond This Level of Debate: A Reply to Rosalind Gill." *Science, Technology, & Human Values* 21:354–357.

Hughes, T. 1987. "The Evolution of Large Technological Systems." In Bijker, Hughes, and Pinch.

Jasanoff, S. 1996. "Beyond Epistemology: Relativism and Engagement in the Politics of Science." *Social Studies of Science* 26:393–418.

Kling, R. 1992. "Audiences, Narratives, and Human Values in Social Studies of Technology." *Science, Technology & Human Values* 17:349–365.

Latour, B. 1987. *Science in Action.* Cambridge, Mass.: Harvard University Press.

Latour, B. 1992. "Where Are the Missing Masses? The Sociology of a Few Mundane Artifacts." In Bijker and Law.

Latour, B. 1995. "A Door Must Be Either Open or Shut: A Little Philosophy of Techniques." In Feenberg and Hannay.

MacKenzie, D. 1990. *Inventing Accuracy: A Historical Sociology of Nuclear Missile Guidance.* Cambridge: MIT Press.

MacKenzie, D., and J. Wajcman. 1985a. "Introduction: The Social Shaping of Technology." In MacKenzie and Wajcman.

MacKenzie, D., and J. Wajcman, eds. 1985b. *The Social Shaping of Technology.* Philadelphia: Open University Press.

Martin, B. 1993. "The Critique of Science Becomes Academic." *Science, Technology & Human Values* 18:247–259.

Mitcham, C., ed. 1995. *Research in Philosophy and Technology,* vol. 15: *Social and Philosophical Constructions of Technology.* Greenwich, Conn.: JAI Press.

Mol, A., and J. Mesman. 1996. "Neonatal Food and the Politics of Theory: Some Questions of Method." *Social Studies of Science* 26:419–444.

Noble, D. 1984. *Forces of Production: A Social History of Industrial Automation.* New York: Knopf.

Pels, D. 1996. "The Politics of Symmetry." *Social Studies of Science* 26:277–304.

Pfaffenberger, B. 1992. "Technological Dramas." *Science, Technology & Human Values* 17:282–312.

Pinch, T., and Bijker, W. 1987. "The Social Construction of Facts and Artifacts: Or How the Sociology of Science and the Sociology of Technology Might Benefit Each Other." In Bijker, Pinch, and Hughes.

Radder, H. 1992. "Normative Reflexions on Constructivist Approaches to Science and Technology." *Social Studies of Science* 22:141–173.

Rosen, P. 1993. "The Social Construction of Mountain Bikes: Technology and Postmodernity in the Cycle Industry." *Social Studies of Science* 23:479–513.

Scott, P., E. Richards, and B. Martin. 1990. "Captives of Controversy: The Myth of the Neutral Social Researcher in Contemporary Scientific Controversies." *Science, Technology & Human Values,* 15:474–494.

Sismondo, S. 1993. "Some Social Constructions." *Social Studies of Science,* 23:515–553.

Soper, K. 1995. "Feminism and Ecology: Realism and Rhetoric in the Discourses of Nature." *Science, Technology & Human Values* 20:311–331.

Staudenmaier, J. 1995. "Problematic Stimulation: Historians and Sociologists Constructing Technology Studies." In Mitcham.

Wajcman, J. 1991. *Feminism Confronts Technology.* University Park: Pennsylvania State University Press.

Winner, L. 1977. *Autonomous Technology.* Cambridge: MIT Press.

Winner, L. 1980. "Do Artifacts Have Politics?" *Daedalus* 109:121–136.

Winner, L. 1991. "Upon Opening the Black Box and Finding It Empty: Social Constructivism and the Philosophy of Technology." In J. Pitt, and E. Lugo, eds., *The Technology of Discovery and the Discovery of Technology.* Blacksburg, Va: Society for Philosophy and Technology.

Winner, L. 1994. "Reply to Mark Elam." *Science, Technology, & Human Values* 19:107–109.

Woolgar, S. 1991. "The Turn to Technology in Social Studies of Science." *Science, Technology, & Human Values* 16:20–50.

Woolgar, S. 1993. "What's at Stake in the Sociology of Technology? A Reply to Pinch and Winner." *Science, Technology, & Human Values* 18:523–529.

8

Women and the Assessment of Technology: To Think, to Be; to Unthink, to Free

Corlann Gee Bush

Everything is what it is, what it isn't, and its direct opposite. That technique, so skillfully executed might help account for the compelling irrationality . . . *double double think is very easy to deal with if we just realize that we have only to double double unthink it.*

—Dworkin 1974, p. 63

Although Andrea Dworkin is here analyzing Pauline Reage's literary style in the *Story of O,* her realization that we can "double double unthink" the mind fetters by which patriarchal thought binds women is an especially useful one. For those of us who want to challenge and change female victimization, it is a compelling concept.

SOMETHING ELSE AGAIN

The great strength of the women's movement has always been its twin abilities to unthink the sources of oppression and to use this analysis to create a new and synthesizing vision. Assertiveness is, for example, something else again: a special, learned behavior that does more than merely combine attributes of passivity and aggressiveness. Assertiveness is an unthinking and a transcendence of those common, control-oriented behaviors.[1]

Similarly, in their books *Against Our Will* and *Rape: The Power of Consciousness,* Susan Brownmiller (1974) and Susan Griffin (1979) unthink rape as a crime of passion and rethink it as a crime of violence, insights which led to the establishment of rape crisis and victim advocacy services. But a good feminist shelter home-crisis service is something else again: it is a place where women are responsible for the safety and security of other women, where women tech self-defense and self-esteem to each other. In like manner, women's spirituality is something else again. Indebted both to Mary Daly for unthinking Christianity in *Beyond God the Father* (1973) and *The Church and the Second Sex* (1968) and to witchcraft for rethinking ritual, women's spirituality is more than a synthesis of those insights, it is a transformation of them.

In other words, feminist scholarship and feminist activism proceed not through a sterile, planar dialectic of thesis, antithesis, synthesis but through a dynamic process of unthinking, rethinking, energizing, and transforming. At its best, feminism creates new life forms out of experiences as common as seawater and insights as electrifying as lightning.

From *Machina Ex Dea*, pp. 151–170, Joan Rothschild, ed. (New York: Teacher's College Press). Copyright © 1983 by Joan Rothschild. Reprinted by permission of Joan Rothschild.

The purpose of this chapter is to suggest that a feminist analysis of technology would be, like assertiveness, something else again. I will raise some of the questions that feminist technology studies should seek to ask, and I will attempt to answer them. Further, I hope to show how scholars, educators, and activists can work together toward a transformation of technological change in our society.

The endeavor is timely not least books such as this, journal issues, articles, and conferences are increasingly devoting time and energy to the subject or because technologically related political issues such as the antinuclear movement and genetic engineering consume larger and larger amounts of both our new space and our consciousness. The most important reason why feminists must unthink and rethink women's relationship to technology is that the *tech-fix* (Weinberg 1966, p. 6) and the public policies on which it is based are no longer working. The tech-fix is the belief that technology can be used to solve all types of problems, even social ones. Belief in progress and the tech-fix has long been used to rationalize inequality: it is only a matter of time until technology extends material benefits to all citizens, regardless of race, sex, class, religion, or nationality.

> Technology has expanded our productive capacity so greatly that even though our distribution is still inefficient, and unfair by Marxian precepts, there is more than enough to go around. Technology has provided a "fix"—greatly expanded production of goods—which enables our capitalistic society to achieve many of the aims of the Marxist social engineer without going through the social revolution Marx viewed as inevitable. Technology has converted the seemingly intractable social problem of *widespread* poverty into a relatively tractable one (Weinberg 1966, p. 7).

While Weinberg himself advocated cooperation among social *and* technical engineers in order to make a better society, and thereby, a better life, for all of us who are part of society" (Weinberg 1966, p. 10), less conscientious philosophers and politicians have seen in the tech-fix a justification for laissez-faire economics and discriminatory public policy. Despite its claim to the contrary, the tech-fix has not worked well for most women or for people of color; recent analyses of the feminization of poverty, for example, indicate that jobs, which have always provided men with access to material goods, do not get women out of poverty.

> Social welfare programs based on the old male model of poverty do not consider the special nature of women's poverty. One fact that is little understood and rarely reflected in public welfare policy is that women in poverty are almost invariably productive workers, participating fully in both the paid and the unpaid work force. The inequalities of present public policies molded by the traditional economic role of women cannot continue. Locked into poverty by capricious programs designed by and for male policymakers . . . women who are young and poor today are destined to grow old and poor as the years pass. Society cannot continue persisting with the male model of a job automatically lifting a family out of poverty . . . (McKee 1982, p. 36).

As this example illustrates, the traditional social policies for dealing with inequality—*get a job*—and traditional technological solutions—*produce more efficiently*—have not worked to make a better society for women. Therefore, it is essential that women begin the unthinking of these traditions and the rethinking of new relationships between social and technical engineering.

UNTHINKING TECH-MYTHS

In the poem, "To An Old House in America," Adrienne Rich describes the attitude that women should take toward the task of unthinking public policy in regard to technology: "I do not want to

simplify/Or: I would simplify/By naming the complexity/It has not made o'er simple all along"
(Rich 1975, p. 240). Partly because it is in their best interest to do so and partly because they
truly see nothing else, most politicians and technocrats paint the canvas of popular opinion about
technology with the broadest possible brushstrokes, rendering it, in pure type, as TOOL, as
THREAT, or as TRIUMPH.[2] From each of these assumptions proceed argument, legislation, public
policy, and, ironically, powerlessness. In order to develop a feminist critique of technology, we
must analyze these assumptions and unthink them, making them simpler by naming their complexity.

   The belief that technology represents the triumph of human intelligence is one of America's
most cherished cultural myths; it is also the easiest to understand, analyze, and disprove. Unfortunately, to discuss it is to resort to clichés: "There's nothing wrong that a little good old American
ingenuity can't fix"; "That's progress"; or "Progress is our most important product." From such
articles of faith in technology stemmed Manifest Destiny, the mechanization of agriculture, the
urbanization of rural and nomadic cultures, the concept of the twentieth as the "American Century," and every World's Fair since 1893. That such faith seems naïve to a generation that lives
with the arms race, acid rain, hazardous waste, and near disasters at nuclear power plants is not to
diminish one *byte* either of Western culture's faith in the tech-fix or its belief that technological
change equals material progress. And, indeed, like all generalizations, this myth is true—at least
partially. Technology *has* decreased hardships and suffering while raising standards of health, living, and literacy throughout the industrialized world.

   But, not without problems, as nay-sayers are so quick to point out. Those who perceive technology as the ultimate threat to life on the planet look upon it as an iatrogenic disease, one created,
like nausea in chemotherapy patients, by the very techniques with which we treat the disease. In
this view, toxic wastes, pollution, urban sprawl, increasing rates of skin cancer, even tasteless tomatoes are all problems created through our desires to control nature through technology. Characterized by their desire to go cold turkey on the addiction to the tech-fix, contemporary critics of
technology participate in a myriad of activities and organizations (Zero Population Growth, Friends
of the Earth, Sierra Club, the Greenpeace Foundation) and advocate a variety of goals (peace, arms
limitation, appropriate technology, etc.). And, once again, their technology-as-threat generalization
is true, or at least as true as its opposite number: in truth, no one, until Rachael Carson (1955), paid
much attention to the effects of technology on the natural world it tried to control; indeed, technology has created problems as it has set out to solve others.

   Fortunately, the inadequacy of such polarized thinking is obvious: technology is neither
wholly good nor wholly bad. "It has both positive and negative effects, and it usually has the two
at the same time and in virtue of each other" (Mesthene 1970, p. 26). Every innovation has both
positive and negative consequences that pulse through the social fabric like waves through water.

   Much harder to unthink is the notion that technologies are merely tools: neither good nor bad
but neutral, moral only to the extent that their user is moral. This, of course, is the old saw "guns
don't kill people, people kill people" writ large enough to include not only guns and nuclear weapons but also cars, televisions, and computer games. And there is truth here, too. Any given person
can use any given gun at any given time either to kill another person for revenge or to shoot a
grouse for supper. The gun is the tool through which the shooter accomplishes his or her objectives.
However, just as morality is a collective concept, so too are guns. As a class of objects, they comprise a technology that is designed for killing in a way that ice picks, hammers, even knives—all
tools that have on occasion been used as weapons—are not. To believe that technologies are neutral
tools subject only to the motives and morals of the user is to miss completely their collective significance. Tools and technologies have what I can only describe as *valence,* a bias or "charge"
analogous to that of atoms that have lost or gained electrons through ionization. A particular tech-

nological system, even an individual tool, has a tendency to interact in similar situations in identi-fiable and predictable ways. In other words, particular tools of technologies tend to be favored in certain situations, tend to perform in a predictable manner in these situations, and tend to bend other interactions to them. Valence tends to seek out or fit in with certain social norms and to ignore or disturb others.

Jacques Ellul (1964) seems to be identifying like valence when he describes "the specific weight" with which technique is endowed:

> It is not a kind of neutral matter, with no direction, quality, or structure. It is a power endowed with its own peculiar force. It refracts in its own specific sense the wills which make use of it and the ends proposed for it. Indeed, independently of the objectives that man pretends to assign to any given technical means, that means always conceals in itself a finality which cannot be evaded (pp. 140–41).

While this seems to be overstating the case a bit—valence is not the atom, only one of its attri-butes—tools and techniques do have tendencies to pull or push behavior in definable ways. Guns, for example, are valenced to violence; the presence of a gun in a given situation raises the level of violence by its presence alone. Television, on the other hand, is valenced to individuation; despite the fact that any number of people may be present in the same room at the same time, there will not be much conversation because the presence of the TV itself pulls against interaction and pushes toward isolation. Similarly, automobiles and microwave ovens are individuating technologies while trains and campfires are accretionary ones.

Unthinking tech-myths and understanding valence also require greater clarity of definition (Winner 1977, pp. 10–12). Several terms, especially *tool, technique,* and *technology,* are often used interchangeably when, in fact, they describe related but distinguishable phenomena. *Tools* are the implements, gadgets, machines, appliances, and instruments themselves. A hammer is, for exam-ple, a tool as is a spoon or an automatic washing machine. *Techniques* are the skills, methods, procedures, and processes that people perform in order to use tools. Carpentry is, therefore, a tech-nique that utilizes hammers, baking is a technique that uses spoons, and laundering a technique that employs washing machines. *Technology* refers to the organized systems of interactions that utilize tools and involve techniques for the performance of tasks and the accomplishment of objec-tives. Hammers and carpentry are some of the tools and techniques of architectural or building technology. Spoons and baking, washing machines and laundering are some of the tools and tech-niques of domestic or household technology.

A feminist critique of the public policy debate over technology should, thus, unthink the tripartite myth that sees technology in simple categories as tool, triumph, or threat. In unthinking it, we can simplify it by naming its complexity:

- A tool is not a simple isolated thing but is a member of a class of objects designed for specific purposes.
- Any given use of tools, techniques, or technologies can have both beneficial and detrimen-tal effects at the same time.
- Both use and effect are expressions of a valence or propensity for tools to function in cer-tain ways in certain settings.
- Polarizing the rhetoric about technology enables advocates of particular points of view to gain adherents and power while doing nothing to empower citizens to understand, discuss, and control technology on their own.

"Making it o'er simple all along" has proven an excellent technique for maintaining social control. The assertion that technology is beneficial lulls people into believing that there is nothing wrong that can't be fixed, so they do nothing. Likewise, the technophobia that sees technology as evil frightens people into passivity and they do nothing. The argument that technology is value-free either focuses on the human factor in technology in order to obscure its valence or else concentrates on the autonomy of technology in order to obscure its human control. In all cases, the result is that people feel they can do nothing. In addition, by encouraging people to argue with and blame each other, rhetoric wars draw public attention away from more important questions such as: Who is making technological decisions? On what basis? What will the effects be?

CONTEXT, CONTEXT, WHITHER ART THOU, CONTEXT?

In unthinking the power dynamics of technological decision making, a feminist critique needs to pay special attention to the social messages whispered in women's ears since birth: mother to daughter, "Don't touch that, you'll get dirty"; father to daughter, "Don't worry your pretty little head about it"; teacher to young girl, "It doesn't matter if you can't do math"; woman to woman, "Boy, a man must have designed this."

Each of these statements is talking about a CONTEXT in which technological decisions are made, technical information is conveyed, and technological innovations are adopted. That such social learning is characterized by sex stereotyping should come as no surprise. What may be surprising is not the depth of women's ignorance—after all, women have, by and large, been encouraged to be ignorant—but the extent to which men in general, inventors, technocrats, even scholars, all share an amazing ignorance about the context in which technology operates. There are four:

1. *The design or development context* which includes all the decisions, materials, personnel, processes, and systems necessary to create tools and techniques from raw materials.
2. *The user context* which includes all the motivations, intentions, advantages, and adjustments called into play by the use of particular techniques or tools.
3. *The environmental context* that describes nonspecific physical surroundings in which a technology or tool is developed and used.
4. *The cultural context* which includes all the norms, values, myths, aspirations, laws and interactions of the society of which the tools or technique is a part.

Of these, much more is known about the design or developmental context of technology than about the other three put together. Western culture's collective lack of knowledge about all but the developmental context of technology springs in part from what Langdon Winner calls technological orthodoxy: a "philosophy of sorts" that has seldom been "subject to the light of critical scrutiny" (Winner 1979, p. 75). Standard tenets of technological orthodoxy include:

* That men know best what they themselves have made.
* That the things men make are under their firm control.
* That technologies are neutral: they are simply tools that can be used one way or another; the benefit or harm they bring depends on how men use them (Winner 1979, p. 76).

If one accepts these assumptions, then there is very little to do except study processes of design and invent ever-newer gadgets. The user and environmental contexts become obscured if not invisible, an invisibility further confirmed by the fact that, since the Industrial Revolution, men have been inventors and designers while women have been users and consumers of technology. By and large, men have created, women have accommodated.

The sex role division of labor that characterizes Western societies has ensured that boys and girls have been brought up with different expectations, experiences, and training, a pattern that has undergone remarkably little change since the nineteenth century.

> Games for girls were carefully differentiated from boys' amusement. A girl might play with a hoop or swing gently, but the "ruder and more daring gymnastics of boys" were outlawed. Competitive play was also anathema: a "little girl should never be ambitious to swing higher than her companions." Children's board games afforded another insidious method of inculcating masculinity and femininity. On a boys' game board the player moved in an upward spiral, past temptations, obstacles, and reverses until the winner reached a pinnacle of propriety and prestige. A girls' playful enactment of her course in life moved via a circular ever-inward path to the "mansion of happiness," a pastel tableau of mother and child. The dice of popular culture were loaded for both sexes and weighted with domesticity for little women. The doctrine of (separate) spheres was thereby insinuated in the personality of the child early in life and even during the course of play (Ryan 1979, p. 92).

It is difficult to invent a better mousetrap if you're taught to be afraid of mice; it is impossible to dream of becoming an engineer if you're never allowed to get dirty.

As compared to women, men do, indeed, know a great deal about what they would call the "design interface" of technology; they know more about how machines work; they discovered the properties of elements and the principles of science. They know math, they develop cost-benefit-risk analyses; they discover, invent, engineer, manufacture, and sell. Collectively, men know almost everything there is to know about the design and development of tools, techniques, and systems; but they understand far less about how their technologies are used—in part because there is less money in understanding than in designing, in part because the burden of adjusting to technological change falls more heavily on women. What is worse, however, is that most men do not know that they do not know anything women and the user context.

> From the preliminary conceptualizations to the final marketing of a product, most decision making about technology is done by men who design, usually subconsciously, a model of the physical world in which they would like to live, using material artifacts which meet the needs of the people—men—they best know. The result [is] technological development based on particular sets of male conditioning, values, and roles . . . (Zimmerman 1981, p. 2).

Ironically, until very recently, most women did not realize that they possessed information of any great significance. With all the cultural attention focused on the activity in the developmental context, it was hard to see beyond the glare of the spotlights into the living rooms and kitchens and laundries where women were working and living out the answers to dozens of unverbalized questions: How am I spending my time here? How is my work different from what I remember my mother doing? Am I really better off? Why does everything seem so out of control? Rephrased, these are the questions that will comprise a feminist assessment of technology: How have women's roles changed as a result of modern technology? Has women's status in society kept pace with the standard of living? Do women today have more opportunities or merely more exceptions? What is the relationship of material possessions to personal freedom?

Think for a moment about washing machines. Almost every family in the United States has access to one; across the country, women spend thousands of hours each day in sorting, washing, drying, folding, and ironing clothes. The automatic washing machine has freed women from the pain and toil described so well by Agnes Smedley (1973) in *Daughter of Earth.* But as washing technology has changed, so too has clothing (it gets dirtier faster) and wardrobes (we own more clothes) and even standards of cleanliness (clothes must be whiter than white), children change clothes more often, there are more clothes to wash. Joann Vanek (1974, p. 118), in her work on time spent in housework, asserts that women spent as much time in household related tasks in 1966 as they did in 1926.

More has changed, however, than just standards of cleanliness. Doing laundry used to be a collective enterprise. When I was a child in the late 1940s and early 1950s, my mother and grandmother washed the family's clothes together. My grandmother owned a semi-automatic machine but she lived 45 minutes away; my mother had hot water, a large sink, and five children. Every Sunday we would dump the dirty clothes in a big wicker basket and drive to my grandmother's house where all the womenfolk would spend the afternoon in the basement, talking and laughing as we worked. By evening, the wicker basket would again be full, but this time we neatly folded, clean smelling piles of socks, sheets, towels, and underwear that would have to last us a week. Crisply ironed dresses and slacks, on hangers, waited to be hung, first on those little hooks over the side doors of the car, then in our closets at home.

Nostalgic as these memories are, doing laundry was not romantic. It was exhausting, repetitious work, and neither my mother nor I would trade in our own automatic washers to go back to it (Armitage 1982, pp. 3–6). Yet, during my childhood, laundry was a communal activity, an occasion for gossip, friendship, and bonding. Laundering was hard work, and everyone in the family and in the society knew it and respected us as laborers. Further, having laundry and a day on which to do it was an organizing principle (Monday, washday; Tuesday, iron; Wednesday . . .) around which women allocated their time and resources. And, finally, there was closure, a sense of completion and accomplishment impossible to achieve today when my sister washes, dries, folds, and irons her family's clothes everyday or when I wash only because I have nothing to wear.

Admittedly, this homey digression into soap opera (One Woman's Wash) is a far cry from the design specification and cost-benefit analyses men use to describe and understand the developmental context of washing machines, but it is equally valid for it describes the user context in the user's terms. Analyzing the user context of technological change is a process of collecting thousands and thousands of such stories and rethinking them into an understanding of the effects of technological change on women's lives.[3] From unthinking the developmental context and rethinking the user context, it is only a short step to studying the environmental and cultural contexts of technological change. Of these, our knowledge of the environmental context is the better developed, partly because we have given it more serious attention but mostly because environmental studies has been legitimate career options for men.

While concern about the effects of technology on the natural environment is an idea that can be traced back to de Crevecoeur (1968 [1782]) and James Fenimore Cooper (1832), Rachael Carson (1955, 1961, 1962) is the person most responsible for our current level of ecological awareness and for the scientific rather than aesthetic basis on which it rests. As we learn more about the fragile reciprocity within ecosystems, we begin to unthink the arrogance of our assumption that we are separate from and superior to nature. In an ecosystem, it is never possible to do only *one* thing; for every action there are chain reactions of causes and effects. The continued survival of the world depends upon developing more precise models of the environment so we can predict and prevent actual catastrophe without being immobilized by the risking of it.

Perhaps no one could have foreseen that the aerosol sprays we used to apply everything from paint to anti-perspirant would degrade the earth's ozone layer, but no one seems to have asked. The drums for burying toxic waste would eventually corrode and leak seems so obvious that millions ought to have been able to predict the risk, yet no one seems to have had the desire or the clout to deal with the problem of hazardous waste before it became a crisis. In pursuit of progress, we have been content to ignore the ecological consequences of our technological decisions because, until it was pointed out to us, we did not realize that there *was* an environmental context surrounding the tools we use.

The environmental impact analysis (EIA) has become the most popular means by which governments and industries attempt to predict and assess the ecological impact of technological change. While most EIAs are long, tedious, and nonconfrontive, the idea behind them and much of the work that has gone into them is sound. In her articles on appropriate technology, Judy Smith (1978, 1981) from the Women and Technology Project in Missoula, Montana, has suggested that sex-role impact reports could be used to improve our understanding of the cultural context of technology in much the same way that the EIA has improved our knowledge of the environmental context.

And we do need something, for we know next to nothing about the interactions of culture and technology, having always seen these as separate phenomena. Most people welcome technological change because it is *material,* believing that it makes things better, but it doesn't make them different. They resist social change because it is *social* and personal; it is seen as making things different . . . and worse. The realization that technological change stimulates social change is not one that most people welcome.

Feminists need to unthink this cultural blindness. Because women are idealized as culture carriers, as havens of serenity in a heartless world (Lasch 1977), women are supposed to remain passive while the rest of the culture is allowed, even encouraged, to move rapidly ahead. Women are like the handles of a slingshot whose relatively motionless support enables the elastic and shot to build up energy and to accelerate past them at incredible speeds. The culture measures its progress by women's stasis. When women do try to move, when they try to make changes rather than accommodations, they are accused of selfishness, of me-ness, of weakening the family, of being disloyal to civilization (Rich 1979, pp. 275–310).

However, it is crucial that feminists continue to unthink and rethink the cultural contexts of technology for a reason more significant than our systematic exclusion from it: it is dangerous not to. Technology always enters into the present culture, accepting and exacerbating the existing norms and values. In a society characterized by a sex-role division of labor, any tool or technique—it has valence, remember—will have dramatically different effects on men than on women.

Two examples will serve to illustrate this point. Prior to the acquisition of horses between the late sixteenth and mid-seventeenth centuries, women and dogs were the beasts of burden for Native American tribes on the Great Plains. Mobility was limited by both the topography and the speed at which people and dogs could walk. Physical labor was women's province in Plains culture, but since wealth in those societies was determined by how many dogs a person "owned" and since women owned the dogs, the status of women in pre-equestrian tribes was relatively high—they owned what men considered wealth (Roe 1955, p. 29). Women were central to the economic and social life of their tribes in more than the ownership of dogs. They controlled the technology of travel and food: they were responsible for the foraging, gathering, and preserving of food for the tribe and, in many cases, determined the time and routes of tribal migration. They had access to important women's societies and played a central part in religious and community celebrations (Liberty 1982, p. 14).

Women's roles in Plains Indian societies changed profoundly and rapidly as horses were acquired and domesticated. In less than two centuries—for some tribes in less than a generation—a new culture evolved. The most immediate changes were technological and economic; horses became the technology for transportation and they were opened by men. Women could still own dogs, but this was no longer the measure of wealth it had been.

With their "currency" debased, women's status slipped further as important economic, social, and religious roles were reassigned to men. As the buffalo became a major source of food and shelter, the value of women's foraging activities decreased. Hunting ranges were expanded, causing more frequent moves with women doing more of the packing up and less of the deciding about when and where to go. As each tribe's hunting range increased, competition for land intensified; and warfare, raiding, and their concomitants for women—rape and slavery—also increased.

Of course, not all effects were negative. Technologies are substitutes for human labor: horses made women's work easier and more effective. Also, several tribes, including the Blackfeet, allowed a woman to retain ownership of her own horse and saddle. However, a woman was seldom allowed to trade or raid for horses, and her rights to her husband's herd usually ended with his death.

Thus, for Native American women, the horse was a mixed blessing. It eased their burdens and made transportation easier. But it also added new tasks and responsibilities without adding authority over those tasks or increasing autonomy. The opposite was true for men; the horse provided new tasks and responsibilities—men had always been responsible for hunting, defense, and warfare—but it did enhance these traditional roles, giving men more decision-making authority, more autonomy, and more access to status. Paradoxically, while a woman's absolute status was greatly improved by the changes from dog to horse culture, her status relative to men actually declined. In this manner, horses changed the nature of Native American culture on the high plains, but women and men were affected in profoundly different ways.

A similar phenomenon occurred at the end of the horse farming era in the Palouse region of Idaho and Washington in the United States. During the 1920s, it was common for a farmer to employ 15 to 25 hired men and to use 25 to 44 horses to harvest his crops; farmers and their hands worked back-breaking, twenty-hour days. On the other hand, women also worked long days during harvest, cooking five meals a day for as many as forty people. During the year, women were responsible for a family's food, nutrition, health, safety, and sanitation. Women's work had economic value. Performing their traditional roles as wives, mothers, and homemakers, women were economically crucial to the survival of the labor intensive family farm (Bush 1982). Unfortunately, in the same manner that the horse made a Plains Indian woman's work easier even as it lowered her status relative to men, so too did the conversion from horses to diesel power and electricity ease the farm wife's hardships while it decreased the economic significance of her labor. In both cases, technological innovation had profoundly different consequences for men's and women's work. In both cases, the innovation was coded or valenced in such a way that it loaded the status of men's roles while eroding status for women.

TECHNOLOGY AND EQUITY

Technology is, therefore, an equity issue. Technology has everything to do with who benefits and who suffers, whose opportunities increase and whose decrease, who creates and who accommodates. If women are to transform or "re-valence" technology, we must develop ways to assess the equity implications of technological development and development strategies for changing social relationships as well as mechanical techniques. To do this, we must have a definition of technology that will allow us to focus on such questions of equity.

Not surprisingly, there are no such empowering definitions in the existing literature. Equity has not been a major concern of either technophobes or technophiles. In fact, most definitions of technology fall short on several counts. The most commonly accessible definitions, those in dictionaries, tell us little: *Webster's* "the science of the industrial arts" and "science used in a practical way," and the *American Heritage Dictionary*'s "the application of science, especially to industrial and commercial objectives" and "the entire body of methods, and materials used to achieve such objectives" are definitions so abstract as to be meaningless. Other attempts clarify function but lose the crucial connection to science, as in James Burke's (1980, p. 23) "the sum total of all the objects and systems used to produce goods and perform services."

Better definitions connect technology to other categories of human behavior and to human motivation:

> A form of cultural activity devoted to the production or transformation of material objects, or the creation or procedural systems, in order to expand the realm of practical human possibility (Hannay and McGinn 1980, p. 27).

On rare occasions, definitions do not raise equity questions as in John McDermott's attempt:

> Technology, in its concrete, empirical meaning, refers fundamentally to systems of rationalized control over large groups of men, events, and machines by small groups of technically skilled men operating through organizational hierarchy (McDermott 1969, p. 29).

However, this definition is really defining *technocracy* rather than *technology*. More often, there are romantic definitions that enmesh us in cotton candy:

> [Technology's task] is to employ the earth's resources and energy income in such a way as to support all humanity while also enabling all people to enjoy the whole earth, all its historical artifacts and its beautiful places without any man enjoying life around earth at the cost of another (Fuller 1969, p. 348).

While no one could argue with such ideas, Buckminster Fuller leaves us where the boon and bane theorists leave us—confounded by double-think. It is impossible to ask tough questions of such a definition or to examine closely why technology does not now support all humanity equally.

More distressing is the tendency of scholars to use the generis "he/man" to represent all of humanity. For example, "without one man interfering with the other, without any man enjoying life around the earth at the cost of another" is a statement that completely disregards the fact that, around the earth, men enjoy their lives at *women's* cost. Similarly, statements such as "because of the autonomy of technique, man cannot choose his means any more than his ends" (Ellul 1964, p. 40) and "the roots of the machine's genealogical tree is in the brain of this conceptual man . . . after all it was he who made the machine" (Usher 1954, p. 22) grossly mislead us because they obscure the historical and contemporary roles that women have played in technological development.[4] Worse, they reinforce the most disabling myth of all, the assumption that men and women are affected similarly by and benefit equally from technological change.

Therefore, because of the oversimplification of some definitions and the exclusion of women from others, feminists need to rethink a definition of technology that both includes women and facilitates an equity analysis. Such a definition might be:

> Technology is a form of human cultural activity that applies the principles of science and mechanics to the solution of problems. It includes the resources, tools, processes, personnel, and systems developed to perform tasks and create immediate particular, and personal and/or competitive advantages in a given ecological, economic, and social context (Bush in *Taking Hold of Technology* 1981, p. 1).

The chief virtue of this definition is its consideration of advantage; people accept and adopt technology to the extent that they see advantage for themselves and, in competitive situations, disadvantage for others. Thus, an equity analysis of an innovation should focus on benefits and risks within the contexts in which the technology operates. An equity analysis of a technology would examine the following:

The Developmental Context:

- the principles of science and mechanics applied by the tool or technique
- the resources, tools, processes, and systems employed to develop it
- the tasks to be performed and the specific problems to be solved

The User Context:

- the current tool, technique, or system that will be displaced at this time
- the interplay of this innovation with others that are currently in use
  the immediate personal advantage and competitive advantage created by the use of technology
- the second and third level consequences for individuals

The Environmental Context:

- the ecological impact of accepting the technology versus the impact of continuing current techniques

The Cultural Context:

- the impact on sex roles
- the social system affected
- the organization of communities
- the economic system involved and the distribution of goods within this system

A specific example will serve to illustrate how an equity analysis might be approached. Refrigeration was "invented" in the 1840s in Apalachicola, Florida, by John Gorrie as a by-product of his work on a cure for malaria (Burke 1980, p. 238). Gorrie's invention was a freezing machine that used a steam-driven piston to compress air in a cylinder that was surrounded by salt water. (As the piston advances, it compresses air in the cylinder; as the piston retracts, the air expands.) An expanding gas draws heat extracted from all the heat available in the surrounding brine. If a flow of continuously cold air is then pumped out of the cylinder into the surrounding air, the result is air conditioning; if the air is continuously allowed to cool the brine solution, the brine itself will draw heat from water, causing it to freeze and make ice. If the gas (air) or brine is allowed to circulate in a closed system, heat will be drawn from the surrounding air or matter (food), causing refrigeration.

The Developmental Context

Thus, refrigeration applies the laws of science (specifically the properties of gases) and the principles of mechanics (thermodynamics and compression) to perform the tasks of making ice, preserving and freezing food, and cooling air. Refrigeration also solves the problems of retarding food spoilage and coping with heat waves, thereby creating personal advantage. The resources and tools used include a gas, a solution, a source of energy, and a piston-driven compressor.

The developmental context is enormously complex and interconnected; however, a general analysis would include all the supply, manufacture, and distribution systems for the refrigeration units themselves—everything from the engineers who design the appliances, to the factory workers who make, inspect, and pack them, to the truckers who transport them, to the clerk who sells them. A truly expansive analysis of the developmental context would also include the food production, packing, and distribution systems required to make available even one box of frozen peas as well as the artists, designers, paper products, and advertisers who package the peas and induce us to buy them.

The User Context

Refrigeration has affected our lives in such a myriad of ways that elaborating on them all would require another paper in itself. Refrigeration has important commercial uses as well as medical ones, and it would not be overstating the case to assert that there is no aspect of modern life that has not been affected by refrigeration. Nonetheless, a more limited analysis of refrigeration as it has affected domestic and family life in the United States is both revealing and instructive.

To the self-sufficient farm family of the early twentieth century, refrigeration meant release from the food production and preservation chores that dominated much of men's and women's lives: canning garden produce to get the family through the winter; butchering, smoking, and drying meat from farm-raised hogs and cattle; milking cows daily and churning butter. The advantages of owning a refrigerator in such a situation were immediate and dramatic: food could be cooked ahead of serving time allowing women to spend far less time in meal preparation; freezing produce and meat was a faster, easier, and more sanitary process than canning or smoking, again saving women time and improving the family's health. The refrigerator thus generated positive changes for women, freeing them from hard, hot work and improving their absolute status. However, the second and third level effects of refrigeration technology were not as benign for women as the primary effects.

Since refrigeration kept food fresh for long periods of time, fresh produce could be shipped across country, thus improving nutrition nationwide. Food processing and preservation moved out of the home, and new industries and services paid workers to perform the duties that had once been almost solely women's domestic responsibility. Within the home, the nature of women's work changed from responsibility for managing food production to responsibility for managing food consumption. Also, farmers stopped growing food for family subsistence and local markets and started growing cash crops for sale on national and international markets. Opportunities for employment shifted from farm labor to industrial labor, and families moved from rural areas to cities and suburbs. Thus, the use of refrigeration changed the work roles of individual women and men and, through them, the economy, the content of work, and the nature of culture and agriculture.

The Environmental Context

An analysis of the environmental context of refrigeration technology would examine the effects of the developmental and user contexts on the environmental by asking such questions as: Since refrigeration affects agriculture, what are the ecological effects of cash crop monoculture on, say, soil erosion or the use of pesticides? Since refrigeration retards the growth of bacteria and preserves blood and pharmaceuticals, what are the consequences for disease control? What are the effects of increased transportation of food on energy supplies and air pollution?

The Cultural Context

Finally, an examination of the sex role impact of refrigeration technology would reveal a disparate effect on men and women. In the United States, men have been largely responsible for food production, women for food preservation and preparation. Refrigerators were a valenced technology that affected women's lives by, generally, removing food preservation from their domestic duties and relocating it in the market economy. Women now buy what they once canned. Women's traditional roles have been eroded, as their lives have been made easier. On the other hand, men, who originally had very little to do with food preservation, canning or cooking, now control the processes by which food is manufactured and sold. Men's roles and responsibilities have been loaded and their opportunities increased, although their work has not necessarily been made easier. Refrigeration has, thus, been adopted and diffused throughout a sexist society; we should not be surprised to learn that its effects have been dissimilar and disequitable.

THE GREAT CHAIN OF CAUSATION

Of course, not one of us thinks about the effects of refrigeration on soil erosion or women's status when we open the fridge to get a glass of milk. We are gadget-rich and assessment-poor in this society, yet each private act connects us to each other in a great chain of causation. Unfortunately, to think about the consequences of one's actions is to risk becoming immobilized; so the culture teaches us to double think rather then think, and lulls us into believing that individual solutions can work for the collective good.

Of course, we can continue to double think such things only as long as we can foist negative effects and disadvantages off onto someone else: onto women if we are men, onto blacks if we are white, onto youth if we are old, onto the aged if we are young. Equity for others need not concern us as long as *we* are immediately advantaged.

Feminists above all, must give the lie to this rationale, to unthink it; for if the women's movement teaches anything, it is that there can be no individual solutions to collective problems. A feminist transformation of technological thought must include unthinking the old myths of technology as threat or triumph and rethinking the attendant rhetoric. A feminist unthinking of technology should strive for a holistic understanding of the contexts in which it operates and should present an unflinching analysis of its advantages and disadvantages. Above all, a feminist assessment of technology must recognize technology as an equity issue. The challenge to feminists is to transform society in order to make technology equitable and to transform technology in order to make society equitable. A feminist technology should, indeed, be something else again.

NOTES

1. I am indebted for this insight to Betsy Brown and the other students in my seminar "The Future of the Female Principle," University of Idaho, Spring 1982.

2. In *Technological Change: Its Impact on Man and Society* (1970), Emmanuel Mesthene identifies "three unhelpful views about technology: technology as blessing, technology as curse, and technology as unworthy of notice." He does not mention the "technology as neutral tool argument," perhaps because he is one of its leading proponents.

3. Obviously, oral history is the only way that scholars can accumulate this data. Oral historians should ask respondents questions about their acquisition of and adaptation to household appliances. Such questions might include: "When did you get electricity?" "What was the first appliance you bought?" "What was your first washing machine like?" "How long did it take you to learn how to use it?" "What was your next machine

like?" "When did you get running water?" "Are you usually given appliances for presents or do you buy them yourself?" et cetera.

4. This situation is slowly changing thanks to much good work by Elise Boulding (1976), Patricia Draper (1975), Nancy Tanner and Adrienne Zihlman (1976), and Autumn Stanley (1984).

REFERENCES

Armitage, Susan. 1982. Wash on Monday: The housework of farm women in transition. *Plainswoman* VI, 2:3–6.

Boulding, Elise. 1976. *The underside of history: A view of women through time.* Boulder, Colo.: Westview Press.

Brownmiller, Susan. 1975. *Against our will: Men, women, and rape.* New York: Simon and Schuster.

Burke, James. 1980. *Connections.* Boston: Little Brown.

Bush, Corlann Gee. 1982. The barn is his; the house is mine: Agricultural technology and sex roles. *Energy and transport.* Eds. George Daniels and Mark Rose. Beverly Hills, Calif.: Sage Publications, 235–59.

Carson, Rachel. 1955. *The edge of the sea.* Boston: Houghton Mifflin.

Carson, Rachel. 1961. *The sea around us.* New York: Oxford University Press.

Carson, Rachel. 1962. *Silent spring.* Boston: Houghton Mifflin.

Cooper, James Fenimore. 1832. *The pioneer.* Philadelphia: Cary & Lea.

Daly, Mary. 1968. *The church and the second sex.* New York: Harper & Row.

Daly, Mary. 1973. *Beyond God the father: Toward a philosophy of women's liberation.* Boston: Beacon Press.

De Crevecoeur, Michael Guillaume St. Jean. 1968. *Letters from an American farmer: Reprint of 1782 edition.* Magnolia, Mass.: Peter Smith.

Draper, Patricia. 1975. !Kung women: Contrasts in sexual egalitarianism in foraging and sedentary contexts. *Toward and anthropology of women,* Ed. Rayna Reiter. New York: Monthly Review Press, 77–109.

Dworkin, Andrea. 1974. *Woman hating.* New York: E. P. Dutton.

Ellul, Jacques. 1964. *The technological society.* New York: Knopf.

Fuller, R. Buckminster. 1969. *Utopia or oblivion: The prospect for humanity.* New York: Bantam Books.

Griffin, Susan. 1979. *Rape: The power of consciousness.* New York: Harper & Row.

Hannay, N. Bruce, and McGinn, Robert. 1980. The anatomy of modern technology: Prolegomenon to an improved public policy for the social management of technology. *Daedalus* 109, 1:25–53.

Lasch, Christopher. 1977. *Haven in a heartless world: The family besieged.* New York: Basic Books.

Liberty, Margot. 1982. Hell came with horses: Plains Indian women in the equestrian era. *Montana: The Magazine of Western History* 32, 3:10–19.

McDermott, John. 1969. Technology: The opiate of the intellectuals. *New York Review of Books XVI,* 2 (July): 25–35.

McKee, Alice. 1982. The feminization of poverty. *Graduate Woman* 76, 4:34–36.

Mesthene, Emmanuel G. 1970. *Technological change: Its impact on man and society.* Cambridge, Mass.: Harvard University Press.

Rich, Adrienne. 1975. *Poems: Selected and new 1950–1974.* New York: W. W. Norton.

Rich, Adrienne. 1979. Disloyal to civilization. *On lies, secrets, and silence: Selected prose.* New York: W. W. Norton.

Roe, Frank Gilbert. 1955. *The Indian and the horse.* Norman: University of Oklahoma Press.

Ryan, Mary P. 1979. *Womanhood in America: From colonial times to the present.* 2nd ed. New York: Franklin Watts.

Smedley, Agnes. 1973. *Daughter of earth.* Old Westbury, N.Y.: Feminist Press.

Smith, Judy. 1978. *Something old, something new, something borrowed, something due: Women and appropriate technology.* Butte, Mont.: National Center for Appropriate Technology.

Smith, Judy. 1981. Women and technology. What is at stake? *Graduate Woman* 75, 1:33–35.

Stanley, Autumn. 1984. *Mothers of invention: Women inventors and innovators through the ages.* Metuchen, N.J.: Scarecrow Press.

Taking hold on technology: Topic guide for 1981–83. 1981. Washington, D.C.: American Association of University Women.

Tanner, Nancy, and Zihlman, Adrienne. 1976. Women in evolution: Part I: Innovation and selection in human origins. *Signs* 1 (Spring):585–608.

Usher, Abbott Payson. 1954. *A history of mechanical inventions.* Cambridge, Mass.: Harvard University Press.

Vanek, Joann. 1974. Time spent in housework. *Scientific American* 231 (November):116–20.

Weinberg, Alvin M. 1966. Can technology replace social engineering? *University of Chicago Magazine* 59:6–10.

Winner, Langdon. 1977. *Autonomous technology: Technics-out-of-control as a theme in political thought.* Cambridge: MIT Press.

Winner, Langdon. 1979. The political philosophy of alternative technology. *Technology in Society* 1:75–86.

Zimmerman, Jan. 1981. Introduction. *Future, technology and woman: Proceedings of the conference.* San Diego, Calif.: Women's Studies Department, San Diego State University.

9

Design Methodology and the Nature of Technical Artifacts

Peter Kroes

The aim of design methodology is to improve design processes; this means that it takes a normative stance towards its object of study. Given this aim it is no surprise that research in design methodology has always had a strong focus on the nature of design processes. The study of the nature of technical artefacts, considered to be the outcome of a design process, has received little attention. For several reasons, however, including its normative standpoint, design methodology cannot avoid a closer analysis of the nature of technical artifacts. Here, an interpretation of technical artifacts in terms of a dual nature—which refers to the fact that they are physical and intentional objects at the same time—is offered and it is argued that this interpretation has far-reaching consequences for the research agenda of design methodology. The paper starts with a comparison of design methodology with the methodology of science. This is followed by an exposition of the dual nature of technical artefacts. Finally, some consequences of this interpretation of technical artefacts for the research agenda of design methodology are discussed.

DESIGN METHODOLOGY VERSUS METHODOLOGY OF SCIENCE

There are two striking differences in orientation between methodological studies of technical design and of scientific research. Design methodology takes a normative stance toward design and is very much process oriented, whereas research methodology is descriptive and strongly product oriented. We will first have a closer look at these differences.

Methodological studies of science, as part of the broader discipline of the philosophy of science, typically concentrate on the outcomes of scientific research processes, such as empirical claims, laws, theories and explanations, and focus on questions about the interpretation of these products and their reliability (or truth). This product orientation in the methodology of science is related to the classic logical positivist's distinction between the context of discovery (how are phenomena, laws, theories, etc., discovered?) and the context of justification (how are phenomena, laws, theories, etc., justified?) and their highly influential idea that there is no "logic" of scientific discovery. Insofar as logical positivists were process oriented, they were interested in a very special

Peter Kroes, "Design Methodology and the Nature of Technical Artefacts," *Design Studies* 23, no. 3 (May 2002): 287–302. Copyright © 2002 Elsevier Science Ltd. Reprinted by permission.

kind of process, namely a "rational reconstruction of science," that is according to Carnap "a schematized description of an imaginary procedure, consisting of rationally prescribed steps, which would lead to essentially the same results as the actual [. . .] process" (p. 16). The real research process, as it was actually performed, was only of minor interest to them or of no interest at all. Admittedly, there has been an empirical turn in the philosophy of science since the work of Kuhn, but although this turn has led to more interest in actual research processes, particularly in experiments, the underlying issues remain, as before, issues about the interpretation and justification of scientific research. It is mainly a descriptive activity.

Design methodology, characterized by one of its leading figures, Nigel Cross, as "the study of principles, practices and procedures of design" (see note 1) aims at improving design practice and is strongly process oriented.

According to Cross's history of design methodology, the founding fathers of this discipline has a strong normative attitude: design methodology should contribute to the improvement of design practice, particularly by exploiting scientific methods. Cross's history also illustrates the strong process orientation of design methodology. He mentions five categories of recent work in this field, four of which are explicitly process (activity) oriented (namely, the development of design methods, the management of design process, the nature of design activity and the philosophy of design method which deals with the philosophical analysis and reflection on design activity); only the work which he classifies under the heading "the nature of design problems" is not process oriented.

A look at the broader field of design research indeed confirms the strong bent towards processes and activities in this field. According to Dorst, two paradigms within design research can be distinguished: design as rational problem solving and design as reflective practice, and both are process oriented. Schon's theory about the reflective practitioner, which has attracted much attention in recent years within design research, approached design as a reflective process. Bucciarelli's work, also well known within this field, analyzes design as an essentially social process. Finally, a quick scan of the contents of volumes 16 (1995) through 22 (2001) of one of the leading journals in this field, *Design Studies,* confirms the strong process orientation. The topics addressed typically concern: (creativity in) design thinking, design progress, communication of design knowledge, managing design information, the role of computers in design, design as a cognitive activity, decision making in design, et cetera. This journal explicitly presents itself as a forum for the discussion and development of the theoretical aspects of design, including its methodology and values, but almost without exception the methodological contributions concern the actual methods and techniques used in solving design tasks, not the methods and techniques used in justifying the outcome of a design process (see note 2).

So, there are two differences between design methodology and research methodology: the former takes a normative stance and is process oriented, the latter is descriptive and product oriented (see figure 9.1). Compared to research methodology, design methodology is interested in a

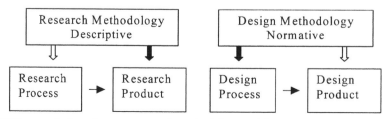

Figure 9.1.   Differences in focus between research methodology and design methodology

rather different kinds of "rational reconstruction," namely in a schematized description of real (not imaginary) design procedures, consisting of rationally prescribed steps, which should lead to essentially better (not the same) results compared with existing design procedure. Not surprisingly, therefore, design methodology has resulted in a variety of schemes for dividing the design process into various phases, varying from the very simple analysis—synthesis—evaluation scheme to, for instance, the rather detailed and elaborate scheme proposed by the Verien Deutscher Ingenieure (VDI).

Because of these differences, design methodology and methodology of science bear little resemblance to each other. It is even confusing to call both "methodology" because that suggests that they address similar kinds of questions for design and research. That is not in fact the case. Suppose we were to construe a field called "design methodology" analogous to the field called "methodology of science." Then it would have to deal with the following kinds of questions, some of which will surface again later on (see note 3).

- What is a design?
- What makes a design a good or a successful design?
- What are the proper criteria for evaluating proposed solutions for a given design problem?
- Is it possible to characterize in a general (logical?) way notions such as the effectiveness and efficiency of design solutions?
- How can a proposed solution for a design problem be rationally justified?
- How can design decisions with regard to trade-offs between conflicting design specifications be rationally justified?

Shifting attention from design solutions to design methods, the following questions crop up.

- Does the correct application of design methods guarantee a successful outcome or make a successful outcome probable to a certain degree?
- If so, is there a "logic of design methods," that is, can we understand this property of design methods from a logical (analytical) point of view?

Furthermore, a technological design (ideally) contains an explanation of how a given physical (chemical, biological) devise realizes a certain function.

- How is such a technological explanation, that is, an explanation of a function in terms of a physical (chemical, etc.) structure, possible?
- What kind of adequacy conditions apply to technological explanations?

These questions concern either the justification of the outcome of the design process or a rational reconstruction of the design process in the Carnapian sense (i.e., in terms of imaginary steps and procedures).

Design methodology, as it has been practiced up till now, has largely neglected these questions. Because it aims at the improvement of design practice, it has focused mainly on the design process. By analyzing in detail the nature of the process, it tries to rationally reconstruct it in the sense described earlier. In my opinion, however, design methodology will have to address some of the issues described above for at least two reasons. The first is that the design process and the design product are so intimately related to each other that an understanding of the nature of the design process requires insight into the nature of the product designed and vice versa. Consider the design of various kinds of artifacts, for example, the steering wheel of a car, an air bag, a car, a car transport system, a police system, a law on traffic regulations. Roughly speaking, these artifacts may be ordered on an axis ranging from technical objects through socio-technical objects to social objects. It is a matter of fact that the design processes which lie at the basis of these various kinds of artefacts differ strongly. It seems implausible that it will be possible to construct a domain-independent theory about design processes, which will cover all these cases (see note 4). An analysis of the design process of technical artifacts should therefore take into account the specific nature of those objects (see note 5). Second, the normative stance taken by design methodology towards the design process implies that it cannot escape questions concerning the quality of the outcome of that process. Since that outcome is the design of a technical artifact, it has to address some of the questions listed above about criteria for success of a design. So, let us now turn to a closer analysis of the nature of technical artifacts.

THE DUAL NATURE OF TECHNICAL ARTIFACTS

According to the view defended below, technical artifacts have a dual nature: on the one hand they are physical objects (man-made constructions) that may be used to perform a certain function, on the other hand they are intentional objects since it is the function of a technical artifact that distinguished it from physical (natural) objects and this function has meaning only within a context of intentional human action. Before presenting this dual nature view of technical artifacts, I will briefly discuss Herbert Simon's theory on artificial things as exposed in his classic, *The sciences of the artificial* (in the following text, page numbers refer to this book). This theory proves to be a useful stepping stone to the dual nature view. For Simon, the science of the artificial will closely resemble the science of engineering because engineering deals with the synthesis of things. In contrast to the scientist, the engineer and more in particular the designer are "concerned with the things *ought* to be—how they ought to be in order to *attain goals,* and to *function*" (pp. 4–5). One of the striking features of (technical) artifacts is precisely that they can be characterized in terms of functions and goals. Functions and goals are analyzed by Simon in the following way (p. 5):

> Let us look a little more closely at the functional or purposeful aspect of artificial things. Fulfillment of purpose or adaptations to a goal involves a relation among three terms: the purpose or goal, the character of the artifact, and the environment in which the artifact performs.

For instance, the purpose of a clock is to tell time and the character of the clock refers to its physical make-up (gears, springs, etc., for a mechanical clock). Finally the environment is important because not every kind of clock is useful in every environment; sundials can only perform their function in sunny climates. Simon's analysis of artifacts is represented in a schematic way in figure 9.2 (see note 6).

According to Simon, the environment of an artifact is very important because it moulds the artifact. He considers the artifact to be a kind of "interface" between "an 'inner' environment, the substance and organization of the artifact itself, and an 'outer' environment, the surroundings in which it operated" (p. 6). The inner environment of the artifact, its character, is shaped in such a way that it realizes the goals set in the outer environment (p. 10). Therefore, the science of the artificial has to focus on this interface, since the "artificial world is centered precisely on this interface between the inner and outer environments; it is concerned with attaining goals by adapting the former to the latter" (p. 113).

Simon's distinction between inner and outer environment points to two different ways of looking at technical artifacts. Looked at from the outer environment, the technical artifact presents itself primarily as something, whatever its inner environment, that fulfills a certain goal, purpose or function. From this perspective the artifact is characterized primarily in a functional way; the inner

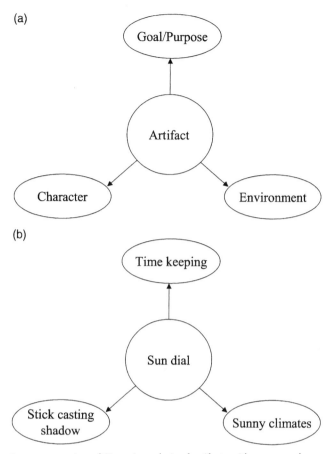

Figure 9.2. Schematic representation of Simon's analysis of artifacts with an example

environment remains a black box. Looked at from the inner environment, the artifact is described as some kind of physical system; from this perspective, the goal that it fulfills in the environment remains a black box (see note 7). As Simon remarks (p. 7) "*Given* an airplane, or *given* a bird, we can analyze them by the methods of natural science without any particular attention to purpose or adaptation, without reference to the interface between what I have called the inner and outer environments." These two different ways of characterizing an artifact, in terms of its inner and outer environment, correspond closely to what we call the dual nature of technical artifacts.

The view that technical artefacts have a dual nature finds its origin in the observation that we employ in our thinking, speaking and doing two basic conceptualizations of the world, and that we do not know how to integrate these two together into one coherent conceptualization (see note 8). On the one hand, we see the world as consisting of physical objects interacting through causal connections. This will be called the "physical" or "structural" conceptualization which is employed and developed by the physical sciences. On the other hand, we see the world as consisting partly of agents (primarily human beings), who intentionally represent the world and act intentionally in it, and whose behaviour is explained partly in terms of reasons (and not causes). This is the "intentional" conceptualization of the world which underlies most of the social sciences. One aspect of this latter conceptualization is that certain activities are interpreted in terms of realizations of goals and that functions are attributed to certain objects or activities. The existence of these two different conceptualizations poses a problem in cases where both offer competing explanations for the same kind of phenomenon, for example, for raising a hand to vote in a meeting: one in terms of physiological causes, the other in terms of reasons. This is the well-known mind-body problem.

The question that concerns us is how technical artifacts fit into these two conceptualizations of the world. Our starting point for exploring these issues will be following characterization of technical artefacts: *technical artifacts are objects with a technical function and with a physical structure consciously designed, produced and used by humans to realize its function* (see note 9). In short, a technical artifact is a physical object with a technical function. This characterization of the technical artifact makes it a hybrid kind of object which does not fit in either the physical or the intentional conceptualization. Looked upon as merely physical objects, technical artefacts fit into the physical conceptualization of the world; the way the artifact works can be explained in terms of causal processes. But as a mere physical object, it is not a technical artifact. This means that technical artifacts cannot be described exhaustively within the physical conceptualization, since it has no place for its functional features. But neither can it be described exhaustively within the intentional conceptualization since its functionality must be realized through an appropriate physical structure and the intentional conceptualization has not place for the physical features of a technical artifact (see note 10). Hence the conclusion that technical artifacts have a dual nature: On the one hand they are physical, on the other intentional objects.

According to the above line of thought, the notion of technical artifact is related to three key notions, namely the notion of a physical structure, of a (technical) function and of a context of intentional human action (see figure 9.3 and note 11).

The inclusion of the context of human action into our analysis of artifacts needs some clarification, since we have characterized technical artifacts earlier as physical objects with technical functions. I have included the context of human action because it makes no sense to speak about technical functions without reference to a context of human action. As remarked earlier, functional discourse is part of the intentional conceptualization of the world; it is meaningless to speak about technical functions without a context of intentional (human) action. This can be expressed in an ontological way by saying that some context of human action is constitutive for a technical function. This is in line with Searle's claim that technical functions are attributed, in or with regard to some context of human action, to objects; they are not intrinsic properties of those objects. In his analysis of the relational ontology of technical artifacts, Meijers also claims that human action is constitutive for functions: "A central part of my argument focuses on functions and functional properties. These are realized by the physical structure of the artifact together with the practice of its design and use" (p. 81). Thus, in figure 9.3 function and context of human action are intimately connected; they both belong to the domain of the intentional. Technical artifacts have a dual nature since they are at the same time part of the domain of the physical and of the intentional.

There are some notable differences between our analysis of technical artifacts and Simon's. Simon's notion of goal or purpose has been replaced by the notion of function. This may seem an insignificant move but it is not, because we may attribute functions to technical artifacts but not goals (in the sense of an aim or an end (telos)). That notion has its place in a context of intentional human action; within such a context a means used to achieve a goal (end, aim) is attributed a function. Thus, Simon's analysis implicitly refers to a context of human action by referring to goals and purposes. Furthermore, the notion of environment is a context of human action. Simon's claim that the artifact has to adapt to its environment then reduces to the, rather obvious, claim that the artifact has to adapt to the context of human action in which it is used. Nevertheless, this is a noticeable change because it brings out the fact that not any kind of environment is relevant for the analysis of technical artifacts; only references to environments comprising a context of human action are appropriate. In his example of the sundial, for instance, Simon interprets the environment

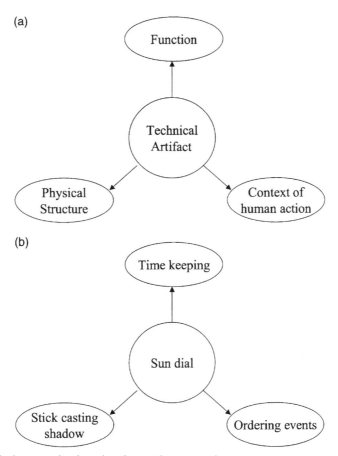

Figure 9.3.   The dual nature of technical artifacts with an example

in a physical way (sunny climates are the required environment for sundials). But this is problematic. It is not this physical environment that turns the object involved, a stick that casts a shadow on a surface, into an artifact of this type of sundial. Only within the context of human action (e.g., of ordering events or comparing time intervals) does this physical object acquire a function and become a technical artifact (a time-keeping device or clock).

The main difference between Simon's analysis and ours is that the latter gives a much more prominent and explicit place to a context of human action in analyzing the nature of technical artifacts. The advantage of this is that it brings much more into the open the dual nature of technical artefacts: we cannot make sense of technical artifacts without taking into consideration their physical structure, but also not without their context of intentional human action. Within Simon's analysis this dual nature stays more implicit and is related to the two different perspectives on technical artifacts, namely the perspective of the inner environment (physical structure) and the perspective of the outer environment (context of human action).

Note that the above characterization of a technical artifact involves processes in an essential way: without some context of human action (activity, processes) the notion of function loses its meaning, and what is left of a technical artifact without its function is just some physical object. In order to arrive at a better understanding of how the design process is involved in characterizing a technical artifact, we have to take a closer look at what we have called "context of human action."

This is a very general and rather vague term. With regard to technical artefacts, at least two significant kinds of context of human action can be distinguished, namely the design context and the user context (see figure 9.4 and note 12).

In these two contexts the technical artifact manifests itself in different ways. In the design context, the main emphasis lies on how to construct a physical system (object) that realizes a certain function. This function is often described in terms of a list of specifications which the object to be designed must meet. Here we encounter what Simon calls the "inner environment" of a technical artifact. In the context of use, the "outer environment" presents itself. There, the function of the artifact in relation to the realization of goals (ends) is of prime importance and the physical constitution of the technical artifact becomes of secondary importance. Note, however, that in the context of design as well as in the context of use it is not only a physical structure, just as it is not only a function in the context of use.

In many cases, there is no continuity between a context of design and a context of use in the sense that the same people who design a technical artifact also use it. This situation creates problems with regard to the communication of functions between designers and users. To what extent is it possible to design a technical artifact so that it will communicate its "proper" function, that is, the function it was designed for, to its potential users? Or is it the case that the technical artifact itself plays no intermediary role at all in the communication of its function, which means that this communication has to be established by other means? It is interesting to note that Dipert has worked out a theory of technical artifacts in which it is a defining feature of artifacts that they are explicitly designed to communicate their artifactuality and functionality to their users. This view presupposed that in principle it is possible that an "artifact itself" communicates its function. However, since the function is not an intrinsic property of the artifact, it is not clear what it is in the "artifact itself" that is the source for the communication of its function. Part of this problem may be solved for the communication of its function. Part of this problem may be solved by taking into consideration the user manual of a technical artifact. A user manual has at least two functions: it is a means to communicate the intended function to the user and to make this function accessible to the user by prescribing which actions have to be performed to realize the intended function. If we assume that a user manual is an integral part of the technical artifact, then part of the communication problem can be solved easily by way of the user manual (see note 13). But even in that case the question as to how much of the function of a technical artifact can be communicated without

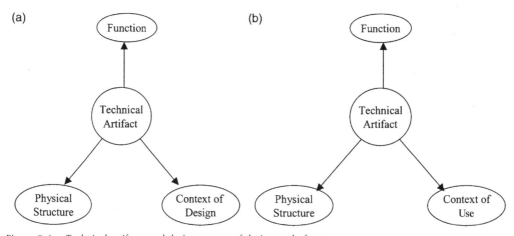

Figure 9.4.   Technical artifacts and their contexts of design and of use

recourse to a user manual is of great importance for design practice. To study this matter, design methodology will have to focus on the nature of a technical artifact, on the various ways technical artefacts may communicate their function, and what kind of theory of communication this presupposes.

At this point I conclude the exposition of the dual nature of technical artifacts. In the next section, I will explore some of the consequences of this interpretation of nature of technical artifacts for the agenda setting of design methodology, particularly from the point of view of the relationship between the design process and the design product.

DISCUSSION: CONSEQUENCES FOR THE RESEARCH AGENDA OF DESIGN METHODOLOGY

Our analysis of technical artifacts as having a dual nature of itself leads to a question that is of crucial importance for understanding the nature of design processes and therefore deserves a prominent place on the research agenda of design methodology. This question is: How can we account for the fact that designers are able to bridge the gap between a functional and a structural description of a technical artifact? That they are able to bridge the gap stands without question. But from a philosophical point of view, we are dealing here with two different conceptualizations of an artifact. It is not clear how these two are related to each other and how it is possible to go from one conceptualization to the other. Schematically (see figure 9.5), a design process may be characterized as starting with a functional description of the desired artifact; this may be considered to be the input of a design process. This functional description is a black box description with regard to the physical structure of the technical artifact. It is precisely the task of the designer to fill this black box with a physical structure such that this structure will realize the intended function. The output of a design process, therefore, is a description of a physical structure which adequately performs the function, that is, with a design of the technical artifact (which may be taken to include the user manual).

Given the interpretation of the design process, two observations may be made. First, designers manage to bridge the gap between functional and structural descriptions of artefacts in a systematic way; they use all kinds of design methods to help them solve their design problems. Second, they are in most cases able to explain why a proposed design will adequately fulfill its function. From the point of view of the dual nature of technical artifacts, these observations raise the following questions:

- What kind of design methods are used by designers to bridge the gap between the two modes of describing technical artifacts?
- How are we to interpret the role of these design methods in bridging the gap between the two conceptualizations of the artifact? In other words, can we provide a rational account

Figure 9.5.   The design process and the gap between functional and structural descriptions

of the use of these design methods, showing why their use is successful, given the conceptual gap?

- How do designers explain the function of an artifact in terms of its structure?
- How can a function be explained in terms of a physical structure, given the conceptual gap between the two kinds of descriptions involved (see note 14)?

In order to answer these questions much empirical and conceptual work still remains to be done. But given the aim of design methodology to improve design practice, it cannot avoid addressing these questions: without clarification of these issues an adequate understanding of the nature of the design of technical artifacts is, to say the least, problematic (see note 15).

The final topic that I would like to draw attention to concerns the quality of a design, in particular the notion of a successful design. It is self-evident that design methodology has to establish some criteria for the quality, the success and the failure of design processes if we are to take its normative stance towards design processes seriously. Otherwise, the notion of an improvement of a design process loses its meaning. These criteria are also necessary to uphold the idea that designing technical artifacts is partly a rational activity (see note 16). Without some criteria for improvement or progress, the notion of rationality becomes problematic.

So, what are the criteria for quality on the basis of which design processes may be evaluated? In line with their process orientation, design methodologists seem to have approached this problem primarily from the point of view of the organization and management of design processes. There are many prescriptive phase diagrams of how to split up the overall design process into various parts. The suggestion is implied, explicitly or implicitly, that following these diagrams will lead to or at least contribute to the quality (success) of the design process. Thus, implementing adequately the prescriptive phase diagram becomes a criterion for success. Without an assumption of this kind, the rationale behind these diagrams becomes problematic. This may be part of the answer, but it is highly questionable whether it addresses the real issues involved. It is not difficult to imagine, and probably has often actually been the case, that a design process follows painstakingly all the required procedure and nevertheless its outcome is deemed a failure by the people involved. In such cases, the design process has to be considered a success, whereas its outcome is a failure (the proverbial successful operation with a dead patient). Conversely, a badly organized and poorly managed design process may lead to an excellent design.

The relationship between the quality of a design process and the quality of its outcome does not seem to be straightforward. Abiding by the rules of procedural rationality is not a sufficient criterion for success (neither does it seem to be a necessary criterion). More is involved, namely the criteria in terms of which the outcome of a design process is evaluated. So, we arrive at the question: What is a good or successful design? That itself is a complicated issue and it is doubtful that there is one set of criteria that is universally valid in every context. In our analysis of technical artifacts we have distinguished so far two different contexts of human action, namely the design context and the use context. It is not all self-evident that the same criteria for quality apply in both contexts. Within a design context, a general criterion for success may be that a proposed design meets all the specifications and constraints. A particular design that satisfies this criterion may nevertheless be considered a failure in the context of use because it does not meet the expectations or satisfy the needs of the users; the latter will be their criterion of success. This situation may be due to, for instance, poor communication between designers and users about the desired functionality. But even if we assume that the communication about needs and functions is flawless, then it is nevertheless doubtful whether the community of designers applies the same criteria of quality as the community of users. For instance, the introduction of an "engineering change," that is, a change in the physical make-up of an artifact that does not affect in any way its technical function, may considerably improve the quality of a design judged in the context of design, whereas in the context of use its quality remains the same.

Apart from the context of design and the context of use, technical artefacts figure in many other contexts of human action, such as the context of production, context of maintenance, context of consumer markets, et cetera. Each of these contexts has its own criteria for quality and success which may be relevant to the way the quality of the design of the artefact is evaluated. Aesthetic criteria pose a problem of their own in evaluating the quality of a design because it is a problem to find objective standards for these criteria (see note 17). Moreover, the importance of these criteria varies strongly over different engineering domains (for instance, in many areas of electrical and mechanical engineering they are almost conclusively to be drawn from the foregoing, namely that a clear insight into the notion of the quality (success) of a design or a design process is lacking.

Given this conclusion, it is rather remarkable that, although design methodology professes to aim at improving design processes, it has, to my knowledge, not addressed these issues systematically. Either the success of a design process depends wholly or in part on the success of the outcome of this process, in which case it appears rather obvious that the above issues about the quality of a design should rank high on the research agenda of design methodology. Or the success of a design process does not depend at all on the success of its outcome, but in that case the rationale for improving the design process, that is, for the aim of design methodology, becomes problematic. Design methodology needs a foundation for its normative point of view on design processes, and it appears plausible that this foundation is partly to be found in criteria for a successful design.

ACKNOWLEDGMENTS

I would like to thank the members of the Department of Philosophy of the Delft University of Technology for their valuable comments on an earlier version of this paper.

NOTES

1. Quotes in Dorst (p. 8). There seems to be some confusion about what design methodology is about and how it related to the wider field of design studies. I will not go into this matter; for my purpose, Cross's description is a good starting point.

2. A notable exception is Galle.

3. For more details about how design methodology in this sense fits into the broader field known as the philosophy of engineering design, see Kroes.

4. This claim does not go undisputed; see for instance, Simon, who remarks: "The intellectual activity that produces material artifacts is no different fundamentally from the one that prescribes remedies for a sick patient or the one that devises a new sales plan for a company or a social welfare policy for a state" (p. 111).

5. In the following analysis, I will consider the outcome of a design process to be a technical artifact. Although a design is not yet itself something that justifiably may be called a (full-blooded) technical artifact, it is an integral part of a process that produces a technical artifact. Moreover, the ultimate validation of a design involves the actual making and use of the technical artifact described in the design. Thus, if the design validation phase is taken to be part of the design process, this process implicitly implies the making of the intended technical artifact.

6. The arrows stand for conceptual implication: the notion of an artifact conceptually implies the notion of a character, a goal or purpose, and an environment.

7. For more details about the black box character of fundamental and physical descriptions of objects, see Kroes.

8. Parts of the following are based on the NOW grant application *The dual nature of technical artifacts,*

1999. Written jointly by Anthonie Meijers, Maarten Franssen, Pieter Vermaas, Wybo Houkes and the author. For the full text of this application, see http://www.dualnature.tudelft.nl

9. Of course, all kinds of demarcation problems arise about software or natural objects used for practical purposes. I will leave those problems aside. This characterization seems to be adequate for technical artifacts which are the result of engineering design and development.

10. This is related to the fact that a functional description is, from a physical point of view, a black box description of an object; in general, a functional description states that something, whatever it may be from a physical point of view, may be used as a means to realize a certain state of affairs.

11. In a more or less similar way, Losonsky analyses the nature of artifacts in terms of the following there features: internal structure, purpose and manner of use.

12. For an action-theoretical account of the design and use context, see W. Houkes, P. Vermaas, K. Dorst and M. de Vries, "Design and use as plans: an action-theoretic account," in this issue of *Design Studies*.

13. "Part of the communication problem," because there is no guarantee that a user will reconstruct from the technical artefact (including its user manual) a function that is identical to the intended function. An argument for including the user manual in the technical artifact is that it strengthens the ties between a technical artifact and a context of action, since the user manual prescribes how the artifact in question has to be used in order to realize its intended function.

14. For a discussion of this issue, see Kroes.

15. The fact that designers have to deal with two conceptualizations of the world has not gone unnoticed in design methodology. Rosenman and Gero (1998), for instance, explicitly characterize design as involving the "transition of concepts from the socio-cultural environment to the description of technical objects" (p. 161).

16. The notion of rationality may be interpreted in this context in various ways. It may be taken in the sense that it is possible to provide arguments why certain design decisions will lead to better results than other ones. If we take designing to be a goal-oriented activity, then the notion of rationality as adaptation of means to an end may be applied (means-end-rationality).

17. For a discussion of the role of aesthetic criteria in design and how this role affects the question whether design is an art or a science, see Kroes.

REFERENCES

Bucciarelli, L. L. *Designing engineers* (Cambridge: MIT Press, 1994).

Carnap, R. "Intellectual autobiography," in P. A. Schilpp ed., *The philosophy of Rudolf Carnap* (La Salle, IL: Open Court, 1963).

Cross, N. A. "History of design methodology," in M. J. De Vries, N. Cross, and D. P. Grant, eds., *Design methodology and relationships with science* (Dordrecht: Kluwer Academic Publishers, 1993), pp. 15–27.

Dipert, R. R. *Artefacts, art works and agency* (Philadelphia: Temple Press, 1993).

Dorst, K. "Describing design; a comparison of paradigms," PhD thesis (The Netherlands: Delft University of Technology, 1997).

Galle, P. "Design rationalization and the logic of design." *Design Studies*. Vol. 17 No. 3 (1996) 253–75.

Kroes, P. A. "Technological explanations: the relation between structure and function of technological objects." *Technè*, Vol. 3, No. 3 (1998). http://scholar.lib.edu/ejournals/SPT/v3n3/html/KROES.htm

Kroes, P. A. "Reflections on technological design as art." *Integrated Design and Process Technology*, Vol. IDPT-3 (1998), 104–109.

Losonsky, M. "The nature of artifacts." *Philosophy*. Vol. 65 (1990) 81–85.

Meijers, A. W. M. "The relational ontology of technical artefact," in P. Kroes and A. Meijers, eds. *Research in philosophy and technology*, Vol. 20 (2000), 81–96.

Rosenman, M. A. and Gero, J. S. "Purpose and function in design: from the socio-cultural to the techno-physical." *Design Studies,* Vol. 19 (1998).

Schön, D. *The reflective practitioner: how professionals think in action* (Aldershot: Ashgate, 1991).

Simon, H, A. *The sciences of the artificial,* 3rd ed. (Cambridge: MIT Press, 1996).

Searle, J. R. *The construction of social reality* (London: Penguin Books, 1995).

Verein Deutscher Ingenieure (VDI), *Systematic approach to the design of technical systems and products: Guideline* (*VDI 2221* Beuth Verlag, Berlin, 1987).

10

Democratic Rationalization:
Technology, Power, and Freedom

Andrew Feenberg

THE LIMITS OF DEMOCRATIC THEORY

Technology is one of the major sources of public power in modern societies. So far as decisions affecting our daily lives are concerned, political democracy is largely overshadowed by the enormous power wielded by the masters of technical systems: corporate and military leaders, and professional associations of groups such as physicians and engineers. They have far more to do with control over patterns of urban growth, the design of dwellings and transportation systems, the selection of innovations, and our experience as employees, patients, and consumers than all the governmental institutions of our society put together.

Marx saw this situation coming in the middle of the nineteenth century. He argued that traditional democratic theory erred in treating the economy as an extrapolitical domain ruled by natural laws such as the law of supply and demand. He claimed that we will remain disenfranchised and alienated so long as we have no say in industrial decision making. Democracy must be extended from the political domain into the world of work. This is the underlying demand behind the idea of socialism.

Modern societies have been challenged by this demand for over a century. Democratic political theory offers no persuasive reason of principle to reject it. Indeed, many democratic theorists endorse it.[1] What is more, in a number of countries, socialist parliamentary victories or revolutions have brought parties to power dedicated to achieving it. Yet today we do not appear to be much closer to democratizing industrialism than in Marx's time.

This state of affairs is usually explained in one of the following two ways.

On the one hand, the common sense view argues that modern technology is incompatible with workplace democracy. Democratic theory cannot reasonably press for reforms that would destroy the economic foundations of society. For evidence, consider the Soviet case: although they were socialists, the communists did not democratize industry, and the current democratization of

A revised version of Andrew Feenberg, "Subversive Rationality: Technology, Power, and Democracy," in *Inquiry* 35/3–4 (1992): 301–22. Copyright © 1992 Taylor & Francis. Reprinted by permission of Taylor & France, Oslo: Norway. (www.tandf.no/inquiry)

Soviet society extends only to the factory gate. At least regarding the ex–Soviet Union, everyone can agree on the need for authoritarian industrial management.

On the other hand, a minority of radical theorists claims that technology is not responsible for the concentration of industrial power. That is a political matter, due to the victory of capitalist and communist elites in struggles with the underlying population. No doubt modern technology lends itself to authoritarian administration, but in a different social context it could just as well be operated democratically.

In what follows, I will argue for a qualified version of this second position, somewhat different from both the usual Marxist and democratic formulations. The qualification concerns the role of technology, which I see as *neither* determining nor neutral. I will argue that modern forms of hegemony are based on the technical mediation of a variety of social activities, whether it be production or medicine, education or the military, and that, consequently, the democratization of our society requires radical technical as well as political change.

This is a controversial position. The common sense view of technology limits democracy to the state. By contrast, I believe that unless democracy can be extended beyond its traditional bounds into the technically mediated domains of social life, its use value will continue to decline, participation will wither, and the institutions we identify with a free society will gradually disappear.

Let me turn now to the background to my argument. I will begin by presenting an overview of various theories that claim that insofar as modern societies depend on technology, they require authoritarian hierarchy. These theories presuppose a form of technological determinism that is refuted by historical and sociological arguments I will briefly summarize. I will then present a sketch of a nondeterministic theory of modern society I call *critical theory of technology*. This alternative approach emphasizes contextual aspects of technology ignored by the dominant view. I will argue that technology is not just the rational control of nature; both its development and impact are also intrinsically social. I will then show that this view undermines the customary reliance on efficiency as a criterion of technological development. That conclusion, in turn, opens broad possibilities of change foreclosed by the usual understanding of technology.

DYSTOPIAN MODERNITY

Max Weber's famous theory of rationalization is the original argument against industrial democracy. The title of this chapter implies a provocative reversal of Weber's conclusions. He defined rationalization as the increasing role of calculation and control in social life, a trend leading to what he called the "iron cage" of bureaucracy.[2] "Democratic" rationalization is thus a contradiction in terms.

Once traditionalist struggle against rationalization has been defeated, further resistance in a Weberian universe can only reaffirm irrational life forces against routine and drab predictability. This is not a democratic program but a romantic antidystopian one, the sort of thing that is already foreshadowed in Dostoyevsky's *Notes from Underground* and various back-to-nature ideologies.

My title is meant to reject the dichotomy between rational hierarchy and irrational protest implicit in Weber's position. If authoritarian social hierarchy is truly a contingent dimension of technical progress, as I believe, and not a technical necessity, then there must be an alternative way of rationalizing society that democratizes rather than centralizes control. We need not go underground or native to preserve threatened values such as freedom and individuality.

But the most powerful critiques of modern technological society follow directly in Weber's footsteps in rejecting this possibility. I am thinking of Heidegger's formulation of "the question of technology" and Ellul's theory of "the technical phenomenon."[3] According to these theories, we have become little more than objects of technique, incorporated into the mechanism we have created. As Marshall McLuhan once put it, technology has reduced us to the "sex organs of machines." The only hope is a vaguely evoked spiritual renewal that is too abstract to inform a new technical practice.

These are interesting theories, important for their contribution to opening a space of reflection on modern technology. I will return to Heidegger's argument in the conclusion to this chapter. But first, to advance my own argument, I will concentrate on the principal flaw of *dystopianism,* the identification of technology in general with the specific technologies that have developed in the last century in the West. These are technologies of conquest that pretend to an unprecedented autonomy; their social sources and impacts are hidden. I will argue that this type of technology is a particular feature of our society and not a universal dimension of "modernity" as such.

TECHNOLOGICAL DETERMINISM

Determinism rests on the assumption that technologies have an autonomous functional logic that can be explained without reference to society. *Technology* is presumably social only through the purpose it serves, and purposes are in the mind of the beholder. Technology would thus resemble science and mathematics by its intrinsic independence of the social world.

Yet unlike science and mathematics, technology has immediate and powerful social impacts. It would seem that society's fate is at least partially dependent on a nonsocial factor that influences it without suffering a reciprocal influence. This is what is meant by *technological determinism.* Such a deterministic view of technology is commonplace in business and government, where it is often assumed that progress is an exogenous force influencing society rather than an expression of changes in culture and values.

The dystopian visions of modernity I have been describing are also deterministic. If we want to affirm the democratic potentialities of modern industrialism, we will therefore have to challenge their deterministic premises. These I will call the *thesis of unilinear progress* and the *thesis of determination by the base.* Here is a brief summary of these two positions.

1) Technical progress appears to follow a unilinear course, a fixed track, from less to more advanced configurations. Although this conclusion seems obvious from a backward glance at the development of any familiar technical object, in fact it is based on two claims of unequal plausibility: first, that technical progress proceeds from lower to higher levels of development; and second, that that development follows a single sequence of necessary stages. As we will see, the first claim is independent of the second and is not necessarily deterministic.

2) Technological determinism also affirms that social institutions must adapt to the "imperatives" of the technological base. This view, which no doubt has its source in a certain reading of Marx, is now part of the common sense of the social sciences.[4] Below, I will discuss one of its implications in detail: the supposed "trade-off" between prosperity and environmental values.

These two theses of technological determinism present decontextualized, self-generating technology as the unique foundation of modern society. Determinism thus implies that our technology and its corresponding institutional structures are universal, indeed, planetary in scope. There may be many forms of tribal society, many feudalisms, and even many forms of early capitalism, but there is only one modernity and it is exemplified in our society for good or ill. Developing societies should take note: as Marx once said, calling the attention of his backward German compatriots to British advances: *De te fabula narratur*—of you the tale is told.[5]

CONSTRUCTIVISM

The implications of determinism appear so obvious that it is surprising to discover that neither of its two theses can withstand close scrutiny. Yet contemporary sociology of technology undermines the first thesis of unilinear progress, whereas historical precedents are unkind to the second thesis of determination by the base.

Recent constructivist sociology of technology grows out of new social studies of science. These studies challenge our tendency to exempt scientific theories from the sort of sociological examination to which we submit nonscientific beliefs. They affirm the *principle of symmetry,* according to which all contending beliefs are subject to the same type of social explanation regardless of their truth or falsity.[6] A similar approach to technology rejects the usual assumption that technologies succeed on purely functional grounds.

Constructivism argues that theories and technologies are underdetermined by scientific and technical criteria. Concretely, this means two things: first, there is generally a surplus of workable solutions to any given problem, and social actors make the final choice among a batch of technically viable options; and second, the problem–definition often changes in the course of solution. The latter point is the more conclusive but also more difficult of the two.

Two sociologists of technology, Pinch and Bijker, illustrate it with the early history of the bicycle.[7] The object we take to be a self-evident "black box" actually started out as two very different devices: a sportsman's racer and a utilitarian transportation vehicle. The high front wheel of the sportsman's bike was necessary at the time to attain high speeds, but it also caused instability. Equal-sized wheels made for a safer but less exciting ride. These two designs met different needs and were in fact different technologies with many shared elements. Pinch and Bijker call this original ambiguity (of the object designated as a "bicycle") "interpretative flexibility."

Eventually the "safety" design won out, and it benefited from all the later advances that occurred in the field. In retrospect, it seems as though the high wheelers were a clumsy and less efficient stage in a progressive development leading through the old "safety" bicycle to current designs. In fact, the high wheeler and the safety shared the field for years, and neither was a stage in the other's development. The high wheeler represents a possible alternative path of bicycle development that addressed different problems at the origin.

Determinism is a species of Whig history that makes it seem as though the end of the story was inevitable from the very beginning by projecting the abstract technical logic of the finished object back into the past as a cause of development. That approach confuses our understanding of the past and stifles the imagination of a different future. Constructivism can open up that future, although its practitioners have hesitated so far to engage the larger social issues implied in their method.[8]

INDETERMINISM

If the thesis of unilinear progress falls, the collapse of the notion of determination by the technological base cannot be far behind. Yet it is still frequently invoked in contemporary political debates.

I shall return to these debates later in this chapter. For now, let us consider the remarkable anticipation of current attitudes in the struggle over the length of the workday and over child labor in mid-nineteenth-century England. The debate on the Factory Bill of 1844 was entirely structured around the deterministic opposition of technological imperatives and ideology. Lord Ashley, the chief advocate of regulation, protested in the name of familial ideology that

> the tendency of the various improvements in machinery is to supersede the employment of adult males, and substitute in its place, the labour of children and females. What will be the effect on future generations, if their tender frames be subjected, without limitation or control, to such destructive agencies.[9]

He went on to deplore the decline of the family consequent upon the employment of women, which "disturbs the order of nature" and deprives children of proper upbringing. "It matters not whether it be prince or peasant, all that is best, all that is lasting in the character of a man, he has learnt at his mother's knees." Lord Ashley was outraged to find that

> females not only perform the labour, but occupy the places of men; they are forming various clubs and associations, and gradually acquiring all those privileges which are held to be the proper portion of the male sex. . . . They meet together to drink, sing, and smoke; they use, it is stated, the lowest, most brutal, and most disgusting language imaginable.

Proposals to abolish child labor met with consternation on the part of factory owners, who regarded the little (child) worker as an "imperative" of the technologies created to employ him. They denounced the "inefficiency" of using full-grown workers to accomplish tasks done as well or better by children, and they predicted all the usual catastrophic economic consequences—increased poverty, unemployment, loss of international competitiveness—from the substitution of more costly adult labor. Their eloquent representative, Sir J. Graham, therefore urged caution:

> We have arrived at a state of society when without commerce and manufactures this great community cannot be maintained. Let us, as far as we can, mitigate the evils arising out of this highly artificial state of society; but let us take care to adopt no step that may be fatal to commerce and manufactures.

He further explained that a reduction in the workday for women and children would conflict with the depreciation cycle of machinery and lead to lower wages and trade problems. He concluded that "in the close race of competition which our manufacturers are now running with foreign competitors . . . such a step would be fatal." Regulation, he and his fellows maintained in words that echo still, is based on a "false principle of humanity, which in the end is certain to defeat itself." One might almost believe that Ludd had risen again in the person of Lord Ashley: the issue is not really the length of the workday, "but it is in principle an argument to get rid of the whole system of factory labour." Similar protestations are heard today on behalf of industries threatened with what they call environmental "Luddism."

Yet what actually happened once the regulators succeeded in imposing limitations on the workday and expelling children from the factory? Did the violated imperatives of technology come back to haunt them? Not at all. Regulation led to an intensification of factory labor that was incompatible with the earlier conditions in any case. Children ceased to be workers and were redefined socially as learners and consumers. Consequently, they entered the labor market with higher levels of skill and discipline that were soon presupposed by technological design. As a result, no one is nostalgic for a return to the good old days when inflation was held down by child labor. That is simply not an option (at least not in the developed capitalist world).

This example shows the tremendous flexibility of the technical system. It is not rigidly constraining but on the contrary can adapt to a variety of social demands. This conclusion should not be surprising given the responsiveness of technology to social redefinition discussed previously. It means that technology is just another dependent social variable, albeit an increasingly important one, and not the key to the riddle of history.

Determinism, I have argued, is characterized by the principles of unilinear progress and determination by the base; if determinism is wrong, then technology research must be guided by the following two contrary principles. In the first place, technological development is not unilinear but branches in many directions, and it could reach generally higher levels along more than one different track. And, secondly, technological development is not determining for society but is overdetermined by both technical and social factors.

The political significance of this position should also be clear by now. In a society where determinism stands guard on the frontiers of democracy, indeterminism cannot but be political. If technology has many unexplored potentialities, no technological imperatives dictate the current social hierarchy. Rather, technology is a scene of social struggle, a *parliament of things* on which civilizational alternatives contend.

INTERPRETING TECHNOLOGY

In the next sections of this chapter, I would like to present several major themes of a nondeterminist approach to technology. The picture sketched so far implies a significant change in our definition of technology. It can no longer be considered as a collection of devices, nor, more generally, as the sum of rational means. These are tendentious definitions that make technology seem more functional and less social than in fact it is.

As a social object, technology ought to be subject to interpretation like any other cultural artifact, but it is generally excluded from humanistic study. We are assured that its essence lies in a technically explainable function rather than a hermeneutically interpretable meaning. At most, humanistic methods might illuminate extrinsic aspects of technology, such as packaging and advertising, or popular reactions to controversial innovations such as nuclear power or surrogate motherhood. Technological determinism draws its force from this attitude. If one ignores most of the connections between technology and society, it is no wonder that technology then appears to be self-generating.

Technical objects have two hermeneutic dimensions that I call their *social meaning* and their *cultural horizon*.[10] The role of social meaning is clear in the case of the bicycle introduced above. We have seen that the construction of the bicycle was controlled in the first instance by a contest of interpretations: was it to be a sportsman's toy or a means of transportation? Design features such as wheel size also served to signify it as one or another type of object.[11]

It might be objected that this is merely an initial disagreement over goals with no hermeneutic significance. Once the object is stabilized, the engineer has the last word on its nature, and the humanist interpreter is out of luck. This is the view of most engineers and managers; they readily grasp the concept of *goal* but they have no place for *meaning*.

In fact, the dichotomy of goal and meaning is a product of functionalist professional culture, which is itself rooted in the structure of the modern economy. The concept of a goal strips technology bare of social contexts, focusing engineers and managers on just what they need to know to do their job.

A fuller picture is conveyed, however, by studying the social role of the technical object and the lifestyles it makes possible. That picture places the abstract notion of a goal in its concrete social context. It makes technology's contextual causes and consequences visible rather than obscuring them behind an impoverished functionalism.

The functionalist point of view yields a decontextualized temporal cross-section in the life of the object. As we have seen, determinism claims implausibly to be able to get from one such momentary configuration of the object to the next on purely technical terms. But in the real world, all sorts of unpredictable attitudes crystallize around technical objects and influence later design changes. The engineer may think these are extrinsic to the device he or she is working on, but they are its very substance as a historically evolving phenomenon.

These facts are recognized to a certain extent in the technical fields themselves, especially in computers. Here we have a contemporary version of the dilemma of the bicycle discussed above. Progress of a generalized sort in speed, power, and memory goes on apace while corporate planners struggle with the question of what it is all for. Technical development does not point definitively toward any particular path. Instead, it opens branches, and the final determination of the "right" branch is not within the competence of engineering because it is simply not inscribed in the nature of the technology.

I have studied a particularly clear example of the complexity of the relation between the technical function and meaning of the computer in the case of French videotext.[12] Called *Teletel,* this system was designed to bring France into the Information Age by giving telephone subscribers access to databases. Fearing that consumers would reject anything resembling office equipment, the telephone company attempted to redefine the computer's social image; it was no longer to appear as a calculating device for professionals but was to become an informational network for all.

The telephone company designed a new type of terminal, the *Minitel,* to look and feel like an adjunct to the domestic telephone. The telephonic disguise suggested to some users that they ought to be able to talk to each other on the network. Soon the *Minitel* underwent a further redefinition at the hands of these users, many of whom employed it primarily for anonymous online chatting with other users in the search for amusement, companionship, and sex.

Thus, the design of the *Minitel* invited communications applications that the company's engineers had not intended when they set about improving the flow of information in French society. Those applications, in turn, connoted the *Minitel* as a means of personal encounter, the very opposite of the rationalistic project for which it was originally created. The "cold" computer became a "hot" new medium.

At issue in the transformation is not only the computer's narrowly conceived technical function but also the very nature of the advanced society it makes possible. Does networking open the doors to the Information Age in which, as rational consumers hungry for data, we pursue strategies of optimization? Or is it a postmodern technology that emerges from the breakdown of institutional and sentimental stability, reflecting, in Lyotard's words, the "atomisation of society into flexible networks of language games?"[13] In this case, technology is not merely the servant of some predefined social purpose; it is an environment within which a way of life is elaborated.

In sum, differences in the way social groups interpret and use technical objects are not merely extrinsic but also make a difference in the nature of the objects themselves. *What* the object *is* for the groups that ultimately decide its fate determines what it *becomes* as it is redesigned and improved over time. If this is true, then we can only understand technological development by studying the sociopolitical situation of the various groups involved in it.

TECHNOLOGICAL HEGEMONY

In addition to the sort of assumptions about individual technical objects that we have been discussing so far, that situation also includes broader assumptions about social values. This is where the study of the cultural horizon of technology comes in. This second hermeneutic dimension of technology is the basis of modern forms of social hegemony; it is particularly relevant to our original question concerning the inevitability of hierarchy in technological society.

As I will use the term, *hegemony* is a form of domination so deeply rooted in social life that it seems natural to those it dominates. One might also define it as that aspect of the distribution of social power that has the force of culture behind it.

The term *horizon* refers to culturally general assumptions that form the unquestioned background to every aspect of life.[14] Some of these support the prevailing hegemony. For example, in feudal societies, the *chain of being* established hierarchy in the fabric of God's universe and protected the caste relations of the society from challenge. Under this horizon, peasants revolted in the name of the king, the only imaginable source of power. Rationalization is our modern horizon, and technological design is the key to its effectiveness as the basis of modern hegemonies.

Technological development is constrained by cultural norms originating in economics, ideology, religion, and tradition. We discussed earlier how assumptions about the age composition of the labor force entered into the design of nineteenth-century production technology. Such assumptions seem so natural and obvious that they often lie below the threshold of conscious awareness.

This is the point of Herbert Marcuse's important critique of Weber.[15] Marcuse shows that the concept of rationalization confounds the control of labor by management with control of nature by technology. The search for control of nature is generic, but management only arises against a specific social background, the capitalist wage system. Workers have no immediate interest in output in this system, unlike earlier forms of farm and craft labor, since their wage is not essentially linked to the income of the firm. Control of human beings becomes all-important in this context.

Through mechanization, some of the control functions are eventually transferred from human overseers and parcelized work practices to machines. Machine design is thus socially relative in a way that Weber never recognized, and the "technological rationality" it embodies is not universal but particular to capitalism. In fact, it is the horizon of all the existing industrial societies, communist as well as capitalist, insofar as they are managed from above. (In a later section, I discuss a generalized application of this approach in terms of what I call the *technical code*.)

If Marcuse is right, it ought to be possible to trace the impress of class relations in the very design of production technology as has indeed been shown by such Marxist students of the labor process as Harry Braverman and David Noble.[16] The assembly line offers a particularly clear instance because it achieves traditional management goals, such as deskilling and pacing work, through technical design. Its technologically enforced labor discipline increases productivity and profits by increasing control. However, the assembly line only appears as technical progress in a specific social context. It would not be perceived as an advance in an economy based on workers' cooperatives in which labor discipline was more self-imposed than imposed from above. In such a society, a different technological rationality would dictate different ways of increasing productivity.[17]

This example shows that technological rationality is not merely a belief, an ideology, but is also effectively incorporated into the structure of machines. Machine design mirrors back the social factors operative in the prevailing rationality. The fact that the argument for the social relativity of modern technology originated in a Marxist context has obscured its most radical implications. We are not dealing here with a mere critique of the property system, but have extended the force of that critique down into the technical "base." This approach goes well beyond the old economic distinction between capitalism and socialism, market and plan. Instead, one arrives at a very different distinction between societies in which power rests on the technical mediation of social activities and those that democratize technical control and, correspondingly, technological design.

DOUBLE ASPECT THEORY

The argument to this point might be summarized as a claim that social meaning and functional rationality are inextricably intertwined dimensions of technology. They are not ontologically distinct, for example, with meaning in the observer's mind and rationality in the technology proper. Rather they are *double aspects* of the same underlying technical object, each aspect revealed by a specific contextualization.

Functional rationality, like scientific-technical rationality in general, isolates objects from their original context in order to incorporate them into theoretical or functional systems. The institutions that support this procedure, such as laboratories and research centers, themselves form a special context with their own practices and links to various social agencies and powers. The notion of "pure" rationality arises when the work of decontextualization is not itself grasped as a social activity reflecting social interests.

Technologies are selected by these interests from among many possible configurations. Guiding the selection process are social codes established by the cultural and political struggles that define the horizon under which the technology will fall. Once introduced, technology offers a material validation of the cultural horizon to which it has been preformed. I call this the *bias* of technology: apparently neutral, functional rationality is enlisted in support of a hegemony. The more technology society employs, the more significant is this support.

As Foucault argues in his theory of "power/knowledge," modern forms of oppression are not so much based on false ideologies as on the specific technical "truths" that form the basis of the dominant hegemony and that reproduce it.[18] So long as the contingency of the choice of "truth" remains hidden, the deterministic image of a technically justified social order is projected.

The legitimating effectiveness of technology depends on unconsciousness of the cultural-political horizon under which it was designed. A recontextualizing critique of technology can uncover that horizon, demystify the illusion of technical necessity, and expose the relativity of the prevailing technical choices.

THE SOCIAL RELATIVITY OF EFFICIENCY

These issues appear with particular force in the environmental movement today. Many environmentalists argue for technical changes that would protect nature and in the process improve human life as well. Such changes would enhance efficiency in broad terms by reducing harmful and costly side effects of technology. However, this program is very difficult to impose in a capitalist society. There is a tendency to deflect criticism from technological processes to products and people, from a priori prevention to a posteriori cleanup. These preferred strategies are generally costly and reduce efficiency under the horizon of the given technology. This situation has political consequences.

Restoring the environment after it has been damaged is a form of collective consumption, financed by taxes or higher prices. These approaches dominate public awareness. This is why environmentalism is generally perceived as a cost involving trade-offs, and not as a rationalization increasing overall efficiency. But in a modern society obsessed by economic well-being, that perception is damning. Economists and businesspeople are fond of explaining the price we must pay in inflation and unemployment for worshipping at Nature's shrine instead of Mammon's. Poverty awaits those who will not adjust their social and political expectations to technology.

This trade-off model has environmentalists grasping at straws for a strategy. Some hold out the pious hope that people will turn from economic to spiritual values in the face of the mounting problems of industrial society. Others expect enlightened dictators to impose technological reform even if a greedy populace shirks its duty. It is difficult to decide which of these solutions is more improbable, but both are incompatible with basic democratic values.[19]

The trade-off model confronts us with dilemmas—environmentally sound technology versus prosperity, workers' satisfaction and control versus productivity, and so on—when what we need are syntheses. Unless the problems of modern industrialism can be solved in ways that both enhance public welfare and win public support, there is little reason to hope that they will ever be solved. But how can technological reform be reconciled with prosperity when it places a variety of new limits on the economy?

The child labor case shows how apparent dilemmas arise on the boundaries of cultural change, specifically, where the social definition of major technologies is in transition. In such situations, social groups excluded from the original design network articulate their unrepresented interests politically. New values the outsiders believe would enhance their welfare appear as mere ideology to insiders who are adequately represented by the existing designs.

This is a difference of perspective, not of nature. Yet the illusion of essential conflict is renewed whenever major social changes affect technology. At first, satisfying the demands of new groups after the fact has visible costs and, if it is done clumsily, will indeed reduce efficiency until better designs are found. But usually better designs can be found and what appeared to be an insuperable barrier to growth dissolves in the face of technological change.

This situation indicates the essential difference between economic exchange and technique. Exchange is all about trade-offs: more of A means less of B. But the aim of technical advance is precisely to avoid such dilemmas by elegant designs that optimize several variables at once. A single cleverly conceived mechanism may correspond to many different social demands, one structure to many functions.[20] Design is not a zero-sum economic game but an ambivalent cultural process that serves a multiplicity of values and social groups without necessarily sacrificing efficiency.

THE TECHNICAL CODE

That these conflicts over social control of technology are not new can be seen from the interesting case of the "bursting boilers."[21] Steamboat boilers were the first technology regulated in the United States. In the early nineteenth century, the steamboat was a major form of transportation similar to the automobile or airlines today. Steamboats were necessary in a big country without paved roads and lots of rivers and canals. But steamboats frequently blew up when the boilers weakened with age or were pushed too hard. After several particularly murderous accidents in 1816, the city of Philadelphia consulted with experts on how to design safer boilers, the first time an American governmental institution interested itself in the problem. In 1837, at the request of Congress, the Franklin Institute issued a detailed report and recommendations based on rigorous study of boiler construction. Congress was tempted to impose a safe boiler code on the industry, but boilermakers and steamboat owners resisted and government hesitated to interfere with private property.

It took from that first inquiry in 1816 to 1852 for Congress to pass effective laws regulating the construction of boilers. In that time, 5,000 people were killed in accidents on steamboats. Is this many casualties or few? Consumers evidently were not too alarmed to continue traveling by riverboat in ever increasing numbers. Understandably, the ship owners interpreted this as a vote of confidence and protested the excessive cost of safer designs. Yet politicians also won votes demanding safety.

The accident rate fell dramatically once technical changes such as thicker walls and safety valves were mandated. Legislation would hardly have been necessary to achieve this outcome had it been technically determined. But, in fact, boiler design was relative to a social judgment about safety. That judgment could have been made on strictly market grounds, as the shippers wished, or politically, with differing technical results. In either case, those results *constitute* a proper boiler. What a boiler *is* was thus defined through a long process of political struggle culminating finally in uniform codes issued by the American Society of Mechanical Engineers.

This example shows just how technology adapts to social change. What I call the *technical code* of the object mediates the process. That code responds to the cultural horizon of the society at the level of technical design. Quite down-to-earth technical parameters such as the choice and processing of materials are *socially* specified by the code. The illusion of technical necessity arises from the fact that the code is thus literally "cast in iron," at least in the case of boilers.[22]

Conservative antiregulatory social philosophies are based on this illusion. They forget that the design process always already incorporates standards of safety and environmental compatibility; similarly, all technologies support some basic level of user or worker initiative. A properly made technical object simply *must* meet these standards to be recognized as such. We do not treat conformity as an expensive add-on, but regard it as an intrinsic production cost. Raising the standards means altering the definition of the object, not paying a price for an alternative good or ideological value as the trade-off model holds.

But what of the much discussed cost–benefit ratio of design changes such as those mandated by environmental or other similar legislation? These calculations have some application to transitional situations, before technological advances responding to new values fundamentally alter the terms of the problem. But, all too often, the results depend on economists' very rough estimates of the monetary value of such things as a day of trout fishing or an asthma attack. If made without prejudice, these estimates may well help to prioritize policy alternatives. But one cannot legitimately generalize from such policy applications to a universal theory of the costs of regulation.

Such fetishism of efficiency ignores our ordinary understanding of the concept that alone is relevant to social decision making. In that everyday sense, efficiency concerns the narrow range of values that economic actors routinely affect by their decisions. Unproblematic aspects of technology are not included. In theory, one can decompose any technical object and account for each of its elements in terms of the goals it meets, whether it be safety, speed, reliability, and the like, but in practice no one is interested in opening the "black box" to see what is inside.

For example, once the boiler code is established, such things as the thickness of a wall or the design of a safety valve appear as essential to the object. The cost of these features is not broken out as the specific "price" of safety and compared unfavorably with a more efficient but less secure version of the technology. Violating the code in order to lower costs is a crime, not a trade-off. And since all further progress takes place on the basis of the new safety standard, soon no one looks back to the good old days of cheaper, insecure designs.

Design standards are only controversial while they are in flux. Resolved conflicts over technology are quickly forgotten. Their outcomes, a welter of taken-for-granted technical and legal standards, are embodied in a stable code and form the background against which economic actors manipulate the unstable portions of the environment in the pursuit of efficiency. The code is not varied in real world economic calculations but treated as a fixed input.

Anticipating the stabilization of a new code, one can often ignore contemporary arguments that will soon be silenced by the emergence of a new horizon of efficiency calculations. This is what happened with boiler design and child labor; presumably, the current debates on environmentalism will have a similar history, and we will someday mock those who object to cleaner air as a "false principle of humanity" that violates technological imperatives.

Noneconomic values intersect the economy in the technical code. The examples we are dealing with illustrate this point clearly. The legal standards that regulate workers' economic activity have a significant impact on every aspect of their lives. In the child labor case, regulation helped to widen educational opportunities with consequences that are not primarily economic in character. In the riverboat case, Americans gradually chose high levels of security, and boiler design came to reflect that choice. Ultimately, this was no trade-off of one good for another, but a noneconomic decision about the value of human life and the responsibilities of government.

Technology is thus not merely a means to an end; technical design standards define major portions of the social environment, such as urban and built spaces, workplaces, medical activities and expectations, life patterns, and so on. The economic significance of technical change often pales beside its wider human implications in framing a way of life. In such cases, regulation defines the cultural framework *of* the economy; it is not an act *in* the economy.

HEIDEGGER'S "ESSENCE" OF TECHNOLOGY

The theory sketched here suggests the possibility of a general reform of technology. But dystopian critics object that the mere fact of pursuing efficiency or technical effectiveness already does inadmissible violence to human beings and nature. Universal functionalization destroys the integrity of all that is. As Heidegger argues, an "objectless" world of mere resources replaces a world of "things" treated with respect for their own sake as the gathering places of our manifold engagements with "being."[23]

This critique gains force from the actual perils with which modern technology threatens the world today. But my suspicions are aroused by Heidegger's famous contrast between a dam on the Rhine and a Greek chalice. It would be difficult to find a more tendentious comparison. No doubt, modern technology is immensely more destructive than any other. And Heidegger is right to argue that means are not truly neutral, and that their substantive content affects society independent of the goals they serve. But I have argued here that this content is not *essentially* destructive; rather, it is a matter of design and social insertion.

However, Heidegger rejects any merely social diagnosis of the ills of technological societies and claims that the source of their problems dates back at least to Plato, that modern societies merely realize a *telos* immanent in Western metaphysics from the beginning. His originality consists in pointing out that the ambition to control being is itself a way of being and hence subordinate at some deeper level to an ontological dispensation beyond human control. But the overall effect of his critique is to condemn human agency, at least in modern times, and to confuse essential differences between types of technological development.

Heidegger distinguishes between the *ontological* problem of technology, which can only be addressed by achieving what he calls "a free relation" to technology, and the merely *ontic* solutions proposed by reformers who wish to change technology itself. This distinction may have seemed more interesting in years gone by than it does today. In effect, Heidegger is asking for nothing more than a change in attitude toward the selfsame technical world. But that is an idealistic solution in the bad sense, and one that a generation of environmental action would seem decisively to refute.

Confronted with this argument, Heidegger's defenders usually point out that his critique of technology is not merely concerned with human attitudes but also with the way being reveals itself. Roughly translated out of Heidegger's language, this means that the modern world has a technological form in something like the sense in which, for example, the medieval world had a religious form. *Form* is no mere question of attitude but takes on a material life of its own: power plants are the gothic cathedrals of our time. But this interpretation of Heidegger's thought raises the expectation that he will offer criteria for a reform of technology. For example, his analysis of the tendency of modern technology to accumulate and store up nature's powers suggests the superiority of another technology that would not challenge nature in Promethean fashion.

Unfortunately, Heidegger's argument is developed at such a high level of abstraction he literally cannot discriminate between electricity and atom bombs, agricultural techniques and the Holocaust. In a 1949 lecture, he asserted: "Agriculture is now the mechanized food industry, in essence the same as the manufacturing of corpses in gas chambers and extermination camps, the same as the blockade and starvation of nations, the same as the production of hydrogen bombs."[24] All are merely different expressions of the identical enframing that we are called to transcend through the recovery of a deeper relation to being. And since Heidegger rejects technical regression while leaving no room for a better technological future, it is difficult to see in what that relation would consist beyond a mere change of attitude.

HISTORY OR METAPHYSICS

Heidegger is perfectly aware that technical activity was not "metaphysical" in his sense until recently. He must therefore sharply distinguish modern technology from all earlier forms of technique, obscuring the many real connections and continuities. I would argue, on the contrary, that what is new about modern technology can only be understood against the background of the traditional technical world from which it developed. Furthermore, the saving potential of modern technology can only be realized by recapturing certain traditional features of technique. Perhaps this is why theories that treat modern technology as a unique phenomenon lead to such pessimistic conclusions.

Modern technology differs from earlier technical practices through significant shifts in emphasis rather than generically. There is nothing unprecedented in its chief features, such as the reduction of objects to raw materials, the use of precise measurement and plans, and the technical control of some human beings by others, large scales of operation. It is the centrality of these features that is new, and of course the consequences of that are truly without precedent.

What does a broader historical picture of technology show? The privileged dimensions of modern technology appear in a larger context that includes many currently subordinated features that were defining for it in former times. For example, until the generalization of Taylorism, technical life was essentially about the choice of a vocation. Technology was associated with a way of life, with specific forms of personal development, virtues, and so on. Only the success of capitalist deskilling finally reduced these human dimensions of technique to marginal phenomena.

Similarly, modern management has replaced the traditional collegiality of the guilds with new forms of technical control. Just as vocational investment in work continues in certain exceptional settings, so collegiality survives in a few professional or cooperative workplaces. Numerous historical studies show that these older forms are not so much incompatible with the *essence* of technology as with capitalist economics. Given a different social context and a different path of technical development, it might be possible to recover these traditional technical values and organizational forms in new ways in a future evolution of modern technological society.

Technology is an elaborate complex of related activities that crystallizes around tool making and using in every society. Matters such as the transmission of techniques or the management of its natural consequences are not extrinsic to technology per se but are dimensions of it. When, in modern societies, it becomes advantageous to minimize these aspects of technology, that too is a way of accommodating it to a certain social demand, not the revelation of its preexisting essence. In so far as it makes sense to talk about an essence of technology at all, it must embrace the whole field revealed by historical study, and not only a few traits ethnocentrically privileged by our society.

There is an interesting text in which Heidegger shows us a jug "gathering" the contexts in which it was created and functions. This image could be applied to technology as well, and in fact there is one brief passage in which Heidegger so interprets a highway bridge. Indeed, there is no reason why modern technology cannot also gather its multiple contexts, albeit with less romantic pathos than jugs and chalices. This is in fact one way of interpreting contemporary demands for such things as environmentally sound technology, applications of medical technology that respect human freedom and dignity, urban designs that create humane living spaces, production methods that protect workers' health and offer scope for their intelligence, and so on. What are these demands if not a call to reconstruct modern technology so that it gathers a wider range of contexts to itself rather than reducing its natural, human, and social environment to mere resources?

Heidegger would not take these alternatives very seriously because he reifies modern technology as something separate from society, as an inherently contextless force aiming at pure power. If this is the essence of technology, reform would be merely extrinsic. But at this point, Heidegger's position converges with the very Prometheanism he rejects. Both depend on the narrow definition of technology that, at least since Bacon and Descartes, has emphasized its destiny to control the world to the exclusion of its equally essential contextual embeddedness. I believe that this definition reflects the capitalist environment in which modern technology first developed.

The exemplary modern master of technology is the entrepreneur, single-mindedly focused on production and profit. The enterprise is a radically decontextualized platform for action, without the traditional responsibilities for persons and places that went with technical power in the past. It is the autonomy of the enterprise that makes it possible to distinguish so sharply between intended and unintended consequences, between goals and contextual effects, and to ignore the latter.

The narrow focus of modern technology meets the needs of a particular hegemony; it is not a metaphysical condition. Under that hegemony, technological design is unusually decontextualized and destructive. It is that hegemony that is called to account, not technology per se, when we point out that today technical means form an increasingly threatening life environment. It is that hegemony, as it has embodied itself in technology, that must be challenged in the struggle for technological reform.

DEMOCRATIC RATIONALIZATION

For generations, faith in progress was supported by two widely held beliefs: that technical necessity dictates the path of development, and that the pursuit of efficiency provides a basis for identifying that path. I have argued here that both these beliefs are false and that, furthermore, they are ideologies employed to justify restrictions on opportunities to participate in the institutions of industrial society. I conclude that we can achieve a new type of technological society that can support a broader range of values. Democracy is one of the chief values a redesigned industrialism could better serve.

What does it mean to democratize technology? The problem is not primarily one of legal rights but of initiative and participation. Legal forms may eventually routinize claims that are asserted informally at first, but the forms will remain hollow unless they emerge from the experience and needs of individuals resisting a specifically technological hegemony.

That resistance takes many forms, from union struggles over health and safety in nuclear power plants to community struggles over toxic waste disposal to political demands for regulation of reproductive technologies. These movements alert us to the need to take technological externalities into account and demand design changes responsive to the enlarged context revealed in that accounting.

Such technological controversies have become an inescapable feature of contemporary political life, laying out the parameters for official "technology assessment."[25] They prefigure the creation of a new public sphere embracing the technical background of social life, and a new style of rationalization that internalizes unaccounted costs born by "nature," in other words, something or somebody exploitable in the pursuit of profit. Here, respect for nature is not antagonistic to technology but enhances efficiency in broad terms.

As these controversies become commonplace, surprising new forms of resistance and new types of demands emerge alongside them. Networking has given rise to one among many such innovative public reactions to technology. Individuals who are incorporated into new types of technical networks have learned to resist through the net itself in order to influence the powers that control it. This is not a contest for wealth or administrative power, but a struggle to subvert the technical practices, procedures, and designs structuring everyday life.

The example of the *Minitel* can serve as a model of this new approach. In France, the computer was politicized as soon as the government attempted to introduce a highly rationalistic information system to the general public. Users "hacked" the network in which they were inserted and altered its functioning, introducing human communication on a vast scale where only the centralized distribution of information had been planned.

It is instructive to compare this case to the movements of AIDS patients.[26] Just as a rationalistic conception of the computer tends to occlude its communicative potentialities, so in medicine, caring functions have become mere side effects of treatment, which is itself understood in exclusively technical terms. Patients become objects of this technique, more or less "compliant" to management by physicians. The incorporation of thousands of incurably ill AIDS patients into this system destabilized it and exposed it to new challenges.

The key issue was access to experimental treatment. In effect, clinical research is one way in which a highly technologized medical system can care for those it cannot yet cure. But until quite recently, access to medical experiments has been severely restricted by paternalistic concern for patients' welfare. AIDS patients were able to open up access because the networks of contagion in which they were caught were paralleled by social networks that were already mobilized around gay rights at the time the disease was first diagnosed.

Instead of participating in medicine individually as objects of a technical practice, they challenged it collectively and politically. They "hacked" the medical system and turned it to new purposes. Their struggle represents a counter tendency to the technocratic organization of medicine, an attempt at a recovery of its symbolic dimension and caring functions.

As in the case of the *Minitel,* it is not obvious how to evaluate this challenge in terms of the customary concept of politics. Nor do these subtle struggles against the growth of silence in technological societies appear significant from the standpoint of the reactionary ideologies that contend noisily with capitalist modernism today. Yet the demand for communication these movements represent is so fundamental that it can serve as a touchstone for the adequacy of our concept of politics to the technological age.

These resistances, like the environmental movement, challenge the horizon of rationality under which technology is currently designed. Rationalization in our society responds to a particular definition of technology as a means to the goal of profit and power. A broader understanding of technology suggests a very different notion of rationalization based on responsibility for the human and natural contexts of technical action. I call this *democratic rationalization* because it requires technological advances that can only be made in opposition to the dominant hegemony. It represents an alternative to both the ongoing celebration of technocracy triumphant and the gloomy Heideggerian counterclaim that "only a God can save us" from technocultural disaster.[27]

Is democratic rationalization in this sense socialist? There is certainly room for discussion of the connection between this new technological agenda and the old idea of socialism. I believe there is significant continuity. In socialist theory, workers' lives and dignity stood for the larger contexts that modern technology ignores. The destruction of their minds and bodies on the workplace was viewed as a contingent consequence of capitalist technical design. The implication that socialist societies might design a very different technology under a different cultural horizon was perhaps given only lip service, but at least it was formulated as a goal.

We can make a similar argument today over a wider range of contexts in a broader variety of institutional settings with considerably more urgency. I am inclined to call such a position socialist and to hope that, in time, it can replace the image of socialism projected by the failed communist experiment.

More important than this terminological question is the substantive point I have been trying to make. Why has democracy not been extended to technically mediated domains of social life despite a century of struggles? Is it because technology excludes democracy, or because it has been used to suppress it? The weight of the argument supports the second conclusion. Technology can support more than one type of technological civilization, and may someday be incorporated into a more democratic society than ours.

NOTES

This chapter expands a presentation of my book *Critical Theory of Technology* (New York: Oxford University Press, 1991), delivered at the American Philosophical Association, December 28, 1991, and first published in an earlier version in *Inquiry* 35, nos. 3–4 (1992): 301–22.

1. See, for example, Joshua Cohen and Joel Rogers, *On Democracy: Toward a Transformation of American Society* (Harmondsworth, UK: Penguin, 1983); and Frank Cunningham, *Democratic Theory and Socialism* (Cambridge: Cambridge University Press, 1987).

2. Max Weber, *The Protestant Ethic and the Spirit of Capitalism,* trans. T. Parsons (New York: Scribners, 1958), 181–82.

3. Martin Heidegger, *The Question Concerning Technology,* trans. W. Lovitt (New York: Harper & Row, 1977); and Jacques Ellul, *The Technological Society,* trans. J. Wilkinson (New York: Vintage, 1964).

4. Richard W. Miller, *Analyzing Marx: Morality, Power and History* (Princeton, N.J.: Princeton University Press, 1984), 188–95.

5. Karl Marx, *Capital* (New York: Modern Library, 1906), 13.

6. See, for example, David Bloor, *Knowledge and Social Imagery* (Chicago: University of Chicago Press, 1991), 175–79. For a general presentation of constructivism, see Bruno Latour, *Science in Action* (Cambridge, Mass.: Harvard University Press, 1987).

7. Trevor Pinch and Wiebe Bijker, "The Social Construction of Facts and Artefacts: or How the Sociology of Science and the Sociology of Technology Might Benefit Each Other," *Social Studies of Science*, no. 14 (1984).

8. See Langdon Winner's blistering critique of the characteristic limitations of the position, entitled, "Upon Opening the Black Box and Finding It Empty: Social Constructivism and the Philosophy of Technology," in *The Technology of Discovery and the Discovery of Technology: Proceedings of the Sixth International Conference of the Society for Philosophy and Technology* (Blacksburg, Va.: Society for Philosophy and Technology, 1991).

9. *Hansard's Debates, Third Series: Parliamentary Debates, 1830–1891* 73 (February 22–April 22, 1844). The quoted passages are found between 1088 and 1123.

10. A useful starting point for the development of a hermeneutics of technology is offered by Paul Ricoeur in "The Model of the Text: Meaningful Action Considered as a Text," in P. Rabinow and W. Sullivan, eds., *Interpretive Social Science: A Reader* (Berkeley: University of California Press, 1979).

11. Michel de Certeau used the phrase "rhetorics of technology" to refer to the representations and practices that contextualize technologies and assign them a social meaning. De Certeau chose the term "rhetoric" because that meaning is not simply present at hand but communicates a content that can be articulated by studying the connotations that technology evokes. See the special issue of *Traverse*, no. 26 (October 1982),

entitled *Les Rhetoriques de la Technologie,* and, in that issue, especially, Marc Guillaume's article, "Telespectres": 22–23.

12. See chapter 7, "From Information to Communication: The French Experience with Videotext," in Andrew Feenberg, *Alternative Modernity* (Berkeley: University of California Press, 1995).

13. Jean-François Lyotard, *La Condition Postmoderne* (Paris: Editions de Minuit, 1979), 34.

14. For an approach to social theory based on this notion (called, however, *doxa,* by the author), see Pierre Bourdieu, *Outline of a Theory of Practice,* trans. R. Nice (Cambridge: Cambridge University Press, 1977), 164–70.

15. Herbert Marcuse, "Industrialization and Capitalism in the Work of Max Weber," in *Negations,* trans. J. Shapiro (Boston: Beacon, 1968).

16. Harry Braverman, *Labor and Monopoly Capital* (New York: Monthly Review, 1974); and David Noble, *Forces of Production* (New York: Oxford University Press, 1984).

17. Bernard Gendron and Nancy Holstrom, "Marx, Machinery and Alienation," *Research in Philosophy and Technology* 2 (1979).

18. Foucault's most persuasive presentation of this view is *Surveiller et Punir* (Paris: Gallimard, 1975).

19. See, for example, Robert Heilbroner, *An Inquiry into the Human Prospect* (New York: W. W. Norton, 1975). For a review of these issues in some of their earliest formulations, see Andrew Feenberg, "Beyond the Politics of Survival," *Theory and Society*, no. 7 (1979).

20. This aspect of technology, called *concretization,* is explained in Gilbert Simondon, *Du Mode d'Existence des Objets Techniques* (Paris: Aubier, 1958), ch. 1.

21. John G. Burke, "Bursting Boilers and the Federal Power," in M. Kranzberg and W. Davenport, eds., *Technology and Culture* (New York: New American Library, 1972).

22. The technical code expresses the "standpoint" of the dominant social groups at the level of design and engineering. It is thus relative to a social position without, for that matter, being a mere ideology or psychological disposition. As I will argue in the last section of this chapter, struggle for sociotechnical change can emerge from the subordinated standpoints of those dominated within technological systems. For more on the concept of standpoint epistemology, see Sandra Harding, *Whose Science? Whose Knowledge?* (Ithaca, N.Y.: Cornell University Press, 1991).

23. The texts by Heidegger discussed here are, in order, "The Question Concerning Technology," "The Thing," and "Building Dwelling Thinking," all in *Poetry, Language, Thought,* trans. A. Hofstadter (New York: Harper & Row, 1971).

24. Quoted in T. Rockmore, *On Heidegger's Nazism and Philosophy* (Berkeley: University of California Press, 1992), 241.

25. Alberto Cambrosio and Camille Limoges, "Controversies as Governing Processes in Technology Assessment," in *Technology Analysis & Strategic Management* 3, no. 4 (1991).

26. For more on the problem of AIDS in this context, see Andrew Feenberg, "On Being a Human Subject: Interest and Obligation in the Experimental Treatment of Incurable Disease," *The Philosophical Forum* 23, no. 3 (Spring 1992).

27. "Only a God Can Save Us Now," Martin Heidegger interviewed in *Der Spiegel*, translated by D. Schendler, *Graduate Philosophy Journal* 6, no. 1 (Winter 1977).

11

A Collective of Humans and Nonhumans: Following Daedalus's Labyrinth

Bruno Latour

The Greeks used to distinguish the straight path of reason and scientific knowledge, *episteme*, from the clever and crooked path of technical know-how, *metis*. Now that we have seen how indirect, devious, mediated, interconnected, vascularized are the paths taken by scientific facts, we may be able to find a different genealogy for technical artifacts as well. This is all the more necessary because so much of science studies relies on the notion of "construction," borrowed from technical action. As we are going to see, however, the philosophy of technology is no more directly useful for defining human and nonhuman connections than epistemology has been, and for the same reason: in the modernist settlement, theory fails to capture practice. Technical action, thus, presents us with puzzles as bizarre as those involved in the articulation of facts. Having grasped how the classical theory of objectivity fails to do any justice to the practice of science, we are now going to see that the notion of "technical efficiency over matter" in no way accounts for the subtlety of engineers. We may then be able, finally, to understand these nonhumans, which are, I have been claiming since the beginning, full-fledged actors in our collective; we may understand at last why we do not live in a society gazing out at a natural world or in a natural world that includes society as one of its components. Now that nonhumans are no longer confused with objects, it may be possible to imagine the collective in which humans are entangled with them.

In the myth of Daedalus, all things deviate from the straight line. After Daedalus's escape from the labyrinth, Minos used a subterfuge worthy of Daedalus himself to find the clever craftsman's hiding place and to take revenge. Minos, in disguise, heralded far and wide his offer of a reward to anyone who could thread the circumvoluted shell of a snail. Daedalus, hidden at the court of King Cocalus and unaware that the offer was a trap, managed the trick by replicating Ariadne's cunning: he attached thread to an ant and, after allowing it to enter the shell through a hole at its apex, he induced the ant to weave its way through this tiny labyrinth. Triumphant, Daedalus claimed his reward, but King Minos, equally triumphant, asked for Daedalus's extradition to Crete. Cocalus abandoned Daedalus; still, this artful dodger managed, with the help of Cocalus's daughters, to divert the hot water from the plumbing system he had installed in the palace, so that it fell, as if by accident, on Minos in his bath. (The king died, boiled like an egg.) Only for a brief while could

Minos outwit his master engineer—Daedalus was always one ruse, one machination ahead of his rivals.

Daedalus embodies the sort of intelligence for which Odysseus (of whom the *Iliad* says that he is *polymetis*, a bag of tricks) is most famed. Once we enter the realm of engineers and craftsmen, no unmediated action is possible. A *daedalion*, the word in Greek that has been used to describe the labyrinth, is something curved, veering from the straight line, artful but fake, beautiful but contrived. Daedalus is an inventor of contraptions: statues that seem to be alive, military robots that watch over Crete, an ancient version of genetic engineering that enables Poseidon's bull to impregnate Pasiphae to conceive the Minotaur—for which he builds the labyrinth, from which, via another set of machines, he manages to escape, losing his son Icarus on the way. Despised, indispensable, criminal, ever at war with the three kings who draw their power from his machinations, Daedalus is the best eponym for technique—and the concept of *daedalion* is the best tool for penetrating the evolution of what I have called so far the collective, which in this chapter I want to define more precisely. Our path will lead us not only through philosophy but through what could be called a *pragmatogony* that is, a wholly mythical "genesis of things," in the fashion of the cosmogonies of the past.

FOLDING HUMANS AND NONHUMANS INTO EACH OTHER

To understand techniques—technical means—and their place in the collective, we have to be as devious as the and to which Daedalus attached his thread. The straight lines of philosophy are of no use when it is *daedalia,* which we have to explore. To cut a hole at the apex of the shell and weave my thread, I need to define, in opposition to Heidegger, what meditation means in the realm of techniques. For Heidegger a technology is never an instrument, a mere tool. Does that mean that technologies mediate action? No, because we have ourselves become instruments for no other end than instrumentality itself (Heidegger 1997). Man—there is no Woman in Heidegger—is possessed by technology, and it is a complete illusion to believe that we can master it. We are, on the contrary, framed by this *Gestell,* which is one way in which Being is unveiled. Is technology inferior to science and pure knowledge? No, because, for Heidegger, far from serving as applied science, technology dominates all, even the purely theoretical sciences. By rationalizing and stockpiling nature, science plays into the hands of technology, whose sole end is to rationalize and stockpile nature without end. Our modern destiny—technology—appears to Heidegger radically different from *poesis,* the kind of "making" that ancient craftsmen knew how to achieve. Technology is unique, insuperable, omnipresent, superior, a monster born in our midst which has already devoured its unwitting midwives. But Heidegger is mistaken. I will try to show why by using a simple, well-known example to demonstrate the impossibility of speaking of any sort of mastery in our relations with nonhumans, *including* their supposed mastery over us.

"Guns kill people" is a slogan of those who try to control the unrestricted sale of guns. To which the National Rifle Association (NRA) replies with another slogan, "Guns don't kill people; *people* kill people." The first slogan is materialist: the gun acts by virtue of *material* components irreducible to the social qualities of the gunman. On account of the gun the law-abiding citizen, a good guy, becomes dangerous. The NRA, meanwhile, offers (amusingly enough, given its political views) a *sociological* version more often associated with the Left: that the gun does nothing in itself or by virtue of its material components. The gun is a tool, a medium, a neutral carrier of human will. If the gunman is a good guy, the gun will be used wisely and will kill only when appropriate. If the gunman is a crook or a lunatic, then, with *no change in the gun itself,* a killing that would in any case occur will be (simply) carried out more efficiently. What does the gun add to the shooting? In the materialist account, *everything:* an innocent citizen becomes a criminal by virtue of the gun in her hand. The gun enables, of course, but also instructs, directs, even pulls the trigger—and who, with a knife in her hand, has not wanted at some time to stab someone or something? Each artifact has its script, its potential to take hold of passersby and force them to play roles in its story. By contract, the sociological version of the NRA renders the gun a *neutral* carrier of will that *adds nothing* to the action, playing the role of a passive conductor, through which good and evil are equally able to flow.

I have caricatured the two positions, of course, in an absurdly diametrical opposition. No materialist claims, more exactly, is that the good citizen is *transformed* by carrying the gun. A good citizen who, without a gun, might simply be angry may become a criminal if he gets his hands on a gun—as if the gun had the power to change Dr. Jekyll into Mr. Hyde. Materialists thus make the intriguing suggestion that our qualities as subjects, our competences, our personalities, depend on what we hold in our hands. Reversing the dogma of moralism, the materialists insist that we are what we have—what we have in our hands, at least.

As for the NRA, its members cannot truly maintain that the gun is so neutral an object that it has no part in the act of killing. They have to acknowledge that the gun *adds* something, though not to the moral state of the person holding it. For the NRA, one's moral state is a Platonic essence: one is born either a good citizen or a criminal. Period. As such, the NRA account is moralist—what matters is what you are, not what you have. The sole contribution of the gun is to speed the act. Killing by fists or knives is simply slower, dirtier, messier. With a gun, one kills better, but at no point does the gun modify one's goal. Thus NRA sociologists make the troubling suggestion that we can master techniques, that techniques are nothing more than pliable and diligent slaves. This simple example is enough to show that artifacts are no easier to grasp than facts: it is going to take us a long time to understand precisely what things make us do.

The First Meaning of Technical Mediation: Interference

Who or what is responsible for the act of killing? Is the gun no more than a piece of mediating technology? The answer to these questions depends on what mediation means. A first sense of mediation (I will offer four) is what I will call the *program of action,* the series of goals and steps and intentions that an agent can describe in a story like the one about the gun and the gunman (see figure 11.1). If the agent is human, is angry, wants to take revenge, and if the accomplishment of the agent's goal is interrupted for whatever reason (perhaps the agent is not strong enough), then the agent makes a *detour*: one cannot speak of techniques any more than of science without speaking of

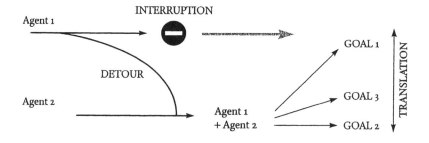

FIRST MEANING OF MEDIATION: GOAL TRANSLATION

Figure 11.1.   We can portray the relation between two agents as a translation of their goals which results in a composite goal that is different from the two original goals.

daedalia. (Although in English the word "technology" tends to replace the word "technique," I will make use of both terms throughout, reserving the tainted term "technoscience" for a very specific stage in my mythical pragmatogony.) Agent 1 falls back on Agent 2, here a gun. Agent 1 enlists the gun or is enlisted by it—it does not matter which—and a third agent emerges from a fusion of the other two.

The question now becomes which goal the new composite agent will pursue. If it returns, after its detour, to Goal 1, then the NRA story obtains. The gun is then a tool, merely an intermediary. If Agent 3 drifts from Goal 1 to Goal 2, then the materialist story obtains. The gun's intent, the gun's will, the gun's script have superseded those of Agent 1; it is human action that is no more than an intermediary. Note that in the figure it makes no difference if Agent 1 and Agent 2 are reversed. The myth of the Neutral Tool under complete human control and the myth of the Autonomous Destiny that no human can master are symmetrical. But a third possibility is more commonly realized: the creation of a new goal that corresponds to neither agent's program of action. (You only wanted to injure but, with a gun now in your hand, you want to kill.) Earlier I called this uncertainty about goals translation. As should be clear by now, translation does not mean a shift from one vocabulary to another, from one French word to one English word, for instance, as if the two languages existed independently. I used translation to mean displacement, drift, invention, mediation, the creation of a link that did not exist before and that to some degree modified the original two.

Which of them, then, the gun or the citizen, is the *actor* in this situation? *Someone else* (a citizen-gun, a gun-citizen). If we try to comprehend techniques while assuming that the psychological capacity of humans is forever fixed, we will not succeed in understanding how techniques are created nor even how they are used. You are a different person with the gun in your hand. Essence is existence and existence is action. If I define you by what you have (the gun), and by the series of associations that you enter into when you use what you have (when you fire the gun), then you are modified by the gun—more so or less so, depending on the weight of the other associations that you carry.

This translation is wholly symmetrical. You are different with a gun in your hand; the gun is different with you holding it. You are another subject because you hold the gun; the gun is another object because it has entered into a relationship with you. The gun is no longer the gun-in-the-armory or the gun-in-the-drawer or the gun-in-the-pocket, but the gun-in-your-hand, aimed at someone who is screaming. What is true of the subject, of the gunman, is as true of the object, of the gun that is held. A good citizen becomes a criminal, a bad guy becomes a worse guy; a silent gun becomes a fired gun, a new gun becomes a used gun, a sporting gun becomes a weapon. The twin mistake of the materialists and the sociologists is to start with essences, those of subjects or those of objects. As we saw earlier, that starting point renders impossible our measurement of the mediating role of techniques as well as those of science. If we study the gun and the citizen as propositions, however, we realize that neither subject nor object (nor their goals) is fixed. When the propositions are articulated, they join into a new proposition. They become "someone, something" else.

It is now possible to shift our attention to this "someone else," the hybrid actor comprising (for instance) gun and gunman. We must learn to attribute—redistribute—actions to many more agents than are acceptable in either the materialist or the sociological account. Agents can be human or (like the gun) nonhuman, and each can have goals (or functions, as engineers prefer to say). Since the word "agent" in the case of nonhumans is uncommon, a better term, as we have seen, is actant. Why is this nuance important? Because, for example, in my vignette of the gun and the gunman, I could replace the gunman with "a class of unemployed loiterers," translating the individual agent into a collective; or I could talk of "unconscious motives," translating it into a subindividual agent. I could redescribe the gun as "what the gun lobby puts in the hands of unsuspecting children," translating it from an object into an institution or a commercial network; or I could call it "the action of a trigger on a cartridge through the intermediary of a spring and a firing-pin," translating it into a mechanical series of causes and consequences. These examples of actor-actant symmetry force us to abandon the subject-object dichotomy, a distinction that prevents the understanding of collectives. It is neither people nor guns that kill. Responsibility for action must be shared among the various actants. And this is the first of the four meanings of mediation.

The Second Meaning of Technical Mediation: Composition

One might object that a basic asymmetry lingers—women make computer chips, but no computer has ever made women. Common sense, however, is not the safest guide here, any more than it is in the sciences. The difficulty we just encountered with the example of the gun remains, and the solution is the same: the prime mover of an action becomes a new, distributed, and nested series of practices whose sum may be possible to add up but only if we respect the mediating role of all the actants mobilized in the series.

To be convincing on this point will require a short inquiry into the way we talk about tools. When someone tells a story about the invention, fabrication, or use of a tool, whether in the animal kingdom or the human, whether in the psychological laboratory or the historical or the prehistoric, the structure is the same. Some agent has a goal or goals; suddenly the access to the goal is interrupted by that breach in the straight path that distinguished *metis* from *episteme.* The detour, a *daedalion,* begins (figure 11.2). The agent, frustrated, turns around in a mad and random search, and then, whether by insight or eureka or by trial and error (there are various psychologies available to account for this moment) the agent seizes upon some other agent—a stick, a partner, an electrical current—and then, so the story goes, returns to the previous task, removes the obstacle, and achieves the goal. Of course, in most tool stories there is not one but two or several *subprograms* nested in one another. A chimpanzee might seize a stick and, finding it too blunt, begin, after another crisis, another subprogram, to sharpen the stick, inventing en route a compound tool. (How far the multiplication of these subprograms can continue raises interesting questions in cognitive psychology and evolutionary theory). Although one can imagine many other outcomes—for instance, the loss of the original goal in the maze of subprograms)—let us suppose that the original task has been resumed.

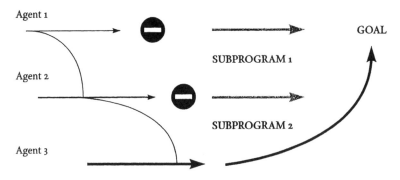

SECOND MEANING OF MEDIATION: COMPOSITION

Figure 11.2.   If the number of subprograms is increased, then the composite goal—here the thick curved line—becomes the common achievement of each of the agents bent by the process of successive translation

What interests me here is the *composition* of action marked by the lines that get longer at each step in Figure 11.2. Who performs the action? Agent 1 plus Agent 2 plus Agent 3. Action is a property of associated entities. Agent 1 is allowed, authorized, enabled, afforded by the others. The chimp plus the sharp stick reach (not reaches) the banana. The attribution to one actor of the role of prime mover in no way weakens the necessity of a composition of forces to explain the action. It is by mistake, or unfairness, that our headlines read "Man flies," "Woman goes into space." Flying is a property of the whole association of entities that includes airports and planes, launch pads and ticket counters. B-52s do not fly, the U.S. Air Force flies. Action is simply not a property of humans *but of an association of actants,* and this is the second meaning of technical mediation. Provisional "actorial" roles may be attributed to actants only because actants are in the process of exchanging competences, offering one another new possibilities, new goals, new functions. Thus symmetry holds in the case of fabrication as it does in the case of use.

But what does symmetry mean? Symmetry is defined by what is conserved through transformations. In the symmetry between humans and nonhumans, I keep constant the series of competences, of properties, that agents are able to swap by overlapping with one another. I want to situate myself at the stage *before* we can clearly delineate subjects and objects, goals and functions, form and matter, before the swapping of properties and competences is observable and interpretable. Full-fledged human subjects and respectable objects out there in the world cannot be my starting point; they may be my point of arrival. Not only does this correspond to the notion of articulation, but it is also consistent with many well-established myths that tell us that we have been made by our tools. The expression *Homo faber* or, better, *Homo faber fabricates* describes, for Hegel and Andre Leroi-Gourhan and Marx and Bergson, a dialectical movement that ends by making us sons and daughters of our own works. As for Heidegger, the relevant myth is that "So long as we represent technology as an instrument, we remain held fast in the will to master it. We press on past the essence of technology" (Heidegger 1977, p. 32). We will see later what can be done with dialectics and the *Gestell,* but if inventing myths is the only way to get on with the job, I shall not hesitate to make up a new one and even to throw in a few more of my diagrams.

The Third Meaning of Technical Mediation: The Folding of Time and Space

Why is it so difficult to measure, with any precision, the mediating role of techniques? Because the action that we are trying to measure is subject to blackboxing, a process that makes the joint production of actors and artifacts entirely opaque. Daedalus's maze shrouds itself in secrecy. Can we open the labyrinth and count what is inside?

Take, for instance, an overhead projector. It is a point in a sequence of action (in a lecture, say), a silent and mute intermediary, taken for granted, completely determined by its function. Now suppose the projector breaks down. The crisis reminds us of the projector's existence. As the repairmen swarm around it, adjusting this lens, tightening that bulb, we remember that the projector is made of several parts, each with its role and function and its relatively independent goals. Whereas a moment before the projector scarcely existed, now even its parts have individual existence, each its own "black box." In an instant our "projector" grew from being composed of zero to one to many. How many actants are really there? The philosophy of technology we need has little use for arithmetic.

The crisis continues. The repairmen fall into a routinized sequence of actions, replacing parts. It becomes clear that their actions are composed of steps in a sequence that integrated several human gestures. We no longer focus on an object but see a group of people gathered *around* an object. A shift has occurred between actant and mediator.

Figures 11.1 and 11.2 showed that goals are redefined by associations with nonhuman actants, and that action is a property of the whole association, not only of those actants called human. However, as figure 11.3 shows, the situation is even more confused, since the *number* of actants varies from step to step. The composition of objects also varies: sometimes objects appear stable, sometimes they appear agitated, like a group of humans around a malfunctioning artifact. Thus the projector may count for one part, for nothing, for one hundred parts, for so many humans, for no humans—and each part itself may count for one, for zero, for many, for an object, for a group. In the seven steps of figure 11.3, each action may proceed toward either the dispersion of actants or their integration into a single punctuated whole (a whole that, soon thereafter, will count for nothing). We need to account for all seven steps.

Look around the room in which you are puzzling over figure 11.3. Consider how many black boxes there are in the room. Open the black boxes; examine the assemblies inside. Each of the parts inside the black box is itself a black box full of parts. If any part were to break, how many humans would immediately materialize around each? How far *back* in time, *away* in space, should we retrace our steps to follow all those silent entities that contribute peacefully to your reading this chapter at your desk? Return each of these entities to step 1; imagine the time when each was disinterested and going its own way, without being bent, enrolled, enlisted, mobilized, folded in any of the others' plots. From which forest should we take our wood? In which quarry should we let the stones quietly rest?

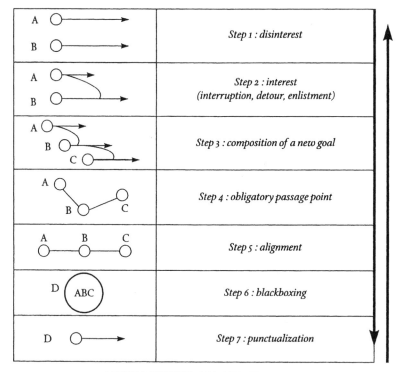

THIRD MEANING OF MEDIATION:
REVERSIBLE BLACKBOXING

Figure 11.3.   Any given assembly of artifacts may be moved up or down this succession of steps depending on the crisis they go through. What we may consider, in routine use, as one agent (step 7) may turn out to be composed of several (step 6) that may not even be aligned (step 4). The history of the earlier translations they had to go through may become visible, until they are freed again from any influence of the others (step 1).

Most of these entities now sit in silence, as if they did not exist, invisible, transparent, mute, bringing to the present scene their force and their action from who knows how many millions of years past. They have a peculiar ontological status, but does this mean that they do not act, that they do not mediate action? Can we say that because we have made all of them—and who is this "we," by the way? Not I, certainly—should they be considered slaves or tools or merely evidence of a *Gestell?* The depth of our ignorance about techniques is unfathomable. We are not even able to count their number, nor can we tell whether they exist as objects or as assemblies or as so many sequences of skilled actions. Yet there remain philosophers who believe there are such things as abject objects. . . . If science studies once believed that relying on the construction of artifacts would help account for facts, it is in for a surprise. Nonhumans escape the strictures of objectivity twice; they are neither objects known by a subject nor objects manipulated by a master (nor, of course, are they masters themselves).

The Fourth Meaning of Technical Mediation: Crossing the Boundary between Signs and Things

The reason for such ignorance is made clearer when we consider the fourth and most important meaning of mediation. Up to this point I have used the terms "story" and "program of action," "goal" and "function," "translation" and "interest," "human" and "nonhuman," as if techniques were dependable denizens that support the world of discourse. But techniques modify the matter of our expression, not only its form. Techniques have meaning, but they produce meaning via a special type of articulation that crosses the common sense boundary between signs and things.

Here is a simple example of what I have in mind: the speed bump that forces drivers to slow down on campus, which in French is called a "sleeping policeman." The driver's goal is translated, by means of the speed bump, from "slow down so as not to endanger students" into "slow down and protect your car's suspension." The two goals are far apart, and we recognize here the same displacement as in our gun story. The driver's first version appeals to morality, enlightened disinterest, and reflection, whereas the second appeals to pure selfishness and reflex action. In my experience, there are many more people who would respond to the second than to the first: selfishness is a trait more widely distributed than respect for law and life—at least in France! The driver modifies his behavior through the mediation of the speed bump: he falls back from morality to force. But from an observer's point of view it does not matter through which channel a given behavior is attained. From her window the chancellor sees that cars are slowing down, respecting her injunction, and for her that is enough.

The transition from reckless to disciplined drivers has been effected through yet another detour. Instead of signs and warnings, the campus engineers have used concrete and pavement. In this context the notion of detour, of translation, should be modified to absorb, not only (as with previous examples) a shift in the definition of goals and functions, but also *a change in the very matter of expression.* The engineers' program of action, "make drivers slow down on campus," is now articulated with concrete. What would the right word be to account for this articulation? I could have said "objectified" or "reified" or "realized" or "materialized" or "engraved," but these words imply an all-powerful human agent imposing his will on shapeless matter, while nonhumans also act, displace goals, and contribute to their definition. As we see, it is not easy to find the right term for the activity of techniques. In the meantime I want to propose yet another term, *delegation* (see figure 11.4).

Not only has one meaning, in the example of the speed bump, been displaced into another, but an action (the enforcement of the speed law) has been translated into another kind of expression. The engineers' program is delegated in concrete, and in considering this shift we leave the

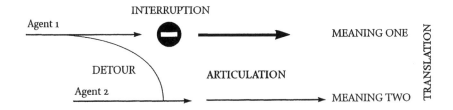

FOURTH MEANING OF MEDIATION: DELEGATION

Figure 11.4.   As in figure 11.1, the introduction of a second agent in the path of a first one implies a process of translation; but here the shift in meaning is much greater, since the very nature of the "meaning" has been modified. The matter of the expression has changed along the way.

relative comfort of linguistic metaphors and enter unknown territory. We have not abandoned meaningful human relations and abruptly entered a world of brute material relations—although this might be the impression of drivers, used to dealing with negotiable signs but now confronted by nonnegotiable speed bumps. The shift is not from discourse to matter because, for the engineers, the speed bump is one *meaningful articulation* within a gamut of propositions from which they are no more free to choose than the syntagms and paradigms we saw earlier. What they can do is to explore the associations and the substitutions that trace a unique trajectory through the collective. Thus *we remain in meaning but no longer in discourse;* yet we do not reside among mere objects. Where are we?

Before we can even begin to elaborate a philosophy of techniques we have to understand delegation as yet another type of shifting.

If I say to you, for instance, "Let us imagine ourselves in the shoes of the campus engineers when they decided to install the speed bumps," I not only transport you into another space and time but translate you into another actor. I shift you out of the scene you now occupy. The point of spatial, temporal, and "actorial" shifting, which is basic to all fiction, is to make the reader travel without moving. You make a detour through the engineers' office, but without leaving your seat. You lend me, for a time, a character who, with the aid of your patience and imagination, travels with me to another place, becomes another actor, then returns to become yourself in your own world again. This mechanism is called identification, by means of which the "enunciator" (I) and the "enunciatee" (you) both invest in the shifting delegates of ourselves within other composite frames of reference.

In the case of the speed bump the shift is "actorial": the "sleeping policeman," as the bump is known, is not a policeman, does not resemble one in the least. The shift is also spatial: on the campus road there now resides a new actant that slows down cars (or damages them). Finally, the shift is temporal: the bump is there night and day. But the enunciator of this technical act has disappeared from the scene—where are the engineers? Where is the policeman?—while someone, something, reliably acts as lieutenant, holding the enunciator's place. Supposedly the co-presence of enunciators and enunciatees is necessary for an act of fiction to be possible, but what we have now is an absent engineer, a constantly present speed bump, and an enunciatee who has become the user of an artifact.

One may object that this comparison between fictional shifting and the shifts of delegation in technical activity is spurious: to be transported in imagination from France to Brazil is not the same as taking a plane from France to Brazil. True enough, but where does the difference reside? With imaginative transportation, you simultaneously occupy all frames of reference, shifting into and out of all the delegated *personae* that the storyteller offers. Through fiction, *ego, hic, nunc* may be shifted, may become other *personae,* in other places, at other times. But aboard the plane I cannot occupy more than one frame of reference at a time (unless, of course, I sit back and read a novel which takes me, say, to Dublin on a fine June day in 1904). I am seated in an object-institution that connects two airports through an airline. The act of transportation has been *shifted down*, not out—down to planes, engines, and automatic pilots, object-institutions to which has been delegated the task of moving while the engineers and managers are absent (or limited to monitoring). The co-presence of enunciators and enunciatees has collapsed, along with their many frames of reference, to a single point in time and space. All the frames or reference of the engineers, air-traffic controllers, and ticket agents have been brought together into the single frame of reference of Air France flight 1107 to São Paulo.

An object *stands in* for an actor and creates an asymmetry between absent markers and occasional users. Without this detour, this shifting down, we would not understand how an enunciator could be absent: either it is there, we would say, or it does not exist. But through shifting down another combination of absence and presence becomes possible. In delegation it is not, as in fiction, that I am here and elsewhere, that I am myself and someone else, but that an action, long past, of an actor, long disappeared, is still active here, today, on me. I live in the midst of technical *delegates;* I am folded into nonhumans.

The whole philosophy of techniques has been preoccupied by this detour. Think of technology as *congealed* labor. Consider the very notion of investment: a regular course of action is suspended, a detour is initiated via several types of actants, and the return is a fresh hybrid that carries past acts into the present and permits its many investors to disappear while also remaining present. Such detours subvert the order of time and space—in a minute I may mobilize forces set into motion hundreds or millions of years ago in faraway places. The relative shapes of actants and their ontological status may be completely reshuffled—techniques act as *shape-changers*, making a cop out of a barrel of wet concrete, lending a policeman the permanence and obstinacy of stone. The relative ordering of presence and absence is redistributed—we hourly encounter hundreds even thousands, of absent makers who are remote in time and space yet simultaneously active and present. And through such detours, finally, the political order is subverted, since I rely on many delegated actions that themselves make me do things on behalf of others who are no longer here, the course of whose existence I cannot even retrace.

A detour of this kind is not easy to understand, and the difficulty is compounded by the accusation of fetishism made by critics of technology. It is us, the human makers (so they say), that you see in those machines, those implements, us under another guise, our own hard work. We should restore the human labor (so they command) that stands behind those idols. We heard this story told, to different effect, by the NRA: guns do not act on their own, only humans do so. A fine story, but it comes centuries too late. Humans are no longer *by themselves.* Our delegation of action to other actants that now share our human existence has developed so far that a program of antifetishism could only lead us to a nonhuman world, a lost, phantasmagoric world *before* the mediation of artifacts. The erasure of delegation by the critical antifetishists would render the shifting *down* to technical artifacts as opaque as the shifting *out* to scientific facts (see figure 11.4).

But we cannot fall back on materialism either. In artifacts and technologies we do not find the efficiency and stubbornness of matter, imprinting chains of cause and effect onto malleable humans. The speed bump is ultimately *not* made of matter; it is full of engineers and chancellors and lawmakers, commingling their wills and their story lines with those of gravel, concrete, paint, and standard calculations. The mediation, the technical translation, that I am trying to understand resides in the blind spot in which society and matter exchange properties. The story I am telling is not a *Homo faber* story, in which the courageous innovator breaks away from the constraints of social order to make contact with hard and inhuman but—at last—objective matter. I am struggling to approach the zone where some, though not all, of the characteristics of pavement become police-men, and some, though not all, of the characteristics of policemen become speed bumps. I have earlier called this zone articulation, and this is not, as I hope is now clear, a sort of golden mean or dialectic between objectivity and subjectivity. What I want to find is another Ariadne's thread to follow how Daedalus folds, weaves, plots, contrives, finds solutions where none are visible, using any expedient at hand, in the cracks and gaps of ordinary routines, swapping properties among inert, animal, symbolic, concrete, and human materials.

Technical Is a Good Adjective, Technique a Lousy Noun

We now understand that techniques do not exist as such, that there is nothing that we can define philosophically or sociologically as an object, as an artifact or a piece of technology. There does not exist, any more in technology than in science, anything to play the role of the foil for the human soul in the modernist scenography. The noun "technique"—or its upgraded version, "technol-ogy"—does not need to be used to separate humans from the multifarious assemblies with which they combine. But there is an *adjective,* technical, that we can use in many different situations, and rightly so.

"Technical" applies, first of all, to a subprogram, or a series of nested subprograms, like the ones discussed earlier. When we say "this is a technical point," it means that we have to *deviate* for a moment from the main task and that we will eventually *resume* our normal course of action, which is the only focus with our attention. A black box opens momentarily, and will soon be closed again, becoming completely invisible in the main sequence of action.

Second, "technical" designates the subordinate role of people, skills, or objects that occupy this secondary function of being present, indispensable, but invisible. It thus indicates a specialized and highly circumscribed task, clearly subordinate in a hierarchy.

Third, the adjective designates a hitch, a snag, a catch, a hiccup in the smooth functioning of the subprograms, as when we say that "there is a technical problem to solve first." Here the devia-tion may not lead us back to the main road, as with the first meaning, but may *threaten* the original goal entirely. Technical is no longer a mere detour, but an obstacle, a roadblock, the beginning of a detour, of a long translation, maybe of a whole new labyrinth. What should have been a means may become an end, at least for a while, or maybe a maze, in which we are lost forever.

The fourth meaning carries the same uncertainty about what is an end and what is a means. "Technical skill" and "technical personnel" apply to those with a unique ability, a knack, a gift, and also to the ability to make themselves *indispensable,* to occupy privileged though inferior posi-tions which might be called, borrowing a military term, obligatory passage points. So technical people, objects, or skills are at once inferior (since the main task will eventually be resumed), indis-pensable (since the goal is unreachable without them), and, in a way, capricious, mysterious, uncer-tain (since they depend on some highly specialized and sketchily circumscribed knack). Daedalus the perverse and Vulcan the limping god are good illustrations of this meaning of technical. So the adjective technical has a useful meaning that agrees in common parlance with the first three types of mediation defined above, interference, composition of goals, and blackboxing.

"Technical" also designates a very specific type of *delegation*, of movement, of shifting down, that crosses over with entities that have a different timing, different spaces, different proper-ties, different ontologies, and that are made to share the same destiny, thus creating a new actant. Hence the noun form is often used as well as the adjective, as when we say "a technique of commu-nication," "a technique for boiling eggs." In this case the noun does not designate a thing, but a *modus operandi,* a chain of gestures and know-how bringing about some anticipated result.

If one ever comes face to face with a technical object, this is never the beginning but the *end* of a long process of proliferating mediators, a process in which all relevant subprograms, nested one into another, meet in a "simple" task. Instead of the legendary kingdom in which subjects meet objects, one generally finds oneself in the realm of the *personne morale,* of what is called the "body corporate" or the "artificial person." Three extraordinary terms! As if the personality became moral by becoming collective, or collective by becoming artificial, or plural by doubling the Saxon word body with a Latin synonym, *corpus.* A *body corporate* is what we and our artifacts have become. We are an object-institution.

The point sounds trivial if applied asymmetrically. "Of course," one might say, "a piece of technology must be seized and activated by a human subject, a purposeful agent." But the point I am making is symmetrical: what is true of the "object" is still truer of the "subject." There is no sense in which humans may be said to exist as humans without entering into commerce with what authorizes and enables them to exist (that is, to act). A forsaken gun is a mere piece of matter, but what would an abandoned gunner be? A human, yes (a gun is only one artifact among many), but not a soldier—and certainly not one of the NRA's law-abiding Americans. Purposeful action and intentionality may not be properties of objects, but they are not properties of humans either. They are the properties of institutions, of apparatuses, of what Foucault called *dispositifs.* Only corporate bodies are able to absorb the proliferation of mediators, to regulate their expression, to redistribute skills, to force boxes to blacken and close. Objects that exist simply as objects, detached from a collective life, are unknown, buried in the ground. Technical artifacts are as far from the status of efficiency as scientific facts are from the noble pedestal of objectivity. Real artifacts are always parts of institutions, trembling in their missed status as mediators, mobilizing faraway lands and people, ready to become people or things, not knowing if they are composed of one or of many, of a black box counting for one or of a labyrinth concealing multitudes. Boeing 747s do not fly, airlines fly.

Part II

TECHNOLOGY AND ETHICS

All technologies raise implicit ethical questions. Anything humans make and do is subject to ethical evaluation about appropriate uses, acceptable consequences, and right or wrong actions. Most often, traditional ethical theories provide an adequate framework for assessing ethical problems associated with technology. We can usually resolve questions easily in terms of either a utilitarian framework of weighing consequences with the aim of maximizing happiness, a deontological framework of rights and responsibilities, or a virtue ethics framework emphasizing good character development and citizenship. In most cases it is only a matter of applying our traditional moral principles to the situations created by technologies. Most of our artifacts and machines are innocuous, anyway. They do not in themselves radically transform our daily lives. But sometimes they do. New technological innovations may test the limits of traditional moral principles. The situations created by new technologies can raise moral questions that are so unusual that we may need to develop new moral frameworks in order to assess them. Technologies allow us to create and extend life in an unprecedented fashion; they equip us to gather, store, and manipulate information in ways that affect our privacy, freedom, and rights; and they can radically alter the life prospects for humans, animals, and the planet itself. Arguably, radical innovations in technology require similar radical innovations in our understanding of morality.

Arguably, technology itself is moral. Ethics might not apply exclusively to humans but to technology as well. That would mean that technologies (and not just people) are subject to moral evaluation. We could then speak of good-or-bad, right-or-wrong technologies and not just good-or-bad, right-or-wrong human actions. It is the difference between ethics *and* technology, and ethics *of* technology. Only the latter treats technologies themselves as having moral properties designed into them. One of the challenges for the philosophy of technology is to evaluate the moral dimensions of artifacts, calling attention to the ways in which things are also moral. Another challenge is to argue for new ways of thinking about how to design morality into things.

Each of the readings in this section examines the ways that ethics and technology are inseparably tied together. The readings address the need to develop new frameworks for an ethics of technology; how technology affects our notions of rights and liberties; the role of moral reasoning in engineering design practice; how artifacts have morality; and the role of feminist care ethics in assessing human-technology relations.

In the first reading, "Technology and Responsibility: Reflections on the New Tasks of Ethics," Hans Jonas argues that the changes caused by modern technology on the very nature of human action require a revision in the way we think about morality. Traditional ethical theories presume four characteristics of human action: 1) action on non-human things (the whole realm of *techné*) is ethically neutral; 2) ethics is anthropocentric, concerned only with human relations; 3) human nature remains unchanged by *techne*; and 4) ethics is concerned with a limited time span and imme-diate circumstances. For traditional ethical theories the domain of moral conduct includes only our contemporaries and extends only to a finite temporal and spatial horizon. The moral knowledge requisite for responsible action is limited in a corresponding fashion; we are not accountable for what we cannot know—that is, the distant future and far-away places. Yet as Jonas notes, modern technology changes everything. Our new power to affect the entire planet opens up a new dimen-sion of responsibility previously inconceivable. Accordingly, ethics must grow to encompass this new dimension.

Each of the four characteristics of human action no longer holds. *Techné*, in the form of modern technology, has made activity in nature ethically significant. Consequently, we need to move beyond anthropocentrism and respect nature as well. In addition, human nature changes as we technologically transform ourselves and our environment, therefore politics has to change to include a longer, broader scope of responsibility. Finally, our moral responsibilities now include obligations to consider and respect future generations, the environment, and the entire planet to ensure a world fit for habitation. Unfortunately, Jonas notes, our technical capacities have out-stripped our moral knowledge, leaving us responsible for technologies with unforeseeable conse-quences. In response to these challenges, Jonas proposes a revision of Kant's duty-based moral philosophy. This revision in Kant's *Categorical Imperative* establishes an obligation to respect the continuation of humanity into the indefinite future. He concludes the chapter by applying this new moral law to technologies that transform human nature (that prolong life and postpone death, con-trol behavior through chemicals and implants, and exercise genetic control over life). The task for the new ethics Jonas proposes is to balance our technological powers, limited knowledge, and broadened scope of responsibility.

In "Technology, Demography, and the Anachronism of Traditional Rights," Robert E. McGinn analyzes the problematic the interplay of technology, individual rights, and increasing numbers of people. As more and more people exercise their right to own or use the optimal technol-ogies available, the result is likely to diminish the quality of life for everyone. There are three components to the problematic pattern McGinn identifies among technology, rights, and popula-tion. First is "technological maximality" (TM)—technology or technology-related phenomena embodying one or more aspect that is the greatest scale or highest degree previously attained. TM can be manifested in devices and systems or in aspects of their production, use, size, speed, and other "maximal" properties. Second is "traditional rights"—natural, universal, and inviolable enti-tlements owed to individuals. Third is the increasing number of rights-bearing citizens engaged in technologically maximalist practices. McGinn argues that the combination of these three factors often puts the quality of life of an entire society at risk. Examples of the "troubling triad" include the right to life-prolonging medical technologies (problematic related effect: drains resources), motorized vehicles (problematic related effect: pollutes the environment), and private property ownership (problematic related effect: over-development threatens the biosphere).

The basic conundrum is that we cannot reduce the number of people and thus diminish the burden on the environment, and limits to individual rights to TM are usually seen as illegal if not immoral. McGinn argues that we need a new *contextualized* theory of human rights based on needs that are vital to all human life. This new moral theory of rights is not absolute; it occasionally can be restricted depending on the circumstances. A contextualized theory attempts to balance the rights and needs of individuals as well as the welfare and happiness of society with the problematic realities of technological development and TM. McGinn identifies six grounds for revising the absolute character of individual rights. He then tests these grounds in cases involving urban planning and medicine to show how a balance between rights and TM can be achieved.

In "NEST-ethics: Patterns of Moral Argumentation About New and Emerging Science and Technology," Tsjalling Swiertsra and Arie Rip examine the typical arguments made for and against controversial new technologies and scientific techniques. They doubt that there is such thing as a unique nano or bio-ethics, dedicated to the unique questions raised by nanotechnology and biotechnology. But they strongly believe there is a unique form of ethical deliberation that appears whenever an emerging technology threatens well-established social conventions and moral routines. Swierstra and Rip offer an inventory of NEST-ethics arguments and demonstrate how proponents and opponents alike rely on recurring tropes and motifs. These argument-patterns have become the standard repertoire, acting like a tool-kit for debaters. Once we have mapped the argumentative terrain, we should have a better understanding of how debates typically unfold and gain an improved ability to make good NEST-ethical arguments.

The authors identify several argumentative tropes: technological versus social determinism; the wow-yuck pattern; the successes and failures of past experiences; and moral habituation versus moral corruption. After examining these patterns, the authors consider the NEST-arguments that rely on familiar moral theories. Utilitarian arguments are the most common, framing the issues in terms of an optimistic view of technological progress or a pessimistic view of technological risks. Deontological arguments are typically used to counter optimistic promises by invoking such things as "autonomy" and "human dignity," but they are just as often used to support new technologies. And arguments from virtue ethics typically frame NEST issues in terms of our visions of a technological good life or the limits we might not want to transgress. Swierstra and Rip invite us to view NEST discussions as rational, consensus-seeking deliberations that aim for temporary stabilizations and workable compromises. Above all, they remind us that ethics and new technologies co-evolve. While there are indeed recurrent patterns of moral argumentation there is also learning and acclimation to technology, changes in the repertoire of tropes, and new issues that continually arise.

In "Moralizing Technology" Peter-Paul Verbeek examines whether things—and not just humans—can be moral agents. Technologies help to shape the quality of our lives in countless ways. More importantly, they also help to shape our daily decisions and actions. The question Verbeek asks is whether things influence human action enough to attribute moral responsibility to things. Even if we conclude that things themselves do not have moral dimensions then what exactly is the impact of technologies on our moral decisions and actions? Can material things provide answers to our moral problems? The key concept for Verbeek is *technological mediation*, a concept borrowed from Ihde and Latour. Verbeek investigates the normative sense of technological mediation to show that artifacts have built-in morality and thus a kind of moral agency we usually reserve only for humans.

Verbeek starts with Idhe's notion of the perceptual mediation of experience by technology. He agrees that things play an active (not neutral) role in the relationship between humans and their world. He then builds on Latour's notion of the "script" of technologies, which suggest specific actions and discourage others. The script of something prescribes how users should act when using a thing. Technologies mediate both perception and action and, therefore, figure into the central concern of morality: how to act? Material things help to answer this basic moral question by inviting or even requiring specific forms of action when they are used. When things tell us how to act then they too can be considered moral agents.

Verbeek argues that artifacts indeed have morality. Things have both intentionality (the intention to act in a specific way) and freedom (the capability to realize an intention). Initially, this seems like a wildly implausible claim. How can an artifact form intentions? How can it possess any kind of autonomy? The answer, according to Verbeek, is that all forms of intentionality and freedom are a joint effort of human beings and technological artifacts. Following Latour, he affirms that neither humanity nor technology make sense without reference to the other. Humans and technologies are co-shaped and co-determined by each other. It makes more sense to speak of "hybrid intentionality" and "hybrid freedom."

This analysis of the moral agency of artifacts has important implications for designers of technology. Designers "materialize morality" into things that inevitably play a mediating role in our lives. Technology design is an inherently moral activity. Verbeek concludes with a discussion of how to augment the design methodology, "Constructive Technology Assessment," by incorporating into it a concern for materialized morality. Designers would then be able to better anticipate the mediating role of technologies in human actions and, hopefully, design (things that act) more responsibly.

In "Technological Ethics in a Different Voice," Diane Michelfelder considers Borgmann's proposed reform of technology from the perspective of *care ethics*. Like Borgmann, Michelfelder believes that technology can fulfill its promise of freedom and the good life but only if we use technology to live a life of engagement. She concurs that ordinary, everyday life (of work, play, family, and friendships) has moral significance. But unlike Borgmann, Michelfelder turns to care ethics to make sense of our everyday, interpersonal relationships including our relationships with technology. Care ethics emphasizes precisely what traditional, Enlightenment moral philosophy (with its emphasis of abstract, universalist, impartial procedures) overlooks: face-to-face encounters, compassion and love, and attention to unique situations and personal life. Care ethics, she claims, is also a better approach for evaluating the moral aspects of everyday life that Enlightenment morality in principle excludes, like the act of mothering or the maintenance of friendships.

Michelfelder then uses the framework of a feminist version of care ethics to analyze our moral relationships with technologies to see if Borgmann's distinction between devices (that disengage us) and focal things (that engage us) is valid. She finds his distinction to be problematic. Rather than classify technologies as either devices or things we should examine the ways in which people—in particular women—actually experience material objects. She agrees with the goal of technological reform but questions the usefulness of his distinction if we can use devices to relate more fully to one another and to the world, hence as focal things. Some devices, like telephones, foster rather than threaten our engagement with each other. When devices are used in a context that builds relationships, they help deliver the promise of technology to make our lives better. Ultimately what is important, Michelfelder argues, is not whether an object is a device or focal thing, or if it is designed in a more democratic process, but whether or not a technology plays a role making our everyday lives more meaningful.

12

Technology and Responsibility

Hans Jonas

All previous ethics—whether in the form of issuing direct enjoinders to do and not to do certain things, or in the form of defining principles for such enjoinders, or in the form of establishing the ground of obligation for obeying such principles—had these interconnected tacit premises in common: that the human condition, determined by the nature of man and the nature of things, was given once for all; that the human good on that basis was readily determinable; and that the range of human action and therefore responsibility was narrowly circumscribed. It will be the burden of my argument to show that these premises no longer hold, and to reflect on the meaning of this fact for our moral condition. More specifically, it will be my contention that with certain developments of our powers the *nature of human action* has changed, and since ethics is concerned with action, it should follow that the changed nature of human action calls for a change in ethics as well: this not merely in the sense that new objects of action have added to the case material on which received rules of conduct are to be applied, but in the more radical sense that the qualitatively novel nature of certain of our actions has opened up a whole new dimension of ethical relevance for which there is no precedent in the standards and canons of traditional ethics.

I

The novel powers I have in mind are, of course, those of modern *technology*. My first point, accordingly, is to ask how this technology affects the nature of our acting, in what ways it makes acting under its dominion *different* from what it has been through the ages. Since throughout those ages man was never without technology, the question involves the human difference of *modern* from previous technology. Let us start with an ancient voice on man's powers and deed which in an archetypal sense itself strikes, as it were, a technological note—the famous Chorus from Sophocles' *Antigone*.

> Many the wonders but nothing more wondrous than man.
> This thing crosses the sea in the winter's storm, making his path through the roaring waves. And

she, the greatest of gods, the Earth—deathless she is, and unwearied—he wears her away as the ploughs go up and down from year to year and his mules turn up the soil.

The tribes of the lighthearted birds he ensnares, and the races of all the wild beasts and the salty brood of the sea, with the twisted mesh of his nets, he leads captive, this clever man.

He controls with craft the beasts of the open air, who roam the hills. The horse with his shaggy mane he holds and harnesses, yoked about the neck, and the strong bull of the mountain.

Speech and thought like the wind and the feelings that make the town, he has taught himself, and shelter against the cold, refuge from rain. Ever resourceful is he. He faces no future helpless. Only against death shall he call for aid in vain. But from baffling maladies has he contrived escape.

Clever beyond all dreams the inventive craft that he has which may drive him one time or another to well or ill.

When he honors the laws of the land the gods' sworn right high indeed in his city; but stateless the man who dares to do what is shameful.

This awestruck homage to man's powers tells of his violent and violating irruption into the cosmic order, the self-assertive invasion of nature's various domains by his restless cleverness; but also of his building—through the self-taught powers of speech and thought and social sentiment—the home for his very humanity, the artifact of the city. The raping of nature and the civilizing of himself go hand in hand. Both are in defiance of the elements, the one by venturing into them and overpowering their creatures, the other by securing an enclave against them in the shelter of the city and its laws. Man is the maker of his life qua human, bending circumstances to his will and needs, and except against death he is never helpless.

Yet there is a subdued and even anxious quality about this appraisal of the marvel that is man, and nobody can mistake it for immodest bragging. With all his boundless resourcefulness, man is still small by the measure of the elements: precisely this makes his sallies into them so daring and allows those elements to tolerate his forwardness. Making free with the denizens of land and sea and air, he yet leaves the encompassing nature of those elements unchanged, and their generative powers undiminished. Them he cannot harm by carving out his little dominion from theirs. They last, while his schemes have their short lived way. Much as he harries Earth, the greatest of gods, year after year with his plough—she is ageless and unwearied; her enduring patience he must and can trust, and must conform. And just as ageless is the sea. With all his netting of the salty brood, the spawning ocean is inexhaustible, nor is it hurt by the plying of ships, nor sullied by what is jettisoned into its deeps. And no matter how many illnesses he contrives to cure, mortality does not bow to cunning.

All this holds because man's inroads into nature, as seen by himself, were essentially superficial, and powerless to upset its appointed balance. Nor is there a hint, in the *Antigone* chorus or anywhere else, that this is only a beginning and that greater things of artifice and power are yet to come—that man is embarked on an endless course of conquest. He had gone thus far in reducing necessity, had learned by his wits to wrest that much from it for the humanity of his life, and there he could stop. The room he had thus made was filled by the city of men—meant to enclose and not to expand—and thereby a new balance was struck within the larger balance of the whole. All the well or ill to which man's inventive craft may drive him one time or another is inside the human enclave and does not touch the nature of things.

The immunity of the whole, untroubled in its depth by the importunities of man, that is, the essential immutability of Nature as the cosmic order, was indeed the backdrop to all of mortal man's enterprises, between the abiding and the changing: the abiding was Nature, the changing his own works. The greatest of these works was the city, and on it he could offer some measure of abidingness by the laws he made for it and undertook to honor. But no long-range certainty pertained to this contrived abidingness. As a precarious artifact, it can lapse or go astray. Not even within its artificial space, with all the freedom it gives to man's determination of self, can the arbitrary ever supersede the basic terms of his being. The very inconstancy of human fortunes assures the constancy of the human condition. Chance and luck and folly, the great equalizers in human affairs, act like an entropy of sorts and make all definite designs in the long run revert to the perennial norm. Cities rise and fall, rules come and go, families prosper and decline; no change is there to stay, and in the end, with all the temporary deflections balancing each other out, the state of man is as it always was. So here too, in his very own artifact, man's control is small and his abiding nature prevails.

Still, in this citadel of his own making, clearly set off from the rest of things and entrusted to him, was the whole and sole domain of man's responsible action. Nature was not an object of human responsibility—she taking care of herself and, with some coaxing and worrying, also of man: not ethics, only cleverness applied to her. But in the city, where men deal with men, cleverness must be wedded to morality, for this is the soul of its being. In this intra-human frame dwells all traditional ethics and matches the nature of action delimited by this frame.

II

Let us extract from the preceding those characteristics of human action which are relevant for a comparison with the state of things today.

1) All dealing with the non-human world, that is, the whole realm of *techne* (with the exception of medicine), was ethically neutral—in respect to both the object and the subject of such action: in respect to the object, because it impinged but little on the self-sustaining nature of things and thus raised no question of permanent injury to the integrity of its object, the natural order as a whole; and in respect to the agent subject it was ethically neutral because *techne* as an activity conceived itself as a determinate tribute to necessity and not as an indefinite, self-validating advance to mankind's major goal, claiming in its pursuit man's ultimate effort and concern. The real vocation of man lay elsewhere. In brief, action on non-human things did not constitute a sphere of authentic ethical significance.

2) Ethical significance belonged to the direct dealing of man with man, including the dealing with himself: all traditional ethics is *anthropocentric*.

3) For action in this domain, the entity "man" and his basic condition was considered constant in essence and not itself an object of reshaping *techne*.

4) The good and evil about which action had to care lay close to the act, either in the praxis itself or in its immediate reach, and were not a matter for remote planning. This proximity of ends pertained to time as well as space. The effective range of action was small, the time-span of foresight, goal-setting and accountability was short, control of circumstances limited. Proper conduct had its immediate criteria and almost immediate consummation. The long run of consequences beyond was left to change, fate or providence. Ethics accordingly was of the here and now, of occasions as they arise between men, of the recurrent, typical situations of private and public life. The good man was he who met these contingencies with virtue and wisdom, cultivating these powers in himself, and for the rest resigning himself to the unknown.

All enjoinders and maxims of traditional ethics, materially different as they may be, show this confinement to the immediate setting of the action. "Love thy neighbor as thyself"; "Do unto others as you would wish them to do unto you"; "Instruct your child in the way of truth"; "Strive for excellence by developing and actualizing the best potentialities of your being qua man"; "Subordinate your individual good to the common good"; "Never treat your fellow man as a means only but always *also* as an end in himself"—and so on. Note that in all those maxims the agent and the "other" of his action are sharers of a common present. It is those alive now and in some commerce with me that have a claim on my conduct as it affects them by deed or omission. The ethical universe is composed of contemporaries, and its horizon to the future is confined by the foreseeable span of their lives. Similarly confined is its horizon of place, within which the agent and the other meet as neighbor, friend or foe, as superior and subordinate, weaker and stronger, and in all the other roles in which humans interact with one another. To this proximate range of action all morality was geared.

III

It follows that the *knowledge* that is required—besides the moral will—to assure the morality of action, fitted these limited terms: it was not the knowledge of the scientist or the expert, but knowledge of a kind readily available to all men of goodwill. Kant went so far as to say that "human reason can, in matters of morality, be easily brought to a high degree of accuracy and completeness even in the most ordinary intelligence";[1] that "there is no need of science or philosophy for knowing what man has to do in order to be honest and good, and indeed to be wise and virtuous. . . . [Ordinary intelligence] can have as good hope of hitting the mark as any philosopher can promise himself";[2] and again: "I need no elaborate acuteness to find out what I have to do so that my willing be morally good. Inexperienced regarding the course of the world, unable to anticipate all the contingencies that happen in it," I can yet know how to out in accordance with the moral law.[3]

Not every thinker in ethics, it is true, went so far in discounting the cognitive side of moral action. But even when it received much greater emphasis, as in Aristotle, where the discernment of the situation and what is fitting for it makes considerable demands on experience and judgment, such knowledge has nothing to do with the science of things. It implies, of course, a general conception of the human good as such, a conception predicated on the presumed invariables of man's nature and condition, which may or may not find expression in a theory of its own. But its translation into practice requires a knowledge of the here and now, and this is entirely non-theoretical. This "knowledge" proper to virtue (of the "where, when, to whom, and how") stays with the immediate issue, in whose defined context the action *as the agent's own* takes its course and within which it terminates. The good or bad of the action is wholly decided within that short-term context. Its moral quality shines forth from it, visible to its witnesses. No one was held responsible for the unintended later effects of his well-intentioned, well-considered, and well-performed act. The short arm of human power did not call for a long arm of predictive knowledge; the shortness of the one is as little culpable as that of the other. Precisely because the human good, known in its generality, is the same for all time, its relation or violation takes place at each time, and its complete locus is always the present.

IV

All this has decisively changed. Modern technology has introduced actions, objects, and consequences of such novel scale that the framework of former ethics can no longer contain them. The *Antigone* chorus on the *deinotes,* the wondrous power, of man would have to read differently now;

and its admonition to the individual to honor the laws of the land would no longer be enough. To be sure, the old prescriptions of the "neighbor" ethics—of justice, charity, honesty, and so on—still hold in their intimate immediacy of the nearest, day by day sphere of human interaction. But this sphere is overshadowed by a growing realm of collective action where doer, deed, and effect are no longer the same as they were in the proximate sphere, and which by the enormity of its powers forces upon ethics a new dimension of responsibility never dreamt of before.

Take, for instance, as the first major change in the inherited picture, the critical *vulnerability* of nature to man's technological intervention—unsuspected before it began to show itself in damage already done. This discovery, whose shock led to the concept and nascent science of ecology, alters the very concept of ourselves as a causal agency in the larger scheme of things. It brings to light, through the effects, that the nature of human action has de facto changed, and that an object of an entirely new order—no less than the whole biosphere of the planet—has been added to what we must be responsible for because of our power over it. And of what surpassing importance an object, dwarfing all previous objects of active man! Nature as a human responsibility is surely a *novum* to be pondered in ethical theory. What kind of obligation is operative in it? Is it more than a utilitarian concern? Is it just prudence that bids us not to kill the goose that lays the golden eggs, or saw off the branch on which we sit? But the "we" that here sits and may fall into the abyss is all future mankind, and the survival of the species is more than a prudential duty of its present members. Insofar as it is the fate of *man,* as affected by the condition of nature, which makes us care about the preservation of nature, such care admittedly still retains the anthropocentric focus of all classical ethics. Even so, the difference is great. The containment of nearness and contemporaneity is gone, swept away by the spatial spread and time-span of the cause-effect trains which technological practice sets afoot, even when undertaken for proximate ends. Their irreversibility conjoined to their aggregate magnitude injects another novel factor into the moral equation. To this take their cumulative character: their effects add themselves to one another, and the situation for later acting and being becomes increasingly different from what it was for the initial agent. The cumulative self-propagation of the technological change of the world thus constantly overtakes the conditions of its contributing acts and moves through none but unprecedented situations, for which the lessons of experience are powerless. And not even content with changing its beginning to the point of unrecognizability, the cumulation as such may consume the basis of the whole series, the very condition of itself. All this would have to be co-intended in the will of the single action if this is to be a morally responsible one. Ignorance no longer provides it with an alibi.

Knowledge, under these circumstances, becomes a prime duty beyond anything claimed for it heretofore, and the knowledge must be commensurate with the causal scale of our action. The fact that it cannot really be thus commensurate, that is, that the predictive knowledge falls behind the technical knowledge which nourishes our power to act, itself assumes ethical importance. Recognition of ignorance becomes the obverse of the duty to know and thus part of the ethics which must govern the ever more necessary self-policing of our out-sized might. No previous ethics had to consider the global condition of human life and the far-off future, even existence, of the race. Their now being an issue demands, in brief, a new concept of duties and rights, for which previous ethics and metaphysics provide not even the principles, let alone a ready doctrine.

And what if the new kind of human action would mean that more than the interest of man alone is to be considered—that our duty extends further and the anthropocentric confinement of former ethics no longer holds? It is at least not senseless anymore to ask whether the condition of extra-human nature, the biosphere as a whole and in its parts, now subject to our power, has become a human trust and has something of a moral claim on us not only for our ulterior sake but for its own and in its own right. If this were the case it would require quite some rethinking in basic principles of ethics. It would mean to seek not only the human good, but also the good of things extra-human, that is, to extend the recognition of "ends in themselves" beyond the sphere of man and make the human good include the care for them. For such a role of stewardship no previous ethics has prepared us—and the dominant, scientific view of *Nature* even less. Indeed, the latter emphatically denies us all conceptual means to think of Nature as something to be honored, having reduced it to the indifference of necessity and accident, and divested it of any dignity of ends. But still, a silent plea for sparing its integrity seems to issue from the threatened plenitude of the living world. Should we heed this plea, should we grant its claim as sanctioned by the nature of things, or dismiss it as a mere sentiment on our part, which we may indulge as far as we wish and can afford to do? If the former, it would (if taken seriously in its theoretical implications) push the necessary rethinking beyond the doctrine of action, that is, ethics, into the doctrine of being, that is, metaphysics, in which all ethics must ultimately be grounded. On this speculative subject I will here say no more than that we should keep ourselves open to the thought that natural science may not tell the whole story about Nature.

V

Returning to strictly intra-human considerations, there is another ethical aspect to the growth of *techne* as a pursuit beyond the pragmatically limited terms of former times. Then, so we found, *techne* was a measured tribute to necessity, not the road to mankind's chosen goal—a means with a finite measure of adequacy to well-defined proximate ends. Now, *techne* in the form of modern technology has turned into an infinite forward-thrust of the race, its most significant enterprise, in whose permanent, self-transcending advance to ever greater things the vocation of man tends to be seen, and whose success of maximal control over things and himself appears as the consummation of his destiny. Thus the triumph of *Homo faber* over his external object means also his triumph in the internal constitution of *Homo sapiens,* of whom he used to be a subsidiary part. In other words, technology, apart from its objective works, assumes ethical significance by the central place it now occupies in human purpose. Its cumulative creation, the expanding artificial environment, continuously reinforces the particular powers in man that created it, by compelling their unceasing inventive employment in its management and further advance, and by rewarding them with additional success—which only adds to the relentless claim. This positive feedback of functional necessity and reward—in whose dynamics pride of achievement must not be forgotten—assures the growing ascendancy of one side of man's nature over all the others, and inevitably at their expense. If nothing succeeds like success, nothing also entraps like success. Outshining in prestige and starving in resources whatever else belongs to the fullness of man, the expansion of his power is accompanied by a contraction of his self-conception and being. In the image he entertains of himself—the potent self-formula which determines his actual being as much as it reflects it—man now is evermore the maker of what he has made and the doer of what he can do, and most of all the preparer of what he will be able to do next. But not you or I: it is the aggregate, not the individual doer or deed that matters here; and the indefinite future, rather than the contemporary context of the action, constitutes the relevant horizon of responsibility. This requires imperatives of a new sort. If the realm of making has invaded the space of essential action, then morality must invade the realm of making, from which it had formerly stayed aloof, and must do so in the form of public policy. With issues of such inclusiveness and such lengths of anticipation public policy has never had to deal before. In fact, the changed nature of human action changes the very nature of politics.

For the boundary between "city" and "nature" has been obliterated: the city of men, once an enclave in the non-human world, spreads over the whole of terrestrial nature and usurps its place. The difference between the artificial and the natural has vanished, the natural is swallowed up in the sphere of the artificial, and at the same time the total artifact, the works of man working on and through himself, generates a "nature" of its own, i.e., a necessity with which human freedom has to cope in an entirely new sense. Once it could be said *Fiat justitia, pereat mundus,* "Let justice be done, and may the world perish"—where "world," of course, meant the renewable enclave in the imperishable whole. Not even rhetorically can the like be said anymore when the perishing of the whole through the doings of man—be they just or unjust—has become a real possibility. Issues never legislated on come into the purview of the laws which the total city must give itself so that there will be a world for the generations of man to come.

That there *ought* to be through all future time such a world fit for human habitation, and that it ought in all future time to be inhabited by a mankind worthy of the human name, will be readily affirmed as a general axiom or a persuasive desirability of speculative imagination (as persuasive and undemonstrable as the proposition that there being a world at all is "better" than there being none): but as a *moral* proposition, namely, a practical *obligation* toward the posterity of a distant future, and a principle of decision in present action, it is quite different from the imperatives of the previous ethics of contemporaneity; and it has entered the moral scene only with our novel powers and range of prescience.

The *presence of man in the world* had been a first and unquestionable given, from which all idea of obligation in human conduct started out. Now it has itself become an *object* of obligation—the obligation namely to ensure the very premise of all obligation, that is, the *foothold* for a moral universe in the physical world—the existence of mere *candidates* for a moral order. The difference this makes for ethics may be illustrated in one example.

VI

Kant's categorical imperative said: "Act so that you *can* will that the maxim of our action be made the principle of a universal law." The "can" here invoked is that of reason and its consistency with itself: *Given* the existence of a community of human agents (acting rational beings), the action must be such that it can without self-contradiction be imagined as a general practice of that community. Mark that the basic reflection of morals here is not itself a moral but a logical one: The "I *can* will" or "I *cannot* will" expresses logical compatibility or incompatibility, not moral approbation or revulsion. But there is no self-contradiction in the thought that humanity would once come to an end, therefore also none in the thought that the happiness of present and proximate generations would be bought with the unhappiness or even non-existence of later ones—as little as, after all, in the inverse thought that the existence or happiness of later generations would be bought with the unhappiness or even partial extinction of present ones. The sacrifice of the future for the present is *logically* no more open to attack than the sacrifice of the present for the future. The difference is only that in the one case the series goes on, and in the other it does not. But that it *ought to go on,* regardless of the distribution of happiness or unhappiness, even with a persistent preponderance of unhappiness over happiness, nay, even of immorality over morality[4]—this cannot be derived from the rule of self-consistency *within* the series, long or short as it happens to be: it is a commandment of a very different kind, lying outside and "prior" to the series as a whole, and its ultimate grounding can only be metaphysical.

An imperative responding to the new type of human action and addressed to the new type of agency that operates it might run thus: "Act so that the effects of your action are compatible with the permanence of genuine human life"; or expressed negatively: "Act so that the effects of your action are not destructive of the future possibility of such life"; or simply: "Do not compromise the conditions for an indefinite continuation of humanity on earth"; or most generally: "In your present choices, include the future wholeness of Man among the objects of your will."

It is immediately obvious that no rational contradiction is involved in the violation of this kind of imperative. I *can* will the present good with sacrifice of the future good. It is also evident that the new imperative addresses itself to public policy rather than private conduct, which is not in the causal dimension to which that imperative applies. Kant's categorical imperative was addressed to the individual, and its criterion was instantaneous. It enjoined each of us to consider what would happen *if* the *maxim* of my present action were made, or at this moment already were, the principle of a universal legislation; the self-consistency or inconsistency of such a *hypothetical* universalization is made the test for my *private* choice. But it was no part of the reasoning that there is any probability of my private choice *in fact* becoming universal law, or that it might contribute to its becoming that. The universalization is a thought-experiment by the private agent not to test the immanent morality of his action. Indeed, real consequences are not considered at all, and the principle is one not of objective responsibility but of the subjective quality of my self-determination. The new imperative invokes a different consistency: not that of the act with itself, but that of its eventual *effects* with the continuance of human agency in times to come. And the "universalization" it contemplates is by no means hypothetical—that is, a purely logical transference from the individual "me" to an imaginary, causally unrelated "all" ("*if* everybody acted like that"); on the contrary, the actions subject to the new imperative—actions of the collective whole—have their universal reference in their actual scope of efficacy: they "totalize" themselves in the progress of their momentum and thus are bound to terminate in shaping the universal dispensation of things. This adds a *time* horizon to the moral calculus which is entirely absent from the instantaneous logical operation of the Kantian imperative: whereas the latter extrapolates into an ever-present order of abstract compatibility, our imperative extrapolates into a predictable real *future* as the open-ended dimension of our responsibility.

VII

Similar comparisons could be made with all the other historical forms of the ethics of contemporaneity and immediacy. The new order of human action requires a commensurate ethics of foresight and responsibility, which is as new as are the issues with which it has to deal. We have seen that these are the issues posed by the works of *Homo faber* in the age of technology. But among those novel works we haven't mentioned yet the potentially most ominous class. We have considered *techne* only as applied to the non-human realm. But man himself has been added to the objects of technology. *Homo faber* is turning upon himself and gets ready to make over the maker of all the rest. This consummation of his power, which may well portend the overpowering of man, this final imposition of art on nature, calls upon the utter resources of ethical thought, which never before has been faced with elective alternatives to what were considered the definite terms of the human condition.

a) Take, for instance, the most basic of these "givens," man's mortality. Whoever before had to make up his mind on its desirable and *eligible* measure? There was nothing to choose about the upper limit, the "three score years and ten, or by reason of strength fourscore." Its inexorable rule was the subject of lament, submission, or vain (not to say foolish) wish-dreams about possible exceptions—strangely enough, almost never of affirmation. The intellectual imagination of a George Bernard Shaw and a Jonathan Swift speculated on the privilege of not having to die, or the curse of not being able to die. (Swift with the latter was the more perspicacious of the two.) Myth and legend toyed with such themes against the acknowledged background of the unalterable, which made the earnest man rather pray "teach us to number our days that we may get a heart of wisdom" (Psalm 90). Nothing of this was in the realm of doing, and effective decision. The question was only how to relate to the stubborn fact.

But lately, the dark cloud of inevitability seems to lift. A practical hope is held out by certain advances in cell biology to prolong, perhaps indefinitely extend the span of life by counteracting biochemical processes of aging. Death no longer appears as a necessity belonging to the nature of life, but as an avoidable, at least in principle tractable and long-delayable, organic malfunction. A perennial yearning of mortal man seems to come nearer fulfillment, and for the first time we have in earnest to ask the question "How desirable is this? How desirable for the individual, and how for the species?" These questions involve the very meaning of our finitude, the attitude toward death, and the general biological significance of the balance of death and procreation. Even prior to such ultimate questions are the more pragmatic ones of who should be eligible for the boon: persons of particular quality and merit? Of social eminence? Those that can pay for it? Everybody? The last would seem the only just course. But it would have to be paid for at the opposite end, at the source. For clearly, on a population-wide scale, the price of extended age must be a proportional slowing of replacement, that is, a diminished access of new life. The result would be a decreasing proportion of youth in an increasingly aged population. How good or bad would that be for the general condition of man? Would the species gain or lose? And how *right* would it be to preempt the place of youth? Having to die is bound up with having been born: mortality is but the other side of the perennial spring of "a natality" (to use Hannah Arendt's term). This had always been ordained; now its meaning has to be pondered in the sphere of decision.

To take the extreme (not that it will ever be obtained): if we abolish death, we must abolish procreation as well, for the latter is life's answer to the former, and so we would have a world of old age with no youth, and of known individuals with no surprises of such that had never been before. But this perhaps is precisely the wisdom in the harsh dispensation of our mortality: that it grants us the eternally renewed promise of the freshness, immediacy and eagerness of youth, together with the supply of otherness as such. There is no substitute for this in the greater accumulation of prolonged experience: it can never recapture the unique privilege of seeing the world for the first time and with new eyes, never relive the wonder which, according to Plato, is the beginning of philosophy, never the curiosity of the child, which rarely enough lives on as thirst for knowledge in the adult, until it wanes there too. This ever renewed beginning, which is only to be had at the price of ever repeated ending, may well be mankind's hope, its safeguard against lapsing into boredom and routine, its chance of retaining the spontaneity of life. Also, the role of the *memento mori* in the individual's life must be considered, and what its attenuation to indefiniteness may do to it. Perhaps a non-negotiable limit to our expected time is necessary for each of us as the incentive to number our days and make them count.

So it could be that what by intent is a philanthropic gift of science to man, the partial granting of his oldest wish—to escape the curse of mortality—turns out to be to the detriment of man. I am not indulging in prediction and, in spite of my noticeable bias, not even in valuation. My point is that already the promised gift raises questions that had never been asked before in terms of practical choice, and that no principle of former ethics, which took the human constants for granted, is competent to deal with them. And yet they must be dealt with ethically and by principle and not merely by the pressure of interest.

b) It is similar with all the other, quasi-utopian powers about to be made available by the advances of biomedical science as they are translated into technology. Of these, *behavior control* is much nearer to practical readiness than the still hypothetical prospect I have just been discussing, and the ethical questions it raises are less profound but have a more direct bearing on the moral conception of man. Here again, the new kind of intervention exceeds the old ethical categories.

They have not equipped us to rule, for example, on mental control by chemical means or by direct electrical action of the brain via implanted electrodes—undertaken, let us assume, for defensible and even laudable ends. The mixture of beneficial and dangerous potentials is obvious, but the lines are not easy to draw. Relief of mental patients from distressing and disabling symptoms seems unequivocally beneficial. But from the relief of the *patient,* a goal entirely in the tradition of the medical art, there is an easy passage to the relief of *society* from the inconvenience of difficult individual behavior among its members: that is, the passage from medical to social application; and this opens up an indefinite field with grave potentials. The troublesome problems of rule and unruliness in modern mass society make the extension of such control methods to non-medical categories extremely tempting for social management. Numerous questions of human rights and dignity arise. The difficult question of preemption care versus enabling care insists on concrete answers. Shall we induce learning attitudes in school children by the mass administration of drugs, circumventing the appeal to autonomous motivation? Shall we overcome aggression by electronic pacification of brain areas? Shall we generate sensations of happiness or pleasure or at least contentment through independent stimulation (or tranquilizing) of the appropriate centers—independent, that is, of the objects of happiness, pleasure, or content and their attainment in personal living and achieving? Candidacies could be multiplied. Business firms might become interested in some of these techniques for performance-increase among their employees.

Regardless of the question of compulsion or consent, and regardless also of the question of undesirable side-effects, each time we thus bypass the human way of dealing with human problems, short-circuiting it by an impersonal mechanism, we have taken away something from the dignity of personal selfhood and advanced a further step on the road from responsible subjects to programmed behavior systems. Social functionalism, important as it is, is only one side of the question. Decisive is the question of what kind of individuals the society is composed of to make its existence valuable as a whole. Somewhere along the line of increasing social manageability at the price of individual autonomy, the question of the worthwhileness of the human enterprise must pose itself. Answering it involves the image of man we entertain. We must think it anew in light of the things we can do to it now and could never do before.

c) This holds even more with respect to the last object of a technology applied on man himself—the genetic control of future men. This is too wide a subject for cursory treatment. Here I merely point to this most ambitious dream of *Homo faber,* summed up in the phrase that man will take his own evolution in hand, with the aim of not just preserving the integrity of the species but of modifying it by improvements of his own design. Whether we have the right to do it, whether we are qualified for that creative role, is the most serious question that can be posed to man finding himself suddenly in possession of such failed powers. Who will be the image-makers, by what standards, and on the basis of what knowledge? Also, the question of the moral right to experiment on future human beings must be asked. These and similar questions, which demand an answer before we embark on a journey into the unknown, show most vividly how far our powers to act are pushing us beyond the terms of all former ethics.

VIII

The ethically relevant common feature in all the examples adduced is what I like to call the inherently "utopian" drift of our actions under the conditions of modern technology, whether it works on non-human or on human nature, and whether the "utopia" at the end of the road be planned or unplanned. By the kind and size of its snowballing effects, technological power propels us into goals of a type that was formerly the preserve of Utopias. To put it differently, technological power has turned what used and ought to be tentative, perhaps enlightening, plays of speculative reason into competing blueprints for projects, and in choosing between them we have to choose between extremes of remote effects. The one thing we can really know of them is their extremism as such—that they concern the total condition of nature on our globe and the very kind of creatures that shall, or shall not, populate it. In consequence of the inevitably "utopian" scale of modern technology, the salutary gap between everyday and ultimate issues, between occasions is closing. Living now constantly in the shadow of unwanted, built-in, automatic utopianism, we are constantly confronted with issues whose positive choice requires supreme wisdom—an impossible, and in particular for contemporary man, who denies the very existence of its object: that is to say, objective value and truth. We need wisdom most when we believe in it least.

If the new nature of our acting then calls for a new ethics of long-range responsibility, coextensive with the range of our power, it calls in the name of that very responsibility also for a new kind of humility—a humility not like former humility, that is, owing to the littleness, but owing to the excessive magnitude of our power, which is the excess of our power to act over our power to foresee and our power to evaluate and to judge. In the face of the quasi-eschatological potentials of our technological processes, ignorance of the ultimate implications becomes itself a reason for responsible restraint—as the second best to the possession of wisdom itself.

One other aspect of the required new ethics of responsibility for and to a distant future is worth mentioning: the insufficiency of representative government to meet the new demands on its normal principles and by its normal mechanics. For according to these, only *present* interests make themselves heard and felt and enforce their condition. It is to them that public agencies are accountable, and this is the way in which concretely the respecting of rights comes about (as distinct from their abstract acknowledgement). But the *future* is not represented, it is not a force that can throw its weight into the scales. The non-existent has no lobby, and the unborn are powerless. Thus accountability to them has no political reality behind it yet in present decision making, and when they can make their complaint, then we, the culprits, will not longer be there.

This raises to an ultimate pitch the old question of the power of the wise, or the force of ideas not allied to self-interest, in the body politic. What *force* shall represent the future in the present? However, before *this* question can become earnest in practical terms, the new ethics must find its theory, on which dos and don'ts can be based. That is: before the question of what *force,* comes the question of what *insight* or value-knowledge shall represent the future in the present.

IX

And here is where I get stuck, and where we all get stuck. For the very same movement which put us in possession of the powers that have now to be regulated by norms—the movement of modern knowledge called science—has by a necessary complementarity eroded the foundations from which norms could be derived; it has destroyed the very idea of norm as such. Not, fortunately, the feeling for norm and even for particular norms. But this feeling, becomes uncertain of itself when contradicted by alleged knowledge or at least denied all sanction by it. Anyway and always does it have a difficult enough time against the loud clamors of greed and fear. Now it must in addition blush before the frown of superior knowledge, as unfounded and incapable of foundation. First, Nature has been "neutralized" with respect to value, then man himself. Now we shiver in the nakedness of a nihilism in which near-omnipotence is paired with near-emptiness, greatest capacity with knowing least what for. With the apocalyptic pregnancy of our actions, that very knowledge which we lack has become more urgently needed than at any other stage in the adventure of mankind. Alas, urgency is no promise of success. On the contrary, it must be avowed that to seek for wisdom today requires a good measure of unwisdom. The very nature of the age which cries out for an ethical theory makes it suspiciously look like a fool's errand. Yet we have no choice in the matter but to try.

It is a question whether without restoring the category of the sacred, the category most thoroughly destroyed by the scientific enlightenment, we can have an ethics able to cope with the extreme powers which we possess today and constantly increase and are almost compelled to use. Regarding those consequences imminent enough still to hit ourselves, fear can do the job—so often the best substitute for genuine virtue or wisdom. But this means fails us towards the more distant prospects, which here matter the most, especially as the beginnings seem mostly innocent in their smallness. Only awe of the sacred with its unqualified veto is independent to fit computations of mundane fear and the solace of uncertainty about distant consequences. But religion as a soul-determining force is no longer there to be summoned to the aid of ethics. The latter must stand on its worldly feet—that is, on reason and its fitness for philosophy. And while of faith it can be said that it either is there or is not, of ethics it holds that it must be there.

It must be there because men act, and ethics is for the reordering of actions and for regulating the power to act. It must be there all the more, then, the greater the powers of acting that are to be regulated; and with their size, the ordering principle must also fit their kind. Thus, novel powers to act require novel ethical rules and perhaps even a new ethics.

"Thou shalt not kill" was enunciated because man has the power to kill and often the occasion and even inclination for it—in short, because killing is actually done. It is only under the *pressure* of real habits of action, and generally of the fact that always action already takes place, without *this* having to be commanded first, that ethics as the ruling of such acting under the standard of the good or the permitted enters the stage. Such a *pressure* emanates from the novel technological powers of man, whose exercise is given with their existence. *If* they really are as novel in kind as here contended, and if by the kind of their potential consequences they really have abolished the moral neutrality which the technical commerce with matter hitherto enjoyed—then their pressure bids to seek for new prescriptions in ethics which are competent to assume their guidance, but which first of all can hold their own theoretically against that very pressure. To the demonstration of those premises this chapter was devoted. If they are accepted, then we who make thinking our business have a task to last us for our time. We must do it in time, for since we act anyway, we shall have some ethic or other in any case, and without a supreme effort to determine the right one, we may be left with a wrong one by default.

NOTES

1. Immanuel Kant, *Groundwork of the Metaphysics of Morals* trans. H. J. Paton. New York: Harper & Row, 1964, preface.
2. *Op. cit.,* chapter I.
3. *Ibid.* (I have followed H. J. Paton's translation with some changes.)
4. On this last point, the biblical God changed his mind to an all-encompassing "yes" after the Flood.

13

Technology, Demography, and the Anachronism of Traditional Rights

Robert E. McGinn

INTRODUCTION

Critics of the influence of technology on society debit the unhappy outcomes they decry to different causal accounts. Some target *specific characteristics or purposes* of technologies that they hold are inherently objectionable. For example, certain critics believe biotechnologies such as human in vitro fertilisation and the genetic engineering of transgenic animals to be morally wrong, regardless of who controls or uses them, and attribute what they see as the negative social consequences of these innovations to their defining characteristics or informing purposes. Others, eschewing the technological determinism implicit in such a viewpoint, find fault with the *social contexts* of technological developments and hold these contexts—more precisely, those who shape and control them—responsible for such unhappy social outcomes as result. For example, some critics have blamed the tragic medical consequences of silicone gel breast implants on a profit-driven rush to market these devices and on lax government regulation. Still other critics focus on *users* of technologies, pointing to problematic aspects of the use and operation of the technics and technical systems at their disposal. For example, some attributed the fatal crash of a DC-10 in Chicago in 1979 and the rash of reports in the mid-1980s of spontaneous acceleration of Audi automobiles upon braking to the alleged carelessness of maintenance workers and consumers.

Questions of the validity and relative value of these theory-laden approaches aside, this chapter identifies and analyses an important source of problematic technology-related influence on society of a quite different nature. Neither wholly technical, nor wholly social, nor wholly individual in nature, the source discussed below combines technical, social, and individual elements.

The source in question is a recurrent pattern of socio-technical practice characteristic of contemporary Western societies. The pattern poses a challenge to professionals in fields as diverse as medicine, city planning, environmental management, and engineering. While not intrinsically problematic—indeed, the pattern sometimes yields beneficial consequences—the pattern is *poten-*

Originally published as "Technology, Demography, and the Anachronism of Traditional Rights," *Journal of Applied Philosophy,* Vol. 11, No. 1, 1994. Copyright © 1994 Society for Applied Philosophy, Blackwell Publishers, 108 Cowley Road, Oxford, OX4 1JF, UK and 3 Cambridge Center, Cambridge, MA 02142, USA.

tially problematic. Its manifestations frequently dilute or jeopardize the quality of life in societies in which they unfold. Unless appropriate changes are forthcoming, the pattern's effects promise to be even more destructive in the future. In what follows, I shall describe and clarify the general pattern, explore its sources of strength, elaborate a conceptual/theoretical change that will be necessary to bring the pattern under control and mitigate its negative effects, and survey some conflicts over recent efforts to do just that in two social arenas: urban planning and medicine.

THE PATTERN: NATURE AND MEANING

The pattern in question involves the interplay of technology, rights, and numbers. It may be characterised thus:

> "technological maximality," unfolding under the auspices of "traditional rights" supposedly held and exercised by a large and increasing number of parties, is apt to dilute or diminish contemporary societal quality of life.

Let us begin by defining the three key expressions in this formulation.

First, in speaking of an item of technology or a technology-related phenomenon as exhibiting "technological maximality" (TM), I mean the *quality of embodying in one or more of its aspects or dimensions the greatest scale or highest degree previously attained or currently possible in that aspect or dimension.*

Thus understood, TM can be manifested in various forms. Some hinge on the characteristics of technological products and systems, while others have to do with aspects of their production, diffusion, use, or operation. Making material artifacts (technics) and sociotechnical systems of hitherto unequalled or unsurpassed scale or performance might be viewed as paradigmatic forms of TM. However, the TM concept is also intended to encompass maximalist phenomena having to do with processes as well as products. Examples of technological maximality of process include producing or diffusing as many units as possible of a technic in a given time interval or domain, and using a technic or system as intensively or extensively as possible in a given domain or situation. It is important to recognise that technological maximality can obtain even where no large-scale or super-powerful technics or technical systems are involved. Technological maximality can be reflected in *how* humans interact with and use their technics and systems as much as in technic and system characteristics proper. TM, one might say, has adverbial as well as substantive modes. In sum, technology can be maximalist in one or more of the following nine senses:

- product size or scale
- product performance (power, speed, efficiency, scope, etc.)
- speed of production of a technic or system
- volume of production of a technic or system
- speed of diffusion of a technic or system
- domain of diffusion of a technic or system
- intensity of use or operation of a technic or system
- domain of use or operation of a technic or system
- duration of use or operation of a technic or system

Secondly, "traditional rights" are entitlements of individuals as traditionally conceived in modern Western societies. For example, in the traditional Western conception individual rights have been viewed as timelessly valid and morally inviolable. Traditional individual rights often interpreted in this absolutist way include the right to life as well as liberty, property, and procreative rights.

Thirdly, the "large and increasing number of parties" factor refers to the presence in most kinds of context in contemporary Western societies of many, indeed also a growing number of, parties—usually individual humans—each of whom supposedly holds rights of the above sort and may exercise them in, among other ways, technologically maximalist behavior.

Before proceeding, I want to stress that this paper is neither a critique of "technological maximality" per se nor a celebration of E. F. Schumacher's "small is beautiful" idea. For, like the above triadic pattern, TM (or, for that matter, technological minimality) per se is not inherently morally objectionable or problematic. Specimens of technological maximality such as the then unprecedentedly large medieval Gothic cathedrals and the mammoth Saturn V rockets of the kind that took Apollo XI toward the moon suffice to refute any such claim. Rather, it is the *conjunction* of the three above-mentioned factors in repeated patterns of sociotechnical practice—*large and increasing numbers of parties engaged in technologically maximalist practices as something that each party supposedly has a morally inviolable right to do*—that is apt to put societal quality of life at risk. With this in mind, in what follows we shall refer to the combustible mixture of these three interrelated factors as "the troubling triad."

The triadic pattern is surprisingly widespread. Consider the following examples:

1) the intensive, often protracted use of life-prolonging technologies or technological procedures in thousands of cases of terminally ill or irreversibly comatose patients, or in the case of those needing an organ transplant or other life-sustaining treatment, such uses supposedly being called for by the inviolable right to life
2) the proliferation of mopeds, all-terrain, snowmobile, and other kinds of versatile transport vehicles in special or fragile environmental areas, such use supposedly being sanctioned by rider mobility rights
3) the erection of growing numbers of high-rise buildings in city centres, as supposedly permitted by owner or developer property rights

As suggested by these examples, our pattern of sociotechnical practice unfolds in diverse spheres of human activity. Problematic phenomena exemplifying the pattern in other arenas include the infestation of American national parks by tens of thousands of small tourist aircraft overflights per year; the depletion of ocean fishing areas through the use of hundreds of enormous, mechanically operated, nylon monofilament nets; and the decimation of old-growth forests in the northwestern United States through the use of myriad potent chain saws. The untoward effects exacted by the unfolding of our triadic pattern include steep financial and psychological tolls, the depletion and degradation of environmental resources, and the dilution and disappearance of urban amenities. In short, the costs of the ongoing operation of the triadic pattern are substantial and increasing.

To this point, the pattern identified above makes reference to a number of individual agents, each of whom engages in or is involved with a specimen of technologically maximal behavior, for example, having life-prolonging technologies applied intensively to herself or himself, using a technic "extensively"—meaning either "in a spatially widespread manner" or "frequently"—in a fragile, limited, or distinctive domain; or erecting a megastructure.

However, as characterised above, our pattern obscures the fact that TM can be present in *aggregative* as well as non-aggregative situations. Each of a large number of individuals, acting under the auspices of a right construed in traditional fashion, can engage in behavior that while not technologically maximal in itself becomes so when aggregated over all relevant agents. Of course, aggregating over a number of cases each of which is *already* technologically maximal compounds the maximality in question, and probably also its effects on society. We may say, therefore, that there is *individual* TM (where the individual behavior in question is technologically maximal) and *aggregative* TM, the latter having two subspecies: *simple-aggregative,* where the individual behavior is *not* technologically maximal, and *compound-aggregative,* where the individual behavior is *already* technologically maximal.

One reason why simple-aggregative TM is troubling is that individual agents may have putative rights to engage in specimens of non-technologically-maximal behavior that, taken individually, seem innocuous or of negligible import. However, contemporary environments or contexts do not automatically become larger or more robust in proportion to technic performance improvements, increasing costs of contemporary technics and systems, or the increasing number of those with access to or affected by these items. Hence, the aggregation of individually permissible behavior over all relevant agents with access to technics can result in substantial harm to societal quality of life. One can therefore speak of "public harms of aggregation." For example, the failure of each of a large number of people to recycle their garbage is technological behavior that when aggregated provides an instance of problematic technological maximality of use. Aggregating the individually innocuous effects of a large number of people driving motor vehicles that emit pollutants yields the same story: individually innocuous behavior can, when aggregated over a large group, yield a significant, noxious outcome.

THE PATTERN: SOURCES OF STRENGTH

How is the power of the pattern under discussion to be accounted for? Put differently, why does the troubling triad come under so little critical scrutiny when it has such untoward effects on individual and societal quality of life? In the case of simple-aggregative TM, the reason is that the effects of the behavior of the individual agent are negligibly problematic. It is difficult to induce a person to restrict her or his behavior when it is not perceivably linkable to the doing of significant harm to some recognized protectable individual or societal interest.

More generally, the strength of the pattern derives from the effects of factors of various sorts that lend impetus to its constituent elements. Let us examine each element in the pattern in turn.

Technological Maximality

The modern drive to achieve increases in efficiency and economies of scale and thereby reap enhanced profits is unquestionably an important factor that fuels various modes of technological maximality. One thinks in this connection of maximalist technics such as the Boeing 747 and the Alaska pipeline as well as the diffusion speed and scope modes of TM for personal technics like the VCR and CD player.

However, economic considerations do not tell the entire causal story. Cultural phenomena also play an important role and help explain the low level of resistance to our pattern. Technological maximality is encouraged by the "technological fix" mentality deeply entrenched in Western countries. Should anything go awry as a result of some technological maximalist practice, one can always, it is assumed, concoct a technological fix to remedy or at least patch up the situation in time. Moreover, there is much individual and group prestige to be garnered in modern Western societies by producing, possessing, or using the biggest, fastest, or more potent technic or technological project; more generally, by being, technologically speaking, "the-firstest-with-the-mostest." Further, influential sectors of Western opinion gauge societal progress and even a society's level of civilization by the degree to which it attains and practises certain forms of technological maximality. Small-scale, appropriate technology may be fine for developing countries but resorting to it would be seen as culturally retrogressive for a technologically "advanced" society such as the United States.

Technological maximality is often associated with construction projects. In 1985, American developer Donald Trump announced what proved to be abortive plans for a 150-story, 1,800-foot-tall Television City on the West Side of Manhattan, a megastructure that he revealingly called "the world's greatest building."[1] The demise in contemporary Western society of shared qualitative standards for making comparative value judgments has created a vacuum often filled by primarily quantitative standards of value. This, in turn, has fuelled technological maximality as a route to invidious distinction. If a building is quantitatively "the greatest," it must surely be qualitatively "the best," a convenient confusion of quality and quantity.

TM in the sense of virtually unrestricted technic use throughout special environments is greatly encouraged by modern Western cultural attitudes toward nature. Unlike in many traditional societies, land and space are typically perceived as homogeneous in character. No domains of land or space are sacred areas, hence possibly off limits to certain technological activity. On the contrary, in the contemporary United States, nature is often regarded more as a playground for technology-intensive human activity. Dune buggy riders were incensed when environmentalists sued to force the National Park Service to ban off-road vehicles from the fragile dunes at Cape Cod National Seashore. The leader of the Massachusetts Beach Buggy Association lamented that "it seems like every year they come up with more ways to deprive people of recreational activities,"[2] a comment that comes close to suggesting that rider rights have been violated.

The United States has no monopoly on TM. For example, France has a long tradition of technological maximality. The country's fascination with "grands travaux," large-scale technical projects conceived by politicians, public engineers, or civil servants, is several centuries old.[3] Encompassing classic projects such as Napoleon's Arc de Triomphe, Hausmann's transformation of central Paris, and Eiffel's Tower, the maximalist trend has also been manifested in the nationwide SNCF electric railroad system and the Anglo-French Concorde supersonic transport airplane. More recent specimens suggesting that TM is alive and well in France include audacious undertakings such as the Channel Tunnel, ever more potent nuclear power stations, the T.G.V. *(très grand vitesse)* train, and the Mitterand government's plan for building the world's largest library, dubbed by critics the "T.G.B." *(très grand bibliothèque).*[4] Such projects are not pursued solely or primarily for economic motives but for reasons of national prestige and grandeur, certification of governmental power and competence, as symbols of cultural superiority, and as monuments to individual politicians.

Another cultural factor that fosters certain modes of TM is the relatively democratic consumer culture established in the U.S. and other Western countries in the twentieth century. For the American people, innovative technics should not be reserved for the competitive advantage and enjoyment of the privileged few. Rather, based on experience with technics such as the automobile, the phone, and the television, it is believed and expected that such items should and will become available to the great mass of the American people. This expectation, cultivated by corporate advertising in order to ensure sufficient demand for what industry has the capacity to produce, in turn greatly facilitates technological maximality of production and diffusion.

Traditional Rights

Many modern Western societies are founded on belief in what were once called the "rights of man," a term that succeeded the earlier phrase "natural rights." Building on Locke's thought about natural rights, the Bill of Rights enacted by the British Parliament in 1689 provided for rights to, among other things, life, liberty, and property. The U.S. Declaration of Independence of 1776 declared that "all men . . . are endowed by their Creator with certain inalienable rights; that among these are life, liberty, and the pursuit of happiness." The French "Declaration des droits de I'homme et du citoyen" of 1789 asserts that "the purpose of all political association is the conservation of the natural and inalienable rights of man: these rights are liberty, property, security and resistance to oppression." In the 1940s Eleanor Roosevelt promoted use of the current expression "human rights" when she determined through her work in the United Nations "that the rights of men were not understood in some parts of the world to include the rights of women."[5] Although later articles of the 1948 U.N. Universal Declaration of Human Rights make reference to novel "economic and social rights" that are more clearly reflections of a particular stage of societal development, the document's Preamble refers to inalienable human rights and its early articles are couched in the language of "the old natural rights tradition."[6] Thus, in the dominant modern Western conception, individual rights are immutable, morally inviolable, and, for many, God-given.

What has this development to do with technological maximality? Things are declared as rights in a society under particular historical circumstances. When the declaration that something is a fundamental right in a society is supported by that society's dominant political-economic forces, it is safe to assume that recognition of and respect for that right is congruent with and adaptive in relation to prevailing social conditions. However, given the millennial history of the perceived close relationship between morality and religion, to get citizens of a society to take a declared "right of man" seriously, it has often seemed prudent to represent rights thus designated as having some kind of transcendental seal of approval: for example, God's blessing, correspondence with the alleged inherent fabric of the universe, or reference to them in some putatively sacred document. The right in question is thereby imbued with an immutable character, as if, although originating in specific historical circumstances, the right was nevertheless timelessly valid. Such a conception of rights can support even technologically maximal exercises of particular rights of this sort.

However, the specific sets of social-historical circumstances that gave birth to such rights eventually changed, whereas, on the whole, the perceived nature of the rights in question has not. Continuing to affirm the same things as categorical rights can become dysfunctional under new, downstream social-historical conditions; in particular, when the technics and systems to which citizens and society have access have changed radically. Endowing traditional rights with a quasi-sacred status to elicit respect for them has made it more difficult to delimit or retire them as rights further down the historical road, for example, in the present context of rampant technological maximality. In essence, the cultural strategy used to legitimate traditional rights has bestowed on them considerable intellectual inertia, something which has proved difficult to alter even though technological and demographic changes have radically transformed the context in which those rights are exercised and take effect.

A recent example of how continued affirmation of traditional rights in the context of unprecedented technological maximality can impede or disrupt societal functioning is that of the automated telephone dialer. These devices can systematically call and leave prerecorded messages at every number in a telephone exchange, including listed and unlisted numbers, cellular telephones, pagers, corporate switchboards, and unattended answering machines. By one estimate, at least 20,000 such machines are likely to be at work each day dialing some 20 million numbers around the United States. As a consequence of the potency and number of autodialers, significant communications breakdowns have already occurred.[7]

When Oregon legislators banned the commercial use of autodialers, two small-business owners who used the devices in telemarketing brought suit to invalidate the legislation on the grounds that, among other things, it violated their right to free speech. One issue here is whether U.S. society should leave its traditional robust right to free speech intact when threats to or violations of other important protectable interests, for example, privacy, emergency preparedness, and efficient organisational operation, result from exercise of this right in revolutionary technological contexts such as those created by the intensive commercial use of autodialers and fax machines for "junk calls" and "junk mail." Significantly, the American Civil Liberties Union, which supported the plaintiffs in the Oregon case, argued for preserving the traditional free speech right unabridged.

In November 1991 Congress passed the Telephone Consumer Protection Act that banned the use of autodialers for calling homes, except for emergency notification or if a party had explicitly agreed to receive such calls. However, the decision to ban turned on the annoying personal experiences of Congressional representatives and their constituents with unsolicited sales calls, not on any principled confrontation with the tension between traditional rights, technological maximality, and increasing numbers.[8] Not surprisingly, in 1993 a U.S. District Court Judge blocked enforcement of the law, ruling that it violated the constitutional right to free speech.[9]

Increasing Numbers

The positive attitude in the U.S. toward an increasing national population was adaptive in the early years of the Republic when more people were needed to settle the country and fuel economic growth. Today, even while evidencing concern over rapid population increases in less developed countries, the U.S. retains strongly pronatalist tax policies and evidences residues of the long-standing belief that when it comes to population "more is better." The "land of unlimited opportunity" myth, belief that America has an unlimited capacity to absorb population increases without undermining its quality of life, and conviction that intergenerational fairness requires that just as America opened its doors widely to earlier generations of impoverished or persecuted peoples so also should it continue to do so today; these and other beliefs militate against taking the difficult steps that might decrease or further slow the rate of increase of the American population, hence of the number of rights claimants.

THE PATTERN AS SELF-REINFORCING

Not only do powerful cultural factors foster each of the three elements of the pattern, it is also self-reinforcing. For example, reproductive freedom, derived from the right of freedom or liberty, is sacrosanct in contemporary Western societies. This belief aids and abets the increasing numbers factor, something that in turn fuels technological maximality (e.g., in technic and system size and production and diffusion rates) to support the resultant growing population. Put differently, the increasing numbers factor intensifies the interaction of the elements of the troubling triad. Under such circumstances the latter can undergo a kind of chain reaction: increases in any of its elements tend to evoke increases in one or both of the other two, and so forth. The rights to life, liberty, property and the pursuit of happiness have traditionally been construed as "negative rights," that is, as entitlements *not to be done to* in certain ways: not to be physically attacked, constrained, deprived of one's property, et cetera. But, in the context of new technologies, some such rights have also taken on a positive facet: entitlement of the individual *to be done to* in certain ways, for example, to be provided with access to various kinds of life-sustaining medical technologies and to be provided with certain kinds of information in possession of another party. A positive-faceted right to life encourages further technological maximality in both development and use, something that in turn increases the number of rights holders.

TOWARD A CONTEXTUALIZED THEORY OF HUMAN RIGHTS

A society that generates an ever more potent technological arsenal and, in the name of democratic consumerism, makes its elements available in ever larger numbers to a growing citizenry whose members believe they have inviolable rights to make, access, and use those items in individually or aggregatively technologically maximalist ways, risks and may even invite progressive impairment of its quality of life. Substantial changes will be necessary if this scenario is to be avoided, especially in the United States.

What changes might help avoid this outcome?

Decrease, stabilise, or at least substantially cut the rate of increase in the number of rights holders.

To think that any such possibility could be achieved in the foreseeable future is utopian at this juncture in Western cultural history. In spite of projections about the environmental consequences of a doubled or trebled world population, no politician of standing has raised the question of population limitation as a desirable goal for the United States or any other Western society. For this possibility to be realisable, it would seem that certain traditional rights, that is to say, those relating to reproductive behaviour and mobility, would have to be significantly reined in, a most unlikely prospect.

Put a tighter leash on individual technologically maximal behaviour.

As with the previous possibility, this option too would seem to require abridging certain traditional exercise rights or changing the underlying, quasi-categorical traditional conception of individual rights to a more conditional one. Alternatively, if one could demonstrate that untrammeled operation of the pattern is producing effects that undermine various intangible individual or societal interests, this might furnish a reason for leash-tightening. However, for various reasons, such demonstrations, even if feasible, are rarely socially persuasive.[10]

This situation suggests that one thing that may be crucial to avoiding the above scenario is elaboration and diffusion of *a new theory of moral rights*. While detailed elaboration and defence of such a theory is not feasible here, an acceptable theory should at least include accounts of the basis, function, status, and grounds for limitation of individual rights. Brief remarks on these components follow.

Basis

Western intellectual development has reached a stage in which individual moral rights can be given a more empirical, naturalistic basis. It should be acknowledged that the epistemological plausibility of rights talk need not, indeed should not, depend upon untestable beliefs in the existence and largesse of a deity interested in protecting the vital interests of individual human beings by endowing them with inalienable rights. Human rights can be plausibly anchored in basic human needs, that is, universal features of human "wiring" that must be satisfied to an adequate degree if the individual is to survive or thrive.[11] The notion then would be that something qualifies as an individual human right if and only if its protection is vital to the fulfilment of one or more underlying basic human needs. This *bottom-up* approach has the virtue of making discourse about moral rights more empirically grounded than traditional top-down theological or metaphysical approaches.

Function

In the new theory I propose, moral rights have a mundane though important function: to serve as conceptual spotlights that focus attention on aspects of human life that are essential to individual survival or thrival. The reason why such searchlights are needed is that such aspects of human life are ever at risk of being neglected because of political or social inequalities, socially conditioned preoccupation with ephemera, or the tendency of human agents to overlook or discount the interests of parties outside of their respective immediate geographical and temporal circles.

Status

Joel Feinberg has distinguished three degrees of absoluteness for individual moral rights:[12]

1) A right can be absolute in the sense of "bounded exceptionlessness," that is, binding without exception in a finite, bounded domain, as with, for example, the right to freedom of speech.
2) A right can be absolute in the (higher) sense of an "ideal directive," that is, always deserving of respectful, favourable consideration, even when, after all things have been considered, it is concluded that the right must regrettably be overridden, as with, for example, the right to privacy.
3) A right can be absolute in the (still higher) sense of "unbounded exceptionlessness" and "non-conflictability," that is, binding without exception in an unbounded domain and not intrinsically susceptible to conflict with itself or another right, in the way that, for example, the right to free speech is conflictable, as exemplified in the hectoring of a speaker. The right not to be subjected to gratuitous torture is a plausible candidate for a right that is absolute in this third sense.

In the theory we propose, individual moral rights will not be absolute in the third, highest degree, only in the first or second degree, depending on technological and demographic circumstances and on the effects on societal quality of life of aggregated maximalist exercise of the right in question.

Grounds for Decreasing the Absoluteness of Individual Rights

There are at least six kinds of circumstantial grounds that may justify restriction or limitation of an individual moral right because of the bearing of its technologically maximal exercise on societal quality of life:

1) If the very existence of society is called into question by the exercise of a putative right, for example, exercise of the right to self-defence by the acquisition of the capability of making and using weapons or other technologies of mass destruction.
2) If continued effective social functioning is threatened by the exercise of a right, for example, the disruption of telecommunication by the operation of automatic phone dialers operated under the auspices of the right of free speech.
3) If some natural resource vital to society is threatened through the exercise of a right, for example, the reduction of fishing areas or forests to non-sustainable conditions by technologically maximal harvesting practices.
4) If a seriously debilitating financial cost is imposed on society by the widespread or frequent exercise of a right, as with mushrooming public health care payments for private kidney dialysis treatment in the name of the right to life.
5) If some phenomenon of significant aesthetic, cultural, historical, or spiritual value to a people is jeopardised by the exercise of a right, for example, the destruction of a recognised architectural landmark by affixing its façade to a newly built, incongruous, mega-structure under the aegis of a private property right.
6) If some highly valued social amenity would be seriously damaged or eliminated through the exercise of a right. For example, between 1981 and 1989 convivial public space at the Federal Plaza in downtown Manhattan was effectively eliminated by the installation of an enormous sculpture (Richard Serra's 120 ft long by 14 ft high "Tilted Arc"). The artist

unsuccessfully sued the government attempting to halt removal of the work as a violation of his First Amendment right of free speech, while many of his supporters cited the right to free artistic expression.[13]

In the case of simple-aggregative TM, the only option to acquiescence is to demonstrate the significant harm done to a protectable societal interest by the aggregated act and attempt to effect an ethical revaluation of putatively harmless individual behavior; in other words, to lower the threshold of individual wrongdoing to reflect the manifest wrong effected by aggregation. With such a revaluation, the individual would have no right to act as he or she once did because of the newly declared immorality of the individual act. This process may be underway vis-à-vis the individual's disposal of home refuse without separation for recyling.

In sum, we need a *contextualized theory of human rights.* An acceptable theory of rights in contemporary technological society must be able to take on board the implications of their exercise in a context in which a rapidly changing, potent technological arsenal is diffused throughout a populous, materialistic, democratic society. Use of such a technological arsenal by a large and growing number of rights holders has considerable potential for diluting or diminishing societal quality of life. Indeed, insistence on untrammeled, entitled use of potent or pervasive technics by a large number of individuals can be self-defeating, for example, by yielding a state of social affairs incompatible with other social goals whose realisation the group also highly values.

At a deeper level, what is called into question here is the viability of modern Western individualism. Can, say, contemporary U.S. society afford to continue to promote technology-based individualism in the context of the diffusion and use of multiple potent technics by a large and ever growing population? Or is the traditional concept of individualism itself in need of revision or retirement? The ideology of individualism in all areas of life may have been a viable one in the early modern era, one with a less potent and diverse technological arsenal and a less populous society. But can contemporary Western societies have their ideological cake and eat it too? Can individualism continue to be celebrated and promoted even as a greater and greater number of citizens have access to powerful technics and systems that they, however technologically unsocialized, believe themselves entitled to use in maximalist ways?

RECENT STRUGGLES TO ADAPT INDIVIDUAL RIGHTS TO TECHNOLOGICAL MAXIMALITY AND INCREASING NUMBERS

In recent years, struggles to adapt individual rights to the realities of technological maximality in populous democratic societies have been waged incessantly on several professional fronts. Let us briefly discuss some pertinent developments in two such fields: urban planning and medicine.

Urban Planning

Two urban planning concerns involving our pattern, over which there were protracted struggles in the 1980s, are building construction and the unrestricted movement of cars. In 1986, the city of San Francisco, California, became the first large city in U.S. history to impose significant limits on the proliferation of downtown high-rise buildings. After several unsuccessful previous efforts, a citizen initiative was finally approved that established a building height limit and a cap on the amount of new high-rise floor space that can be added to the downtown area each year. The majority of San Francisco voters came to believe that the aggregate effects of the continued exercise of essentially unrestricted individual property rights by land owners and developers in technologically maximal ways—entitled erection of numerous highrises—was undermining the quality of city life.

In 1990, voters of Seattle, Washington, reached the same conclusion and approved a similar citizen initiative.

As for cars, the 1980s saw the adoption in a few Western countries of substantial limits on their use in cities. For example, to combat air pollution and enhance the quality of urban social life, citizens of Milan and Florence voted overwhelmingly in the mid-1980s to impose limits on the use of cars. In Milan, they are prohibited from entering the *centro storico* between 7:30 a.m. and 6:30 p.m., while in Florence much of the *centro storico* has been turned into a pedestrians-only zone. In California, the cities of Berkeley and Palo Alto installed barriers to prevent drivers from traversing residential streets in the course of cross-town travel. Revealingly, in a debate in the California State Senate over legislation authorising Berkeley to keep its barriers, one senator argued that "We should be *entitled* to use all roadways. . . . Certain individuals think they're too good to have other people drive down their streets" (emphasis added).[14] The phenomenon combated by the road barriers is a clear instance of aggregative TM of use unfolding under the auspices of traditional mobility rights exercised by large numbers of car-drivers. The senator's mind reading notwithstanding, it would seem that citizens, perceiving this pattern as jeopardising the safety of children and diluting the neighbourhood's residential character (read: quality of social life), prevailed on authorities to diminish the long-established domain of driver mobility rights.

Efforts to restrict individual property and mobility rights in urban settings in light of the quality-of-life consequences of their aggregated, technologically maximalist exercise have initiated a high-stakes struggle that promises to grow in importance and be vigorously contested for the foreseeable future.

Medicine

An important issue in the area of medicine that involves our pattern is the ongoing tension between the right to life and the widespread intensive use of life-prolongation technology. Following World War II, the change in the locus of dying from the home to the technology-intensive hospital enabled the full arsenal of modern medical technology to be mobilised in service of the right to life. However, the quality of the prolonged life was often so abysmal that efforts to pull back from application of technologically maximal life-extending medical care eventually surfaced.

The Karen Ann Quinlin case (1975–1985) was a landmark in the United States. The Quinlins asked their comatose daughter's doctor to disconnect her respirator. He refused, as did the New Jersey Court of Appeals.[15] The latter argued, significantly for our purposes, that "the right to life and the preservation of it are 'interests of the highest order.'" In other words, in the Appeals Court's view, respecting the traditional individual right to life was held to require ongoing provision of technologically maximal medical care. The New Jersey Supreme Court eventually found for the Quinlins, not by revoking this idea but by finding that a patient's privacy interest grows in proportion to the invasiveness of the medical care to which the patient is subjected, and that that interest can be exercised in a proxy vein by the patient's parents.[16]

The equally celebrated Nancy Cruzan case (1983–1990) was essentially an extension of Quinlin, except that the technological means of life extension that Nancy's parents sought to terminate were her food and hydration tubes. Many who opposed the Cruzans believed that removal of these tubes was tantamount to killing their comatose daughter, that is, to violating her right to life. In their view, respect for Nancy's right to life required continued application of these technological means without limitation of time, regardless of the quality of life being sustained. The Missouri Supreme Court concluded that the state's interest in the preservation of life is "unqualified," that is, that the right to life is inviolable. The Court held that in the absence of "clear and convincing evidence" that a patient would not want to be kept alive by machines in the state into which he or she had fallen, that is, would not wish to exercise her or his right to life under such circumstances, the perceived absoluteness of the right to life drove continued application of the life-prolonging technology.[17]

The 1990 case of Helga Wanglie, seemingly commonplace at the outset, took on revolutionary potential. Hospitalized after fracturing her hip, Mrs. Wanglie suffered a respiratory attack that cut off oxygen to her brain. By the time she could be resuscitated, the patient had incurred severe brain damage and lapsed into a vegetative state believed irreversible by hospital doctors. Despite this prognosis and after extensive consultation with the doctors, Mrs. Wanglie's family refused to authorize disconnection of the respirator that prolonged her life, asserting that the patient "want[ed] everything done." According to Mr. Wanglie, "she told me, 'Only He who gave life has the right to take life.'"[18]

Unprecedentedly, believing that further medical care was inappropriate, the hospital brought suit in court to obtain authorisation to disconnect the patient's respirator against her family's wishes. Predictably, this suit was unsuccessful, but Mrs. Wanglie died shortly thereafter.[19, 20] Had the suit succeeded, it would have marked a significant departure from traditional thinking and practice concerning the right to life. Care would have been terminated not at the behest of patient or guardians, something increasingly familiar in recent years, but rather as the result of a conclusion by a care-providing institution that further treatment was "futile." Projected quality of patient life would have taken precedence over the patient's inviolable right to life as asserted by guardians and the absoluteness of the right to life would have been diminished. Consensus that further treatment, however intensive or extensive, offered no reasonable chance of restoring cognitive functioning would have been established as a sufficient condition for mandatory cessation of care.

There are thousands of adults and children in the U.S. and other Western societies whose lives of grim quality are sustained by technological maximality in the name of the right to life, understood by many as categorically binding.[21] The financial and psychological tolls exacted by this specimen of compound-aggregative TM are enormous and will continue to grow until the right to life—its nature and limits—is adapted to the individual and aggregative implications of the technologies used on its behalf.

The troubling triadic pattern should be of concern to many kinds of professional practitioners, not just public officials. Professionals such as urban designers, environmental managers, engineers, and physicians are increasingly confronted in their respective practices with problematic consequences of the continued operation of the troubling triad. Each such individual must decide whether to conduct her or his professional practice—processing building permits, managing natural resource use, designing technics and sociotechnical systems, and treating patients—on the basis of traditional individualistic conceptions of rights unmodified by contemporary technological capabilities and demographic realities, or to alter the concepts and constraints informing her or his practice to reflect extant forms of technological maximality. The fundamental reason why the triadic pattern should be of concern to practising professionals is that failure to combat it is tantamount to acquiescing in the increasingly serious individual and societal harms apt to result from its predictable repeated manifestations. Professionals have an important role to play in raising societal consciousness about the costs of continuing to rely on anachronistic concepts of individual rights in contemporary technological societies. To date, doctors have made some progress in this effort but other professional groups have not even begun to rise to the challenge.

CONCLUSION

In the coming years U.S. citizens and other Westerners will face some critical choices. If we persist in gratifying our seemingly insatiable appetite for technological maximality, carried out under the auspices of anachronistic conceptions of rights claimed by ever increasing numbers of people, we shall pay an increasingly steep price in the form of a diminishing societal quality of life. Consciousness-raising, through education and responsible activism, though maddeningly slow, seems the most viable route to developing the societal ability to make discriminating choices about technological practices and their aggregated effects. However accomplished, developing that ability is essential if we are to secure a future of quality for our children and theirs. Taming the troubling triadic pattern would be an excellent place to begin this quest. The technodemographic anachronism of selected traditional rights should be recognised and a new, naturalistic, non-absolutist theory of human rights should be elaborated, one that stands in dynamic relationship to evolving technological capabilities and demographic trends. Whether or not such a new theory of rights emerges, becomes embodied in law, and alters the contours of professional practice in the next few decades will be critically important to society in the twenty-first century and beyond.

NOTES

1. *New York Times,* 19 November, 1985, p. 1.

2. *Newsweek,* 25 July, 1983, p. 22.

3. See for example: Cecil O. Smith Jr. (1990) The Longest Run: Public Engineers and Planning in France, *American Historical Review,* 95, No. 3, pp. 657–692.

4. *New York Times,* Section II, 22 December, 1991, p. 36.

5. Maurice Cranston (1983) Are There Any Human Rights?, *Daedalus,* 112, No. 4, p. 1.

6. Maurice Cranston (1973) *What Are Human Rights?* (New York, Taplinger), pp. 53–54.

7. *New York Times,* 30 October, 1991, p. A1.

8. *New York Times,* 28 November, 1991, pp. D1 and D3.

9. *New York Times,* 23 May, 1993, I, p. 26.

10. Robert E. McGinn (1979) In Defense of Intangibles: the Responsibility-Feasibility Dilemma in Modern Technological Innovation, *Science, Technology, and Human Values,* No. 29, pp. 4–10.

11. See, for example, David Braybrooke (1968) Let Needs Diminish That Preferences May Prosper, in *Studies in Moral Philosophy* (Oxford: Blackwell), pp. 86–107, and the same author's (1987) *Meeting Needs* (Princeton, N.J.: Princeton University Press) for careful analysis of the concept of basic human needs. For discussion of the testability of claims that something is a bona fide basic human need, see Amatai Etzioni (1968), Basic Human Needs, Alienation, and Inauthenticity, *American Sociological Review,* 33, pp. 870–885. On the relationship between human rights and human needs, see also C. B. Macpherson, quoted in D. D. Raphael (1967), *Political Theory and the Rights of Man* (London, Macmillan), p. 14.

12. Joel Feinberg (1973) *Social Philosophy* (Englewood Cliffs, N.J.: Prentice-Hall), pp. 85–88.

13. See, for example, J. Hitt (ed.) The Storm in the Plaza, *Harper's Magazine,* July 1985, pp. 27–33.

14. *San Francisco Chronicle,* 2 July, 1983, p. 6.

15. *In re Quinlin,* 137 N.J. super 227 (1975).

16. *In re Quinlin,* 70 N.J. 10,335 A. 2d 647 (1976).

17. After the U.S. Supreme Court decision upholding the Missouri Supreme Court was handed down, three of the patient's friends provided new evidence of her expressed wish to be spared existence in a technologically sustained vegetative state. This led, in a lower court rehearing, to a judgment permitting parental exercise of the patient's recognised privacy interest through the withdrawal of her food and hydration tubes. Nancy Cruzan expired twelve days after this decision was announced. See *New York Times,* 15 December, 1990, A1 and A9, and 27 December, 1990, A1 and A13.

18. *New York Times,* 10 January, 1991, A16.

19. Ibid., 2 July, 1991, A12.

20. Ibid., 6 July, 1991, I, 8.

21. U.S. Congress, Office of Technology Assessment (1987) *Technology-Dependent Children: Hospital v. Home Care Sustaining Technologies and the Elderly* (Washington, D.C.: U.S. Government Printing Office).

14

Technological Ethics in a Different Voice

Diane P. Michelfelder

The rapid growth of modern forms of technology has brought both a threat and a promise for liberal democratic society. As we grapple to understand the implications of new techniques for extending a woman's reproductive life or the spreading underground landscape of fiber-optic communication networks or any of the other developments of contemporary technology, we see how these changes conceivably threaten the existence of a number of primary goods traditionally associated with democratic society, including social freedom, individual autonomy, and personal privacy. At the same time, we recognize that similar hopes and promises have traditionally been associated with both technology and democracy. Like democratic society itself, technology holds forth the promise of creating expanded opportunities and a greater realm of individual freedom and fulfillment. This situation poses a key question for the contemporary philosophy of technology. How can technology be reformed to pose more promise than threat for democratic life? How can technological society be compatible with democratic values?

One approach to this question is to suggest that the public needs to be more involved with technology not merely as thoughtful consumers but as active participants in its design. We can find an example of this approach in the work of Andrew Feenberg. As he argues, most notably in his recent book, *Alternative Modernity: The Technical Turn in Philosophy and Social Theory,* the advantage of technical politics, of greater public participation in the design of technological objects and technologically mediated services such as health care, is to open up this process to the consideration of a wider sphere of values than if the design process were to be left up to bureaucrats and professionals, whose main concern is with preserving efficiency. Democratic values such as personal autonomy and individual agency are part of this wider sphere. For Feenberg, the route to technological reform and the preservation of democracy thus runs directly through the intervention of nonprofessionals in the early stages of the development of technology (Feenberg 1995).

By contrast, the route taken by Albert Borgmann starts at a much later point. His insightful explorations into the nature of the technological device—that "conjunction of machinery and commodity" (Borgmann 1992b, 296)—do not take us into a discussion of how public participation in the design process might result in a device more reflective of democratic virtues. Borgmann's interest in technology starts at the point where it has already been designed, developed, and ready for our consumption. Any reform of technology, from his viewpoint, must first pass through a serious examination of the moral status of material culture. But why must it start here, rather than earlier, as Feenberg suggests? In particular, why must it start here for the sake of preserving democratic values?

In taking up these questions in the first part of this paper, I will form a basis for turning in the following section to look at Borgmann's work within the larger context of contemporary moral theory. With this context in mind, in the third part of this paper I will take a critical look from the perspective of feminist ethics at Borgmann's distinction between the thing and the device, a distinction on which his understanding of the moral status of material culture rests. Even if from this perspective this distinction turns out to be questionable, it does not undermine, as I will suggest in the final part of this paper, the wisdom of Borgmann's starting point in his evaluation of technological culture.

PUBLIC PARTICIPATION AND TECHNOLOGICAL REFORM

One of the developments that Andrew Feenberg singles out in *Alternative Modernity* to back up his claim that public involvement in technological change can further democratic culture is the rise of the French videotext system known as Teletel (Feenberg 1995, 144–66). As originally proposed, the Teletel project had all the characteristics of a technocracy-enhancing device. It was developed within the bureaucratic structure of the French government-controlled telephone company to advance that government's desire to increase France's reputation as a leader in emerging technology. It imposed on the public something in which it was not interested: convenient access from home terminals (Minitels) to government-controlled information services. However, as Feenberg points out, the government plan for Teletel was foiled when the public (thanks to the initial assistance of computer hackers) discovered the potential of the Minitels as a means of communication. As a result of these interventions, Feenberg reports, general public use of the Minitels for sending messages eventually escalated to the point where it brought government use of the system to a halt by causing it to crash. For Feenberg, this story offers evidence that the truth of social constructivism is best seen in the history of the computer.

Let us imagine it does offer this evidence. What support, though, does this story offer regarding the claim that public participation in technical design can further democratic culture? In Feenberg's mind, there is no doubt that the Teletel story reflects the growth of liberal democratic values. The effect generated by the possibility of sending anonymous messages to others over computers is, according to Feenberg, a positive one, one that "enhances the sense of personal freedom and individualism by reducing the 'existential' engagement of the self in its communications" (Feenberg 1995, 159). He also finds that in the ease of contact and connection building fostered by computer-mediated communication, any individual or group of individuals who is a part of building these connections becomes more empowered (Feenberg 1995, 160).

But as society is strengthened in this way, in other words, as more and more opportunities open up for electronic interaction among individuals, do these opportunities lead to a more meaningful social engagement and exercise of individual freedom? As Borgmann writes in *Technology and the Character of Contemporary Society* (or *TCCL*): "The capacity for significance is where human freedom should be located and grounded" (Borgmann 1984, 102). Human interaction without significance leads to disengagement; human freedom without significance leads to banality of agency. If computer-mediated communications take one where Feenberg believes they do (and there is little about the more recent development of Internet-based communication to raise doubts about this), toward a point where personal life increasingly becomes a matter of "staging . . . personal performances" (Feenberg 1995, 160), then one wonders what effect this has on other values important for democratic culture: values such as self-respect, dignity, community, and personal responsibility.

The Teletel system, of course, is just one example of technological development, but it provides an illustration through which Borgmann's concern with the limits of public participation in the design process as a means of furthering the democratic development of technological society can be understood. Despite the philosophical foundations of liberal democracy in the idea that the state should promote equality by refraining from supporting any particular idea of the human good, in practice, he writes, "liberal democracy is enacted as technology. It does not leave the question of the good life open but answers it along technological lines" (Borgmann 1984, 92). The example we have been talking about illustrates this claim. Value-neutral on its surface with respect to the good life, Feenberg depicts the Teletel system as encouraging a play of self-representation and identity that develops at an ever-intensifying pace while simultaneously blurring the distinction between private and public life. The value of this displacement, though, in making life more meaningful, is questionable.

To put it in another way, for technology to be designed so that it offers greater opportunities for more and more people, what it offers has to be put in the form of a commodity. But the more these opportunities are put in the form of commodities, the more banal they threaten to become. This is why, in Borgmann's view, technical politics cannot lead to technical reform.

For there truly to be a reform of technological society, Borgmann maintains, it is not enough only to think about preserving democratic values. One also needs to consider how to make these values meaningful contributors to the good life without overly determining what the good life is. "The good life," he writes, "is one of engagement, and engagement is variously realized by various people" (Borgmann 1984, 214). While a technical politics can influence the design of objects so that they reflect democratic values, it cannot guarantee that these values will be more meaningfully experienced. While a technical politics can lead to more individual freedom, it does not necessarily lead to an enriched sense of freedom. For an object to lead to an enriched sense of freedom, it needs, according to Borgmann, to promote unity over dispersement, and tradition over instantaneity. Values such as these naturally belong to objects, or can be acquired by them, but cannot be designed into them.

To take some of Borgmann's favorite examples, a musical instrument such as a violin can reflect the history of its use in the texture of its wood (Borgmann 1992b, 294); with its seasonal variations, a wilderness area speaks of the natural belonging together of time and space (Borgmann 1984, 191). We need to bring more things like these into our lives, and use technology to enhance our direct experience of them (as in wearing the right kinds of boots for a hike in the woods), for technology to deliver on its promise of bringing about a better life. As Borgmann writes toward the end of *TCCL,* "So counterbalanced, technology can fulfill the promise of a new kind of freedom and richness" (Borgmann 1984, 248).

Thus for Borgmann the most critical moral choices that one faces regarding material culture are "material decisions" (Borgmann 1992a, 112): decisions regarding whether to purchase or adopt a technical device or to become more engaged with things. These decisions, like the decisions to participate in the process of design of an artifact, tend to be inconspicuous. The second type of decision, as Wiebe E. Bijker, Thomas P. Hughes, and Trevor Pinch have shown (1987), fades from public memory over time. The end result of design turns into a "black box" and takes on the appearance of having been created solely by technical experts. The moral decisions Borgmann describes are just as inconspicuous because of the nature of the context in which they are discussed and made. This context is called domestic life. "Technology," he observes, "has step by step stripped the household of substance and dignity" (Borgmann 1984, 125). Just as Borgmann recalls our attention to the things of everyday life, he also makes us remember the importance of the household as a locus for everyday moral decision making. Thus Borgmann's reflections on how technology might be reformed can also be seen as an attempt to restore the philosophical significance of ordinary life.

BORGMANN AND THE RENEWAL OF PHILOSOPHICAL INTEREST IN ORDINARY LIFE

In this attempt, Borgmann does not stand alone. Over the course of the past two decades or so in North America, everyday life has been making a philosophical comeback. Five years after the publication of Borgmann's *TCCL* appeared Charles Taylor's *Sources of the Self,* a fascinating and ambitious account of the history of the making of modern identity. Heard throughout this book is the phrase "the affirmation of everyday life," a life characterized in Taylor's understanding by our nonpolitical relations with others in the context of the material world. As he sees it, affirming this life is one of the key features in the formation of our perception of who we are (Taylor 1989, 13). Against the horizons of our lives of work and play, friendship and family, we raise moral concerns that go beyond the questions of duties and obligations familiar to philosophers. What sorts of lives have the character of good lives, lives that are meaningful and worth living? What does one need to do to live a life that would be good in this sense? What can give my life a sense of purpose? In raising these questions, we affirm ordinary life. This affirmation is so deeply woven into the fabric of our culture that its very pervasiveness, Taylor maintains, serves to shield it from philosophical sight (Taylor 1989, 498).

Other signs point as well to a resurgence of philosophical interest in the moral dimensions of ordinary life. Take, for example, two fairly recent approaches to moral philosophy. In one of these approaches, philosophers such as Lawrence Blum, Christina Hoff Summers, John Hartwig, and John Deigh have been giving consideration to the particular ethical problems triggered by interpersonal relationships, those relationships among persons who know each other as friends or as family members or who are otherwise intimately connected. As George Graham and Hugh LaFollette note in their book *Person to Person,* these relationships are ones that almost all of us spend a tremendous amount of time and energy trying to create and sustain (Graham and LaFollette 1989, 1). Such activity engenders a significant amount of ethical confusion. Creating new relationships often means making difficult decisions about breaking off relationships in which one is already engaged. Maintaining interpersonal relationships often means making difficult decisions about what the demands of love and friendship entail. In accepting the challenge to sort through some of this confusion in a philosophically meaningful way, those involved with the ethics of interpersonal relationships willingly pay attention to ordinary life. In the process, they worry about the appropriateness of importing the standard moral point of view and standard moral psychology used for our dealings with others in larger social contexts—the Kantian viewpoint of impartiality and the distrust of emotions as factors in moral decision making—into the smaller and more intimate settings of families and friendships.

Another, related conversation about ethics includes thinkers such as Virginia Held, Nel Noddings, Joan Tronto, Rita Manning, Marilyn Friedman, and others whose work has been influenced by Carol Gilligan's research into the development of moral reasoning among women. I will call the enterprise in which these theorists are engaged feminist ethics, since I believe that description would be agreeable to those whom I have just mentioned, all of whom take the analysis of women's moral experiences and perspectives to be the starting point from which to rethink ethical theory.[1] Like interpersonal ethics, feminist ethics (particularly the ethics of care) places particular value on our relationships with those with whom we come into face-to-face contact in the context of familial and friendly relations. Its key insight lies in the idea that the experience of looking out for those immediately around one, an experience traditionally associated with women, is morally significant, and needs to be taken into account by anyone interested in developing a moral theory that would be a satisfactory and useful guide to the moral dilemmas facing us in all areas of life. Thus this approach to ethics also willingly accepts the challenge of paying philosophical attention to ordinary life. This challenge is summed up nicely by Virginia Held: "Instead of importing into the household principles derived from the marketplace, perhaps we should export to the wider society the relations suitable for mothering persons and children" (Held 1987, 122).

On the surface, these three paths of ethical inquiry—Borgmann's ethics of modern technology, the ethics of interpersonal relationships, and feminist ethics—are occupied with different ethical questions. But they are united, it seems to me, in at least two ways. First, they are joined by their mutual contesting of the values upon which Kantian moral theory in particular and the Enlightenment in general are based. Wherever the modernist project of submitting public institutions and affairs to one's personal scrutiny went forward, certain privileges were enforced: that of reason over emotion, the "naked self" over the self in relation to others, impartiality over partiality, the public realm over the private sphere, culture over nature, procedural over substantive reasoning, and mind over body. In addition to the critique of Kantian ethics already mentioned by philosophers writing within a framework of an ethics of interpersonal relationships, feminist ethics has argued that these privileges led to the construction of moral theories insensitive to the ways in which women represent their own moral experience. Joining his voice to these critiques, Borgmann has written (while simultaneously praising the work of Carol Gilligan), "Universalism neglects . . . ways of empathy and care and is harsh toward the human subtleties and frailties that do not convert into the universal currency. . . . The major liability of moral universalism is its dominance; the consequence of dominance is an oppressive impoverishment of moral life" (Borgmann 1992a, 54–55).

A second feature uniting these relatively new forms of moral inquiry is a more positive one. Each attempts to limit further increases in the "impoverishment of moral life" by calling attention to the *moral* aspects of typical features of ordinary life that have traditionally been overlooked or even denied. The act of mothering (for Virginia Held), the maintenance of friendships (for Lawrence Blum) and the loving preparation of a home-cooked meal (for Borgmann) have all been defended, against the dominant belief to the contrary, as morally significant events.[2]

Despite the similarities and common concerns of these three approaches to moral philosophy, however, little engagement exists among them. Between feminist ethics and the ethics of interpersonal relationships, some engagement can be found: for instance, the "other-centered" model of friendship discussed in the latter is of interest to care ethicists as part of an alternative to Kantian ethics. However, both of these modes of ethical inquiry have shown little interest in the ethical dimensions of material culture. Nel Noddings, for example, believes that while caring can be a moral phenomenon when it is directed toward one's own self and that of others, it loses its moral dimension when it is directed toward things. In her book *Caring,* she defends the absence of discussion of our relations to things in her work: "as we pass into the realm of things and ideas, we move entirely beyond the ethical. . . . My main reason for setting things aside is that we behave ethically only through them and not toward them" (Noddings 1984, 161–62).

And yet in ordinary life ethical issues of technology, gender, and interpersonal relationships overlap in numerous ways. One wonders as a responsible parent whether it is an act of caring to buy one's son a Mighty Morphin Power Ranger. If I wish to watch a television program that my spouse cannot tolerate, should I go into another room to watch it or should I see what else is on television so that we could watch a program together? Is a married person committing adultery if he or she has an affair with a stranger in cyberspace? Seeing these interconnections, one wonders what might be the result were the probing, insightful questioning initiated by Borgmann into the moral significance of our material culture widened to include the other voices mentioned here. What would we learn, for instance, if Borgmann's technological ethics were explored from the perspective of feminist ethics?

In the context of this paper I can do no more than start to answer this question. With this in mind, I would like to look at one of the central claims of *TCCL:* the claim that the objects of material culture fall into the category of either things or devices.

FEMINISM AND THE DEVICE PARADIGM

As Borgmann describes them, things are machines that, in a manner of speaking, announce their own narratives and as a result are generous in the effects they can produce. For example, we can see the heat of the wood burning in the fireplace being produced in front of our eyes—the heat announces its own story, its own history, in which its relation to the world is revealed. In turn, fireplaces give us a place to focus our attention, to regroup and reconnect with one another as we watch the logs burn. In this regard, Borgmann speaks compellingly not only of the fireplace but also of wine: "Technological wine no longer bespeaks the particular weather of the year in which it grew since technology is at pains to provide assured, i.e. uniform, quality. It no longer speaks of a particular place since it is a blend of raw materials from different places" (Borgmann 1984, 49).

Devices, on the other hand, hide their narratives by means of their machinery and as a result produce only the commodity they were intended to produce. When I key the characters of the words I want to write into my portable computer they appear virtually simultaneously on the screen in front of me. I cannot see the connection between the one event and the other, and the computer does not demand that I know how it works in order for it to function. The commodity we call "processed words" is the result. While things lead to "multi-sided experiences," devices produce "one-sided experiences."[3] Fireplaces provide warmth, the possibility of conviviality, and a closer tie to the natural world; a central heating system simply provides warmth.

What thoughts might a philosopher working within the framework of feminist ethics have about this distinction? To begin with, I think she would be somewhat uneasy with the process of thinking used to make decisions about whether a particular object would be classified as a thing or a device. In this process, Borgmann abstracts from the particular context of the object's actual use and focuses his attention directly on the object itself. The view that some wine is "technological," as the example described above shows, is based on the derivation of the wine, the implication being that putting such degraded wine on the table would lead to a "one-sided experience" and further thwart, albeit in a small way, technology's capability to contribute meaningfully to the good life. In a feminist analysis of the moral significance of material culture, a different methodology would prevail. The analysis of material objects would develop under the assumption that understanding people's actual experiences of these objects, and in particular understanding the actual experiences of women who use them, would be an important source of information in deciding what direction a technological reform of society should take.

The attempt to make sense of women's experience of one specific technological innovation is the subject of communication professor Lana Rakow's book *Gender on the Line: Women, the Telephone, and Community Life* (1992). As its title suggests, this is a study of the telephone practices of the women residents of a particular community, a small midwestern town she called, to protect its identity, Prospect.

Two features of Rakow's study are of interest with regard to our topic. One relates to the discrepancy between popular perceptions of women's use of the telephone, and the use revealed in her investigation. She was well aware at the beginning of her study of the popular perception, not just in Prospect but widespread throughout American culture, of women's use of the telephone. In the popular perception, characterized by expressions such as "Women just like to talk on the phone" and "Women are on the phone all the time," telephone conversations among women appear as "productivity sinks," as ways of wasting time. Understandably from this perception the telephone could appear as a device used for the sake of idle chatter that creates distraction from the demands of work and everyday life. This is how Borgmann sees it:

> The telephone network, of course, is an early version of hyperintelligent communication, and we know in what ways the telephone has led to disconnectedness. It has extinguished the seemingly austere communication via letters. Yet this austerity was wealth in disguise. To write a letter one needed to sit down, collect one's thoughts and world, and commit them laboriously to paper. Such labor was a guide to concentration and responsibility. (Borgmann 1992a, 105)

Rakow's study, however, did not support the popular perception. She found that the "women-talk" engaged in by her subjects was neither chatter nor gossip. Rather, it was a means to the end of producing, affirming, and reinforcing the familial and community connections that played a very large role in defining these women's lives. Such "phone work," very often consisting of exchanges of stories, was the stuff of which relations were made: "Women's talk holds together the fabric of the community, building and maintaining relationships and accomplishing important community relations" (Rakow 1992, 34).

Let me suggest some further support for this view from my own experience. While I was growing up, I frequently witnessed this type of phone work on Sunday afternoons as my mother would make and receive calls from other women to discuss "what had gone on at church." Although these women had just seen each other at church several hours before, their phone calls played exactly the role that Rakow discovered they played in Prospect. At the time, they were not allowed to hold any positions of authority within the organizational structure of this particular church. The meaning of these phone calls would be missed by calling them idle talk; at least in part, these phone visits served to strengthen and reinforce their identity within the gendered community to which these women belonged.

Another interesting feature of Rakow's study was its discovery of how women used the telephone to convey care:

> Telephoning functions as a form of care-giving. Frequency and duration of calls . . . demonstrate a need for caring or to express care (or a lack of it). Caring here has the dual implication of caring *about* and caring *for*—that is, involving both affection and service. . . . While this [care-giving role] has been little recognized or valued, the caring work of women over the telephone has been even less noted. (Rakow 1992, 57)

As one of the places where the moral status of the care-giving role of women has been most clearly recognized and valued, feminist ethics is, of course, an exception to this last point. Rakow's recognition of the telephone as a means to demonstrate one's caring for speaks directly to Nel Noddings's understanding of why giving care can be considered a moral activity (Noddings 1984). In caring one not only puts another's needs ahead of one's own, but, in reflecting on how to take care of those needs, one sees oneself as being related to, rather than detached from, the self of the other. In commenting that not only checking on the welfare of another woman or phoning her on her birthday but "listening to others who need to talk is also a form of care" (Rakow 1992, 57), Rakow singles out a kind of caring that well reflects Noddings's description. More often one needs to listen to others who call one than one needs to call others; and taking care of the needs of those who call often involves simply staying on the phone while the other talks. As Rakow correctly points out, this makes this particular practice of telephone caring a form of work. Those who criticize the ethics of care for taking up too much of one's time with meeting the needs of individual others might also be critical of Rakow's subjects who reported that

> they spend time listening on the phone when they do not have the time or interest for it. . . . One elderly woman . . . put a bird feeder outside the window by her telephone so she can watch the birds when she has to spend time with these phone calls. "I don't visit; I just listen to others," she said. (Rakow 1992, 57)

As these features of telephone conversations came to light in the interviews she conducted with the women of Prospect, Rakow began to see the telephone as "a gendered, not a neutral, technology" (Rakow 1992, 33). As a piece of gendered technology, the telephone arguably appears more like a thing than like a device, allowing for, in Borgmann's phrase, the "focal practice" of caring to take place. Looking at the telephone from this perspective raises doubts about Borgmann's assessment of the telephone. Has the telephone in fact become a substitute for the thing of the letter, contributing to our widespread feelings of disconnectedness and to our distraction? Rakow's fieldwork provides support for the idea that phone work, much like letter writing, can be "a guide to concentration and responsibility." By giving care over the phone, the development of both these virtues is supported. Thus, on Borgmann's own terms—"The focal significance of a mental activity should be judged, I believe, by the force and extent with which it gathers and illuminates the tangible world and our appropriation of it" (Borgmann 1984, 217)—it is difficult to see how using the telephone as a means of conveying care could not count as a focal concern.

Along with the question of whether a particular item of our material culture is or is not a device, looking at the device paradigm from a feminist point of view gives rise to at least two other issues. One is connected to an assumption on which this paradigm rests: that the moral significance of an object is directly related to whether or not that object is a substitute for the real thing. This issue is also connected to the idea that because technological objects are always substitutes for the real thing, the introduction of new technology tends to be a step forward in the impoverishment of ordinary life.

Certainly technological objects are always substitutes for *something or another*. A washing machine is a substitute for a washing board, dryers are substitutes for the line out back, krab [*sic*] is often found these days on salad bars, and so forth. In some cases, the older object gradually fades from view, as happened with the typewriter, which (but only as of fairly recently) is no longer being produced. In other cases, however, the thing substituted for is not entirely replaced, but continues to coexist alongside the substitute. In these cases, it is harder to see how the technological object is a substitute *for the real thing,* and thus harder to see how the introduction of the new object threatens our sense of engagement with the world. While it is true that telephones substitute for letter writing, as Borgmann observes, the practice of letter writing goes on, even to the point of becoming intertwined with the use of the telephone. Again, from Rakow:

> The calls these women make and the letters they send literally call families into existence and maintain them as a connected group. A woman who talks daily to her two nearby sisters demonstrated the role women play in keeping track of the well-being of family members and changes

in their lives. She said, "If we get a letter from any of them (the rest of the family) we always call and read each other the letters." (Rakow 1992, 64)

Perhaps, though, the largest question prompted by Rakow's study has to do with whether Borgmann's distinction itself between things and devices can hold up under close consideration of the experiences and practices of different individuals. There are many devices that can be and are used as the women in this study used the telephone. Stereos, for example, can be a means for someone to share with someone else particular cuts on a record or songs from a CD to which she or he attaches a great deal of personal significance. In this way, stereos can serve as equipment that aid the development of mutual understanding and relatedness, rather than only being mechanisms for disengagement. The same goes for the use of the computer as a communicative device. Empirical investigations into the gendered use of computer-mediated communications suggest that while women do not necessarily use this environment like the telephone, as a means of promoting care, they do not "flame" (send electronic messages critical of another individual) nearly as much as do men, and they are critical of men who do engage in such activity.[4]

In particular, from a feminist perspective one might well wonder whether, in Borgmann's language, the use of those "conjunctions of machinery and commodity" inevitably hamper one's efforts at relating more to others and to the world. Borgmann argues that because devices hide their origins and their connections to the world, they cannot foster our own bodily and social engagement with the world. But as I have tried to show here, this is arguably not the case. Whether or not a material object hides or reveals "its own story" does not seem to have a direct bearing on that object's capacity to bind others together in a narrative web. For instance, older women participating in Rakow's study generally agreed that telephones improved in their ability to serve as a means of social support and caregiving once their machinery became more hidden: when private lines took the place of party lines and the use of an operator was not necessary to place a local call. To generalize, the machinery that clouds the story of a device does not appear to prevent that device from playing a role in relationship building.

DEVICES AND THE PROMISE OF TECHNOLOGY

While a child growing up in New Jersey, I looked forward on Friday evenings in the summer to eating supper with my aunt and uncle. I would run across the yard separating my parents' house from theirs to take my place at a chair placed at the corner of the kitchen table. The best part of the meal, I knew, would always be the same, and that was why I looked forward to these evenings. While drinking lemonade from the multicolored aluminum glasses so popular during the 1950s, we would eat Mrs. Paul's fish sticks topped with tartar sauce. With their dubious nutritional as well as aesthetic value, fish sticks are to fresh fish as, in a contrast described eloquently by Borgmann, Cool Whip is to fresh cream (Borgmann 1987, 239–42). One doesn't know the seas in which the fish that make up fish sticks swim. Nearly anyone can prepare them in a matter of minutes. Still, despite these considerations, these meals were marked by family sociability and kindness, and were not hurried affairs.

I recall these meals now with the following point in mind. One might be tempted by the course of the discussion here to say that the objects of material culture should not be divided along the lines proposed in *TCCL* but divided in another manner. From the perspective of feminist ethics, one might suggest that one needs to divide up contemporary material culture between relational things, things open up the possibility of caring relations to others, and nonrelational things: things that open up the possibility of experience but not the possibility of relation. Telephones, on this way of looking at things, would count as relational things. Virtual reality machines, such as the running simulator Borgmann imagines in *CPD,* or golf simulators that allow one to move from the green of the seventeenth hole at Saint Andrews to the tee of the eighteenth hole at Pebble Beach, would be nonrelational things. One can enjoy the experiences a virtual golf course makes possible, but one cannot in turn, for example, act in a caring manner toward the natural environment it so vividly represents. But the drawback of this distinction seems similar to the drawback of the distinction between things and devices: the possibility of using a thing in a relational and thus potentially caring manner seems to depend more on the individual using that thing and less on the thing itself. Depending on who is playing it, a match of virtual golf has the potential of strengthening, rather than undoing, narrative connections between oneself, others and the world.

But if our discussion does not lead in this direction, where does it lead? Let me suggest that although it does not lead one to reject the device paradigm outright, it does lead one to recognize that while any device does use machinery to produce a commodity, the meaning of one's experience associated with this device does not necessarily have to be diminished. And if one can use technology (such as the telephone) to carry out focal practices (such as caregiving), then we might have cause to believe that there are other ways to recoup the promise of technology than Borgmann sees. As mentioned earlier, his hope is that we will give technology more of a supporting role in our lives than it has at present (Borgmann 1984, 247), a role he interprets as meaning that it should support the focal practices centered around focal things. But if devices can themselves support focal practices, then the ways in which technology can assume a supporting role in our lives are enhanced.

But if the idea that devices can support focal practices is in one way a challenge to the device paradigm, in another way it gives additional weight to the notion that there are limits to reforming technology through the process of democratic design. When they are used in a context involving narrative and tradition, devices can help build engagement and further reinforce the cohesiveness of civil society. Robert Putnam has pointed out the importance of trust and other forms of "social capital" necessary for citizens to interact with each other in a cooperative manner. As social capital erodes, democracy itself, he argues, is threatened (Putnam 1995, 67). While this paper has suggested that devices can under some conditions further the development of social capital, it is difficult to see how they can be deliberately designed to do so. In thinking about how to reform technology from a democratic perspective, we need to remember the role of features of ordinary life such as narrative and tradition in making our experience of democratic values more meaningful. Borgmann's reminder to us of this role is, it seems to me, one of the reasons why *TCCL* will continue to have a significant impact in shaping the field of the philosophy of technology.

NOTES

1. I am not using "feminist ethics" in a technical sense, but as a way of referring to the philosophical approach to ethics that starts from a serious examination of the moral experience of women. For philosophers such as Alison Jaggar, the term feminist ethics primarily means an ethics that recognizes the patriarchal domination of women and the need for women to overcome this system of male domination. Thus she and others might disagree that the ethics of care, as I take it here, is an enterprise of feminist ethics.

2. For example, Virginia Held has written: "[Feminist moral inquiry] pays attention to the neglected experience of women and to such a woefully neglected though enormous area of human moral experience as that of mothering. . . . That this whole vast region of human experience can have been dismissed as 'natural' and thus as irrelevant to morality is extraordinary" (Held 1995, 160).

3. The term "multi-sided experiences" is used by Mihaly Csikszentmihalyi and Eugene Rochbert-Halton in their work *The Meaning of Things,* discussed in Borgmann 1992b.

4. See, for example, Susan Herring, "Gender Differences in Computer-Mediated Communication: Bringing Familiar Baggage to the New Frontier" (unpublished paper).

15

NEST-ethics: Patterns of Moral Argumentation About New and Emerging Science and Technology

Tsjalling Swierstra and Arie Rip

INTRODUCTION

Can there be a dedicated nano-ethics, just as there is, by now, a bio-ethics? And should there be? The title of the journal *Nanoethics* is careful in that it creates some distance to nanotechnology in its sub-title: *Ethics of technologies that converge at the nano-scale.* While there are an increasing number of articles and comments that call for nano-ethics these are mostly calls for more ethical reflection in general, or focus on the utopian and doomsday scenarios that have been put forward. (e.g., Mnyusiwalla et al., 2003, Gordijn 2003, Lin and Allhoff, 2006) There is little specific to nanotechnology that would warrant the prefix "nano." (Grunwald,2005) This is different from the case of bio-ethics, where aspects and issues derived from living creatures are a shared starting point. Nanotechnology has no such common referent other than that phenomena and manipulations occur at the nano-scale. It is an umbrella term covering a host of heterogeneous technologies, from electronics to materials and on to medical use of nanoparticles. At the same time, there are calls for nano-ethics, and working on nano-ethics is a business proposition for organizations like the Centre for Responsible Nanotechnology (http://CRNano.org) and The Nanoethics Group (www.nanoethics.org).

More important for our query about the status of nano-ethics is the fact that nanotechnologies are *enabling* technologies. By making existing technologies smaller and faster, well-known ethical issues, say privacy and new ICT, or point-of-care diagnostics and professional-medical responsibilities, can become more pressing, but not necessarily different in kind. Quite a number of reports reflect this by taking sector issues (medical, environment, military) or moral principles (equity, privacy, safety, sustainability, security) as their starting point, rather than specific features of nanotechnology. (For a good example, see Van Est et al., 2004.)

Still, there are calls for nano-ethics. And actors present issues about nanotechnology as ethical issues. A striking example (which we will use again later on) is this quote from Philip J. Bond, U. S. Under-Secretary of Commerce, "Responsible nanotechnology development," in the SwissRe workshop of December 2004 (SwissRe, 2004b, p. 7):

> Given nanotechnology's extraordinary economic and societal potential, it would be unethical, in my view, to attempt to halt scientific and technological progress in nanotechnology. Nanotechnology offers the potential for improving people's standard of living, healthcare, and nutrition; reducing or even eliminating pollution through clean production technologies; repairing existing environmental damage; feeding the world's hungry; enabling the blind to see and the deaf to hear; eradicating diseases and offering protection against harmful bacteria and viruses; and even extending the length and the quality of life through the repair or replacement of failing organs. Given this fantastic potential, how can our attempt to harness nanotechnology's power at the earliest opportunity—to alleviate so many earthly ills—be anything other than ethical? Conversely, how can a choice to halt be anything other than unethical?

In this quote one sees how actors tend to use the qualifiers "ethical" and "unethical" to indicate what is good (must be done) and bad (must not be done). The key feature for our discussion, however, is that it makes a general point about progress thanks to new technology, rather than saying anything specific about nanotechnology other than that it is wonderful and will enable the blind to see and the deaf to hear (phrases with a biblical ring to them). Clearly, there are ethics involved, but these are not nano-ethics, but ethics of progress and/or negative impact through new technology.

Similarly, the recent exercises in public engagement with nanotechnology and its promises and possible concerns, like focus groups and a citizen jury in the United Kingdom, or nano-dialogue projects funded by the European Union, tend to come up with reports which are quite general and could apply to any new or emerging technology. This has led to critical comments, by nanotechnology actors as well as analysts: why do these exercises at all if nothing specific to nanotechnology comes out of the discussions and reflections?

What we will do in this article is to turn this criticism around, and see it as a finding, and starting point for further analysis. There appear to be certain patterns of moral argumentation about new and emerging technology, which are then applied to nanotechnology. In other words, while there may not be a nano-ethics, there definitely is a NEST-ethics. The prefix NEST stands for New and Emerging Science and Technology. Our contention is that most ethical questions presently raised about nanotechnology belong to NEST-ethics. In the next two sections we will show that there are indeed typical argumentative patterns that together constitute a NEST-ethics. Such a NEST-ethics is not given once and for all. It evolves with further experiences with new science and technology, like stem cells and now also promises and concerns about nanotechnology. But there are also strong continuities.

NEST-ethics typically exists as a set of recurring tropes and argumentative patterns. By "trope" we understand a recurring motif or argument that is supposed to have particular force. By argumentative "pattern" we understand two or more ethical arguments that hang together in the sense that they provoke each other into existence. The tropes and the "storylines" in the argumentative patterns have become a repertoire that is available in late-modern societies, both as a framing of how actors view issues and expect others to view them, and as a kind of toolkit that can be drawn upon in concrete debates.

The to-and-fro of moral argumentation about NEST is actually played out at two levels. There are what one might call meta-ethical issues, addressing the relation between technology and morality, with a particular focus on the "new and emerging" feature. We will discuss these issues first. The remaining tropes and patterns in the repertoire can subsequently be clustered according to the dominant moral standard referred to implicitly or explicitly: collective utility (utilitarianism, consequentialism more broadly), duties and rights (deontology), the just distribution of costs and benefits (theories of justice), or conceptions of the good life (virtue ethics, or as we prefer to say: good life ethics—which merges into "good society" ethics).

This turns out to be more than making an inventory. Most often, there is a pattern: the debate starts with seemingly obvious consequentialist arguments. These are then criticized, and this provokes reference to equity, basic values and aspects of the good life. In response, there are attempts to blackbox these references, and return to consequentialist arguments which are easier to handle discussion and management (in a broad sense) of new technology in society. And this allows a simple division of moral labour where scientists and other introducers of new and emerging science and technology are justified in pushing on as long as they are willing to consider side-effects.

TECHNOLOGY, MORALITY AND ETHICS: PRELIMINARY REFLECTIONS

New and emerging science and technology constitute novelties, already within the world of science and technology. Of course, some novelties are more novel than others. A distinction is often made between incremental and radical (or disruptive) innovation. But they are still innovations, and some existing alignments will be threatened, or at least opened up. Then, a process of re-alignment starts which runs more or less smoothly. In science and technology, the creation of novelty is actively pursued, and the new findings and options and proofs of principle are expected to be taken up. There is also resistance to change, however, as the history of science and technology amply shows.

Responses of society to new and emerging science and technology and its societal embedding also vary. One can hypothesize that when a new technology can be fitted to existing artefacts, routines and strategies, and/or when it appears to address existing or newly articulated needs and desires (when new experiences are offered as with Sony's Walkman, cf. Du Gay et al., 1997), its embedding will go smoothly. There are also general cultural patterns in the response to novelty. It can be seen as the hero who shall conquer, overcome the barriers of the existing order which will soon become obsolete; or the deviant, the wayward, and if it persists, it becomes the sinner that must be punished for going against the existing order. The moral flavour of the terms is not accidental.

The link with moral argumentation and ethics requires some reconsideration of ethics as well. We will build on philosophical pragmatism, especially in John Dewey's version (Dewey, 1994; Keulartz et al., 2002; 2004). In the pragmatics of everyday life, morals exist mainly as routines which are considered to be self-evident so that people are hardly aware of their existence. These moral routines once started their existence as conscious solutions for conflicting stakeholder interests/rights or as answers to the question: what would be a good life to lead, as an individual and/or as a community? But afterwards, we unthinkingly obey these tacit norms and unthinkingly pursue these tacit values. For example, as Bernard Williams (1985) pointed out, "normal" people do not consciously decide that it is immoral to kill an obnoxious colleague. The thought should not even cross their minds. And if it does, this indicates abnormality.

We become aware of moral routines when people disobey them, when conflicts between routines emerge and a moral dilemma arises, or when they are no longer able to provide satisfactory responses to new problems. To put it strongly: Whereas morality is characterized by unproblematic acceptance, ethics is marked by explicitness and controversy. Ethics is "hot" morality; morality is "cold" ethics. We perform ethics when we put up moral routines for discussion. For example: in discussions about emerging technologies, values like health, safety, sustainability and economic growth are usually "cold"; the use of embryonic stem cells or the possibility of human enhancement are "hot."

Emerging technologies, and the accompanying promises and concerns, can rob moral routines of their self-evident invisibility and turn them into topics for discussion, deliberation, modification, reassertion. This is also an effect of promoters of a new technology who stress its novelty to attract the attention of parties who are needed for their financial, political or moral support.

Nanotechnology is not just about new phenomena at the nano-scale and their manipulation, it is about new possibilities for diagnosis and drug delivery, about a third industrial revolution, about human enhancement, up to a heaven on earth where the blind will see and the deaf will hear (as in Under-Secretary Bond's quote).

Working with a novelty necessarily means venturing into the unknown. The extent of the unknown may be large, as in the case of genetic manipulation, or small as when an improved ingredient to toothpaste is advertised as "New! Better!" A principle problem is that it is never known what the extent of the unknown is. We may think that the new ingredient of toothpaste is harmless, but it might turn out to create a new allergy (as has happened occasionally; tests can only test for known allergies). Thus, NEST-ethics will have to address two issues at the same time: there is ignorance about what the new technology might become and do, and moral routines cannot be relied upon unquestioningly. The newly emerging technology robs morals of their self-evident invisibility, and transforms them into ethics.

The friction between established moral routines and new technology is a well-known issue, with the advent of the contraceptive pill as a canonical example (and one where behaviors changed and morality adapted). And instead of friction, the new possibilities may open up spaces for reflection. A small but interesting example from micro-technology to be further enabled by nano-miniaturization is the Verichip™: a passive RFID chip with a person's identification that is implanted under the skin, and is used (on a voluntary basis) by regular visitors to nightclubs in Barcelona and Rotterdam. They don't need to carry an identity card anymore, and if the chip includes a money deposit, they can also pay with it, and don't need to carry a wallet. Just come as themselves—with their identity enhanced by the implanted chip. Many more options are possible for nightclub visitors, and the company also pushes other uses. The ethical question here is what sort of identity we want to construct (with the active support of the company) for ourselves. A new good life is being articulated stimulated by technical ability to construct a variety of such lives.

Our use of morality as "cold" and ethics as "hot" helps us to highlight an important phenomenon. And it is this phenomenon of opening up of existing moral routines and moral orders which is important, not our specific use of the terms—even if this helps us to make our point, and write this article. Terminology is difficult anyway, because actors use the label "ethics," and particularly "ethical" to refer to what is good to do, and should be done (or refrained from doing), rather than the reflexive discussion about what might be good to do that we would highlight.

We acknowledge that our conception of "ethics" takes us some distance from how actors use this label. But this is necessary for understanding what happens. In NEST-debates "ethic" is often positioned as a brake on technology (like technology assessment used to be labelled as technology "harassment"). But positions promoting technology are every inch as ethical as positions harassing or limiting technology in the name of some higher value. This implies that Under-Secretary Bond's simple contrast of ethical and unethical cannot be kept up. Instead, "ethical" is the articulation (including contestation) of what used to be morally self-evident.

Our approach to ethics is also broader compared with how ethical arguments in NEST-discussions are contrasted with economic, environmental, social, political, medical, or metaphysical arguments. The use of the acronym ELSA for Ethical, Legal and Social Aspects, introducing three different aspects, is a case in point. We argue that there is no principle difference. Presumably "non-ethical" arguments in the end refer to stakeholders' interests/rights and/or conceptions of the good life—thus, ethics. For example, economic arguments in favor of, or opposed to, an emerging technology usually follow a clearly utilitarian logic, with a focus on maximizing collective happiness. Or metaphysical considerations on human-machine interactions refer to ethical conceptions of what constitutes a good life for humans.

For health and environmental risk issues, this point is particularly important, because these are often treated as technical questions. As has been shown in detail for recombinant DNA, genetic modification, and telecommunication standards (Von Schomberg, 1997; Rip, 2001; Schmidt and Werle, 1998), the focus on technical questions is only possible when some closure of the open-ended ethical (or normative, or political, or foundational) debate has occurred, and further discussion can be delegated to technical-analytical work. Conversely, the technical discussion can be opened up again to ethical discussion when the assumptions protecting the technical approach are ·questioned.

The evolution of the debate on health and environmental risks of nano-particles can be understood in these terms. In the late 1990s, there were some early warnings, based on very limited evidence and on the analogy with risks of asbestos. When precautionary approaches were advocated, and particularly when (in 2003) the ETC group proposed a moratorium on nano-particle production, the debate became acrimonious, and the right of ETC and other critics to raise their voice was contested. Explicit and implicit normative questions about the sort of life we should lead: avoiding risks, or experimenting and learning, or even embracing risks, were at issue. The closure of this debate can be located in time, linked to the wide acceptance of reinsurance company Swiss-Re's report *Nanotechnology: Small Matter, Many Unknowns* (SwissRe, 2004). Risks of nano-particles became a legitimate question: further research was pushed (and funded), and government agencies started to investigate ways to regulate such risks. Thus, it was uncertainty about the extent of such risks that was at issue, rather than ignorance about the nature of potential hazards. By now, some actors, concerned nano-scientists as well as NGOs like Greenpeace UK, realize that this focus on handling the risks technically and in terms of new regulation is providing wider considerations about the desirability of developing and using nano-particles.

We see a pattern here. In addition, the recourse to the technical is itself a normative position, and thus a meta-ethical issue.

NEST ETHICS: META-ETHICAL ISSUES

NEST-ethics starts with the opening up an existing order by a scientific or technological novelty that undermines the self-evidence of existing moral routines, in combination with the additional challenge of our ignorance about the nature and effects of this novelty. We can then identify and characterize patterns of moral argumentation as they occur (and with examples from nanotechnology). Interestingly, part of the argumentation is at a meta-level: about our background understanding of the issues and how to approach them, rather than about substantial questions about good action and the good life. This is linked to the "new and emerging" aspect of NEST, where it is too early to reach conclusions about concrete ethical issues, but the prospect of having to do so induces discussion of how to go about it—which raises meta-ethical questions.

The first meta-ethical issue derives from the prima facie presupposition of NEST-ethics that one can influence the development of new technology, and so has to discuss desirability as well as feasibility. This leads into a long-standing discussion about technological determinism, and its more recent counterpoint, social determination, or at least social construction, of technological development. In the technological determinist view, emerging technologies will materialize anyhow, independent of what people think, deliberate or decide. The problem of how to act under conditions of ignorance is thus "solved" by denying human agency. Technological determinism might be justified by appealing to a transcendent technological reason, unfolding/materializing itself like a Hegelian idea. Actually, transcendent reason tends to be replaced by immanent strategic games, an equally unyielding, superhuman international competition: if we don't do it, our competitors will, so the new technology will happen anyway. Human agency is not completely denied, but delegated to the strategic games that actors continue to play, and depend on. The self-fulfilling prophecy of Moore's Law for semiconductors is a clear example. Conversely, those who are not part of the strategic games experience another lack of agency, that of outsiders. In recent focus groups and other public engagement exercises on nanotechnology, particularly in Britain, members of the public voiced their experience of not having any agency, and were then joined, to the surprise of both parties, by nano-scientists being involved but unable to make a difference either.

More central actors, like firms developing new technological options, and government agencies enabling new technology development and constraining it through regulation, might be seen as carriers of agency. Even then, there is the fact of non-malleability of technological developments, not because of inherent technological determinism but because directions and path dependencies emerge at the collective level, in a sense behind the back of the actors. The emergence of paradigms and dominant designs are examples. As one of us (AR) has argued, such non-malleability is itself a societal construction, but once in place, it cannot be easily undermined. Human agency, so dear to classical ethics, has to be replaced by distributed and collective agency, and a time dimension has to be introduced. Human agency can make some difference at an early stage (even if the issues and directions are still unclear), but much less so at a later stage, when alignments have solidified.

Actual moral argumentation patterns are divided, depending on the situation and audience. When addressing external audiences, promoters of new technology use the deterministic metaphor of a train that cannot be stopped so as to enrol funders and publics. These are then painted as fatalistic, and experience themselves that way. Internally, however, the determinism is leavened by the possibility to do better. Illustrative is Vicki Colvin's testimony before U. S. Congress, April 2003, on the "wow-yuck pattern" of public appreciation of new and emerging science and technology. She presents this pattern as a recurrent phenomenon, but then adds that we can counteract it if "we" understand what "we" (the promoters of NEST) did wrong. So nanotechnology actors need not repeat the mistakes made by biotechnology actors. Thus, determinism is repositioned as a contingent result of actors' behaviors and interactions leading to unintended outcomes at the collective level. Understanding of such processes then enables agency, in the sense of making a bit of difference rather than forcing one's way.

An interesting further aspect is the subterraneous link between how the promoters (or enactors, or insiders) position new technology, and how outsiders, publics, critics do so. Enactors position the technology as promising as such, independently of the efforts that actors must make. They let the promising technology speak for them, and so give it agency. Critics and publics see technology as exogenous, entering society from somewhere outside. For the critics, Franklin (2006) notes:

> "This view [of Fukuyama and Habermas] of genetic manipulation as *a force unto itself*, hostile to social order and integration. . . . Here 'biotechnology' is attributed a sinister agency." (p. 87).

For the public, a similar response was visible in the reports of a focus group, and their discussion of new technologies, including nanotechnology, and their perceived ability to transform society and nature. One participant says: "It'll get out of the cage I'm sure"—so it is a wild beast that has to be contained (Kearnes et al. 2006, p. 53). Thus, "outsiders" also picture new technology as an independent force. In other words, there is an unholy alliance with the insiders, and this perpetuates the myth of exogenous technology.

There are other patterns of second-order moral argumentation. One such pattern derives from the dual way that past experiences can be drawn upon. The past is mobilized to give credibility to arguments favoring promoting NEST but also to arguments pleading for prudence and precaution.

There are general arguments in favor: the new technology will bring us all kinds of good, because technologies have done so in the past; mankind has progressed because our forbears did not shrink away from their duties. Prometheus is invoked here (and sometimes there are second thoughts about such progress being a Faustian bargain—by now, the two tropes are often linked). It does not matter that first only technological "haves" profit, because eventually the benefits will trickle down to the lower strata. Even the poorest person is nowadays materially better off than kings were in the Middle Ages.

And then there are arguments cautioning against the emerging technology: technologies always have unintended, and quite often unwanted, side-effects; there are always bad people misusing technology; new technology makes the rich richer and the poor often more powerless; scientists and technologists are always promising more than they can make come true; and so on.

The trope that humans (some humans) end up misusing new technologies for destructive purposes can be called the inverse King Midas trope: whereas the mythical Greek king turned everything he touched into gold, modern (Western) civilization turns everything into a means of destruction (and both Midas and civilization got into trouble). This bleak view of mankind can lead to the conclusion that we should not go for more and more "technological toys." The ethics are more complex, however, as is clear in the debate about guns (in the USA): do guns kill people (so no more guns), or do people kill people (so people must do better)?

We have used a simple dichotomy between promotion and caution here, but there is more at play than contentions between proponents and opponents. There is a sequence of actions and inter-actions, which creates a specific pattern of moral argumentation. (Swierstra et al., 2002a) First, there is recognition and announcement of a novel technological option and its promise. In response, those pleading for prudence and precaution stress the novelty of the new technology as well, but now to communicate the message that there is not just uncertainty, but ignorance about effects of the new technology. The promoters now face a quandary. They had started the (NEST-ethical) discussion by stressing the novelty of the emerging technology, so as to attract attention and enrol allies. This move then creates opposition that cannot simply be negated. One strategy that is used often is to play down the novelty of the NEST, presenting it as nothing unusual. What was first introduced as a "revolution" is now toned down to business as usual." In the case of nanotechnology, the slogan then is: we're just making things smaller and faster. Or in the discussion about health and environmental risks of nanoparticles: we've had nano-sized particles around all the time, in soot from fires and in exhausts of diesel engines.

The message thus shifts: the new technology is in fact not new at all—the past contains all kinds of precedents for the emerging technology. Haven't we been genetically modifying animals since the first breeding experiments? Are twins not living proofs that clones are willed by God and/ or in accordance with natural order? Is education not a basic form of human enhancement already? Is writing itself not a technology that produced texts, an external medium to which we delegate part of our cognitive power and autonomy, no different from when we will interface our brains with computers? If we see these earlier technologies as being in accordance with our present moral intuitions, we should now be consistent and see the new technologies as similarly acceptable. This is not only an argument from *precedent*: the new technology is nothing new; but also an injunction to stay with the moral intuitions that have evolved in our interaction with earlier technological developments.

This last point creates an opening for technology critics to turn the argument from precedent around, creating an argument from *consequent*. Instead of legitimizing future developments in terms of criteria of the past and present, they de-legitimize the past and present by applying criteria derived from a desirable future. For example, if possible cloning of farm animals is seen as an unacceptable form of commodification of living organisms, the concern about commodification should be used as well to reconsider our current acceptance of the bio industry. And if we are really worried about toxicity of nano tubes, should we not be consistent and worry about all the fine dust currently produced by exhaust fumes as well?

There can be further rebuttals; the pattern continues. The steps in this pattern have become expected in our late-modern risk society, as it were as moves in a game. In that way, the pattern will create the positions of proponent and opponent: an inquiry into possible side-effects will be treated as an indication of opposition to the new technology, and thus call up further arguments legitimating the original inquiry—turning the innocent inquirer into an actual opponent.

The third main pattern of meta-ethical argumentation is linked to a basic characteristic of NEST-ethics, that is to say, the possibility that emerging technologies may change our morals and ethical considerations. This gives rise to two mirroring arguments. Technologically induced moral change can be projected as almost inevitable—the habituation argument—or depicted as a threat—moral corruption. The pattern of argumentation is now about the general relation between morals and technology. Such arguments do not aim to win the match by gaining the most points, but by revaluing the whole match. As a strategy, such a revaluation occurs late in the game, when first-round arguments seem unable to win the day.

The first argument is the *habituation* argument. Its basic tenet is that although at present the new technology is in conflict with established morals, there will be reconsideration of the morals when people become used to the new technology and its possibilities and limitations. In time morality will adapt. Precedents are quoted, ranging from overcoming fright (as for the first trains, which might even frighten the cows in the meadow so that their milk would turn sour) to more explicit changes in morals. People called contraceptives immoral because these would severe sex from procreation (and lead to wanton sex); many are now quite happy to accept that it did so indeed. Louise Brown, the first child created by IVF, was greeted as a miracle or a monster, depending on your view. Nowadays the excitement seems distant, and IVF is accepted while recognizing that it is invasive and a psychological burden. Similarly, so the argument runs, people might now feel uncomfortable about interfacing the body and the brain with silicon-based implants, to enhance the human; in another ten years they will no longer understand what the fuzz was all about.

The habituation argument can become part of an action plan, for example, of promoters of a new technology who sit out the flak until people have become used to the new technology. This is sometimes part of explicit policies. A balanced example is the Dutch Embryo Act, which prohibits the creation of embryos for scientific research in section 24, but in section 33.2 explicitly states that this prohibition has to be reassessed after five years to see whether prevailing moral insights might have evolved by that time.

The second argument is the argument of moral corruption. It comes in two forms: the *slippery slope* argument stresses the temporal dimension of this corruption; the *colonisation* argument the spatial dimension. The argument can be deployed in its own right, taking as its starting point that humans have to be protected against their own bent towards the immoral. As a strategy, it comes into play when (parts of) public opinion seems to favor the emerging technology and no convincing moral arguments against the emerging technology itself have turned up. In such a situation opponents can argue that the new technology, although seemingly innocuous or even beneficial now, will inevitably invoke further technological steps that will later result in applications that are blatantly immoral. The only way to stop this from happening is to prohibit the emerging technology from the start. For example: if implanting a chip in the brains of paralyzed patients will enable them to communicate with the world, who in her right mind would want to deny that this is a good thing? But that same technology, once developed, will be marketed for other less deserving consumers and for less legitimate, manipulative or hedonistic, purposes. The implanted chips, for example, can and will then also be used for manipulative mind-control or like a new kind of drugs.

In its spatial form, the moral corruption argument leads to the same conclusion: better stop now before the new technology can spread and be taken up for the wrong goals. The new technology might indeed address legitimate needs of a minority, but it is impossible to stop others making less legitimate use of the technology once this is developed. The technology will spread out. Nuclear proliferation and the attempts to contain it is a case in point. Nano technology will develop ultra small bio-sensors that will permanently monitor our body processes. This can be important in hospitals. No longer confined to the laboratory and the hospital, however, these devices will result in the complete medicalisation of our everyday lives.

The two types of moral corruption argument lead to proposals for moratoriums and other ways of self- and other-containment. The call for a voluntary moratorium on recombinant DNA research in 1974 and 1975, by molecular biologists themselves, is a well-known example (Krimsky, 1982). Bans on cloning, because of the risk that boys from Brazil will be cloned, are a current example (there can also be deontological arguments for such a ban, see next section). Such proposals quickly turn into debates about practicalities, and the question of feasibility of containment. Promoters of the new technology will use infeasibility of global containment as an argument to allow them to continue—because somebody elsewhere will certainly do so, and we (perhaps with higher moral standards) had better be in as well. Such debates overlap with what we call "consquentialist contestation" and will discuss in the next section.

NEST-ETHICS: PATTERNS OF ETHICAL ARGUMENTATION

In practice, NEST-ethics starts with a consequentialist pattern of ethical argumentation: the new and emerging technology is deemed desirable, or not, because its consequences are desirable, or not. Since such consequences are still speculative, they have the form of promises, or warnings and concerns when put forward in an action-oriented context. NEST-ethical discussion typically starts with the promises made by scientists and technologists and those who identify with their message about the new options. (See the Philip Bond quote in the introduction.) These promises reflect the passion and confidence of those who make them, but they are also a way to attract attention, and thus financial, political and moral support for the new ventures.

While promises can enroll allies, they can also raise doubts and critical questions, already from actors pushing other promises which compete for the same scarce resources. Such discussions occur, for example, around the promise of fuel cells and the hydrogen economy (Avadikyan et al., 2003). The feasibility and desirability of a hydrogen economy is questioned by those who push other energy futures and other technologies to carry them. For nanotechnology, such contestation remains subdued because nanotechnology is relevant for all sorts of applications, and thus unspecific in terms of what it is competing with.

Critical reactions can also focus on the new technology itself, independent of alternative technological options. In the consequentialist pattern of moral argumentation, critics then have to identify undesirable consequences to get a hearing. A struggle ensues about the nature and plausibility of the various consequences. Such consequentialist contestation is further fuelled by a cultural expectation, in late-modern societies, that there will be proponents as well as opponents of a new technology. (Rip and Talma, 1998), somewhat independent of its specific features. In fact, by now there are NGOs like Greenpeace with a professional opponent role. When new technologies emerge, they will try and identify the negative consequences. That is their business model. Or sometimes, as in the case of Greenpeace UK for nanotechnology, come up with a balanced appraisal—which is then not believed by proponents because they project the stereotypical opponent role on Greenpeace.*

Other patterns of argumentation can be characterized as deontological, as focusing on justice, or as drawing on "good life" ethics, as we will show below. In practice, they are most often additional to the consequentialist pattern.

Consequentialist arguments

Consequentialist contestation follows a distinctive pattern, which is fuelled by two general perspectives on technology, which are linked to the meta-ethical discussion of agency. There is the optimistic view that technological progress is basically beneficial, and a pessimistic view of technology as inherently risky and dangerous. The optimistic belief in technological progress short-circuits the problem of uncertainty and ignorance by arguing that there may be small mishaps, but all in all, and in the long run, the new technology will benefit us. As we discussed already, this optimism gets extra "muscle" by combining it with determinism: you should not want to stop this technological advance, but you cannot, either. Resistance is bad as well as well as futile.

A priori pessimism about the effects of new technology gets rid of the uncertainty and ignorance just as well; you may not know exactly what will go wrong, but go wrong it will. The critical stance that goes with pessimism might lead to attempts at changing the course of events, that is, some voluntarism. But just as often we see pessimism and determinism combining into fatalism. Resistance against fate is then undertaken in a spirit of duty, not of hope. There is more to say about the issues of uncertainty and agency, as these are linked to basic views of nature and society. For example, as Mary Douglas and others have argued, a view of nature as resilient goes together with a conviction that we can, and thus should, go for technological progress, there is no need to bother about side-effects until they appear. The alternative view of nature as vulnerable is linked to a view of technology as a "monster" that might have to be banned, or at least contained from the beginning. (Douglas and Wildavsky, 1982)

Consequentialist contestation is inevitable in late-modern societies. The pattern of moral argumentation starts with promises which have the form: if we invest in this new and emerging science and technology, this will increase our knowledge as well as our scope in manipulating the natural world, which will eventually result in increasing general happiness when application of such knowledge and manipulation leads to positive effects x, y and z. Such a claim can and will be challenged along three axes.

The first axis concerns the basis of the promises made, that is, the questioning of their plausibility. Because promises are based on assumptions about, or projections on, the future, one can demand that we get our "facts" straight before taking these promises seriously. Some optimists predict that nanotechnology will help to interface human brains and computers, so that we can "learn" French by simply implanting a chip. Highly improbable, the objection goes, because learn-

---

* Greenpeace UK (2003) introduced a more "moderate" NGO stand in the nano risk debate: they argued that one should not just focus on risks, but also on nanotech opportunities. They did propose a moratorium on the release of engineered NPs in the environment, but simultaneously recognized this moratorium to be "impractical and probably damaging." They called on industry to be responsible and to significantly increase funding for environmental risk research. However, this complex message was not received by the proponents of nano technology. Rob Atkinson, director of the Progressive Policy Institute, and Mark Modzelewski, executive director of the Nano Business Alliance accused the Greenpeace report of being misleading propaganda and a form of "industrial terrorism." This rejection rested more on Greenpeace's reputation than on the actual contents of its report. (The quotes are from Tim Harper's *TNT Weekly* at the time.)

ing a language is extremely complex—as is shown by the sometimes hilarious results of translation software programs. Clearly, there are attempts to check how speculative such predictions are, even if these will be inconclusive because they are themselves part of the consequentialist contestation. This is visible in the prolonged debate between (the late) Richard Smalley and Eric Drexler about the principle possibility of molecular assembly. Smalley's objections ("fat and sticky fingers") were not directly countered by Drexler, who tended to refer to the occurrence of molecular assembly in living cells to make his position plausible. The pressure to assess the so-called realism of the Drexlerian scenario is also visible in the stipulation in the U. S. twenty-first century Nanotechnology Act of 2003 to do just that.

The second axis along which promises can be contested, is not the plausibility of the benefits, but the ratio of benefits and costs. Do the latter not outweigh the former? Skeptics will stress the danger of not acknowledging our *cognitive limits*. Transgressing these limits, as the *topos* of the sorcerer's apprentice teaches us, means sowing seeds for future disaster. This line of argument can lead to the demand to first "get the facts straight," for example, by first assessing health, environment and safety risks of nano-particles before going into wholesale production and use. Proponents of nanotechnology first contested this demand ("there is no risk"), then grudgingly took it up while production continued, and now recognize it as a real concern. They fear a backlash if the public finds out in a later stage that the risks were "underestimated." Simply acknowledging risks, however, might not be enough to prevent such a backlash. Social studies have shown that experts generally perceive risks in quantitative terms, whereas the general public perceives risk in more qualitative or narrative terms (cf. also, Wiedemann et al., 2003). As a result, there is a real chance of miscommunication between these two parties.

The third axis of consequentialist contestation consists of questioning whether the benefits promised are really benefits. This will shift the discussion to another level, because this no longer is a factual question but an explicitly normative one. Promises of benefits imply views and criteria about what is beneficial, even if these remain implied. Such views and criteria can be unproblematic, when all participants agree that health, absence of hunger, economic growth, and cheaper products are desirable and that hunger, sickness, and poverty are not. In the case of cochlear implants, however, the promise of allowing the deaf to hear again was contested by the deaf community, with its own culture, and now officially recognized language. The utilitarian criterion of "maximizing happiness" has been shown by philosophers to be inadequate. In the case of cochlear implants for the deaf, the whole notion of happiness, in the sense of what is considered to be beneficial, can shift according to the culture from within one is viewing happiness.

Underlying most consequentialist arguments is a utilitarian ethics, with its moral drive to reduce pain and to maximize happiness. In modern times, avoiding or reducing pain (*non-maleficence, primum non nocere*) is taken to have priority over maximizing happiness (*beneficence*). The underlying idea is that suffering is not only more pressing than sub-optimal happiness, but also a somehow more objective or uncontested criterion than happiness. (Popper, 1946) Ideas about what makes a person happy vary, whereas people tend to agree about what counts as suffering. Few would deny that hunger and sickness are harms that need mending. Thus, one can understand why those consequences of an emerging technology are foregrounded which reduce hunger and disease. The facile way in which agricultural biotechnology was (and continues to be) linked with reducing hunger in developing countries is visible again for nanotechnology.

Minimizing suffering and reducing harm are phrased as positive goals. In practice they often also function as it were negatively: as long as a new technology does not harm anyone, it does not need ethical discussion. This laissez-faire attitude can itself be formulated as a political and ethical principle (cf. Mill, 1989). It does raise questions about the burden of evidence, which are taken up as patterns of moral argumentation. Often, it requires critics to argue that the new technology might cause harm to some stakeholders and thus cannot be pursued freely. Those favoring an emerging technology do not have (or do not see) a duty to check for possible harms (except when regulation requires them to do so, as with the registration of new medical drugs). Furthermore, a new technological option tends to be developed with certain concrete stakeholders in mind, so at least some of the benefits will be clearly defined. In contrast, possible harms are often speculative, lie farther away in the future and/or space, and concern as yet anonymous, collective stakeholders. The asymmetry of benefits and harms is almost unavoidable, structures not just argumentation but also action, and has given rise to increasing recognition of the need for early warning (Harremoës, 2001; Swierstra et al, 2002b).

There are three recurring rhetorical tropes in this consequentialist cluster. The first is about upstream solutions for downstream problems. This trope is very visible in promises about genetic therapy, and in human enhancement debates generally. No longer, so the argument goes, will we have to muddle through by fighting symptoms; genetic therapy and enhancement technologies will finally enable us to go to the (biological, molecular) root of the (medical and socio-economic) problems and solve them there. In nanotechnology, upstream solutions are pushed when nanotechnology enables enhancement technologies, but also in relation to drug delivery and to problems of developing countries.

Secondly, a sceptic might first allow that the emerging technology will indeed plausibly deliver some of its promises, but then proceed to deny that this makes the emerging technology necessary. Here the first trope about "upstream solutions" gives way to a second trope about the (un)desirability of "technological fixes" and "social fixes."** There may well exist alternatives that address the problems, say of environment or poverty, as they appear here and now. These alternatives are argued for by labelling the proposed upstream solution a technological fix, with its pejorative connotation of pushing through a technological approach with all sorts of harmful side effects. The assumption here is that social problems deserve social solutions, not technical ones that only address the symptoms anyway. Proponents can open up this trope by arguing that the technological solution is much more feasible and realistic than a cumbersome social one. In that case it is narrow minded and irresponsible—in the light of the pressing problems—to cling to a dogma of social problems deserving social solutions.

A third trope is precaution, that is, precautionary approaches in general and the specific precautionary principle that is now part of EU regulations (Jones et al., 2006). In terms of Mary Douglas's cultural theory, this trope of precaution belongs with the hierarchists and bureaucrats, not with collectivists/sectists whose precautionary concern is to ban the monster of new technology. Thus, in the formulation of the European Union, there must be "reasonable grounds for concern for the possibility of adverse effects" before there can be measures "to ensure the chosen high level of protection in the [European] Community based on a broad cost-benefit analysis whereby priority will be given to human health and the environment" (Rip, 2006a).

Presently for nanotechnology, the focus is on risks of nano-particles. There, precautionary approaches have been narrowed to health and environmental risks, and wider concerns about the need for nano-particle based products are backgrounded. We discussed this example already at the end of section 2. What is interesting here is how some actors, for example, Greenpeace UK, are concerned about the narrowing of the agenda, question the benefits of nano-particles (cf., the fourth-hurdle argument as discussed for biotechnology), and start offering good-life ethical arguments. In other words, consequentialist contestation has led to partial resolutions, that is, of the issue (here, health and environmental risks of nano-particles) that was foregrounded in the debate, but there are residual concerns which cannot be addressed in the consequentialist pattern of argu-

---

** These are analyst's terms, not necessarily actor's terms. They help us to identify the tropes in the debate.

mentation. This creates openings for deontological and good-life ethical arguments, for the next step in the evolution of the debate. Sometimes, as with stem cells, deontological arguments are present at an early stage already.

Deontological arguments

Deontological (i.e. right- and duty based) arguments are expected to be up front when the new technology touches upon deeply felt convictions and existential interests. They can also function as a check on consequentialism, because deontological principles, in our societies, appear to have a right of way before consequences. Even if the principles can be contested by referring to benefits that we now forego, or risks we have to suffer—an example of the latter would be that individual choice and autonomy, as a principle in medical ethics, has to be modified, for example because of the possibility of community genetics.

Technologies may appear to produce desirable over-all consequences, but they can still conflict with deeply seated moral convictions about duties and rights, often but not necessarily protecting the interests of individuals or minorities that are threatened by the majority interests favored in consequentialism (because of embedded utilitarianism). A good example is medical experimentation on humans. Here deontological principles protect *individual* patients—or embryos—from being subjected to cruel experiments that in fact could benefit *public* health.

In NEST-debates deontological arguments are often introduced to counter optimistic promises. But deontological principles are not only called upon to frustrate emerging technologies. Common moral principles *supporting* new technologies are: a duty to further human progress, a duty to help diminish suffering, a duty to acquire knowledge, and last but certainly not least: the right to choose freely whether or not to use a particular technology (as long as this does not harm others, of course).

There are three main ways along which deontological principles can be contested. First by invoking another principle with a higher priority, for example, by stressing that the principle of nonmaleficence (*primum non nocere*) outweighs the principle of beneficence. An example is the claim: "Although miniaturized surveillance techniques might increase security, this does not make the accompanying infringement of privacy rights acceptable." A second way is by arguing that the principle does not apply in the case of this specific technology. "Of course it would be wrong to kill human beings, but you cannot seriously consider a human embryo of less than two weeks old, a human being."

The third way to counter a concrete deontological argument is by interpreting and applying the principles differently. The same principle is mobilized to prohibit a new technology and to endorse it. For example, we all endorse the principle that people should have a right to choose freely whether or not to use a technology. But thinking through what will happen in a competitive world full of inequalities, the same principle entails that human enhancement should be forbidden. When some individuals exercise their right and start to technically enhance their offspring, this in practice forces other parents to follow suit. Allowing enhancement techniques to be available therefore effectively infringes upon the right of the other parents to choose freely *not* to use these techniques.

This is recurrent argument, and there are further moves, like emphasizing that the other parents are still free to choose, only the effects of their choice may hinder their offspring to compete with the kids who were enhanced. We note that the structure of this pattern of argumentation is the same as the argument about new technology as an unstoppable train, because "if we don't do it, our competitors will," which we offered as part of a meta-ethical issue in the beginning of section

3. The meta-ethical issue is visible in the reference to our present competitive world, and to forces felt in practice, which are both treated as given, and to be accepted.

Justice arguments

Distributive justice, in the immediate sense of how the benefits and the risks will be distributed, is an important issue, even if it gets only passing reference in NEST-discussions. The low prominence in these discussions has to do with the mostly speculative nature of the impacts. One can still project, and hope that inequities will be mitigated, somehow. For technologies closer to implementation than nanotechnology, for example biotechnology, distributive justice is higher on the agenda. Still, there are common patterns in the moral argumentation.

There are contrasting views of what constitutes distributive justice, depending on the distributive criterion that is used: equality, need, merit, effort, or a combination of these [as with Rawls (1971) on principles of justice]. For NEST, the paradigmatic issue in the discussions is a techno-divide, the gap between rich and poor countries, and between poor and rich strata of the population within a country. And the basic tenet, accepted by most of the discussants (even if the reasons are not clear), seems to be Rawls's "maximum" rule; the new technology will only advance justice when it will benefit those who are now worst off: the poor (countries).

Arguments supporting developing the new technology in rich countries, and with affluent consumers as the first target group, must then include a trickle-down effect. The new technology will create more goods/value, and therefore everyone can have a larger piece—in *absolute* terms—of the expanded cake. Although the new technology might at first benefit the rich countries who had the resources to develop it, in the end the poor (countries) might be the ones profiting most:

> what at first appears to be very "high-tech" and costly and therefore perhaps irrelevant for developing countries, in the end might come to be of most value for those same developing countries. Thus NT, were it to develop in the way it ought, might ultimately be of most value for the poor and sick in the developing world (Mnyusiwalla et al, 2003).

There is a further move in this pattern of argumentation. Even if the new technology does make the majority better off in absolute terms, it might still widen the (nano) divide between those reaping most benefits and those left to pick up the crumbs. The relative position of the latter group will worsen as a result of the emerging technology:

> The transition from a pre-nano to a post-nano world could be very traumatic and could exacerbate the problem of haves versus have-nots. Have-nots do not easily obtain access to new technologies: the difference between the lives of the nano-rich and the nano-poor will likely be striking (Smith, 2001, 204).

This latter argument need not, however, lead to denunciating the new technology. The conclusion, most often, is a plea for developing the technology in directions that specifically address the needs of the poor developing countries. (Meridian Institute, 2006) Thus, for some proponents, the issue of distributive justice is more than putting trust in the trickledown effect of new technology.

Arguments from "good life" ethics

What sort of good life can be achieved thanks to new and emerging science and technology? The promises of enactors about new options tend to short-circuit this question by projecting wonderful new possibilities without reflecting on how "good" this kind of "life" actually might be. In contrast, commentators and critical groups will sometimes outline a "good life" and use this as a reference when discussing and assessing new science and technology. This is particularly clear in the environmental movement (up to "deep ecology"). The ETC group, which now focuses on critical evaluation of ongoing developments in nanotechnology, is a good example because it had started with a view on the good life, emphasizing ecology (E), better use of technology (T) and reluctance towards concentration (C), and this is still visible in its arguments about nanotechnology in society.

One framing of a "good life" occurs through culturally shaped identities and aspirations: who are we and who do we want to be? Indicative are references to archetypical figures and myths. Those promoting new technology typically draw upon a Promethean identity, mixed with some frontiers rhetoric, "Boldly go where no man went before." Conversely, sceptics and adversaries warn against Faustian bargains, and against "hubris"; proud Icarus soaring high in the skies like the Gods—and then plummeting to his death.

Another framing of what is a "good life" is visible in discourses of limits. In the biotechnology debate, a recurrent *motif* was that humans should not play God. The concrete reference was to the possibility of recreating nature. God's Creation would then be shorthand, somewhat independent of theistic religious connotations, for respect towards what has evolved, instead of it being objectified, instrumentalized, commodified, subjected and manipulated.

Further limits are derived from what is deemed to be natural. There exists a (hidden, but to be explicated) moral order in nature that should be followed. If not we will create monsters, as Victor Frankenstein did. Mary Shelley's novel is more complex, as it includes the experience and feelings of the monster, and suggests that monsters are the result of lack of care and love, rather than because of the technology that went into their creation. In the debate about genetically modified food, especially in the United Kingdom, the term "Frankenstein food" or even "Frankenfood" has become shorthand for what is inadmissible. "This is not what we want to be on the shelves of our supermarkets."

The focus on limits is often conservative: do not transgress what is already there. Another way of viewing what is out there, and what might put limits on the aims of control that are associated with technology, is the idea that human beings need "otherness" and cannot flourish in a completely controlled and manipulated, human built, "brave new world" that plies itself obediently to our every desire. This *motif* is visible when people start to extol unspoilt nature, human imperfection, suffering, death and fate. We want the world to put up resistance to our touch, to show robustness, to surprise and provoke us. We do not want the world to become a mirror in which we only see our own image reflected.

The debate about the good life follows the lines laid out by the short-cuts, rather than discussing the good life as such. Proponents of a new technology will offer more technology-friendly interpretations of what God wants us to do, or even argue that God means us to play Him (and others of course will question His existence). The moral order implied in nature is queried by arguing that it is our ability to create technology what truly constitutes our human nature. Others flatly deny that there is any moral order hidden in nature (apart from the well-known naturalistic fallacy). And while mankind has a bad track-record as to wielding our technological powers wisely, we are learning and making progress. And finally, even when we will one day live in a complete 'technotope', this will do nothing to diminish our experience of "otherness" because technology is every inch as capricious, surprising and different as nature is (Rip, forthcoming).

Two final comments. Good life arguments can lead to clashes between incommensurable worldviews. Instead, and to avoid such clashes, their persuasive force is drained by treating the arguments as private beliefs. For proponents, it suffices that the new technology will help realize the wants and preferences of at least some interested parties. They are not much interested in shared conceptions of the good life, and position conceptions of the good life as private. They argue that although the belief in limits might be a perfectly respectable private opinion, it does not constitute a legitimate argument in a public discussion. This (liberal) argument gives the "good life" part of NEST-ethics a peculiarly asymmetric and somewhat slippery character: arguments are not met by counterarguments, but relocated from the public to the private domain (Swierstra, 2002).

Secondly, it is not always possible to draw a sharp line between good life ethics and deontology. The former envisages substantive, thick, conceptions of the good, whereas the latter concentrates on "thin" conceptions of the good or on what is "right." What belongs to the "good" and what belongs to the "right" is always a matter of contention. To illustrate this: in a society where most people believe in God and in a moral order hidden in nature, the limits are all accepted as belonging to the domain of the right, of what is neutral. In a modern, secular, pluralist society the same limits would have to be qualified as belonging to good life ethics because they rest on non-neutral, and therefore substantive, conceptions of the good.

CONCLUSION

One might view NEST discussions as rational, consensus-seeking deliberations. The idealized paradigm for this type of deliberation is the classical Athenian *agora*: the market place where the free citizens gathered to decide, solely on the strength of arguments, the good of their polis. Instead, one should start with actor strategies, serving particular interests. Not an agora where Rousseau's *volonté generale* takes form, but an *arena* where some win and others lose. In an arena consensus is never reached, although a workable compromise is sometimes achieved. Consensus-seeking models can at best provide temporary stabilizations.

It is not a fight of all against all, however. In the public sphere in our societies, characterisized as functioning democracies, all players are forced to seek legitimacy for their standpoints. Such legitimacy can only be acquired by participating in deliberation. To win in the arena, participants have to act as if they were in an agora. In other words: even if the agora is an illusion, it is a necessary one, and it is productive (Swierstra, 1998).

Since Machiavelli, political theorists have pointed out that struggle among an irreducible plurality of perspectives can be productive. Diversity, heterogeneity, incommensurability, and antagonism—they can tear the fabric apart but they can also help to keep it vital and vigorous. Probing each other's worlds goes together with competition for primacy in a universe of discourse with others who cannot beforehand be branded as unreasonable. Such reflexive awareness rejects the naivety of dogmatic beliefs, recognizes its own fallibility, and leaves room for reasonable dissensus. Pragmatist ethicists can contribute by helping develop different tools for "conflict" and "dilemma" management to enhance mutual respect (Keulartz et al., 2004).

As an instance of NEST-ethics, nano-ethics will reproduce the general patterns to some extent, but also modify them. An important point, which remained implicit in our discussion of NEST-ethics, is the co-evolution of ethics and new technologies: while there are recurrent patterns of moral argumentation, there is also learning, shifts in repertoires, new issues coming up. The presently widespread acceptance of precautionary approaches (definitely in Europe) is an example of such a shift. What one now sees happen with nanotechnology is a further kind of precaution: promoters do not want impasses to occur as happened with green biotechnology, and go out of their way to communicate with publics and politicians (Rip, 2006b), and want to discuss ethics and societal aspects at an early stage. While the debate still often follows the lines of the patterns of moral argumentation we outlined, there are now openings for further articulation.

REFERENCES

Avadikyan. A., P. Cohendet, J.-A. Heraud (eds.) (2003). *The Economic Dynamics of Fuel Cell Technologies*. Berlin: Springer Verlag.

Caspary, W. R. (2000). *Dewey on Democracy*. Ithaca, NY: Cornell University Press.

Collingridge, David (1980). *The Social Control of Technology*. London: Frances Pinter.

DEFRA (Department for Environment, Food and Rural Affairs) (2006), *UK Voluntary Reporting Scheme for Engineered Nanoscale Materials*. London: DEFRA, September 2006. http://www.defra.gov.uk/ENVI RONMENT/nanotech/policy

Dewey, J. (1954). *The Later Works*. Athens, OH: Swallow Press Books.

Dewey, J. (1994). *The Moral Writings of John Dewey. Revised Edition*. J. Gouinlock (ed.) Amherst, New York: Prometheus Books.

Douglas, Mary, and Aaron Wildavsky (1982). *Risk and Culture. An Essay on the Selection of Technological and Environmental Dangers*. Berkeley: University of California Press.

Du Gay, Paul, Stuart Hall, Linda Janes, Hugh Mackay and Keith Negus (1997). *Doing Cultural Studies. The Story of the Sony Walkman*. Milton Keynes: Open University, in association with Sage Publications.

ETC (2003). *No Small Matter II: The Case for a Global Moratorium—Size Matters!* Occasional Paper Series 7(1).

Fisher, Elizabeth, Judith Jones and Rene von Schomberg (eds.) (2006). *Implementing the Precautionary Principle. Perspectives and Prospects*. Edward Elgar.

Fisher, E. and R.L. Mahajan (2006). Midstream modulation of nanotechnology in an academic research laboratory, *Proceedings of IMECHE2006* (ASME International Mechanical Engineering Congress and Exposition, Chicago, November 5–10, 2006).

Franklin, Sarah (2006). Better by design?, pp. 86–94, in Paul Miller and James Wilsdon (eds.), *Better Humans? The Politics of Human Enhancement and Life Extension*. London: DEMOS.

Gordijn, B. (2005). Nanoethics: From utopian dreams and apocalyptic nightmares towards a more balanced view. *Science and Engineering Ethics* (11), 4, 521–533.

Greenpeace UK (2003). Future Technologies, today's choices. Nanotechnology, artificial intelligence and robotics; a technical, political and institutional map of emerging technologies. 24 July 2003, http://www .greenpeace.org.uk/MultimediaFiles/Live/FullReport/5886.pdf

Grunwald, Amin (2005). Nanotechnology. A new field of ethical inquiry?. *Science and Engineering Ethics*, 11, 2: 187–201.

Harremoës, Poul et al. (2001). *Late Lessons from Early Warnings: The Precautionary Principle 1896–2000*. Copenhagen: European Environmental Agency.

Kearnes, Matthew, Phil Macnaghten and James Wilsdon (2006). *Governing at the Nanoscale. People, Policies and Emerging Technologies*. London: DEMOS.

Keulartz, J., M. Schermer, M. Korthals, T. Swierstra (Eds.) (2002). *Pragmatist Ethics for a Technological Culture*. Deventer: Kluwer Academic Publishers.

Keulartz, J., M. Schermer, M. Korthals & T. Swierstra (2004). Ethics in a technological culture. A programmatic proposal for a pragmatist approach. *Science, Technology and Human Values*), (29) 1, 3–29.

Krimsky, Sheldon (1982). *Genetic Alchemy. The Social History of the Recombinant DNA Controversy*. Cambridge, Mass.: MIT Press.

Lin, Patrick, and Fritz Allhoff (2006). Nanoethics and human enhancement. Originally published in *Nanotechnology Perceptions: A Review of Ultraprecision Engineering and Nanotechnology,* Volume 2, No. 1, March 27 2006.

Mill, J.S. (1989). *On Liberty and Other Writings*. Cambridge: Cambridge University Press.

Mnyusiwalla A, Daar AS, Singer PA. Mind the gap: Science and ethics in nanotechnology. *Nanotechnology* 2003; 14: 9–13.

Popper, K.R. (1966). *The Open Society and its Enemies*. London, New York: Routledge & Kegan Paul.

Rawls, John (1971). *A Theory of Justice*, Cambridge, Mass.: Harvard University Press.

Renn, Ortwin, and Mike Roco (2006). *Nanotechnology Risk Governance*. Geneva: International Risk Governance Council. IRGC White Paper No. 2.

Rip, Arie (2001). Contributions from social studies of science and constructive technology assessment, in Andrew Stirling (ed.) *On Science and Precaution in the Management of Technological Risk. Volume II.*

Case Studies. Sevilla: Institute for Prospective Technology Studies (European Commission Joint Research Centre), November 2001, pp. 94–122.

———(2006a). The tension between fiction and precaution in nanotechnology, in Fisher, Jones and Von Schomberg (eds.), *Implementing the Precautionary Principle. Perspectives and Prospects*, Edward Elgar, pp. 423–448.

———(2006b). Folk Theories of Nanotechnologists, *Science as Culture 15*(4) (December) 349–365.

Rip, Arie, and Johan W. Schot (2002). Identifying *loci* for influencing the dynamics of technological development, in Knut Sørensen and Robin Williams (eds), *Shaping Technology, Guiding Policy; Concepts, Spaces and Tools*. Cheltenham: Edward Rip, Arie and Siebe Talma, Antagonistic patterns and new technologies, in C. Disco and B.J.R. van der Meulen (eds.) *Getting New Technologies Together* (Berlin: Walter de Gruyter, 1998), pp. 285–306.

Smith, R.H. (2001). Social, ethical, and legal implications of nanotechnology, in M.C.Roco & W.S.Bainbridge, *Societal Implications of Nanoscience and Nanotechnology*.

Swierstra, Tsjalling E. (1998). *De sofocratische verleiding. Het ondemocratische karakter van een aantal moderne rationaliteitsconcepties*. Kampen: Kok Agora.

———(2002a). Moral vocabularies and public debate: The cases of cloning and new reproductive technologies, in T. E. Swierstra, J. Keulartz, J. M. Korthals & M. Schermer (Eds.), *Pragmatist Ethics for a Technological Culture*. Deventer: Kluwer Academic Publishers, 223–240.

Swierstra, T., Keulartz, J., & Korthals, M. (2002). *You only live twice. De maatschappelijke component van het genomics onderzoek*. 26–09–2002.

Swierstra, T. E., & Tonkens, E. (2002). Klakkeloze keuzevrijheid. Kanttekeningen bij de dominantie van keuzevrijheid in hedendaags beleid. *Filosofie en praktijk (23) 4*, 3–19.

Swiss Re (2004). *Nanotechnology: Small Matter, Many Unknowns*. http://www.swissre.com/INTERNET/pws filpr.nsf/vwFilebyIDKEYLu/ULUR-5YNGET/$FILE/Publ04_Nanotech_en.pdf

Swiss Re Centre for Global Dialogue (2005). *Nanotechnology. Small size–large impact?* Zürich: Swiss Reinsurance Company. Risk Dialogue Series.

Van Est, R., I. Malsch, and A.Rip (2004). *Om het kleine te waarderen. . . . Een schets van nanotechnologie, publiek debat, toepassingsgebieden en maatschappelijke aandachtspunten*. Den Haag: Rathenau Instituut, [Working documents 93].

van Lente, Harro, and Arie Rip, Expectations in technological developments: An example of prospective structures to be filled in by agency, in C. Disco and B.J.R. van der Meulen (eds.), *Getting New Technologies Together* (Berlin: Walter de Gruyter, 1998), pp. 195–220.

Von Schomberg, Rene (1997). *Argumentatie in de context van een wetenschappelijke controverse. Een analyse van de discussie over de introductie van genetisch gemodificeerde organismen in het milieu*, Delft: Eburon. PhD thesis, University of Twente, 27 March 1997.

Williams, Bernard (1985). *Ethics and the Limits of Philosophy*. Cambridge: Cambridge University Press.

16

Moralizing Technology: On the Morality of Technological Artifacts and Their Design

Peter-Paul Verbeek

INTRODUCTION: MORALITY AND MATERIALITY

Ethics appears to be at the eve of a new Copernican revolution. A few centuries ago, the Enlightenment, with Kant as its major representative, brought about a turnover hitherto unequalled by moving the source of morality from God to humans. But currently there seem to be good reasons to move the source of morality one place further. It increasingly becomes clear that we should not consider morality as a solely human affair, but also as a matter of *things*. Just like human beings, material objects appear to be able to provide answers to moral questions. The artifacts we deal with in our daily lives help to determine our actions and decisions in myriad ways. And answering the question how to act is the ethical activity *par excellence*. This "material turn" in ethics raises many questions, though. Is the conclusion that things influence human actions reason enough to actually attribute morality to materiality? Can things be considered moral agents, and if so, to what extent? And is it morally right to go even one step further and try to explicitly shape this morality of things, by consciously steering human behavior with the help of the material environment?

The ethics of engineering design is perhaps the best place to start analyzing the moral dimension of technological artifacts, since this is also the place where human beings can take responsibility for the moral aspects of their products. In its current form, though, the ethics of engineering design tends to follow a somewhat externalist approach to technology. It mainly focuses on the importance of taking individual responsibility ("whistle blowing") to prevent technological disasters, and on methods to assess and balance the risks accompanying new technologies. Favorite case studies concern technologies which have caused a lot of problems that could have been prevented by responsible actions of engineers, like the exploding space shuttle *Challenger*, or the Ford Pinto with its rupturing gas tank in crashes over 25 miles per hour. Case studies like these merely address technologies in terms of their functionality: technologies are designed to do something, and if they fail to do so properly, they were badly designed. What such case studies fail to take into account are the impacts of technologies on our moral decisions and actions, and on the quality of our lives.

When technologies are used, they always help to shape the context in which they fulfill their function. Technological artifacts help to shape human actions and perceptions, and create new practices and ways of living. Cell phones, for example, contribute explicitly to the nature of our communications and interactions. And technologies like ultrasound play active roles in our decisions regarding unborn life. Functionality is too limited a concept for engineering ethics. Mediations transcend functionality: they form a surplus to it, which occurs once the technology is functioning. When technologies fulfill their functions, they also help to shape the actions and experiences of their users. This phenomenon has been analyzed as "technological mediation": technologies mediate the experiences and practices of their users (Latour 1992; Ihde 1990; Verbeek 2005).

Up to now, the concept of mediation has mainly functioned in *descriptive* settings: in analyses of the role of technologies in their use contexts. In this article, I investigate how it can be deployed in a *normative* setting. The concept of mediation lays bare ethical questions regarding technology design that transcend the common sense idea that technologies only need to be morally evaluated in terms of the goals for which they are designed or of the quality of their functioning. Technological mediations have at least as much moral relevance as technological risks and disaster prevention, which are dominating the ethics of technology. By mediating human experiences and practices, technologies help to shape the quality of our lives and, more importantly, our moral actions and decisions. In order to address this moral dimension of technologies adequately, therefore, the ethics of technology should expand its approach to include technological mediation and its moral relevance, enabling designers to take responsibility not only for the quality of the functioning of their designs, but also of their built-in morality.

In order to cover all relevant aspects of the role of technological artifacts in their use contexts and to provide a vocabulary for analyzing these aspects, I will first elaborate the notion of "technological mediation." After this, I will investigate the implications of this mediation approach for ethics. First, I will investigate to what extent the moral relevance of mediation can be reason to attribute (a specific form of) moral agency to technology. Moral agency requires at least some form of intentionality and some degree of freedom, which artifacts form of intentionality, and to rethink the role of freedom in moral agency. I will show that this makes it possible to elaborate on a specific notion of moral agency, which does justice to the moral relevance of technological mediation.

Second, I will investigate how the concept of mediation can be made fruitful for design ethics. Integrating mediation in engineering ethics is a complex task, however. Firstly, the ambition to design technologies with the explicit aim to influence human actions raises moral questions itself. It is not self-evident, after all, that all attempts to steer human behavior are morally justified, and steering human beings with the help of technology raises associations with the totalitarian technocracy of Orwell's Big Brother. Moreover, if some forms of behavior-steering technologies can be seen as morally acceptable—and I think such technologies do exist—it is very complicated to design them, since there is no linear connection between the activities of designers and the mediating role of the artifacts they are designing. As I will make clear, this mediating role also depends on the unpredictable ways in which the technologies are used. For this reason, I will suggest some ways to cope with this unpredictability.

TECHNOLOGICAL MEDIATION

For analyzing the role of technologies in the daily lives of human beings, the concept of *technological mediation* is a helpful tool; especially in the way it was developed in "postphenomenological" philosophy of technology (cf. Verbeek 2005). Phenomenology—in my elementary definition—is the philosophical analysis of the structure of the relations between humans and their lifeworld. From such a phenomenological perspective, the influence of technology on human behavior can be analyzed systematically, in terms of the role technology plays in human-world relations. Technological mediation then concerns the role of technology in human action (conceived as the ways in which human beings are present in their world), and human experience (conceived as the ways in which their world is present to them).

Human-Technology Relations

A good starting point for understanding technological mediation is Martin Heidegger's classical analysis of the role of tools in the everyday relation between humans and their world. According to Heidegger (1927), tools should be understood as "connections" or "linkages" between humans and reality. Heidegger indicates the way in which tools are present to human beings when they are used as "readiness-to-hand." Tools that are used for doing something typically withdraw from people's attention; the attention of, for example, a person who drives a nail into a wall, is not directed at the hammer, but at the nail. People's involvement with reality takes place *through* the ready-to-hand artifact. Only when it breaks down, it asks attention for itself again. The artifact is then, in Heidegger's words, "present-at-hand" and is not able to facilitate a relationship between a user and his or her world anymore.

Even though ready-to-hand artifacts withdraw from people's attention, they do play a constitutive role in the human-world relation that arises around them. When a technological artifact is used, it facilitates people's involvement with reality, and in doing so it coshapes how humans can be present in their world and their world for them. In this sense, things-in-use can be understood as *mediators* of human-world relationships. Technological artifacts are not neutral intermediaries, but actively co-shape people's being in the world: their perceptions and actions, experience and existence.

The positions of the North American philosopher Don Ihde and the French philosopher and anthropologist Bruno Latour offer concepts for building a vocabulary to gain a closer understanding of this mediating role of technologies. In order to build this vocabulary, I discern two perspectives on mediation: one that focuses on perception and another one on praxis. Each of these perspectives approached the human-world relationship from a different side. The hermeneutic or "experience-oriented" perspective starts from the side of world, and directs itself at the ways reality can be interpreted and be present for people. The main category here is *perception*. The pragmatic or "praxis-oriented" perspective approached human-world relations from the human side. Its central question is how human beings act in their world and shape their existence. The main category here is *action*.

Mediation of Perception

The central hermeneutic question for a "philosophy of mediation" is how artifacts mediate human experiences and interpretations of reality. Don Ihde's philosophy of technology is a good starting point for answering this question, because of its focus on the technological mediation of perception. Ihde elaborated Heidegger's tool-analysis into analysis of the relationships between humans and technological artifacts (Ihde 1990). He discerns several relationships human beings can have with technologies. Two of these can be indicated as relations of mediation.[1]

Firstly, Ihde discerns the "embodiment relation," which is his equivalent to Heidegger's "readiness-to-hand." In the embodiment relation, technologies are "incorporated" by their users, establishing a relationship between humans and their world "through" the technological artifact. This embodiment relation, for instance, occurs when looking through a pair of glasses; the artifact is not perceived itself, but it helps to perceive the environment. Technological artifacts become extensions of the human body here, as it were. Secondly, Ihde discerns the "hermeneutic relation."

In this relation, technologies do not provide access to reality because they are "incorporated," but because they provide a representation of reality, which requires interpretation (hence the name "*hermeneutic* relation"). A thermometer, for instance, established a relationship between humans and reality in terms of temperature. Reading off a thermometer does not result in a direct sensation of heat or cold, but gives a value which requires interpretation in order to tell something about reality.

Ihde shows that technologies, when mediating our sensory relationship with reality, transform what we perceive. According to Ihde, the transformation of perception always has a structure of amplification and reduction. Mediating technologies amplify specific aspects of reality while reducing other aspects. When looking at a tree with an infrared camera, for instance, most aspects of the tree that are visible for the naked eye get lost, but at the same time a new aspect of the tree becomes visible: one can now see whether it is healthy or not. Ihde calls this transforming capacity of technology "technological intentionality": technologies have "intentions," they are not neutral instruments but play an active role in the relationship between humans and their world.

These intentionalities are not fixed properties of artifacts, however. They get shaped within the relationship humans have with these artifacts. Within different relationships technologies can have a different "identity." The telephone and the typewriter, for instance, were not developed as communication and writing technologies, but as equipment for the blind and the hard of hearing to help them hear and write. In their use context they were interpreted quite differently, however. This phenomenon Ihde calls *multistability*: a technology can have several "stabilities," depending on the way it is embedded in a use context. Technological intentionalities, therefore, are always dependent on the specific stabilities that come about.

Ihde's analysis of the transformation of perception has important hermeneutic implications. In fact, it shows that mediating artifacts help to determine how reality can be present for and interpreted by people. Technologies help to shape what counts as "real." This hermeneutic role of things has important ethical consequences, since it implies that technologies can actively contribute to the moral decisions human beings make. Medical imaging technologies, like MRI and ultrasound, are good examples of this. Obstetrical ultrasound makes visible aspects of a living fetus in the womb, which cannot be seen without them, and which inform us about the health of the unborn child. But the specific way in which ultrasound scanners represent what they "see" helps to shape how the unborn child is perceived and interpreted, and what decisions are made. In this way, technologies fundamentally shape people's experience of disease, pregnancy, or their unborn child. The very fact of having an ultrasound scan made lets the fetus be present in terms of health and disease, and in terms of our ability to prevent children with this disease from being born (cf. Verbeek 2002).

Mediation of Action

Within the praxis-perspective, the central question is how artifacts mediate people's actions and the way they live their lives. While perception, from a phenomenological point of view, consists in the way the world is present for humans, praxis can be seen as the way humans are present in their world. The work of Bruno Latour offers many interesting concepts for analyzing how artifacts mediate action (cf. Latour 1992, 1994). Latour points out that what humans do is in many cases co-shaped by the things they use. Actions are not only the result of individual intentions and the social structures in which human beings find themselves (the classical agency-structure dichotomy), but also of people's material environment. The concept introduced by Latour and Akrich to describe the influence of artifacts on human actions is "script." Like the script of a movie or a theater play, artifacts prescribe their users how to act when they use them. A speed bump, for instance, has the script "slow down when you approach me"; a plastic coffee cup "throw me away after use."

This influence of artifacts on human actions is of a specific nature. When scripts are at work, things mediate action as material things, not as immaterial signs. A traffic sign makes people slow down because of what it signifies, not because of its material presence in the relation between humans and world. And we do not discard a plastic coffee cup because its user's manual tells us to do so, but because it simply is physically not able to survive being cleaned several times. The influence of technological artifacts on human actions can be non-linguistic. Things are able to exert influence as *material things,* not only as *signs* or *carriers of meaning.*

As is the case with perception, in the mediation of action *transformations* occur. Following Latour, within the domain of action these transformations can be indicated as "translations" of "programs of action." Latour attributes programs of actions to all entities—human and nonhuman. When an entity enters a relationship with another entity, the original programs of action of both are translated into a new one. When somebody's action program is to "prepare meals quickly," and this program is added to that of a microwave oven ("heating food quickly"), the action program of the resulting "composite actor" might be "regularly eating instant meals individually."

In the translation of action, a similar structure can be discerned as in the transformation of perception. Just as in the mediation of perception some aspects of reality are amplified and others are reduced, in the mediation of action one could say that specific actions are "invited," while others are "inhibited." The scripts of artifacts suggest specific actions and discourage others.

The nature of this invitation-inhibition structure is as context-dependent as the amplification-reduction structure of perception. Ihde's concept of multistability also applies within the context of the mediation of action. The telephone has had a major influence on the separation of people's geographical and social context, by making it possible to maintain social relationships outside our immediate living environment. But it could only have this influence because it is used as a communication technology, not as the hearing aid it was originally supposed to be.

An important difference with respect to the mediation of perception, however, is the way in which action-mediating artifacts are present. Artifacts do not only mediate action from a ready-to-hand position but also from a present-at-hand position. A gun, to mention an unpleasant example, mediates action from a ready-to-hand position, translating "express my anger" or "take revenge" into "kill that person." A speed bump, however, cannot be embodied. It will never be ready-to-hand; it exerts influence on people's actions from a present-at-hand position.

Vocabulary

The science-technology studies (STS) concept of "scripts," indicating the influence of technological artifacts on human actions, can be seen as part of a more encompassing framework for understanding the role of technologies in the relation between humans and reality. The main concepts of this framework together form a "vocabulary for technological mediation," which could be helpful to analyze the role of technologies in their use context. Artifacts mediate perception by means of technological *intentionalities:* the active and intentional influence of technologies. They mediate action by means of *scripts,* which prescribe how to act when using the artifact. This latter form of meditation is most important for the ethics of engineering design, since it concerns human actions, and ethics concerns the moral question "how to act?" Technological mediation appears to be context-dependent, and always entails a *translation* of action and a *transformation* of perception. The translation of action has a structure of *invitation* and *inhibition;* the transformation of perception a structure of *amplification* and *reduction.*

Table 16.1: A vocabulary for technological mediation

experience	praxis
mediation of perception	mediation of action
technological intentionality	script
transformation of perception	translation of action
amplification and reduction	imitation and inhibition
delegation: *deliberate inscription*	
multistability: *context-dependency*	

DO ARTIFACTS HAVE MORALITY?

The phenomenon of technological mediation has important implications for ethical theory and for the ethics of engineering design. I will address these implications separately. In this section, I will investigate to what extent the technological mediation of human actions and interpretations of reality can be reason to attribute a specific form of moral agency to technological artifacts. In the next section, I will investigate how engineering could benefit from incorporating the notion of mediation, and of "material moral agency."

The question of the moral significance of technological artifacts has been playing a role on the backbenches of the philosophy of technology for quite some time now. Already in 1986 Langdon Winner asked himself "Do artifacts have politics?" This question was grounded in his analysis of a number of "racist" overpasses in New York, which were deliberately built so low that only cars could pass beneath them, but not buses, thus preventing the dark-skinned population—unable to afford a car—from accessing the beach (Winner 1986). Bruno Latour (1992) subsequently argued that artifacts are bearers of morality as they are constantly making all kinds of moral decisions for people. For example, he shows that the moral decision of how fast one drives is often delegated to a speed bump in the road with the script "slow down before reaching me." Anyone complaining about deteriorating morality, according to Latour, should use their eyes better, as the objects around us are crammed with morality.

As elaborated above, many of our actions and interpretations of the world are co-shaped by the technologies we use. Telephones mediate the way we communicate with others, cars help to determine the acceptable distance from home to work, thermometers co-shape our experience of health and disease, and prenatal diagnostic technologies generate difficult questions regarding pregnancy and abortion. This mediating role of technologies also pertains to actions and decisions we usually call "moral"—ranging from the speed we find morally acceptable to our decisions about unborn life. If ethics is about the question "how to act," and technologies help to answer this question, technologies appear to do ethics, or at least help us to do so. Analogously to Winner's claim that artifacts have politics, therefore, the conclusion seems justified that artifacts have morality: technologies play an active role in moral action and decision-making.

But how to understand this material morality? Does it actually imply that artifacts can be considered moral agents? In ethical theory, to qualify as a moral agent at least requires the possession of *intentionality* and some degree of *freedom*. In order to be held morally accountable for an action, an agent needs to have the intention to act in a specific way, and the freedom to actually realize this intention. Both requirements seem problematic with respect to artifacts—at least, at first sight. Artifacts, after all, do not seem to be able to form intentions, and neither do they possess any form of autonomy. Yet, both requirements for moral agency deserve further analysis.

Technological Intentionality

At a first glance, it might seem absurd to speak about artifacts in terms of intentionality. A closer inspection of what we mean by "intentionality" in relation to what artifacts actually "do," however, makes it possible to attribute a specific form of intentionality to artifacts. In order to show that, it is important to make a distinction here between two aspects of "intentionality." Firstly, intentionality entails the ability to *form intentions,* and secondly, this forming of intentions can be considered something *original* or *spontaneous* in the sense that it literally "springs from" or is "originated by" the agent possessing intentionality. Both aspects of intentionality will appear not to be as alien to technological artifacts as they might seem.

First of all, the "mediation approach" to technology, as elaborated above, makes it possible to attribute to artifacts the ability to actually form intentions. In this approach, technologies are analyzed in terms of their mediating roles in relations between human and reality. The core idea is that technologies, when used, always establish a relation between users and their environment. Technologies do not only enable us to shape *how* we act and experience things. They are not neutral instruments or intermediaries, but active mediators that help shape the relation between people and reality. This "technological intentionality" can be illustrated by further elaborating the example of obstetrical ultrasound, which was introduced in the previous section. Ultrasound is not simply a functional means to make visible an unborn child in the womb. It actively helps to shape the way the unborn child is given in human experience, and in doing so it informs the choices his or her expecting parents make. Because of the ways in which ultrasound mediated the relations between the fetus and the future parents, it constitutes both fetus and parents in specific ways.

Ultrasound brings about a number of "translations" of the relations between parents and the fetus, while mediating their visual contact. First of all, ultrasound isolates the fetus from the female body. In doing so, it creates a new ontological status of the fetus, as a separate living being rather than forming a unity with his or her mother. This creates the space to make decisions about the fetus apart from the pregnant woman in whose body it is growing. Secondly, ultrasound places the fetus in a context of medical norms. It makes visible defects of the neural tube, and makes it possible to measure the thickness of the fetal neck fold, which forms an indication of the risk that the child will suffer from Down's Syndrome. In doing so, ultrasound translates pregnancy into a medical process; the fetus into a possible patient; and congenital defects into preventable suffering. As a result, pregnancy becomes a process of choices: the choice to have tests like neck fold measurements done at all, and the choice of what to do if anything is "wrong." Moreover, parents are constituted as decision-makers regarding the life of their unborn child. To be sure, the role of ultrasound is ambivalent here: on the one hand it may encourage abortion, making it possible to prevent suffering; on the other hand it may discourage abortion, enhancing emotional bonds between parents and the unborn child by visualizing "fetal personhood."

In all of these examples, artifacts are active: they help to shape human actions, interpretations, and decisions, which would have been different without the artifact. To be sure, artifacts do not have intentions like human beings do, because they cannot *deliberately* do something. But their lack of consciousness does not take away the fact that artifacts can have intentions in the literal sense of the Latin word "intendere," which means "to direct," "to direct one's course," "to direct one's mind." The intentionality of artifacts is to be found in their directing role in the actions and experiences of human beings. Technological mediation, therefore, can be seen as a specific, material form of intentionality.

With regard to the second aspect of intentionality, the "originality" of intentions, a similar argumentation can be given. For even though artifacts evidently cannot form intentions entirely on their own (again because of their lack of consciousness) their mediating roles, however, cannot be entirely reduced to the intentions of their designers and users. Otherwise, the intentionalities of artifacts would be a variant of what Searle indicated as "derived intentionality" (Searle 1983), entirely reducible to human intentionalities. Quite often, technologies mediate human actions and experiences without human beings having told them to do so. Some technologies, for instance, are used differently than their designers had envisaged. The first cars—which only made 15 km/h—were used primarily for sports, and for medical purposes; driving at a speed of 15 km/h was considered to create an environment of "thin air," which was supposed to be healthy for people with lung diseases. Only after it got interpreted as a means for long distance transport could the car get to play its current role in the division between labor and leisure (Baudet 1986). In this case, unexpected mediations come about in specific use contexts. But unforeseen mediations can also emerge when technologies are used as intended. The very fact that the introduction of mobile phones has led to changes in youth culture—such as the fact that young people appear to make ever less appointments with each other, since everyone can call and be called at any time and place—was not intended by the designers of the cell phone, even though it is used here in precisely the context the designers had envisaged.

It seems plausible, then, to attribute a specific form of intentionality to artifacts. This "material" form of intentionality is quite different from human intentionality, in that it cannot exist without human intentionalities supporting it. Only within the relations between humans and reality, artifacts help to constitute both the objects in reality that are experiences or acted upon and the subjects that are experiencing and acting. This implies that the subjects who act or make decisions about actions are never purely human, but rather a complex blend of humanity and technology. When making a decision about abortion on the basis of technologically mediated knowledge about the chances that the child will suffer from a serious disease, this decision is not "purely" human, but neither is it entirely induced by technology. The very situation of having to make this decision and the very ways in which the decision is made, were co-shaped by technological artifacts. Without these technologies, either there would not be a situation of choice, or the decision would be made on the basis of a different relation to the situation. At the same time, the technologies involved do not *determine* human decisions here. Moral decision-making is a joint effort of human beings and technological artifacts.

Strictly speaking, then, there is no such thing as "technological intentionality"; intentionality is always a hybrid affair, involving both human and nonhuman intentions, or, better, "composite intentions" with intentionality distributed over the human and the nonhuman elements in human-technology-world relationships. Rather than being "derived" from human agents, this intentionality comes about in associations between human and nonhumans. For that reason, it could be called "hybrid intentionality."

Technology and Freedom

But what about the second requirement for moral agency we discerned at the beginning of this paper: freedom, or even autonomy? Now that we have concluded that artifacts may have some form of intentionality, can we also say that they have freedom? Obviously not. Again, freedom requires the possession of a mind, which artifacts do not have. Technologies, therefore, cannot be free agents like human beings are. But nevertheless there are good arguments not to exclude artifacts entirely from the realm of freedom that is required for moral agency. In order to show this, I will first elaborate that human freedom in moral decision-making is never absolute, but always bound to the specific situations in which decisions are to be made, including their material infrastructure. Second, I will argue that in the human-technology associations that embody hybrid intentionality, freedom should also be seen as distributed over the human and nonhuman elements in the associations.

Even though freedom is obviously needed in order to be accountable for one's actions, the thoroughly technologically mediated character of our daily lives makes it difficult to take freedom as an absolute criterion for moral agency. After all, as became clear above, in virtually every moral decision we make, technologies play an important role. The decision of how fast to drive and therefore how much risk to run harming other people is always mediated by the lay-out of the road, the power of the engine of the car, the presence or absence of speed bumps and speed cameras, et cetera. And the decision to have surgery or not is most often mediated by all kinds of imaging technologies, blood tests, et cetera, which help constitute the body in specific ways, thus organizing specific situations of choice.

To be sure, moral agency does not necessarily require complete autonomy. Some degree of freedom can be enough to be held morally accountable for an action. And not all freedom is taken away by technological mediations, as the examples of abortion and driving speed made clear. In these examples, human behavior is not determined by technology, but rather co-shaped by it, with humans still being able to reflect on their behavior and making decisions about it. This does not take away the fact, however, that most mediations, like those provided by speed bumps and by the presence of ultrasound scanners as a common option in medical practice, occur in a pre-reflexive manner, and can in no way be escaped in moral decision-making. The moral dilemmas of whether or not to have an abortion and of how fast to drive would not exist in the same way without the technologies involved in these practices—such dilemmas are rather *shaped* by these technologies. Technologies cannot be defined away from our daily lives. The concept of freedom presupposes a form of sovereignty with respect to technology that human beings simply do not possess.

This conclusion can be read in two distinct ways. The first is that mediation has nothing to do with morality whatsoever. If moral agency requires freedom and technological mediation limits or even annihilates human freedom, only non-technologically mediated situations leave room for morality. Not only are technological artifacts unable to make moral decisions, but also does technology-induced human behavior have a non-moral character. A good example of this criticism is down the often-heard negative reactions to explicit behavior-steering technologies like speed limiters in cars. Usually, the resistance against such technologies is supported with two kinds of arguments. First, there is the fear that human freedom is threatened and that democracy is exchanged for technocracy. Should all human actions be guided by technology, the criticism goes, the outcome would be a technocratic society in which moral problems are solved by machines instead of people. Second, there is the charge of immorality or, at best, amorality. Actions not the product of our own free will but induced by technology can not be described as "moral." And, which is worse, behavior-steering technologies might create a form of moral laziness that is fatal to the moral abilities of citizens.

Yet, these criticisms are deeply problematic. After all, the analyses of technological mediation given above show that human actions are *always* mediated. To phrase it in Latour's words: "Without technological detours, the properly human cannot exist. Morality is no more human than technology, in the sense that it would originate from an already constituted human who would be master of itself as well as of the universe. Let us say that it traverses the world and, like technology, that it engenders in its wake forms of humanity, choices of subjectivity, modes of objectification, and various types of attachment" (Latour 2002). And this is precisely what opponents of speed limitation forget. Also without speed limiters, the actions of drivers are continually mediated: indeed, as cars can easily exceed speed limits and as our roads are so wide and the bends so gentle as to permit driving fast, we are consistently being invited to further explore the space between the accelerator and the floor. Therefore, giving the inevitable technological mediations a desirable form rather than rejecting outright the idea of a "moralized technology" in fact attests to a sense of responsibility.

The conclusion that mediation and morality are at odds with each other, therefore, is not satisfying. It is virtually impossible to think of any morally relevant situation in which technology does not play a role. And it would be throwing out the child with the bathwater to conclude that there is no room for morality and moral judgments in all situations in which technologies play a role. Therefore, an alternative solution is needed of the apparent tension between technological mediation and ethics. Rather than taking absolute freedom as a prerequisite for moral agency, we need to reinterpret freedom as an agent's ability to relate to what determines him or her. Human actions always take place in a stubborn reality, and for this reason, absolute freedom can only be attained by ignoring reality, and therefore by giving up the possibility to act at all. Freedom is not a lack of forces and constraints; it rather is the existential space human beings have to realize their existence. Humans have a relation to their own existence and to the ways in which this is co-shaped by the material culture in which it takes place. The material situatedness of human existence *creates* specific forms of freedom, rather than impeding them. Freedom consists in the possibilities that are opened up for human beings to have a relation to the environment in which they live and to which they are bound.

This redefinition of freedom, to be sure, still leaves no room to attribute freedom to technological artifacts. But it does take them back into the realm of freedom, rather than excluding them from it altogether. On the one hand, after all, they help to *constitute* freedom, by providing the material environment in which human existence takes place and gets its shape. And on the other hand, artifacts can enter associations with human beings, while these associations—consisting partly of material artifacts—are the places where freedom is to be located. For even though freedom is never absolute but always gets shape by technological and contextual mediations, these very mediations also create the space for moral decision-making. Just like intentionality, freedom too appears to be a hybrid affair, most often located in associations of humans and artifacts.

MORALIZING TECHNOLOGY

This analysis of the moral agency of technological artifacts has important implications for the ethics of technology and technology design. First of all, the mediation approach to technology makes clear that moral issues regarding technology development comprise more than weighing technological risks and preventing disasters, however important these activities in fact are. What is at stake when technologies are introduced in society are also the ways in which these technologies will mediate human actions and experiences, thus helping to shape our moral decisions and our quality of life. The ethics of technology design, therefore, should also occupy itself with taking responsibility for the future mediating roles of technologies-in-design.

Moreover, the analysis of technological mediation shows that, even without explicit moral reflection, technology design is inherently a moral activity. By designing artifacts that will inevitably play a mediating role in people's actions and experience, thus helping to shape (moral) decisions and practices, designers "materialize morality"; they are "doing ethics by other means" (cf. Verbeek 2006). This conclusion makes it even more urgent to expand the scope of the ethics of technology in order to include the moral dimensions of the artifacts themselves, and to try and give shape to these dimensions in a responsible way.

Taking Mediation into Ethics

There are two ways to take mediation analysis into the ethics of technology and design. First of all, they can be used to develop moral assessments of technologies in terms of their mediating roles in human practices and experiences. Secondly, the conclusion that artifacts do have a specific form of morality also shifts ethics from the domain of language to that of materiality. When artifacts have moral relevance and even embody a specific form of moral agency, ethics cannot only occupy itself with developing conceptual frameworks for moral reflection, but should also engage itself with the actual development of the material environments that help shape moral action and decision-making. Hans Achterhuis has called this the "moralization of technology" (Achterhuis 1995).

The first way to take mediation into ethics is closest to common practices in the ethics of technology. In fact, it comes down to an augmentation of the current focus on risk assessment and disaster prevention. Rather than focusing on the acceptability and preventability of negative consequences of the introduction of new technologies, it aims to assess the impact of the mediating capacities of technologies-in-design for human practices and experiences. When an action-ethical approach is followed here, moral reflection is directed at the question whether the actions resulting from specific technological mediation can be morally justified. This reflection can take place along deontological or consequentialist lines. But in many cases, a virtue-ethical life-ethical approach is at least as fruitful to assess technological mediations, focusing on the quality of the *practices* that are introduced by the mediating technologies, and their implications for the kind of life we are living. Not only the impact of mediation on specific human actions is important then, but also the ways in which mediating technologies help to constitute human beings and the world they are experiencing and in which they are acting. To return to the example of ultrasound again: rather than merely assessing the impact of routine ultrasound scans in obstetrical health care in terms of safety and abortion rates, a life-ethical approach would try to assess the quality of the practices that arise around ultrasound scanning, in which the fetus and its expecting parents are constituted in specific ways (as possible patients versus decision-makers) and in specific relations to each other (situations of choice).

The second way to augment the ethics of technology with the approach of technological mediation is not only to assess mediations, but also to try to help *shape* them. Rather than working from an external standpoint vis-à-vis technology, aiming at rejecting or accepting new technologies, the ethics of technology then aims to *accompany* technological developments (Hottois), experiencing with mediations and finding ways to discuss and assess how one could deal with these mediations, and what kinds of living-with-technology are to be preferred. This direction was taken by the Dutch philosopher Hans Achterhuis (1995; 1998), who translated Latour's analysis of scripts into a plea for an explicit "moralization of technology." Instead of only moralizing other *people* ("do not shower too long"; "buy a ticket before you enter the subway"), humans should also moralize their *material environment*. To a water-saving showerhead the task could be delegated to see to it that not too much water is used when showering, and to a turnstyle the task to make sure that only people with a ticket can enter the train.

Achterhuis' plea for a moralization of technology received severe criticism (cf. Achterhuis 1998, 28–31). In the debate that arose around this issue in The Netherlands, two types of arguments were brought in against his ideas. Firstly, human freedom was thought to be attacked when human actions are explicitly and consciously steered with the help of technology. This reduction of human freedom was even perceived as a threat to human dignity; if human actions are not a result from deliberate decisions but from steering technologies, people were thought to be deprived from what makes them human. Moreover, if they are not acting in freedom, their actions cannot be called "moral." Human beings then simply show a type of behavior that was desired by the designers of the technology, instead of explicitly choosing to act this way. Secondly, Achterhuis was accused of jettisoning the democratic principles of our society, because his plea for developing behavior-steering technology was considered an implicit propagation of technocracy. When moral issues are solved by the technological activities of designers instead of democratic activities of politicians, these critics hold, not humans but technology will be in control.

These arguments can be countered, though. First of all, human dignity is not necessarily attacked when limitations of freedom occur. Our legal constitution implies a major limitation of freedom, after all, but this does not make it a threat to our dignity. Human behavior is determined in many ways, and human freedom is limited in many ways. Few people will protest against the legal prohibition of murder, so why protest to the material inhibition imposed by a speed bump to drive too fast at places where children are often playing on the pavement? Secondly, the analysis of technological mediation made clear that technologies *always* help to shape human actions. Therefore, paying explicit attention to the mediating role of technologies should be seen as taking influencing human actions, we had better try and give this influence a desirable form. Besides, as will become clear below in the example of a Dutch industrial design initiative, the "moralizing" role of technologies does not necessarily have the form of exerting *force* on human beings to act in specific ways. Technologies can also *seduce* people to do certain things; they can invite specific actions without forcefully exacting them.

These counterarguments, however, do not take away the anxiety that a technocracy would come about when technologies are explicitly moralized. It might be true that technologies do not differ from laws in limiting human freedom, but laws come about in a democratic way, and the moralization of technology does not. Yet, this does not justify the conclusion that it is better to refrain from paying explicit attention to technological mediation during the design process. If technologies are not moralized explicitly, after all, the responsibility for technological mediation is left to the designers only. Precisely this would amount to a form of technocracy. A better conclusion would be that it is important to find a democratic way to "moralize technology." In the following, I will elaborate a way to do this.

Designing Mediations

The moral impediments to the moralization of technology can be countered much easier than the practical impediments. The moralization of technological artifacts is not as easy as it might seem to be. In order to "build in" specific forms of mediation in technologies, designers need to anticipate the future mediating role of the technologies they are designing. And this is a complex task, since there is no direct relationship between the activities of designers and the mediating role of the technologies they are designing. As became clear above, the mediating role of technologies comes about in a complex interplay between technologies and their users.

Technologies are "multistable," as Don Ihde calls it. They have no fixed identitiy, but only get defined in their context of use. If this were not the case, accepting the idea of technological mediation would take us back to technological determinism; technologies would then be able to determine the behavior of their users all by themselves instead of being part of a sociotechnical network. This multistability makes it difficult to predict the ways in which technologies will influence human actions, and accordingly to evaluate this influence in ethical terms. Technologies can be used in unforeseen ways, and therefore have an unforeseen influence on human actions. The energy-saving light bulb is a good example here, having actually resulted in an increased energy consumption since such bulbs often appear to be used in places previously left unlit, such as in the garden or on the façade, thereby canceling out of their economizing effect (Steg 1999; Weegink 1996). Moreover, unintentional and unexpected forms of mediation can arise when technologies do get used in the way their designers intended. A good example is the revolving door which keeps out not only cold air but also wheelchair users. In short, designers play a seminal role in realizing particular forms of mediation, but not the only role. Users with their interpretations and forms of appropriation also have a part to play; and so do technologies, which give rise to unintended and unanticipated forms of mediation.

Designers thus help to shape the mediating roles of technologies, but these roles also depend on the ways in which the technologies are used and on the ways in which the technologies in question allow unforeseen mediations to emerge. The suggestion that "scripts" are a result of "inscriptions" (Akrich) or "delegations" (Latour), therefore, does not do enough justice to the complex way in which mediation comes about. Designers cannot simply "inscribe" a desired form of morality into an artifact. The mediating role of technologies is not only the result of the activities of the designers, who inscribe scripts or delegate responsibilities, but also depends on the users, who interpret and appropriate technologies, and on the technologies themselves, which can evoke "emergent" forms of mediation.

In all human actions and all interpretations informing moral decisions, three forms of agency are at work: (1) the agency of the human being performing the action or making the moral decision (in interaction with the technology), but also appropriating the technological artifact in a specific way; (2) the agency of the artifact mediating these actions and decisions, sometimes in unforeseen ways; and (3) the agency of the designer who—either implicitly or in explicit delegations—gives a specific shape to the artifact used, and thus helps to shape the eventual mediating role of the artifact. Taking responsibility for technological mediation, therefore, comes down to entering into an interaction with the agency of future users and the artifact-in-design, rather than acting as a "prime mover" (cf. Smith 2003).

The unpredictability of the mediating role of technology that follows from this does not imply, however, that designers are by definition unequipped to deal with it. In order to cope with the unpredictability and complexity of technological mediation, it is important to seek links between the design context and the future use context. Design specifications should be derived not only from the product's intended function but also from an informed prediction of the product's mediating roles and a moral assessment of these roles. A key tool to bring about this coupling of design context and use context, however trivial it may sound, is the designer's moral imagination. By trying to imagine the ways technology-in-design could be used and by shaping user operations and interpretations from that perspective, a designer can include the product's mediating role in his or her moral assessment back during the design phase. Performing a mediation analysis (cf. Verbeek 2006) can be a good basis for making an *informed prediction* of the future mediating role of a technology. As an example of this approach I will briefly discuss the work done by the Dutch industrial designers collective *Eternally Yours*. A second way to formulate an informed prediction of the future mediating role of technologies is a more systematic one. It consists in an augmentation of the existing design methodology of *Constructive Technology Assessment* in such a way, that it becomes an instrument for a democratically organized moralization of technology.

Anticipation by Imagination: Eternally Yours

An interesting example of anticipating mediation by imagination is the work of the Dutch industrial designers collective *Eternally Yours. Eternally Yours* engages in eco-design, but in an unorthodox way (cf. Van Hinte 1997; Verbeek 2005). It does not want to address the issue of sustainability only in the usual terms of reducing pollution in production, consumption, and waste. The actual problem, *Eternally Yours* holds, is that most of our products are thrown away far before actually being worn out. Meeting this problem could be way more effective than reducing pollution in the different stages of products' life-cycles. For this reason, *Eternally Yours* focuses on developing ways to create product longevity. It does so by investigating how the coming about of attachment between products and their users could be stimulated and enhanced.

In order to stimulate longevity, *Eternally Yours* seeks to design things that invite people to use and cherish them as long as possible. *"It's time for a new generation of products, that can age slowly and in a dignified way, become our partners in life and support our memories,"* as *Eternally Yours* approvingly quoted the Italian designer Ezio Manzini in its letterhead. *Eternally Yours* investigates what characteristics of products are able to evoke a bond with their users. According to *Eternally Yours*, three dimensions can be discerned in the lifespan of products. Things have a technical, an economical, and a psychological lifespan. Products can turn into waste because they simply are broken and cannot be repaired anymore; because they are outdated by newer models that have appeared in the market; and because they do not fit people's preferences and taste anymore. For *Eternally Yours*, the psychological lifespan is the most important. The crucial question for sustainable design is therefore: how can the psychological lifetime of products be prolonged?

Eternally Yours developed many ideas to answer this question. For instance, it searched for forms and materials that could stimulate longevity. Materials were investigated that do not get unattractive when aging but have "quality of wear." Leather, for instance, is mostly found more beautiful when it has been used for some time, whereas a shiny polished chromium surface looks worn out with the first scratch. An interesting example of a design in this context is the upholstery of a couch that was designed by Sigrid Smits. In the velour that was used for it, a pattern was stitched that is initially invisible. When the couch has been used for a while, the pattern gradually becomes visible. Instead of aging in an unattractive way, this couch renews itself when getting old. *Eternally Yours* does not only pay attention to materials and product surfaces, however. It also investigates the ways in which services around products can influence their lifespan. The availability of repair and upgrading services can prevent people from discarding products prematurely.

The most important way to stimulate longevity that should be mentioned in the context of this article, however, consists in designing products that evoke a bond with their users by engaging users in their functioning. Most technologies ask as little attention for themselves as possible when people are using them. Technologies, after all, are often designed to disburden people: a central heating system liberates us from the necessity to gather wood, chop it, fill the hearth, clean it, et cetera. We only need to switch a button and our house gets warm. But this disburdening character also creates a loss of "engagement" with technological products. Ever fewer interactions are needed to use them (cf. Borgmann 1995). One of the downsides of this development is that this also affects the attachment between human beings and technological products. The product as a material entity has become less important than the function it fulfills. In many cases, human beings are not invited to interact with the technological artifact they are using, but only to consume the commodity it procures.

The work of *Eternally Yours* shows that this loss of engagement can be countered in a playful way. Technological products could invite users to interact with them without being so demanding that nobody would be prepared to use them. An interesting example in this direction is an engaging "electric/ceramic heater" that was designed by Sven Adolph. It consists of a heating element with several concentric, cylindrically shaped ceramic shells of different height around it, that all have a vertical aperture. The shells can be arranged in several ways, so that they radiate their warmth in specific directions. This artifact is not a purely functional heater that withdraws into pure functionality like common radiators, which are hidden under the windowsill and are only turned on and off. It is an engaging product that asks for attention and involvement in its functioning, much like a campfire. You cannot hide it under the windowsill but have to put it in the middle of the room. You cannot escape it if you need warmth: you have to sit around it. Its shells have to be arranged if we want it to function. Simply turning the heater on and off is not enough: you actually have to be involved in its functioning if you want it to work.

The activities of *Eternally Yours* can be seen as a form of "anticipating mediation by imagination." Sigrid Smiths' couch and Sven Adolph's heater were designed explicitly from the perspective of their possible mediating role in the interactions and affective relationships their owners will have with them. They mediate the behavior of their users in such a way that they are likely to get attached more to these artifacts than to other couches or heaters. These products were not only designed as functional objects, but as artifacts that actively mediate the behavior of their users. The products of *Eternally Yours* embody an "environmental ethics:" they seduce their users to cherish them rather than throwing them away prematurely.

Augmenting Constructive Technology Assessment

A second way to make an "informed prediction" about the mediating role of a technology-in-design is a more systematic one. To establish a connection between the context of use and the context of design, designers could also employ a method that was developed precisely for making such a connection: the method of Constructive Technology Assessment (CTA) (cf. Schot 1992; Rip, Misa and Schot 1995). CTA creates a link between the contexts of design and use in a practical way: it aims to involve all relevant stakeholders in the design of technologies. In order to make use of the CTA methodology within the context of technological mediation, it needs to be augmented, though.

CTA is based on an evolutionary view of technology development. The process of technology development is seen as generating "variations" that are exposed to a "selection environment," which is formed by entities like the market and government regulations. In this selection environment, only the "fittest" variations will survive. There is an important difference between the generation of technologies and the generation of biological species, though. Contrary to biological evolution, in technology development there is a connection or "nexus" between variation and selection. After all, designers can anticipate the selection environment when they are designing technologies, in order to prevent that much effort is put in developing technologies which will not be accepted by consumers or by government regulations.

CTA is a method to employ this nexus in a systematical way, by feeding back assessments of the technology-on-design by all relevant actors—like users, pressure groups, designers, companies et cetera—into the design process. It does so by organizing meetings of all relevant actors in which the aim is to reach consensus about the design of the technology that is "constructively assessed." This form of technology assessment is called "constructive" because it does not assess technologies after they have been developed, but during their development, so that these assessments can be used to modify the original design. Besides this, CTA can be seen as a democratization of the designing process. When a CTA design methodology is followed, not only designers determine what a technology will look like, but all relevant social actors. Following this method, therefore, could take away the fear for technocracy that was discussed above.

Seen from the perspective of technological mediation, however, CTA also has limitations that need to be overcome. CTA primarily focuses on *human* actors, and pays too little attention to the actively mediating role of the *nonhuman* actor that is at the center of all activity: the technology-in-design. CTA claims to open the black box of technology by analyzing the complex dynamics of technology development. It bases itself on the constructivist notion that technologies are not "given" but the outcome of a process in which many actors are involved. Other interactions between the actors might have resulted in a different technology. But by analyzing the dynamics of *technology development* the black box of technology is only opened half way. It reveals how technologies emerge from their *design context,* but their role in their *use context* remains black-boxed. Therefore, organizing a democratically, domination-free discussion between all relevant actors is not enough to lay bare all relevant aspects of the technology in question. The mediating role of the technology-in-design is likely to remain hidden during the entire CTA process if it is not put explicitly and systematically on the agenda.

For this reason, participants in the CTA process should not only be invited to integrate assessments of users and pressure groups in product specifications, but also to anticipate possible mediating roles of the technology-in-design. The vocabulary for analyzing mediation, as presented in section 2 of this paper, could be helpful for doing this. Approaching the artifact-in-design in terms of mediation offers a perspective that can be used when creating a nexus between the contexts of design and use.

When the CTA method is augmented in this way, the method of "anticipation by imagination" is given a more systematic character. Creating space for all relevant stakeholders to anticipate the possible mediating role of the technology-in-design enhances the chance that as many possible mediating roles are taken into account. To be sure, this augmentation of the CTA methodology does not guarantee that all mediating roles of the technology in design will be predicted. It creates a connection between the "inscriptions" within the context of design and the "interpretations" or "appropriations" within the context of use, but this cannot possibly cover all "emergent" mediating roles of the technology. Yet, it might be a fruitful way to give shape to the responsibility of designers that becomes visible from the analysis of technical mediation.

CONCLUSION

The analyses of technological mediation, which have been elaborated over the past years in STS and philosophy of technology, have major implications for the ethics of engineering design. The insight that technologies inevitably play a mediating role in the actions of users, makes the work of designers an inherently moral activity. Ethics is about the question how to act, and technologies appear to be able to give material answers to this question by inviting or even exacting specific form of action when they are used. This implies that technological mediation could play an important role in the ethics of engineering design. Designers should not only focus on the functionality of technologies but also on their mediating roles. The fact that technologies always mediate human actions charges designers with the responsibility to anticipate these mediating roles.

This anticipation is a complex task, however, since the mediating role of technologies is not entirely predictable. But even though the future cannot be predicted with full accuracy, ways do exist to develop well-informed and rationally grounded conjectures. In order to cope with the uncertainty regarding the future role of technologies in their use contexts, designers should try to bridge the gap between the context of use and the context of design.

One way to do so is by carrying out a "mediation analysis" with the help of the designer's imagination, which can be facilitated by the vocabulary developed in this article. Such an analysis will not allow designers to predict entirely how the technology they are designing will actually be used, but it will help to identify possible use practices and the forms of mediation that might emerge alongside it.

Designers could also make use of an augmented form of constructive technology assessment, in which the connection between design and use if not only made in imagination but also in practice. In this case, a mediation analysis is carried out not only by the designer individually, but by all stakeholders together, who engage in a democratically organized debate in order to decide how to feed back the outcomes of this analysis into the design process. Following this method could take away part of the fear that deliberately designing behavior-steering technology would lead to technocracy, since the inevitable mediating role of technology is made subject to democratic decision-making here.

To be sure, this anticipation of technological mediation introduces new complexities in the design process. Designers, for instance, might have to deal with trade-offs: in some cases, designing a product with specific desirable mediating characteristics might have negative consequences for the usefulness or attractiveness of the product. Introducing automatic speed-influencing in cars will make sure that drivers keep to the speed limit, but at the cost of the experience of freedom—which appears to be rather important to some car drivers, judging by the fierce societal resistance against speed-limiting measures. Also, when anticipating the mediating role of technologies, prototypes might be developed and rejected because they are likely to bring about undesirable mediations. Dealing with such trade-offs and undesirable spin-offs require a separate moral decision-making process.

Technology design appears to entail more than inventing functional products. The perspective of technological mediation, which has been developed in STS and in the philosophy of technology, reveals that designing should be regarded as a form of materializing morality. This implies that the ethics of engineering design should take more seriously the moral charge of technological products and rethink the moral responsibility of designers accordingly.

NOTES

1. Ihde also distinguished two relations which do not directly concern mediation. Firstly, he identifies the "alterity relations," in which technologies are the terminus of our experience. This relation—which mirrors Heidegger's "presence at hand"—occurs when interacting with a device as if it were another living being, for instance when buying a train ticket at an automatic ticket dispenser. Secondly, Ihde discerns the "background relation." In this relation, technologies play a role at the background of our experience, creating a context for it. An example of this relation is the automatic switching on and off of the refrigerator.

REFERENCES

Achterhuis, H. (1995). "De moralisering van de apparaten." In: *Socialisme en Democratie* 52, 1:3–12.
Achterhuis, H. (1998). *De erfenis van de utopia.* Amsterdam: Ambo.
Akrich, M. (1992). "The De-scription of Technical Objects." In W. E. Bijiker and J. Law, *Shaping Technology/Building Society.* Cambridge: MIT Press, 205–224.
Baudet, H. (1986). *Een vertrouwde wereld: 100 jaar innovatie in Nederland.* Amsterdam: Bert Bakker.
Borgmann, A. (1995). "The Moral Significance of the Material Culture." In A. Feenberg and A. Hannay (eds.), *Technology and the Politics of Knowledge.* Bloomington: Indiana University Press.
Heidegger, M. (1927). *Sein und Zeit.* Tuberingen: Max Niemeyer Verlag.
Hinte, E. van (1997). *Eternally Yours: Visions on Product Endurance.* Rotterdam: 010 Publishers.
Ihde, D. (1990). *Technology and the Lifeworld.* Bloomington: Indiana University Press.
Ihde, D. (1993a). *Philosophy of Technology—An Introduction.* New York: Paragon House.
Ihde, D. (1993b). *Postphenomenology.* Evanston, Ill: Northwestern Univeristy Press.
Ihde, D. (1998). *Expanding Hermeneutics.* Evanston, Ill: Northwestern University Press.
Kuijk, L. (2004). "Prenataal Onderzoek: Abortus als logisch vervolg." In *Trouw,* 3 January 2004. Amsterdam: PCM Publishers.

Latour, B. (1998). "Veiligheidsgordel—de verloren massa van de moraliteit." In M. Schwartz and R. Jansma, *De technologische cultuur,* Amsterdam: De Balie.

Latour, B. (1992). "Where are the Missing Masses?—The Sociology of a Few Mundane Artifacts." In W.E. Bijker and J. Law (ed.), *Shaping Technology/Building Society.* Cambridge: MIT Press.

Latour, B. (1993). *We Have Never Been Modern* (trans. C. Porter). Cambridge: Harvard University Press. (translation of: *Nous n'avons jamais été moderns,* Paris: La Découverte, 1991).

Latour, B. (1994). "On Technical Mediation—Philosophy, Sociology, Genealogy." In *Common Knowledge* 3, 29–64.

Latour, B. (2002). "Morality and Technology: The End of the Means." In *Theory, Culture & Society,* Col. 19, no. 5-6: 247–260.

Latour, B. (2005). "From *Realpolitik* to *Dingpolitik*—or How to Make Things Public." In B. Latour and P. Weibel (eds.), *Making Things Public: Atmospheres of Democracy.* Cambridge: MIT Press.

Rip, A., T. Misa, and J. Schot, eds. (1995). *Managing Technology in Society—The approach of constructive technology assessment.* London: Pinter.

Schot, J. (1992). "Constructive Technology Assessment and Technology Dynamics: The Case of Clean Technologies." *Science, Technology and Human Values* 17, 1:36–56.

Searle, J. R. (1983). *Intentionality: An Essay in the Philosophy of Mind.* Cambridge: Cambridge University Press.

Smith, A. (2003). "Do You Believe in Ethics? Latour and Ihde in the Trenches of the Sciences Wars." In Don Ihde and Evan Selinger (eds.), *Chasing Technoscience: Matrix for Materiality.* Bloomington and Indianapolis: Indiana University Press.

Steg, L. (1999). *Verspilde energie? Wat doen en laten Nederlanders voor het milieu.* The Hague: Sociaal en Cultureel Planbureau (SCP Cahier no. 156).

Swierstra, T. (1999). Moeten artefacten morel gerehabiliteerd? *K&M—tijdschrift voor empirishe filosofie* 4: 317–326.

Verbeek, P. P. (2002). "Pragmatism and Pragmata—bioethics and the technological mediation of experience." In J. Keulartz, *Pragmatist Ethics for a Technological Culture.* Dordrecht: Kluwer.

Verbeek, P. P. (2006). "Materializing Morality—Design Ethics and Technological Mediation." In *Science, Technology, and Human Values*, 3.

Vries, G. de (1999). *Zeppelins—over filosofie, technologie en cultuur.* Amsterdam: Van Gennep.

Weegink, R. J. (1996). *Basisonderzoek elektriciteitsverbruik kleinverbruikers BEK'95.* Arnhem: EnergieNed

Winner, L. (1986). "Do Artifacts Have Politics?" In L. Winner, *The Whale and the Reactor.* Chicago: University of Chicago Press.

Part III

TECHNOLOGY AND POLITICS

Technology plays a vital role in the organization of social life. It affects the way we live, the way we work, the way we get our information, the nature of our healthcare, and countless other basic features of social life. Decisions about the design, development, and administration of technologies often have lasting (even irreversible) effects on a society. Technology may bring opportunity and advantages to some members of society but risks and disadvantages to other members. It may give some members of society power and control that other members do not enjoy. Or it may expose some members to hazards and risks from which others are free. Technology is always political. It helps shape our lives as citizens, and thus is bound up with questions of freedom, democracy, social justice, and our vision of the good life. Like the ethical dimension, a political dimension is intrinsic to making and using technology. The challenging questions are not just about the political effects or consequences of technology but about the political dimensions of technological practice itself. It is not about politics *and* technology, but politics *of* technology.

The common thread running through the chapters in this section is the conviction that no firm distinction can be made between technological and political concerns. Technology policy is bound up with political problems regarding distributive justice, social equality, and other political considerations; and public policy is bound up with technology policy insofar as artifacts mediate and influence the character of our lives as citizens. The readings here address this interplay of technology and politics by examining the idea that technologies have political qualities designed into them; the relationship of democracy to technology policy; the conflict between surveillance technologies and our right to privacy; the relationship between technological change and the unchanging nature of the Constitution; and the interplay of globalization and technology.

In "Do Artifacts Have Politics?" Langdon Winner explores the ideas that technical things have political qualities built into them and that these qualities embody specific forms of power and authority. Winner examines two ways in which artifacts have political qualities. First are instances in which a technological device or system is used to settle a social or political problem in a community. In these cases, the technology is far from neutral but designed to produce results that structure social relations, reinforce vested interests, or engineer human relations. Examples include highways with low-overpasses designed to prevent bus traffic (thus excluding poor people from certain areas), riot-proof campuses designed to prevent large crowds from amassing, and curb cuts in sidewalks designed to accommodate people in wheelchairs. In each case, the "technological deck is stacked" in advance to favor certain social and political interests. Winner says that some technological devices and systems function like laws that build order into our world. They constitute a "form of life," as Wittgenstein would call it, forming the background and context of our lives.

The second manner in which artifacts have political qualities is when they are "inherently political technologies." These are rigid and inflexible technologies, which require (or are at least compatible with) particular political relationships. They unavoidably shape and pattern political relationships, for example as centralized or decentralized, egalitarian or unegalitarian, repressive or liberating. The strong version of the idea that technology has political qualities holds that artifacts themselves require specific social conditions in order to operate. The weaker version holds that a given technology is merely compatible with particular social and political relationships. Examples of political technologies include systems of energy, communication, and transportation, all of which are large scale, centralized, hierarchical organizations administered by highly skilled managers. The very nature of these systems opposes more democratic and decentralized forms of social-political organization.

Winner invites us to weigh claims of practical necessity and efficiency against the moral-political ideals of democratic self-management, justice, and community. If the politics internal to technological management cannot be separated so neatly from public policy, why should we sacrifice our political rights for the sake of technical efficiency? If the technologies themselves have intractable properties then it doesn't matter what kind of political system they function in; their internal political qualities will remain unchanged. Winner urges us to be mindful of the political qualities and contexts of technology, and to opt for democracy over efficiency.

Michel Foucault, in "Panopticism," excerpted from his *Discipline and Punish*: *The Birth of the Prison* (1977), examines the role played by surveillance structures commonly found in prisons, hospitals, and military barracks. The *panopticon* (an actual design that still is in use today) not only detains and imprisons but functions as a laboratory for studying people used in "normalizing detention." The panopticon was developed alongside techniques for creating a conception of the human body that can be examined and regulated in the smallest details of life. What Foucault calls "disciplinary power" is the way that power is subtly exercised on people for the sake of order and control. The effectiveness of power is enhanced by organizing, measuring, supervising, and correcting what is considered abnormal. The very idea of what makes a person normal or abnormal was furthered through the study of individuals in panoptic environments; experts were given the opportunity to examine, label, and experiment on people in a controlled setting. The "normal person" became defined in relation to the madman, the leper, the plague victim, the criminal, and beggars/vagabonds. The panopticon, Foucault proposes, is the machinery and set of techniques that enable a society to control its citizens by discovering and inventing new psychological and behavioral categories to be controlled subsequently by power structures, as well as internalized by people themselves. The panopticon automates and de-individualizes power; its very architecture assures its continual functioning. It is a quasi-autonomous form of power that functions as a political technology.

The panopticon led to new forms of power and knowledge as it normalized and integrated people into society. There is nothing necessarily insidious about disciplinary practices; they are merely techniques for ordering humans. Disciplinary practices include techniques that aim to exercise power at the lowest cost, maximize the extent of power, and increase the docility and utility of entire system of people and institutions. The disciplinary society described by Foucault evolved alongside Capitalism and the Enlightenment. Even democracies depend on disciplinary mechanisms that classify and order people according to a norm or scale. Power and knowledge reinforce one another to form the foundation of social life. More power creates more categories of knowledge; more knowledge refines and extends the scope of power. This is how the liberatory Enlightenment project is at the same time a process of increased discipline: we are achieving greater freedom as power and knowledge becomes more detailed and more controlling. Foucault's contribution to a philosophy of technology is to focus on the overlooked, small, hidden forms of panoptic power that lead to the various disciplinary techniques critical to the formation of identity.

In "Strong Democracy and Technology," Richard Sclove argues that citizens have the right to participate in decisions about a society's basic organization, structure, and evolution. On this model of "strong democracy," citizens should be included in decisions about more than just legal and political matters; they should be included in decisions about technological design and practice, as well. Something as important as the technical organization of society should be established by democratic procedures in a way that satisfies the common interests of citizens. Sclove's argument is strictly moral. If a decision is legitimate, it must have the informed, free consent of those affected by it. In the United States decisions about the design, management, and uses of technological systems are made by elected officials and market forces, often influenced by small groups of technically skilled people, who we can only hope have our best interests in mind. Sclove maintains that what is at stake in having such important decisions about our lives made by other people is nothing less than our rights and liberties. If we have little or no say in decisions that shape and pattern our collective fate then our autonomy is unacceptably compromised.

The implication for public policy is to create mechanisms that would enable people to participate in technological design and management, as well as to contest or reject a technology wherever they determine that rights, liberties, and collective well-being is threatened. Such decisions should be made in a democratic process that would include representatives from grassroots organizations, public interest groups, academic scientists, and community organizations. Yet Sclove takes this proposal even further and argues that strong democracy requires democratic background conditions that foster among citizens a sense of civic responsibility and readiness to participate fully in decision-making processes. These background conditions include: *democratic politics* to ensure full representation and participation at both the local and national levels; *democratic community* to nurture the bonds of mutual understanding and respect among citizens; and *democratic work* to satisfy the needs and interests of everyone in a way that fosters the realization of our talents and capabilities. Strong democracy implies not only a democratic politics of technology but the democratic institutions, communities, and workplaces that would reflect, support, and enable such a politics of technology.

"Bigger Monster, Weaker Chains," authored by Jay Stanley and Barry Steinhardt, is an American Civil Liberties Union (ACLU) report from 2003 that criticizes the increasing surveillance of people's private lives undertaken by the United States government in the aftermath of 9/11. The report argues that the notion of a *surveillance society*, where every facet of our private lives is monitored and recorded, is far from a paranoid delusion. Granted, the federal government, law enforcement, and private sector rely on information gathering in order to protect the general welfare, fight crime, and conduct business. But the surveillance of citizens has risen so sharply in recent years that the ACLU warns against the potential for abuse. The U.S. Patriot Act, for example, expands the government's authority to spy on citizens while reducing the checks and balances on those powers. The U.S. Pentagon's 2003 "Total Information Awareness Program" (later renamed "Terrorist Information Awareness Program") expands surveillance by providing the government unified access to every possible government and commercial database. Such a massive project is made possible by the convergence of surveillance technologies, laws that make such activities permissible, and an amenable political climate. The authors of the ACLU report, however, argue that the trend toward increasing the scope of surveillance, coupled with changes in the law that weaken the limitations of the government, threatens our personal privacy and undermines the liberties guaranteed to us in the Constitution and Bill of Rights. Mass surveillance, they argue, is not the best way to prevent terrorist attacks.

The report identifies several threats to privacy, including video surveillance, data surveillance (collecting information on individuals to construct a character portrait), the commodification of information (the gathering and sale of information by private corporations), DNA scanning, and financial information mining. As new technologies become more sophisticated, more prevalent, and cheaper to use the only barriers that remain between the government and people are legal and political. To prevent us from becoming a surveillance society the authors recommend new, comprehensive privacy laws, new regulations for new technologies, and a revival of the Fourth Amendment to the Constitution to guarantee protection from the government. In their follow-up piece, "Even Bigger, Even Weaker," (2007) the authors argue that checks and balances against the government's surveillance powers continue to be eroded. They ask, how much information do you need to know that a person is *not* a terrorist?

In "The Constitution in Cyberspace," Laurence Tribe inquires whether the changed physical and temporal context of cyberspace and other new technologies requires changes in the Constitution itself. Are we in danger of losing core values like freedom, privacy, and equality because the notions of space and time in cyberspace are very different from when the Constitution was written? Or can the core values of the Constitution be applied to new technologies that the Framers never could have even imagined? Tribe believes that the Constitution—in particular, the Fourth Amendment—continues be relevant regardless of how new technologies change our lives. He examines five basic assumptions underlying Constitutional interpretation to show how they can be adapted to suit the current technological landscape.

The first axiom of the Constitution is that it regulates actions by the government not actions undertaken by individuals and groups. New technologies do not change the function of the Constitution to limit the powers of the government and to protect private groups. The second axiom is that a person's mind, body, and property belong to that person, not to the public as a whole. The Constitution, however, only regulates some private, commercial activity. Most questions about new technologies (e.g., copying software, patent protection, and other issues of cyber-property) are political, not constitutional. The third axiom is that the government should remain neutral as to the value or content of information regardless of its physical or virtual status. The fourth axiom is that the Constitution is founded on normative principles that are not affected by developments in science and technology. Morality is concerned only with what *should be*, not with *what is*. The fifth axiom is that the Constitution's norms must be invariant despite technological transformations. At its core the Constitution protects people, not places or things. Tribe concludes that the Constitution "must be read through technologically transparent lenses." New technologies may raise new moral, legal, and political challenges but they do not change the core values of the Constitution.

In "Technology Transfer and Globalization," Evan Selinger applies the insights of a phenomenologically informed philosophy of technology to development ethics, a branch of applied ethics concerned with evaluating the moral dimensions of socioeconomic change primarily in poor countries and regions. Both development ethics and the philosophy of technology are concerned with understanding and judging the ways that human creations transform and affect the worlds we inhabit. Yet, despite significant overlap, the literatures in development ethics and philosophy of technology rarely refer to one another. Selinger seeks to bridge this gap in the literature by examining how technology plays a role in one aspect of globalization: the transfer of technology from affluent nations to poor nations. He endeavors to give a detailed account of specific technological practices not only to determine their impact on developing countries but also to uncover the ways in which technologies and technological practices themselves can be value-laden.

Selinger illustrates the dynamics of globalization and technology transfer by recounting the fate of the Village Phone (VP) program in Bangladesh. The program was created by the 2006 Nobel Peace Prize recipient Muhammad Yunus, who founded the Grameen Bank, a community development organization that provides "microfinancing" (small loans) to people with no collateral in impoverished areas to assist them in their entrepreneurial projects. The VP program furnished Bangladeshi women with the resources to start a business that provides wireless (cell phone) pay phone services. The program has been a mixed success. On one hand, it has empowered women who have been disenfranchised by religious fundamentalism; on the other hand, it has also disempowered women who still find themselves subjected to chauvinistic cultural norms. Yet, Selinger notes that the VP champions and critics alike fail to pay sufficient attention to the actual experience—what it is like from the perspective of the participants—to engage in this new technological practice. Both sides overlook the crucial role that the technology plays in the lives these women and in the lives of the community. Such a phenomenological approach highlights the paradoxical nature of technological practices that are often simultaneously good and bad, or in the case of the VP, which instill relations of gender independence and gender dependence. Selinger's phenomenological approach calls attention to the ambiguous experience of the transfer of technology and globalization.

17

Do Artifacts Have Politics?

Langdon Winner

No idea is more provocative in controversies about technology and society than the notion that technical things have political qualities. At issue is the claim that the machines, structures, and systems of modern material culture can be accurately judged not only for their contributions to efficiency and productivity and their positive and negative environmental side effects, but also for the ways in which they can embody specific forms of power and authority. Since ideas of this kind are a persistent and troubling presence in discussions about the meaning of technology, they deserve explicit attention.

It is no surprise to learn that technical systems of various kinds are deeply interwoven in the conditions of modern politics. The physical arrangements of industrial production, warfare, communications, and the like have fundamentally changed the exercise of power and the experience of citizenship. But to go beyond this obvious fact and to argue that certain technologies *in themselves* have political properties seems, at first glance, completely mistaken. We all know that people have politics; things do not. To discover either virtues or evils in aggregates of steel, plastic, transistors, integrated circuits, chemicals, and the like seems just plain wrong, a way of mystifying human artifice and of avoiding the true sources, the human sources of freedom and oppression, justice and injustice. Blaming the hardware appears even more foolish than blaming the victims when it comes to judging conditions of public life.

Hence, the stern advice commonly given those who flirt with the notion that technical artifacts have political qualities: What matters is not technology itself, but the social or economic system in which it is embedded. This maxim, which in a number of variations is the central premise of a theory that can be called the social determination of technology, has an obvious wisdom. It serves as a needed corrective to those who focus uncritically upon such things as "the computer and its social impacts" but who fail to look behind technical devices to see the social circumstances of their development, deployment, and use. This view provides an antidote to naïve technological determinism—the idea that technology develops as the sole result of an internal dynamic and then, unmediated by any other influence, molds society to fit its patterns. Those who have not recognized the ways in which technologies are shaped by social and economic forces have not gotten very far.

But the corrective has its own shortcomings; taken literally, it suggests that technical *things* do not matter at all. Once one has done the detective work necessary to reveal the social origins—

power holders behind a particular instance of technological change—one will have explained everything of importance. This conclusion offers comfort to social scientists. It validates what they had always suspected, namely, that there is nothing distinctive about the study of technology in the first place. Hence, they can return to their standard models of social power—those of interest-group politics, bureaucratic politics, Marxist models of class struggle, and the like—and have everything they need. The social determination of technology is, in this view, essentially no different from the social determination of, say, welfare policy or taxation.

There are, however, good reasons to believe that technology is politically significant in its own right, good reasons why the standard models of social science only go so far in accounting for what is most interesting and troublesome about the subject. Much of modern social and political thought contains recurring statements of what can be called a theory of technological politics, an odd mongrel of notions often crossbred with orthodox liberal, conservative, and socialist philosophies.[1] The theory of technological politics draws attention to the momentum of large-scale sociotechnical systems, to the response of modern societies to certain technological imperatives, and to the ways human ends are powerfully transformed as they are adapted to technical means. This perspective offers a novel framework of interpretation and explanation for some of the more puzzling patterns that have taken shape in and around the growth of modern material culture. Its starting point is a decision to take technical artifacts seriously. Rather than insist that we immediately reduce everything to the interplay of social forces, the theory of technological politics suggests that we pay attention to the characteristics of technical objects and the meaning of those characteristics. A necessary complement to, rather than a replacement for, theories of the social determination of technology, this approach identifies certain technologies as political phenomena in their own right. It points us back, to borrow Edmund Husserl's philosophical injunction, *to the things themselves.*

In what follows I will outline and illustrate two ways in which artifacts can contain political properties. First are instances in which the invention, design, or arrangement of a specific technical device or system becomes a way of settling an issue in the affairs of a particular community. Seen in the proper light, examples of this kind are fairly straightforward and easily understood. Second are cases of what can be called "inherently political technologies," man-made systems that appear to require or to be strongly compatible with particular kinds of political relationships. Arguments about cases of this kind are much more troublesome and closer to the heart of the matter. By the term "politics" I mean arrangements of power and authority in human associations as well as the activities that take place within those arrangements. For my purposes here, the term "technology" is understood to mean all of modern practical artifice, but to avoid confusion I prefer to speak of "technologies" plural, smaller or larger pieces or systems of hardware of a specific kind.[2] My intention is not to settle any of the issues here once and for all, but to indicate their general dimensions and significance.

TECHNICAL ARRANGEMENTS AND SOCIAL ORDER

Anyone who has traveled the highways of America and has gotten used to the normal height of overpasses may well find something a little odd about some of the bridges over the parkways on Long Island, New York. Many of the overpasses are extraordinarily low, having as little as nine feet of clearance at the curb. Even those who happened to notice this structural peculiarity would not be inclined to attach any special meaning to it. In our accustomed way of looking at things such as roads and bridges, we see the details of form as innocuous and seldom give them a second thought.

It turns out, however, that some two hundred or so low-hanging overpasses on Long Island are there for a reason. They were deliberately designed and built that way by someone who wanted to achieve a particular social effect. Robert Moses, the master builder of roads, parks, bridges, and other public works of the 1920s to the 1970s in New York, built his overpasses according to specifications that would discourage the presence of buses on his parkways. According to evidence provided by Moses' biographer, Robert A. Caro, the reasons reflect Moses' social class bias and racial prejudice. Automobile-owning whites of "upper" and "comfortable middle" classes, as he called them, would be free to use the parkways for recreation and commuting. Poor people and blacks, who normally used public transit, were kept off the roads because the twelve-foot-tall buses could not handle the overpasses. One consequence was to limit access of racial minorities and low-income groups to Jones Beach, Moses' widely acclaimed public park. Moses made doubly sure of this result by vetoing a proposed extension of the Long Island Railroad to Jones Beach.

Robert Moses' life is a fascinating story in recent U.S. political history. His dealings with mayors, governors, and presidents; his careful manipulation of legislatures, banks, labor unions, the press, and public opinion could be studied by political scientists for years. But the most important and enduring results of his work are his technologies, the vast engineering projects that give New York much of its present form. For generations after Moses' death and the alliances he forged have fallen apart, his public works, especially the highways and bridges he built to favor the use of the automobile over the development of mass transit, will continue to shape that city. Many of his monumental structures of concrete and steel embody a systematic social inequality, a way of engineering relationships among people that, after a time, became just another part of the landscape. As New York planner Lee Koppleman told Caro about the low bridges on Wantagh Parkway, "The old son of a gun had made sure that buses would *never* be able to use his goddamned parkways."[3]

Histories of architecture, city planning, and public works contain many examples of physical arrangements with explicit or implicit political purposes. One can point to Baron Haussmann's broad Parisian thoroughfares, engineered at Louis Napoleon's direction to prevent any recurrence of street fighting of the kind that took place during the revolution of 1848. Or one can visit any number of grotesque concrete buildings and huge plazas constructed on university campuses in the United States during the late 1960s and early 1970s to defuse student demonstrations. Studies of industrial machines and instruments also turn up interesting political stories, including some that violate our normal expectations about why technological innovations are made in the first place. If we suppose that new technologies are introduced to achieve increased efficiency, the history of technology shows that we will sometimes be disappointed. Technological change expresses a panoply of human motives, not the least of which is the desire of some to have dominion over others even though it may require an occasional sacrifice of cost savings and some violation of the normal standard of trying to get more from less.

One poignant illustration can be found in the history of nineteenth-century industrial mechanization. At Cyrus McCormick's reaper manufacturing plant in Chicago in the middle 1880s, pneumatic molding machines, a new and largely untested innovation, were added to the foundry at an estimated cost of $500,000. The standard economic interpretation would lead us to expect that this step was taken to modernize the plant and achieve the kind of efficiencies that mechanization brings. But historian Robert Ozanne has put the development in a broader context. At the time, Cyrus McCormick II was engaged in a battle with the National Union of Iron Molders. He saw the addition of the new machines as a way to "weed out the bad element among the men," namely, the skilled workers who had organized the union local in Chicago.[4] The new machines, manned by unskilled laborers, actually produced inferior castings at a higher cost than the earlier process. After three years of use the machines were, in fact, abandoned, but by that time they had served their purpose—the destruction of the union. Thus, the story of these technical developments at the

McCormick factory cannot be adequately understood outside the record of workers' attempts to organize, police repression of the labor movement in Chicago during that period, and the events surrounding the bombing at Haymarket Square. Technological history and U.S. political history were at that moment deeply intertwined.

In the examples of Moses' low bridges and McCormick's molding machines, one sees the importance of technical arrangements that precede the *use* of the things in question. It is obvious that technologies can be used in ways that enhance the power, authority, and privilege of some over others, for example, the use of television to sell a candidate. In our accustomed way of thinking technologies are seen as neutral tools that can be used well or poorly, for good, evil, or something in between. But we usually do not stop to inquire whether a given device might have been designed and built in such a way that it produces a set of consequences logically and temporally *prior to any of its professed uses.* Robert Moses' bridges, after all, were used to carry automobiles from one point to another; McCormick's machines were used to make metal castings; both technologies, however, encompassed purposes far beyond their immediate use. If our moral and political language for evaluating technology includes only categories having to do with tools and uses, if it does not include attention to the meaning of the designs and arrangements of our artifacts, then we will be blinded to much that is intellectually and practically crucial.

Because the point is most easily understood in the light of particular intentions embodied in physical form, I have so far offered illustrations that seem almost conspiratorial. But to recognize the political dimensions in the shapes of technology does not require that we look for conscious conspiracies or malicious intentions. The organized movement of handicapped people in the United States during the 1970s pointed out the countless ways in which machines, instruments, and structures of common use—buses, buildings, sidewalks, plumbing fixtures, and so forth—made it impossible for many handicapped persons to move freely about, a condition that systematically excluded them from public life. It is safe to say that designs unsuited for the handicapped arose more from long-standing neglect than from anyone's active intention. But once the issue was brought to public attention, it became evident that justice required a remedy. A whole range of artifacts have been redesigned and rebuilt to accommodate this minority.

Indeed, many of the most important examples of technologies that have political consequences are those that transcend the simple categories "intended" and "unintended" altogether. These are instances in which the very process of technical development is so thoroughly biased in a particular direction that it regularly produces results heralded as wonderful breakthroughs by some social interests and crushing setbacks by others. In such cases it is neither correct nor insightful to say, "Someone intended to do somebody else harm." Rather one must say that the technological deck has been stacked in advance to favor certain social interests and that some people were bound to receive a better hand than others.

The mechanical tomato harvester, a remarkable device perfected by researchers at the University of California from the late 1940s to the present offers an illustrative tale. The machine is able to harvest tomatoes in a single pass through a row, cutting the plants from the ground, shaking the fruit loose, and (in the newest models) sorting the tomatoes electronically into large plastic gondolas that hold up to twenty-five tons of produce headed for canning factories. To accommodate the rough motion of these harvesters in the field, agricultural researchers have bred new varieties of tomatoes that are hardier, sturdier, and less tasty than those previously grown. The harvesters replace the system of handpicking in which crews of farm workers would pass through the fields three or four times, putting ripe tomatoes in lug boxes and saving immature fruit for later harvest.[5] Studies in California indicate that the use of the machine reduces costs by approximately five to seven dollars per ton as compared to hand harvesting.[6] But the benefits are by no means equally divided in the agricultural economy. In fact, the machine in the garden has in this instance been the occasion for a thorough reshaping of social relationships involved in tomato production in rural California.

By virtue of their very size and cost of more than $50,000 each, the machines are compatible only with a highly concentrated form of tomato growing. With the introduction of this new method of harvesting, the number of tomato growers declined from approximately 4,000 in the early 1960s to about 600 in 1973, and yet there was a substantial increase in tons of tomatoes produced. By the late 1970s an estimated 32,000 jobs in the tomato industry had been eliminated as a direct consequence of mechanization.[7] Thus, a jump in productivity to the benefit of very large growers has occurred at the sacrifice of other rural agricultural communities.

The University of California's research on and development of agricultural machines such as the tomato harvester eventually became the subject of a lawsuit filed by attorneys for California Rural Legal Assistance, an organization representing a group of farm workers and other interested parties. The suit charged that university officials are spending tax monies on projects that benefit a handful of private interests to the detriment of farm workers, small farmers, consumers, and rural California generally and asks for a court injunction to stop the practice. The university denied these charges, arguing that to accept them "would require elimination of all research with any potential practical application."[8]

As far as I know, no one argued that the development of the tomato harvester was the result of a plot. Two students of the controversy, William Friedland and Amy Barton, specifically exonerate the original developers of the machine and the hard tomato from any desire to facilitate economic concentration in that industry.[9] What we see here instead is an ongoing social process in which scientific knowledge, technological invention, and corporate profit reinforce each other in deeply entrenched patterns, patterns that bear the unmistakable stamp of political and economic power. Over many decades agricultural research and development in U.S. land-grant colleges and universities has tended to favor the interests of large agribusiness concerns.[10] It is in the face of such subtly ingrained patterns that opponents of innovations such as the tomato harvester are made to seem "antitechnology" or "antiprogress." For the harvester is not merely the symbol of a social order that rewards some while punishing others; it is in a true sense an embodiment of that order.

Within a given category of technological change there are, roughly speaking, two kinds of choices that can affect the relative distribution of power, authority, and privilege in a community. Often the crucial decision is a simple "yes or no" choice—are we going to develop and adopt the thing or not? In recent years many local, national, and international disputes about technology have centered on "yes or no" judgments about such things as food additives, pesticides, the building of highways, nuclear reactors, dam projects, and proposed high-tech weapons. The fundamental choice about an antiballistic missile or supersonic transport is whether or not the thing is going to join society as a piece of its operating equipment. Reasons given for and against are frequently as important as those concerning the adoption of an important new law.

A second range of choices, equally critical in many instances, has to do with specific features in the design or arrangement of a technical system after the decision to go ahead with it has already been made. Even after a utility company wins permission to build a large electric power line, important controversies can remain with respect to the placement of its route and the design of its towers; even after an organization has decided to institute a system of computers, controversies can still arise with regard to the kinds of components, programs, modes of access, and other specific features the system will include. Once the mechanical tomato harvester had been developed in its basic form, a design alteration of critical social significance—the addition of electronic sorters, for example—changed the character of the machine's effects upon the balance of wealth and power in California agriculture. Some of the most interesting research on technology and politics at present focuses upon the attempt to demonstrate in a detailed, concrete fashion how seemingly innocuous design features in mass transit systems, water projects, industrial machinery, and other technologies actually mask social choices of profound significance. Historian David Noble has studied two kinds of automated machine tool systems that have different implications for the relative power of management and labor in the industries that might employ them. He has shown that although the basic electronic and mechanical components of the record/playback and numerical control systems are similar, the choice of one design over another has crucial consequences for social struggles on the shop floor. To see the matter solely in terms of cost cutting, efficiency, or the modernization of equipment is to miss a decisive element in the story.[11]

From such examples I would offer some general conclusions. These correspond to the interpretation of technologies as "forms of life" presented earlier, filling in the explicitly political dimensions of that point of view.

The things we call "technologies" are ways of building order in our world. Many technical devices and systems important in everyday life contain possibilities for many different ways of ordering human activity. Consciously or unconsciously, deliberately or inadvertently, societies choose structures for technologies that influence how people are going to work, communicate, travel, consume, and so forth over a very long time. In the processes by which structuring decisions are made, different people are situated differently and possess unequal degrees of power as well as unequal levels of awareness. By far the greatest latitude of choice exists the very first time a particular instrument, system, or technique is introduced. Because choices tend to become strongly fixed in material equipment, economic investment, and social habit, the original flexibility vanishes for all practical purposes once the initial commitments are made. In that sense technological innovations are similar to legislative acts or political foundings that establish a framework for public order that will endure over many generations. For that reason the same careful attention one would give to the rules, roles, and relationships of politics must also be given to such things as the building of highways, the creation of television networks, and the tailoring of seemingly insignificant features on new machines. The issues that divide or unite people in society are settled not only in the institutions and practices of politics proper, but also, and less obviously, in tangible arrangements of steel and concrete, wires and semiconductors, nuts and bolts.

INHERENTLY POLITICAL TECHNOLOGIES

None of the arguments and examples considered thus far addresses a stronger, more troubling claim often made in writings about technology and society—the belief that some technologies are by their very nature political in a specific way. According to this view, the adoption of a given technical system unavoidably brings with it conditions for human relationships that have a distinctive political cast—for example, centralized or decentralized, egalitarian or inegalitarian, repressive or liberating. This is ultimately what is at stake in assertions such as those of Lewis Mumford that two traditions of technology, one authoritarian, the other democratic, exist side by side in Western history. In all the cases cited above the technologies are relatively flexible in design and arrangement and variable in their effects. Although one can recognize a particular result produced in a particular setting, one can also easily imagine how a roughly similar device or system might have been built or situated with very much different political consequences. The idea we must now examine and evaluate is that certain kinds of technology do not allow such flexibility, and that to choose them is to choose unalterably a particular form of political life.

A remarkably forceful statement of one version of this argument appears in Friedrich Engels's little essay "On Authority," written in 1872. Answering anarchists who believed that authority is an evil that ought to be abolished altogether, Engels launches into a panegyric for authoritarianism, maintaining, among other things, that strong authority is a necessary condition in modern industry. To advance his case in the strongest possible way, he asks his readers to imagine that the revolution has already occurred. "Supposing a social revolution dethroned the capitalists, who now exercise their authority over the production and circulation of wealth. Supposing, to adopt entirely the point of view of the anti-authoritarians, that the land and the instruments of labour had become the collective property of the workers who use them. Will authority have disappeared or will it have only changed its form?"[12]

His answer draws upon lessons from three sociotechnical systems of his day, cotton-spinning mills, railways, and ships at sea. He observes that on its way to becoming finished thread, cotton moves through a number of different operations at different locations in the factory. The workers perform a wide variety of tasks, from running the steam engine to carrying the products from one room to another. Because these tasks must be coordinated and because the timing of the work is "fixed by the authority of the steam," laborers must learn to accept a rigid discipline. They must, according to Engels, work at regular hours and agree to subordinate their individual wills to the persons in charge of factory operations. If they fail to do so, they risk the horrifying possibility that production will come to a grinding halt. Engels pulls no punches. "The automatic machinery of a big factory," he writes, "is much more despotic than the small capitalists who employ workers ever have been."[13]

Similar lessons are adduced in Engels's analysis of the necessary operating conditions for railways and ships at sea. Both require the subordination of workers to an "imperious authority" that sees to it that things run according to plan. Engels finds that far from being an idiosyncrasy of capitalist social organization, relationships of authority and subordination arise "independently of all social organization, [and] are imposed upon us together with the material conditions under which we produce and make products circulate." Again, he intends this to be stern advice to the anarchists who, according to Engels, thought it possible simply to eradicate subordination and superordination at a single stroke. All such schemes are nonsense. The roots of unavoidable authoritarianism are, he argues, deeply implanted in the human involvement with science and technology. "If man, by dint of his knowledge and inventive genius, has subdued the forces of nature, the latter avenge themselves upon him by subjecting him, insofar as he employs them, to a veritable despotism independent of all social organization."[14]

Attempts to justify strong authority on the basis of supposedly necessary conditions of technical practice have an ancient history. A pivotal theme in the *Republic* is Plato's quest to borrow the authority of *technē* and employ it by analogy to buttress his argument in favor of authority in the state. Among the illustrations he chooses, like Engels, is that of a ship on the high seas. Because large sailing vessels by their very nature need to be steered with a firm hand, sailors must yield to their captain's commands; no reasonable person believes that ships can be run democratically. Plato goes on to suggest that governing a state is rather like being captain of a ship or like practicing medicine as a physician. Much the same conditions that require central rule and decisive action in organized technical activity also create this need in government.

In Engels's argument, and arguments like it, the justification for authority is no longer made by Plato's classic analogy, but rather directly with reference to technology itself. If the basic case is as compelling as Engels believed it to be, one would expect that as a society adopted increasingly complicated technical systems as its material basis, the prospects for authoritarian ways of life would be greatly enhanced. Central control by knowledgeable people acting at the top of a rigid social hierarchy would seem increasingly prudent. In this respect his stand in "On Authority" appears to be at variance with Karl Marx's position in Volume I of *Capital*. Marx tries to show that increasing mechanization will render obsolete the hierarchical division of labor and the relationships of subordination that, in his view, were necessary during the early stages of modern manufacturing. "Modern Industry," he writes, "sweeps away by technical means the manufacturing division of labor, under which each man is bound hand and foot for life to a single detail operation. At the same time, the capitalistic form of that industry reproduces this same division of labour in a still more monstrous shape; in the factory proper, by converting the workman into a living appendage of the machine."[15] In Marx's view the conditions that will eventually dissolve the capitalist division of labor and facilitate proletarian revolution are conditions latent in industrial technology itself. The differences between Marx's position in *Capital* and Engels's in his essay raise an important question for socialism: What, after all, does modern technology make possible or necessary in political life? The theoretical tension we see here mirrors many troubles in the practice of freedom and authority that had muddied the tracks of socialist revolution.

Arguments to the effect that technologies are in some sense inherently political have been advanced in a wide variety of contexts, far too many to summarize here. My reading of such notions, however, reveals there are two basic ways of stating the case. One version claims that the adoption of a given technical system actually requires the creation and maintenance of a particular set of social conditions as the operating environment of that system. Engels's position is of this kind. A similar view is offered by a contemporary writer who holds that "if you accept nuclear power plants, you also accept a techno-scientific-industrial-military elite. Without these people in charge, you could not have nuclear power."[16] In this conception some kinds of technology require their social environments to be structured in a particular way in much the same sense that an automobile requires wheels in order to move. The thing could not exist as an effective operating entity unless certain social as well as material conditions were met. The meaning of "required" here is that of practical (rather than logical) necessity. Thus, Plato thought it a practical necessity that a ship at sea have one captain and an unquestionably obedient crew.

A second, somewhat weaker, version of the argument holds that a given kind of technology is strongly compatible with, but does not strictly require, social and political relationships of a particular stripe. Many advocates of solar energy have argued that technologies of that variety are more compatible with a democratic, egalitarian society than energy systems based on coal, oil, and nuclear power; at the same time they do not maintain that anything about solar energy requires democracy. Their case is, briefly, that solar energy is decentralizing in both a technical and political sense: technically speaking, it is vastly more reasonable to build solar systems in a disaggregated, widely distributed manner than in large-scale centralized plants; politically speaking, solar energy accommodates the attempts of individuals and local communities to manage their affairs effectively because they are dealing with systems that are more accessible, comprehensible, and controllable than huge centralized sources. In this view solar energy is desirable not only for its economic and environmental benefits, but also for the salutary institutions it is likely to permit in other areas of public life.[17]

Within both versions of the argument there is a further distinction to be made between conditions that are internal to the workings of a given technical system and those that are external to it. Engels's thesis concerns internal social relations said to be required within cotton factories and railways, for example; what such relationships mean for the condition of society at large is, for him, a separate question. In contrast, the solar advocate's belief that solar technologies are compatible with democracy pertains to the way they complement aspects of society removed from the organization of those technologies as such.

There are, then, several different directions that arguments of this kind can follow. Are the social conditions predicated said to be required by, or strongly compatible with, the workings of a given technical system? Are those conditions internal to that system or external to it (or both)? Although writings that address such questions are often unclear about what is being asserted, arguments in this general category are an important part of modern political discourse. They enter into many attempts to explain how changes in social life take place in the wake of technological innovation. More important, they are often used to buttress attempts to justify or criticize proposed courses of action involving new technology. By offering distinctly political reasons for or against the adoption of a particular technology, arguments of this kind stand apart from more commonly employed, more easily quantifiable claims about economic costs and benefits, environmental impacts, and possible risks to public health and safety that technical systems may involve. The issue here does not concern how many jobs will be created, how much income generated, how many pollutants added, or how many cancers produced. Rather, the issue has to do with ways in which choices about technology have important consequences for the form and quality of human associations.

If we examine social patterns that characterize the environments of technical systems, we find certain devices and systems almost invariably linked to specific ways of organizing power and authority. The important question is: Does this state of affairs derive from an unavoidable social response to intractable properties in the things themselves, or is it instead a pattern imposed independently by a governing body, ruling class, or some other social or cultural institution to further its own purposes?

Taking the most obvious example, the atom bomb is an inherently political artifact. As long as it exists at all, its lethal properties demand that it be controlled by a centralized, rigidly hierarchical chain of command closed to all influences that might make its workings unpredictable. The internal social system of the bomb must be authoritarian; there is no other way. The state of affairs stands as a practical necessity independent of any larger political system in which the bomb is embedded, independent of the type of regime or character of its rulers. Indeed, democratic states must try to find ways to ensure that the social structures and mentality that characterize the management of nuclear weapons do not "spin off" or "spill over" into the polity as a whole.

The bomb is, of course, a special case. The reasons very rigid relationships of authority are necessary in its immediate presence should be clear to anyone. If, however, we look for other instances in which particular varieties of technology are widely perceived to need the maintenance of a special pattern of power and authority, modern technical history contains a wealth of examples.

Alfred D. Chandler in *The Visible Hand,* a monumental study of modern business enterprise, presents impressive documentation to defend the hypothesis that the construction and day-to-day operation of many systems of production, transportation, and communication in the nineteenth and twentieth centuries require the development of particular social form—a large-scale centralized, hierarchical organization administered by highly skilled managers. Typical of Chandler's reasoning is his analysis of the growth of the railroads.[18]

> Technology made possible fast, all-weather transportation; but safe, regular, reliable movement of goods and passengers, as well as the continuing maintenance and repair of locomotives, rolling stock, and track, roadbed, stations, roundhouses, and other equipment, required the creation of a sizable administrative organization. It meant the employment of a set of managers to supervise these functional activities over an extensive geographical area; and the appointment of an administrative command of middle and top executives to monitor, evaluate, and coordinate the work of managers responsible for the day-to-day operations.

Throughout his book Chandler points to ways in which technologies used in the production and distribution of electricity, chemicals, and a wide range of industrial goods "demanded" or "required" this form of human association. "Hence, the operational requirements of railroads demanded the creation of the first administrative hierarchies in American business."[19]

Were there other conceivable ways of organizing these aggregates of people and apparatus? Chandler shows that a previously dominant social form, the small traditional family firm, simply could not handle the task in most cases. Although he does not speculate further, it is clear that he believes there is, to be realistic, very little latitude in the forms of power and authority appropriate within modern sociotechnical systems. The properties of many modern technologies—oil pipelines and refineries, for example—are such that overwhelmingly impressive economies of scale and speed are possible. If such systems are to work effectively, efficiently, quickly, and safely, certain requirements of internal social organization have to be fulfilled; the material possibilities that modern technologies make available could not be exploited otherwise. Chandler acknowledges that as one compares sociotechnical institutions of different nations, one sees "ways in which cultural attitudes, values, ideologies, political systems, and social structure affect these imperatives."[20] But the weight of argument and empirical evidence in *The Visible Hand* suggests that any significant departure from the basic pattern would be, at best, highly unlikely.

It may be that other conceivable arrangements of power and authority, for example, those of decentralized, democratic worker self-management, could prove capable of administering factories, refineries, communications systems, and railroads as well as or better than the organizations Chandler describes. Evidence from automobile assembly teams in Sweden and worker-managed plants in Yugoslavia and other countries is often presented to salvage these possibilities. Unable to settle controversies over this matter here, I merely point to what I consider to be their bone of contention. The available evidence tends to show that many large, sophisticated technological systems are in fact highly compatible with centralized, hierarchical managerial control. The interesting question, however, has to do with whether or not this pattern is in any sense a requirement of such systems, a question that is not solely empirical. The matter ultimately rests on our judgments about what steps, if any, are practically necessary in the workings of particular kinds of technology and what, if anything, such measures require of the structure of human associations. Was Plato right in saying that a ship at sea needs steering by a decisive hand and that this could only be accomplished by a single captain and an obedient crew? Is Chandler correct in saying that the properties of large-scale systems require centralized, hierarchical managerial control?

To answer such questions, we would have to examine in some detail the moral claims of practical necessity (including those advocated in the doctrines of economics) and weigh them against moral claims of other sorts, for example, the notion that it is good for sailors to participate in the command of a ship or that workers have a right to be involved in making and administering decisions in a factory. It is characteristic of societies based on large, complex technological systems, however, that moral reasons other than those of practical necessity appear increasingly obsolete, "idealistic," and irrelevant. Whatever claims one may wish to make on behalf of liberty, justice, or equality can be immediately neutralized when confronted with arguments to the effect, "Fine, but that's no way to run a railroad" (or steel mill, or airline, or communication system, and so on). Here we encounter an important quality in modern political discourse and in the way people commonly think about what measures are justified in response to the possibilities technologies make available. In many instances, to say that some technologies are inherently political is to say that certain widely accepted reasons of practical necessity—especially the need to maintain crucial technological systems as smoothly working entities—have tended to eclipse other sorts of moral and political reasoning.

One attempt to salvage the autonomy of politics from the bind of practical necessity involves the notion that conditions of human association found in the internal workings of technological systems can easily be kept separate from the polity as a whole. Americans have long rested content in the belief that arrangements of power and authority inside industrial corporations, public utilities, and the like have little bearing on public institutions, practices, and ideas at large. That "democracy stops at the factory gates" was taken as a fact of life that had nothing to do with the practice of political freedom. But can the internal politics of technology and the politics of the whole community be so easily separated? A recent study of business leaders in the United States, contemporary exemplars of Chandler's "visible hand of management," found them remarkably impatient with such democratic scruples as "one man, one vote." If democracy doesn't work for the firm, the most critical institution in all of society, American executives ask, how well can it be expected to work for the government of a nation—particularly when that government attempts to interfere with the achievements of the firm? The authors of the report observe that patterns of authority that work effectively in the corporation become for businessmen "the desirable model against which to compare political and economic relationships in the rest of society."[21] While such findings are far from conclusive, they do reflect a sentiment increasingly common in the land: what dilemmas such as the energy crisis require is not a redistribution of wealth or broader public participation but, rather, stronger, centralized public and private management.

An especially vivid case in which the operational requirements of a technical system might influence the quality of public life is the debates about the risks of nuclear power. As the supply of uranium for nuclear reactors runs out, a proposed alternative fuel is the plutonium generated as a by-product in reactor cores. Well-known objections to plutonium recycling focus on its unacceptable economic costs, its risks of environmental contamination, and its dangers in regard to the international proliferation of nuclear weapons. Beyond these concerns, however, stands another less widely appreciated set of hazards—those that involve the sacrifice of civil liberties. The widespread use of plutonium as a fuel increases the chance that this toxic substance might be stolen by terrorists, organized crime, or other persons. This raises the prospect, and not a trivial one, that extraordinary measures would have to be taken to safeguard plutonium from theft and to recover it should the substance be stolen. Workers in the nuclear industry as well as ordinary citizens outside could well become subject to background security checks, covert surveillance, wiretapping, informers, and even emergency measures under martial law—all justified by the need to safeguard plutonium.

Russell W. Ayres's study of the legal ramifications of plutonium recycling concludes: "With the passage of time and the increase in the quantity of plutonium in existence will come pressure to eliminate the traditional checks the courts and legislatures place on the activities of the executive and to develop a powerful central authority better able to enforce strict safe-guards." He avers that "once a quantity of plutonium had been stolen, the case for literally turning the country upside down to get it back would be overwhelming." Ayres anticipates and worries about the kinds of thinking that, I have argued, characterize inherently political technologies. It is still true that in a world in which human beings make and maintain artificial systems nothing is "required" in an absolute sense. Nevertheless, once a course of action is under way, once artifacts such as nuclear power plants have been built and put in operation, the kinds of reasoning that justify the adaptation of social life to technical requirements pop up as spontaneously as flowers in the spring. In Ayres's words, "Once recycling begins and the risks of plutonium theft become real rather than hypothetical, the case for governmental infringement of protected rights will seem compelling."[22] After a certain point, those who cannot accept the hard requirements and imperatives will be dismissed as dreamers and fools.

*   *   *

The two varieties of interpretation I have outlined indicate how artifacts can have political qualities. In the first instance we noticed ways in which specific features in the design or arrangement of a device or system could provide a convenient means of establishing patterns of power and authority in a given setting. Technologies of this kind have a range of flexibility in the dimensions of their material form. It is precisely because they are flexible that their consequences for society must be understood with reference to the social actors able to influence which designs and arrangements are chosen. In the second instance we examined ways in which the intractable properties of certain kinds of technology are strongly, perhaps unavoidably, linked to particular institutionalized patterns of power and authority. Here the initial choice about whether or not to adopt something is decisive in regard to its consequences. There are no alternative physical designs or arrangements that would make a significant difference; there are, furthermore, no genuine possibilities for creative intervention by different social systems—capitalist or socialist—that could change the intractability of the entity or significantly alter the quality of its political effects.

To know which variety of interpretation is applicable in a given case is often what is at stake in disputes, some of them passionate ones, about the meaning of technology for how we live. I have argued a "both/and" position here, for it seems to me that both kinds of understanding are applicable in different circumstances. Indeed, it can happen that within a particular complex of technology—a system of communication or transportation, for example—some aspects may be flexible in their possibilities for society, while other aspects may be (for better or worse) completely intractable. The two varieties of interpretation I have examined here can overlap and intersect at many points.

These are, of course, issues on which people can disagree. Thus, some proponents of energy from renewable resources now believe they have at last discovered a set of intrinsically democratic, egalitarian, communitarian technologies. In my best estimation, however, the social consequences of building renewable energy systems will surely depend on the specific configurations of both hardware and the social institutions created to bring that energy to us. It may be that we will find ways to turn this silk purse into a sow's ear. By comparison, advocates of the further development of nuclear power seem to believe that they are working on a rather flexible technology whose adverse social effects can be fixed by changing the design parameters of reactors and nuclear waste disposal systems. For reasons indicated above, I believe them to be dead wrong in that faith. Yes, we may be able to manage some of the "risks" to public health and safety that nuclear power brings. But as society adapts to the more dangerous and apparently indelible features of nuclear power, what will be the long-range toll in human freedom?

My belief that we ought to attend more closely to technical objects themselves is not to say that we can ignore the contexts in which those objects are situated. A ship at sea may well require, as Plato and Engels insisted, a single captain and obedient crew. But a ship out of service, parked at the dock, needs only a caretaker. To understand which technologies and which contexts are important to us, and why, is an enterprise that must involve both the study of specific technical systems and their history as well as a thorough grasp of the concepts and controversies of political theory. In our times people are often willing to make drastic changes in the way they live to accommodate technological innovation while at the same time resisting similar kinds of changes justified on political grounds. If for no other reason than that, it is important for us to achieve a clearer view of these matters than has been our habit so far.

NOTES

1. Langdon Winner, *Autonomous Technology: Technics-Out-of-Control as a Theme in Political Thought* (Cambridge: MIT Press, 1977).

2. The meaning of "technology" I employ in this essay does not encompass some of the broader definitions of that concept found in contemporary literature, for example, the notion of "technique" in the writings of Jacques Ellul. My purposes here are more limited. For a discussion of the difficulties that arise in attempts to define "technology," see *Autonomous Technology*, 8–12.

3. Robert A. Caro, *The Power Broker: Robert Moses and the Fall of New York* (New York: Random House, 1974), 318, 481, 514, 546, 951–958, 952.

4. Robert Ozanne, *A Century of Labor-Management Relations at McCormick and International Harvester* (Madison: University of Wisconsin Press, 1967), 20.

5. The early history of the tomato harvester is told in Wayne D. Rasmussen, "Advances in American Agriculture: The Mechanical Tomato Harvester as a Case Study," *Technology and Culture* 9:531–543, 1968.

6. Andrew Schmitz and David Seckler, "Mechanized Agriculture and Social Welfare: The Case of the Tomato Harvester," *American Journal of Agricultural Economics* 52:569–577, 1970.

7. William H. Friedland and Amy Barton, "Tomato Technology," *Society* 13:6, September/October 1976. See also William H. Friedland, *Social Sleepwalkers: Scientific and Technological Research in California Agriculture,* University of California, Davis, Department of Applied Behavioral Sciences, Research Monograph No. 13, 1974.

8. *University of California Clip Sheet* 54:36, May 1, 1979.

9. "Tomato Technology."

10. A history and critical analysis of agricultural research in the land-grant colleges is given in James Hightower, *Hard Tomatoes, Hard Times* (Cambridge: Schenkman, 1978).

11. David F. Noble, *Forces of Production: A Social History of Machine Tool Automation* (New York: Alfred A. Knopf, 1984).

12. Friedrich Engels, "On Authority," in *The Marx-Engels Reader,* ed. 2, Robert Tucker (ed.) (New York: W. W. Norton, 1978), 731.

13. Ibid.

14. Ibid., 732, 731.

15. Karl Marx, *Capital,* vol. 1, ed. 3, translated by Samuel Moore and Edward Aveling (New York: Modern Library, 1906), 530.

16. Jerry Mander, *Four Arguments for the Elimination of Television* (New York: William Morrow, 1978), 44.

17. See, for example, Robert Argue, Barbara Emanuel, and Stephen Graham, *The Sun Builders: A People's Guide to Solar, Wind and Wood Energy in Canada* (Toronto: Renewable Energy in Canada, 1978). "We think decentralization is an implicit component of renewable energy; this implies the decentralization of energy systems, communities and of power. Renewable energy doesn't require mammoth generation sources of disruptive transmission corridors. Our cities and towns, which have been dependent on centralized energy supplies, may be able to achieve some degree of autonomy, thereby controlling and administering their own energy needs" (16).

18. Alfred D. Chandler, Jr., *The Visible Hand: The Managerial Revolution in American Business* (Cambridge: Belknap, 1977), 244.

19. Ibid.

20. Ibid., 500.

21. Leonard Silk and David Vogel, *Ethics and Profits: The Crisis of Confidence in American Business* (New York: Simon and Schuster, 1976), 191.

22. Russell W. Ayres, "Policing Plutonium: The Civil Liberties Fallout," *Harvard Civil Rights—Civil Liberties Law Review* 10 (1975):443, 413–414, 374.

18

Panopticism

Michel Foucault

Bentham's *Panopticon* is the architectural figure of this composition. We know the principle on which it was based: at the periphery, an annular building; at the centre, a tower; this tower is pierced with wide windows that open onto the inner side of the ring; the peripheric building is divided into cells, each of which extends the whole width of the building; they have two windows, one on the inside, corresponding to the windows of the tower; the other, on the outside, allows the light to cross the cell from one end to the other. All that is needed, then, is to place a supervisor in a central tower and to shut up in each cell a madman, a patient, a condemned man, a worker or a schoolboy. By the effect of backlighting, one can observe from the tower, standing out precisely against the light, the small captive shadows in the cells of the periphery. They are like so many cages, so many theatres, in which each actor is alone, perfectly individualized and constantly visible. The panoptic mechanism arranges spatial unities that make it possible to see constantly and to recognize immediately. In short, it reverses the principle of the dungeon; or rather of its three functions—to enclose, to deprive of light and to hide—it preserves only the first and eliminates the other two. Full lighting and the eye of a supervisor capture better than darkness, which ultimately protected. Visibility is a trap.

To begin with, this made it possible—as a negative effect—to avoid those compact, swarming, howling masses that were to be found in places of confinement, those painted by Goya or described by Howard. Each individual, in his place, is securely confined to a cell from which he is seen from the front by the supervisor; but the side walls prevent him from coming into contact with his companions. He is seen, but he does not see; he is the object of information, never a subject in communication. The arrangement of his room, opposite the central tower, imposes on him an axial visibility; but the divisions of the ring, those separated cells, imply a lateral invisibility. And this invisibility is a guarantee of order. If the inmates are convicts, there is no danger of a plot, an attempt at collective escape, the planning of new crimes for the future, bad reciprocal influences; if they are patients, there is no danger of contagion; if they are madmen there is no risk of their committing violence upon one another; if they are schoolchildren, there is no copying, no noise,

From Michel Foucault, *Discipline and Punish: The Birth of the Prison*, trans. Alan Sheridan, (New York: Pantheon, 1977. Originally published in French as *Surveiller et Punir.* Copyright © 1975 by Editions Gallimard. Reprinted by permission of Georges Borchardt, Inc., for Editions Gallimard.

no chatter, no waste of time; if they are workers, there are no disorders, no theft, no coalitions, none of those distractions that slow down the rate of work, make it less perfect or cause accidents. The crowd, a compact mass, a locus of multiple exchanges, individualities merging together, a collective effect, is abolished and replaced by a collection of separated individualities. From the point of view of the guardian, it is replaced by a multiplicity that can be numbered and supervised; from the point of view of the inmates, by a sequestered and observed solitude (Bentham, 60–64).

Hence the major effect of the Panopticon: to induce in the inmate a state of conscious and permanent visibility that assures the automatic functioning of power. So to arrange things that the surveillance is permanent in its effects, even if it is discontinuous in its action; that the perfection of power should tend to render its actual exercise unnecessary; that this architectural apparatus should be a machine for creating and sustaining a power relation independent of the person who exercises it; in short, that the inmates should be caught up in a power situation of which they are themselves the bearers. To achieve this, it is at once too much and too little that the prisoner should be constantly observed by an inspector: too little, for what matters is that he knows himself to be observed; too much, because he has no need in fact of being so. In view of this, Bentham laid down the principle that power should be visible and unverifiable. Visible: the inmate will constantly have before his eyes the tall outline of the central tower from which he is spied upon. Unverifiable: the inmate must never know whether he is being looked at at any one moment; but he must be sure that he may always be so. In order to make the presence or absence of the inspector unverifiable, so that the prisoners, in their cells, cannot even see a shadow, Bentham envisaged not only venetian blinds on the windows of the central observation hall, but, on the inside, partitions that intersected the hall at right angles and, in order to pass from one quarter to the other, not doors but zig-zag openings; for the slightest noise, a gleam of light, a brightness in a half-opened door would betray the presence of the guardian. The Panopticon is a machine for dissociating the see/being seen dyad: in the peripheric ring, one is totally seen, without ever seeing; in the central tower, one sees everything without ever being seen.

It is an important mechanism, for it automatizes and disindividualizes power. Power has its principle not so much in a person as in a certain concerted distribution of bodies, surfaces, lights, gazes; in an arrangement whose internal mechanisms produce the relation in which individuals are caught up. The ceremonies, the rituals, the marks by which the sovereign's surplus power was manifested are useless. There is a machinery that assures dissymmetry, disequilibrium, difference. Consequently, it does not matter who exercises power. Any individual, taken almost at random, can operate the machine: in the absence of the director, his family, his friends, his visitors, even his servants (Bentham, 45). Similarly, it does not matter what motive animates him: the curiosity of the indiscreet, the malice of a child, the thirst for knowledge of a philosopher who wishes to visit this museum of human nature, or the perversity of those who take pleasure in spying and punishing. The more numerous those anonymous and temporary observers are, the greater the risk for the inmate of being surprised and the greater his anxious awareness of being observed. The Panopticon is a marvellous machine which, whatever use one may wish to put it to, produces homogeneous effects of power.

A real subjection is born mechanically from a fictitious relation. So it is not necessary to use force to constrain the convict to good behaviour, the madman to calm, the worker to work, the schoolboy to application, the patient to the observation of the regulations. Bentham was surprised that panoptic institutions could be so light: there were no more bars, no more chains, no more heavy locks; all that was needed was that the separations should be clear and the openings well arranged. The heaviness of the old "houses of security," with their fortress-like architecture, could be replaced by the simple, economic geometry of a "house of certainty." The efficiency of power, its constraining force have, in a sense, passed over to the other side—to the side of its surface of application. He who is subjected to a field of visibility, and who knows it, assumes responsibility for the constraints of power; he makes them play spontaneously upon himself; he inscribes in himself the power relation in which he simultaneously plays both roles; he becomes the principle of his own subjection. By this very fact, the external power may throw off its physical weight; it tends to the non-corporal; and, the more it approaches this limit, the more constant, profound and permanent are its effects: it is a perpetual victory that avoids any physical confrontation and which is always decided in advance.

Bentham does not say whether he was inspired, in his project, by Le Vaux's menagerie at Versailles: the first menagerie in which the different elements are not, as they traditionally were, distributed in a park (Loisel, 104–7). At the centre was an octagonal pavilion which, on the first floor, consisted of only a single room, the king's *salon;* on every side large windows looked out onto seven cages (the eighth side was reserved for the entrance), containing different species of animals. By Bentham's time, this menagerie had disappeared. But one finds in the programme of the Panopticon a similar concern with individualizing observation, with characterization and classification, with the analytical arrangement of space. The Panopticon is a royal menagerie; the animal is replaced by man, individual distribution by specific grouping and the king by the machinery of a furtive power. With this exception, the Panopticon also does the work of a naturalist. It makes it possible to draw up differences: among patients, to observe the symptoms of each individual, without the proximity of beds, the circulation of miasmas, the effects of contagion confusing the clinical tables; among schoolchildren, it makes it possible to observe performances (without there being any imitation or copying), to map aptitudes, to assess characters, to draw up rigorous classifications and, in relation to normal development, to distinguish 'laziness and stubbornness' from "incurable imbecility"; among workers, it makes it possible to note the aptitudes of each worker, compare the time he takes to perform a task, and if they are paid by the day, to calculate their wages (Bentham, 60–64).

So much for the question of observation. But the Panopticon was also a laboratory; it could be used as a machine to carry out experiments, to alter behaviour, to train or correct individuals. To experiment with medicines and monitor their effects. To try out different punishments on prisoners, according to their crimes and character, and to seek the most effective ones. To teach different techniques simultaneously to the workers, to decide which is the best. To try out pedagogical experiments—and in particular to take up once again the well-debated problem of secluded education, by using orphans. One would see what would happen when, in their sixteenth or eighteenth year, they were presented with other boys or girls; one could verify whether, as Helvetius thought, anyone could learn anything; one would follow "the genealogy of every observable idea"; one could bring up different children according to different systems of thought, making certain children believe that two and two do not make four or that the moon is a cheese, then put them together when they are twenty or twenty-five years old; one would then have discussions that would be worth a great deal more than the sermons or lectures on which so much money is spent; one would have at least an opportunity of making discoveries in the domain of metaphysics. The Panopticon is a privileged place for experiments on men, and for analysing with complete certainty the transformations that may be obtained from them. The Panopticon may even provide an apparatus for supervising its own mechanisms. In this central tower, the director may spy on all the employees that he has under his orders: nurses, doctors, foremen, teachers, warders; he will be able to judge them continuously, alter their behaviour, impose them upon them the methods he thinks best; and it will even be possible to observe the director himself. An inspector arriving unexpectedly at the centre of the Panopticon will be able to judge at a glance, without anything being concealed from him, how the entire establishment is functioning. And, in any case, enclosed as he is in the middle of this architectural mechanism, is not the director's own fate entirely bound up with it? The incompetent physician who has allowed contagion to spread, the incompetent prison governor or workshop manager will be the first victims of an epidemic or a revolt. "By every tie I could devise," said the master of the Panopticon, "my own fate had been bound up by me with theirs"' (Bentham, 177). The Panopticon functions as a kind of laboratory of power. Thanks to its mechanisms of observation, it gains in efficiency and in the ability to penetrate into men's behaviour; knowledge follows the advances of power, discovering new objects of knowledge over all the surfaces on which power is exercised.

The plague-stricken town, the panoptic establishment—the differences are important. They mark, at a distance of a century and a half, the transformations of the disciplinary programme. In the first case, there is an exceptional situation: against an extraordinary evil, power is mobilized; it makes itself everywhere present and visible; it invents new mechanisms; it separates, it immobilizes, it partitions; it constructs for a time what is both a counter-city and the perfect society; it imposes an ideal functioning, but one that is reduced, in the final analysis, like the evil that it combats, to a simple dualism of life and death: that which moves brings death, and one kills that which moves. The Panopticon, on the other hand, must be understood as a generalizable model of functioning; a way of defining power relations in terms of the everyday life of men. No doubt Bentham presents it as a particular institution, closed in upon itself. Utopias, perfectly closed in upon themselves, are common enough. As opposed to the ruined prisons, littered with mechanisms of torture, to be seen in Piranese's engravings, the Panopticon presents a cruel, ingenious cage. The fact that it should have given rise, even in our own time, to so many variations, projected or realized, is evidence of the imaginary intensity that it has possessed for almost two hundred years. But the Panopticon must not be understood as a dream building: it is the diagram of a mechanism of power reduced to its ideal form; its functioning, abstracted from any obstacle, resistance or friction, must be represented as a pure architectural and optical system: it is in fact a figure of political technology that may and must be detached from any specific use.

It is polyvalent in its applications; it serves to reform prisoners, but also to treat patients, to instruct schoolchildren, to confine the insane, to supervise workers, to put beggars and idlers to work. It is a type of location of bodies in space, of distribution of individuals in relation to one another, of hierarchical organization, of disposition of centres and channels of power, of definition of the instruments and modes of intervention of power, which can be implemented in hospitals, workshops, schools, prisons. Whenever one is dealing with a multiplicity of individuals on whom a task or a particular form of behaviour must be imposed, the panoptic schema may be used. It is—necessary modifications apart—applicable "to all establishments whatsoever, in which, within a space not too large to be covered or commanded by buildings, a number of persons are meant to be kept under inspection" (Bentham, 40; although Bentham takes the penitentiary house as his prime example, it is because it has many different functions to fulfil—safe custody, confinement, solitude, forced labour and instruction).

In each of its applications, it makes it possible to perfect the exercise of power. It does this in several ways: because it can reduce the number of those who exercise it, while increasing the number of those on whom it is exercised. Because it is possible to intervene at any moment and because the constant pressure acts even before the offences, mistakes or crimes have been committed. Because, in these conditions, its strength is that it never intervenes, it is exercised spontaneously and without noise, it constitutes a mechanism whose effects follow from one another. Because, without any physical instrument other than architecture and geometry, it acts directly on individuals; it gives "power of mind over mind." The panoptic schema makes any apparatus of power more intense: it assures its economy (in material, in personnel, in time); it assures its efficacity by its preventative character, its continuous functioning and its automatic mechanisms. It is a way of obtaining from power "in hitherto unexampled quantity," "a great and new instrument of government . . . ; its great excellence consists in the great strength it is capable of giving to *any* institution it may be thought proper to apply it to" (Bentham, 66).

It's case of "it's easy once you've thought of it" in the political sphere. It can in fact be integrated into any function (education, medical treatment, production, punishment); it can increase the effect of this function, by being linked closely with it; it can constitute a mixed mechanism in which relations of power (and of knowledge) may be precisely adjusted, in the smallest detail, to the processes that are to be supervised; it can establish a direct proportion between "surplus power" and "surplus production." In short, it arranges things in such a way that the exercise of power is not added on from the outside, like a rigid, heavy constraint, to the functions it invests, but is so subtly present in them as to increase their efficiency by itself increasing its own points of contact. The panoptic mechanism is not simply a hinge, a point of exchange between a mechanism of power and a function; it is a way of making power relations function in a function, and of making a function function through these power relations. Bentham's Preface to *Panopticon* opens with a list of the benefits to be obtained from his "inspection-house": "*Morals reformed—health preserved—industry invigorated—instruction diffused—public burthens lightened*—Economy seated, as it were, upon a rock—the gordian knot of the Poor-Laws not cut, but united—all by a simple idea in architecture!" (Bentham, 39).

Furthermore, the arrangement of this machine is such that its enclosed nature does not preclude a permanent presence from the outside: we have seen that anyone may come and exercise in the central tower the functions of surveillance, and that, this being the case, he can gain a clear idea of the way in which the surveillance is practised. In fact, any panoptic institution, even if it is as rigorously closed as a penitentiary, may without difficulty be subjected to such irregular and constant inspections: and not only by the appointed inspectors, but also by the public; any member of society will have the right to come and see with his own eyes how the schools, hospitals, factories, prisons function. There is no risk, therefore, that the increase of power created by the panoptic machine may degenerate into tyranny; the disciplinary mechanism will be democratically controlled, since it will be constantly accessible "to the great tribunal committee of the world." This Panopticon, subtly arranged so that an observer may observe, at a glance, so many different individuals, also enables everyone to come and observe any of the observers. The seeing machine was once a sort of dark room into which individuals spied; it has become a transparent building in which the exercise of power may be supervised by society as a whole.

The panoptic schema, without disappearing as such or losing any of its properties, was destined to spread throughout the social body; its vocation was to become a generalized function. The plague-stricken town provided an exceptional disciplinary model: perfect, but absolutely violent; to the disease that brought death, power opposed its perpetual threat of death; life inside it was reduced to its simplest expression; it was, against the power of death, the meticulous exercise of the right of the sword. The Panopticon, on the other hand, has a role of amplification; although it arranges power, although it is intended to make it more economic and more effective, it does so not for power itself, nor for the immediate salvation of a threatened society: its aim is to strengthen the social forces—to increase production, to develop the economy, spread education, raise the level of public morality; to increase and multiply.

How is power to be strengthened in such a way that, far from impeding progress, far from weighing upon it with its rules and regulations, it actually facilitates such progress? What intensificator of power will be able at the same time to be a multiplicator of production? How will power, by increasing its forces, be able to increase those of society instead of confiscating them or impeding them? The Panopticon's solution to this problem is that the productive increase of power can be assured only if, on the one hand, it can be exercised continuously in the very foundations of society, in the subtlest possible way, and if, on the other hand, it functions outside these sudden, violent, discontinuous forms that are bound up with the exercise of sovereignty. The body of the king, with its strange material and physical presence, with the force that he himself deploys or transmits to some few others, is at the opposite extreme of this new physics of power represented by panopticism; the domain of panopticism is, on the contrary, that whole lower region, that region of irregular bodies, with their details, their multiple movements, their heterogeneous forces, their spatial relations; what are required are mechanisms that analyse distributions, gaps, series, combinations, and which use instruments that render visible, record, differentiate and compare: a physics of a relational and multiple power, which has its maximum intensity not in the person of the king, but in the bodies that can be individualized by these relations. At the theoretical level, Bentham defines another way of analysing the social body and the power relations that traverse it; in terms of practice, he defines a procedure of subordination of bodies and forces that must increase the utility of power while practising the economy of the prince. Panopticism is the general principle of a new "political anatomy" whose object and end are not the relations of sovereignty but the relations of discipline.

The celebrated, transparent, circular cage, with its high tower, powerful and knowing, may have been for Bentham a project of a perfect disciplinary institution; but he also set out to show how one may "unlock" the disciplines and get them to function in a diffused, multiple, polyvalent way throughout the whole social body. These disciplines, which the classical age had elaborated in specific, relatively enclosed places—barracks, schools, workshops—and whose total implementation had been imagined only at the limited and temporary scale of a plague-stricken town, Bentham dreamt of transforming into a network of mechanisms that would be everywhere and always alert, running through society without interruption in space or in time. The panoptic arrangement provides the formula for this generalization. It programmes, at the level of an elementary and easily transferable mechanism, the basic functioning of a society penetrated through and through with disciplinary mechanisms.

There are two images, then, of discipline. At one extreme, the discipline-blockade, the enclosed institution, established on the edges of society, turned inwards towards negative functions: arresting evil, breaking communications, suspending time. At the other extreme, with panopticism, is the discipline-mechanism: a functional mechanism that must improve the exercise of power by making it lighter, more rapid, more effective, a design of subtle coercion for a society to come. The movement from one project to the other, from a schema of exceptional discipline to one of a generalized surveillance, rests on a historical transformation: the gradual extension of the mechanisms of discipline throughout the seventeenth and eighteenth centuries, their spread throughout the whole social body, the formation of what might be called in general the disciplinary society.

"Discipline" may be identified neither with an institution nor with an apparatus; it is a type of power, a modality for its exercise, comprising a whole set of instruments, techniques, procedures, levels of application, targets; it is a "physics" or an "anatomy" of power, a technology. And it may be taken over either by "specialized" institutions (the penitentiaries or 'houses of correction' of the nineteenth century), or by institutions that use it as an essential instrument for a particular end (schools, hospitals), or by pre-existing authorities that find in it a means of reinforcing or reorganizing their internal mechanisms of power (one day we should show how intra-familial relations, essentially in the parents–children cell, have become 'disciplined', absorbing since the classical age external schemata, first educational and military, then medical, psychiatric, psychological, which have made the family the privileged locus of emergence for the disciplinary question of the normal and the abnormal); or by apparatuses that have made discipline their principle of internal functioning (the disciplinarization of the administrative apparatus from the Napoleonic period), or finally by state apparatuses whose major, if not exclusive, function is to assure that discipline reigns over society as a whole (the police).

On the whole, therefore, one can speak of the formation of a disciplinary society in this movement that stretches from the enclosed disciplines, a sort of social "quarantine," to an indefinitely generalizable mechanism of "panopticism." Not because the disciplinary modality of power has replaced all the others; but because it has infiltrated the others, sometimes undermining them, but serving as an intermediary between them, linking them together, extending them and above all making it possible to bring the effects of power to the most minute and distant elements. It assures an infinitesimal distribution of the power relations.

A few years after Bentham, Julius gave this society its birth certificate (Julius, 384–86). Speaking of the panoptic principle, he said that there was much more there than architectural ingenuity: it was an event in the "history of the human mind." In appearance, it is merely the solution of a technical problem; but, through it, a whole type of society emerges. Antiquity had been a civilization of spectacle. "To render accessible to a multitude of men the inspection of a small number of objects": this was the problem to which the architecture of temples, theatres and circuses responded. With spectacle, there was a predominance of public life, the intensity of festivals, sensual proximity. In these rituals in which blood flowed, society found new vigour and formed for a moment a single great body. The modern age poses the opposite problem: "To procure for a small number, or even for a single individual, the instantaneous view of a great multitude." In a society in which the principal elements are no longer the community and public life, but, on the one hand, private individuals and, on the other, the state, relations can be regulated only in a form that is the exact reverse of the spectacle: "It was to the modern age, to the ever-growing influence of the state, to its ever more profound intervention in all the details and all the relations of social life, that was reserved the task of increasing and perfecting its guarantees, by using and directing towards that great aim the building and distribution of buildings intended to observe a great multitude of men at the same time."

Julius saw as a fulfilled historical process that which Bentham had described as a technical programme. Our society is one not of spectacle, but of surveillance; under the surface of images, one invests bodies in depth; behind the great abstraction of exchange, there continues the meticulous, concrete training of useful forces; the circuits of communication are the supports of an accumulation and a centralization of knowledge; the play of signs defines the anchorages of power; it is not that the beautiful totality of the individual is amputated, repressed, altered by our social order, it is rather that the individual is carefully fabricated in it, according to a whole technique of forces and bodies. We are much less Greeks than we believe. We are neither in the amphitheatre, nor on the stage, but in the panoptic machine, invested by its effects of power, which we bring to ourselves since we are part of its mechanism. The importance, in historical mythology, of the Napoleonic character probably derives from the fact that it is at the point of junction of the monarchical, ritual exercise of sovereignty and the hierarchical, permanent exercise of indefinite discipline. He is the individual who looms over everything with a single gaze which no detail, however minute, can escape: "You may consider that no part of the Empire is without surveillance, no crime, no offence, no contravention that remains unpunished, and that the eye of the genius who can enlighten all embraces the whole of this vast machine, without, however, the slightest detail escaping his attention" (Treilhard, 14). At the moment of its full blossoming, the disciplinary society still assumes with the Emperor the old aspect of the power of spectacle. As a monarch who is at one and the same time a usurper of the ancient throne and the organizer of the new state, he combined into a single symbolic, ultimate figure the whole of the long process by which the pomp of sovereignty, the necessarily spectacular manifestations of power, were extinguished one by one in the daily exercise of surveillance, in a panopticism in which the vigilance of intersecting gazes was soon to render useless both the eagle and the sun.

The formation of the disciplinary society is connected with a number of broad historical processes—economic, juridico-political and, lastly, scientific—of which it forms part.

1) Generally speaking, it might be said that the disciplines are techniques for assuring the ordering of human multiplicities. It is true that there is nothing exceptional or even characteristic in this: every system of power is presented with the same problem. But the peculiarity of the disciplines is that they try to define in relation to the multiplicities a tactics of power that fulfils three criteria: firstly, to obtain the exercise of power at the lowest possible cost (economically, by the low expenditure it involves; politically, by its discretion, its low exteriorization, its relative invisibility, the little resistance it arouses); secondly, to bring the effects of this social power to their maximum intensity and to extend them as far as possible, without either failure or interval; thirdly, to link this "economic" growth of power with the output of the apparatuses (educational, military, industrial or medical) within which it is exercised; in short, to increase both the docility and the utility of all the elements of the system. This triple objective of the disciplines corresponds to a well-known historical conjuncture. One aspect of this conjuncture was the large demographic thrust of the eighteenth century; an increase in the floating population (one of the primary objects of discipline is to fix; it is an anti-nomadic technique); a change of quantitative scale in the groups to be supervised or manipulated (from the beginning of the seventeenth century to the eve of the French Revolution, the school population had been increasing rapidly, as had no doubt the hospital population; by the end of the eighteenth century, the peace-time army exceeded 200,000 men). The other aspect of the conjuncture was the growth in the apparatus of production, which was becoming more and more extended and complex; it was also becoming more costly and its profitability had to be increased. The development of the disciplinary methods corresponded to these two processes, or rather, no doubt, to the new need to adjust their correlation. Neither the residual forms of feudal power nor the structures of the administrative monarchy, nor the local mechanisms of supervision, nor the unstable, tangled mass they all formed together could carry out this role: they were hindered from doing so by the irregular and inadequate extension of their network, by their often conflicting functioning, but above all by the "costly" nature of the power that was exercised in them. It was costly in several senses: because directly it cost a great deal to the Treasury; because the system of corrupt offices and farmed-out taxes weighed indirectly, but very heavily, on the population; because the resistance it encountered forced it into a cycle of perpetual reinforcement; because it proceeded essentially by levying (levying on money or products by royal, seigniorial, ecclesiastical taxation; levying on men or time by *corvées* of press-ganging, by locking up or banishing vagabonds). The development of the disciplines marks the appearance of elementary techniques belonging to a quite different economy: mechanisms of power which, instead of proceeding by deduction, are integrated into the productive efficiency of the apparatuses from within, into the growth of this efficiency and into the use of what it produces. For the old principle of "levying-violence," which governed the economy of power, the disciplines substitute the principle of "mildness-production-profit." These are the techniques that make it possible to adjust the multiplicity of men and the multiplication of the apparatuses of production (and this means not only 'production' in the strict sense, but also the production of knowledge and skills in the school, the production of health in the hospitals, the production of destructive force in the army).

In this task of adjustment, discipline had to solve a number of problems for which the old economy of power was not sufficiently equipped. It could reduce the inefficiency of mass phenomena: reduce what, in a multiplicity, makes it much less manageable than a unity; reduce what is opposed to the use of each of its elements and of their sum; reduce everything that may counter the advantages of number. That is why discipline fixes; it arrests or regulates movements; it clears up confusion; it dissipates compact groupings of individuals wandering about the country in unpredictable ways; it establishes calculated distributions. It must also master all the forces that are formed from the very constitution of an organized multiplicity; it must neutralize the effects of counter-power that spring from them and which form a resistance to the power that wishes to dominate it: agitations, revolts, spontaneous organizations, coalitions—anything that may establish horizontal conjunctions. Hence the fact that the disciplines use procedures of partitioning and verticality, that they introduce, between the different elements at the same level, as solid separations as possible, that they define compact hierarchical networks, in short, that they oppose to the intrinsic, adverse force of multiplicity the technique of the continuous, individualizing pyramid. They must also increase the particular utility of each element of the multiplicity, but by means that are the most rapid and the least costly, that is to say, by using the multiplicity itself as an instrument of this growth. Hence, in order to extract from bodies the maximum time and force, the use of those overall methods known as time-tables, collective training, exercises, total and detailed surveillance. Furthermore, the disciplines must increase the effect of utility proper to the multiplicities, so that each is made more useful than the simple sum of its elements: it is in order to increase the utilizable effects of the multiple that the disciplines define tactics of distribution, reciprocal adjustment of bodies, gestures and rhythms, differentiation of capacities, reciprocal coordination in relation to apparatuses or tasks. Lastly, the disciplines have to bring into play the power relations, not above but inside the very texture of the multiplicity, as discreetly as possible, as well articulated on the other functions of these multiplicities and also in the least expensive way possible: to this correspond anonymous instruments of power, coextensive with the multiplicity that they regiment, such as hierarchical surveillance, continuous registration, perpetual assessment and classification. In short, to substitute for a power that is manifested through the brilliance of those who exercise it, a power that insidiously objectifies those on whom it is applied; to form a body of knowledge about these individuals, rather than to deploy the ostentatious signs of sovereignty. In a word, the disciplines are the ensemble of minute technical inventions that made it possible to increase the useful size of multiplicities by decreasing the inconveniences of the power which, in order to make them useful, must control them. A multiplicity, whether in a workshop or a nation, an army or a school, reaches the threshold of a discipline when the relation of the one to the other becomes favorable.

If the economic take-off of the West began with the techniques that made possible the accumulation of capital, it might perhaps be said that the methods for administering the accumulation of men made possible a political take-off in relation to the traditional, ritual, costly, violent forms of power, which soon fell into disuse and were superseded by a subtle, calculated technology of subjection. In fact, the two processes—the accumulation of men and the accumulation of capital—cannot be separated; it would not have been possible to solve the problem of the accumulation of men without the growth of an apparatus of production capable of both sustaining them and using them; conversely, the techniques that made the cumulative multiplicity of men useful accelerated the accumulation of capital. At a less general level, the technological mutations of the apparatus of production, the division of labour and the elaboration of the disciplinary techniques sustained an ensemble of very close relations (cf. Marx, *Capital,* vol. I, chapter XIII and the very interesting analysis in Guerry and Deleule). Each makes the other possible and necessary; each provides a model for the other. The disciplinary pyramid constituted the small cell of power within which the separation, coordination and supervision of tasks was imposed and made efficient; and analytical partitioning of time, gestures and bodily forces constituted an operational schema that could easily be transferred from the groups to be subjected to the mechanisms of production; the massive projection of military methods onto industrial organization was an example of this modelling of the division of labour following the model laid down by the schemata of power. But, on the other hand, the technical analysis of the process of production, its "mechanical" breaking-down, were projected onto the labour force whose task it was to implement it: the constitution of those disciplinary machines in which the individual forces that they bring together are composed into a whole and therefore increased is the effect of this projection. Let us say that discipline is the unitary technique by which the body is reduced as a "political" force at the least cost and maximized as a useful force. The growth of a capitalist economy gave rise to the specific modality of disciplinary power, whose general formulas, techniques of submitting forces and bodies, in short, "political anatomy," could be operated in the most diverse political régimes, apparatuses or institutions.

2) The panoptic modality of power—at the elementary, technical, merely physical level at which it is situated—is not under the immediate dependence or a direct extension of the great juridico-political structures of a society; it is nonetheless not absolutely independent. Historically, the process by which the bourgeoisie became in the course of the eighteenth century the politically dominant class was masked by the establishment of an explicit, coded and formally egalitarian juridical framework, made possible by the organization of a parliamentary, representative régime. But the development and generalization of disciplinary mechanisms constituted the other, dark side of these processes. The general juridical form that guaranteed a system of rights that were egalitarian in principle was supported by these tiny, everyday, physical mechanisms, by all those systems of micro-power that are essentially non-egalitarian and asymmetrical that we call the disciplines. And although, in a formal way, the representative régime makes it possible, directly or indirectly, with or without relays, for the will of all to form the fundamental authority of sovereignty, the disciplines provide, at the base, a guarantee of the submission of forces and bodies. The real, corporal disciplines constituted the foundation of the formal, juridical liberties. The contract may have been regarded as the ideal foundation of law and political power; panopticism constituted the technique, universally widespread, of coercion. It continued to work in depth on the juridical structures of society, in order to make the effective mechanisms of power function in opposition to the formal framework that it had acquired. The "Enlightenment," which discovered the liberties, also invented the disciplines.

In appearance, the disciplines constitute nothing more than an infra-law. They seem to extend the general forms defined by law to the infinitesimal level of individual lives; or they appear as methods of training that enable individuals to become integrated into these general demands. They seem to constitute the same type of law on a different scale, thereby making it more meticulous and more indulgent. The disciplines should be regarded as a sort of counterlaw. They have the precise role of introducing insuperable asymmetries and excluding reciprocities. First, because discipline creates between individuals a "private" link, which is a relation of constraints entirely different from contractual obligation; the acceptance of a discipline may be underwritten by contract; the way in which it is imposed, the mechanisms it brings into play, the non-reversible subordination of one group of people by another, the "surplus" power that is always fixed on the same side, the inequality of position of the different "partners" in relation to the common regulation, all these distinguish the disciplinary link from the contractual link, and make it possible to distort the contractual link systematically from the moment it has as its content a mechanism of discipline. We know, for example, how many real procedures undermine the legal fiction of the work contract: workshop discipline is not the least important. Moreover, whereas the juridical systems define juridical subjects according to universal norms, the disciplines characterize, classify, specialize; they distribute along a scale, around a norm, hierarchize individuals in relation to one another and, if necessary, disqualify and invalidate. In any case, in the space and during the time in which they exercise their control and bring into play the asymmetries of their power, they effect a suspension of the law that is never total, but is never annulled either. Regular and institutional as it may be, the discipline, in its mechanism, is a "counter-law." And, although the universal juridicism of modern society seems to fix limits on the exercise of power, its universally widespread panopticism enables it to operate, on the underside of the law, a machinery that is both immense and minute, which supports, reinforces, multiplies the asymmetry of power and undermines the limits that are traced around the law. The minute disciplines, the panopticisms of every day may well be below the level of emergence of the great apparatuses and the great political struggles. But, in the genealogy of modern society, they have been, with the class domination that traverses it, the political counterpart of the juridical norms according to which power was redistributed. Hence, no doubt, the importance that has been given for so long to the small techniques of discipline, to those apparently insignificant tricks that it has invented, and even to those "sciences" that give it a respectable face; hence the fear of abandoning them if one cannot find any substitute; hence the affirmation that they are at the very foundation of society, and an element in its equilibrium, whereas they are a series of mechanisms for unbalancing power relations definitively and everywhere; hence the persistence in regarding them as the humble, but concrete form of every morality, whereas they are a set of physico-political techniques.

To return to the problem of legal punishments, the prison with all the corrective technology at its disposal is to be resituated at the point where the codified power to punish turns into a disciplinary power to observe; at the point where the universal punishments of the law are applied selectively to certain individuals and always the same ones; at the point where the redefinition of the juridical subject by the penalty becomes a useful training of the criminal; at the point where the law is inverted and passes outside itself, and where the counter-law becomes the effective and institutionalized content of the juridical forms. What generalizes the power to punish, then, is not the universal consciousness of the law in each juridical subject; it is the regular extension, the infinitely minute web of panoptic techniques.

3) Taken one by one, most of these techniques have a long history behind them. But what was new, in the eighteenth century, was that, by being combined and generalized, they attained a level at which the formation of knowledge and the increase of power regularly reinforce one another in a circular process. At this point, the disciplines crossed the "technological" threshold. First the hospital, then the school, then, later, the workshop were not simply "reordered" by the disciplines; they became, thanks to them, apparatuses such that any mechanism of objectification could be used in them as an instrument of subjection, and any growth of power could give rise in them to possible branches of knowledge; it was this link, proper to the technological systems, that made possible within the disciplinary element the formation of clinical medicine, psychiatry, child psychology, educational psychology, the rationalization of labour. It is a double process, then: an epistemological "thaw" through a refinement of power relations; a multiplication of the effects of power through the formation and accumulation of new forms of knowledge.

The extension of the disciplinary methods is inscribed in a broad historical process: the development at about the same time of many other technologies—agronomical, industrial, economic. But it must be recognized that, compared with the mining industries, the emerging chemical industries or methods of national accountancy, compared with the blast furnaces or the steam engine, panopticism has received little attention. It is regarded as not much more than a bizarre little utopia, a perverse dream—rather as though Bentham had been the Fourier of a police society, and the Phalanstery had taken on the form of the Panopticon. And yet this represented the abstract formula of a very real technology, that of individuals. There were many reasons why it received little praise; the most obvious is that the discourses to which it gave rise rarely acquired, except in the academic classifications, the status of sciences; but the real reason is no doubt that the power that it operates and which it augments is a direct, physical power that men exercise upon one another. An inglorious culmination had an origin that could be only grudgingly acknowledged. But it would be unjust to compare the disciplinary techniques with such inventions as the steam engine or Amici's microscope. They are much less; and yet, in a way, they are much more. If a historical equivalent or at least a point of comparison had to be found for them, it would be rather in the "inquisitorial" technique.

But the penitentiary Panopticon was also a system of individualizing and permanent documentation. The same year in which variants of the Benthamite schema were recommended for the building of prisons, the system of "moral accounting" was made compulsory: an individual report of a uniform kind in every prison, on which the governor or head-warder, the chaplain and the instructor had to fill in their observations on each inmate: "It is in a way the *vade mecum* of prison administration, making it possible to assess each case, each circumstance and, consequently, to know what treatment to apply to each prisoner individually" (Ducpétiaux, 56–57). Many other, much more complete, systems of recording were planned or tried out (cf., for example, Gregory, 199ff; Grellet-Wammy, 23–25 and 199–203). The overall aim was to make the prison a place for the constitution of a body of knowledge that would regulate the exercise of penitentiary practice. The prison has not only to know the decision of the judges and to apply it in terms of the established regulations: it has to extract unceasingly from the inmate a body of knowledge that will make it possible to transform the penal measure into a penitentiary operation; which will make of the penalty required by the offence a modification of the inmate that will be of use to society. The autonomy of the carceral régime and the knowledge that it creates make it possible to increase the utility of the penalty, which the code had made the very principle of its punitive philosophy: "The governor must not lose sight of a single inmate, because in whatever part of the prison the inmate is to be found, whether he is entering or leaving, or whether he is staying there, the governor must also justify the motives for his staying in a particular classification or for his movement from one to another. He is a veritable accountant. Each inmate is for him, in the sphere of individual education, a capital invested with penitentiary interest" (Lucas, II, 449–50). As a highly efficient technology, penitentiary practice produces a return on the capital invested in the penal system and in the building of heavy prisons.

Similarly, the offender becomes an individual to know. This demand for knowledge was not, in the first instance, inserted into the legislation itself, in order to provide substance for the sentence and to determine the true degree of guilt. It is as a convict, as a point of application for punitive mechanisms, that the offender is constituted himself as the object of possible knowledge.

But this implies that the penitentiary apparatus, with the whole technological programme that accompanies it, brings about a curious substitution: from the hands of justice, it certainly receives a convicted person; but what it must apply itself to is not, of course, the offence, nor even exactly the offender, but a rather different object, one defined by variables which at the outset at least were not taken into account in the sentence, for they were relevant only for a corrective technology. This other character, whom the penitentiary apparatus substitutes for the convicted offender, is the *delinquent*.

The delinquent is to be distinguished from the offender by the fact that it is not so much his act as his life that is relevant in characterizing him. The penitentiary operation, if it is to be a genuine re-education, must become the sum total existence of the delinquent, making of the prison a sort of artificial and coercive theatre in which his life will be examined from top to bottom. The legal punishment bears upon an act; the punitive technique on a life; it falls to this punitive technique, therefore, to reconstitute all the sordid detail of a life in the form of knowledge, to fill in the gaps of that knowledge and to act upon it by a practice of compulsion. It is a biographical knowledge and a technique for correcting individual lives. The observation of the delinquent "should go back not only to the circumstances, but also to the causes of his crime; they must be sought in the story of his life, from the triple point of view of psychology, social position and upbringing, in order to discover the dangerous proclivities of the first, the harmful predispositions of the second and the bad antecedents of the third. This biographical investigation is an essential part of the preliminary investigation for the classification of penalties before it becomes a condition for the classification of moralities in the penitentiary system. It must accompany the convict from the court to the prison, where the governor's task is not only to receive it, but also to complete, supervise and rectify its various factors during the period of detention" (Lucas, II, 440–42). Behind the offender, to whom the investigation of the facts may attribute responsibility for an offence, stands the delinquent whose slow formation is shown in a biographical investigation. The introduction of the "biographical" is important in the history of penalty. Because it establishes the "criminal" as existing before the crime and even outside it. And, for this reason, a psychological causality, duplicating the juridical attribution of responsibility, confuses its effects. At this point one enters the 'criminological' labyrinth from which we have certainly not yet emerged: any determining cause, because it reduces responsibility, marks the author of the offence with a criminality all the more formidable and demands penitentiary measures that are all the more strict. As the biography of the criminal duplicates in penal practice the analysis of circumstances used in gauging the crime, so one sees penal discourse and psychiatric discourse crossing each other's frontiers; and there, at their point of junction, is formed the notion of the "dangerous" individual, which makes it possible to draw up a network of causality in terms of an entire biography and to present a verdict of punishment-correction.

The penitentiary technique and the delinquent are in a sense twin brothers. It is not true that it was the discovery of the delinquent through a scientific rationality that introduced into our old prisons the refinement of penitentiary techniques. Nor is it true that the internal elaboration of penitentiary methods has finally brought to light the "objective" existence of a delinquency that the abstraction and rigidity of the law were unable to perceive. They appeared together, the one extending from the other, as a technological ensemble that forms and fragments the object to which it applies its instruments. And it is this delinquency, formed in the foundations of the judicial apparatus, among the *"basses œuvres,"* the servile tasks, from which justice averts its gaze, out of the shame it feels in punishing those it condemns, it is this delinquency that now comes to haunt the untroubled courts and the majesty of the laws; it is this delinquency that must be known, assessed, measured, diagnosed, treated when sentences are passed. It is now this delinquency, this anomaly, this deviation, this potential danger, this illness, this form of existence, that must be taken into account when the codes are rewritten. Delinquency is the vengeance of the prison on justice. It is a revenge formidable enough to leave the judge speechless. It is at this point that the criminologists raise their voices.

But we must not forget that the prison, that concentrated and austere figure of all the disciplines, is not an endogenous element in the penal system as defined at the turn of the eighteenth and nineteenth centuries. The theme of a punitive society and of a general semio-technique of punishment that has sustained the "ideological" codes—Beccarian or Benthamite—did not itself give rise to the universal use of the prison. This prison came from elsewhere—from the mechanisms proper to a disciplinary power. Now, despite this heterogeneity, the mechanisms and effects of the prison have spread right through modern criminal justice; delinquency and the delinquents have become parasites on it through and through. One must seek the reason for this formidable "efficiency" of the prison. But one thing may be noted at the outset: the penal justice defined in the eighteenth century by the reformers traced two possible but divergent lines of objectification of the criminal: the first was the series of "monsters," moral or political, who had fallen outside the social pact; the second was that of the juridical subject rehabilitated by punishment. Now the "delinquent" makes it possible to join the two lines and to constitute under the authority of medicine, psychology or criminology, an individual in whom the offender of the law and the object of a scientific technique are superimposed—or almost—one upon the other. That the grip of the prison on the penal system should not have led to a violent reaction of rejection is no doubt due to many reasons. One of these is that, in fabricating delinquency, it gave to criminal justice a unitary field of objects, authenticated by the "sciences," and thus enabled it to function on a general horizon of "truth."

The prison, that darkest region in the apparatus of justice, is the place where the power to punish, which no longer dares to manifest itself openly, silently organizes a field of objectivity in which punishment will be able to function openly as treatment and the sentence be inscribed among the discourses of knowledge. It is understandable that justice should have adopted so easily a prison that was not the offspring of its own thoughts. Justice certainly owed the prison this recognition.

19

Strong Democracy and Technology

Richard E. Sclove

> In West Central Minnesota, local farmers have been opposing an electrical transmission line for over four years . . . The public relations man for the utility said . . . , "You should be proud to have the biggest powerline in the world in your country," but the farmers felt differently.
>
> To people who love and care for the land, a transmission line of this size is a desecration. People who once felt they lived in a democratic society feel they have been betrayed and no longer control their own lives.
>
> —*Minnesota farmer and protester Alice Tripp*[1]

How does the key insight that technologies represent a species of social structure bear on the relationship between technology and democracy? The answer depends partly on one's concept of democracy. One common view is that, as a matter of justice, people should be able to influence the basic social circumstances of their lives. This view implies organizing society along relatively egalitarian and participatory lines, a vision that Benjamin Barber has labeled "strong democracy."[2]

Historic examples approaching this ideal include New England town meetings, the confederation of self-governing Swiss villages and cantons, and the English and American tradition of trial by a jury of peers. Strong democracy is apparent also in the methods or aspirations of various social movements such as the late-19th-century American Farmers Alliance, the 1960s U.S. civil rights movement, and the 1980s Polish Solidarity movement. In each of these cases ordinary people claimed the rights and responsibilities of active citizenship concerning basic social issues.

The strong democratic tradition contrasts with more passive or inegalitarian models of democracy that in practice tend to prevail today, so-called thin democracy.[3] Here the focus shifts from a core concern with substantive political equality and with citizens' active engagement in political discourse, or in seeking their common good, to a preoccupation with representative institutions, periodic elections, and competition among conflicting private interests, elites, and power blocs. Within thin democracies power is less evenly distributed; citizens can vote for representatives but ordinarily have little direct influence on important public decisions.

The contest—both in theory and in practice—between the strong and thin democratic traditions is long-standing and unlikely to be resolved soon. Rather than stopping now to compare and contrast the two, I propose initially to suspend judgment and simply posit a specific, strong democratic model of how societies ought to be organized.

TECHNOLOGY AND DEMOCRACY

The strong democratic ideal envisions extensive opportunities for citizens to participate in important decisions that affect them. A decision qualifies as important particularly insofar as it bears on a society's basic organization or structure. The commitment to egalitarian participation does not preclude continued reliance on some representative institutions, but these should be designed to support and incorporate, rather than to replace, participatory processes.

Complementing this procedural standard of strong democracy is a substantive standard: in their political involvements citizens ought, whatever else they do, to grant precedence to respecting any important concerns or interests common to everyone. Above all, they should perpetuate their society's basic character as a strong democracy. Apart from this one substantive moral obligation, citizens are free to attend as they wish to their diverse and perhaps conflicting personal concerns.

This model of democracy, even in schematic form, is sufficient for deriving a prescriptive theory of democracy and technology: *If citizens ought to be empowered to participate in determining their society's basic structure, and technologies are an important species of social structure, it follows that technological design and practice should be democratized.* Strong democracy's complementary procedural and substantive components entail, furthermore, that technological democratization incorporate two corresponding elements. Procedurally, people from all walks of life require expanded opportunities to shape their evolving technological order. And substantively, the resulting technologies should be compatible with citizens' common interests and affinities—to whatever extent such exist—and particularly with their fundamental interest in strong democracy itself.

Democratic Evaluation, Choice, and Governance

The preceding argument suggests that processes of technological development that are today guided by market forces, economic self-interest, distant bureaucracies, or international rivalry should be subordinated to democratic prerogatives. Only in this way can technologies begin actively to support, rather than to coerce or constrict, people's chosen ways of life. For example, residents of many American cities have grown resigned to daily traffic jams, sprawling shopping malls, the stress associated with combining careers with parenthood, and the television as babysitter. This pattern of sociotechnological organization is largely haphazard.[4]

At other times, an existing technological order, or its process of transformation, reflects the direct intentions of powerful organizations or elites. For instance, this chapter's epigraph alludes to an electric utility consortium that proceeded, despite adamant local opposition, to construct a huge transmission line across prime Minnesota farmland. That outcome was not haphazard or unplanned, but neither did it reflect democratic preferences.

Technological evolution can thus encompass social processes ranging from the haphazard to the bitterly contested or blatantly coercive. None of these processes is strongly democratic. This is not to say that every particular technology must suddenly be subjected to formal political review. Each time one is moved to buy a fork or to sell a pencil sharpener, one should not have to defend the decision before a citizens' tribunal or a congressional committee. Not all technologies exert an equal structural influence. However, consider a modern society's treatment of another genus of social structure: various kinds of law. The rules that parents create for their children are subject to relatively little social oversight. But rule making by federal agencies is governed by extensive formal procedure, and even more stringent procedure is required to amend a national constitution. Why should the treatment of technology be so different?

Whether a technology requires political scrutiny and, if so, where and how exhaustively, should correspond roughly to the degree to which it promises, fundamentally or enduringly, to affect social life. This implies the need for a graduated set of democratic procedures for reviewing existing technological arrangements, monitoring emerging ones, and ensuring that the technological order is compatible with informed democratic wishes.

Within such a system, citizens or polities that believe that a set of technologies may embody significant structural potency ought always to have the opportunity to make that case in an appropriate political forum. Beyond this, there should be a system of ongoing democratic oversight of the entire technological order, scanning for the unanticipated emergence of undemocratic technological consequences or dynamics, and prepared when necessary to intervene remedially in the interest of democratic norms.

This does not mean, however, that everyone has to participate in each technological decision that becomes politicized. Logistical nightmares aside, there is more to life than politics. But in contrast to the present state of affairs, there should be abundant opportunity for widespread and effective participation. Ideally, each citizen would at least occasionally exercise that opportunity, particularly on technological matters significant to him or her.

For example, in the early 1970s the cry rang out that there was natural gas beneath the frigid and remote northwest corner of Canada. Eager to deliver the fuel to urban markets, energy companies began planning to build a high-pressure, chilled pipeline across thousands of miles of wilderness, the traditional home of the Inuit (Eskimos) and various Indian tribes. At that point, a Canadian government ministry, anticipating significant environmental and social repercussions, initiated a public inquiry under the supervision of a respected Supreme Court justice, Thomas R. Berger.

The MacKenzie Valley Pipeline Inquiry (also called the Berger Inquiry) began with preliminary hearings open to participation by any Canadian who felt remotely affected by the pipeline proposal. Responding to what they heard, Berger and his staff then developed a novel format to encourage a thorough, open, and accessible inquiry process. One component involved formal, quasi-judicial hearings comprising conventional expert testimony with cross-examination. But Berger also initiated a series of informal "community hearings." Travelling 17,000 miles to thirty-five remote villages, towns, and settlements, the Berger Inquiry took testimony from nearly 1,000 native witnesses. The familiarity of a local setting and the company of family and neighbors encouraged witness spontaneity and frankness. One native commented: "It's the first time anybody bothered asking us how we felt."[5]

Disadvantaged groups received funding to support travel and other needs related to competent participation. The Canadian Broadcasting Company carried daily radio summaries of both the community and the formal hearings, in English as well as in six native languages. Thus each community was aware of evidence and concerns that had previously been expressed. Moreover, by interspersing formal hearings with travel to concurrent community hearings, Berger made clear his intention to weigh respectfully the testimony of both Ph.D. scientists and teenaged subsistence fishermen.

Berger's final report quoted generously from the full range of witnesses and became a national bestseller. Based on testimony concerning environmental, socioeconomic, cultural, and other issues, the judge recommended a ten-year delay in any decision to build a pipeline through the MacKenzie Valley, as well as a host of more specific steps (including a major new wilderness park and a whale sanctuary). Within months the original pipeline proposal was rejected, and the Canadian Parliament instead approved an alternate route paralleling the existing Alaska Highway.[6]

Some might fault the MacKenzie Valley Inquiry for depending so much on the democratic sensibilities and good faith of one man—Judge Berger—rather than empowering the affected native groups to play a role in formulating the inquiry's conclusions. Nevertheless, the process was vastly more open and egalitarian than is the norm in industrial societies. It contrasts sharply with the steps forced on those Minnesota farmers, mentioned earlier, who were loathe to see a transmission line strung across their fields:

> The farmers have tried to use every legitimate legal and political channel to make known to the utility company, the government and the public their determination to save the land and to maintain safety in their workplaces. The farmers and their urban supporters have been met with indifference and arrogance by both the utility and the government. Turned away at the state capitol, they have taken their case to the courts again and again, only to be rebuffed.[7]

The Berger Inquiry represents just one example of a more democratic means of technological decision making.

Democratic Technologies

Besides fostering democratic procedures for technological decision making, we must seek technological outcomes that are substantively democratic. The purpose of democratic procedures is, most obviously, to help ensure that technologies structurally support popular aspirations, whatever they may be. The alternative is to continue watching aspirations tacitly conform themselves to haphazardly generated technological imperatives or to authoritarian decisions.

However, according to strong democratic theory, citizens and their representatives should grant precedence here to two kinds of aspirations. First and most importantly, technologies should—independent of their diverse focal purposes—structurally support the social and institutional conditions necessary to establish and maintain strong democracy itself. (These conditions are discussed later in this chapter.) Second, technologies should structurally respect any other important concerns common to all citizens.

This does not necessarily mean shifting social resources to the design of technologies that focally support democracy or other common goods. That is the instinct of many strong democrats, and some such efforts may be appropriate. For example, there might be a constructive role within strong democracy for electronically mediated "town meetings."

However, a preoccupation with certain technologies' focal functions, if it excludes commensurate attention to their nonfocal functions and to those of other technologies, is apt to prove disappointing or counterproductive. It might, for example, do little good to televise more political debates without first inquiring whether a nonfocal consequence of watching television is to induce passivity rather than critical engagement.[8] And is it obviously more urgent to seek any new technologies that are focally democratic before contemplating the redesign of existing technologies that, nonfocally, are antidemocratic? How can one know that the adverse effect of the latter is not sufficient to override any beneficial effect intended by the former? For instance, it may be fruitless to try to foster civic engagement via interactive telecommunications unless communities are prepared at the same time to promote convivially designed town and city centers; neighborhood parks, greenhouses, workshops, and daycare centers; technologies compatible with democratically managed workplaces and flexible work schedules; more democratically governed urban technological infrastructures; and other steps toward constituting democratic communities.

In other words, societies do not require a special subset of technologies that are focally democratic as a complement to the remaining majority of technologies that are inconsequential to politics, because the remaining technologies are not, in fact, inconsequential. The overall objective ought to be a technological order that structurally manifests a democratic design style. Considering the entirety of a society's disparate technologies—both their focal and nonfocal aspects—is the technological order strongly democratic? That is the first question.

Owing in part to modern societies' persistent neglect of their structural potency, technologies have never systematically been evaluated from the standpoint of their bearing on democracy. Therefore, upon scrutiny, many existing technologies may prove structurally undemocratic. Furthermore, from a dynamic perspective, they may erect obstacles to efforts intended to further democratization. For example, the declining interest in political participation observed within most industrial democracies might be partly attributable to latent subversion of democracy's necessary conditions by technologies. We can start testing such conjectures after formulating criteria for distinguishing structurally democratic technologies from their less democratic counterparts.

Contestable Democratic Design Criteria

If democratic theory can specify that technologies ought above all to be compatible with strong democracy, does that prescription preempt the most important questions that democratic procedures for technological decision making might otherwise address? No. In the first place, this leaves many important questions to the discretion of democratic judgment. These involve debating shared and personal concerns and then striving to ensure that technologies structurally support them. But even on the prior question of seeking a technological order that structurally supports strong democracy, there is a broad and critical role for democratic involvement.

The simple idea that technologies ought to be compatible with strong democracy is entirely abstract. To become effective, it must be expanded into a sequence of successively more specific guidelines for technological design, what I call democratic design criteria. But to specify such criteria with greater precision and content, and then to use them, one must adduce and interpret a progressively wider selection of evidence and exercise judgment. Thus as democratic design criteria become more specific and are applied, the grounds upon which they might reasonably be contested increase.

Moreover, even an expanded system of design criteria will always remain essentially incomplete. For instance, as social circumstances shift or as novel technologies are developed, new criteria will be needed and old ones will have to be reevaluated. In addition, no finite set of criteria can ever fully specify an adequate technological design. Democratic design is ultimately a matter of art and judgment.[9]

This guarantees an ongoing central role for democratic procedure. Democratic theory and its theorists—or anyone else—can help initiate the process of formulating and using design guidelines. However, self-selected actors have neither the knowledge nor the right to make determinative discretionary judgments on behalf of other citizens. Individuals cannot, for example, possibly know what their common interests and preferred democratic institutions are until after they have heard others express their hopes and concerns, and listened to comments on their own, and until everyone has had some chance to reflect on their initial desires and assumptions. Also, individuals cannot trust themselves, pollsters, or scientists to make objective judgments on behalf of others, because invariably each person's, professional's, or group's interests are at stake in the outcome, subtly influencing perception and reasoning. Only democratic forums can supply impartiality born of the balance among multiple perspectives, the opportunity for reflection, and the full range of social knowledge needed to reach legitimate determinations.

Hence it makes sense to seek democratic procedures for formulating and applying rationally contestable design criteria for democratic technologies. These will be "contestable" because the process of generating and refining design criteria cannot be finalized. As technology, social knowledge, and societies and their norms change, one can expect shifts in these design criteria. However, the criteria will be "rationally" or democratically contestable because such shifts need not be arbitrary. They should reflect citizens' current best assessment of the conditions required to realize strong democracy and other shared values. (See figure 19.1 for the basic ingredients of a strong democratic politics of technology.)

Contrast

The theory of democracy and technology developed here contrasts with predecessor theories that emphasize either broadened participation in decision making or else evolving technologies that support democratic social relations, but that do not integrate these procedural and substantive concerns.[10]

The theory also contrasts with a prevalent view, one that arose during the 19th century, that American mass-production technology was democratic because it made consumer goods widely and cheaply available. Democracy thus became equated with a perceived tendency toward equality of opportunity in economic consumption. This earlier view was insensitive to the structural social consequences associated with production technologies themselves and with the goods and services they produced. Furthermore, as a consequence of this blind spot, the theory foresaw no need to complement the market mechanism for making technological choices with any type of political oversight.

FREEDOM: THE MORAL BASIS OF STRONG DEMOCRACY

This chapter opened by simply positing a strongly democratic model of democracy; let us now briefly consider a moral argument supporting the model's desirability. Among human goods or

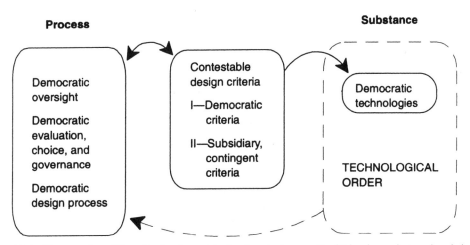

Figure 19.1   Democratic politics of technology. A technology is democratic if it has been designed and chosen with democratic participation or oversight and—considering its focal and nonfocal aspects—is structurally compatible with strong democracy and with citizens' other important common concerns. Within a democratic politics of technology, reflection on existing and proposed technologies plays a role in generating democratic design criteria. Use of these criteria then mediates between democratic procedures and the evolution of a substantively democratic technological order.

The figure distinguishes between two categories of design criteria: (I) priority goes to criteria that help ensure that technologies are compatible with democracy's necessary conditions; (II) subsidiary criteria can then reflect technologies' structural bearing on citizens' other concerns and interests. The dashed line indicates that the entire existing technological order exerts a structural influence on politics generally, including (in this instance) the possibility of a democratic politics of technology.

values, freedom is widely regarded as preeminent. Freedom is a fundamental precondition of all our willful acts, and hence of pursuing all other goods. But under what conditions is one free? Normally, people consider themselves free when no one is interfering with what they want to do. However, this familiar view is not entirely adequate. Suppose a woman is externally free to pursue her desires, but her desires are purely and directly a product of social conditioning or compulsively self-destructive (e.g., heroin addiction)? How truly free is she?

Such considerations suggest that actions are fully free when guided by something in addition to external incentive, social compulsion, or even a person's own instinctive psychological inclinations. That "something" Immanuel Kant identified as morality—specifically, compliance with moral principles that individuals prescribe to themselves. Morality expresses freedom in ways that cannot otherwise occur, even when one chooses among one's own competing psychological inclinations. With the freedom that morality secures, one acquires the dignity of being autonomously self-governing, an "end unto oneself." [11]

But what should the content of moral self-prescriptions be? Kant envisioned one overarching moral principle, what he called the "categorical imperative." One can think of it as a formal restatement of the Golden Rule: always treat others with the respect that you would wish them to accord you, including your fundamental interest in freedom. In Kant's words, "Act so that you treat humanity, whether in your own person or in that of another, always as an end and never as a means only." [12] Thus, in Kantian philosophy the concept of autonomy connotes moral community and readiness to act on behalf of the common good, rather than radical individualism.

However, suppose I behave morally, but nobody else reciprocates? There would be small freedom for me in that kind of society. Living in interdependent association with others (as people do and must) provides innumerable opportunities that could not otherwise exist, including the opportunity to develop moral autonomy. But it also subjects each person to the consequences of others' actions. Should these consequences seem arbitrary or contrary to their interests, people might well judge their freedom diminished.

As a solution, suppose each person's actions were governed by regulative structures, such as laws and government institutions, that they participated in choosing (hence strong democratic procedure) and that respected any important common concerns, particularly their preeminent interest in freedom (hence strong democratic substance). In a society of this sort, laws and other social structures would each stand, in effect, as explicit expressions of mutual agreement to live in accord with Kant's categorical imperative (i.e., to respect oneself and others as ends). [13]

Strong democracy asks that citizens grant priority to commonalities not for their society's own sake, independent of its individual members, but because it is on balance best for each individual member. Strong democratic procedure expresses and develops individual moral freedom, while its structural results constitute conditions requisite to perpetuating maximum equal freedom. Insofar as it envisions democratic procedures for evolving and governing democratic structures, let us call this a model of "democratic structuration."* In other words, democratic structuration represents strong democracy's basic principle of collective self-organization.

---

*The basic concept of structuration is that people's thoughts and behavior are invariably shaped by structures that—through ordinary activities (or sometimes extraordinary ones, such as revolution or constitutional convention)—they participate collectively and continuously in generating, reproducing, or transforming. A rough analogy from the natural world can help convey the idea. Consider a river as a process shaped and guided by a structure: its banks. As the river flows, it is continuously modifying its banks, here through erosion, there through deposition of sediment. Over time the river cuts deep gorges, meanders back and forth across broad floodplains, crafts oxbows and bypasses, and establishes at its mouth a complex deltaic formation. Hence the river is a vibrant example of structuration: a process conditioned by enduring structures that it nonetheless helps continuously to reconstitute.

Combine the preceding normative argument with the conventionally slighted insight that technologies function as an important species of social structure. It follows that evolving a democratic technological order is a moral responsibility of the highest order. A democratic politics of technology—one comprising democratic means for cultivating technologies that structurally support democracy—is needed to transform technology from an arbitrary, irrational, or undemocratic social force into a substantive constituent of human freedom (see figure 19.2).

Of course, democracy is by no means the only issue that needs to be considered when making decisions about technology. Citizens might well wish to make technological decisions based partly

I. Philosophical Case for Strong Democracy:

A. Freedom is a highest order human value. Respecting other people's freedom is a moral duty, necessary for realizing one's own freedom. (Kanitian moral theory)

B. Given the inalterable fact of real-world social interdependence, the opportunity to fully develop and express individual freedom can best be secured within a context of democratic structuration:

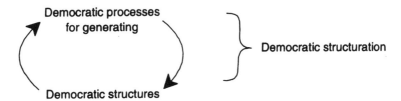

Under these circumstances both social processes and their structural results, by respecting people's freedom, embody Kantian morality. Here structures support rather than constrain people's highest order interest in freedom. (Neo-Rousseauian, or strong democratic, political theory)

II. Philosophical Case for a Democratic Politics of Technology:

Applying the preceding argument (I) to technology:

C. Technologies are a species of social structure. Therefore, it is morally vital that they, like other social structures, be generated and governed via democratic structuration. (The content of democratic structuration, as it applies to technology, is elaborated in figure 19.1.)

Figure 19.2   Philosophical argument for strong democracy and for a democratic politics of technology.

In social life what we do and who we are (or may become) is similarly guided by our society's basic structures: its laws, major political and economic institutions, cultural beliefs, and so on. But our activities nevertheless produce cumulative material and psychological results, not fully determined by structure, that in turn are woven back into our society's evolving structural complex. Hence at every moment we contribute marginally—or, upon occasion, dramatically—to affirming or transforming our society's basic structures.

The word "structuration"—introduced by Giddens (1979, chap. 2)—is not aesthetically pleasing, but it has achieved wider currency than any synonym. I propose the term "democratic structuration" to embed this explanatory concept in a normative context, suggesting that the means and the ends of structuration should be guided by an overarching respect for moral freedom.

on practical, economic, cultural, environmental, religious, or other grounds. But among these diverse considerations, priority should go to the question of technologies' bearing on democracy. This is because democracy is fundamental, establishing the necessary background circumstance for us to be able to decide fairly and effectively what other issues to take into account in both our technological and nontechnological decision making. (Granting priority to democracy within technological decisions would be somewhat analogous to ensuring compatibility with the U.S. Constitution when drafting or debating proposed laws or regulations.)

It would be presumptuous, however, to insist that the case supporting strong democracy is entirely conclusive. This implies that the contestability attributed earlier to democratic technological design criteria stretches logically back into the supporting theory's philosophical core.

DEMOCRATIC BACKGROUND CONDITIONS

Establishing democratic structuration depends on a number of background conditions. These need to be elaborated here in just enough detail to permit subsequent derivation of democratic design criteria for technologies. The requisite background conditions include: (1) some commonality of purpose, attachment, or outlook among citizens (at a minimum, general recognition of a preeminent interest in living in a strong democracy); (2) some general readiness on the part of citizens to accord higher political priority to advancing important common purposes than to narrower personal concerns; and (3) institutions that foster these circumstances. These background conditions, in turn, incorporate three organizing principles: democratic politics, democratic community, and democratic work (see figure 19.3).

Democratic Politics

A strong democracy, by definition, affords citizens roughly equal, and maximally extensive, opportunities to guide their society's evolution. What kinds of political institutions does this imply?

Participation and Representation

Is there a middle ground between the present systems of representation, in which a few people participate in deciding most important issues, and obsessive participation, in which *all* people are expected to participate actively in *all* important issues. Barber characterizes the middle ground as a democracy

> in which all of the people govern themselves in at least some public matters at least some of the time. . . . Active citizens govern themselves directly . . . , not necessarily . . . in every instance, but frequently enough and in particular when basic policies are being decided and when significant power is being deployed.[14]

Moreover, on issues in which individuals choose not to participate, they should know that generally others with a similar point of view are participating competently, in effect on their behalf. This may entail, among other things, institutional mechanisms to ensure that the views of socially disadvantaged groups are fully represented and that their needs and rights are respected.[15]

Broadened and equalized opportunities for participation are more than a matter of formal legal rights. They must be supported by relatively equal access to the resources required for efficacy, including time and money. Today, for example, politicians, government functionaries, soldiers, and jurors are paid to perform their civic duties. Why not, when necessary, pay citizens to perform theirs, as did the ancient Athenians? Fairness and equality may also be served by increasing the ratio of representatives chosen by lot to those chosen by vote.[16]

Political Decentralization and Federation

What can help prevent representation from gradually usurping the role of an active citizenry? A partial answer lies in some sort of devolution of centralized political institutions (in which the population's large size renders meaningful participation by all impossible) in favor of a plurality of more autonomous, local political units. By means of small-scale local politics, more voices can be heard and each can carry more weight than in a larger polity. Decisions can be more responsive to individuals, thereby increasing citizens' incentive to participate. There is also potential for small polities to be able to govern themselves somewhat more consensually than can the larger society.

However, various considerations—such as the importance of protecting minorities from local repression—suggest the need to embed decentralization within a larger, federated democratic system. The detailed form of federation must be decided contextually, but its thrust should be toward (1) subsidiarity (i.e., decisions should be made at the lowest political level competent to make them), (2) egalitarianism within and among polities, and (3) global awareness or nonparochialism

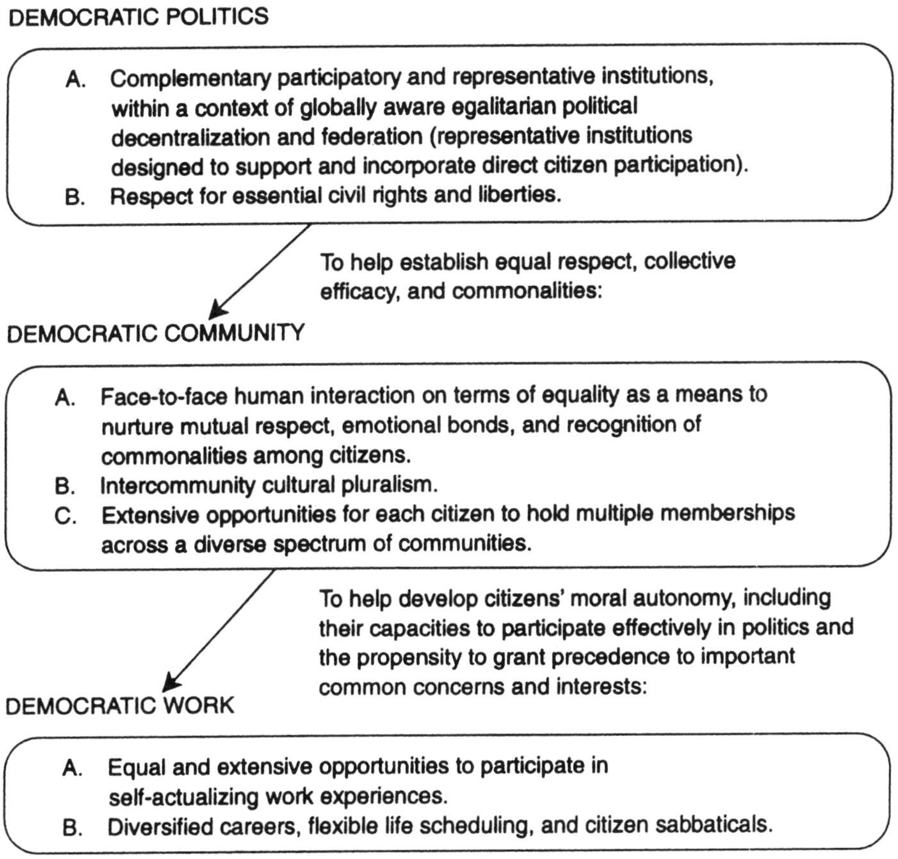

DEMOCRATIC POLITICS

A. Complementary participatory and representative institutions, within a context of globally aware egalitarian political decentralization and federation (representative institutions designed to support and incorporate direct citizen participation).
B. Respect for essential civil rights and liberties.

To help establish equal respect, collective efficacy, and commonalities:

DEMOCRATIC COMMUNITY

A. Face-to-face human interaction on terms of equality as a means to nurture mutual respect, emotional bonds, and recognition of commonalities among citizens.
B. Intercommunity cultural pluralism.
C. Extensive opportunities for each citizen to hold multiple memberships across a diverse spectrum of communities.

To help develop citizens' moral autonomy, including their capacities to participate effectively in politics and the propensity to grant precedence to important common concerns and interests:

DEMOCRATIC WORK

A. Equal and extensive opportunities to participate in self-actualizing work experiences.
B. Diversified careers, flexible life scheduling, and citizen sabbaticals.

Figure 19.3   Some of strong democracy's principal necessary conditions.

(i.e., ideally, everyone manifests a measure of knowledge and concern with the entire federated whole—or even beyond it).

In short, power should be relatively diffuse and equal. Political interaction and accountability should be multidirectional—flowing horizontally among local polities, vertically back and forth between local polities and more comprehensive political units, and cross-cut in less formal ways by nonterritorially based groups, voluntary associations, and social movements.

Agenda Setting and Civil Rights

What if citizens are widely empowered to participate in societal choices, but the menu of choice is so restricted that they cannot express their true wishes? Numerous political theorists agree that decision-making *processes* are democratically inadequate, even spurious, unless they are combined with relatively equal and extensive opportunities for citizens, communities, and groups to help shape decision-making *agendas*. Various civil liberties and protections are also democratically essential either because they are intrinsic to respecting people as moral agents or because we require them in order to function as citizens.[17]

Democratic Community

A strong democracy requires local communities composed of free and equal members that are substantially self-governing. Such communities help constitute the foundation of a decentrally federated democratic polity, and hence of political participation, freedom, and efficacy. They do this in part by establishing a basis for individual empowerment within collectivities that, as such, are much more able than individual citizens to contest the emergence of democratically unaccountable power elsewhere in society (e.g., in neighboring communities, private corporations, nonterritorial interest groups, or higher echelons of federative government). Local communities also provide a key site for coming to know oneself and others fully and contextually as moral agents. The defining features of a democratic community include social structures and practices that nurture collective efficacy, mutual respect, and moral and political equality, and, if possible, help sustain a measure of communitywide commonality.[18]

Strong democratic theory does not envision a perfect societal harmony of interest, sentiment, or perspective. Rather, the central aims of strong democratic practice include seeking amid the fray for any existing areas of commonness; striving to invent creative solutions that, in a just manner, enhance the ratio of concordance to that of conflict; and balancing the search for common purpose against respect for enduring differences and against coercive pressures toward conformity.

Of course, countless forms of community and human association are not locally based. However, strong democracy places a special weight on local community as a foundation (but not a culmination) because of the distinctive and inescapable physical and moral interdependencies that arise at the local level; the territorial grounding of political jurisdictions; and the distinctive quality of mutual understanding, learning, and personal growth that can take place through sustained, contextually situated, face-to-face discourse and interaction.

Cultural Pluralism

If one next considers an entire society's overall pattern of kinds of communities and associations, one discovers that strong democracy does more than *permit* diversity among them: it *requires* diversity. Specifically, democracies should manifest a certain kind of institutional and cultural pluralism: equal respect and protection for all cultures, communities, traditions, and ways of life whose practices can reasonably be construed to affirm equal respect and freedom for all. (Cultures that fail to meet this standard may not warrant unqualified respect, but neither do they warrant determined intervention—unless, that is, they seriously threaten the viability of other, democratic cultures or oppress their own members involuntarily.[19])

There are two principal reasons for this requirement. First, equal respect for people entails respecting their cultural heritage. To undermine a culture corrodes the social bases of its members' sense of self and purpose. Second, all people share an interest in living in a society and a world comprised of many cultures. Cultural pluralism supplies alternative viewpoints from which individuals can learn to see their own culture's strengths and limitations, thereby enriching their lives, understanding, and even survival prospects. Moreover, it provides alternative kinds of communities to which people can travel or move if they become sufficiently dissatisfied with their own.

Democratic Macrocommunity

Democratic politics beyond the level of a single community or group requires generally accepted means of addressing disagreements, and ideally a measure of societywide mutual respect or commonality. The alternatives can include authoritarianism, civil violence, or even genocide—as modern history vividly demonstrates. How, then, can local or association-specific solidarity, together with translocal cultural pluralism, possibly be reconciled with the conditions needed for societywide democracy?

Cultures and groups invariably disagree on fundamental matters sometimes. Nonetheless, there is reason to believe that local democratic communities represent a promising foundation for cultivating societywide respect or commonality. For one thing, ethnic hatred and violence are frequently associated with longstanding political-economic inequalities, not with extant approximations to strong democracy. Moreover, often it is probably harder to escape acknowledging and learning to accommodate differences when engaged in local democracy than in translocal association or politics, where there can be more leeway to evade, deny, or withdraw from differences.

The alternative notion of forging a macropolitical culture at the expense of local democratic communities risks coercion or a mass society, in which people relate abstractly rather than as concrete, multidimensional moral agents. Members of a mass society cannot feel fully respected as whole selves, and furthermore, they are vulnerable to self-deception concerning other citizens' needs and to manipulation by those feigning privileged knowledge of the common good.[20]

One method of nurturing local nonparochialism is to pursue cooperative relations among communities that are distantly located and culturally distinct. There is a good model in those modern U.S. and European cities that have established collaborative relations with communities in other nations regarding matters of peace, international justice, environmental protection, or economic development.[21]

Another route to nonparochialism is to ensure that people have extensive opportunities to experience life in a variety of different kinds of communities. Generally, such opportunities should involve experience both in (1) a culturally diverse array of small face-to-face communities or groups and (2) socially comprehensive, nonterritorially based communities.* The former would encourage concrete understanding of the lives and outlooks of different kinds of people and communities; the latter would provide practical opportunities to generalize and apply what one has learned from these diverse experiences to the problems and well-being of society as a whole.

---

*The latter are communities or organizations that manifest a multifaceted concern with the well-being of a wide range of kinds of people, if not the entire society or world. Examples include broad-based political parties or movements; federation-level government agencies; and translocal, nongovernmental social service organizations.

Nonterritorial associations that are not socially comprehensive—such as ethnic associations, labor unions, churches, single-issue political organizations, and so on—can obviously function as one kind of rewarding community for their members. However, they seem less likely to provoke deep, multiculturally informed comprehension of an entire society.

Democratic Work

People often think of "work" as something they do primarily to earn a living. Here, however, work is interpreted as a lifelong process whose central functions include individual self-development as well as social maintenance (both biological and cultural). "Democratic work" thus denotes (1) work activity through which one can discover, develop, and express one's creative powers, strengthen one's character, and enhance one's self-esteem, efficacy, and moral growth (including readiness to act on behalf of common interests and concerns); (2) a work setting that permits one to help choose the product, intermediate activities, and social conditions of one's labor, thereby developing political competence within a context of democratic self-governance; and (3) the creation of material or other cultural products that are consistent with democracy's necessary conditions, that are useful or pleasing to oneself or to others, and that thus contribute to social maintenance and mutual and self-respect.[22]

Democratic work contributes richly to individual autonomy and democratic society. Hence, societies cannot be considered strongly democratic if there is involuntary unemployment or if, for example, many people are compelled to work in social environments that are tedious and hierarchically structured, while a few elite managers make important decisions that affect many other citizens. To the extent that good jobs are scarce or societal maintenance requires a certain amount of drudgery or other unpleasant work, vigorous efforts should be made to ensure the sharing of both unpleasant and pleasurable activities.[23]

Diversified Careers and Citizen Sabbaticals

Numerous social thinkers have suggested that people should be able to work in a variety of different careers—either in linear sequence or, preferably, in fluidly alternating succession (sometimes called "flexible life scheduling").[24] However, one reason that is often overlooked concerns the cultivation of citizens' readiness to respect people everywhere as ends in themselves and to act on behalf of societywide interests.

To capture this benefit might require a societywide system, analogous to faculty sabbaticals or the U.S. Peace Corps, that would encourage each person to occasionally take a leave of absence from his or her home community, to live and work for perhaps a month each year or a year each decade in another community, culture, or region. This sabbatical system should include opportunities for the broadest possible number and range of people to take turns within translocal government and administration.

Citizens could then return to their home communities with a deeper appreciation of the diverse needs of other communities, a broader experiential basis from which to conceive of their society's general interest, and lingering emotional attachments to the other communities. (Note, for instance how increased contact between white and African-American soldiers in the U.S. armed services has generally reduced racial prejudice there, thus increasing receptivity to societywide racial integration and equality.[25]) Citizen sabbaticals would thus provide one concrete means of implementing the earlier proposal that citizens have the opportunity for lived experience in a culturally diverse array of communities.

Guiding Principles

Do the three seemingly distinct social domains—formal politics, community, and work—mean that different kinds of basic institutions each contribute to democracy in an essentially different but complementary way? Suppose, instead, one conceives of democratic politics, community, and work as three guiding principles that should each to some extent be active within every basic institutional setting or association (recall figure 19.3). An actual workplace, for example, may be conceived primarily as a locus of self-actualizing experience and production (work), but it should also ordinarily be governed democratically (politics) and help nurture mutual regard (community).

Failure to embody, within each of a society's many settings and associations, all three principles will tend to result in a whole society much less than the sum of its institutional parts. When, for example, each of a society's basic institutions is merely monoprincipled not only does each fall short of constituting a democratic microcosm, but each in addition tends to stress and overtax the capacities of the others.

Democratic Knowledge

Widespread political participation and the experience of diverse cultures and forms of work amount to an experientially based program of civic education. Living this way, one could hardly help but acquire extensive knowledge of one's world and society. This is not only positive; it is also democratically vital and a civil right. Competent citizenship, moral development, self-esteem, and cultural maintenance all depend on extensive opportunities, available to both individuals and cultural groups, to participate in producing, contesting, disseminating, and critically appropriating social knowledge, norms, and cultural meaning.[26]

Formal politics, in particular, must incorporate procedures that support collective self-education and deliberation. The means might include ensuring multiple independent sources of information with effective representation of minority perspectives; open and diverse means of participatory political communication and deliberation (including subsidies to disadvantaged groups that would otherwise be excluded); and extensive and convenient means of monitoring government performance.

NOTES

1. Tripp (1980, pp. 19, 33).
2. Barber (1984). I find the term "strong democracy" congenial and I agree with many of Barber's ideas. However, I also draw on other theoretical treatments that complement or qualify his model, such as Rawls (1971), Mansbridge (1980), Bowles and Gintis (1986), Gould (1988), Young (1989), and Cohen et al. (1992). These works also include extensive criticism of thin democracy or its underlying, classical liberal presuppositions.
3. Barber (1984).
4. Social struggles, the influence of business interests and social elites, and city planning have all, of course, left their mark on contemporary cities. But often planning has done little but help establish the infrastructure under which anarchic growth could proceed. On the technologically influenced evolution of American urban and suburban areas, see, for example, Mumford (1961), Cowan (1983), Hayden (1984), Jackson (1985), and Rose and Tarr (1987).
5. Berger (1977, vol. I, p. 22).
6. Sources on the Berger Inquiry include Berger (1977), Gamble (1978), and OECD (1979, pp. 61–77).
7. Tripp (1980, p. 19).
8. E.g., Mander (1978).

9. On the art and process of design, see Pye (1982), O'Cathain (1984), Ferguson (1992), and Thomas (1994).

10. Examples of conceptions of technology and democracy that are either purely procedural or purely substantive include, respectively, OECD (1979) and Mumford (1964). Notable prior steps toward synthesizing these two components include Illich (1973), Bookchin (1982), Goldhaber (1986), and Feenberg (1991).

11. Kant (1959, pp. 46–54).

12. Ibid., p. 47.

13. In its essential logic, this solution is Rousseau's in *The Social Contract* (1968).

14. Barber (1984, pp. xiv, 151).

15. Young (1989).

16. Barber (1984, pp. 290–293). On Athenian democracy, see Finley (1983, pp. 70–84).

17. E.g., Cobb and Elder (1972), Rawls (1971).

18. Small communities can sometimes be parochial or oppressive, although—by definition—not if they are strongly democratic. For evidence and argument contrary to the view that local communities are invariably parochial or oppressive, see Mansbridge (1980), Taylor (1982), and Morgan (1984). Various social observers (e.g., Lee 1976) argue that the dangers of enforced social conformity are greater in a mass society than in one composed of smaller, differentiated, democratic social units. On communities as mediating structures that empower individuals, see Frug (1980), Bowles and Gintis (1986), and Berry (1993). The potential power inherent in local democratic community is made manifest, for example, in Old Order Amish people's success in resisting the joint coercive force toward conformity resulting from external market relations, persistent social ridicule, and one-time government opposition to various Amish practices (such as refusal to participate in Social Security, military service, and the public school system).

19. See Rawls (1971, sect. 35) and Gutmann (1987, pp. 41–47).

20. Calhoun (1991).

21. Brugmann (1989).

22. For one reasonable attempt at synthesizing a number of contrasting modern views on individual moral development, see Kegan (1982). Empirical research has not decisively confirmed or refuted the thesis that self-governed workplaces encourage political participation beyond the workplace (e.g., compare Elden 1981 with Greenberg 1986).

23. Schor (1992). For examples of contemporary workplaces in which menial tasks are shared, see Rothschild and Whitt (1986) and Holusha (1994). Feminist writing on the importance of sharing childcare responsibilities are also salient here (e.g., Hochschild 1989).

24. E.g., Sirianni (1988).

25. Lovell and Stiehm (1989).

26. For perspectives on democratic knowledge production and civic education, see, for example, Schutz (1946), Adams and Horton (1975), Goodwyn (1978), Foucault (1980), Freire (1980), Guess (1981), Barber (1984), Belenky (1986), Gutmann (1987), and Harding (1993).

20

Bigger Monster, Weaker Chains: The Growth of an American Surveillance Society

Jay Stanley and Barry Steinhardt

INTRODUCTION

Privacy and liberty in the United States are at risk. A combination of lightning-fast technological innovation and the erosion of privacy protections threatens to transform Big Brother from an oft-cited but remote threat into a very real part of American life. We are at risk of turning into a Surveillance Society.

The explosion of computers, cameras, sensors, wireless communication, GPS, biometrics, and other technologies in just the last ten years is feeding a surveillance monster that is growing silently in out midst. Scarcely a month goes by in which we don't read about some new high-tech way to invade people's privacy, from face recognition to implantable microchips, data-mining, DNA chips, and even "brain wave fingerprinting." The fact is, there are no longer any *technical* barriers to the Big Brother regime portrayed by George Orwell.

Even as this surveillance monster grows in power, we are weakening the legal chains that keep it from trampling our lives. We should be responding to intrusive new technologies by building stronger restraints to protect our privacy; instead, we are doing the opposite—loosening regulations on government surveillance, watching passively as private surveillance grows unchecked, and contemplating the introduction of tremendously powerful new surveillance infrastructures that will tie all this information together.

A gradual weakening of our privacy rights has been underway for decades, but many of the most startling developments have come in response to the terrorist attacks of September 11. But few of these hastily enacted measures are likely to increase our protection against terrorism. More often than not, September 11 has been used as a pretext to loosen constraints that law enforcement has been chafing under for years.

It doesn't require some apocalyptic vision of American democracy being replaced by dictatorship to worry about a surveillance society. There is a lot of room for the United States to become a meaner, less open and less just place without any radical change in government. All that's required is the continued construction of new surveillance technologies and the simultaneous erosion of privacy protections.

"Bigger Monster, Weaker Chains," by Jay Stanley and Barry Steinhard, January 2003. American Civil Liberties Union Technology and Liberty Program. Reprinted by permission of the American Civil Liberties Union.

It's not hard to imagine how in the near future we might see scenarios like the following:

- An African-American man from the central city visits an affluent white suburb to attend a co-worker's barbeque. Later that night, a crime takes place elsewhere in the neighborhood. The police review surveillance camera images, use face recognition to identify the man, and pay him a visit at home the next day. His trip to the suburbs where he "didn't belong" has earned him an interrogation from suspicious police.
- A tourist walking through an unfamiliar city happens upon a sex shop. She stops to gaze at several curious items in the store's window before moving along. Unbeknownst to her, the store has set up the newly available "Customer Identification System," which detects a signal being emitted by a computer chip in her driver's license and records her identity and the date, time, and duration of her brief look inside the window. A week later, she gets a solicitation in the mail mentioning her "visit" and embarrassing her in front of her family.

Such possibilities are only the tip of the iceberg. The media faithfully reports the latest surveillance gadgets and the latest moves to soften the rules on government spying, but rarely provides the big picture. That is unfortunate, because each new threat to our privacy is much more significant as part of the overall trend than it seems when viewed in isolation. When these monitoring technologies and techniques are combined, they can create a surveillance network far more powerful than any single one would create on its own.

The good news is that these trends can be stopped. As the American people realize that each new development is part of this larger story, they will give more and more weight to protecting privacy, and support the measures we need to preserve our freedom.

THE GROWING SURVEILLANCE MONSTER

In the film *Minority Report,* which takes place in the United States in the year 2050, people called "Pre-cogs" can supposedly predict future crimes, and the nation has become a perfect surveillance society. The frightening thing is that except for the psychic Pre-cogs, the technologies of surveillance portrayed in the film already exist or are in the pipeline. Replace the Pre-cogs with "brain fingerprinting"—the supposed ability to ferret out dangerous tendencies by reading brain waves—and the film's entire vision no longer lies far in the future. Other new privacy invasions are coming at us from all directions, from video and data surveillance to DNA scanning to new data-gathering gadgets.

Video Surveillance

Surveillance video cameras are rapidly spreading throughout the public arena. A survey of surveillance cameras in Manhattan, for example, found that it is impossible to walk around the city without being recorded nearly every step of the way. And since September 11 the pace has quickened, with new cameras being placed not only in some of our most sacred public spaces, such as the National Mall in Washington and the Statue of Liberty in New York harbor, but on ordinary public streets all over America.

As common as video cameras have become, there are strong signs that, without public action, video surveillance may be on the verge of a revolutionary expansion in American life. There are three factors propelling this revolution:

1. **Improved technology.** Advances such as the digitization of video mean cheaper cameras, cheaper transmission of far-flung video feeds, and cheaper storage and retrieval of images.
2. **Centralized surveillance.** A new centralized surveillance center in Washington, D.C., is an early indicator of what technology may bring. It allows officers to view images from video cameras across the city—public buildings and streets, neighborhoods, Metro stations, and even schools. With the flip of a switch, officers can zoom in on people from cameras a half mile away.[1]
3. **Unexamined assumptions that cameras provide security.** In the wake of the September 11 attacks, many embraced surveillance as the way to prevent future attacks and prevent crime. But it is far from clear how cameras will increase security. U.S. government experts on security technology, noting that "monitoring video screens is both boring and mesmerizing," have found in experiments that after only 20 minutes of watching video monitors, "the attention of most individuals has degenerated to well below acceptable levels."[2] In addition, studies of cameras' effect on crime in Britain, where they have been extensively deployed, have found no conclusive evidence that they have reduced crime.[3]

These developments are creating powerful momentum toward pervasive video surveillance of our public spaces. If centralized video facilities are permitted in Washington and around the nation, it is inevitable that they will be expanded—not only in the number of cameras but also in their power and ability. It is easy to foresee inexpensive, one-dollar cameras being distributed throughout our cities and tied via wireless technology into a centralized police facility where the life of the city can be monitored. Those video signals could be stored indefinitely in digital form in giant but inexpensive databases, and called up with the click of a mouse at any time. With face recognition, the video records could even be indexed and searched based on who the systems identify—correctly, or all too often, incorrectly.

Several airports around the nation, a handful of cities, and even the National Park Service at the Statue of Liberty have installed face recognition. While not nearly reliable enough to be effective as a security application,[4] such a system could still violate the privacy of a significant percentage of the citizens who appeared before it, as well as those who do not appear before it but are falsely identified as having done so. Unlike, say, an iris scan, face recognition doesn't require the knowledge, consent, or participation of the subject; modern cameras can easily view faces from over 100 yards away.

Further possibilities for the expansion of video surveillance lie with unmanned aircraft, or drones, which have been used by the military and the CIA overseas for reconnaissance, surveillance, and targeting. Controlled from the ground, they can stay airborne for days at a time. Now there is talk of deploying them domestically. Senate Armed Services Committee Chairman John Warner (R-VA) said in December 2002 that he wants to explore their use in Homeland Security, and a number of domestic government agencies have expressed interest in deploying them. Drones are likely to be just one of many ways in which improving robotics technology will be applied to surveillance.[5]

The bottom line is that surveillance systems, once installed, rarely remain confined to their original purpose. Once the nation decides to go down the path of seeking security through video surveillance, the imperative to make it work will become overwhelming, and the monitoring of citizens in public places will quickly become pervasive.

Data Surveillance

An insidious new type of surveillance is becoming possible that is just as intrusive as video surveillance—what we might call "data surveillance." Data surveillance is *the collection of information about an identifiable individual, often from multiple sources, that can be assembled into a portrait of that person's activities.*[6] Most computers are programmed to automatically store and track usage data, and the spread of computer chips in our daily lives means that more and more of our activities leave behind "data trails." It will soon be possible to combine information from different sources to re-create an individual's activities with such detail that it becomes no different than being followed around all day by a detective with a video camera.

Some think comprehensive public tracking will make no difference, since life in public places is not "private" in the same way as life inside the home. This is wrong; such tracking would represent a radical change in American life. A woman who leaves her house, drives to a store, meets a friend for coffee, visits a museum, and then returns home may be in public all day, but her life is still private in that she is the only one who has an overall view of how she spent her day. In America, she does not expect that her activities are being watched or tracked in any systematic way—she expects to be left alone. But if current trends continue, it will be impossible to have any contact with the outside world that is not watched and recorded.

The Commodification of Information

A major factor driving the trend toward data surveillance forward is the commodification of personal information by corporations. As computer technology exploded in recent decades, making it much easier to collect information about what Americans buy and do, companies came to realize that such data is often very valuable. The experience of marketing efforts gives businesses a strong incentive to know as much about consumers as possible so they can focus on the most likely new customers. Surveys, sweepstakes questionnaires, loyalty programs and detailed product registration forms have proliferated in American life—all aimed at gathering information about consumers. Today, any consumer activity that is *not* being tracked and recorded is increasingly being viewed by businesses as money left on the table.

On the Internet, where every mouse click can be recorded, the tracking and profiling of consumers is even more prevalent. Web sites can not only track what consumers buy, but what they *look at*—and for how long, and in what order. With the end of the Dot Com era, personal information has become an even more precious source of hard cash for those Internet ventures that survive. And of course Americans use the Internet not just as a shopping mall, but to research topics of interest, debate political issues, seek support for personal problems, and many other purposes that can generate deeply private information about their thoughts, interests, lifestyles, habits, and activities.

Genetic Privacy

The relentless commercialization of information has also led to the breakdown of some longstanding traditions, such as doctor-patient confidentiality. Citizens share some of their most intimate and embarrassing secrets with their doctors on the old-fashioned assumption that their conversations are confidential. Yet those details are routinely shared with insurance companies, researchers, marketers, and employers. An insurance trade organization called the Medical Information Bureau even keeps a centralized medical database with records on millions of patients. Weak new medical privacy rules will do little to stop this behavior.

An even greater threat to medical privacy is looming: genetic information. The increase in DNA analysis for medical testing, research, and other purposes will accelerate sharply in coming years, and will increasingly be incorporated into routine health care.

Unlike other medical information, genetic data is a unique combination: both difficult to keep confidential and extremely revealing about us. DNA is very easy to acquire because we constantly slough off hair, saliva, skin cells, and other samples of our DNA (household dust, for example, is made up primarily of dead human skin cells). That means that no matter how hard we strive to keep our genetic code private, we are always vulnerable to other parties' secretly testing samples of our DNA. The issue will be intensified by the development of cheap and efficient DNA chips capable of reading parts of our genetic sequences.

Already, it is possible to send away a DNA sample for analysis. A testing company called Genelex reports that it has amassed 50,000 DNA samples, many gathered surreptitiously for paternity testing. "You'd be amazed," the company's CEO told *U.S. News & World Report.* "Siblings have sent in mom's discarded Kleenex and wax from her hearing aid to resolve the family rumors."[7]

Not only is DNA easier to acquire than other medical information, revealing it can also have more profound consequences. Genetic markers are rapidly being identified for all sorts of genetic diseases, risk factors, and other characteristics. None of us knows what time bombs are lurking in our genomes.

The consequences of increased genetic transparency will likely include:

- **Discrimination by insurers.** Health and life insurance companies could collect DNA for use in deciding who to insure and what to charge them, with the result that a certain proportion of the population could become uninsurable. The insurance industry has already vigorously opposed efforts in Congress to pass meaningful genetic privacy and discrimination bills.
- **Employment discrimination.** Genetic workplace testing is already on the rise, and the courts have heard many cases. Employers desiring healthy, capable workers will always have an incentive to discriminate based on DNA—an incentive that will be even stronger as long as health insurance is provided through the workplace.
- **Genetic spying.** Cheap technology could allow everyone from schoolchildren to dating couples to nosy neighbors to routinely check out each other's genetic codes. A likely high-profile example: online posting of the genetic profiles of celebrities or politicians.

Financial privacy

Like doctor-patient confidentiality, the tradition of privacy and discretion by financial institutions has also collapsed; financial companies today routinely put the details of their customers' financial lives up for sale.

A big part of the problem is the Gramm-Leach-Bliley Act passed by Congress in 1999. Although Gramm-Leach is sometimes described as a "financial privacy law," it created a very weak privacy standard—so weak, in fact, that far from protecting Americans' financial privacy, the law has had the effect of ratifying the increasing abandonment of customer privacy by financial companies.

Gramm-Leach effectively gives financial institutions permission to sell their customers' financial data to anyone they choose. That includes the date, amount, and recipient of credit card charges or checks a customer has written; account balances; and information about the flow of deposits and withdrawals through an account. Consumers provide a tremendous amount of information about themselves when they fill our applications to get a loan, buy insurance, or purchase securities, and companies can also share that information. In fact, the only information a financial company may NOT give out about you is your account number.

Under Gramm-Leach, you get no privacy unless you file complex paperwork, following a financial institution's precise instructions before a deadline they set, and repeating the process for each and every financial service provider who may have data about you. And it is a process that many companies intentionally make difficult and cumbersome; few let consumers "opt out" of data sharing through a Web site or phone number, or even provide a self-addressed envelope.

Gramm-Leach is an excellent example of the ways that privacy protections are being weakened even as the potential for privacy invasion grows.

New Data-Gathering Technologies

The discovery by business of the monetary value of personal information and the vast new project of tracking the habits of consumers has been made possible by advances in computers, databases, and the Internet. In the near future, other new technologies will continue to fill out the mosaic of information it is possible to collect on every individual. Examples include:

- **Cell phone location data.** The government has mandated that manufacturers make cell phones capable of automatically reporting their location when an owner dials 911. Of course, those phones are capable of tracking their location at other times as well. And in applying the rules that protect the privacy of telephone records to this location data, the government is weakening those rules in a way that allows phone companies to collect and share data about the location and movements of their customers.
- **Biometrics.** Technologies that identify us by unique bodily attributes such as our fingerprints, faces, iris patterns, or DNA are already being proposed for inclusion on national ID cards and to identify airline passengers. Face recognition is spreading. Fingerprint scanners have been introduced as security or payment mechanisms in office buildings, college campuses, grocery stores and even fast-food restaurants. And several companies are working on DNA chips that will be able to instantly identify individuals by the DNA we leave behind everywhere we go.
- **Black boxes.** All cars built today contain computers, and some of those computers are being programmed in ways that are not necessarily in the interest of owners. An increasing number of cars contain devices akin to the "black boxes" on aircraft that record details about a vehicle's operation and movement. Those devices can "tattle" on car owners to the police or insurance investigators. Already, one car rental agency tried to charge a customer for speeding after a GPS device in the car reported the transgression back to the company. And cars are just one example of how products and possessions can be programmed to spy and inform on their owners.
- **RFID chips.** RFID chips, which are already used in such applications as toll-booth speed passes, emit a short-range radio signal containing a unique code that identifies each chip. Once the cost of these chips falls to a few pennies each, plans are underway to affix them to products in stores, down to every can of soup and tube of toothpaste. They will allow everyday objects to "talk" to each other—or to anyone else who is listening. For example, they could let market researchers scan the contents of your purse or car from five feet away, or let police officers scan your identification when they pass you on the street.
- **Implantable GPS chips.** Computer chips that can record and broadcast their location have also been developed. In addition to practical uses such as building them into shipping containers, they can also serve as location "bugs" when, for example, hidden by a suspicious husband in a wife's purse. And they can be implanted under the skin (as can RFID chips).

If we do not act to reverse the current trend, data surveillance—like video surveillance—will allow corporations or the government to constantly monitor what individual Americans do every day. Data surveillance would cover *everyone,* with records of every transaction and activity squir-reled away until they are sucked up by powerful search engines, whether as part of routine security checks, a general sweep for suspects in an unsolved crime, or a program of harassment against some future Martin Luther King.

GOVERNMENT SURVEILLANCE

Data surveillance is made possible by the growing ocean of privately collected personal data. But who would conduct that surveillance? There are certainly business incentives for doing so; compa-nies called data aggregators (such as Acxiom and ChoicePoint) are in the business of compiling detailed databases or individuals and then selling that information to others. Although these compa-nies are invisible to the average person, data aggregation is an enormous, multi-billion-dollar indus-try. Some databases are even "co-ops" where participants agree to contribute data about their customers in return for the ability to pull our cross-merchant profiles of customers' activities.

The biggest threat to privacy, however, comes from the government. Many Americans are naturally concerned about corporate surveillance, but only the government has the power to take away liberty—as has been demonstrated starkly by the post-September 11 detention of suspects without trial as "enemy combatants."

In addition, the government has unmatched power to centralize all the private sector data that is being generated. In fact, the distinction between government and private-sector privacy invasions is fading quickly. The Justice Department, for example, reportedly has an $8 million contract with data aggregator ChoicePoint that allows government agents to tap into the company's vast database of personal information on individuals.[8] Although the Privacy Act of 1974 banned the government from maintaining information on citizens who are not the targets of investigations, the FBI can now evade that requirement by simply purchasing information that has been collected by the private sector. Other proposals—such as the Pentagon's "Total Information Awareness" project and airline passenger profiling programs—would institutionalize government access to consumer data in even more far-reaching ways (see below).

Government Databases

The government's access to personal information begins with the thousands of databases it main-tains on the lives of Americans and others. For instance:

- The FBI maintains a giant database that contains millions of records covering everything from criminal records to stolen boats and databases with millions of computerized finger-prints and DNA records.
- The Treasury Department runs a database that collects financial information reported to the government by thousands of banks and other financial institutions.
- A "new hires" database maintained by the Department of Health and Human Services, which contains the name, address, social security number, and quarterly wages of every working person in the United States.
- The federal Department of Education maintains an enormous information bank holding years worth of educational records on individuals stretching from their primary school years through higher education. After September 11, Congress gave the FBI permission to access the database without probable cause.
- State departments of motor vehicles of course possess millions of up-to-date files contain-

ing a variety of personal data, including photographs of most adults living in the United States.

Communications Surveillance

The government also performs an increasing amount of eavesdropping on electronic communications. While technologies like telephone wiretapping have been around for decades, today's technologies cast a far broader net. The FBI's controversial "Carnivore" program, for example, is supposed to be used to tap into the e-mail traffic of a particular individual. Unlike a telephone wiretap, however, it doesn't cover just one device but (because of how the Internet is built) filters through *all* the traffic on the Internet Service Provider to which it has been attached. The only thing keeping the government from trolling through all this traffic is software instructions that are written by the government itself. (Despite that clear conflict of interest, the FBI has refused to allow independent inspection and oversight of the device's operation.)

Another example is the international eavesdropping program code-named Echelon. Operated by a partnership consisting of the United States, Britain, Canada, Australia, and New Zealand, Echelon reportedly grabs e-mail, phone calls, and other electronic communications from its far-flung listening posts across most of the earth. (U.S. eavesdroppers are not supposed to listen in on the conversations of Americans, but the question about Echelon has always been whether the intelligence agencies of participating nations can set up reciprocal, back-scratching arrangements to spy on each others' citizens.) Like Carnivore, Echelon may be used against particular targets, but to do so its operators must sort through massive amounts of information about potentially millions of people. That is worlds away from the popular conception of the old wiretap where an FBI agent listens to one line. Not only the volume of intercepts but the potential for abuse is now exponentially higher.

The "Patriot" Act

The potential for the abuse of surveillance powers has also risen sharply due to a dramatic post-9/11 erosion of legal protections against government surveillance of citizens. Just six weeks after the September 11 attacks, a panicked Congress passed the "USA PATRIOT Act," an overnight revision of the nation's surveillance laws that vastly expanded the government's authority to spy on its own citizens and reduced checks and balances on those powers, such as judicial oversight. The government never demonstrated that restraints on surveillance had contributed to the attack, and indeed much of the new legislation had nothing to do with fighting terrorism. Rather, the bill represented a successful use of the terrorist attacks by the FBI to roll back unwanted checks on its power. The most powerful provisions of the law allow for:

- **Easy access to records.** Under the PATRIOT Act, the FBI can force anyone to turn over records on their customers or clients, giving the government unchecked power to rifle through individuals' financial records, medical histories, Internet usage, travel patterns, or any other records. Some of the most invasive and disturbing uses permitted by the Act involve government access to citizens' reading habits from libraries and bookstores. The FBI does not have to show suspicion of a crime, can gag the recipient of a search order from disclosing the search to anyone, and is subject to no meaningful judicial oversight.
- **Expansion of the "pen register" exception in wiretap law.** The PATRIOT Act expands exceptions to the normal requirement for probable cause in wiretap law.[9] As with its new

power to search records, the FBI need not show probable cause or even reasonable suspicion of criminal activity, and judicial oversight is essentially nil.

- **Expansion of the intelligence exception in wiretap law.** The PATRIOT Act also loosens the evidence needed by the government to justify an intelligence wiretap or physical search. Previously the law allowed exceptions to the Fourth Amendment for these kinds of searches only if "the purpose" of the search was to gather foreign intelligence. But the Act changes "the purpose" to "a significant purpose," which lets the government circumvent the Constitution's probably cause requirement even when its main goal is ordinary law enforcement.[10]
- **More secret searches.** Except in rare cases, the law has always required that the subject of a search be notified that a search is taking place. Such notice is a crucial check on the government's power because it forces the authorities to operate in the open and allows the subject of searches to challenge their validity in court. But the PATRIOT Act allows the government to conduct searches without notifying the subjects until long after the search has been executed.

Under these changes and other authorities asserted by the Bush administration, U.S. intelligence agents could conduct a secret search of an American citizen's home, use evidence found there to declare him an "enemy combatant," and imprison him without trial. The courts would have no chance to review these decisions—indeed, they might never even find out about them.[11]

The "TIPS" Program

In the name of fighting terrorism, the Bush administration has also proposed a program that would encourage citizens to spy on each other. The administration initially planned to recruit people such as letter carriers and utility technicians, who, the White House said, are "well-positioned to recognize unusual events." In the face of fierce public criticism, the administration scaled back the program, but continued to enlist workers involved in certain key industries. In November 2002 Congress included a provision in the Homeland Security Act prohibiting the Bush administration from moving forward with TIPS.

Although Congress killed TIPS, the fact that the administration would pursue such a program reveals a disturbing disconnect with American values and a disturbing lack of awareness of the history of governmental abuses of power. Dividing citizen from citizen by encouraging mutual suspicion and reporting to the government would dramatically increase the government's power by extending surveillance into every nook and cranny of American society. Such a strategy was central to the Soviet Union and other totalitarian regimes.

Loosened Domestic Spying Regulations

In May 2002, Attorney General John Ashcroft issued new guidelines on domestic spying that significantly increase the freedom of federal agents to conduct surveillance on American individuals and organizations. Under the new guidelines, FBI agents can infiltrate "any event that is open to the public," from public meetings and demonstrations to political conventions to church services to 12-step programs. This was the same basis upon which abuses were carried out by the FBI in the 1950s and 1960s, including surveillance of political groups that disagreed with the government, anonymous letters sent to the spouses of targets to try to ruin their marriages, and the infamous campaign against Martin Luther King, who was investigated and harassed for decades. The new guidelines are purely for spying on Americans; there is a separate set of foreign guidelines that cover investigations inside the United States of foreign powers and terrorist organizations such as Al Qaeda.

Like the TIPS program, Ashcroft's guidelines sow suspicion among citizens and extend the government's surveillance power into the capillaries of American life. It is not just the reality of government surveillance that chills free expression and the freedom that Americans enjoy. The same negative effects come when we are constantly forced to wonder whether we *might* be under observation—whether the person sitting next to us is secretly informing the government that we are "suspicious."

THE SYNERGIES OF SURVEILLANCE

Multiple surveillance techniques added together are greater than the sum of their parts. One example is face recognition, which combines the power of computerized software analysis, cameras, and databases to seek matches between facial images. But the real synergies of surveillance come into play with data collection.

The growing piles of data being collected on Americans represent an enormous invasion of privacy, but our privacy has actually been protected by the fact that all this information still remains scattered across many different databases. As a result, there exists a pent-up capacity for surveillance in American life today—a capacity that will be fully realized if the government, landlords, employers, or other powerful forces gain the ability to *draw together* all this information. A particular piece of data about you—such as the fact that you entered your office at 10:29 AM on July 5, 2001—is normally innocuous. But when enough pieces of that kind of data are assembled together, they add up to an extremely detailed and intrusive picture of an individual's life and habits.

Data Profiling and "Total Information Awareness"

Just how real this scenario is has been demonstrated by another ominous surveillance plan to emerge from the effort against terrorism: the Pentagon's "Total Information Awareness" program. The aim of this program is to give officials easy, unified access to every possible government and commercial database in the world.[12] According to program director John Poindexter, the program's goal is to develop "ultra-large-scale" database technologies with the goal of "treating the world-wide, distributed, legacy databases as if they were one centralized database." The program envisions a "full-coverage database containing all information relevant to identifying" potential terrorists and their supporters. As we have seen, the amount of available information is mushrooming by the day, and will soon be rich enough to reveal much of our lives.

The TIA program, which is run by the Defense Advanced Research Projects Agency (DARPA), not only seeks to bring together the oceans of data that are already being collected on people, but would be designed to afford what DARPA calls "easy future scaling" to embrace new sources of data as they become available. It would also incorporate other work being done by the military, such as their "Human Identification at a Distance" program, which seeks to allow identification and tracking of people from a distance, and therefore without their permission or knowledge.[13]

Although it has not received nearly as much media attention, a close cousin of TIA is also being created in the context of airline security. This plan involves the creation of a system for conducting background checks on individuals who wish to fly and then separating out either those who appear to be the most trustworthy passengers (proposals known as "trusted traveler") or flagging the least trustworthy (a proposal known as CAPS II, Computer Assisted Passenger Screening) for special attention.

The *Washington Post* has reported that work is being done on CAPS II with the goal of creating a "vast air security screening system designed to instantly pull together every passenger's travel history and living arrangements, plus a wealth of other personal and demographic information" in the hopes that the authorities will be able to "profile passenger activity and intuit obscure clues about potential threats." The government program would reportedly draw on enormous stores of personal information from data aggregators and other sources, including travel records. Plans call for using complex computer algorithms, including highly experimental technologies such as "neural networks," to sort through the reams of new personal information and identify "suspicious" people.[14]

The dubious premise of programs like TIA and CAPS II—that "terrorist patterns" can be ferreted out from the enormous mass of American lives, many of which will inevitably be quirky, eccentric, or riddled with suspicious coincidences—probably dooms them to failure. But failure is not likely to lead these programs to be shut down—instead, the government will begin feeding its computers more and more personal information in a vain effort to make the concept work. We will then have the worst of both worlds: poor security and a super-charged surveillance tool that would destroy Americans' privacy and threaten our freedom.

It is easy to imagine these systems being expanded in the future to share their risk assessments with other security systems. For example, CAPS could be linked to a photographic database and surveillance cameras equipped with face recognition software. Such a system might sound an alarm when a subject who has been designated as "suspicious" appears in public. The Suspicious Citizen could then be watched from a centralized video monitoring facility as he moves around the city.

In short, the government is working furiously to bring disparate sources of information about us together into one view, just as privacy advocates have been warning about for years. That would represent a radical branching off from the centuries-old Anglo-American tradition that the police conduct surveillance only where there is evidence of involvement in wrongdoing. It would seek to protect us by monitoring *everyone* for signs of wrongdoing—in short, by instituting a giant dragnet capable of sifting through the personal lives of Americans in search of "suspicious" patterns. The potential for abuse of such a system is staggering.

The massive defense research capabilities of the United States have always involved the search for ways of outwardly defending our nation. Programs like TIA[15] involve turning those capabilities inward and applying them to the American people—something that should be done, if at all, only with extreme caution and plenty of public input, political debate, checks and balances, and Congressional oversight. So far, none of those things have been present with TIA or CAPS II.

National ID Cards

If Americans allow it, another convergence of surveillance technologies will probably center around a national ID card. A national ID would immediately combine new technologies such as biometrics and RFID chips along with an enormously powerful database (possibly distributed among the 50 states). Before long, it would become an overarching means of facilitating surveillance by allowing far-flung pools of information to be pulled together into a single, incredibly rich dossier or profile of our lives. Before long, office buildings, doctors' offices, gas stations, highway tolls, subways, and buses would incorporate the ID card into their security or payment systems for greater efficiency, and data that is currently scattered and disconnected will get organized around the ID and lead to the creation of what amounts to a national database of sensitive information about American citizens.

History has shown that databases created for one purpose are almost inevitable expanded to other uses; Social Security, which was prohibited by federal law from being used as an identifier when it was first created, is a prime example. Over time, a national ID database would inevitably contain a wider and wider range of information and become accessible to more and more people for more and more purposes that are further and further removed from its original justification.

The most likely route to a national ID is through our driver's licenses. Since September 11, the American Association of Motor Vehicle Administrators has been forcefully lobbying Congress for funds to establish nationwide uniformity in the design and content of driver's licenses—and more importantly, for tightly interconnecting the databases that lie behind the physical licenses themselves.

An attempt to retrofit driver's licenses into national ID cards will launch a predictable series of events bringing us toward a surveillance society:

- Proponents will promise that the IDs will be implements in limited ways that won't devastate privacy and other liberties.
- Once a limited version of the proposals is put in place, its limits as an anti-terrorism measure will quickly become apparent. Like a dam built halfway across a river, the IDs cannot possibly be effective unless their coverage is total.
- The scheme's ineffectiveness—starkly demonstrated, perhaps, by a new terrorist attack—will create an overwhelming imperative to "fix" and "complete" it, which will turn it into the totalitarian tool that proponents promised it would never become.

A perfect example of that dynamic is the requirement that travelers present driver's licenses when boarding airplanes, instituted after the explosion (now believed to have been mechanical in cause) that brought down TWA Flight 800 in 1996. On its own, the requirement was meaningless as a security measure, but after September 11 its existence quickly led to calls to begin tracking and identifying citizens on the theory that "we already have to show ID, we might as well make it mean something."

Once in place, it is easy to imagine how national IDs could be combined with an RFID chip to allow for convenient, at-a-distance verification of ID. The IDs could then be tied to access control points around our public places, so that the unauthorized could be kept out of office buildings, apartments, public transit, and secure public buildings. Citizens with criminal records, poor CAPS ratings or low incomes could be barred from accessing airports, sports arenas, stores, or other facilities. Retailers might add RFID readers to find out exactly who is browsing their aisles, gawking at their window displays from the sidewalk or passing by without looking. A network of automated RFID listening posts on the sidewalks and roads could even reveal the location of all citizens at all times. Pocket ID readers could be used by FBI agents to sweep up the identities of everyone at a political meeting, protest march, or Islamic prayer service.

CONCLUSION

If we do not take steps to control and regulate surveillance to bring it into conformity with our values, we will find ourselves being tracked, analyzed, profiled, and flagged in our daily lives to a degree we can scarcely imagine today. We will be forced into an impossible struggle to conform to the letter of every rule, law, and guideline, lest we create ammunition for enemies in the government or elsewhere. Our transgressions will become permanent Scarlet Letters that follow us throughout our lives, visible to all and used by the government, landlords, employers, insurance companies and other powerful parties to increase their leverage over average people. Americans will not be able to engage in political protest or go about their daily lives without the constant awareness that we are—or could be—under surveillance. We will be forced to constantly ask of even the smallest action taken in public, "Will this make me look suspicious? Will this hurt my chances for future employment? Will this reduce my ability to get insurance?" The exercise of free speech will be chilled as Americans become conscious that their every word may be reported to the government by FBI infiltrators, suspicious fellow citizens, or an Internet Service Provider.

Many well-known commentators like Sun Microsystems CEO Scott McNealy have already pronounced privacy dead. The truth is that a surveillance society does loom over us, and privacy, while not yet dead, is on life support.

Heroic Measures Are Required to Save It

Four main goals need to be attained to prevent this dark potential from being realized: a change in the terms of the debate, passage of comprehensive privacy laws, passage of new laws to regulate the powerful and invasive new technologies that have and will continue to appear, and a revival of the Fourth Amendment to the U.S. Constitution.

1) Changing the Terms of the Debate

In the public debates over every new surveillance technology, the forest too often gets lost for the trees, and we lose sight of the larger trend: the seemingly inexorable movement toward a surveillance society. It will always be important to understand and publicly debate every new technology and every new technique for spying on people. But unless each new development is also understood as just one piece of the larger surveillance mosaic that is rapidly being constructed around us, Americans are not likely to get excited about a given incremental loss of privacy like the tracking of cars through toll booths or the growing practice of tracking consumers' supermarket purchases.

We are being confronted with fundamental choices about what sort of society we want to live in. But unless the terms of the debate are changed to focus on the forest instead of individual trees, too many Americans will never even recognize the choice we face, and a decision against preserving privacy will be made by default.

2) Comprehensive Privacy Laws

Although broad-based protections against government surveillance, such as the wiretap laws, are being weakened, at least they exist. But surveillance is increasingly being carried out by the private sector—frequently at the behest of government—and the laws protecting Americans against nongovernmental privacy invasions are pitifully weak.

In contrast to the rest of the developed world, the U.S. has no strong, comprehensive law protecting privacy—only a patchwork of largely inadequate protections. For example, as a result of many legislators' discomfort over the disclosure of Judge Robert Bork's video rental choices during his Supreme Court confirmation battle, video records are now protected by a strong privacy law. Medical records are governed by a separate, far weaker law that allows for widespread access to extremely personal information. Financial data is governed by yet another "privacy" law—Gramm-Leach—which as we have seen really amounts to a license to share financial information.

Another law protects only the privacy of children under age thirteen on the Internet. And layered on top of this sectoral approach to privacy by the federal government is a geographical patchwork of constitutional and statutory privacy protections in the states.

The patchwork approach to privacy is grossly inadequate. As invasive practices grow, Americans will face constant uncertainty about when and how these complex laws protect them, contributing to a pervasive sense of insecurity. With the glaring exception of the United States, every advanced industrialized nation in the world has enacted overarching privacy laws that protect citizens against private-sector abuses. When it comes to this fundamental human value, the United States is an outlaw nation. For example, the European Union bars companies from evading privacy rules by transferring personal information to other nations whose data-protection policies are "inadequate." That is the kind of law that is usually applied to Third World countries, but the EU counts the United States in this category.

We need to develop a baseline of simple and clear privacy protections that crosses all sectors of our lives and give it the force of law. Only then can Americans act with a confident knowledge of when they can and cannot be monitored.

3) New Technologies and New Laws

The technologies of surveillance are developing at the speed of light, but the body of law that protects us is stuck back in the Stone Age. In the past, new technologies that threatened our privacy, such as telephone wiretapping, were assimilated over time into our society. The legal system had time to adapt and reinterpret existing laws, the political system had time to consider and enact new laws or regulations, and the culture had time to absorb the implications of the new technology for daily life. Today, however, change is happening so fast that none of this adaptation has time to take place—a problem that is being intensified by the scramble to enact unexamined anti-terrorism measures. The result is a significant danger that surveillance practices will become entrenched in American life that would never be accepted if we had more time to digest them.

Since a comprehensive privacy law may never be passed in the United States—and certainly not in the near future—law and legal principles must be developed or adapted to rein in particular new technologies such as surveillance cameras, location-tracking devices, and biometrics. Surveillance cameras, for example, must be subject to force-of-law rules covering important details like when they will be used, how long images will be stored, and when and with whom they will be shared.

4) Reviving the Fourth Amendment

> The right of the people to be secure in their persons, houses, papers, and effects, against unreasonable searches and seizures, shall not be violated, and no warrants shall issue, but upon probable cause, supported by oath or affirmation, and particularly describing the place to be searched, and the persons or things to be seized.
>
> —Fourth Amendment to the U.S. Constitution

The Fourth Amendment, the primary Constitutional bulwark against government invasion of our privacy, was a direct response to the British authorities' use of "general warrants" to conduct broad searches of the rebellious colonists.

Historically, the courts have been slow to adapt the Fourth Amendment to the realities of developing technologies. It took almost 40 years for the U.S. Supreme Court to recognize that the Constitution applies to the wiretapping of telephone conversations.[16]

In recent years—in no small part as the result of the failed "war on drugs"—Fourth Amendment principles have been steadily eroding. The circumstances under which police and other government officials may conduct warrantless searches has been rapidly expanding. The courts have allowed for increased surveillance and searches on the nation's highways and at our "borders" (the legal definition of which actually extends hundreds of miles inland from the actual border). And despite the Constitution's plain language covering "persons" and "effects," the courts have increasingly allowed for warrantless searches when we are outside of our homes and "in public." Here the courts have increasingly found we have no "reasonable expectation" of privacy and that therefore the Fourth Amendment does not apply.

But like other Constitutional provisions, the Fourth Amendment needs to be understood in contemporary terms. New technologies are endowing the government with the twenty-first century equivalent of Superman's X-ray vision. Using everything from powerful video technologies that can literally see in the dark, to biometric identification techniques like face recognition, to "brain fingerprinting" that can purportedly read our thoughts, the government is now capable of conducting broad searches of our "persons and effects" while we are going about our daily lives—even while we are in "public."

The Fourth Amendment is in desperate need of a revival. The reasonable expectation of privacy cannot be defined by the power that technology affords the government to spy on us. Since that power is increasingly limitless, the "reasonable expectation" standard will leave our privacy dead indeed.

But all is not yet lost. There is some reason for hope. In an important pre-9/11 case, *Kyllo v. U.S.,*[17] the Supreme Court held that the reasonable expectation of privacy could not be determined by the power of new technologies. In a remarkable opinion written by conservative Justice Antonin Scalia, the Court held that without a warrant the police could not use a new thermal imaging device that searches for heat sources to conduct what was the functional equivalent of a warrantless search for marijuana cultivation in Danny Kyllo's home.

The Court specifically declined to leave Kyllo "at the mercy of advancing technology." While Kyllo involved a search of a home, it enunciates an important principle: the Fourth Amendment must adapt to new technologies. That principle can and should be expanded to general use. The Framers never expected the Constitution to be read exclusively in terms of the circumstances of 1791.

NOTES

1. Jess Bravin, "Washington Police to Play 'I Spy' with Cameras, Raising Concerns," *Wall Street Journal,* Feb. 13, 2002.

2. See http://www.ncjrs.org/school/ch2a_5.html.

3. See http://www.scotcrim.u-net.com/researchc2.htm.

4. The success rate of face recognition technology has been dismal. The many independent findings to that effect include a trial conducted by the U.S. military in 2002, which found that with a reasonably low false-positive rate, the technology had less than a 20 percent chance of successfully identifying a person in its database who appeared before the camera. See http://www.aclu.org/issues/privacy/FINAL_1_Final_Steve_King.pdf, 17th slide.

5. Richard H. P. Sia, "Pilotless Aircraft Makers Seek Role for Domestic Uses," *Congress Daily,* December 17, 2002

6. Data surveillance is often loosely referred to as "data mining." Strictly speaking, however, data mining refers to the search for hidden patterns in large, pre-existing collections of data (such as the finding that sales of both beer and diapers rise on Friday nights). Data mining need not involve personally identifiable information. Data surveillance, on the other hand, involves the collection of information about an identifiable individ-

ual. Note, however, that when data surveillance is carried out on a mass scale, a search for patterns in people's activities—data mining—can then be conducted as well. This is what appears to be contemplated in the Total Information Awareness and CAPS II programs (see below).

7. Dana Hawkins, "As DNA Banks Quietly Multiply, Who Is Guarding the Safe?" *U.S. News & World Report,* Dec. 2, 2002.

8. Glenn R. Simpson, "Big Brother-in-Law: If the FBI Hopes to Get The Goods on You, It May Ask ChoicePoint," *Wall Street Journal,* April 13, 2001.

9. The expanded exception involves what are called "pen register/trap and trace" warrants that collect "addressing information" but not the content of a communication. Those searches are named after devices that were used on telephones to show a list of telephone numbers dialed and received (as opposed to tapping into actual conversations). The PATRIOT Act expands the pen register exception onto the Internet in ways that will probably be used by the government to collect the actual content of communications and that allow nonspecific "nationwide" warrants in violation of the Fourth Amendment's explicit requirement that warrants "must specify the place to be searched."

10. In August, the secret "FISA" court that oversees domestic intelligence spying released an opinion rejecting a Bush Administration attempt to allow criminal prosecutors to use intelligence warrants to evade the Fourth Amendment entirely. The court noted that agents applying for warrants had regularly filed false and misleading information. In November 2002, however, the FISA appeals court (three judges chosen by Supreme Court Chief Justice William Rehnquist), meeting for the first time ever, ruled in favor of the government.

11. See Charles Lane, "In Terror War, 2nd Track for Suspects," *Washington Post,* December 1, 2002. Online at http://www.washingtonpost.com/wp-dyn/articles/A58308-2002Nov30.html.

12. See "Pentagon Plans a Computer System That Would Peek at Personal Data of Americans," *New York Times,* Nov. 9, 2002; "U.S. Hopes to Check Computers Globally," *Washington Post,* Nov. 12, 2002; "The Poindexter Plan," *National Journal,* Sept. 7, 2002.

13. Quotes are from the TIA homepage at http://www.darpa.mil/iao/index.htm and from public 8/2/02 remarks by Poindexter, online at http://www.fas.org/irp/agency/dod/poindexter/html.

14. Robert O'Harrow Jr., "Intricate Screening of Fliers In Works," *Washington Post,* Feb. 1, 2002, p. A1.

15. The TIA is just one part of a larger post-9/11 expansion of federal research and development efforts. The budget for military R&D spending alone has been increased by 18 percent in the current fiscal year to a record $58.8 billion. Bob Davis, "Massive Federal R&D Initiative to Fight Terror Is Under Way," *Wall Street Journal,* November 25, 2002

16. In 1967 the Supreme Court finally recognized the right to privacy in telephone conversations in the case *Katz v. U.S.* (380 US 347), reversing the 1928 opinion *Olmstead v. U.S.* (277 US 438) 17190 F.3d 1041, 2001.

21

The Constitution in Cyberspace: Law and Liberty Beyond the Electronic Frontier

Laurence H. Tribe

My topic is how to "map" the text and structure of our Constitution onto the texture and topology of *cyberspace.* That's the term coined by cyberpunk novelist William Gibson, which many now use to describe the "place"—a place without physical walls or even physical dimensions—where ordinary telephone conversations "happen," where voice-mail and e-mail messages are stored and sent back and forth, and where computer-generated graphics are transmitted and transformed, all in the form of interactions, some real-time and some delayed, among countless users, and between users and the computer itself.

Some use the cyberspace concept to designate fantasy worlds or *virtual realities* of the sort Gibson described in his novel *Neuromancer,* in which people can essentially turn their minds into computer peripherals capable of perceiving and exploring the data matrix. The whole idea of virtual reality, of course, strikes a slightly odd note. As one of Lily Tomlin's most memorable characters once asked, "What's reality, anyway, but a collective hunch?" Work in this field tends to be done largely by people who share the famous observation that reality is overrated!

However that may be, cyberspace connotes to some users the sorts of technologies that people in Silicon Valley work on when they try to develop "virtual racquetball" for the disabled, computer-aided design systems that allow architects to walk through "virtual buildings" and remodel them *before* they are built, "virtual conferencing" for business meetings, or maybe someday even "virtual day care centers" for latchkey children. The user snaps on a pair of goggles hooked up to a high-powered computer terminal, puts on a special set of gloves (and perhaps other gear) wired into the same computer system, and, looking a little bit like Darth Vader, pretty much steps into a computer-driven, drug-free, three-dimensional, interactive, infinitely expandable hallucination complete with sight, sound, and touch—allowing the user literally to move through, and experience, information.

I'm using the term cyberspace much more broadly, as many have lately. I'm using it to encompass the full array of computer-mediated audio and/or video interactions that are already widely dispersed in modern societies—from things as ubiquitous as the ordinary telephone to

From Laurence H. Tribe, "The Constitution in Cyberspace," *The Humanist* (Sept./Oct., 1991) 51:5, pp. 15–21.

things that are still coming online like computer bulletin boards and networks like Prodigy or the WELL (Whole Earth 'Lectronic Link), based in San Francisco. My topic, broadly put, is the implications of that rapidly expanding array for our constitutional order. It is a constitutional order that tends to carve up the social, legal, and political universe along lines of *physical place* or *temporal proximity.* The critical thing to note is that these very lines, in cyberspace, either get bent out of shape or fade out altogether. The question, then, becomes: when the lines along which our Constitution is drawn warp or vanish, what happens to the Constitution itself?

SETTING THE STAGE

To set the stage with a perhaps unfamiliar example, consider a decision handed down in 1990, *Maryland v. Craig,* in which the U.S. Supreme Court upheld the power of a state to put an alleged child abuser on trial with the defendant's accuser testifying not in the defendant's presence but by one-way, closed-circuit television. The Sixth Amendment, which of course antedated television by a century and a half, says: "In all criminal prosecutions, the accused shall enjoy the right . . . to be confronted with the witnesses against him." Justice O'Connor wrote for a bare majority of five justices that the state's procedures nonetheless struck a fair balance between costs to the accused and benefits to the victim and to society as a whole. Justice Scalia, joined by the three "liberals" then on the Court (Justices Brennan, Marshall, and Stevens), dissented from that cost–benefit approach to interpreting the Sixth Amendment. He wrote:

> The Court has convincingly proved that the Maryland procedure serves a valid interest, and gives the defendant virtually everything the Confrontation Clause guarantees (everything, that is, except confrontation). I am persuaded, therefore, that the Maryland procedure is virtually constitutional. Since it is not, however, actually constitutional I [dissent].

Could it be that the high-tech, closed-circuit TV context, almost as familiar to the Court's youngest justice as to his even younger law clerks, might've had some bearing on Justice Scalia's sly invocation of "virtual" constitutional reality? Even if Justice Scalia wasn't making a pun on virtual reality, and I suspect he wasn't, his dissenting opinion about the Confrontation Clause requires *us* to "confront" the recurring puzzle of how constitutional provisions written two centuries ago should be construed and applied in ever-changing circumstances.

Should contemporary society's technology-driven cost–benefit fixation be allowed to water down the old-fashioned value of direct confrontation that the Constitution seemingly enshrined as basic? I would hope not. In that respect, I find myself in complete agreement with Justice Scalia.

But new technological possibilities for seeing your accuser clearly without having your accuser see you at all—possibilities for sparing the accuser any discomfort in ways that the accuser couldn't be spared before one-way mirrors or closed-circuit TVs were developed—*should* lead us at least to ask ourselves whether *two*-way confrontation, in which your accuser is supposed to be made uncomfortable, and thus less likely to lie, really *is* the core value of the Confrontation Clause. If so, virtual confrontation should be held constitutionally insufficient. If not—if the core value served by the Confrontation Clause is just the ability to *watch* your accuser say that you did it— then "virtual" confrontation should suffice. New technologies should lead us to look more closely at just what *values* the Constitution seeks to preserve. New technologies should *not* lead us to react reflexively *either* way—either by assuming that technologies the Framers didn't know about make their concerns and values obsolete, or by assuming that those new technologies couldn't possibly provide new ways out of old dilemmas and therefore should be ignored altogether.

The one-way mirror yields a fitting metaphor for the task we confront. As the Supreme Court said in a different context several years ago, "The mirror image presented [here] requires us to step through an analytical looking glass to resolve it" (*NCAA v. Tarkanian,* 109 S. Ct. at 462). The world in which the Sixth Amendment's Confrontation Clause was written and ratified was a world in which "being confronted with" your accuser *necessarily* meant a simultaneous physical confrontation so that your accuser had to *perceive* you being accused by him. Closed-circuit television and one-way mirrors changed all that by *decoupling* those two dimensions of confrontation, marking a shift in the conditions of information transfer that is in many ways typical of cyberspace.

What does that sort of shift mean for constitutional analysis? A common way to react is to treat the pattern as it existed *prior* to the new technology (the pattern in which doing "A" necessarily *included* doing "B") as essentially arbitrary or accidental. Taking this approach, once the technological change makes it possible to do "A" *without* "B"—to see your accuser without having him or her see you, or to read someone's mail without her knowing it, to switch examples—one concludes that the "old" Constitution's inclusion of "B" is irrelevant; one concludes that it is enough for the government to guarantee "A" alone. Sometimes that will be the case; but it's vital to understand that, sometimes, it won't be.

A characteristic feature of modernity is the subordination of purpose to accident—an acute appreciation of just how contingent and coincidental the connections we are taught to make often are. We understand, as moderns, that many of the ways we carve up and organize the world reflect what our social history and cultural heritage, and perhaps our neurological wiring, bring to the world, and not some irreducible "way things are." A wonderful example comes from a 1966 essay by Jorge Luis Borges, "Other Inquisitions." There, the essayist describes the following taxonomy of the animal kingdom, which he purports to trace to an ancient Chinese encyclopedia entitled *The Celestial Emporium of Benevolent Knowledge:*

On those remote pages it is written that animals are divided into:

(a) those belonging to the Emperor
(b) those that are embalmed
(c) those that are trained
(d) suckling pigs
(e) mermaids
(f) fabulous ones
(g) stray dogs
(h) those that are included in this classification
(i) those that tremble as if they were mad
(j) innumerable ones
(k) those drawn with a very fine camel's hair brush
(l) others
(m) those that have just broken a water pitcher
(n) those that, from a great distance, resemble flies

Contemporary writers from Michel Foucault, in *The Archaeology of Knowledge,* through George Lakoff, in *Women, Fire, and Dangerous Things,* use Borges's Chinese encyclopedia to illustrate a range of different propositions, but the *core* proposition is the supposed arbitrariness—the political character, in a sense—of all culturally imposed categories.

At one level, that proposition expresses a profound truth and may encourage humility by combating cultural imperialism. At another level, though, the proposition tells a dangerous lie: it suggests that we have descended into the nihilism that so obsessed Nietzsche and other thinkers—a world where *everything* is relative, all lines are up for grabs, and all principles and connections are just matters of purely subjective preference or, worse still, arbitrary convention. Whether we believe that killing animals for food is wrong, for example, becomes a question indistinguishable from whether we happen to enjoy eating beans, rice, and tofu.

This is a particularly pernicious notion in a era when we pass more and more of our lives in cyberspace, a place where, almost by definition, our most familiar landmarks are rearranged or disappear altogether—because there is a pervasive tendency, even (and perhaps especially) among the most enlightened, to forget that the human values and ideals to which we commit ourselves may indeed be universal and need not depend on how our particular cultures, or our latest technologies, carve up the universe we inhabit. It was my very wise colleague from Yale, the late Art Leff, who once observed that, even in a world without an agreed-upon God, we can still agree—even if we can't "prove" mathematically—that "napalming babies is wrong."

The Constitution's core values, I'm convinced, need not be transmogrified, or metamorphosed into oblivion, in the dim recesses of cyberspace. But to say that they *need* not be lost there is hardly to predict that they *will* not be. On the contrary, the danger is clear and present that they *will* be.

The "event horizon" against which this transformation might occur is already plainly visible:

Electronic trespassers like Kevin Mitnick don't stop with cracking pay phones, but break into NORAD—the North American Defense Command computer in Colorado Springs—not in a *WarGames* movie, but in real life.

Less challenging to national security but more ubiquitously threatening, computer crackers download everyman's credit history from institutions like TRW, start charging phone calls (and more) to everyman's number, set loose "worm" programs that shut down thousands of linked computers, and spread "computer viruses" through everyman's work or home PC.

It is not only the government that feels threatened by "computer crime"; both the owners and the users of private information services, computer bulletin boards, gateways, and networks feel equally vulnerable to this new breed of invisible trespasser. The response from the many who sense danger has been swift, and often brutal, as a few examples illustrate.

In March 1990, U.S. Secret Service agents staged a surprise raid on Steve Jackson Games, a small games manufacturer in Austin, Texas, and seized all paper and electronic drafts of its newest fantasy role-playing game, *GURPS Cyberpunk,* calling the game a "handbook for computer crime."

By the spring of 1990, up to one quarter of the U.S. Treasury Department's investigators had become involved in a project of eavesdropping on computer bulletin boards, apparently tracking notorious hackers like "Acid Phreak" and "Phiber Optik" through what one journalist dubbed "the dark canyons of cyberspace."

In May 1990, in the now famous (or infamous) "Operation Sun Devil," more than 150 secret service agents teamed up with state and local law enforcement agencies, and with security personnel from AT&T, American Express, U.S. Sprint, and a number of the regional Bell telephone companies, armed themselves with over two dozen search warrants and more than a few guns, and seized 42 computers and 23,000 floppy discs in fourteen cities from New York to Texas. Their target: a looseknit group of people in their teens and twenties, dubbed the "Legion of Doom."

I am not describing an Indiana Jones movie. I'm talking about America in the 1990s.

THE PROBLEM

The Constitution's architecture can too easily come to seem quaintly irrelevant, or at least impossible to take very seriously, in the world as reconstituted by the microchip. I propose today to canvass five axioms of our constitutional law—five basic assumptions that I believe shape the way American constitutional scholars and judges view legal issues—and to examine how they can adapt to the cyberspace age. My conclusion (and I will try not to give away too much of the punch line here) is that the Framers of our Constitution were very wise indeed. They bequeathed us a framework for all seasons, a truly astonishing document whose principles are suitable for all times and all technological landscapes.

Axiom 1: There Is a Vital Difference between Government and Private Action

The first axiom I will discuss is the proposition that the Constitution, with the sole exception of the Thirteenth Amendment prohibiting slavery, regulates action by the *government* rather than the conduct of *private* individuals and groups. In an article I wrote in the *Harvard Law Review* in November 1989, "The Curvature of Constitutional Space," I discussed the Constitution's metaphormorphosis from a Newtonian to an Einsteinian and Heisenbergian paradigm. It was common, early in our history, to see the Constitution as "Newtonian in design with its carefully counterpoised forces and counterforces, its [geographical and institutional] checks and balances."

Indeed, in many ways contemporary constitutional law is still trapped within and stunted by that paradigm. But today at least, some postmodern constitutionalists tend to think and talk in the language of relativity, quantum mechanics, and chaos theory. This may quite naturally suggest to some observers that the Constitution's basic strategy of decentralizing and diffusing power by constraining and fragmenting governmental authority in particular has been rendered obsolete.

The institutional separation of powers among the three federal branches of government, the geographical division of authority between the federal government and the fifty state governments, the recognition of national boundaries, and, above all, the sharp distinction between the public and private spheres become easy to deride as relics of a simpler, precomputer age. Thus Eli Noam, in the First Ithiel de Sola Pool Memorial Lecture, delivered in October 1990 at MIT, notes that computer networks and network associations acquire quasi-governmental powers as they necessarily take on such tasks as mediating their members' conflicting interests, establishing cost shares, creating their own rules of admission and access and expulsion, even establishing their own de facto taxing mechanisms. In Professor Noam's words, "Networks become political entities," global nets that respect no state or local boundaries. Restrictions on the use of information in one country (to protect privacy, for example) tend to lead to export of that information to other countries, where it can be analyzed and then used on a selective basis in the country attempting to restrict it. *Data havens* reminiscent of the role played by the Swiss in banking may emerge, with few restrictions on the storage and manipulation of information.

A tempting conclusion is that, to protect the free speech and other rights of *users* in such private networks, judges must treat these networks not as associations that have rights of their own *against* the government but as virtual "governments" in themselves—as entities against which individual rights must be defended in the Constitution's name. Such a conclusion would be misleadingly simplistic. There are circumstances, of course, when nongovernmental bodies like privately owned "company towns" or even huge shopping malls should be subjected to legislative and administrative controls by democratically accountable entities, or even to judicial controls as though they were arms of the state—but that may be as true (or as false) of multinational corporations or foundations, or transnational religious organizations, or even small-town communities, as it is of computer-mediated networks. It's a fallacy to suppose that just because a computer bulletin board or network or gateway is something *like* a shopping mall, government has as much constitutional duty—or even authority—to guarantee open public access to such a network as it has to guarantee open public access to a privately owned shopping center like the one involved in the U.S.

Supreme Court's famous *PruneYard Shopping Center* decision of 1980, arising from San Jose, California.

The rules of law, both statutory and judge-made, through which each state *allocates* private powers and responsibilities themselves represent characteristic forms of government action. That's why a state's rules for imposing liability on private publishers, or for deciding which private contracts to enforce and which ones to invalidate, are all subject to scrutiny for their consistency with the federal Constitution. But as a general proposition, it is only what *governments* do, either through such rules or through the actions of public officials, that the U.S. Constitution constrains. And nothing about any new technology suddenly erases the Constitution's enduring value of restraining *government* above all else, and of protecting all private groups, large and small, from government.

It's true that certain technologies may become socially indispensable—so that equal or at least minimal access to basic computer power, for example, might be as significant a constitutional goal as equal or at least minimal access to the franchise, or to dispute resolution through the judicial system, or to elementary and secondary education. But all this means (or should mean) is that the Constitution's constraints on government must at times take the form of imposing *affirmative duties* to assure access rather than merely enforcing *negative prohibitions* against designated sorts of invasion or intrusion.

Today, for example, the government is under an affirmative obligation to open up criminal trials to the press and the public, at least where there has not been a particularized finding that such openness would disrupt the proceedings. The government is also under an affirmative obligation to provide free legal assistance for indigent criminal defendants, to assure speedy trials, to underwrite the cost of counting ballots at election time, and to desegregate previously segregated school systems. But these occasional affirmative obligations don't, or shouldn't, mean that the Constitution's axiomatic division between the realm of public power and the realm of private life should be jettisoned.

Nor would the "indispensability" of information technologies provide a license for government to impose strict content, access, pricing, and other types of regulation. *Books* are indispensable to most of us, for example—but it doesn't follow that government should therefore be able to regulate the content of what goes onto the shelves of *bookstores*. The right of a private bookstore owner to decide which books to stock and which to discard, which books to display openly and which to store in limited access areas, should remain inviolate. And note, incidentally, that this needn't make the bookstore owner a *publisher* who is liable for the words printed in the books on her shelves. It's a common fallacy to imagine that the moment a computer gateway or bulletin board begins to exercise powers of selection to control who may be online, it must automatically assume the responsibilities of a newscaster, a broadcaster, or an author. For computer gateways and bulletin boards are really the "bookstores" of cyberspace; most of them organize and present information in a computer format, rather than generating more information content of their own.

Axiom 2: The Constitutional Boundaries of Private Property and Personality Depend on Variables Deeper Than Social Utility and Technological Feasibility

The second constitutional axiom, one closely related to the private-public distinction of the first axiom, is that a person's mind, body, and property belong *to that person* and not to the public as a whole. Some believe that cyberspace challenges that axiom because its entire premise lies in the existence of computers tied to electronic transmission networks that process digital information. Because such information can be easily replicated in series of 1s and 0s, anything that anyone has come up with in virtual reality can be infinitely reproduced. I can log on to a computer library, copy a "virtual book" to my computer disk, and send a copy to your computer without creating a gap on anyone's bookshelf. The same is true of valuable computer programs, costing hundreds of dollars, creating serious piracy problems. This feature leads some, like Richard Stallman of the Free Software Foundation, to argue that in cyberspace everything should be free—that information can't be owned. Others, of course, argue that copyright and patent protections of various kinds are needed in order for there to be incentives to create *cyberspace property* in the first place.

Needless to say, there are lively debates about what the optimal incentive package should be as a matter of legislative and social policy. But the only *constitutional* issue, at bottom, isn't the utilitarian or instrumental selection of an optimal policy. Social judgments about what ought to be subject to individual appropriation, in the sense used by John Locke and Robert Nozick, and what ought to remain in the open public domain, are first and foremost *political* decisions.

To be sure, there are some constitutional constraints on these political decisions. The Constitution does not permit anything and everything to be made into a private commodity. Votes, for example, theoretically cannot be bought and sold. Whether the Constitution itself should be read (or amended) so as to permit all basic medical care, shelter, nutrition, legal assistance, and, indeed, computerized information services to be treated as mere commodities, available only to the highest bidder, are all terribly hard questions—as the Eastern Europeans discovered in the early 1990s as they drafted their own constitutions. But these are not questions that should ever be confused with issues of what is technologically possible, about what is realistically enforceable, or about what is socially desirable.

Similarly, the Constitution does not permit anything and everything to be *socialized* and made into a public good available to whoever needs or "deserves" it most. I would hope, for example, that the government could not use its powers of eminent domain to "take" live body parts like eyes or kidneys or brain tissue for those who need transplants and would be expected to lead particularly productive lives. In any event, I feel certain that whatever constitutional right each of us has to inhabit his or her own body and to hold onto his or her own thoughts and creations should not depend solely on cost–benefit calculations, or on the availability of technological methods for painlessly effecting transfers or for creating good artificial substitutes.

Axiom 3: Government May Not Control Information Content

A third constitutional axiom, like the first two, reflects a deep respect for the integrity of each individual and a healthy skepticism toward government. The axiom is that, although information and ideas have real effects in the social world, it's not up to government to pick and choose for us in terms of the *content* of that information or the *value* of those ideas.

This notion is sometimes mistakenly reduced to the naïve child's ditty that "sticks and stones may break my bones, but words can never hurt me." Anybody who's ever been called something awful by children in a schoolyard knows better than to believe any such thing. The real basis for First Amendment values isn't the false premise that information and ideas have no real impact, but the belief that information and ideas are too *important* to entrust to any government censor or overseer.

If we keep that in mind, and *only* if we keep that in mind, will we be able to see through the tempting argument that, in the Information Age, free speech is a luxury we can no longer afford. That argument becomes especially tempting in the context of cyberspace, where sequences of 0s and 1s may become virtual life forms. Computer "viruses" roam the information nets, attaching themselves to various programs and screwing up computer facilities. Creation of a computer virus involves writing a program; the program then replicates itself and mutates. The electronic code involved is very much like DNA. If information content is *speech,* and if the First Amendment is to apply in cyberspace, then mustn't these viruses be speech—and mustn't their writing and dissemination be constitutionally protected? To avoid that nightmarish outcome, mustn't we say that the First Amendment is *inapplicable* to cyberspace?

The answer is no. Speech is protected, but deliberately yelling "Boo!" at a cardiac patient may still be prosecuted as murder. Free speech is a constitutional right, but handing a bank teller a holdup note that says, "Your money or your life," may still be punished as robbery. Stealing someone's diary may be punished as theft—even if you intend to publish it in book form. And the Supreme Court, over the past fifteen years, has gradually brought advertising within the ambit of protected expression without preventing the government from protecting consumers from deceptive advertising. The lesson, in short, is that constitutional principles are subtle enough to bend to such concerns. They needn't be broken or tossed out.

Axiom 4: The Constitution Is Founded on Normative Conceptions of Humanity That Advances in Science and Technology Cannot "Disprove"

A fourth constitutional axiom is that the human spirit is something beyond a physical information processor. That axiom, which regards human thought processes as not fully reducible to the operations of a computer program, however complex, must not be confused with the silly view that, because computer operations involve nothing more than the manipulation of "on" and "off" states of myriad microchips, it somehow follows that government control or outright seizure of computers and computer programs threatens no First Amendment rights because human thought processes are not directly involved. To say that would be like saying that government confiscation of a newspaper's printing press and tomorrow morning's copy has nothing to do with speech but involves only a taking of metal, paper, and ink. Particularly if the seizure or the regulation is triggered by the content of the information being processed or transmitted, the First Amendment is of course fully involved. Yet this recognition that information processing by computer entails something far beyond the mere sequencing of mechanical or chemical steps still leaves a potential gap between what computers can do internally and in communication with one another—and what goes on within and between human minds. It is that gap to which this fourth axiom is addressed; the very existence of any such gap is, as I'm sure you know, a matter of considerable controversy.

What if people like the mathematician and physicist Roger Penrose, author of *The Emperor's New Mind,* are wrong about human minds? In that provocative recent book, Penrose disagrees with those artificial intelligence, or AI, gurus who insist that it's only a matter of time until human thought and feeling can be perfectly simulated or even replicated by a series of purely physical operations—that it's all just neurons firing and neurotransmitters flowing, all subject to perfect modeling in suitable computer systems. Would an adherent of that AI orthodoxy, someone whom Penrose fails to persuade, have to reject as irrelevant for cyberspace those constitutional protections that rest on the anti-AI premise that minds are *not* reducible to really fancy computers?

Consider, for example, the Fifth Amendment, which provides that "no person shall be . . . compelled in any criminal case to be a witness against himself." The Supreme Court has long held that suspects may be required, despite this protection, to provide evidence that is not "testimonial" in nature—blood samples, for instance, or even exemplars of one's handwriting or voice. In 1990, in a case called *Pennsylvania v. Muniz,* the Supreme Court held that answers to even simple questions like "When was your sixth birthday?" are testimonial because such a question, however straightforward, nevertheless calls for the product of mental activity and therefore uses the suspect's mind against him or her. But what if science could eventually describe thinking as a process no more complex than, say, riding a bike or digesting a meal? Might the progress of neurobiology and computer science eventually overthrow the premises of the *Muniz* decision?

I would hope not. For the Constitution's premises, properly understood, are *normative* rather than descriptive. The philosopher David Hume was right in teaching that no "ought" can ever be logically derived from an "is." If we should ever abandon the Constitution's protection for the distinctively and universally human, it won't be because robotics or genetic engineering or computer science has led us to deeper truths, but rather because they have seduced us into more profound confusions. Science and technology open options, create possibilities, suggest incompatibilities, and generate threats. They do not alter what is "right" or what is "wrong." The fact that those notions are elusive and subject to endless debate need not make them totally contingent on contemporary technology.

Axiom 5: Constitutional Principles Should Not Vary with Accidents of Technology

In a sense, that's the fifth and final constitutional axiom I would urge upon this gathering: that the Constitution's norms, at their deepest level, must be invariant under merely *technological* transformations. Our constitutional law evolves through judicial interpretation, case by case, in a process of reasoning by analogy from precedent. At its best, that process is ideally suited to seeing beneath the surface and extracting deeper principles from prior decisions. At its worst, though, the same process can get bogged down in superficial aspects of preexisting examples, fixating upon unessential features while overlooking underlying principles and values.

When the Supreme Court in 1928 first confronted wiretapping and held in *Olmstead v. United States* that such wiretapping involved no "search" or "seizure" within the meaning of the Fourth Amendment's prohibition of "unreasonable searches and seizures," the majority of the Court reasoned that the Fourth Amendment "itself shows that the search is to be of material things—the person, the house, his papers or his effects," and said that "there was no searching" when a suspect's phone was tapped because the Constitution's language "cannot be extended and expanded to include telephone wires reaching to the whole world from the defendant's house or office." After all, said the Court, the intervening wires "are not part of his house or office any more than are the highways along which they are stretched." Even to a law student in the 1960s, as you might imagine, that "reasoning" seemed amazingly artificial. Yet the *Olmstead* doctrine still survived.

It would be illuminating at this point to compare the Supreme Court's initial reaction to new technology in *Olmstead* with its initial reaction to new technology in *Maryland v. Craig,* the 1990 closed-circuit television case with which we began this discussion. In *Craig,* a majority of the justices assumed that, when the eighteenth-century Framers of the Confrontation Clause included a guarantee of two-way *physical* confrontation, they did so solely because it had not yet become technologically feasible for the accused to look his or her accuser in the eye without having the accuser simultaneously watch the accused. Given that this technological obstacle has been removed, the majority assumed, one-way confrontation is now sufficient. It is enough that the accused not be subject to criminal conviction on the basis of statements made outside his presence.

In *Olmstead,* a majority of the justices assumed that, when the eighteenth-century authors of the Fourth Amendment used language that sounded "physical" in guaranteeing against invasions of a person's dwelling or possessions, they did so not solely because physical invasions were at that time the only serious threats to personal privacy, but for the separate and distinct reason that *intangible* invasions simply would not threaten any relevant dimension of Fourth Amendment privacy.

In a sense, *Olmstead* mindlessly read a new technology *out* of the Constitution, while *Craig* absentmindedly read a new technology *into* the Constitution. But both decisions—*Olmstead* and *Craig*—had the structural effect of withholding the protections of the Bill of Rights from threats made possible by new information technologies. *Olmstead* did so by implausibly reading the Constitution's text as though it represented a deliberate decision not to extend protection to threats that eighteenth-century thinkers simply had not foreseen. *Craig* did so by somewhat more plausibly—but still unthinkingly—treating the Constitution's seemingly explicit coupling of two analytically distinct protections as reflecting a failure of technological foresight and imagination, rather than a deliberate value choice.

The *Craig* majority's approach appears to have been driven in part by an understandable sense of how a new information technology could directly protect a particularly sympathetic group, abused children, from a traumatic trial experience. The *Olmstead* majority's approach probably reflected both an exaggerated estimate of how difficult it would be to obtain wiretapping warrants even where fully justified, and an insufficient sense of how a new information technology could directly threaten all of us. Although both *Craig* and *Olmstead* reveal an inadequate consciousness about how new technologies interact with old values, *Craig* at least seems defensible even if misguided, and *Olmstead* seems just plain wrong.

Around twenty-three years ago, as a then-recent law school graduate serving as law clerk to Supreme Court Justice Potter Stewart, I found myself working on a case involving the government's electronic surveillance of a suspected criminal—in the form of a tiny device attached to the outside of a public telephone booth. Because the invasion of the suspect's privacy was accomplished without physical trespass into a "constitutionally protected area," the federal government argued, relying on *Olmstead,* that there had been no "search" or "seizure," and therefore that the Fourth Amendment "right of the people to be secure in their persons, houses, papers, and effects, against unreasonable searches and seizures," simply did not apply.

At first, there were only four votes to overrule *Olmstead* and to hold the Fourth Amendment applicable to wiretapping and electronic eavesdropping. I'm proud to say that, as a twenty-six-year-old kid, I had at least a little bit to do with changing that number from four to seven—and with the argument, formally adopted by a seven-justice majority in December 1967, that the Fourth Amendment "protects people, not places" (389 U.S. at 351). In that decision, *Katz v. United States,* the Supreme Court finally repudiated *Olmstead* and the many decisions that had relied upon it and reasoned that, given the role of electronic telecommunications in modern life, the First Amendment purposes of protecting *free speech* as well as the Fourth Amendment purposes of protecting privacy require treating as a "search" any invasion of a person's confidential telephone communications, with or without physical trespass.

Sadly, nine years later, in *Smith v. Maryland* (1976), the Supreme Court retreated from the *Katz* principle by holding that no search occurs and therefore no warrant is needed when police, with the assistance of the telephone company, make use of a "pen register," a mechanical device placed on someone's phone line that records all numbers dialed from the phone and the times of dialing. The Supreme Court, over the dissents of Justices Stewart, Brennan, and Marshall, found no legitimate expectation of privacy in the numbers dialed, reasoning that the digits one dials are routinely recorded by the phone company for billing purposes. As Justice Stewart, the author of Katz, aptly pointed out,

> That observation no more than describes the basic nature of telephone calls. . . . It is simply not enough to say, after Katz, that there is no legitimate expectation of privacy in the numbers dialed because the caller assumes the risk that the telephone company will expose them to the police. (442 U.S. at 746–747)

Today, the logic of *Smith* is being used to say that people have no expectation of privacy when they use their cordless telephones since they know or should know that radio waves can be easily monitored!

It is easy to be pessimistic about the way in which the Supreme Court has reacted to techno-logical change. In many respects, *Smith* is unfortunately more typical than *Katz* of the way the Court has behaved. For example, when movies were invented, and for several decades thereafter, the Court held that movie exhibitions were not entitled to First Amendment protection. When com-munity access cable TV was born, the Court hindered municipal attempts to provide it at low cost by holding that rules requiring landlords to install small cable boxes on their apartment buildings amounted to a compensable taking of property. And in *Red Lion v. FCC,* decided in 1969 but still not repudiated today, the Court ratified government control of TV and radio broadcast content with the dubious logic that the scarcity of the electromagnetic spectrum justified not merely government policies to auction off, randomly allocate, or otherwise ration the spectrum according to neutral rules, but also much more intrusive and content-based government regulation in the form of the so-called fairness doctrine.

Although the Supreme Court and the lower federal courts have taken a somewhat more enlightened approach in dealing with cable television, these decisions for the most part reveal a curious judicial blindness, as if the Constitution had to be reinvented with the birth of each new technology. Judges interpreting a late-eighteenth-century Bill of Rights tend to forget that, unless its *terms* are read in an evolving and dynamic way, its *values* will lose even the *static* protection they once enjoyed. Ironically, *fidelity* to original values requires *flexibility* of textual interpretation. It was Judge Robert Bork, not famous for his flexibility, who once urged this enlightened view upon then-Judge (now Justice) Scalia, when the two of them sat as colleagues on the U.S. Court of Appeals for the D.C. Circuit.

Judicial error in this field tends to take the form of saying that, by using modern technology ranging from the telephone to the television to computers, we "assume the risk." But that typically begs the question. Justice Harlan, in a dissent penned three decades ago, wrote: "Since it is the task of the law to form and project, as well as mirror and reflect, we should not . . . merely recite . . . risks without examining the *desirability* of saddling them upon society." (*United States v. White*, 1971 401 U.S. at 786). And, I would add, we should not merely recite risks without examining how imposing those risks comports with the Constitution's fundamental values of freedom, privacy, and equality.

Failing to examine just that issue is the basic error I believe federal courts and Congress have made: in regulating radio and TV broadcasting without adequate sensitivity to First Amendment values; in supposing that the selection and editing of video programs by cable operators might be less than a form of expression; in excluding telephone companies from cable and other information markets; in assuming that the processing of 0s and 1s by computers as they exchange data with one another is something less than speech; and in generally treating information processed electroni-cally as though it were somehow less entitled to protection for that reason.

The lesson to be learned is that these choices and these mistakes are not dictated by the Constitution. They are decisions for us to make in interpreting that majestic charter, and in imple-menting the principles that the Constitution establishes.

CONCLUSION

If my own life as a lawyer and legal scholar could leave just one legacy, I'd like it to be the recogni-tion that the Constitution *as a whole* "protects people, not places." If that is to come about, the Constitution as a whole must be read through a technologically transparent lens. That is, we must embrace, as a rule of construction or interpretation, a principle one might call the *cyberspace corol-lary.* It would make a suitable twenty-seventh amendment to the Constitution, one befitting the

200th anniversary of the Bill of Rights. Whether adopted all at once as a constitutional amendment, or accepted gradually as a principle of interpretation that I believe should obtain even without any formal change in the Constitution's language, the corollary I would propose would do for *technology* in the twenty-first century what I believe the Constitution's Ninth Amendment, adopted in 1791, was meant to do for *text*.

The Ninth Amendment says: "The enumeration in the Constitution, of certain rights, shall not be construed to deny or disparage others retained by the people." That amendment provides added support for the long-debated, but now largely accepted, "right of privacy" that the Supreme Court recognized in such decisions as the famous birth control case of 1965, *Griswold v. Connecticut*. The Ninth Amendment's simple message is that the *text* used by the Constitution's authors and ratifiers does not exhaust the values our Constitution recognizes. Perhaps a twenty-seventh amendment could convey a parallel and equally simple message: the *technologies* familiar to the Constitution's authors and ratifiers similarly do not exhaust the *threats* against which the Constitution's core values must be protected.

The most recent amendment, the twenty-sixth, adopted in 1971, extended the vote to eighteen-year-olds. It would be fitting, in a world where youth has been enfranchised, for a twenty-seventh amendment to spell a kind of "childhood's end" for constitutional law. The twenty-seventh amendment, to be proposed for at least serious debate, would read simply:

> This Constitution's protections for the freedoms of speech, press, petition, and assembly, and its protections against unreasonable searches and seizures and the deprivation of life, liberty, or property without due process of law, shall be construed as fully applicable without regard to the technological method or medium through which information content is generated, stored, altered, transmitted, or controlled.

22

Technology Transfer and Globalization

Evan Selinger

INTRODUCTION

Despite the social, political, ethical, and epistemic importance of globalization and technology transfer, philosophers tend to be prioritizing other areas of inquiry. In order to clarify the strengths and weaknesses found in the dominant assessments of these topics, I begin this chapter with a meta-philosophical analysis that reviews representative forms of inquiry. The remainder of the chapter clarifies a vision of how philosophers of technology can pursue a new wave of socially significant investigation. In order to exposit this vision in concrete terms, I turn to the example of the Village Phone program in Bangladesh. While advocates tout this endeavor as a new development paradigm that empowers impoverished and mistreated women by providing them with micro-credit and mobile phones, detractors can find the program's implementation reproducing and augmenting insidious patriarchical forces. By questioning what considerations economic and ethnographic analyses occlude, I not only hope to shed light on the Village Phone program and the underlying trends that drive it, but I further hope to clarify how philosophers of technology can enter into meaningful dialogue with a range of development theorists and practitioners.

ARE PHILOSOPHERS AWARE OF GLOBALIZATION?

For activists, citizens, and theorists alike, "globalization" remains a contested term. In popular and academic discussions, "globalization" is evoked to explain and contextualize many of the extreme and contradictory outcomes that have come to be associated with integrated changes in culture, economics, the environment, politics, and technology. To highlight but a few salient topics, globalization discourse extends to views on: the relations between capitalism, technology, and historical change; the extent to which cultural diversity is a desirable end; the best ways to understand the differences between secular and religious perspectives, and the most useful ways to ameliorate their tensions; the identities of and relations between developed and developing countries; potent environmental changes; phenomenological transformations in how time, space, and place are experienced, and the institutional mechanisms that accommodate these changes and promote additional alterations; and, the putatively declining authority of the nation-state.

Despite living amidst globalization's contentious changes, many philosophers continue to concentrate on other topics; in so doing, they perpetuate the long-standing and rather unfortunate stereotype that philosophy is an esoteric and other-worldly enterprise. To crystallize this point, consider the following results, obtained during a recent search of the *Philosopher's Index*:

- "ethics" (57,845 entries)
- "metaphysics" (55,135 entries
- "aesthetics" (18,528 entries)
- "phenomenology" (9,376 entries)
- "bioethics" (2,510 entries)
- "globalization" (682 entries)
- "development ethics" (12 entries)
- "technology transfer" (5 entries)
- "digital development" (0 entries)
- "microloans" (0 entries)
- "microcredit" (0 entries)
- "Grameen Bank" (2 entries)
- "Grameen Phone" (0 entries)

Globalization and Normative Ethics

When philosophical analysis is given to globalization, the topic of technology is typically reduced to an analytic framing device, a springboard for addressing issues of responsibility that do not stretch or dissolve the conceptual parameters which permit the standard forms of normative analysis (e.g., cosmopolitanism, utilitarianism, the capabilities approach, communitarianism, Habermasian Critical Theory, etc.) to clarify how human agency and human action can be judged in a coherent and potentially systematic fashion. While some discussions about justice, well-being, and moral duty do refer to technology explicitly, the paradigm cases of environmental ethics, labor ethics, cultural ethics, and military ethics, remain more the exception than the rule. And even these analyses scarcely emphasize the concrete dimensions of material culture—what it is, how it can be reproduced and altered, and how it can participate in the organization and disruption of public and private projects—or integrate insights from phenomenology and the cognitive sciences about how embodied human beings respond to artifacts. Such occlusions testify to the fact that the technologies which Bruno Latour (1992) characterizes as "the missing masses" continue to remain largely invisible from the otherwise discerning philosophical eye.

Globalization and Development Ethics

Since the standard philosophical approaches to globalization insufficiently address core problems in development theory and practice, it might be hoped that development ethicists would present robust analyses of technology and technique. David Crocker, Senior Research Scholar at the Institute for Philosophy and Public Policy, defines the relation between development work and development ethics as follows:

> Development—conceived generally as desired or desirable social change—is the work of policy-makers, project managers, grassroots communities, and international aid donors, all whom confront daily moral questions in their work with poor countries. Seeking explicit and reasoned answer to these questions is the work of development philosophers and other ethicists (59).

Crocker further identifies five issues as development ethicists' central concerns (60–63).

1) What is the best way to define the parameters of "development" and the means for achieving it?
2) Who is morally responsible for promoting development? Is it a nation's government, civil society, the market, international institutions, or collaboration between some or all of these actors?
3) Do affluent nations, states, corporations, or individuals have obligations to the poor? If so, what are they?
4) How should the impact and potential of globalization be understood and ethically assessed?
5) How should the moral issues that emerge in development policymaking and practice be addressed and resolved?

Although technological issues are central to all five of these questions, development ethicists mostly address them through other foci. For example, in the wonderful book, *The Ethical Dimensions of Globalization* (2007), contributors focus on the following questions:

- Can general philosophical principles concerning the nature of punishment and justice be brought to bear on the problems of "retribution" and "reconciliation" as they are arising in South African contexts?
- Can practices of "cultural reenactment," such as William Kentridge's animated film study of apartheid, shed light on instances in which democratic citizens are "complicit" in acts of starvation and mass violence occurring far from home that they normally don't feel responsible for?
- Can the appeal to general moral principles be justified and rendered pragmatically useful in instances where human rights and local cultural norms conflict, for example, cases concerning female genital mutilation and child labor?
- Which view of responsibility is more defensible, the view of cosmopolitanism, according to which we are not entitled to treat those near and dear to us as more morally valuable than those human beings with whom we have no ties of family, ethnicity, nationality, or citizenship, or the particularlist view of internationalization, according to which citizens who share in political institutions and a common destiny can justifiably, in some respects, privilege themselves?
- Do advocates of globalization routinely hold views about free trade and migration that are deeply incompatible?

While all of these topics are significant, the development ethicists who consider them nevertheless offer sparse consideration to technology, and this is due to their reliance upon the concepts and methods found in the standard normative philosophical literatures on globalization.

Globalization and the Philosophy of Technology

While the limits of development ethics are somewhat predictable given their sources of intellectual inspiration, it is all the more unfortunate that matters are not much better within the mainstream philosophy of technology. As recently as 1995, the German philosopher Friedrich Rapp could conclude a review essay by identifying the "globalization of technology" as a "new horizon" of philosophical debate that had only "recently" begun to occupy the "center" of discussion. Furthermore, the two main philosophy of technology anthologies that were published after this observation was made, *Philosophy of Technology: The Technological Condition* (2003) and *Readings in the Philosophy of Technology* (2004), scarcely address globalization or technology transfer, even though they engage with many of the philosophical concepts that should be brought to bear upon these issues. And while contemporary analyses of postmodern warfare (e.g., Jean Baudrillard, Paul Virillio, and Slavoj Žižek) do concentrate upon the relation between technology and globalization, they focus mostly on abuses of power that occur during conflict; no consideration is given to technology transfer during peacetime.

Given these trends, it can be lamented that the Society for Philosophy and Technology waited until 2007 to make globalization its central conference theme. Even then, despite laudably including presenters from around the world and a plenary session on Thomas Friedman's *The World Is Flat*, there was less discussion of globalization than one might have hoped for. As an instructive contrast, we can note my home institute—Rochester Institute of Technology, a college that focuses on training students in matters of applied technology, but which lacks a philosophy major—has long incorporated into its mission statements the discourse of preparing students to become "caring and productive members of global society."

When philosophers of technology actually do address globalization, their attention typically remains on the problems and hopes of the West. A countertrend that addresses technological issues in non-Western cultures does exist, and such theorists as Don Ihde, Andrew Feenberg, Carl Mitcham, Hans Poser, Val Dusek, and Aidan Davison have done commendable work. Again, their interventions are exceptional.

The Western bias under discussion here has not gone completely unrecognized. Many of the insights expressed in the philosophy of technology originate in phenomenology, and eminent globalization theorist, Niklas Luhmann, argues that Edmund Husserl's reflections on the crisis of history are tainted by Eurocentrism:

> Most conspicuous is perhaps a Eurocentrism that one rarely finds elsewhere in the twentieth century. European humankind is in crisis, European humankind is in need of salvation—and this by itself. This has certainly nothing to do with imperialism, colonialism, and exploitation, but obviously only with a spiritual consciousness of superiority that not only excludes "the gypsies who constantly vagabond around Europe," but also considers a Europeanization of all other humans "whereas we, if we understand ourselves correctly, will, for instance, never Indianize ourselves." No consideration of the political and economic relations around the globe, no thought of the possibility that the European tradition could slowly be dissolved into other, differently structured relations in a world society. The emphasis on crisis and salvation, autonomously achieved, is owing to these blind spots, which at that time were already non-credible and which would become obviously even less so after the Second World War (38).

Updating this criticism, Trish Glazebrook, a Heidegger scholar, uses her contribution to *Globalization, Technology, and Philosophy* as an opportunity to comment on the Western bias currently found in the philosophy of technology:

> Technology theorists are remarkably silent on the topic of globalization. Although philosophy of technology is burgeoning as a discipline, its proponents have little to say about technology transfer to developing nations, and the impact of the global human condition on technology outside the West, or, as it is also called, the North (143).

To concretize this observation, consider some of the topics reviewed in that volume:

- In "On Globalization, Technology, and the New Justice," globalization protestors are characterized as "profoundly conservative" and unduly "fearful of change." The author's main point is that given the pace of technological and scientific innovation, it is useless nostalgia to look for conceptions of "planetary justice" that are not firmly embedded in contexts where technical systems promote ever-increasing efficiency.

- In "Democracy in the Age of Globalization," globalization is characterized as a desubjectivized postmodern culture that challenges the commitment to virtue. The author's main point is that strategies of resistance need to be cultivated to combat the fact that consumerism and atavistic tribalism have become the dominant horizons for thinking and acting.

- In "Globalization, Technology, and the Authority of Philosophy," globalization is characterized as a movement toward "barbarism." The author's main point is that the speed of technology has created a culture that no longer cultivates the slow and careful patience required for attaining genuine "wisdom."

- In "Communication versus Obligation: The Moral Status of Virtual Community," we are informed that virtual relationships rarely can rise to the level of genuine community. The author's main point is that online interaction routinely fails to provide a context for participants to exhibit the quality of regard and obligation that face-to-face communities can inspire.

- In "The Problem with 'The Problem of Technology,'" we are informed that unless distance from technology can be achieved, the pre-digital past will continue to be looked down upon as a primitive period. The author's main point is that as a consequence of demeaning history, we deprive ourselves of critical resources for understanding and pursuing happiness.

- In "The Human Condition in the Age of Technology," globalization is characterized as a period where "our liberation from the materiality of the world is purchased at the price of inhabiting a parallel world of incomparably less depth and density." The author's main point is that the "ascendancy of the virtual over the real" has undercut "the very reality of human existence."

While all of these issues are worth considering, even if the conclusions are contentious (and, in some cases, perhaps even wrong), it is important to acknowledge that the main problems addressed are difficulties concerning general attitudes toward technology—specifically, Western attitudes that tend to be presented in terms of dilemmas about declining civics faced by a monolithic community, the universal "we." On the rare occasions in which concrete attention is given to specific technological practices, it is solely to determine their moral impact on developed nations. For example, it is noteworthy that the chapter which examines online virtual communities does not question the benefits and detriments that arise (and which can be expected to arise) when laptop computers are exported to developing countries.

Finally, given the persistent appeals to Martin Heidegger throughout the philosophy of technology, it is beneficial to end this meta-philosophical section by noting that in exemplary cases, such evocations diminish, rather than enhance analysis. For example, in *The Creation of the World or Globalization* Jean-Luc Nancy presents a bombastic indictment of globalization that places technology at the center of a dystopian polemic. On the assumption that Heidegger's account of technology and Foucault's account of "biopower" are both, more or less, accurate, Nancy feels justified in articulating dire proposals, such as the claim that "technological and economic planetary domination" are leading to the "disintegration of the world," including, "unprecedented geopolitical, economic, and ecological catastrophe," without referring to any empirical case studies, or even examples (3, 50). Succinctly put, Nancy opposes two possible human "destinies" by contrasting "globalization," which designates a uniform economic and technological logic, with *mondialisation*, which designates the possibility of "authentic world-forming." Embedded in this binary distinction are a variety of other overly reductive contrasts, including demarcations between: creativity and nihilism, immanence and transcendence, unworld and habituation, representation and practice, and principle and mystery. Technology is demonized at the level of metaphysics because Nancy associates "metaphysical history" with "denaturation," and characterizes the history of philosophy as a horizon that limits thought through the technological manipulation of *logos* (77–90). As the translators of the text, Francois Raffoul and David Pettigrew, approvingly note: "The technology of *logos* thus reveals the denaturation of history, of the human being and of life itself. Life, Nancy insists, is no longer pure or bare, but rather produced according to technology. On Nancy's account, life becomes *techne*, and politics the management of *ecotechnology*" (13). *Given this monolithic and essentializing reductivism, it is not surprising that Nancy does not closely consider any empirical examples, or acknowledge benefits, much less ambiguities, that attend to globalized technological practices*!

VILLAGE PHONE: PRELIMINARY CONSIDERATIONS

With the meta-philosophical analysis of globalization and technology transfer completed, the time has come to turn our attention to a specific case of globalized technology transfer and my sense of how philosophers can contribute to extant discussions of it. To establish sufficient context, this transition necessitates a few comments on current affairs.

In recognition of how his microcredit projects "advance democracy and human rights" by creating "economic and social development from below," the 2006 Nobel Committee chose Muhammad Yunus to be the first economist to receive the Peace Prize, an award traditionally bestowed upon politicians and statesmen.[1] Such prestigious and unprecedented recognition suggests that while diverse approaches to microcredit exist, many people view Yunus's methods and aspirations to be the paradigmatic alternatives to "top-down" government-sponsored and NGO-run development initiatives—initiatives that are often equated with a "Western" approach to addressing global poverty.

Yunus founded the Grameen Bank, an institution that achieved international acclaim for offering small "entrepreneurial" loans to impoverished Bangladeshis who lack collateral. He also helped start the Village Phone program (henceforth, VP), an initiative that provides Bangladeshi women with an opportunity to become entrepreneurs by renting calling time on mobile telephones to mostly illiterate villagers who cannot afford to obtain their own telecommunications devices.[2] Inspired by the success of this program and the replicated versions instantiated around the world (e.g., the Philippines, Rwanda, Uganda, and Cameroon), mobile phones have become elevated to symbols of effective digital development; they are routinely characterized as "weapons against poverty."

According to *New York Times* foreign correspondent Celia Dugger, Yunus's economic programs qualify as genuine contributions to peace because microcredit can *empower* women who have been disfranchised by religious fundamentalism:

> [Microcredit] offers hope. It offers, very importantly, *empowerment to women*. Overwhelmingly these microcredit loans are provided to women who are often quite financially powerless in their families. They often don't have rights to inherit property, they don't have bank accounts of their own. So the fact that the woman suddenly has the power to obtain a loan, even a very small loan, can be very important in giving her power and a counterbalance to the appeal of Islamic fundamentalism, which subsumes often the role of women (Sims 2006; emphasis added).

While Dugger conveys a widely held opinion, unanimous agreement does not exist on the matter. In light of reliable ethnographic observations and reasonable views on political agency, some detractors view the Grameen Bank's reforms as *disempowering*.

At first glance, it can be difficult to appreciate why so much importance is given to the matter of whether women are, in fact, empowered by programs such as VP. After all, other issues are pressing. Is VP an effective program for bringing mobile phones to rural villages in Bangladesh? Is it justifiable to use economic reform, technology transfer, or economic reform that employs technology transfer as a means for challenging traditional cultural norms? Is the very notion of "empowerment" so thoroughly Western that it is a chauvinist act to apply it to Bangladeshis?

Although there is no easy answer to the problem of chauvinism, the fact remains that "empowerment" is the dominant concept that assessors use when judging the impact of microcredit programs on women, a population deemed vulnerable, marginalized, and deserving of prioritized attention. Given the prevalence of "empowerment" in both advocacy and critical development literatures, development ethicists have a responsibility to examine whether or not it is the most appropriate term to use for making sense of and evaluating how women's lives change as a consequence of gaining new access to capital and technology. In other words, since "empowerment" has become the primary "talking point" for framing discussions of microcredit programs, philosophers ought to reflect on the underlying presuppositions governing its use. In so doing, the questions raised above will in fact be addressed, even if only indirectly.

The purpose of this essay is to advance discussions of VP by suggesting that the empowerment debate may be based on a poorly posed problem. Contrary to the prevailing accounts that present us with the choice of judging Bangladeshi women to be either fundamentally empowered or else fundamentally disempowered by microcredit, I will contend that the loan recipients should be understood as *embodied subjects who are embedded in conditions in which relations of independence and dependence exist simultaneously*. In arguing that the Grameen Bank's assessors would benefit from reflecting on this *ambiguity*, I will appeal to *applied phenomenological* insights. Although traditional phenomenology has been criticized for being subjectivist, apolitical, insensitive to gender, and reductivist with respect to material culture, I will demonstrate that the phenomenological approach to "lived experience" can shed crucial light on the culturally contingent, value-laden form of labor that a particular group of Bangladeshi women routinely engage in. In this context, I am seeking to bring Marxism, feminism, and the postphenomenology of technology into better dialogue. This endeavor can be considered an exercise in *postphenomenology* because it is written in a "middle-voice" that aims for a subtle equipoise between critique and endorsement of innovative technological practice. Such a position builds upon and therefore is indebted to the dystopian attitude towards technology expressed in previous phenomenological inquiry.[3]

MICROCREDIT EMPOWERS

The women who participate in VP are called "phone ladies," and they have the potential to earn a salary that exceeds the daily income of three-quarters of Bangladeshis (Murphy 2002, 163). When phone ladies are characterized as empowered, the following seven reasons are cited (Yunus 2003; Aminuzzaman, et al. 2003).[4]

First, in targeting women, VP is praised for recognizing the potential of the "poorest of the poor," a marginalized population that routinely has been denied access to credit and exploited by moneylenders. Second, VP is credited for providing women with employment opportunities that traditional Muslim customs of *purdah* inhibit. Under these customs, women are restricted to home-based domestic work; they are discouraged from speaking with males who are not relatives. What

VP facilitates, therefore, is a socially permissible opportunity for phone ladies to speak with male phone clients. Third, as a consequence of the economic opportunities that VP generates, women gain authority and "respect" from their spouses and communities. Fourth, by earning increased income through VP, women are able to take a more active role in their children's futures. For example, they can convey a positive image about women to their daughters, and they have the resources to provide their children, both boys and girls, with better educational opportunities. Fifth, due to the Grameen Bank's social agenda (conveyed in its "Sixteen Resolutions), women who participate in VP are praised for embracing modern values. For example, in order to qualify for loans phone ladies need to eschew the repressive custom of dowry and learn skills that instill self-discipline and appreciation for wellness (e.g., nutrition, sanitation, and family planning are emphasized). Sixth, women are taught to appreciate the virtue of solidarity; in fulfilling the requirements for obtaining loans, phone ladies make pledges to look after one another. Seventh, by promoting "entrepreneurialism," VP is said to do something that charity cannot; it instills pride and confidence, characteristics that ostensibly form the psychological foundation for enhanced civic participation.

For some economists, the empowerment narrative reviewed above risks idealizing microcredit. In response, several moderate criticisms have been offered, including the following from Jayati Ghosh (2006):

> It is a mistake to view microcredit as the universal development panacea which it seems to have become for the international development industry. It can at best be a part of a wider process that also includes working towards reducing asset inequalities, better and more egalitarian access to health and education services, more productive employment opportunities.

Beyond this general judgment, Ghosh highlights four contentious issues.

1) Microcredit operations, including the Grameen Bank's, "depend substantially on subsidies . . . because of high costs of transaction and monitoring." Such subsidies may "imply a transfer of public resources from other public spending, leading to cuts in public health, sanitation and education expenditure."
2) Because microcredit provides small amounts of money and requires borrowers to repay their loans quickly, microcredit may merely function "as a consumption stabilizer, reducing the adverse effects of shocks such as natural calamities or seasonal fluctuations, and provides means for taking advantage of very small business opportunities." As a consequence, microcredit may "amount to no more than a redistribution of incomes among the relatively poor, rather than an overall increase in incomes of the poor."
3) Microcredit borrowers can find themselves in a state of "microcredit dependency" in which they rely on loans for "consumption" rather than "productive use." In some instances, "peer pressure has forced women borrowers to take on expensive loans from moneylenders" to repay their bank loans.
4) Because microcredit institutions require high repayment rates to remain sustainable, they can enact policies that function as "instruments" of "stratification." For example, there "have been cases of women from the most destitute or socially deprived groups being excluded from membership of groups containing better off members, because of fears that their inability to repay will damage the prospects of other members."

While these are all provocative criticisms, none address the following *phenomenological question*:

- What aspects of practice are occluded when quantified analysis, obtained through survey studies and neo-classical economic theories, only foreground selective consequences of participating in VP, notably the "satisfied preferences" that can be detected from a "bird's eye" perspective?

This question focuses on the implications that follow from the fact that advocates of the empowerment position scarcely addressing the social and cultural constraints that many Bangladeshi women experience when they apply for microcredit and maintain the behaviors required for being a borrower in good standing. Since a preliminary answer to this question concerning "lived experience" can be found by consulting *qualitative anthropological* inquiry that does not *disembody* or *disembed* the subjects it studies, we will begin the next section by discussing an influential anthropological text.

MICROCREDIT DISEMPOWERS: ANTHROPOLOGICAL PERSPECTIVE

Aminur Rahman's *Women and Microcredit in Rural Bangladesh* is perhaps the most well-known anthropological critique of the empowerment narratives reviewed in the last section. Rahman's indictment of how traditional Bangladeshi culture absorbs microcredit programs is so scathing that it poses a challenge to numerous "Women in Development" initiatives. These initiatives try to improve the quality of women's lives in developing countries by calling upon "governments, development agencies and international financial institutions to provide aid and resources specifically for women, who would then be able to contribute substantively towards family welfare and national development" (Chowdhry 2001).

Through participant observation and unstructured interviews, Rahman, a "native" of Bangladesh, provides a qualitative *"worm's eye"* study of the day-to-day lives of 120 women from the Tangail district in Bangladesh who borrowed money from the Grameen Bank (22). While Rahman's initial goal was to better understand "the dynamics of the empowerment of women," his up-close observations of "women borrower's lack of power" led him to change course (Rahman 1999, 24).

Rahman discounts the empowerment narratives provided by indigenous theorists and functionaries because he views their judgment as compromised by national pride and personal ambition: "The academics, researchers, and bureaucrats in Bangladesh . . . produce and maintain the hegemonic discourse of the Grameen Bank to establish it as a development 'icon' and to enhance their own reputations" (50). Contrarily, he depicts his own iconoclastic perspective as well-founded because it is informed by four critical ideas.

First, Rahman creates a theory of "disentitlement"; it is a modification of economist and philosopher Amartya Sen's notion of "entitlement" and anthropologist Arjun Appadurai's concept of "enfranchisement" (40–42). Second, Rahman appropriates the distinction between "public" and "hidden" transcripts that political scientist James Scott articulates in *Weapons of the Weak* (42–44). Third, Rahman appeals to aspects of Pierre Bourdieu's sociological practice theory: "habitus," "field," and "capital" (44–48). Finally, Rahman makes use of political theorist Antonio Gramsci's concept of "hegemony" (52). For present purposes, it is not necessary to discuss these ideas in detail. What matters is simply that we recognize that these tools of *ideology critique* incline Rahman to be suspicious of the typical testimony of Grameen Bank employees and borrowers, and the typical analyses of the Grameen Bank's programs.

On the basis of his fieldwork, Rahman comes to see the Grameen Bank's accomplishments as being partially attributable "to its ability to successfully utilize patriarchal structures in facilitating its goals and agendas" (Chowdhry 2001). Rahman concludes:

> Most women borrowers are not the direct benefactors of the credit extended to them. Instead, these women appear to be mediators between their male household members and the bank. Thus the lending institution invests loans in the village to generate profit, but it uses the prevailing patriarchical norms of the village society and the positional vulnerability of women (immobile, shy, passive) for timely repayment and distribution of loans (23).

In support of this distressing outlook, Rahman emphasizes three salient problems.

First, since the Grameen Bank targets its loans to women, some of the husbands who have been excluded by this policy have forced their wives to sign up for loans, only to forcibly appropriate the funds from them. Rather than addressing these loans-by-proxy, the Grameen Bank proclaims that women are empowered by becoming loan recipients (Rahman, 40–41). By touting its commitment to providing opportunities for the most disfranchised Bangladeshi population without acknowledging what actually happens to that population when it pursues these opportunities, Rahman claims that the Grameen Bank generates an "ideology" that obscures the connection between microloan practice and "the larger structure of patriarchy." Worse, because patriarchy becomes intertwined with the lending mechanisms, Rahman insists that the Grameen Bank is guilty of inaugurating "new forms of domination over women in society" (Rahman, 51).

Second, although orthodox narratives emphasize the Grameen Bank's success at instilling empowerment in women by teaching them to assist one another and to abandon detrimental domestic behaviors, Rahman contends that Bangladeshi women tend to present corroborating testimony about these reforms only as a "strategic pose"; their goal is not to tell the truth, but to placate authorities and ensure that they can continue to qualify for funds (Rahman 43). Contrary to the overt testimony, Rahman claims that there are ample instances in which women do not follow through with their commitment to refrain from giving dowry, upgrade sanitation, or engage in substantive changes in how they eat (Rahman 94–96). Additionally, he insists that vicious, but underreported interactions have occurred amongst females themselves in the "lending circles." In these instances, "power hierarchies" underwritten by classicism have prevented genuine solidarity from arising (Rahman, 124–127). Again, Rahman contends that some of women who witness these confrontations stay silent in order to avoid jeopardizing their own loans.

Third, Rahman claims that the Grameen Bank presents an ideological justification that hides the true basis for why it almost exclusively provides loans to women. Officially, the Grameen Bank makes two claims: (1) it promotes social justice by redressing Bangladesh's history of depriving women access to credit; and (2) it respects sexual difference by acknowledging the fact that Bangladeshi women are more fiscally responsible than men (Rahman, 71–72). For example, Yunus maintains that empirical observation establishes that men are inclined to waste their income on frivolous experiences and unnecessary commodities, while women typically prioritize their children's welfare and household necessities.

Contrary to this rationale, Rahman proclaims that the Grameen Bank really targets women because they are an easily manipulated population. More specifically, the Grameen Bank views women as more likely to repay their loans than men because they can be disciplined through *purdah*—cultural norms that emphasize the "virtues" of "submissiveness," "modesty," "purity," "respectability," and "humility" (Rahman, 73–75). Rahman notes that in some cases the Grameen Bank minimizes its transaction costs by threatening women with "shame," and in others it uses or condones violence (both physical and verbal) as a mechanism for pressuring women to make timely payments (Rahman, 123–124). Beyond these scenarios, Rahman insists that the Grameen Bank does not effectively deal with the fact that some husbands will resort to violence to force reluctant wives to sign up for loans.

Although these claims seem extreme when compared with the depictions of microlending espoused in the empowerment narratives, Rahman's central thesis resonates with historical precursors. In this context, it is instructive to recall the uproar that occurred over Bangladesh's reliance on child labor.

In the mid-1990s, discussions took place in the United States about boycotting Bangladeshi goods produced by child labor. In response to congressional consideration of the Child Labor Deterrence Bill and public outcry over a documentary on Wal-Mart importing clothing made by underage laborers, "nervous" Bangladeshi factory owners fired 50,000 children, "75% of the total then employed" (Pierik 2007, 48). Contrary to the "dramatically naïve" perception that these kids would return to school, none of them actually did (Pierik 2007, 48–49). Instead, some remained unemployed despite looking for work; others took jobs—including prostitution—at reduced pay and settled for less adequate nutritional and healthcare conditions (Pierik, 49).

The lesson to be learned from this incident and Rahman's observations is that long-standing social norms in Bangladesh cannot effectively be challenged without first addressing the primary cultural forces which give rise to and sustain them. Thus, economic reform cannot liberate Bangladeshi women until the patriarchal structures that disempower them are directly confronted. Unfortunately, this lesson is obscured by the implicit convictions about technology that underwrite narratives about VP.

VP: TECHNOLOGICAL SCRIPT AND *PURDAH*

Although Rahman's analysis focuses on how patriarchical power permeates the Grameen Bank's microloan iniatives, he does not examine any of the *forms of labor* that women perform after they invest their newly acquired capital into businesses endeavors. Given Rahman's goal of calling attention to the general ways that patriarchy taints microcredit initiatives in Bangladesh, it made sense for him to restrict his focus. What I find disappointing is that none of the critical studies of the Grameen Bank, including ones written after Rahman's book was published, examine the *constraints that women experience when they do their job of renting calling time on mobile phones.*[5] Such occlusion is indicative of how deeply embedded the "instrumental" conception of technology is in both the popular and scholarly imaginations. The persistence of this view suggests that while it is easy to grasp how technology can be put to moral and immoral uses, it can be difficult to appreciate how *technologies and technological practices themselves can be value-laden.*

When mobile phone use is instrumentally analyzed in the context of VP, emphasis is typically given to the salutary ends that the indigenous customers use the technology to pursue (Bayes 2001). For example, because Bangladeshi merchants use mobile phones to gain access to the price of commodities, they can avoid being exploited by middlemen. Additionally, when illiterate Bangladeshis use mobile phones to contact expatriated relatives, they can avoid having their exchanges mediated by religious *imams*. Also, mobile phones provide an opportunity for people who are ill (or who own sick livestock) to obtain medical advice without losing valuable working time or experiencing the hindrance of inefficient transportation systems.

In the rare instances when problems with mobile phone use are addressed, the framework remains instrumentalist; emphasis stays on the salutary consequences that follow when normative human decision-making has more authority than technological or economic influence. For example, the popular coverage that was given to the matter of Bangladeshi parents complaining that their children were being corrupted by the conversations occurring at night (when free calling time was available) focused on how the Bangladesh Telecom Regulatory Commission was petitioning phone vendors to cease from providing this service.

While these scenarios are significant, it is a mistake to treat how VP customers use phones as the only technologically relevant consideration. For if Rahman's analysis is accurate, we should *expect to find patriarchy tainting all of the major opportunities that the Grameen Bank provides*, including opportunities for women to work with mobile phones. Indeed, given Rhaman's reliance on ideology critique, it is surprising that none of the theorists who have been influenced by his views have discussed how the hybrid human-technology phrase "phone lady" evokes Karl Marx's insight that the material conditions which constitute forms of labor can impose identities upon the laborers whose consciousnesses are shaped by the work they perform. While interminable debates continue over Marx's so-called economic and technological determinisms, it is harder to reject his *phenomenology* of what it was like to be a typical nineteenth century factory worker. Marx provides this phenomenology when he discusses the ontological implications that followed from people working under alienating conditions—conditions where one performed a job that was regulated by a standardized regime that reduced human behavior to a functional extension of outputs provided by machines.

To go beyond Rahman's analysis and determine if influence of *purdah* is present in VP, the following research questions need to be answered. Given the centrality of lived experience, *phenomenological considerations* are crucial:

- What opportunities for engaging with customers does VP facilitate as well as inhibit?
- What opportunities for engaging with technology does VP facilitate as well as inhibit?
- What opportunities for engaging with technical professionals does VP facilitate as well as inhibit?

To address these questions concerning VP's *techno-economic script*, it helps to begin by considering the *embodied dynamics of phone use*. In order for phone ladies to present their customers with optimal conditions for conversation, they need to be *silent* and *unobtrusive*. For if the phone ladies speak while their customers engage in discourse, disappointment will likely result and the prospect of repeat business will be compromised. Customers probably will be disappointed because they will find it difficult to concentrate on the very task that motivated them to rent phones in the first place. Additionally, customers can become disappointed if phone ladies engage in extended conversation with them about their calls after the calls are completed. In this instance, the violation of social etiquette may provide a disincentive for customers to return.

The next consideration to address is the matter of *skill* and *judgment*. Unlike merchants who offer multiple goods and services, and who can provide skilled, if not expert judgment concerning different consumerist options, phone ladies do not operate in a context where they can cultivate perspectives that their clientele will value. Rather, the Grameen Bank provides them with the opportunity to offer only one type of service, and that service invariantly requires a default protocol to be followed—a simple, yet strict script in which a phone is traded for a fee. By contrast, customers who enter mobile phone shops in developed nations can talk with employees about the advantages and disadvantages of procuring different phones, different phone peripherals, and different calling plans. Thus, the *very practice of renting mobile phone time is so restrictive* that if phone ladies deserve to be considered "entrepreneurs," it is only in a qualified sense.

Further insight into VP's script can be obtained if we broaden our considerations so as to reflect upon the *available options* that phone ladies have with respect to *working with the technology* they loan out. The mobile phones that Grameen Telecom provides to the phone ladies are devices designed to fulfill *one single function*; they allow people to communicate with one another in real time over potentially vast geographic differences. Although the history of technology is replete with instances in which technologies come to be used in ways that have little relation to what an artifact's designers initially intended, those instances are occasions in which emergent practices could arise because the technologies came to be used in contexts that are less restrictive than VP. In this sense, it is instructive to recall that while the telephone's history can be traced back to visions of the device being used as a prosthetic by the hard of hearing, it became inserted into contexts that rewarded innovation; as a consequence, the telephone transformed into a device that revolutionized how people who do not have hearing disabilities communicate.

Thus, while it must be admitted that phone ladies could, in principle, put their phones to use in innovative ways, it is hard to imagine that such uses would be sustainable, given economic constraints. Because phone ladies are primarily interested in using mobile phones to earn income, and because it is difficult to conceive of impoverished Bangladeshi customers paying to use mobile phones as paperweights, jewelry, or any other non-traditional service or item, it is hard to imagine that phone ladies can do anything other than hand the phone over, without modification, to a customer who is paying to use it to place a call. The techno-economic script simply restricts the phone lady's degrees of freedom to such an extent that they cannot take creative liberties with the artifact they spend considerable time each day with.

Additionally, given the phone ladies' pervasive illiteracy and lack of advanced formal education, they cannot be expected to understand the scientific and engineering principles that underlie mobile telecommunications. Nor can they, given the limited resources at their disposal, be expected to have opportunity to learn about these principles, should they so desire. In light of these limitations, phone ladies need to *rely upon and fully defer to the skilled technicians* who are charged with keeping the phone systems operative and fixing malfunctioning equipment. Since almost all professions rely upon some division and labor, deference to other peoples' expertise is not objectionable in itself. But what is important to note, here, is that the particular web of dependence at issue places phone ladies in an especially vulnerable position. Although phone ladies are depicted by the empowerment narrative as essentially being self-employed entrepreneurs, the fact remains that they can exert little authority when dealing with their business "partners" who are consistently more highly educated. In this context, it is important to acknowledge that the limited authority phone ladies have at home, and the limited authority they have when they meet the requirement of addressing the predominantly male staff of Grameen Bank as "sir," is extended to their limited authority at work.

Furthermore, even if phone ladies run a business that attracts a sizable clientele, *none of their work will lead to skills that can enable them to minimize their future dependence on technical professionals*. Even the prospect of adding additional technological services presently requires external authorities to contribute further input. To this end, when Yunus boasts that in the future phone ladies will have a chance to become "Internet ladies," what he has in mind is the idea that, in *top-down fashion*, *his staff* will find a way to provide impoverished women living in rural areas with working computers that have voice-operated functions that illiterate populations can find user-friendly (2003, 254).

Final insight into the restrictions that limit how the phone ladies can use phones can be obtained if we compare their labor with traditional craft labor—a topic that, admittedly, has been the subject of overly romanticized accounts. Romanticism aside, it remains the case that laborers who produce traditional crafts typically transform raw materials into goods by skillfully using tools. Because such skilled action tends to require discipline to cultivate, it is a form of engagement that humans can be *proud of*. As Marx's philosophical predecessor G. W. F. Hegel notes, the creation of tangible goods can be rewarding because crafts contain an imprint of the artisan's handiwork; this reflection of the human in the thing lessens the gap that, most of the time, separates subject from object. And yet, as the previous remarks in this section suggest, the script that VP provides is not conducive to phone ladies cultivating skill. Consequently, it does not provide an environment for them to view their professional activity as a *personal achievement worthy of pride.*

All these considerations suggest that when VP is understood as a concrete practice, it turns out to be a profession that is predicated upon female laborers embodying many of the characteristics that *purdah* requires. *When phone ladies do their job, they are passive, invisible, deferential, and unremarkable.* The respect they gain is not accorded to them because they are viewed as peers. Rather, since the value of being a phone lady is associated only with an instrumental utility, they are viewed *more as a service than a human being.* And since mobile phones are predominantly rented by men, phone ladies essentially provide a service that is synonymous with male consumption (Aminuzzaman, et al. 2003, 335). Any account that only considers what phone ladies can do with enhanced income, but which glosses over the experience they endure in order to obtain this income is, therefore, incomplete.

THE POLITICS OF DISEMPOWERMENT

Having extended Rahman's views on patriarchy to the *experience of being a phone lady,* the question remains as to whether VP is in principle disempowering. While nobody appears to be advancing a position that strong, some have come close, at least with respect to the underlying microcredit issues. For example, in his essay "The Micro-Credit Cult," libertarian theorist Jeffrey Tucker claims the mandatory changes in lifestyle that the Grameen Bank imposes are tantamount to a cult's demands:

> So let's say you're a borrower in Bangladesh . . . your private life is gone. The Grameen staff is in charge of your family size and the workings of your latrines. Your friends must be Grameenites. You chant the Sixteen Decisions *ad nauseam* and attend tedious exercise sessions and parades. If you're single, the prohibition on dowries limits your marital prospects. If you're married with children, your children are farmed out to Grameen Day Care. You can't have any more if you want to. Plus, you must periodically abandon your primary occupation to dig around in the dirt planting tree seedlings to please international agencies.

Perhaps the most indicting political argument can be found in Aradhana Parmar's essay, "Microcredit, Empowerment, and Agency: Re-Evaluating the Discourse." Although Parmar does not address VP, her interpretation of Rahman coupled with her commitment to a particular version of feminism, leads her to express concern that the Grameen Bank's microlending practices risk disempowering women by "co-opting" their struggles and leaving them "disserviced" and deprived of *political agency* (Parmar 2003, 466–477).

Parmar offers several premises to support her case, beginning by noting that since the Grameen Bank's microloan iniatives are founded by men and predominantly run by male employees, they reduce women to "welfare objects" of reform (Parmar, 465). Under such patronizing conditions, Parmar claims that institutional norms provide women with little "ownership" over the programs they participate in. As we have already discussed, when husbands are not controlling their wives' lives, male development workers and other professionals are authorized to be domineering over female borrowers. Given these patriarchal constraints, Parmar insists that the Grameen Bank's policies are predicated upon viewing women as "incapable" of "identifying their own needs and priorities," and as unable to exercise their own "rationality" for the purpose of developing positive "strategies" and "visions" for combating oppression (471).

Parmar further insists that the Grameen Bank engages in questionable acts of discipline by treating social reform as a matter to be addressed primarily through practices that equate capitalist values, such as individualism and consumerism, with moral values. Because women only obtain new familial and social opportunities by participating in competitive commercial practices, their emancipation is not truly rights-based; instead, it remains contingent on continued financial success. By rewarding Bangladeshi women for believing that they are worthy of respect because they can earn money, the *intrinsic value of human dignity goes unrecognized* and the capacity for women to *experience a form of solidarity that is based on a principled commitment to combating injustice goes unnurtured*. Indeed, from a practical perspective Parmar observes that as women socially advance for reasons related to "labor and capital," they come to seek better material conditions, but do not feel motivated to examine the systematic structures of their oppression (465).

Ultimately, in contrast with the Grameen Bank's imposition of an externally imposed conception of the good life, Parmar argues that a proper empowerment program would assist women to discover their own capacity to create the conditions under which they can act as agents who make "principled choices" (Parmar, 473–474). With empowerment defined in this way, Parmar admonishes the Grameen Bank for failing to appreciate that when properly understood, empowerment is about inner strength, inner conviction, and the inner motivation to create a world where everyone regardless of race, class, or sex has the capacity to exercise autonomous agency. "Empowerment," Parmar concludes, "is based not on 'power over,' but on 'power with' or 'power within'" (474). "Power within" is the gold standard for political agency because it increases "confidence" and "assertiveness" and thereby motivates agents to eliminate "all exploitive structures" (475).

CONCLUSION: DOES VP DISEMPOWER?

How should we interpret Parmar's claims and all the supporting evidence that she draws from (and, as per section 5, could draw from)? Is VP an empowering program, a disempowering program, or is the empowerment-disempowerment debate predicated upon a poorly posed problem?

To answer this question, let us review the two main critical points discussed so far. First, when phone ladies are characterized as empowered, such depictions are primarily based upon survey datum that measure how well individual "preferences" are "satisfied." Such surveys are constructed from a "bird's eye" perspective that does not adequately register *several aspects of lived experience*, including:

- *Whether the women are at liberty to provide honest answers to the questions they are given, or whether patriarchical constraints bias the responses they can provide;*
- *Whether the women need to endure oppressive encounters in order to remain borrowers in good standing, and in order to obtain income by renting calling time;*
- *Whether the women achieve better social and familial standing at the expense of having their intrinsic dignity respected; and*
- *Whether the women achieve a comparatively better quality of life at the expense of developing the characteristics that political agency requires.*

Second, when VP is characterized as an empowering program, *technocratic assumptions* about *technique* and *technology* are typically made. The Grameen Bank's approach to microcredit can be understood as a technique for instilling social change, and the empowerment narratives erroneously suggest that such a technique is culturally transcendent—that it can be imposed on traditional Bangladeshi culture without becoming complicit in its patriarchical norms. Similarly, when the empowerment narratives depict mobile phones as weapons against poverty that challenge patriarchy, they tend to ignore the ways in which certain uses of phones, such as renting them out, become complicit in patriarchical norms. In this case, the mistake consists of viewing technology as culturally transcendent, and not as what Don Ihde (1990) calls a "cultural instrument."

While these considerations give us good reason to be skeptical of the idealizations present in the empowerment narratives, they do not justify the conclusion that phone ladies should be characterized as fundamentally disempowered. If that conclusion were justified, then the only relevant accounts of women's lived experience that future analysts should provide are ones that capture the persistent—if not augmented—presence of patriarchy. In that hypothetical context in which hegemony is all-encompassing, phenomenology would remain a slavish adjunct to ideology critique: only predictable and repetitive patterns of oppression would be emphasized; and gains in independence would consistently be treated as less consequential than the adverse effects of techno-economic scripts that induce relations of dependence.

Ultimately, in order for the unqualified disempowerment conclusion to be valid, three premises would need to be true: *(1) the women who believe they have gained significant independence as a consequence of accruing the benefits reported in the empowerment narratives would have to be experiencing "false consciousness"; (2) subversions of disempowerment (that differ from the behaviors detailed in the empowerment narratives) could not be occurring at present; and (3) the future would need to be closed.* From my perspective, each of these premises is contestable.

With respect to the first premise, even if some of the phone ladies suffer in ways that the empowerment narratives fail to acknowledge, and even if, in some instances, phone ladies are unaware of the extent to which their behavior is compromised, the fact remains that new possibilities for enhancing agency are arising due to access to credit and mobile phones. Without, as Bruno Latour might say, enrolling phone as "allies," these opportunities would not exist. For even if Parmar is right, and "power within" can be meaningfully distinguished from "power over," such a differentiation remains tenuous. Over time, the latter can, as Yunus suggests, become a catalyst for the former. Of course, the latter can also, as Parmar suggests, inhibit the former. But that outcome is an empirical matter; without the assurance of technological or economic determinism, it can only come to light as history unpredictably unfolds.

With respect to the second premise, forms of solidarity between phone ladies may already be occurring, even if they have escaped the attention of analysts. For example, it is possible that in striking a "strategic pose" by pretending to adhere to the "Sixteen Resolutions," Bangladeshi women are cultivating solidarity around their partial subversion of top-down authority. In order to determine if collective consciousness is being formed in this way, anthropologists cannot be content to follow Rahman's lead and treat each instance of women breaking a promise to the Grameen Bank as merely proof of the bank's hypocrisy.

With respect to the third premise, the techno-economic script discussed in section 5 is temporally bounded. Unlike the enduring values embedded in material artifacts, such as speed bumps ("slow down") and disposable coffee cups ("throw me out"), the form of phone lady labor can readily change given shifts in a number of conditions, including alterations to supply-and-demand (Verbeek 2006). For example, if mobile phones proliferate and become more sophisticated, and if competition arises in villages, phone ladies may have the opportunity to engage in certain forms of skilled behavior. They would need to create incentives for customers to use their phones, and this goal could inspire them to arrange their homes in inviting ways, prepare interesting food, etc.

In the final analysis, the problem with "empowerment" and "disempowerment" is that they are modern terms that evoke strong cognitive and emotional responses. They readily conjure images of autonomy and servitude and incline the analysts who use them, even in a qualified sense, to tilt their inquiry in an extreme direction. As a consequence, techno-utopian and techno-dystopian images and rhetoric abound.

What programs like VP do is instill *simultaneous relations of independence and dependence*. As techno-economic reforms, they can create independence only by capitalizing on, and possibly perpetuating, a variety of dependency relations. Indeed, at present, phone ladies only acquire some independence because of a double-dependency; they are dependent on the VP script, and their villages are dependent on their services. In order to create better metrics for assessing these *hybrid relations*, more nuanced accounts of lived experience are necessary—accounts that are sensitive to the impulses toward idealization and ideology critique, but which place *ambiguous experience* in the foreground of the analysis. Attention to this matter will not only improve understanding of VP, but considerations of this sort can provide a new wave for philosophers of technology to interface their analyses with a ranger of development theorists and practitioners.

NOTES

1. The foundation for many of the issues addressed here can be found in three previous articles: Selinger forthcoming a, Selinger forthcoming b, and Selinger 2007.
2. The Village Phone program began in 1997 as a collaborative venture between the Grameen Bank and two companies, a private for-profit company, Grameen Phone Ltd. and a not-for-profit one, Grameen Telecom.
3. For more on postphenomenology, see Selinger 2006.
4. Phone ladies are also referred to as "mobile calling offices."
5. Having restricted my attention to articles and books written in English, I may be overlooking relevant inquiry in other languages. I also may be overlooking sources that fell outside the scope of my searches. These caveats are important for two reasons. First, insofar as I am relying upon secondary literature and not a personally conducted case study, I do not want to overstate the strength of my conclusions. Second, insofar as I am relying upon phenomenological concepts that were developed by Western thinkers, the analysis risks distorting non-Western lifeworlds. This risk is amplified by reliance upon studies that were written in English, for Western audiences.

REFERENCES

Aminuzzaman, S., Baldersheim, H., and Jamil, I. (2003). Talking Back! Empowerment and Mobile Phones in rural Bangaldesh: A Study of the Village Phone Scheme of Grameen Bank. *Contemporary South Asia* 12 (3): 327–348.

Appiah, K. A. (2006). *Cosmopolitanism: Ethics in a World of Strangers*. Princeton, NJ: Princeton University Press.

Baudrillard, J. (2002). *The Spirit of Terrorism*. New York: Verso.

Bayes, A. (2001). Infrastructure and Rural Development: insights from a Grameen Bank Village Phone Initiative in Bangladesh. *Agricultural Economics* 25: 261–272.

Borgmann, A. (2006a). *Real American Ethics*. Chicago: University of Chicago Press.

Borgmann, A. (2006b). Review of *Globalization, Technology, and Philosophy*. *The Canadian Journal of Sociology* 31 (1): 155–157.

Chowdhry, G. (2001). Review of *Women and Microcredit in Rural Bangladesh*. *Journal of Political Ecology: Case Studies in History and Society*. Retrieved on December 16, 2006, from http://jpe.library.arizona.edu/volume_7/Rahman1200.html.

Crocker, D. (2007). Development Ethics and Globalization. In V. Gehring (Ed.) *The Ethical Dimensions of Global Development*. Lanham, MD: Rowman & Littlefield Press, 59–72.

Davison, A. (2001). *Technology and the Contested Meanings of Sustainability*. Albany: SUNY Press.

Dusek, V. (2005). *Philosophy of Technology: An Introduction*. Oxford: Blackwell Publishing.

Feenberg, A. (1995). *Alternative Modernity: The Technical Turn in Philosophy and Social Theory*. Berkeley: University of California Press.

Friedman, T. (2006). *The World Is Flat*. New York: Farrar, Straus, and Giroux.

Gehring, V. (Ed.) (2006). *The Ethical Dimensions of Global Development*. Lanham, MD: Rowman & Little-field Press.

Ghosh, J. (2006). Development as a Nobel Cause. *One World South Asia*. Retrieved October 4, 2007 from www.southasia.oneworld.net/article/view142067/1/2235.

Glazebrook, T. (2004). Global Technology and the Promise of Control. In D. Tabachnik and T. Koivukoski (Eds.) *Globalization, Technology and Philosophy*. Albany: SUNY Press, 143–158.

Ihde, D. (1990). *Technology and the Lifeworld*. Bloomington: Indiana University Press.

Kaplan, D. (Ed.) (2004). *Readings in the Philosophy of Technology*. Lanham, MD: Rowman & Littlefield.

Kellner, D. (2002). Theorizing Globalization. *Sociological Theory* 20 (3): 285–305.

Latour, B. (1993). Where Are the Missing Masses? The Sociology of a Few Mundane Artifacts. In W. Bijker and J. Law (Eds.) *Shaping Technology/Building Society: Studies in Sociotechnical Change*. Cambridge: MIT Press, 225–258.

Latour, B. (1992). *We Have Never Been Modern*. Cambridge, MA: Harvard University Press.

Luhman, N. (2002). *Theories of Distinction*. Palo Alto, CA: Stanford University Press.

McLuhan, M. (1968). *War and Peace in the Global Village*. New York: Simon and Schuster Inc.

Mendieta, E. (2001). Invisible Cities: A Phenomenology of Globalization from Below. *City* 5 (1): 7–26.

Mitcham, C. (1994). Engineering Design Research and Social Responsibility. In K. Schrader-Frechette (Ed.). *Ethics of Scientific Research*. Lanham, MD: Rowman & Littlefield Publishers, 153–168.

Murphy, C. (2002). The Hunt for Globalization that Works. *Fortune* 146 (8): 163–176.

Nancy, J. L. (2007). *The Creation of the World or Globalization*. Trans. Francois Raffoul and David Pettigrew. Albany: SUNY Press.

Nussbaum, M., and Glover, J. (Eds.) (2001). *Women and Human Development*. Cambridge, MA: Cambridge University Press.

Parmar, A. (2003). Microcredit, Empowerment, and Agency—Re-Evaluating the Discourse. *Canadian Journal of Development Studies* 24 (3): 461–476. Retrieved on December 16, 2006, from http://southasia.one world.net/article/view/142067/1/5339.

Pensky, M. (Ed.) (2005). *Globalizing Critical Theory*. Lanham, MD: Rowman and Littlefield.

Pierik, R. (2007). Fighting Child Labor Abroad: Conceptual Problems and Practical Solutions. In V. Gehring (Ed.) *The Ethical Dimensions of Global Development*. New York: Rowman and Littlefield Press, 33–46.

Pogge, T. (2002). *World Poverty and Human Rights: Cosmopolitan Responsibilities and Reforms*. Cambridge, U.K.: Polity Press.

Posser, H. (1991). Technology Transfer and Cultural Background. In W. Konig, H.

Poser, W. Radtke, and Schnell, W. H. (Eds.) *Technological Development, Society, and State*. NJ: World Scientific, 73–81.

Rahman, A. (1999). *Women and Microcredit in Rural Bangladesh*. Boulder, CO: Westview Press.

Rapp, F. (1995). Philosophy of Technology after Twenty Years: A German Perspective. *Techné* 1–2. Retrieved on December 16, 2006 from http://scholar.lib.vt.edu/ejournals/SPT/v1n1n2/rapp1.html

Scharff, R., and Dusek, V. (Eds.) (2002). *Philosophy of Technology: The Technological Condition*. London: Blackwell Publishers.

Scheuerman, W. (2006). Globalization. *Stanford Encyclopedia of Philosophy*. Retrieved on December 16, 2006 from http://plato.stanford.edu/entries/globalization/.

Selinger, E. (forthcoming, a). Does Microcredit Empower? Reflections on the Grameen Bank Debate. *Human Studies*.

Selinger, E. (forthcoming, b). Towards a Reflexive Framework for Development: Technology Transfer after the Empirical Turn. *Synthese*.

Selinger, E. (2007). Technology Transfer: What Can Philosophers Contribute? *Philosophy and Public Affairs Quarterly* 27 1/2: 12–17.

Selinger, E. (Ed.) (2006). *Postphenomenology: A Critical Companion to Ihde*. Albany: SUNY Press.

Sims, C. (2006). Worldview Podcast. *NY Times*. Retrieved on October 23, 2006 at: http://www.nytimes.com/2006/10/21/weekinreview/22worldview.html.

Singer, P., and Mason, J. (2006). *The Way We Eat*. New York: Rodale.

Singer, P. (2004). *One World*. New Haven, CT: Yale University Press.

Sinha, M. (2005). Technology Transfer. In C. Mitcham (Ed.) *Encyclopedia of Science, Technology, and Ethics*, vol. 4, pp. 1912–1914. USA: Macmillan Reference.

Steger, M. (2003). *Globalization: A Very Short Introduction*. New York: Oxford University Press.

Tabachnick, D., and Koivukoski, T. (Eds.) (2004). *Globalization, Technology, and Philosophy*. Albany: SUNY Press.

Tucker, J. (1995). The Micro-Credit Cult. *The Free Market*. Retrieved on February 8, 2007 at: http://www.mises.org/freemarket_detail.asp?control = 215&sortorder = article date.

Verbeek, P.P. (2006). Materializing Morality. *Science, Technology, and Human Values* 31 (3): 361–380.

Verbeek, P. P. (2005). *What Things Do*. University Park: Pennsylvania State University Press.

Virilio, P. (2002). *Ground Zero*. New York: Verso.

Yunus M. (2005). Halving Poverty by 2015—We Can Actually Make It Happen. *The Round Table* 370: 363–375.

Yunus, M. (2003). *Banker to the Poor*. New York: Public Affairs.

Žižek, S. (2002). *Welcome to the Desert of the Real*. New York: Verso.

Part IV

TECHNOLOGY AND HUMAN NATURE

Technology plays a role in shaping human nature. Minimally, technologies mediate our lives and form the basic conditions that shape our identities as individuals and citizens. It is the presence or absence of technological devices and systems that, in large part, forms the differences between life on a farm and life in a city, or life in an advanced industrialized society and life in an underdeveloped society. Technologies also figure into the composition of our identities as part of the stories we tell about who we are. We identify ourselves as, for example, basketball players, home owners, cooks, knitters, music lovers, and a host of other identity constructions that are inseparable from having and using technologies.

But some technologies do more than just influence the construction of personal identity. Some may alter (or promise to alter) the very essence of what it is to be a human being. Technologies may someday make us immune from disease, double or triple our lifespan, surpass us in intelligence or capability, or perhaps even transform us into *posthumans*—human-technology hybrids. These current and future technologies challenge our ideas as to what distinguishes the human from the non-human—that is, what is natural for us to do and to be and what is artificial.

The chapters in this section explore the various ways that technologies shape, transform, and call into question our very idea of human nature. The authors address such subjects as transhumanism and the use of technology to improve human life; the possibilities of nano-biotechnology that supplements human minds and bodies; artificial intelligence and how similar and different computers are from human beings; the role of the human body in knowledge, communication, and computing; and the pros and cons of enhancement technologies that are designed less to treat illness than to make us live longer and better than we currently do. How exactly are humans different from machines? What is it that makes us human? What are the limits (if any) of artificial extensions of human life? When do we stop being humans and start becoming cyborgs and androids? These are the kinds of questions addressed in the following section.

The first chapter in this section is "The Transhumanist FAQ," written by Nick Bostrom (with others) as an attempt to develop the basic principles of the "transhumanism," a movement that advocates the ethical use of technology to expand human capacities. The full version of this document can be found at the homepage of *The World Transhumanist Association* (www.transhumanism.org), an international nonprofit organization, which supports the development of and access to new technologies that enable everyone to enjoy "better minds, better bodies and better lives." Transhumanism is an extension of Enlightenment humanism, the belief that individuals have unconditional worth and the rational capacity to think, to choose, and to act for oneself. Not to be confused with "posthumanism," the belief that humanism is a dated concept and that humans can be technologically redesigned beyond recognition, transhumanism affirms traditional humanist aspirations to better the human condition and improve the world around us.

Bostrom assesses the transformative potentials of several contemporary technologies, including biotechnology (such as genetic engineering, stem cells, and cloning); nanotechnology (intelligent, atomic-level machines); "superintelligence" (computational machines that will outstrip human intelligence); "uploading" (the process of transferring up-and-back from a biological brain to a computer); and the possibility of "the singularity," the theoretical point in the future of incredibly rapid technological development caused by the creation of self-conscious machines. Bostrom evokes Moore's Law (named after the co-founder of *Intel*, Gordon Moore) to argue that the singularity is the inevitable outcome of evolution. Moore's Law states every two years you get twice as much computer power and capacity for the same amount of money. It is, therefore, only a matter of time before we develop computational machines that can replicate the complicated activities of the human brain. Eventually, artificial intelligence will surpass human intelligence and speed up evolution through non-human machines.

Bostrom admits that no one knows what this future would be like for human beings. Will the intelligent machines make our lives better or worse? Will they enhance us or enslave us? "The Transhumanist FAQ" asks and answers questions like these and more, such as, whether new technologies will only benefit the rich and powerful; if transhumanists advocate eugenics; if new technologies will lead to the extinction of humans; and if new technologies are so risky they should be banned. Bostrom makes a compelling case for the continued development of new technologies to improve the quality of life for everyone.

In "Nanotechnology," excerpted from *The Age of Spiritual Machines* (2000), Ray Kurzweil details what it might be like to enhance our bodies at the molecular-cellular level with *nanobot technology*—minute machines (measuring approximately five carbon atoms thick) capable of transforming matter at the atomic level, which could replicate themselves so as to number in the trillions. Although *nanobots* do not yet exist, once they do we should have a thorough and inexpensive means for changing the very structure of matter. Kurzweil believes that intelligent nanotechnology is inevitable. Just as human intelligence is the result of evolution, the next stage is the evolution of computer intelligence that will surpass our cognitive speed, capacity, and accuracy. Kurzweil is a pioneering computer designer and successful entrepreneur, which gives his predictions an air of authority.

The possibilities of nanotechnology are the stuff of science fiction. Kurzweil describes a future in which microscopic machines will be able to travel through the circulatory system, cleaning arteries, destroying cancer cells and tumors, repairing injured tissue, damaged organs, even replacing missing limbs; they could enter our brains and provide sensory stimulation to create any imaginable experience; swarms of nanobots could create virtual fog that would simulate real environments; they could clean our water and air, removing any hazardous matter from the environment; they would be so life-like we might be tempted to have intimate relations with them. Or, the machines could take over and destroy us all. No one really knows. Kurzweil imagines a future where machines fuse with humans to form new entities, new relationships, and create new environments.

Hubert Dreyfus and Stuart Dreyfus would flatly disagree with Kurzweil and the transhumanists. In "Why Computers May Never Think Like People," the Dreyfuses argue that computers are merely rule-governed machines that can never attain the expert skill level demonstrated by human beings. Computers are basically complicated structures of on-off switches that can be used to manipulate symbols. Once researchers in the 1950s saw that one could use symbols to represent facts about the world and rules to represent relationships among facts, they started to program computers that could apply these rules to make logical inferences about facts. In the 1960s, more and more researchers tried to design computers that replicated human problem-solving but with only moderate success. According to the Dreyfuses, this project was doomed to fail from the beginning. Artificial intelligence, they hold, is completely different from the "intuitive intelligence" we have. Computers will never be able to equal (much less exceed) human intelligence because they do not think the way we do. Human beings acquire knowledge based on our bodily capacity for understanding contextual meaning. It has little to do with formal rule-governed behavior. Rather, we acquire know-how through practice and experience.

The authors describe five levels of skill acquisition, ranging from novice to expert, in order to show how acquiring competence to act appropriately in situations involves increasingly less rational and analytic thought, but more intuitive and spontaneous actions. The more competent we are the less we think, deliberate, and choose. The implication for artificial intelligence research is that any attempt to reverse engineer the human brain to model a machine that exceeds our computational abilities misses the point. Human experience is not based on calculations, rules, and information processing. No machine could ever simulate the kind of practical, embodied, contextual, and intuitive understanding humans possess. Hubert Dreyfus and Stuart Dreyfus uncover the underlying assumptions of artificial intelligence research and suggest an alternative model of human experience that is very different from machine logic.

"Interactional Expertise and Embodiment" is a four-part exchange among Evan Selinger, Hubert Dreyfus, and Harry Collins on the relationship between language, knowledge, and artificial intelligence (AI). The issue is whether something needs to possess a body in order to acquire "interactional expertise," or the ability to communicate successfully. The answer to this question will have bearing on how we might design artificial intelligences: with bodies that experience and feel, or without them.

Selinger starts by criticizing Collins's notion of "linguistic socialization," the process of acquiring fluency in a language according to the standards of a given discursive community. The question is how much of a language or discourse do you need in order to "pass" in the eyes of others who already have that competence? The question with respect to AI is whether computers will ever have human-like fluency to pass the Turing Test (to make it impossible for a person to tell if he or she is talking to another person or a computer). Selinger argues that Collins understates the role of embodiment in everyday experience and, especially, in linguistic skill acquisition. Selinger maintains that linguistic socialization can only take place by engaging in practices and by using our bodies and our senses, not just our minds. Once we appreciate the role of embodiment even in language learning, we won't be tempted, as Collins is, to treat perception as if it only occurs in the brain. Selinger challenges Collins on each of his examples that supposedly show how people without sensory experiences can still manage to acquire interactional expertise. In each case (a severely brain damaged woman who learned to converse, colorblind people who can talk about colors, and a sociologist trying to master the discourse of gravitational wave physics to pass as a physicist) Selinger argues that bodily activity—however minimal—always plays a role in linguistic skill acquisition because we can extrapolate from the experiences that even the most minimal body provides.

Collins responds by noting that the experiments conducted on interactional expertise show that a person can actually achieve considerable discursive expertise without any practical involvement in that domain. The right research question should not be whether or not a person can extrapolate from minimal sensory input, but how extrapolation works, how it is accomplished under different circumstances, and how much of a body is actually needed to acquire linguistic fluency in different kinds of domains. Collins sticks to his "minimal embodiment thesis," which is the claim that an individual requires only a minimal body to engage fully with that linguistic community. The experimental data confirms his thesis.

Next, Dreyfus responds to Selinger and Collins by noting that intelligence is not just a matter of demonstrating linguistic skill but involves the capability to make the same discriminations an intelligent person would make. That involves not merely having knowledge but having a mastery of the activity. Full understanding requires bodily involvement not mere secondhand knowledge: it takes *know-how*. Collins responds by noting that his theory of interactional expertise falls in-between the two conventional models of knowledge: formal (disembodied, rule-following) and informal (embodied, situational) but lies much closer to the informal side. He believes that intelligence involves informal and situated knowledge, but the informal situations we find ourselves in can be conversations as well as the kind of embodied activities Selinger and Dreyfus describe.

In "Genetic Interventions and the Ethics of Enhancement of Human Beings," Julian Savulescu examines the use of science and technology not just to prevent or treat disease but to help people to live a longer and/or better life than normal. Some common enhancements include cosmetic surgery (on noses, lips, teeth, breasts, and skin), athletic performance enhancing drugs (anabolic steroids, human growth hormone, blood doping), antidepressants to make people less shy and more outgoing, and other drugs and procedures used to enhance physical, intellectual, and behavioral characteristics. Savulescu focuses on the possibilities of genetic enhancement—that it may one day make it possible to make people more intelligent, to slow down the aging process, to limit aggressiveness and violence, to eliminate alcoholism or anxiety, or to control whatever human features are controlled by genes.

He offers three arguments in favor of genetic enhancement. 1) Choosing not to enhance is wrong; 2) biological manipulations are not morally different from social/environmental manipulations; and 3) there is no difference between the treatment and prevention of disease and enhancement. The aim of medical interventions, he argues, is to help people lead a good life (whatever that might be). If genetic enhancements can improve our moral character, make us more fair-minded or less anti-social, then we have an obligation to pursue those enhancements, as well. Basically, if we have an obligation to treat and prevent disease then we have an obligation to manipulate the characteristics in people that will give them the best opportunity to live the best possible life.

The author considers several objections to genetic enhancement and finds all of them lacking. He believes that we need to shift our frame of reference from healthcare to life enhancement, and that we should use whatever technological means are available to improve how we live. Savulescu argues that enhancement is something we should aspire to achieve and offers guidelines for what would constitute an ethical enhancement for an individual, a child, and for a mentally challenged adult.

In "What's Wrong with Enhancement Technology?" Carl Elliot takes a more cautious approach to the issue and highlights several morally problematic aspects of technological interventions designed to improve our lives. Some of the worrisome issues have to do with blurring the lines between treatment and enhancement and thus challenging the proper goals of medicine. Other issues have to do with what is considered normal and abnormal (and permissible) to change in people thus challenging our accepted social goals. Other problems with enhancement, according to Elliot, include the problem of "cultural complicity" (i.e., reinforcing cultural stereotypes about desirable and undesirable traits), "relative ends" (i.e., others must be un-enhanced for enhancement to exist), and "authenticity" (i.e., am I the same person after enhancement?). Ultimately what's wrong with enhancement technologies may be what Elliot describes as "the drive to mastery" to control every aspect of being human.

23

The Transhumanist FAQ

Nick Bostrom

GENERAL QUESTIONS ABOUT TRANSHUMANISM

What is transhumanism?

Transhumanism is a way of thinking about the future that is based on the premise that the human species in its current form does not represent the end of our development but rather a comparatively early phase. We formally define it as follows:

1) The intellectual and cultural movement that affirms the possibility and desirability of fundamentally improving the human condition through applied reason, especially by developing and making widely available technologies to eliminate aging and to greatly enhance human intellectual, physical, and psychological capacities.
2) The study of the ramifications, promises, and potential dangers of technologies that will enable us to overcome fundamental human limitations, and the related study of the ethical matters involved in developing and using such technologies.

Transhumanism can be viewed as an extension of humanism, from which it is partially derived. Humanists believe that humans matter, that individuals matter. We might not be perfect, but we can make things better by promoting rational thinking, freedom, tolerance, democracy, and concern for our fellow human beings. Transhumanists agree with this but also emphasize what we have the potential to become. Just as we use rational means to improve the human condition and the external world, we can also use such means to improve ourselves, the human organism. In doing so, we are not limited to traditional humanistic methods, such as education and cultural development. We can also use technological means that will eventually enable us to move beyond what some would think of as "human."

It is not our human shape or the details of our current human biology that define what is valuable about us, but rather our aspirations and ideals, our experiences, and the kinds of lives we lead. To a transhumanist, progress occurs when more people become more able to shape themselves, their lives, and the ways they relate to others, in accordance with their own deepest values. Transhumanists place a high value on autonomy: the ability and right of individuals to plan and choose their own lives. Some people may of course, for any number of reasons, choose to forgo the opportunity to use technology to improve themselves. Transhumanists seek to create a world in which autonomous individuals may choose to remain unenhanced or choose to be enhanced and in which these choices will be respected.

Through the accelerating pace of technological development and scientific understanding, we are entering a whole new stage in the history of the human species. In the relatively near future, we may face the prospect of real artificial intelligence. New kinds of cognitive tools will be built that combine artificial intelligence with interface technology. Molecular nanotechnology has the potential to manufacture abundant resources for everybody and to give us control over the biochemical processes in our bodies, enabling us to eliminate disease and unwanted aging. Technologies such as brain-computer interfaces and neuropharmacology could amplify human intelligence, increase emotional well-being, improve our capacity for steady commitment to life projects or a loved one, and even multiply the range and richness of possible emotions. On the dark side of the spectrum, transhumanists recognize that some of these coming technologies could potentially cause great harm to human life; even the survival of our species could be at risk. Seeking to understand the dangers and working to prevent disasters is an essential part of the transhumanist agenda.

What is a posthuman?

It is sometimes useful to talk about possible future beings whose basic capacities so radically exceed those of present humans as to be no longer unambiguously human by our current standards. The standard word for such beings is "posthuman." (Care must be taken to avoid misinterpretation. "Posthuman" does not denote just anything that happens to come after the human era, nor does it have anything to do with the "posthumous." In particular, it does *not* imply that there are no humans anymore.)

Many transhumanists wish to follow life paths which would, sooner or later, require growing into posthuman persons: they yearn to reach intellectual heights as far above any current human genius as humans are above other primates; to be resistant to disease and impervious to aging; to have unlimited youth and vigor; to exercise control over their own desires, moods, and mental states; to be able to avoid feeling tired, hateful, or irritated about petty things; to have an increased capacity for pleasure, love, artistic appreciation, and serenity; to experience novel states of consciousness that current human brains cannot access. It seems likely that the simple fact of living an indefinitely long, healthy, active life would take anyone to posthumanity if they went on accumulating memories, skills, and intelligence.

Posthumans could be completely synthetic artificial intelligences, or they could be enhanced uploads, or they could be the result of making many smaller but cumulatively profound augmentations to a biological human. The latter alternative would probably require either the redesign of the human organism using advanced nanotechnology or its radical enhancement using some combination of technologies such as genetic engineering, psychopharmacology, anti-aging therapies, neural interfaces, advanced information management tools, memory enhancing drugs, wearable computers, and cognitive techniques.

Some authors write as though simply by changing our self-conception, we have become or could become posthuman. This is a confusion or corruption of the original meaning of the term. The changes required to make us posthuman are too profound to be achievable by merely altering some aspect of psychological theory or the way we think about ourselves. Radical technological modifications to our brains and bodies are needed.

It is difficult for us to imagine what it would be like to be a posthuman person. Posthumans may have experiences and concerns that we cannot fathom, thoughts that cannot fit into the three-pound lumps of neural tissue that we use for thinking. Some posthumans may find it advantageous to jettison their bodies altogether and live as information patterns on vast super-fast computer networks. Their minds may be not only more powerful than ours but may also employ different cognitive architectures or include new sensory modalities that enable greater participation in their virtual reality settings. Posthuman minds might be able to share memories and experiences directly, greatly increasing the efficiency, quality, and modes in which posthumans could communicate with each other. The boundaries between posthuman minds may not be as sharply defined as those between humans.

Posthumans might shape themselves and their environment in so many new and profound ways that speculations about the detailed features of posthumans and the posthuman world are likely to fail.

What is a transhuman?

In its contemporary usage, "transhuman" refers to an intermediary form between the human and the posthuman. One might ask, given that our current use of, for example, medicine and information technology enable us to routinely do many things that would have astonished humans living in ancient times, whether we are not already transhuman? The question is a provocative one, but ultimately not very meaningful; the concept of the transhuman is too vague for there to be a definite answer.

A *transhumanist* is simply someone who advocates transhumanism. It is a common error for reporters and other writers to say that transhumanists "claim to be transhuman" or "call themselves transhuman." To adopt a philosophy which says that someday everyone ought to have the chance to grow beyond present human limits is clearly not to say that one is better or somehow currently "more advanced" than one's fellow humans.

TECHNOLOGIES AND PROJECTIONS

Biotechnology, genetic engineering, stem cells, and cloning—what are they and what are they good for?

Biotechnology is the application of techniques and methods based on the biological sciences. It encompasses such diverse enterprises as brewing, manufacture of human insulin, interferon, and human growth hormone, medical diagnostics, cell cloning and reproductive cloning, the genetic modification of crops, bioconversion of organic waste and the use of genetically altered bacteria in the cleanup of oil spills, stem cell research, and much more. Genetic engineering is the area of biotechnology concerned with the directed alteration of genetic material.

Biotechnology already has countless applications in industry, agriculture, and medicine. It is a hotbed of research. The completion of the human genome project—a "rough draft" of the entire human genome was published in the year 2000—was a scientific milestone by anyone's standards. Research is now shifting to decoding the functions and interactions of all these different genes and to developing applications based on this information.

The potential medical benefits are too many to list; researchers are working on every common disease, with varying degrees of success. Progress takes place not only in the development of drugs and diagnostics but also in the creation of better tools and research methodologies, which in turn accelerates progress. When considering what developments are likely over the long-term, such improvements in the research process itself must be factored in. The human genome project was completed ahead of schedule, largely because the initial predictions underestimated the degree to which instrumentation technology would improve during the course of the project. At the same time, one needs to guard against the tendency to hype every latest advance. (Remember all those breakthrough cancer cures that we never heard of again?) Moreover, even in cases where the early promise is borne out, it usually takes ten years to get from proof-of-concept to successful commercialization.

Genetic therapies are of two sorts: somatic and germ-line. In somatic gene therapy, a virus is typically used as a vector to insert genetic material into the cells of the recipient's body. The effects of such interventions do not carry over into the next generation. Germ-line genetic therapy is performed on sperm or egg cells, or on the early zygote, and can be inheritable. (Embryo screening, in which embryos are tested for genetic defects or other traits and then selectively implanted, can also count as a kind of germ-line intervention.) Human gene therapy, except for some forms of embryo screening, is still experimental. Nonetheless, it holds promise for the prevention and treatment of many diseases, as well as for uses in enhancement medicine. The potential scope of genetic medicine is vast: virtually all disease and all human traits—intelligence, extroversion, conscientiousness, physical appearance, etc.—involve genetic predispositions. Single-gene disorders, such as cystic fibrosis, sickle cell anemia, and Huntington's disease are likely to be among the first targets for genetic intervention. Polygenic traits and disorders, ones in which more than one gene is implicated, may follow later (although even polygenic conditions can sometimes be influenced in a beneficial direction by targeting a single gene).

Stem cell research, another scientific frontier, offers great hopes for regenerative medicine. Stem cells are undifferentiated (unspecialized) cells that can renew themselves and give rise to one or more specialized cell types with specific functions in the body. By growing such cells in culture, or steering their activity in the body, it will be possible to grow replacement tissues for the treatment of degenerative disorders, including heart disease, Parkinson's, Alzheimer's, diabetes, and many others. It may also be possible to grow entire organs from stem cells for use in transplantation. Embryonic stem cells seem to be especially versatile and useful, but research is also ongoing into adult stem cells and the "reprogramming" of ordinary cells so that they can be turned back into stem cells with pluripotent capabilities.

The term "human cloning" covers both therapeutic and reproductive uses. In therapeutic cloning, a preimplantation embryo (also known as "blastocyst"—a hollow ball consisting of 30–150 undifferentiated cells) is created via cloning, from which embryonic stem cells could be extracted and used for therapy. Because these cloned stem cells are genetically identical to the patient, the tissues or organs they would produce could be implanted without eliciting an immune response from the patient's body, thereby overcoming a major hurdle in transplant medicine. Reproductive cloning, by contrast, would mean the birth of a child who is genetically identical to the cloned parent: in effect, a younger identical twin.

Everybody recognizes the benefit to ailing patients and their families that come from curing specific diseases. Transhumanists emphasize that, in order to seriously prolong the healthy life span, we also need to develop ways to slow aging or to replace senescent cells and tissues. Gene therapy, stem cell research, therapeutic cloning, and other areas of medicine that have the potential to deliver these benefits deserve a high priority in the allocation of research monies.

What is molecular nanotechnology?

Molecular nanotechnology is an anticipated manufacturing technology that will make it possible to build complex three-dimensional structures to atomic specification using chemical reactions directed by nonbiological machinery. In molecular manufacturing, each atom would go to a selected place, bonding with other atoms in a precisely designated manner. Nanotechnology promises to give us thorough control of the structure of matter.

Since most of the stuff around us and inside us is composed of atoms and gets its characteristic properties from the placement of these atoms, the ability to control the structure of matter on the atomic scale has many applications. As K. Eric Drexler wrote in *Engines of Creation*, the first book on nanotechnology (published in 1986):

> Coal and diamonds, sand and computer chips, cancer and healthy tissue: throughout history, variations in the arrangement of atoms have distinguished the cheap from the cherished, the diseased from the healthy. Arranged one way, atoms make up soil, air, and water; arranged another, they make up ripe strawberries. Arranged one way, they make up homes and fresh air; arranged another, they make up ash and smoke.

Nanotechnology, by making it possible to rearrange atoms effectively, will enable us to transform coal into diamonds, sand into supercomputers, and to remove pollution from the air and tumors from healthy tissue.

Central to Drexler's vision of nanotechnology is the concept of the *assembler*. An assembler would be a molecular construction device. It would have one or more submicroscopic robotic arms under computer control. The arms would be capable of holding and placing reactive compounds so as to positionally control the precise location at which a chemical reaction takes place. The assembler arms would grab a molecular (but *not* necessarily individual atoms) and add it to a workpiece, constructing an atomically precise object step by step. An advanced assembler would be able to make almost any chemically stable structure. In particular, it would be able to make a copy of itself. Since assemblers could replicate themselves, they would be easy to produce in large quantities.

Mature nanotechnology will transform manufacturing into a software problem. To build something, all you will need is a detailed design of the object you want to make and a sequence of instructions for its construction. Rare or expensive raw materials are generally unnecessary; the atoms required for the construction of most kinds of nanotech devices exist in abundance in nature. Dirt, for example, is full of useful atoms.

By working in large teams, assemblers and more specialized nanomachines will be able to build large objects quickly. Consequently, while nanomachines may have features on the scale of a billionth of a meter—a nanometer—the products could be as big as space vehicles or even, in a more distant future, the size of planets.

While it seems fairly well established that molecular nanotechnology is in principle possible, it is harder to determine how long it will take to develop. A common guess among the cognoscenti is that the first assembler may be built around the year 2018, give or take a decade, but there is a large scope for diverging opinion on the upper side of that estimate.

Because the ramifications of nanotechnology are immense, it is imperative that serious thought be given to this topic now. If nanotechnology were to be abused the consequences could be devastating. Society needs to prepare for the assembler breakthrough and do advance planning to minimize the risks associated with it.

What is superintelligence?

A superintelligent intellect (a superintelligence, sometimes called "ultraintelligence") is one that has the capacity to radically outperform the best human brains in practically every field, including scientific creativity, general wisdom, and social skills.

Sometimes a distinction is made between weak and strong superintelligence. *Weak superintelligence* is what you would get if you could run a human intellect at an accelerated clock speed, such as by uploading it to a fast computer. If the upload's clock-rate were a thousand times that of a biological brain, it would perceive reality as being slowed down by a factor of a thousand. It would think a thousand times more thoughts in a given time interval than its biological counterpart.

Strong superintelligence refers to an intellect that is not only faster than a human brain but also smarter in a qualitative sense. No matter how much you speed up your dog's brain, you're not going to get the equivalent of a human intellect. Analogously, there might be kinds of smartness that wouldn't be accessible to even very fast human brains given their current capacities. Something as simple as increasing the size or connectivity of our neuronal networks *might* give us some of these capacities. Other improvements may require wholesale reorganization of our cognitive architecture or the addition of new layers of cognition on top of the old ones.

However, the distinction between weak and strong superintelligence may not be clear-cut. A sufficiently long-lived human who didn't make any errors and had a sufficient stack of scrap paper at hand could in principle compute any Turing computable function. (According to Church's thesis, the class of Turing computable functions is identical to the class of physically computable functions.)

Many but not all transhumanists expect that superintelligence will be created within the first half of this century. Superintelligence requires two things: hardware and software. Chip-manufacturers planning the next generation of microprocessors commonly rely on a well-known empirical regularity known as Moore's Law. In its original 1965-formulation by Intel co-founder Gordon Moore, it stated that the number of components on a chip doubled every year. In contemporary use, the "law" is commonly understood as referring more generally to a doubling of computing power, or of computing power per dollar. For the past couple of years, the doubling time has hovered between eighteen months and two years.

Most experts, Moore included, think that computing power will continue to double about every eighteen months for at least another two decades. This expectation is based in part on extrapolation from the past and in part on consideration of developments currently underway in laboratories. Thus it appears quite likely that human-equivalent hardware will have been achieved within not much more than a couple of decades.

How long it will take to solve the software problem is harder to estimate. One possibility is that progress in computational neuroscience will teach us about the computational architecture of the human brain and what learning rules it employs. We can then implement the same algorithms on a computer. In this approach, the superintelligence would not be completely specified by the programmers but would instead have to grow by learning from experience the same way a human infant does. An alternative approach would be to use genetic algorithms and methods from classical AI. This might result in a superintelligence that bears no close resemblance to a human brain. At the opposite extreme, we could seek to create a superintelligence by uploading a human intellect and then accelerating and enhancing it. The outcome of this might be a superintelligence that is a radically upgraded version of one particular human mind.

The arrival of superintelligence will clearly deal a heavy blow to anthropocentric worldviews. Much more important than its philosophical implications, however, would be its practical effects. Creating superintelligence may be the last invention that humans will ever need to make, since superintelligences could themselves take care of further scientific and technological development. They would do so more effectively than humans. Biological humanity would no longer be the smartest life form on the block. The prospect of superintelligence raises many big issues and concerns that we should think deeply about in advance of its actual development. The paramount question is: What can be done to maximize the chances that the arrival of superintelligence will benefit rather than harm us? The range of expertise needed to address this question extends far beyond the community of AI researchers. Neuroscientists, economists, cognitive scientists, computer scientists, philosophers, ethicists, sociologists, science-fiction writers, military strategists, politicians, legislators, and many others will have to pool their insights if we are to deal wisely with what may be the most important task our species will ever have to tackle.

Many transhumanists would like to become superintelligent themselves. This is obviously a long-term and uncertain goal, but it might be achievable either through uploading and subsequent enhancement or through the gradual augmentation of our biological brains, by means of future nootropics (cognitive enhancement drugs), cognitive techniques, IT tools (e.g., wearable computers, smart agents, information filtering systems, visualization software, etc.), neural-computer interfaces, or brain implants.

What is uploading?

Uploading (sometimes called "downloading," "mind uploading," or "brain reconstruction") is the process of transferring an intellect from a biological brain to a computer.

One way of doing this might be by first scanning the synaptic structure of a particular brain and then implementing the same computations in an electronic medium. A brain scan of sufficient resolution could be produced by disassembling the brain atom for atom by means of nanotechnology. Other approaches, such as analyzing pieces of the brain slice by slice in an electron microscope with automatic image processing have also been proposed. In addition to mapping the connection pattern among the 100 billion-or-so neurons, the scan would probably also have to register some of the functional properties of each of the synaptic interconnections, such as the efficacy of the connection and how stable it is over time (e.g., whether it is short-term or long-term potentiated). Non-local modulators such as neurotransmitter concentrations and hormone balances may also need to be represented, although such parameters likely contain much less data than the neuronal network itself.

In addition to a good three-dimensional map of a brain, uploading will require progress in neuroscience to develop functional models of each species of neuron (how they map input stimuli to outgoing action potentials, and how their properties change in response to activity in learning). It will also require a powerful computer to run the upload, and some way for the upload to interact with the external world or with a virtual reality. (Providing input/output or a virtual reality for the upload appears easy in comparison to the other challenges.)

An alternative hypothetical uploading method would proceed more gradually: one neuron could be replaced by an implant or by a simulation in a computer outside of the body. Then another neuron, and so on, until eventually the whole cortex has been replaced and the person's thinking is implemented on entirely artificial hardware. (To do this for the whole brain would almost certainly require nanotechnology.)

A distinction is sometimes made between destructive uploading, in which the original brain is destroyed in the process, and non-destructive uploading, in which the original brain is preserved intact alongside the uploaded copy. It is a matter of debate under what conditions personal identity would be preserved in destructive uploading. Many philosophers who have studied the problem think that at least under some conditions, an upload of your brain would be you. A widely accepted position is that you survive so long as certain information patterns are conserved, such as your memories, values, attitudes, and emotional dispositions, and so long as there is causal continuity so that earlier stages of yourself help determine later stages of yourself. Views differ on the relative importance of these two criteria, but they can *both* be satisfied in the case of uploading. For the continuation of personhood, on this view, it matters little whether you are implemented on a silicon chip inside a computer or in that gray, cheesy lump inside your skull, assuming both implementations are conscious.

Tricky cases arise, however, if we imagine that several similar copies are made of your uploaded mind. Which one of them is you? Are they all you, or are none of them you? Who owns your property? Who is married to your spouse? Philosophical, legal, and ethical challenges abound. Maybe these will become hotly debated political issues later in this century.

A common misunderstanding about uploads is that they would necessarily be "disembodied" and that this would mean that their experiences would be impoverished. Uploading according to this view would be the ultimate escapism, one that only neurotic body-loathers could possibly feel tempted by. But an upload's experience could in principle be identical to that of a biological human. An upload could have a virtual (simulated) body giving the same sensations and the same possibilities for interaction as a non-simulated boy. With advanced virtual reality, uploads could enjoy food and drink, and upload sex could be as gloriously messy as one could wish. And uploads wouldn't have to be confined to virtual reality: they could interact with people on the outside and even rent robot bodies in order to work in or explore physical reality.

Personal inclinations regarding uploading differ. Many transhumanists have a pragmatic attitude: whether they would like to upload or not depends on the precise conditions in which they would live as uploads and what the alternative use. (Some transhumanists may also doubt whether uploading will be possible.) Advantages of being an upload would include:

- Uploads would not be subject to biological senescence.
- Back-up copies of uploads could be created regularly so that you could be rebooted if something bad happened. (Thus your lifespan would potentially be as long as the universe's.)
- You could potentially live much more economically as an upload since you wouldn't need physical food, housing, transportation, etc.
- If you were running on a fast computer, you would think faster than in a biological implementation. For instance, if you were running on a computer a thousand times more powerful than a human brain, then you would think a thousand times faster (and the external world would appear to you as if it were slowed down by a factor of a thousand). You would thus get to experience more subjective time, and live more, during any given day.
- You could travel at the speed of light as an information patterns, which could be convenient in a future age of large-scale space settlements.
- Radical cognitive enhancements would likely be easier to implement in an upload than in an organic brain.

A couple of other points about uploading:

- Uploading should work for cryonics patients provided their brains are preserved in a sufficiently intact state.
- Uploads could reproduce extremely quickly (simply by making copies of themselves). This implies that resources could very quickly become scarce unless reproduction is regulated.

What is the singularity?

Some thinkers conjecture that there will be a point in the future when the rate of technological development becomes so rapid that the progress-curve becomes nearly vertical. Within a very brief time (months, days, or even just hours), the world might be transformed almost beyond recognition. This hypothetical point is referred to as the singularity. The most likely cause of a singularity would be the creation of some form of rapidly self-enhancing greater-than-human intelligence.

The concept of the singularity is often associated with Vernor Vinge, who regards it as one of the more probable scenarios for the future. Provided that we manage to avoid destroying civilization, Vinge thinks that a singularity is likely to happen as a consequence of advances in artificial intelligence, large systems of networked computers, computer-human integration, or some other form of intelligence amplification. Enhancing intelligence will, in this scenario, at some point lead to a positive feedback loop: smarter systems can design systems that are even more intelligent, and can do so more swiftly than the original human designers. This positive feedback effect would be powerful enough to drive an intelligence explosion that could quickly lead to the emergence of a superintelligent system of surpassing abilities.

The singularity-hypothesis is sometimes paired with the claim that it is impossible for us to predict what comes after the singularity. A post-singularity society might be so alien that we can know nothing about it. One exception might be the basic laws of physics, but even there it is sometimes suggested that there may be undiscovered laws (for instance, we don't yet have an accepted theory of quantum gravity) or poorly understood consequences of known laws that could be exploited to enable things we would normally think of as physically impossible, such as creating traversable wormholes, spawning new "basement" universes, or traveling backward in time. However, unpredictability is logically distinct from abruptness of development and would need to be argued for separately.

Transhumanists differ widely in the probability they assign to Vinge's scenario. Almost all of those who do think that there will be a singularity believe it will happen in this century, and many think it is likely to happen within several decades.

SOCIETY AND POLITICS

Will new technologies only benefit the rich and powerful?

It is clear that everybody can benefit greatly from improved technology. Initially, however, the greatest advantages will go to those who have the resources, the skills, and the willingness to learn to use new tools. One can speculate that some technologies may cause social inequalities to widen. For example, if some form of intelligence amplification becomes available, it may at first be so expensive that only the wealthiest can afford it. The same could happen when we learn how to genetically enhance our children. Those who are already well off would become smarter and make even more money. This phenomenon is not new. Rich parents send their kids to better schools and provide them with resources such as personal connections and information technology that may not be available to the less privileged. Such advantages lead to greater earnings later in life and serve to increase social inequalities.

Trying to ban technological innovation on these grounds, however, would be misguided. If a society judges existing inequalities to be unacceptable, a wiser remedy would be progressive taxation and the provision of community-funded services such as education, IT access in public libraries, genetic enhancements covered by social security, and so forth. Economic and technological progress is not a zero sum game; it's a positive sum game. Technological progress does not solve the hard old political problem of what degree of income redistribution is desirable, but it can greatly increase the size of the pie that is to be divided.

Aren't these future technologies very risky? Could they even cause our extinction?

Yes, and this implies an urgent need to analyze the risks before they materialize and to take steps to reduce them. Biotechnology, nanotechnology, and artificial intelligence pose especially serious risks of accidents and abuse.

One can distinguish between, on the one hand, endurable or limited hazards, such as car crashes, nuclear reactor meltdowns, carcinogenic pollutants in the atmosphere, floods, volcano eruptions, and so forth, and, on the other hand, *existential risks*—events that would cause the extinction of intelligent life or permanently and drastically cripple its potential. While endurable or limited risks can be serious—and may indeed be fatal to the people immediately exposed—they are recoverable; they do not destroy the long-term prospects of humanity as a whole. Humanity has long experience with endurable risks and a variety of institutional and technological mechanisms have been employed to reduce their incidence. Existential risks are a different kind of beast. For most of human history, there were no significant existential risks, or at least none that our ancestors could do anything about. By definition of course, no existential disaster has yet happened. As a species we may therefore be less well prepared to understand and manage this new kind of risk. Furthermore, the reduction of existential risk is a global public good (everybody by necessity benefits from such safety measures, whether or not they contribute to their development), creating a potential free-rider problem, that is, a lack of sufficient selfish incentives for people to make sacrifices to reduce an existential risk. Transhumanists therefore recognize a moral duty to promote efforts to reduce existential risks.

The gravest existential risks facing us in the coming decades will be of our own making. These include:

Destructive uses of nanotechnology. The accidental release of a self-replicating nanobot into the environment, where it would proceed to destroy the entire biosphere, is known as the "gray goo scenario." Since molecular nanotechnology will make use of positional assembly to create nonbiological structures and to open new chemical reaction pathways, there is no reason to suppose that the ecological checks and balances that limit the proliferation of organic self-replicators would also contain nano-replicators. Yet, while gray goo is certainly a legitimate concern, relatively simple engineering safeguards have been described that would make the probability of such a mishap almost arbitrarily small. Much more serious is the threat posed by nanobots deliberately designed to be destructive. A terrorist group or even a lone psychopath, having obtained access to this technology, could do extensive damage or even annihilate life on earth unless effective defensive technologies had been developed beforehand. An unstable arms race between nanotechnic states could also result in our eventual demise. Anti-proliferation efforts will be complicated by the fact that nanotechnology does not require difficult-to-obtain raw materials or large manufacturing plants, and by the dual-use functionality of many of the basic components of destructive nanomachinery. While a nanotechnic defense system (which would act as a global immune system capable of identifying and neutralizing rogue replicators) appears to be possible in principle, it could turn out to be more difficult to construct than a simple destructive replicator. This could create a window of global vulnerability between the potential creation of dangerous replicators and the development of an effective immune system. It is critical that nano-assemblers do not fall into the wrong hands during this period.

Biological warfare. Progress in genetic engineering will lead not only to improvements in medicine but also to the capability to create more effective bioweapons. It is chilling to consider what would have happened if HIV had been as contagious as the virus that causes the common cold. Engineering such microbes might soon become possible for increasing numbers of people. If the RNA sequence of a virus is posted on the Internet, then anybody with some basic expertise and access to a lab will be able to synthesize the actual virus from this description. A demonstration of this possibility was offered by a small team of researchers from New York University at Stony Brook in 2002, who synthesized the polio virus (whose genetic sequence is on the Internet) from scratch and injected it into mice that subsequently became paralyzed and died.

Artificial intelligence. No threat to human existence is posed by today's AI systems or their near-term successors. But if and when superintelligence is created, it will be of paramount importance that it be endowed with human-friendly values. An imprudently or maliciously designed superintelligence, with goals amounting to indifference or hostility to human welfare, could cause our extinction. Another concern is that the first superintelligence, which may become very powerful because of its superior planning ability and because of the technologies it could swiftly develop, would be built to serve only a single person or a small group (such as its programmers or the corporation that commissioned it). While this scenario may not entail the extinction of literally all intelligent life, it nevertheless constitutes an existential risk because the future that would result would be one in which a great part of humanity's potential had been permanently destroyed and in which at most a tiny fraction of all humans would get to enjoy the benefits of posthumanity.

Nuclear war. Today's nuclear arsenals are probably not sufficient to cause the extinction of all humans, but future arms races could result in even larger build-ups. It is also conceivable that an all-out nuclear war would lead to the collapse of modern civilization, and it is not completely certain that the survivors would succeed in rebuilding a civilization capable of sustaining growth and technological development.

Something unknown. All the above risks were unknown a century ago and several of them have only become clearly understood in the past two decades. It is possible that there are future threats of which we haven't yet become aware.

Evaluating the total probability that some existential disaster will do us in before we get the opportunity to become posthuman can be done by various direct or indirect methods. Although any estimate inevitably includes a large subjective factor, it seems that to set the probability to less than 20 percent would be unduly optimistic, and the best estimate may be considerably higher. But depending on the actions we take, this figure can be raised or lowered.

If these technologies are so dangerous, should they be banned? What can be done to reduce the risks?

The position that we ought to relinquish research into robotics, genetic engineering, and nanotechnology has been advocated in an article by Bill Joy (2000). Joy argued that some of the future applications of these technologies are so dangerous that research in those fields should be stopped now. Partly because of Joy's previously technophiliac credentials (he was a software designer and a cofounder of Sun Microsystems), his article, which appeared in *Wired* magazine, attracted a great deal of attention.

Many of the responses to Joy's article pointed out that there is no realistic prospect of a worldwide ban on these technologies; that they have enormous potential benefits that we would not want to forgo; that the poorest people may have a higher tolerance for risk in developments that could improve their condition; and that a ban may actually increase the dangers rather than reduce them, both by delaying the development of protective applications of these technologies, and by weakening the position of those who choose to comply with the ban relative to less scrupulous groups who defy it.

A more promising alternative than a blanket ban is *differential technological development*, in which we would seek to influence the *sequence* in which technologies developed. On this approach, we would strive to retard the development of harmful technologies and their applications, while accelerating the development of beneficial technologies, especially those that offer protection against the harmful ones. For technologies that have decisive military applications, unless they can be verifiably banned, we may seek to ensure that they are developed at a faster pace in countries we regard as responsible than in those that we see as potential enemies. (Whether a ban is verifiable and enforceable can change over time as a result of developments in the international system or in surveillance technology.)

In the case of nanotechnology, the desirable sequence of development is that nanotech immune systems and other defensive measures be deployed before offensive capabilities become available to many independent powers. Once a technology is shared by many, it becomes extremely hard to prevent further proliferation. In the case of biotechnology, we should seek to promote research into vaccines, anti-viral drugs, protective gear, sensors, and diagnostics, and to delay as long as possible the development and proliferation of biological warfare agents and the means of their weaponization. For artificial intelligence, a serious risk will emerge only when capabilities approach or surpass those of humans. At that point one should seek to promote the development of friendly AI and to prevent unfriendly or unreliable AI systems.

Superintelligence is an example of a technology that seems especially worth promoting because it can help reduce a broad range of threats. Superintelligent systems could advise us on policy and make the progress curve for nanotechnology steeper, thus shortening the period of vulnerability between the development of dangerous nanoreplicators and the deployment of effective defenses. If we have a choice, it seems preferable that superintelligence be developed before advanced nanotechnology, as superintelligence could help reduce the risks of nanotechnology but not vice versa. Other technologies that have wide risk-reducing uses include intelligence augmentation, information technology, and surveillance. These can make us smarter individually and collectively or make enforcement of necessary regulation more feasible. A strong prima facie case therefore exists for pursuing these technologies as vigorously as possible. Needless to say, we should also promote non-technological developments that are beneficial in almost all scenarios, such as peace and international cooperation.

In confronting the hydra of existential, limited, and endurable risks glaring at us from the future, it is unlikely that any one silver bullet will provide adequate protection. Instead, an arsenal of countermeasures will be needed so that we can address the various risks on multiple levels.

The first step to tackling a risk is to recognize its existence. More research is needed, and existential risks in particular should be singled out for attention because of their seriousness and because of the special nature of the challenges they pose. Surprisingly little work has been done in this area. The strategic dimensions of our choices must be taken into account, given that some of the technologies in question have important military ramifications. In addition to scholarly studies of the threats and their possible countermeasures, public awareness must be raised to enable a more informed debate of our long-term options.

Some of the lesser existential risks, such as an apocalyptic asteroid impact or the highly speculative scenario involving something like the upsetting of a metastable vacuum state in some future particle accelerator experiment, could be substantially reduced at relatively small expense. Programs to accomplish this—for example, an early detection system for dangerous near-earth objects on potential collation course with Earth, or the commissioning of advance peer review of planned high-energy physics experiments—are probably cost-effective. However, these lesser risks must not deflect attention from the more serious concern raised by more probable existential disasters.

In light of how superabundant the human benefits of technology can ultimately be, it matters less that we obtain all of these benefits in their precisely most optimal form, and more that we obtain them at all. For many practical purposes, it makes sense to adopt the rule of thumb that we should act so as to maximize the probability of an *acceptable* outcome, one in which we attain some (reasonably broad) realization of our potential; or, to put it in negative terms, that we should act so as to minimize net existential risk.

Shouldn't we concentrate on current problems such as improving the situation of the poor, rather than putting our efforts into planning for the "far" future?

Many of the technologies and trends that transhumanists discuss are already reality. Biotechnology and information technology have transformed large sectors of our economies. The relevance of transhumanist ethics is manifest in such contemporary issues as stem cell research, genetically modified crops, human genetic therapy, embryo screening, end of life decisions, enhancement medicine, information markets, and research funding priorities. The importance of transhumanist ideas is likely to increase as the opportunities for human enhancement proliferate.

An argument can be made that the most efficient way of contributing to making the world better is by participating in the transhumanist project. This is so because the stakes are enormous— humanity's entire future may depend on how we manage the coming technological transitions—and because relatively few resources are at the present time being devoted to transhumanist efforts. Even one extra person can still make a significant difference here.

Will extended life worsen overpopulation problems?

Population increase is an issue we would ultimately have to come to grips with even if healthy life-extension were not to happen. Leaving people to die is an unacceptable solution.

A large population should not be viewed simply as a problem. Another way of looking at the same fact is that it means that many persons now enjoy lives that would not have been lived if the population had been smaller. One could ask those who complain about overpopulation exactly which people's lives they would have preferred should not have been led. Would it really have been better if billions of the world's people had never existed and if there had been no other people in their place? Of course, this is not to deny that too-rapid population growth can cause crowding, poverty, and the depletion of natural resources. In this sense there can be real problems that need to be tackled.

How many people the Earth can sustain at a comfortable standard of living is a function of technological development (as well as of how resources are distributed). New technologies, from simple improvements in irrigation and management, to better mining techniques and more efficient power generation machinery, to genetically engineered crops, can continue to improve world resource and food output, while at the same time reducing environmental impact and animal suffering.

Environmentalists are right to insist that the status quo is unsustainable. As a matter of physical necessity, things cannot stay as they are today indefinitely, or even for very long. If we continue to use up resources at the current pace, without finding more resources or learning how to use novel kinds of resources, then we will run into serious shortages sometime around the middle of this century. The deep greens have an answer to this: they suggest we turn back the clock and return to an idyllic pre-industrial age to live in sustainable harmony with nature. The problem with this view is that the pre-industrial age was anything but idyllic. It was a life of poverty, misery, disease, heavy manual toil from dawn to dusk, superstitious fears, and cultural parochialism. Nor was it environmentally sound—as witness the deforestation of England and the Mediterranean region, desertification of large parts of the middle east, soil depletion by the Anasazi in the Glen Canyon area, destruction of farm land in ancient Mesopotamia through the accumulation of mineral salts from irrigation, deforestation and consequent soil erosion by the ancient Mexican Mayas, overhunting of big game almost everywhere, and the extinction of the dodo and other big featherless birds in the South Pacific. Furthermore, it is hard to see how more than a few hundred million people could be maintained at a reasonable standard of living with pre-industrial production methods, so some 90 percent of the world population would somehow have to vanish in order to facilitate this nostalgic return.

Transhumanists propose a much more realistic alternative: not to retreat to an imagined past, but to press ahead as intelligently as we can. The environmental problems that technology creates are problems of intermediary, inefficient technology, of placing insufficient political priority on environmental protection as well as of a lack of ecological knowledge. Technologically less advanced industries in the former Soviet-bloc pollute much more than do their advanced Western counterparts. High-tech industry is typically relatively benign. Once we develop molecular nano-technology, we will not only have clean and efficient manufacturing of almost any commodity, but we will also be able to clean up much of the mess created by today's crude fabrication methods. This would set a standard for a clean environment that today's traditional environmentalists could scarcely dream of.

Nanotechnology will also make it cheaper to colonize space. From a cosmic point of view, Earth is an insignificant speck. It has sometimes been suggested that we ought to leave space untouched in its pristine glory. This view is hard to take seriously. Every hour, through entirely natural processes, vast amounts of resources—millions of times more than the sum total of what the human species has consumed throughout its career—are transformed into radioactive substances or wasted as radiation escaping into intergalactic space. Can we not think of some more creative way of using all this matter and energy?

Even with full-blown space colonization, however, population growth can continue to be a problem, and this is so even if we assume that an unlimited number of people could be transported from Earth into space. If the speed of light provides an upper bound on the expansion speed then the amount of resources under human control will grow only polynomially ($\sim t^3$). Population, on the other hand, can easily grow exponentially ($\sim e^t$). If that happens, then, since a factor that grows exponentially will eventually overtake any factor that grows polynomially, average income will ultimately drop to subsistence levels, forcing population growth to slow. How soon this would happen depends primarily on reproduction rates. A change in average life span would not have a big effect. Even vastly improved technology can only postpone this inevitably for a relatively brief time. The only long-term method of assuring continued growth of average income is some form of population control, whether spontaneous or imposed, limiting the number of new persons created per year. This does not mean that population could not grow, only that the growth would have to be polynomial rather than exponential.

Some additional points to consider:

- In technologically advanced countries, couples tend to have fewer children, often below the replacement rate. As an empirical generalization, giving people increased rational control over their lives, especially through women's education and participation in the labor market, causes couples to have fewer children.
- If one took seriously the idea of controlling population by limiting life span, why not be more active about it? Why not encourage suicide? Why not execute anyone reaching the age of seventy-five?
- If slowing aging were unacceptable because it might lead to there being more people, what about efforts to cure cancer, reduce traffic deaths, or improve worker safety? Why use double standards?
- When transhumanists say they want to extend lifespans, what they mean is that they want to extend healthspans. This means that the extra person-years would be productive and

would add economic value to society. We can all agree that there would be little point in living an extra ten years in a state of dementia.

- The world population growth rate has been declining for several decades. It peaked in 1970 at 2.1 percent. In 2003, it was 1.2 percent; and it is expected to fall below 1.0 percent around 2015 (United Nations 2002). The doomsday predictions of the so-called "Club of Rome" from the early 1970s have consistently turned out to be wrong.
- The more people there are, the more brains there will be working to invent new ideas and solutions.
- If people can look forward to a longer healthy, active life, they will have a personal stake in the future and will hopefully be more concerned about the long-term consequences of their actions.

How does transhumanism relate to religion?

Transhumanism is a philosophical and cultural movement concerned with promoting responsible ways of using technology to enhance human capacities and to increase the scope of human flourishing.

While not a religion, transhumanism might serve a few of the same functions that people have traditionally sought in religion. It offers a sense of direction and purpose and suggests a vision that humans can achieve something greater than our present condition. Unlike most religious believers, however, transhumanists seek to make their dreams come true in *this* world, by relying not on supernatural powers or divine intervention but on rational thinking and empiricism, through continued scientific, technological, economic, and human development. Some of the prospects that used to be the exclusive thunder of the religious institutions, such as very long lifespan, unfading bliss, and godlike intelligence, are being discussed by transhumanists as hypothetical future engineering achievements.

Transhumanism is a naturalistic outlook. At the moment, there is no hard evidence for supernatural forces or irreducible spiritual phenomena, and transhumanists prefer to derive their understanding of the world from rational modes of inquiry, especially the scientific method. Although science forms the basis for much of the transhumanist worldview, transhumanists recognize that science has its own fallibilities and imperfections, and that critical ethical thinking is essential for guiding our conduct and for selecting worthwhile aims to work toward.

Religious fanaticism, superstition, and intolerance are not acceptable among transhumanists. In many cases, these weaknesses can be overcome through a scientific and humanistic education, training in critical thinking, and interaction with people from different cultures. Certain other forms of religiosity, however, may well be compatible with transhumanism.

It should be emphasized that transhumanism is not a fixed set of dogmas. It is an evolving worldview, or rather, a family of evolving worldviews—for transhumanists disagree with each other on many issues. The transhumanist philosophy, still in its formative stages, is meant to keep developing in the light of new experiences and new challenges. Transhumanists want to find out where they are wrong and to change their views accordingly.

Won't it be boring to live forever in a perfect world?

Why not try it and see?

"Perfection" is a vague and treacherous word. There is considerable disagreement among transhumanists about what kind of perfection is attainable and desirable, either in theory or in practice. It is probably wiser to speak of improving the world, rather than making it "perfect." Would it be boring to live for an indefinitely long time in a greatly improved world? The world could surely be improved over the way it is now, including becoming less boring. If you got rid of the pain and stress associated with, say, filling out annual tax returns, people would probably not sit around afterward saying: "Life feels meaningless now that I no longer have income tax forms to fill out."

Admittedly, material improvements to the environment may not, in themselves, be sufficient to bring about lasting happiness. If your accustomed fare is bread and water, then a box of cookies can be a feast. But if every night you eat out at fancy restaurants, such fine fare will soon seem ordinary and normal; and any lesser feast, such as a box of cookies, would be insulting by comparison. Some cognitive scientists speculate that we each have a "set point" of happiness, to which we soon return regardless of changes in the environment. There may be considerable truth to the folk wisdom that an expensive new car does not make you happier (or rather, it makes you happier, but only temporarily). In some ways, human minds and brains are just not designed to be happy. Fortunately, there are several potential viewpoints from which to go about addressing this challenge.

Apes engage in activities that we, as humans, would find repetitive and dull. In the course of becoming smarter, we have become bored by things that would have interested our ancestors. But at the same time we have opened up a vast new space of possibilities for having fun—and the new space is much larger than the previous one. Humans are not simply apes who can obtain more bananas using our intelligence as a tool. Our intelligence enables us to desire new things, such as art, science, and mathematics. If at any point in your indefinitely long life you become bored with the greatly improved world, it may only indicate that the time has come to bump up your intelligence another increment.

If the human brain has a "set point" of happiness to which it returns, maybe this is a design flaw and should be fixed—one of those things that we will end up defining as human, but not humane. It would probably be unwise to eliminate boredom entirely, since boredom can serve to prevent us from wasting too much time on monotonous and meaningless activities. But if we're doing new things, learning, growing more intelligent, and we still aren't happy, for no better reason than that our cognitive architecture is badly designed, then perhaps it is time to redesign it. Present clinical mood-drugs are crude, but nonetheless they can sometimes restore interest and enthusiasm for life—sometimes tiredness and despair has no interesting reason behind it and is simply an imbalance of brain chemistry. Only by compartmentalizing our thinking to a high degree can we imagine a world where there is mature molecular nanotechnology and superhuman artificial intelligence, but the means are still lacking to control the brain circuitry of boredom. Fundamentally, there is no reason why pleasure, excitement, profound well-being and simple joy at being alive could not become the natural, default state of mind for all who desire it.

We should also consider some of the following points:

1) Ordinary life is sometimes boring. So what?
2) Eternal life will be as boring or as exciting as you make it.
3) Is being dead more exciting?
4) If eternal life becomes boring, you will have the option of ending it at any time.

Transhumanism is not about a fancier car, more money, or clever gadgetry, even though this is what the media presents to us as "science" and "advanced technology": transhumanism is about genuine changes to the human condition, including increased intelligence and minds better suited to the achievement of happiness.

24

Twenty-First Century Bodies

Ray Kurzweil

Humankind's first tools were found objects: sticks used to dig up roots and stones used to break open nuts. It took our forebears tens of thousands of years to invent a sharp blade. Today we build machines with finely designed intricate mechanisms, but viewed on an atomic scale, our technology is still crude. "Casting, grinding, milling, and even lithography move atoms in great thundering statistical herds," says Ralph Merkle, a leading nanotechnology theorist at Xerox's Palo Alto Research Center. He adds that current manufacturing methods are "like trying to make things out of Legos with boxing gloves on. . . . In the future, nanotechnology will let us take off the boxing gloves."[1]

Nanotechnology is technology built on the atomic level: building machines one atom at a time. "Nano" refers to a billionth of a meter, which is the width of five carbon atoms. We have one existence proof of the feasibility of nanotechnology: life on Earth. Little machines in our cells called ribosomes build organisms such as humans one molecule, that is, one amino acid, at a time, following digital templates coded in another molecule, called DNA. Life on Earth has mastered the ultimate goal of nanotechnology, which is self-replication.

But as mentioned above, Earthly life is limited by the particular molecular building block it has selected. Just as our human-created computational technology will ultimately exceed the capacity of natural computation (electronic circuits are already millions of times faster than human neural circuits), our twenty-first-century physical technology will also greatly exceed the capabilities of the amino acid–based nanotechnology of the natural world.

The concept of building machines atom by atom was first described in a 1959 talk at Cal Tech titled "There's Plenty of Room at the Bottom," by physicist Richard Feynman, the same guy who first suggested the possibility of quantum computing.[2] The idea was developed in some detail by Eric Drexler twenty years later in his book *Engines of Creation*.[3] The book actually inspired the cryonics movement of the 1980s, in which people had their heads (with or without bodies) frozen in the hope that a future time would possess the molecule-scale technology to overcome their mortal diseases, as well as undo the effects of freezing and defrosting. Whether a future generation would be motivated to revive all these frozen brains was another matter.

After publication of *Engines of Creation*, the response to Drexler's ideas was skeptical and he had difficulty filling out his MIT Ph.D. committee despite Marvin Minsky's agreement to supervise it. Drexler's dissertation, published in 1992 as a book titled *Nanosystems: Molecular Machinery, Manufacturing, and Computation,* provided a comprehensive proof of concept, including detailed analyses and specific designs.[4] A year later, the first nanotechnology conference attracted only a few dozen researchers. The fifth annual conference, held in December 1997, boasted 350 scientists who were far more confident of the practicality of their tiny projects. Nanothinc, an industry think tank, estimated in 1997 that the field already produces $5 billion in annual revenues for nanotechnology-related technologies, including micromachines, microfabrication techniques, nanolithography, nanoscale microscopes, and others. This figure has been more than doubling each year.[5]

THE AGE OF NANOTUBES

One key building material for tiny machines is, again, nanotubes. Although built on an atomic scale, the hexagonal patterns of carbon atoms are extremely strong and durable. "You can do anything you damn well want with these tubes and they'll just keep on truckin'," says Richard Smalley, one of the chemists who received the Nobel Prize for discovering the buckyball molecule.[6] A car made of nanotubes would be stronger and more stable than a car made with steel, but would weigh only fifty pounds. A spacecraft made of nanotubes could be of the size and strength of the U.S. space shuttle, but weigh no more than a conventional car. Nanotubes handle heat extremely well, far better than the fragile amino acids that people are built out of. They can be assembled into all kinds of shapes: wirelike strands, sturdy girders, gears, et cetera. Nanotubes are formed of carbon atoms, which are in plentiful supply in the natural world.

As I mentioned earlier, the same nanotubes can be used for extremely efficient computation, so both the structural and computational technology of the twenty-first century will likely be constructed from the same stuff. In fact, the same nanotubes used to form physical structures can also be used for computation, so future nanomachines can have their brains distributed throughout their bodies.

The best-known examples of nanotechnology to date, while not altogether practical, are beginning to show the feasibility of engineering at the atomic level. IBM created its corporate logo using individual atoms as pixels.[7] In 1996, Texas Instruments built a chip-sized device with half a million moveable mirrors to be used in a tiny high-resolution projector.[8] TI sold $100 million worth of their nanomirrors in 1997.

Chih-Ming Ho of UCLA is designing flying machines using surfaces covered with microflaps that control the flow of air in a similar manner to conventional flaps on a normal airplane.[9] Andrew Berlin at Xerox's Palo Alto Research Center is designing a printer using microscopic air valves to move paper documents precisely.[10]

Cornell graduate student and rock musician Dustin Carr built a realistic-looking but microscopic guitar with strings only fifty nanometers in diameter. Carr's creation is a fully functional musical instrument, but his fingers are too large to play it. Besides, the strings vibrate at 10 million vibrations per second, far beyond the twenty-thousand-cycles-per-second limit of human hearing.[11]

THE HOLY GRAIL OF SELF-REPLICATION:
LITTLE FINGERS AND A LITTLE INTELLIGENCE

Tiny fingers represent something of a holy grail for nanotechnologists. With little fingers and computation, nanomachines would have in their Lilliputian world what people have in the big world: intelligence and the ability to manipulate their environment. Then these little machines could build replicas of themselves, achieving the field's key objective.

The reason that self-replication is important is that it is too expensive to build these tiny machines one at a time. To be effective, nanometer-sized machines need to come in the trillions. The only way to achieve this economically is through combinatorial explosion: let the machines build themselves.

Drexler, Merkle (a coinventor of public key encryption, the primary method of encrypting messages), and others have convincingly described how such a self-replicating nanorobot— *nanobot*—could be constructed. The trick is to provide the nanobot with sufficiently flexible manipulators—arms and hands—so that it is capable of building a copy of itself. It needs some means for mobility so that it can find the requisite raw materials. It requires some intelligence so that it can solve the little problems that will arise when each nanobot goes about building a complicated little machine like itself. *Finally, a really important requirement is that it needs to know when to stop replicating.*

MORPHING IN THE REAL WORLD

Self-replicating machines built at the atomic level could truly transform the world we live in. They could build extremely inexpensive solar cells, allowing the replacement of messy fossil fuels. Since solar cells require a large surface area to collect sufficient sunlight, they could be placed in orbit, with the energy beamed down to Earth.

Nanobots launched into our bloodstreams could supplement our natural immune system and seek out and destroy pathogens, cancer cells, arterial plaque, and other disease agents. In the vision that inspired the cryonics enthusiasts, diseased organs can be rebuilt. We will be able to reconstruct any or all of our bodily organs and systems, and do so at the cellular level. I talked in the last chapter about reverse engineering and emulating the salient computational functionality of human neurons. In the same way, it will become possible to reverse engineer and replicate the physical and chemical functionality of any human cell. In the process we will be in a position to greatly extend the durability, strength, temperature range, and other qualities and capabilities of our cellular building blocks.

We will then be able to grow stronger, more capable organs by redesigning the cells that constitute them and building them with far more versatile and durable materials. As we go down this road, we'll find that some redesign of the body makes sense at multiple levels. For example, if our cells are no longer vulnerable to the conventional pathogens, we may not need the same kind of immune system. But we will need new nanoengineered protections for a new assortment of nanopathogens.

Food, clothing, diamond rings, buildings could all assemble themselves molecule by molecule. Any sort of product could be instantly created when and where we need it. Indeed, the world could continually reassemble itself to meet our changing needs, desires, and fantasies. By the late twenty-first century, nanotechnology will permit objects such as furniture, buildings, clothing, even people, to change their appearance and other characteristics—essentially to change into something else—in a split second.

These technologies will emerge gradually. There is a clear incentive to go down this path. Given a choice, people will prefer to keep their bones from crumbling, their skin supple, their life systems strong and vital. Improving our lives through neural implants on the mental level, and nanotechnology-enhanced bodies on the physical level, will be popular and compelling. It is another one of those slippery slopes—there is no obvious place to stop this progression until the human race has largely replaced the brains and bodies that evolution first provided.

A CLEAR AND FUTURE DANGER

Without self-replication, nanotechnology is neither practical nor economically feasible. And therein lies the rub. What happens if a little software problem (inadvertent or otherwise) fails to halt the self-replication? We may have more nanobots than we want. They could eat up everything in sight.

The movie *The Blob* (of which there are two versions) was a vision of nanotechnology run amok. The movie's villain was this intelligent self-replicating gluttonous stuff that fed on organic matter. Recall that nanotechnology is likely to be built from carbon-based nanotubes, so, like the Blob, it will build itself from organic matter, which is rich in carbon. Unlike mere animal-based cancers, an exponentially exploding nanomachine population would feed on any carbon-based matter. Tracking down all of these bad nanointelligences would be like trying to find trillions of microscopic needles—rapidly moving ones at that—in at least as many haystacks. There have been proposals for nanoscale immunity technologies: good little antibody machines that would go after the bad little machines. The nanoantibodies would, of course, have to scale up at least as quickly as the epidemic of marauding nanomiscreants. There could be a lot of collateral damage as these trillions of machines battle it out.

Now that I have raised this specter, I will try, unconvincingly perhaps, to put the peril in perspective. I believe that it will be possible to engineer self-replicating nanobots in such a way that an *inadvertent*, undesired population explosion would be unlikely. I realize that this may not be completely reassuring, coming from a software developer whose products (like those of my competitors) crash once in a while (but rarely—and when they do, it's the fault of the operating system!). There is a concept in software development of "mission critical" applications. These are software programs that control a process on which people are heavily dependent. Examples of mission-critical software include life-support systems in hospitals, automated surgical equipment, autopilot flying and landing systems, and other software-based systems that affect the well-being of a person or organization. It is feasible to create extremely high levels of reliability in these programs. There are examples of complex technology in use today in which a mishap would severely imperil public safety. A conventional explosion in an atomic power plant could spray deadly plutonium across heavily populated areas. Despite a near meltdown at Chernobyl, this apparently has only occurred twice in the decades that we have had hundreds of such plants operating, both incidents involving recently acknowledged reactor calamities in the Chelyabinsk region of Russia.[12] There are tens of thousands of nuclear weapons, and none has ever exploded in error.

I admit that the above paragraph is not entirely convincing. But the bigger danger is the intentional hostile use of nanotechnology. Once the basic technology is available, it would not be difficult to adapt it as an instrument of war or terrorism. It is not the case that someone would have to be suicidal to use such weapons. The nanoweapons could easily be programmed to replicate only against an enemy; for example, only in a particular geographical area. Nuclear weapons, for all their destructive potential, are at least relatively local in their effects. The self-replicating nature of nanotechnology makes it a far greater danger.

VIRTUAL BODIES

We don't always need real bodies. If we happen to be in a virtual environment, then a virtual body will do just fine. Virtual reality started with the concept of computer games, particularly ones that provided a simulated environment. The first was Space War, written by early artificial-intelligence researchers to pass the time while waiting for programs to compile on their slow 1960s computers.[13] The synthetic space surroundings were easy to render on low-resolution monitors: Stars and other space objects were just illuminated pixels.

Computer games and computerized video games have become more realistic over time, but you cannot completely immerse yourself in these imagined worlds, not without some imagination. For one thing, you can see the edges of the screen, and the all too real world that you have never left is still visible beyond these borders.

If we're going to enter a new world, we had better get rid of traces of the old. In the 1990s the first generation of virtual reality has been introduced in which you don a special visual helmet that takes over your entire visual field. The key to visual reality is that when you move your head, the scene instantly repositions itself so that you are now looking at a different region of a three-dimensional scene. The intention is to simulate what happens when you turn your real head in the real world: The images captured by your retinas rapidly change. Your brain nonetheless understands that the world has remained stationary and that the image is sliding across your retinas only because your head is rotating.

Like most first generation technologies, virtual reality has not been fully convincing. Because rendering a new scene requires a lot of computation, there is a lag in producing the new perspective. Any noticeable delay tips off your brain that the world you're looking at is not entirely real. The resolution of virtual reality displays has also been inadequate to create a fully satisfactory illusion. Finally, contemporary virtual reality helmets are bulky and uncomfortable.

What's needed to remove the rendering delay and to boost display resolution is yet faster computers, which we know are always on the way. By 2007, high-quality virtual reality with convincing artificial environments, virtually instantaneous rendering, and high-definition displays will be comfortable to wear and available at computer game prices.

That takes care of two of our senses—visual and auditory. Another high-resolution sense organ is our skin, and "haptic" interfaces to provide a virtual tactile interface are also evolving. One available today is the Microsoft force-feedback joystick, derived from 1980s research at the MIT Media Lab. A force-feedback joystick adds some tactile realism to computer games, so you feel the rumble of the road in a car-driving game or the pull of the line in a fishing simulation. Emerging in late 1998 is the "tactile mouse," which operates like a conventional mouse but allows the user to feel the texture of surfaces, objects, even people. One company that I am involved in, Medical Learning Company, is developing a simulated patient to help train doctors, as well as enable nonphysicians to play doctor. It will include a haptic interface so that you can feel a knee joint for a fracture or a breast for lumps.[14]

A force-feedback joystick in the tactile domain is comparable to conventional monitors in the visual domain. The force-feedback joystick provides a tactile interface, but it does not totally envelop you. The rest of your tactile world is still reminding you of its presence. In order to leave the real world, at least temporarily, we need a tactile environment that takes over your sense of touch.

So let's invent a virtual tactile environment. We've seen aspects of it in science fiction films (always a good source for inventing the future). We can build a body suit that will detect your own movements as well as provide high resolution tactile stimulation. The suit will also need to provide sufficient force-feedback to actually prevent your movements if you are pressing against a virtual obstacle in the virtual environment. If you are giving a virtual companion a hug, for example, you don't want to move right through his or her body. This will require a force-feedback structure outside the suit, although obstacle resistance could be provided by the suit itself. And since your body inside the suit is still in the real world, it would make sense to put the whole contraption in a booth so that your movements in the virtual world don't knock down lamps and people in your "real" vicinity. Such a suit could also provide a thermal response and thereby allow the simulation of feeling a moist surface—or even immersing your hand or your whole body in water—which is indicated by a change in temperature and a decrease in surface tension. Finally, we can provide a platform consisting of a rotating treadmill device for you to stand (or sit or lie) on, which will allow you to walk or move around (in any direction) in your virtual environment.

So with the suit, the outer structure, the booth, the platform, the goggles, and the earphones, we just about have the means to totally envelop your senses. Of course, we will need some good virtual reality software, but there's certain to be hot competition to provide a panoply of realistic and fantastic new environments as the requisite hardware becomes available.

Oh yes, there is the sense of smell. A completely flexible and general interface for our fourth sense will require a reasonably advanced nanotechnology to synthesize the wide variety of molecules that we can detect with our olfactory sense. In the meantime, we could provide the ability to diffuse a variety of aromas in the virtual reality booth.

Once we are in a virtual reality environment, our own bodies—at least the virtual versions—can change as well. We can become a more attractive version of ourselves, a hideous beast, or any creature real or imagined as we interact with the other inhabitants in each virtual world we enter.

Virtual reality is not a (virtual) place you need go to alone. You can interact with your friends there (who would be in other virtual reality booths, which may be geographically remote). You will have plenty of simulated companions to choose from as well.

DIRECTLY PLUGGING IN

Later in the twenty-first century, as neural implant technologies become ubiquitous, we will be able to create and interact with virtual environments without having to enter a virtual reality booth. Your neural implants will provide the simulated sensory inputs of the virtual environment—and your virtual body—directly in your brain. Conversely, your movements would not move your "real" body, but rather your perceived virtual body. These virtual environments would also include a suitable selection of bodies for yourself. Ultimately, your experience would be highly realistic, just like being in the real world. More than one person could enter a virtual environment and interact with each other. In the virtual world, you will meet other real people and simulated people—eventually, there won't be much difference.

This will be the essence of the Web in the second half of the twenty-first century. A typical "web site" will be a perceived virtual environment, with no external hardware required. You "go there" by mentally selecting the site and then entering that world. Debate Benjamin Franklin on the war powers of the presidency at the history society site. Ski the Alps at the Swiss Chamber of Commerce site (while feeling the cold spray of snow on your face). Hug your favorite movie star at the Columbia Pictures site. Get a little more intimate at the Penthouse or Playgirl site. Of course, there may be a small charge.

REAL VIRTUAL REALITY

In the late twenty-first century, the "real" world will take on many of the characteristics of the virtual world through the means of nanotechnology "swarms." Consider, for example, Rutgers University computer scientist J. Storrs Hall's concept of "Utility Fog."[15] Hall's conception starts with a little robot called a Foglet, which consists of a human-cell-sized device with twelve arms pointing in all directions. At the end of the arms are grippers so that the Foglets can grasp one another to form larger structures. These nanobots are intelligent and can merge their computational capacities with each other to create a distributed intelligence. A space filled with Foglets is called Utility Fog and has some interesting properties.

First of all, the Utility Fog goes to a lot of trouble to simulate its not being there. Hall describes a detailed scenario that lets a real human walk through a room filled with trillions of Foglets and not notice a thing. When desired (and it's not entirely clear who is doing the desiring), the Foglets can quickly simulate any environment by creating all sorts of structures. As Hall puts it, "Fog city can look like a park, or a forest, or ancient Rome one day and Emerald City the next."

The Foglets can create arbitrary wave fronts of light and sound in any direction to create any imaginary visual and auditory environment. They can exert any pattern of pressure to create any tactile environment. In this way, Utility Fog has all the flexibility of a virtual environment, except it exists in the real physical world. The distributed intelligence of the Utility Fog can simulate the minds of scanned (Hall calls them "uploaded") people who are re-created in the Utility Fog as "Fog people." In Hall's scenario, "a biological human can walk through Fog walls, and a Fog (uploaded) human can walk through dumb-matter walls. Of course Fog people can walk through Fog walls, too."

The physical technology of Utility Fog is actually rather conservative. The Foglets are much bigger machines than most nanotechnology conceptions. The software is more challenging, but ultimately feasible. Hall needs a bit of work on his marketing angle: Utility Fog is a rather dull name for such versatile stuff.

There are a variety of proposals for nanotechnology swarms, in which the real environment is constructed from interacting multitudes of nanomachines. In all of the swarm conceptions, physical reality becomes a lot like virtual reality. You can be sleeping in your bed one moment, and have the room transform into your kitchen as you awake. Actually, change that to a dining room as there's no need for a kitchen. Related nanotechnology will instantly create whatever meal you desire. When you finish eating, the room can transform into a study, or a game room, or a swimming pool, or a redwood forest, or the Taj Mahal. You get the idea.

Mark Yim has built a large-scale model of a small swarm showing the feasibility of swarm interaction.[16] Joseph Michael has actually received a U.K. patent on his conception of a nanotechnology swarm, but it is unlikely that his design will be commercially realizable in the twenty-year life of his patent.[17]

It may seem that we will have too many choices. Today, we have only to choose our clothes, makeup, and destination when we go out. In the late twenty-first century, we will have to select our body, our personality, our environment—so many difficult decisions to make! But don't worry— we'll have intelligent swarms of machines to guide us.

THE SENSUAL MACHINE

> *Made double by his lust*
> *he sounds a woman's groans.*
> *A figment of his flesh.*
>
> —from Barry Spacks's poem "The Solitary at Seventeen"

> *I can predict the future by assuming that money and male hormones are the driving*
> *forces for new technology. Therefore, when virtual reality gets cheaper than dating,*
> *society is doomed.*
>
> —Dogbert

The first book printed from a moveable type press may have been the Bible, but the century following Gutenberg's epochal invention saw a lucrative market for books with more prurient topics.[18] New communication technologies—the telephone, motion pictures, television, videotape—have always been quick to adopt sexual themes. The Internet is no exception, with 1998 market estimates of adult online entertainment ranging from $185 million by Forrester Research to $1 billion by Inter@active Week. These figures are for customers, mostly men, paying to view and interact with performers—live, recorded, and simulated. One 1998 estimate cited 28,000 web sites that offer sexual entertainment.[19] These figures do not include couples who have expanded their phone sex to include moving pictures via online video conferencing.

CD-ROMs and DVD disks constitute another technology that has been exploited for erotic entertainment. Although the bulk of adult-oriented disks are used as a means for delivering videos with a bit of interactivity thrown in, a new genre of CD-ROM and DVD provides virtual sexual companions that respond to some mouse-administered fondling.[20] Like most first-generation technologies, the effect is less than convincing, but future generations will eliminate some of the kinks, although not the kinkiness. Developers are also working to exploit the force-feedback mouse so that you can get some sense of what your virtual partner feels like.

Late in the first decade of the twenty-first century, virtual reality will enable you to be with your lover—romantic partner, sex worker, or simulated companion—with full visual and auditory realism. You will be able to do anything you want with your companion except touch, admittedly an important limitation.

Virtual touch has already been introduced, but the all-enveloping, highly realistic, visual-auditory-tactile virtual environment will not be perfected until the second decade of the twenty-first century. At this point, virtual sex becomes a viable competitor to the real thing. Couples will be able to engage in virtual sex regardless of their physical proximity. Even when proximate, virtual sex will be better in some ways and certainly safer. Virtual sex will provide sensations that are more intense and pleasurable than conventional sex, as well as physical experiences that currently do not exist. Virtual sex is also the ultimate in safe sex, as there is no risk of pregnancy or transmission of disease.

Today, lovers may fantasize their partners to be someone else, but users of virtual sex communication will not need as much imagination. You will be able to change the physical appearance and other characteristics of both yourself and your partner. You can make your lover look and feel like your favorite star without your partner's permission or knowledge. Of course, be aware that your partner may be doing the same to you.

Group sex will take on a new meaning in that more than one person can simultaneously share the experience of one partner. Since multiple real people cannot all control the movements of one virtual partner, there needs to be a way of sharing the decision making of what the one virtual body is doing. Each participant sharing a virtual body would have the same visual, auditory, and tactile experience, with shared control of their shared virtual body (perhaps the one virtual body will reflect a consensus of the attempted movements of the multiple participants). A whole audience of people—who may be geographically dispersed—could share one virtual body while engaged in a sexual experience with one performer.

Prostitution will be free of health risks, as will virtual sex in general. Using wireless, very-high-bandwidth communication technologies, neither sex workers nor their patrons need leave their homes. Virtual prostitution is likely to be legally tolerated, at least to a far greater extent than real prostitution is today, as the virtual variety will be impossible to monitor or control. With the risks of disease and violence having been eliminated, there will be far less rationale for proscribing it.

Sex workers will have competition from simulated—computer generated—partners. In the early stages, "real" human virtual partners are likely to be more realistic than simulated virtual partners, but that will change over time. Of course, once the simulated virtual partner is as capable, sensual, and responsive as a real human virtual partner, who's to say that the simulated virtual partner isn't a real, albeit virtual, person?

Is virtual rape possible? In the purely physical sense, probably not. Virtual reality will have a means for users to immediately terminate their experience. Emotional and other means of persuasion and pressure are another matter.

How will such an extensive array of sexual choices and opportunities affect the institution of marriage and the concept of commitment in a relationship? The technology of virtual sex will introduce an array of slippery slopes, and the definition of a monogamous relationship will become far less clear. Some people will feel that access to intense sexual experiences at the click of a mental button will destroy the concept of a sexually committed relationship. Others will argue, as proponents of sexual entertainment and services do today, that such diversions are healthy outlets and serve to maintain healthy relationships. Clearly, couples will need to reach their own understandings, but drawing clear lines will become difficult with the level of privacy that this future technology affords. It is likely that society will accept practices and activities in the virtual arena that it frowns on in the physical world, as the consequences of virtual activities are often (although not always) easier to undo.

In addition to direct sensual and sexual contact, virtual reality will be a great place for romance in general. Stroll with your lover along a virtual Champs-Élysées, take a walk along a virtual Cancún beach, mingle with the animals in a simulated Mozambique game reserve. Your whole relationship can be in Cyberland.

Virtual reality using an external visual-auditory-haptic interface is not the only technology that will transform the nature of sexuality in the twenty-first century. Sexual robots—sexbots—will become popular by the beginning of the third decade of the new century. Today, the idea of intimate relations with a robot or doll is not generally appealing because robots and dolls are so, well, inanimate. But that will change as robots gain the softness, intelligence, pliancy, and passion of their human creators. (By the end of the twenty-first century, there won't be a clear difference between humans and robots. What, after all, is the difference between a human who has upgraded her body and brain using new nanotechnology and computational technologies, and a robot who has gained an intelligence and sensuality surpassing her human creators?)

By the fourth decade, we will move to an era of virtual experiences through internal neural implants. With this technology, you will be able to have almost any kind of experience with just about anyone, real or imagined, at any time. It's just like today's online chat rooms, except that you don't need any equipment that's not already in your head, and you can do a lot more than just chat. You won't be restricted by the limitations of your natural body as you and your partners can take on any virtual physical form. Many new types of experiences will become possible: A man can feel what it is like to be a woman, and vice versa. Indeed, there's no reason why you can't be both at the same time, making real, or at least virtually real, our solitary fantasies.

And then, of course, in the last half of the century, there will be the nanobot swarms—good old sexy Utility Fog, for example. The nanobot swarms can instantly take on any form and emulate any sort of appearance, intelligence, and personality that you or it desires—the human form, say, if that's what turns you on.

THE SPIRITUAL MACHINE

We are not human beings trying to be spiritual. We are spiritual beings trying to be human.

—Jacquelyn Small

Body and soul are twins. God only knows which is which.

—Charles A. Swinburne

We're all lying in the gutter, but some of us are gazing at the stars.

—Oscar Wilde

Sexuality and spirituality are two ways that we transcend our everyday physical reality. Indeed, there are links between our sexual and our spiritual passions, as the ecstatic rhythmic movements associated with some varieties of spiritual experience suggest.

MIND TRIGGERS

We are discovering that the brain can be directly stimulated to experience a wide variety of feelings that we originally thought could be gained only from actual physical or mental experience. Take humor, for example. In the journal *Nature,* Dr. Itzhak Fried and his colleagues at UCLA tell how they found a neurological trigger for humor. They were looking for possible causes for a teenage girl's epileptic seizures and discovered that applying an electric probe to a specific point in the supplementary motor area of her brain caused her to laugh. Initially, the researchers thought that the laughter must be just an involuntary motor response, but they soon realized they were triggering the genuine perception of humor, not just forced laughter. When stimulated in just the right spot of her brain, she found everything funny. "You guys are just so funny—standing around" was a typical comment.[21]

Triggering a perception of humor without circumstances we normally consider funny is perhaps disconcerting (although personally, I find it humorous). Humor involves a certain element of surprise. Blue elephants. The last two words were intended to be surprising, but they probably didn't make you laugh (or maybe they did). In addition to surprise, the unexpected event needs to make sense from an unanticipated but meaningful perspective. And there are some other attributes that humor requires that we don't understand just yet. The brain apparently has a neural net that detects humor from our other perceptions. If we directly stimulate the brain's humor detector, then an otherwise ordinary situation will seem pretty funny.

The same appears to be true of sexual feelings. In experiments with animals, stimulating a specific small area of the hypothalamus with a tiny injection of testosterone causes the animals to engage in female sexual behavior, regardless of gender. Stimulating a different area of the hypothalamus produces male sexual behavior.

These results suggest that once neural implants are commonplace, we will have the ability to produce not only virtual sensory experiences but also the feelings associated with these experiences. We can also create some feelings not ordinarily associated with the experience. So you will be able to add some humor to your sexual experiences, if desired (of course, for some of us humor may already be part of the picture).

The ability to control and reprogram our feelings will become even more profound in the late twenty-first century when technology moves beyond mere neural implants and we fully install our thinking processes into a new computational medium—that is, *when we become software.*

We work hard to achieve feelings of humor, pleasure, and well-being. Being able to call them up at will may seem to rob them of their meaning. Of course, many people use drugs today to create and enhance certain desirable feelings, but the chemical approach comes bundled with many undesirable effects. With neural implant technology, you will be able to enhance your feelings of pleasure and well-being without the hangover. Of course, the potential for abuse is even greater than with drugs. When psychologist James Olds provided rats with the ability to press a button and directly stimulate a pleasure center in the limbic system of their brains, the rats pressed the button endlessly, as often as five thousand times an hour, to the exclusion of everything else, including eating. Only falling asleep caused them to stop temporarily.[22]

Nonetheless, the benefits of neural implant technology will be compelling. As just one example, millions of people suffer from an inability to experience sufficiently intense feelings of sexual pleasure, which is one important aspect of impotence. People with this disability will not pass up the opportunity to overcome their problem through neural implants, which they may already have in place for other purposes. Once a technology is developed to overcome a disability, there is no way to restrict its use from enhancing normal abilities, nor would such restrictions necessarily be desirable. The ability to control our feelings will be just another one of those twenty-first-century slippery slopes.

SO WHAT ABOUT SPIRITUAL EXPERIENCES?

The spiritual experience—a feeling of transcending one's everyday physical and mortal bounds to sense a deeper reality—plays a fundamental role in otherwise disparate religions and philosophies. Spiritual experiences are not all of the same sort but appear to encompass a broad range of mental phenomena. The ecstatic dancing of a Baptist revival appears to be a different phenomenon from the quiet transcendence of a Buddhist monk. Nonetheless, the notion of the spiritual experience has been reported so consistently throughout history, and in virtually all cultures and religions, that it represents a particularly brilliant flower in the phenomenological garden.

Regardless of the nature and derivation of a mental experience, spiritual or otherwise, once we have access to the computational processes that give rise to it, we have the opportunity to understand its neurological correlates. With the understanding of our mental processes will come the opportunity to capture our intellectual, emotional, and spiritual experiences, to call them up at will, and to enhance them.

SPIRITUAL EXPERIENCE THROUGH BRAIN GENERATED MUSIC

There is already one technology that appears to generate at least one of aspect of a spiritual experience. This experimental technology is called Brain Generated Music (BGM), pioneered by Neuro-Sonics, a small company in Baltimore, Maryland, of which I am a director. BGM is a brain-wave biofeedback system capable of evoking an experience called the Relaxation Response, which is associated with deep relaxation.[23] The BGM user attaches three disposable leads to her head. A personal computer then monitors the user's brain waves to determine her unique alpha wavelength. Alpha waves, which are in the range of eight to thirteen cycles per second (cps), are associated with a deep meditative state, as compared to beta waves (in the range of thirteen to twenty-eight cps), which are associated with routine conscious thought. Music is then generated by the computer, according to an algorithm that transforms the user's own brain-wave signal.

The BGM algorithm is designed to encourage the generation of alpha waves by producing pleasurable harmonic combinations upon detection of alpha waves, and less pleasant sounds and sound combinations when alpha detection is low. In addition, the fact that the sounds are synchronized to the user's own alpha wavelength to create a resonance with the user's own alpha rhythm also encourages alpha production.

Dr. Herbert Benson, formerly the director of the hypertension section of Boston's Beth Israel Hospital and now at New England Deaconess Hospital in Boston, and other researchers at the Harvard Medical School and Beth Israel, discovered the neurological-physiological mechanism of the Relaxation Response, which is described as the opposite of the "fight or flight," or stress response.[24] The Relaxation Response is associated with reduced levels of epinephrine (adrenaline) and norepinephrine (noradrenaline), blood pressure, blood sugar, breathing, and heart rates. Regular elicitation of this response is reportedly able to produce permanently lowered blood-pressure levels (to the extent that hypertension is caused by stress factors) and other health benefits. Benson and his colleagues have catalogued a number of techniques that can elicit the Relaxation Response, including yoga and a number of forms of meditation.

I have had experience with meditation, and in my own experience with BGM, and in observing others, BGM does appear to evoke the Relaxation Response. The music itself feels as if it is being generated from inside your mind. Interestingly, if you listen to a tape recording of your own brain-generated music when you are not hooked up to the computer, you do not experience the same sense of transcendence. Although the recorded BGM is based on your personal alpha wavelength, the recorded music was synchronized to the brain waves that were produced by your brain when the music was first generated, not to the brain waves that are produced while listening to the recording. You need to listen to "live" BGM to achieve the resonance effect.

Conventional music is generally a passive experience. Although a performer may be influenced in subtle ways by her audience, the music we listen to generally does not reflect our response. Brain Generated Music represents a new modality of music that enables the music to evolve continually based on the interaction between it and our own mental responses to it.

Is BGM producing a spiritual experience? It's hard to say. The feelings produced while listening to "live" BGM are similar to the deep transcendent feelings I can sometimes achieve with meditation, but they appear to be more reliably produced by BGM.

THE GOD SPOT

Neuroscientists from the University of California at San Diego have found what they call the God module, a tiny locus of nerve cells in the frontal lobe that appears to be activated during religious experiences. They discovered this neural machinery while studying epileptic patients who have intense mystical experiences during seizures. Apparently the intense neural storms during a seizure stimulate the God module. Tracking surface electrical activity in the brain with highly sensitive skin monitors, the scientists found a similar response when very religious nonepileptic persons were shown words and symbols evoking their spiritual beliefs.

A neurological basis for spiritual experience has long been postulated by evolutionary biologists because of the social utility of religious belief. In response to reports of the San Diego research, Richard Harries, the Bishop of Oxford, said through a spokesman that "it would not be surprising if God had created us with a physical facility for belief."[25]

When we can determine the neurological correlates of the variety of spiritual experiences that our species is capable of, we are likely to be able to enhance these experiences in the same way that we will enhance other human experiences. With the next stage of evolution creating a new generation of humans that will be trillions of times more capable and complex than humans today, our ability for spiritual experience and insight is also likely to gain in power and depth.

Just being—experiencing, being conscious—is spiritual, and reflects the essence of spirituality. Machines, derived from human thinking and surpassing humans in their capacity for experience, will claim to be conscious, and thus to be spiritual. They will believe that they are conscious. They will believe that they have spiritual experiences. They will be convinced that these experiences are meaningful. And given the historical inclination of the human race to anthropomorphize the phenomena we encounter, and the persuasiveness of the machines, we're likely to believe them when they tell us this.

Twenty-first-century machines—based on the design of human thinking—will do as their human progenitors have done—going to real and virtual houses of worship, meditating, praying, and transcending—to connect with their spiritual dimension.

NOTES

1. Ralph Merkle's comments on nanotechnology can be found in an overview at his web site at the Xerox Palo Alto Research Center http://sandbox.xerox.com/nano. His site contains links to important publications

on nanotechnology, such as Richard Feynman's 1959 talk and Eric Drexler's dissertation, as well as links to various research centers that focus on nanotechnology.

2. Richard Feynman presented these ideas on December 29, 1959, at the annual meeting of the American Physical Society at the California Institute of Technology (Cal Tech). His talk was first published in the February 1960 issue of Cal Tech's *Engineering and Science*. This article is available online at http://nano.xerox .com/nanotech/feynman.html.

3. Eric Drexler, *Engines of Creation* (New York: Anchor Press/Doubleday, 1986). The book is also accessible online from the Xerox nanotechnology site http://sandbox.xerox.com/nano and also from Drexler's web site at the Foresight Institute http://www.foresight.org/EOC/.index.html.

4. Eric Drexler, *Nanosystems: Molecular Machinery, Manufacturing, and Computation* (New York: John Wiley and Sons, 1992).

5. According to Nanothinc's web site http://www.nanothinc.com/, "Nanotechnology, broadly defined to include a number of nanoscale-related activities and disciplines, is a global industry in which more than 300 companies generate over $5 billion in annual revenues today—and $24 billion in 4 years." Nanothinc includes a list of companies and revenues upon which the figure is based. Some of the nanoapplications generating revenues are micromachines, microelectromechanical systems, autofabrication, nanolithography, nanotechnology tools, scanning probe microscopy, software, nanoscale materials, and nanophase materials.

6. Richard Smalley's publications and work on nanotechnology can be found at the web site for the Center for Nanoscale Science and Technology at Rice University http://cnst.rice.edu/.

7. For information on the use of nanotechnology in creating IBM's corporate logo, read Faye Flam, "Tiny Instrument Has Big Implications." *Knight-Ridder/Tribune News Service,* August 11, 1997, p. 811K7204.

8. Dr. Jeffrey Sampsell at Texas Instruments has written a white paper summarizing research on micromirrors, available at http://www.ti.com/dlp/docs/it/resources/white/overview/over.shtml.

9. A description of the flying machines can be found at the web site of the MEMS (MicroElectroMechanical Systems) and Fluid Dynamics Research Group at the University of California at Los Angeles (UCLA) http://ho.seas.ucla.edu/new/main.htm.

10. Xerox's nanotechnology research is described in Brian Santo, "Smart Matter Program Embeds Intelligence by Combining Sensing, Actuation, Computation—Xerox Builds on Sensor Theory for Smart Materials." *EETimes* (March 23, 1998): 129. More information on this research can be found at the web site for the Smart Matter Research Group at Xerox's Palo Alto Research Center at http://www.parc.xerox.com/spl/proj ects/smart-matter/.

11. For information on the use of nanotechnology in creating the nanoguitar, read Faye Flam, "Tiny Instrument Has Big Implications." *Knight-Ridder/Tribune News Service.*

12. Learn more about the Chelyabinsk region by visiting the web site dedicated to helping the people living in that area at http://www.logtv.com/chelya/chel.html.

13. For more about the story behind Space War, see "A History of Computer Games," *Computer Gaming World* (November 1991): 16–26; and Eric S. Raymond, ed., *New Hacker's Dictionary* (Cambridge, MA: MIT Press, 1992). Space War was developed by Steve Russell in 1961 and implemented by him on the PDP-1 at MIT a year later.

14. Medical Learning Company is a joint venture between the American Board of Family Practice (an organization that certifies the sixty thousand family practice physicians in the United States) and Kurzweil Technologies. The goal of the company is to develop educational software for continuing medical education of physicians as well as other markets. A key aspect of the technology will include an interactive simulated patient that can be examined, interviewed, and treated.

15. Hall's Utility Fog concept is described in J. Storrs Hall, "Utility Fog Part 1," *Extropy,* issue no. 13 (vol. 6, no. 2), third quarter 1994; and J. Storrs Hall, "Utility Fog Part 2," *Extropy,* issue no. 14 (vol. 7, no. 1), first quarter 1995. Also see Jim Wilson, "Shrinking Micromachines: A New Generation of Tools Will Make Molecule-Size Machines a Reality." *Popular Mechanics* 174, no. 11 (November 1997): 55–58.

16. Mark Yim, "Locomotion with a Unit-Modular Reconfigurable Robot," Stanford University Technical Report STAN-CS-TR-95–1536.

17. Joseph Michael, UK Patent #94004227.2.

18. For examples of early "prurient" text publications, see *A History of Erotic Literature* by Patrick J. Kearney (Hong Kong, 1982); and *History Laid Bare* by Richard Zachs (New York: HarperCollins, 1994).

19. *Upside Magazine,* April 1998.

20. For example, the "TFUI" (Touch-and-Feel User Interface) from pixis, as used in their Diva and Space Sirens series of CD-ROMs.

21. From "Who Needs Jokes? Brain Has a Ticklish Spot," Malcolme W. Browne, *New York Times,* March 10, 1998. Also see. I. Fried (with C. L. Wilson, K. A. MacDonald, and E. J. Behnke), "Electric Current Stimulates Laughter," *Scientific Correspondence* 391: 650, 1998.

22. K. Blum et al., "Reward Deficiency Syndrome," *American Scientist,* March–April, 1996.

23. Brain Generated Music is a patented technology of NeuroSonics, a small company in Baltimore, Maryland. The founder, CEO, and principal developer of the technology is Dr. Geoff Wright, who is head of computer music at Peabody Institute of Johns Hopkins University.

24. For details about Dr. Benson's work, see his book *The Relaxation Response* (New York: Avon, 1990).

25. "'God Spot' Is Found in Brain," *Sunday Times* (Britain), November 2, 1997.

25

Why Computers May Never Think Like People

Hubert Dreyfus and Stuart Dreyfus

Scientists who stand at the forefront of artificial intelligence (AI) have long dreamed of autonomous "thinking" machines that are free of human control. And now they believe we are not far from realizing that dream. As Marvin Minsky, a well-known AI professor at MIT, recently put it: "Today our robots are like toys. They do only the simple things they're programmed to. But clearly they're about to cross the edgeless line past which they'll do the things we are programmed to."

Patrick Winston, Minsky's successor as head of the MIT AI Laboratory, agrees: "Just as the Wright Brothers at Kitty Hawk in 1903 were on the right track to the 747s of today, so artificial intelligence, with its attempt to formalize common-sense understanding, is on the way to fully intelligent machines."

Encouraged by such optimistic pronouncements, the US Department of Defense (DOD) is sinking millions of dollars into developing fully autonomous war machines that will respond to a crisis without human intervention. Business executives are investing in "expert" systems whose wisdom they hope will equal, if not surpass, that of their top managers. And AI entrepreneurs are talking of "intelligent systems" that will perform better than we can—in the home, in the classroom, and at work.

But no matter how many billions of dollars the Defense Department or any other agency invests in AI, there is almost no likelihood that scientists can develop machines capable of making intelligent decisions. After twenty-five years of research, AI has failed to live up to its promise, and there is no evidence that it ever will. In fact, machine intelligence will probably never replace human intelligence simply because we ourselves are not "thinking machines." Human beings have an intuitive intelligence that "reasoning" machines simply cannot match.

Military and civilian managers may see this obvious shortcoming and refrain from deploying such "logic" machines. However, once various groups have invested vast sums in developing these machines, the temptation to justify this expense by installing questionable AI technologies will be enormous. The dangers of turning over the battlefield completely to machines are obvious. But it would also be a mistake to replace skilled air-traffic controllers, seasoned business managers, and master teachers with computers that cannot come close to their level of expertise. Computers that "teach" and systems that render "expert" business decisions could eventually produce a generation of students and managers who have no faith in their own intuition and expertise.

We wish to stress that we are not Luddites. There are obvious tasks for which computers are appropriate and even indispensable. Computers are more deliberate, more precise, and less prone to exhaustion and error than the most conscientious human being. They can also store, modify, and tap vast files of data more quickly and accurately than humans can. Hence, they can be used as valuable tools in many areas. As word processors and telecommunication devices, for instance, computers are already changing our methods of writing and our notions of collaboration.

However, we believe that trying to capture more sophisticated skills within the realm of electronic circuits—skills involving not only calculation but also judgment—is a dangerously misguided effort and ultimately doomed to failure.

ACQUIRING HUMAN KNOW-HOW

Most of us know how to ride a bicycle. Does that mean we can formulate specific rules to teach someone else how to do it? How would we explain the difference between the feeling of falling over and the sense of being slightly off-balance when turning? And do we really know, until the situation occurs, just what we would do in response to a certain wobbly feeling? No, we don't. Most of us are able to ride a bicycle because we possess something called "know-how," which we have acquired from practice and sometimes painful experience. That know-how is not accessible to us in the form of facts and rules. If it were, we could say we "know that" certain rules produce proficient bicycle riding.

There are innumerable other aspects of daily life that cannot be reduced to "knowing that." Such experiences involve "knowing how." For example, we know how to carry on an appropriate conversation with family, friends, and strangers in a wide variety of contexts—in the office, at a party, and on the street. We know how to walk. Yet the mechanics of walking on two legs are so complex that the best engineers cannot come close to reproducing them in artificial devices.

This kind of know-how is not innate, as is a bird's skill at building a nest. We have to learn it. Small children learn through trial and error, often by imitating those who are proficient. As adults acquire a skill through instruction and experience, they do not appear to leap suddenly from "knowing that"—a knowledge guided by rules—to experience-based know-how. Instead, people usually pass through five levels of skill: novice, advanced beginner, competent, proficient, and expert. Only when we understand this dynamic process can we ask how far the computer could reasonably progress.

During the novice stage, people learn facts relevant to a particular skill and rules for action that are based on those facts. For instance, car drivers learning to operate a stick shift are told at what speed to shift gears and at what distance—given a particular speed—to follow other cars. These rules ignore context, such as the density of traffic or the number of stops a driver has to make.

Similarly, novice chess players learn a formula for assigning pieces point values independent of their position. They learn the rule: "Always exchange your pieces for the opponent's if the total value of the pieces captured exceeds that of pieces lost." Novices generally do not know that they should violate this rule in certain situations.

After much experience in real situations, novices reach the advanced-beginner stage. Advanced-beginner drivers pay attention to situational elements, which cannot be defined objectively. For instance, they listen to engine sounds when shifting gears. They can also distinguish between the behavior of a distracted or drunken driver and that of the impatient but alert driver. Advanced-beginner chess players recognize and avoid overextended positions. They can also spot situational clues such as a weakened king's side or a strong pawn structure. In all these cases, experience is immeasurably more important than any form of verbal description.

Like the training wheels on a child's first bicycle, initial rules allow beginners to accumulate experience. But soon they must put the rules aside to proceed. For example, at the competent stage, drivers no longer merely follow rules; they drive with a goal in mind. If they wish to get from point A to point B very quickly, they choose their route with an eye to traffic but not much attention to passenger comfort. They follow other cars more closely than they are "supposed" to, enter traffic more daringly, and even break the law. Competent chess players may decide, after weighing alternatives, that they can attack their opponent's king. Removing pieces that defend the enemy king becomes their overriding objective, and to reach it these players will ignore the lessons they learned as beginners and accept some personal losses.

A crucial difference between beginners and more competent performers is their level of involvement. Novices and advanced beginners feel little responsibility for what they do because they are only applying learned rules; if they foul up, they blame the rules instead of themselves. But competent performers, who choose a goal and a plan for achieving it, feel responsible for the result of their choices. A successful outcome is deeply satisfying and leaves a vivid memory. Likewise, disasters are not easily forgotten.

THE INTUITION OF EXPERTS

The learner of a new skill makes conscious choices after reflecting on various options. Yet in our everyday behavior, this model of decision-making—the detached, deliberate, and sometimes agonizing selection among alternatives—is the exception rather than the rule, Proficient performers do not rely on detached deliberation in going about their tasks. Instead, memories of similar experiences in the past seem to trigger plans like those that worked before. Proficient performers recall whole situations from the past and apply them to the present without breaking them down into components or rules.

For instance, a boxer seems to recognize the moment to begin an attack not by following rules and combining various facts about his body's position and that of his opponent. Rather, the whole visual scene triggers the memory of similar earlier situations in which an attack was successful. The boxer is using his intuition, or know-how.

Intuition should not be confused with the re-enactment of childhood patterns or any of the other unconscious means by which human beings come to decisions. Nor is guessing what we mean by intuition. To guess is to reach a conclusion when one does not have enough knowledge or experience to do so. Intuition or know-how is the sort of ability that we use all the time as we go about our everyday tasks. Ironically, it is an ability that our tradition has acknowledged only in women and judged inferior to masculine rationality.

While using their intuition, proficient performers still find themselves thinking analytically about what to do. For instance, when proficient drivers approach a curve on a rainy day, they may intuitively realize they are going too fast. They then consciously decide whether to apply the brakes, remove their foot from the accelerator, or merely reduce pressure on the accelerator. Proficient marketing managers may intuitively realize that they should reposition a product. They may then begin to study the situation, taking great pride in the sophistication of their scientific analysis while overlooking their much more impressive talent—that of recognizing, without conscious thought, the simple existence of the problem.

The final skill level is that of expert. Experts generally know what to do because they have a mature and practiced understanding. When deeply involved in coping with their environment, they do not see problems in some detached way and consciously work at solving them. The skills of experts have become so much a part of them that they need be no more aware of them than they are of their own bodies. Airplane pilots report that as novices they felt they were flying their planes, but as experienced pilots they simply experience flying itself. Grand masters of chess, engrossed in a game, are often oblivious to the fact that they are manipulating pieces on a board. Instead, they see themselves as participants in a world of opportunities, threats, strengths, weaknesses, hopes, and fears. When playing rapidly, they sidestep dangers as automatically as teenagers avoid missiles in a familiar video game.

One of us, Stuart, knows all too well the difference between expert and merely competent chess players; he is stuck at the competent level. He took up chess as an outlet for his analytic talent in mathematics, and most of the other players on his college team were also mathematicians. At some point, a few of his teammates who were not mathematicians began to play fast five- or ten-minute games of chess, and also began eagerly to replay the great games of the grand masters. But Stuart and his mathematical colleagues resisted because fast chess didn't give them the time to *figure out* what to do. They also felt that they could learn nothing from the grand master games, since the record of those games seldom if ever provided specific rules and principles.

Some of his teammates who played fast chess and studied grand master games absorbed a great deal of concrete experience and went on to become chess masters. Yet Stuart and his mathematical friends never got beyond the competent level. Students of math may predominate among chess enthusiasts, but a truck driver is as likely as a mathematician to be among the world's best players. Stuart says he is glad that his analytic approach to chess stymied his progress because it helped him to see that there is more to skill than reasoning.

When things are proceeding normally, experts do not solve problems by reasoning; they do what normally works. Expert air-traffic controllers do not watch blips on a screen and deduce what must be going on in the sky. Rather, they "see" planes when they look at their screens and they respond to what they see, not by using rules but as experience has taught them to. Skilled outfielders do not take the time to figure out where a ball is going. Unlike novices, they simply run to the right spot. In *The Brain*, Richard Restak quotes a Japanese martial artist as saying, "There can be no thought, because if there is thought, there is a time of thought and that means a flaw. . . . If you take the time to think, 'I must use this or that technique', you will be struck while you are thinking."

We recently performed an experiment in which an international chess master, Julio Kaplan, had to add numbers at the rate of about one per second while playing five-second-a-move chess against a slightly weaker but master-level player. Even with his analytical mind apparently jammed by adding numbers, Kaplan more than held his own against the master in a series of games. Deprived of the time necessary to see problems or construct plans, Kaplan still produced fluid and coordinated play.

As adults acquire skills, what stands out is their progression *from* the analytic behavior of consciously following abstract rules *to* skilled behavior based on unconsciously recognizing new situations as similar to remembered ones. Conversely, small children initially understand only concrete examples and gradually learn abstract reasoning. Perhaps it is because this pattern in children is so well known that adult intelligence is so often misunderstood.

By now it is evident that there is more to intelligence than calculative rationality. In fact, experts who consciously reason things out tend to regress to the level of a novice or, at best, a competent performer. One expert pilot described an embarrassing incident that illustrates this point. Once he became an instructor, his only opportunity to fly the four-jet KC-135s at which he had once been expert was during the return flights he made after evaluating trainees. He was approaching the landing strip on one such flight when an engine failed. This is technically an emergency, but an experienced pilot will effortlessly compensate for the pull to one side. Being out of practice, our pilot thought about what to do and then overcompensated. He then consciously corrected himself, and the plane shuddered violently as he landed. Consciously using rules, he had regressed to flying like a beginner.

This is not to say that deliberative rationality has no role in intelligence. Tunnel vision can sometimes be avoided by a type of detached deliberation. Focussing on aspects of a situation that seem relatively unimportant allows another perspective to spring to mind. We once heard an Israeli fighter pilot recount how deliberative rationality may have saved his life by rescuing him from tunnel vision. Having just vanquished an expert opponent, he found himself taking on another member of the enemy squadron who seemed to be brilliantly eluding one masterful ploy after another. Things were looking bad until he stopped following his intuition and deliberated. He then realized that his opponent's surprising maneuvers were really the predictable, rule-following behavior of a beginner. This insight enabled him to vanquish the pilot.

IS INTELLIGENCE BASED ON FACTS?

Digital computers, which are basically complicated structures of simple on-off switches, were first used for scientific calculation. But by the end of the 1950s, researchers such as Allen Newell and Herbert Simon, working together at the Rand Corp., began to exploit the idea that computers could manipulate general symbols. They saw that one could use symbols to represent elementary facts about the world and rules to represent relationships between the facts. Computers could apply these rules and make logical inferences about the facts. For instance, a programmer might give a computer rules about how cannibals like to eat missionaries, and facts about how many cannibals and missionaries must be ferried across a river in one boat that carries only so many people. The computer could then figure out how many trips it would take to get both the cannibals and the missionaries safely across the river.

Newell and Simon believed that computers programmed with such facts and rules could, in principle, solve problems, recognize patterns, understand stories, and indeed do anything that an intelligent person could do. But they soon found that their programs were missing crucial aspects of problem-solving, such as the ability to separate relevant from irrelevant operations. As a result, the programs worked in only a very limited set of cases, such as in solving puzzles and proving theorems of logic.

In the late 1960s, researchers at MIT abandoned Newell and Simon's approach, which was based on imitating people's reports of how they solved problems, and began to work on any processing methods that could give computers intelligence. They recognized that to solve "real-world" problems the computer had to somehow simulate real-world understanding and intuition. In the introduction to *Semantic Information Processing*, a collection of his students' Ph.D. theses, Marvin Minsky describes the heart of the MIT approach:

> If we . . . ask . . . about the common everyday structures—that which a person needs to have ordinary common sense—we will find first a collection of indispensable categories, each rather complex: geometrical and mechanical properties of things and of space; uses and properties of a few thousand objects; hundreds of "facts" about hundreds of people; thousands of facts about tens of people; tens of facts about thousands of people; hundreds of facts about hundreds of organizations . . . I therefore feel that a machine will quite critically need to acquire on the order of a hundred thousand elements of knowledge in order to behave with reasonable sensibility in

ordinary situations. A million, if properly organized, should be enough for a very great intelligence.

However, Minsky's students encountered the same problem that had plagued Newell and Simon: each program worked only in its restricted specialty and could not be applied to other problems. Nor did the programs have any semantics—that is, any understanding of what their symbols meant. For instance, Daniel Bobrow's STUDENT program, which was designed to understand and solve elementary algebraic story problems, interpreted the phrase "the number of times I went to the movies" as the product of the two variables "number of" and "I went to the movies." That's because, as far as the program knew, "times" was a multiplicative operator linking the two phrases.

The restricted, ad hoc character of such work is even more striking in a program called ELIZA, written by MIT computer science professor Joseph Weizenbaum. Weizenbaum set out to show just how much apparent intelligence one could get a computer to exhibit without giving it any real understanding at all. The result was a program that imitated a therapist using simple tricks such as turning statements into questions: it responded to "I'm feeling sad" with "Why are you feeling sad?" When the program couldn't find a stock response, it printed out statements such as "Tell me about your father." The remarkable thing was that people were so easily fooled by these tricks. Weizenbaum was appalled when some people divulged their deepest feelings to the computer and asked others to leave the room while they were using it.

One of us, Hubert, was eager to see a demonstration of the notorious program, and he was delighted when Weizenbaum invited him to sit at the console and interact with ELIZA. Hubert spoiled the fun, however. He unintentionally exposed how shallow the trickery really was by typing, "I'm feeling happy," and then correcting himself by typing, "No, elated." At that point, the program came back with the remark, "Don't be so negative." Why? Because it had been programmed to respond with that rebuke whenever there was a "no" in the input.

MICROWORLDS VERSUS THE REAL WORLD

It took about five years for the shallowness of Minsky's students' programs to become apparent. Meanwhile, Hubert published a book, *What Computers Can't Do,* which asserted that AI research had reached a dead end since it could not come up with a way to represent general common-sense understanding. But just as *What Computers Can't Do* went to press in 1970, Minsky and Seymour Papert, also a professor at MIT, developed a new approach to AI. If one could not deal systematically with common-sense knowledge all at once, they asked, then why not develop methods for dealing systematically with knowledge in isolated subworlds and build gradually from that?

Shortly after that, MIT researchers hailed a computer program by graduate student Terry Winograd as a "major advance" in getting computers to understand human language. The program, called SHRDLU, simulated on a TV screen a robot arm that could move a set of variously shaped blocks. The program allowed a person to engage in a dialogue with the computer, asking questions, making statements, and issuing commands within this simple world of movable blocks. The program relied on grammatical rules, semantics, and facts about blocks. As Winograd cautiously claimed, SHRDLU was a "computer program which 'understands' language in a limited domain."

Winograd achieved success in this restricted domain, or "microworld," because he chose a simple problem carefully. Minsky and Papert believed that by combining a large number of these microworlds, programmers could eventually give computers real-life understanding.

Unfortunately, this research confuses two domains, which we shall distinguish as "universe" and "world." A set of interrelated facts may constitute a "universe" such as the physical universe, but it does not constitute a "world" such as the world of business or theater. A "world" is an organized body of objects, purposes, skills, and practices that make sense only against a background of common human concerns. These "sub-worlds" are not isolable physical systems. Rather, they are specific elaborations of a whole, without which they could not exist.

If Minsky and Papert's microworlds *were* true sub-worlds, they would not have to be extended and combined to encompass the everyday world, because each one would already incorporate it. But since microworlds are only isolated, meaningless domains, they cannot be combined and extended to reflect everyday life. Because scientists failed to ask what a "world" is, another five-year period of AI research ended in stagnation.

Winograd himself soon gave up the attempt to generalize the techniques SHRDLU used. "The AI programs of the late sixties and early seventies are much too literal," he acknowledged in a report for the National Institute of Education. "They deal with meaning as if it were a structure to be built up of the bricks and mortar provided by the words."

From the late 1970s to the present, AI has been wrestling unsuccessfully with what is called the common-sense knowledge problem: how to store and gain access to all the facts human beings seem to know. This problem has kept AI from even beginning to fulfill the predictions Minsky and Simon made in the mid-1960s: that within 20 years computers would be able to do everything humans can.

CAN COMPUTERS COPE WITH CHANGE?

If a machine is to interact intelligently with people, it has to be endowed with an understanding of human life. What we understand simply by virtue of being human—that insults make us angry, that moving physically forward is easier than moving backward—all this and much more would have to be programmed into the computer as facts and rules. As AI workers put it, they must give the computer our belief system. This, of course, presumes that human understanding is made up of beliefs that can be readily collected and stored as facts.

Even if we assume that this is possible, an immediate snag appears: we cannot program computers for context. For instance, we cannot program a computer to know simply that a car is going "too fast." The machine must be programmed in a way free of interpretation—we must stipulate that the car is going "20 miles an hour," for example. Also, computers know what to do only by reference to precise rules, such as "shift to second at 20 miles an hour." Computer programmers cannot use common-sense rules, such as "under normal conditions, shift to second at about 20 miles an hour."

Even if all the facts were stored in a context-free form, the computer still couldn't use them because it would be unable to draw on just the facts or rules that are relevant in each particular context. For example, a general rule of chess is that you should trade material when you're ahead in the value of the pieces on the board. However, you should not apply that rule if the opposing king is much more centrally located than yours, or when you are attacking the enemy king. And there are exceptions to each of these exceptions. It is virtually impossible to include all the possible exceptions in a program and do so in such a way that the computer knows which exception to use in which case.

In the real world, any system of rules has to be incomplete. The law, for instance, always strives for completeness but never achieves it. "Common law" helps, for it is based more on precedents than on a specific code. But the sheer number of lawyers in business tells us that it is impossible to develop a code of law so complete that all situations are unambiguously covered.

To explain our own actions and rules, humans must eventually fall back on everyday practices and simply say, "This is what one does." In the final analysis, all intelligent behavior must hark back to our sense of what we *are*. We can never explicitly formulate this in clear-cut rules and facts; therefore, we cannot program computers to possess that kind of know-how.

Nor can we program them to cope with changes in everyday situations. AI researchers have tried to develop computer programs that describe a normal sequence of events as they unfold. One such script, for instance, details what happens when someone goes to a restaurant. The problem is that so many unpredictable events can occur—one can receive an emergency telephone call or run into an acquaintance—that it's virtually impossible to predict how different people will respond. It all depends on what else is going on and what their specific purpose is. Are these people there to eat, to hobnob with friends, to answer phone calls, or to give the waiters a hard time? To make sense of behavior in restaurants, one has to understand not only what people typically do in eating establishments but why they do it. Thus, even if programmers could manage to list all that is *possibly* relevant in typical restaurant dining, computers could not use the information because they would have no understanding of what is *actually* relevant to specific customers.

THINKING WITH IMAGES, NOT WORDS

Experimental psychologists have shown that people actually use images, not descriptions as computers do, to understand and respond to some situations. Humans often think by forming images and comparing them holistically. This process is quite different from the logical, step-by-step operations that logic machines perform.

For instance, human beings use images to predict how certain events will turn out. If people know that a small box is resting on a large box, they can imagine what would happen if the large box were moved. If they see that the small box is tied to a door, they can also imagine what would result if someone were to open the door. A computer, however, must be given a list of facts about boxes, such as their size, weight, and frictional coefficients, as well as information about how each is affected by various kinds of movements. Given enough precise information about boxes and strings, the computer can deduce whether the small box will move with the large one under certain conditions. People also reason things out in this explicit, step-by-step way—but only if they must think about relationships they have never seen and therefore cannot imagine.

At present, computers have difficulty recognizing images. True, they can store an image as a set of dots and then rotate the set of dots so that a human designer can see the object from any perspective. But to know what a scene depicts, a computer must be able to analyze it and recognize every object. Programming a computer to analyze a scene has turned out to be very difficult. Such programs require a great deal of computation, and they work only in special cases with objects whose characteristics the computer has been programmed to recognize in advance.

But that is just the beginning of the problem. The computer can make inferences only from lists of facts. It's as if to read a newspaper you had to spell out each word, find its meaning in the dictionary, and diagram every sentence, labeling all the parts of speech. Brains do not seem to decompose either language or images this way, but logic machines have no choice. They must break down images into the objects they contain—and then into descriptions of those objects' features—before drawing any conclusions. However, when a picture is converted into a description, much information is lost. In a family photo, for instance, one can see immediately which people are between, behind, and in front of which others. The programmer must list all these relationships for the computer, or the machine must go through the elaborate process of deducing these relationships each time the photo is used.

Some AI workers look for help from parallel processors, machines that can do many things at once and hence make millions of inferences per second. But this appeal misses the point: that human beings seem to be able to form and compare images in a way that cannot be captured by any number of procedures that operate on descriptions.

Take, for example, face recognition. People can not only form an image of a face, but they can also see the similarity between one face and another. Sometimes the similarity will depend on specific shared features, such as blue eyes and heavy beards. A computer, if it has been programmed to abstract such features from a picture of a face, could recognize this sort of similarity.

However, a computer cannot recognize emotions such as anger in facial expressions, because we know of no way to break down anger into elementary symbols. Therefore, logic machines cannot see the similarity between two faces that are angry. Yet human beings can discern the similarity almost instantly.

Many AI theorists are convinced that human brains unconsciously perform a series of computations to perceive such subtleties. While no evidence for this mechanical model of the brain exists, these theorists take it for granted because it is the way people proceed when they are reflecting consciously. To such theorists, any alternative explanation appears mystical and therefore anti-scientific.

But there is another possibility. The brain, and therefore the mind, could still be explained in terms of something material. But it does not have to be an information processing machine. Other physical systems can detect similarity without using any descriptions or rules at all. These systems are known as holograms.

IS THE MIND LIKE A HOLOGRAM?

An ordinary hologram works by taking a picture of an object using two beams of laser light, one of which is reflected off the object and one of which shines directly onto film. When the two beams meet, they create an interference pattern like that produced by the waves from several pebbles thrown into a pond. The light waves form a specific pattern of light and dark regions. A photographic plate records this interference pattern, thus storing a representation of the object.

In ordinary light, the plate just looks blurry, a uniform silvery gray. But if the right frequency of light is projected on to it, the recorded pattern of light and dark shapes the light into a replica of the object. This replica appears three-dimensional: we can view different sides of it as we change position.

What first attracted neuropsychologists to the hologram was that it really is holistic: any small piece of the blur on the photographic plate contains the whole scene. For example, if you cut one corner off a hologram of a table and shine a laser beam through what remains, you do not see an image of a table with a corner missing. The whole table is still there but with fuzzier edges.

Certain areas of the brain also have this property. When a piece is cut out, a person may lose nothing specific from vision, for example. Instead, that person may see everything less distinctly. Holograms have another mindlike property: they can be used for associative memory. If one uses a single hologram to record two different scenes and then bounces laser light off one of the scenes, an image of the other will appear.

In our view, the most important property of holograms is their ability to detect similarity. For example, if we made a hologram of this page and then made a hologram of one of the letters on the page, say the letter *F*, shining a light through the two holograms would reveal an astonishing effect: a black field with bright spots wherever the letter *F* occurs on the page. Moreover, the brightest spots would indicate the *F*s with the greatest similarity to the *F* we used to make our hologram. Dimmer spots would appear where there are imperfect or slightly rotated versions of the *F*. Thus, a hologram can not only identify objects; it can also recognize similarity between them. Yet it employs no descriptions or rules.

The way a hologram can instantly pick out a specific letter on a page is reminiscent of the way people pick out a familiar face from a crowd. It is possible that we distinguish the familiar face from all the other faces by processing rules about objectively identifiable features. But we would have to examine each face in the crowd, detect its features, and compare them with lists of our acquaintances' features. It is much more plausible that our minds work on some variation of the holistic model. While the brain obviously does not contain lasers or use light beams, some scientists have suggested that neurons could process incoming stimuli using interference patterns like those of a hologram.

However, the human mind seems to have an ability that far transcends current holographic techniques: the remarkable ability to recognize whole meaningful patterns without decomposing them into features. Unlike holography, our mind can sometimes detect faces in a crowd that have expressions unlike any we have previously seen on those faces. We can also pick out familiar faces that have changed dramatically because of the growth of a beard or the ravages of time.

We take no stand on the question of whether the brain functions holographically. We simply want to make clear that the information processing computer is not the only physical system that can exhibit mindlike properties. Other devices may provide closer analogies to the way the mind actually works.

Given the above considerations, what level of skill can we expect logic machines to reach? Since we can program computers with thousands of rules combining hundreds of thousands of features, the machines can become what might be thought of as expert novices in any well-structured and well-understood domain. As long as digital computers' ability to recognize images and reason by analogy remains a vague promise, however, they will not be able to approach the way human beings cope with everyday reality. Despite their failure to capture everyday human understanding in computers, AI scientists have developed programs that seem to reproduce human expertise within a specific, isolated domain. The programs are called expert systems. In their narrow areas, such systems perform with impressive competence.

In his book on "Fifth Generation" computers, Edward Feigenbaum, a professor at Stanford, spells out the goal of expert systems: "In the kind of intelligent systems envisioned by the designers of the Fifth Generation, speed and processing power will be increased dramatically. But more important, the machines will have reasoning power: they will automatically engineer vast amounts of knowledge to serve whatever purpose human beings propose, from medical diagnosis to product design, from management decisions to education."

The knowledge engineers claim to have discovered that all a machine needs to behave like an expert in restricted domains are some general rules and lots of very specific knowledge. But can these systems really be expert? If we agree with Feigenbaum that "almost all thinking that professionals do is done by reasoning," and that each expert builds up a "repertory of working rules of thumb," the answer is yes. Given their speed and precision, computers should be as good as or better than people at following rules for deducing conclusions. Therefore, to build an expert system, a programmer need only extract those rules and program them into a computer.

JUST HOW EXPERT ARE EXPERT SYSTEMS?

However, human experts seem to have trouble articulating the principles on which they allegedly act. For example, when Arthur Samuel at IBM decided to write a program for playing checkers in 1947, he tried to elicit "heuristic" rules from checkers masters. But nothing the experts told him allowed him to produce master play. So Samuel supplemented these rules with a program that relies blindly on its memory of past successes to improve its current performance. Basically, the program chooses what moves to make based on rules and a record of all past positions.

This checkers program is one of the best expert systems ever built. But it is no champion. Samuel says the program "is quite capable of beating any amateur player and can give better players a good contest." It did once defeat a state champion, but the champion turned around and defeated the program in six mail games. Nonetheless, Samuel still believes that chess champions rely on heuristic rules. Like Feigenbaum, he simply thinks that the champions are poor at recollecting their compiled rules: "The experts do not know enough about the mental processes involved in playing the game."

INTERNIST-1 is an expert system highly touted for its ability to make diagnoses in internal medicine. Yet according to a recent evaluation of the program published in *The New England Journal of Medicine*, this program misdiagnosed eighteen out of a total of forty-three cases, while clinicians at Massachusetts General Hospital misdiagnosed fifteen. Panels of doctors who discussed each case misdiagnosed only eight. (Biopsies, surgery, and post-mortem autopsies were used to establish the correct diagnosis for each case.) The evaluators found that "the experienced clinician is vastly superior to INTERNIST-1, in the ability to consider the relative severity and independence of the different manifestations of disease and to understand the . . . evolution of the disease process." The journal also noted that this type of systematic evaluation was "virtually unique in the field of medical applications of artificial intelligence."

In every area of expertise, the story is the same: the computer can do better than the beginner and can even exhibit useful competence, but it cannot rival the very experts whose facts and supposed rules it is processing with incredible speed and accuracy.

Why? Because the expert is not following any rules! While a beginner makes inferences using rules and facts just like a computer, the expert intuitively sees what to do without applying rules. Experts must regress to the novice level to state the rules they still remember but no longer use. No amount of rules and facts can substitute for the know-how experts have gained from experience in tens of thousands of situations. We predict that in no domain in which people exhibit such holistic understanding can a system based on rules consistently do as well as experts. Are there any exceptions?

At first glance, at least one expert system seems to be as good as human specialists. Digital Equipment Corp. developed R1, now called XCON, to decide how to combine components of VAX computers to meet consumers' needs. However, the program performs as well as humans only because there are so many possible combinations that even experienced technical editors depend on rule-based methods of problem-solving and take about ten minutes to work out even simple cases. It is no surprise, then, that this particular expert system can rival the best specialists.

Chess also seems to be an exception to our rule. Some chess programs, after all, have achieved master ratings by using "brute force." Designed for the world's most powerful computers, they are capable of examining about ten million possible positions in choosing each move.

However, these programs have an Achilles' heel: they can see only about four moves ahead for each piece. So fairly good players, even those whose chess rating is somewhat lower than the computers, can win by using long-range strategies such as attacking the king side. When confronted by a player who knows its weakness, the computer is not a master-level player.

In every domain where know-how is required to make a judgment, computers cannot deliver expert performance, and it is highly unlikely that they ever will.

Those who are most acutely aware of the limitations of expert systems are best able to exploit their real capabilities. Sandra Cook, manager of the Financial Expert Systems Program at the consulting firm SRI International, is one of these enlightened practitioners. She cautions prospective clients that expert systems should not be expected to perform as well as human experts, nor should they be seen as simulations of human expert thinking.

Cook lists some reasonable conditions under which expert, or rather "competent," systems can be useful. For instance, such systems should be used for problems that can be satisfactorily solved by human experts at such a high level that somewhat inferior performance is still acceptable. Processing of business credit applications is a good example, because rules can be developed for this task and computers can follow them as well as and sometimes better than inexperienced humans. Of course, there are some exceptions to the rules, but a few mistakes are not disastrous. On the other hand, no one should expect expert systems to make stock market predictions because human experts themselves cannot always make such predictions accurately.

Expert systems are also inappropriate for use on problems that change as events unfold. Advice from expert systems on how to control a nuclear reactor during a crisis would come too late to be of any use. Only human experts could make judgments quickly enough to influence events.

It is hard to believe some AI enthusiasts' claim that the companies who use expert systems dominate all competition. In fact, a company that relies too heavily on expert systems faces a genuine danger. Junior employees may come to see expertise as a function of the large knowledge bases and masses of rules on which these programs must rely. Such employees will fail to progress beyond the competent level of performance, and business managers may ultimately discover that their wells of true human expertise and wisdom have gone dry.

COMPUTERS IN THE CLASSROOM

Computers pose a similar threat in the classroom. Advertisements warn that a computer deficiency in the educational diet can seriously impair a child's intellectual growth. As a result, frightened parents spend thousands of dollars on home computers and clamor for schools to install them in the classroom. Critics have likened computer salespeople to the encyclopedia peddlers of a generation ago, who contrived to frighten insecure parents into spending hundreds of dollars for books that contributed little to their offsprings' education.

We feel that there is a proper place for computers in education. However, most of today's educational software is inappropriate, and many teachers now use computers in ways that may eventually produce detrimental results.

Perhaps the least controversial way computers can be used is as tools. Computers can sometimes replace teaching aids ranging from paintbrushes, typewriters, and chalkboards to lab demonstrations. Computer simulations, for instance, allow children to take an active and imaginative role in studying subjects that are difficult to bring into the classroom. Evolution is too slow, nuclear reactions are too fast, factories are too big, and much of chemistry is too dangerous to reproduce realistically. In the future, computer simulations of such events will surely become more common, helping students of all ages in all disciplines to develop their intuition. However, since actual skills can be learned only through experience, it seems only common sense to stick to the world of real objects. For instance, basic electricity should be taught with batteries and bulbs.

Relying too heavily on simulations has its pitfalls. First of all, the social consequences of decisions are often missing from simulations. Furthermore, the appeal of simulations could lead disciplines outside the sciences to stress their formal, analytic side at the expense of lessons based on informal, intuitive understanding. For example, political science departments may be tempted to emphasize mathematical models of elections and neglect the study of political philosophies that question the nature of the state and of power. In some economics departments, econometrics—

which relies heavily on mathematical models—has already pushed aside study of the valuable lessons of economic history. The truth is that no one can assess the dynamic relationships that underlie election results or economies with anything like the accuracy of the laws of physics. Indeed, every election campaign or economic swing offers vivid reminders of how inaccurate predictions based on simulation models can be.

On balance, however, the use of the computer as a tool is relatively unproblematic. But that is not the case with today's efforts to employ the computer as tutor or tutee. Behind the idea that computers can aid, or even replace, teachers is the belief that teachers' understanding of the subject being taught and their profession consists of knowing facts and rules. In other words, the teacher's job is to convey specific facts and rules to students by drill and practice or by coaching.

Actually, if our minds were like computers, drill and practice would be completely unnecessary. The fact that even brilliant students need to practice when learning subtraction suggests that the human brain does not operate like a computer. Drill is required simply to fix the rule in human memory. Computers, by contrast, remember instantly and perfectly. Math students also have to learn that some features such as the physical size and orientation of numbers are irrelevant while others such as position are crucial. In this case, they must learn to "decontextualize," whereas computers have no context to worry about.

There is nothing wrong with using computers as drill sergeants. As with simulation, the only danger in this use stems from the temptation to overemphasize some skills at the expense of others. Mathematics might degenerate into addition and subtraction, English into spelling and punctuation, and history into dates and places.

AI enthusiasts believe that computers can play an even greater role in teaching. According to a 1984 report by the National Academy of Sciences, "Work in artificial intelligence and the cognitive sciences has set the stage for qualitatively new applications of technology to education."

Such claims should give us pause. Computers will not be first-rate teachers unless researchers can solve four basic problems: how to get machines to talk, to listen, to know, and to coach. "We speak as part of our humanness, instinctively, on the basis of past experience," wrote Patrick Suppes of Stanford University, one of the pioneers in computer-aided instruction, in a 1966 *Scientific American* article. "But to get a computer to talk appropriately, we need an explicit theory of talking."

Unfortunately, there is no such theory, and if our analysis of human intelligence is correct, there never will be. The same holds true for the problem of getting computers to listen. Continuous speech recognition seems to be a skill that resists decomposition into features and rules. What we hear does not always correspond to the features of the sound stream. Depending on the context and our expectations, we hear a stream of sound as "I scream," or "ice cream." We assign the space or pause in one of two places, although there is no pause in the sound stream. One expert came up with a sentence that illustrates the different ways we can hear the same stream of sound: "It isn't easy to wreck a nice beach." (Try reading that sentence out loud.)

Without the ability to coach, a computer could hardly substitute for an inexperienced teacher, let alone a Socrates. "Even if you can make the computer talk, listen, and adequately handle a large knowledge data base, we still need to develop an explicit theory of learning and instruction," Suppes writes. "In teaching a student, young or old, a given subject matter, a computer-based learning system can record anything the student does. It can know cognitively an enormous amount of information about the student. The problem is how to use this information wisely, skillfully, and efficiently to teach the student. This is something that the very best human tutors do well, even though they do not understand at all how they do it."

While he recognizes how formidable these obstacles are, Suppes persists in the hope that we can program computers to teach. However, in our view, expertise in teaching does not consist of knowing complicated rules for deciding what tips to give students, when to keep silent, when to intervene—although teachers may have learned such rules in graduate school. Rather, expert teachers learn from experience to draw intuitively and spontaneously on the common-sense knowledge and experience they share with their students to provide the tips and examples they need.

Since computers can successfully teach only novice or, at best, competent performance, they will only produce the sort of expert novices many feel our schools already graduate. Computer programs may actually prevent beginning students from passing beyond competent analysis to expertise. Instead of helping to improve education, computer-aided instruction could easily become part of the problem.

In the air force, for instance, instructors teach beginning pilots a rule for how to scan their instruments. However, when psychologists studied the eye movements of the instructors during simulated flight, the results showed that the instructors were not following the rule they were teaching. In fact, as far as the psychologists could determine, the instructors were not following any rules at all.

Now suppose that the instrument-scanning rule goes into a computer program. The computer monitors eye movements to make sure novices are applying the rule correctly. Eventually, the novices are ready, like the instructors, to abandon the rules and respond to whole situations they perceive as similar to others. At this point, there is nothing more for the computer to teach. If it is still used to check eye movements, it would prevent student pilots from making the transition to intuitive proficiency and expertise.

This is no mere bogeyman. Expert systems are already being developed to teach doctors the huge number of rules that programmers have "extracted" from experts in the medical domain. One can only hope that someone has the sense to disconnect doctors from the system as soon they reach the advanced-beginner stage.

CAN CHILDREN LEARN BY PROGRAMMING?

The concept of using computers as tutees also assumes the information-processing model of the mind. Adherents of this view suppose that knowledge consists of using facts and rules, and that therefore students can acquire knowledge in the very act of programming. According to this theory, learning and learning to program are the same thing.

Seymour Papert is the most articulate exponent of this theory. He is taking his LOGO program into Boston schools to show that children will learn to think more rigorously if they teach a literal-minded but patient and agreeable student—the computer. In Papert's view, programming a computer will induce children to articulate their own program by naming the features they are selecting from their environment, and by making explicit the procedures they are using to relate these features to events. Says Papert: "I have invented ways to take educational advantage of the opportunities to master the art of *deliberately* thinking like a computer, according, for example, to the stereotype of a computer program that proceeds in a step-by-step, literal, mechanical fashion."

Papert's insistence that human know-how can be analyzed has deep roots in our "rationalistic" Western tradition. We can all probably remember a time in school when we knew something perfectly well but our teacher claimed that we didn't know it because we couldn't explain how we got the answer.

Even Nobel laureates face this sort of problem. Physicist Richard Feynman had trouble getting the scientific community to accept his theories because he could not explain how he got his answers. In his book *Disturbing the Universe*, physicist and colleague Freeman Dyson wrote:

> The reason Dick's physics were so hard for the ordinary physicists to grasp was that he did not use equations. . . . He had a physical picture of the way things happen, and the picture gave him the solutions directly with a minimum of calculation. It was no wonder that people who spent their lives solving equations were baffled by him. Their minds were analytical; his was pictorial.

While Papert tries to create a learning environment in which learners constantly face new problems and need to discover new rules, Timothy Gallwey, the author of *Inner Tennis*, encourages learners to achieve mastery by avoiding analytic thinking from the very start. He would like to create a learning environment in which there are no problems at all and so there is never any need for analytic reflection.

Our view lies in between. At any stage of learning, some problems may require rational, analytic thought. Nonetheless, skill in any domain is measured by the performer's ability to act appropriately in situations that might once have been problems but are no longer problems and so do not require analytic reflection. The risk of Gallwey's method is that it leaves the expert without the tools to solve new problems. But the risk of Papert's approach is far greater: it would leave the learner a perpetual beginner by encouraging dependence on rules and analysis.

AI ON THE BATTLEFIELD

The Department of Defense is pursuing a massive Strategic Computing Plan (SCP) to develop completely autonomous land, sea, and air vehicles capable of complex, far-ranging reconnaissance and attack missions. SCP has already spent about $145 million and received approval to spend $150 million in fiscal 1986. To bolster support for this effort, the DOD's Defense Advanced Research Projects Agency (DARPA) points to important advances in AI—expert systems with common sense and systems that can understand natural language. However, no such advances have occurred.

Likewise, computers are no more able today to deal intelligently with "uncertain data" than they were a few years ago when our computerized ballistic-missile warning system interpreted radar reflections from a rising moon as an enemy attack. In a report evaluating the SCP, the congressional Office of Technology Assessment cautioned, "Unlike the Manhattan Project or the Manned Moon Landing Mission, which were principally engineering problems, the success of the DARPA program requires basic scientific breakthroughs, neither the timing nor the nature of which can be predicted."

Even if the Defense Department invests billions of dollars in AI, there is almost no likelihood that this state of affairs will change. Yet once vast sums of money have been spent, there will be a great temptation to install questionable AI-based technologies in a variety of critical areas—from battle management to "data reduction" (figuring out what is really going on given noisy, contradictory data).

Military commanders now respond to a battlefield situation using common sense, experience, and whatever data are available. The frightening prospect of a fully computerized and autonomous defense system is that the expert's ability to use intuition will be replaced by merely competent decision-making. In a crisis, competence is just not good enough.

Furthermore, to justify its expenditures to the public, the military may feel compelled to encourage the civilian sector to adopt similar technologies. Full automation of air-traffic control systems and of skilled factory labor are both real possibilities.

Unless illusions concerning AI are dispelled, we are risking a future in which computers make crucial military and civilian decisions that are best left to human judgment. Knowledgeable

AI practitioners have learned from bitter experience that the development of fully autonomous war machines is unlikely. We hope that military decision-makers or the politicians who fund them will see the light and save US taxpayers' money by terminating this crash program before it is too late.

THE OTHER SIDE OF THE STORY

At this point the reader may reasonably ask: If computers used as logic machines cannot attain the skill level of expert human beings, and if the "Japanese challenge in fifth-generation systems" is a false gauntlet, then why doesn't the public know that? The answer is that AI researchers have a great deal at stake in making it appear that their science and its engineering offspring—expert systems—are on solid ground. They will do whatever is required to preserve this image.

When public television station KCSM in Silicon Valley wanted to do a program on AI to be aired nationally, Stanford AI expert John McCarthy was happy to take part. So was a representative of IntelliCorp, a company making expert systems that wished to air a promotional film. KCSM also invited one of us, Hubert, to provide a balanced perspective. After much negotiating, an evening was finally agreed upon for taping the discussion.

That evening the producer and technicians were standing by at the studio and Hubert had already arrived in San Mateo when word came that McCarthy would not show up because Hubert was to be on the program. A fourth participant, expert-systems researcher Michael Genesereth of Stanford University, also backed out.

All of us were stunned. Representatives from public TV's NOVA science series and CBS news had already interviewed Hubert about AI, and he had recently appeared on a panel with Minsky, Papert, philosopher John Searle of Berkeley, and McCarthy himself at a meeting sponsored by the New York Academy of Sciences. Why not on KCSM? It seems the "experts" wanted to give the impression that they represented a successful science with marketable products and didn't want to answer any potentially embarrassing questions.

The shock tactic worked. The station's executive producer, Stewart Cheifet, rescheduled the taping with McCarthy as well as the demo from IntelliCorp, and he decided to drop the discussion with Hubert. The viewers were left with the impression that AI is a solid, ongoing science which, like physics, is hard at work solving its quite manageable current problems. The public's chance to hear both sides was lost and the myth of steady progress in AI was maintained. The real story remained to be told, and that is what we have tried to do here.

26

Interactional Expertise and Embodiment

Evan Selinger, Hubert Dreyfus, and Harry Collins

Conversational Computers and Interactional Experts: Collins on Human Embodiment

Evan Selinger

INTRODUCTION

In an extensive body of work, Harry Collins applies sociological tools to a problem that he calls "linguistic socialization." Because complete linguistic socialization is said to lead to fluency in a language according to the standards of a given discursive community, it is central to debates about artificial intelligence (AI) and expertise. While Collins is not the only theorist to contribute to these issues, he pursues a sustained research program on how much knowledge language alone can convey.

With respect to AI, there is a debate over whether computers will ever be able to pass the Turing Test. Will machines ever have human-like fluency in natural languages and how might this come about? As Collins contends, for computers to succeed they will have to acquire not only information, but also tacit knowledge. Without tacit knowledge, semantic and referential problems will be insurmountable (1990).

Collin's concept of "interactional expertise" stipulates that one can acquire all of the linguistic understanding of a domain by immersing oneself in the language of the domain without actually engaging in its practices (2004a). It thus gives a positive answer to the question of how much understanding sociologists, activists, and journalists can obtain about a field through the medium of conversation alone. Not everyone agrees that it is possible to learn to talk competently about aspects of a technical field (for example, pass on technical information, assume a sound devil's advocate position on a technical matter, and even make authoritative judgments on a peer review committee) solely by immersing oneself in the talk of the field's "contributory experts"—that is, with people who are physically immersed in the field and who are capable of advancing it.[1] The purpose of this

Evan Selinger, Hubert Dreyfus, and Harry Collins, "Interactional Expertise and Embodiment," *Studies In History and Philosophy of Science Part A* 38, no. 4 (December 2007): 722–40. Copyright © 2007 Elsevier Ltd. Reprinted by permission.

essay is to demonstrate that while Collins furthers philosophical and sociological discussions on how knowledge and language relate, his overall position remains predicted upon a misunderstanding of the phenomenology of embodiment. By omitting developmental consideration of how humans develop linguistic competence or skill itself, Collins misrepresents how knowledge is acquired as well as what kinds of people expert knowers truly are. On the issue of AI, Collins misinterprets the empirical evidence that he introduces for the purpose of supporting his anti-phenomenological claims.

Before establishing these critical points, it will be useful to begin with an overview of Collins's position. For purposes of exposition it is useful to examine how Collins contrasts his position with the phenomenological perspective exemplified by Hubert Dreyfus. I will then present a phenomenological position on embodiment and explain why several of Collins's claims are invalid.[2]

To avoid potential confusion for the reader, some caveats are in order. Although I will be discussing views on the Turing Test throughout the essay, my aim is not to speculate on whether or not machines will ever pass a generalized test of this kind—that is a test that is not domain-restricted. Instead, the analysis that follows is deflationary. I intend to demonstrate that Collins does not illuminate the topics of AI and interactional expertise in the manner that he thinks he does. While I will address how humans typically develop their cognitive architecture I will not discuss whether it is possible for that architecture (or some analogue to it) to be instantiated in a computer. Indeed, I am agnostic about whether a "minimally embodied computer" (Collins's term) could one day produce behavior indistinguishable from successful linguistic socialization. My objective is simply to establish the reasons why the data Collins has gathered does not bear on this question either.

Put in more positive terms, by undermining Collins's examples, I will reintroduce the possibility that human embodiment may be important not only for understanding how humans actually accomplish linguistic socialization but also for figuring out how we might design AI. This is not to say that human embodiment is necessary for such socialization but rather that it is worth more consideration than Collins gives it.

COLLINS ON COMPUTERS AND LINGUISTIC SOCIALIZATION: THE BASIC POSITION

Inquiry into whether computers can be intelligent runs the risk of essentializing the notion of intelligence. This is because there is no single metric for intelligence. Not only is human performance routinely assessed according to different scales that measure different types of intelligences, but it may be reasonable to deem computers intelligent (in some respects) by judging them according to computation standards of performance, or else by assessing the kinds of results they can produce when they collaborate with human beings. Nevertheless, an optimally designed Turing Test demonstrates how much of an understanding of human social life, as revealed by linguistic competence, computers can imitate, or, perhaps, acquire. The question that Collins's notion of interactional expertise raises for the phenomenologist is whether computers will have to possess human or human-like bodies to acquire enough linguistic competence to pass the Turing Test.

Collins states that artificial intelligences do not need bodies even though he also states that they will not be constructed in the foreseeable future because we do not know how to socialize them. Linguistic socialization, he argues, is not crucially dependent on the body. From his perspective, phenomenologists' claims about the significance of embodiment are overstatements (1990, 1992, 1996, 2000, 2004a, b; Collins and Kusch, 1998; Collins et al., 2006). Collins insists that, given the means to communicate and appropriate programming, an immobile box could acquire the "common sense" required to discuss anything pertaining to a typical human form of life and pass the Turing Test. It would somehow have to be connected to the social "form of life" in which the discourse was spoken but this would not require more than a minimal body. Collins agrees that we do not have any idea what the proper programming would consist of nor do we know how to create the means of communication. The crucial point is that the shape or mobility of the body would not be the limiting factor.

To clarify this proposition, Collins sets out two interrelated views on embodiment: the "social embodiment thesis" and the "minimal embodiment thesis" (2000). According to the social embodiment thesis, any particular language that develops can only be completely understood in terms of the bodies of the agents in that culture. For example, human culinary practices cannot be exhaustively analyzed without discussing biology (for example, the need for organic life to acquire nutrition) and human physiology (for example, the contours and sensory dimensions of our hands, mouths, and tongues, our olfactory proclivities, and so on). While this seems to argue for the importance of the body, in Collins's treatment the body is important only at the collective level. Turning to individuals, he claims that it is possible for someone who lacks the type of embodiment that is prevalent in a given society (and which has given rise to the prevalent language of that society) to be linguistically socialized as a member of that society. This is the minimal embodiment thesis— minimal embodiment being just enough embodiment to engage in successful conversation.

Collins concedes, then, that linguistic socialization cannot occur in the absence of a being receiving aural inputs and emitting linguistic outputs. A socializable agent must possess ears (or their equivalent), a larynx (or its equivalent), and enough of the human-like brain—or computational analogue of it—to decode what comes through the ears and reproduce the outputs of the human-like larynx. Even an immobile box would need these features to become linguistically socialized into our world.

Given the diverse forms of embodiment that can fulfill such minimal conditions, Collins does not limit his views on linguistic socialization to the future potential of intelligent computers. "In principle," Collins writes, "if one could find a lion cub that had the potential to have conversations, one could bring it up in human society to speak about chairs as we do in spite of its funny legs" (Collins, 1996, p. 104).

THE COLLINS-DREYFUS DEBATE: COLLINS'S PERSPECTIVE

Collins characterizes his account of linguistic socialization as a position that stands in contrast to the phenomenological critique of artificial intelligence. He takes Hubert Dreyfus as his example of this phenomenological critique as represented by works such as *What Computers Can't Do*. According to Collins's interpretation, Dreyfus advances the following claim: Until computers become embodied like humans, they will, in principle, be unable to make sense of the perspectives that humans take when perceiving, acting, and judging: lacking human perspectives, they will be incapable of passing the Turing Test.

In contrast, Collins believes that (the right kind of) talking computers and the (right kind) of talking lions would be examples of minimally embodied beings who are, in principle, capable of passing the Turing Test. No human topic would be off-limits to them, not even discussions of intimate human experiences, such as love.[3] Collins believes, furthermore, that it is possible to refute Dreyfus by reference to already existing empirical data concerning successful instances of minimally embodied persons becoming linguistically socialized. In this context, he describes three cases—a severely handicapped woman named Madeleine, colorblind people, and a sociologist (himself) trying to master the discourse of gravitational wave physics.

The case of Madeleine

In *The Man Who Mistook His Wife for a Hat,* Oliver Sacks describes one of his patients, Madeleine, as a "congenitally blind woman with cerebral palsy," who, for most of her life, experienced her hands as "useless godforsaken lumps" (1998, p. 59). Douglas Lenat, a computer scientist, appropriates this description as an empirical counter-example that putatively disproves Dreyfus's position on the cognitive importance of human embodiment. From Lenat's perspective, because it was possible for Madeleine to acquire commonsense knowledge from books that were read to her, she is living proof that human embodiment is not decisive for learning natural language.[4] Indeed, despite her severe limitations, Sacks describes Madeleine as an engaging conversationalist.

Dreyfus dismisses Lenat's argument on the grounds that Lenat's disembodied characterization of Madeleine is a distortion of the type of person she is. Dreyfus insists that although Madeleine is disabled, she still shares many of core features of phenomenological embodiment that able-bodied people experience:

> She has feelings, both physical and emotional, and a body that has an inside and an outside and can be moved around in the world. Thus, she can empathize with others and to some extent share the skillful way they encounter the world. (1992, p. xx)

Collins sides with Lenat on this issue—though not on the matter of whether Lenat's favored method of instilling language in a computer would work—and he extends the conversation by providing reasons why Dreyfus's reply should be understood as an inconsistent position. According to Collins, the only thing that Dreyfus's response establishes is that Dreyfus himself actually fails to grasp what embodiment is—that is, whether embodiment is something "physical" or whether it is something "conceptual":

> But under this argument [i.e., the Madeleine example] a body is not so much a physical thing as a conceptual structure. If you can have a body unlike the norm and as unable to use tools, chairs, blind persons' canes and so forth as Madeleine's, yet still have common sense knowledge, then something like today's computers . . . might also acquire common sense given the right programming. It is no longer necessary for machines to move around the world like robots in order to be aware of their situation and exhibit "intelligence." (1996, p. 104)

As Collins sees it, not only is Madeleine mostly a "brain" endowed with "some sensory inputs," but, and also contrary to Dreyfus, either no body or not much of a body is required in order to experience empathy and imagination (2000, p. 188; 2004a, p. 125). Given this depiction, Collins treats the case of Madeleine as empirical proof that a minimally embodied being can become socialized by means of language alone.

The cases of colorblindness and gravitational wave physics

Because Sacks's account comes to us as a secondhand testimony for a popular audience, Collins attempts to gather firsthand empirical evidence from his own experiments. Since he cannot find any more "Madeleines," he tries to make the point of principle in the case of people with lesser deficiencies. In one case, Collins attempts to determine if colorblind people are capable of passing a Turing Test in which conversation focuses on color discourse. According to his recent published study, it turns out that they can (Collins et al., 2006).

The results that Collins obtains seem to confirm his views that bodily ability is not a necessary condition for linguistic ability; the colorblind can talk about color as fluently as color perceivers. On the face of it, it seems that colorblind people acquire their ability to speak as though there were color in their world from talk within the society of color perceiving people; they do not seem to learn much from direct experience with color sensations.

Since this study revolves around human subjects, it is worth putting emphasis upon the parallels that Collins intends to draw between it and the prospects for AI. According to Collins, if colorblind people can pass a Turing Test on color discourse even though their visual systems do not process all of the colors that they can talk about, then additional empirical evidence has been obtained that establishes the possibility that, someday, linguistically socialized computers might be capable of passing a Turing Test on color discourse (or some other perceptual discourse), even if these computers cannot perceive color (or experience some other perception) in the manner that color-perceiving people (or some other group of human perceivers) can.

Collins's third case arises from the fact that despite being trained as a sociologist (and not a natural scientist), he successfully developed such an extensive understanding of gravitational wave physics that, under Turing Test conditions, practicing gravitational wave physicists were unable to distinguish between him and their colleague practicing gravitational wave physicists (ibid.). Collins takes this discursive success as confirmation that he has acquired considerable interactional expertise in gravitational wave physics. In other words, Collins contends that just as colorblind people learn to converse about colors (in the absence of being immersed in the practice of color perception), and just as Madeleine learned to converse about a range of human affairs (in the absence of being immersed in their practices). So too did he learn to converse about gravitational wave physics. All of these accomplishments are possible, according to Collins, because linguistic socialization alone can convey complete linguistic fluency.

This view on linguistic socialization appears, then, to be a radical departure from the philosophical position represented by Dreyfus. Dreyfus argues that one cannot become linguistically socialized into an expert practice without being a contributing expert practitioner oneself. For Dreyfus, in order to be capable of saying everything that can be said about a phenomenon, domain, or experience, one must immerse oneself fully in the corresponding physical activities. Using the practice of surgery as a paradigm case, Dreyfus writes:

> There is surely a way that two expert surgeons can use language to point out important aspects of a situation to each other during a delicate operation. Such authentic language would presuppose a shared background understanding and only make sense to experts currently involved in a shared situation. (2000a, p. 308)

On the Dreyfusian view, only a surgeon can have the appropriate background understanding to take part in the full range of surgical discourse—a discourse that is said to include special "authentic" conversational terms and norms. Even though Dreyfus's quote refers to a cooperation performed operation and not to a Turning Test, use of the qualifier "only" places austere restrictions on the type of people who qualify as candidates for linguistic socialization. To claim that "only . . . [the] experts currently involved in a shared situation" can appreciate the full range of contributory discourse, is to restrict the class of "authentic language" users to the set of people who have a "shared background understanding"—an understanding that, in the Dreyfusian view, can only be acquired if there is physical activity in a "shared situation." Given these restrictions, medical journalists, medical sociologists, computers programmed with social medical software, and talking lions that lack opposable human thumbs would be unable to acquire the full linguistic proficiency in surgical discourse of practicing surgeons.[5]

On this last issue, the empirical evidence on interactional expertise that Collins introduces suggests that his position on what someone can say in the absence of direct experience is tenable. Based on the colorblind and gravitational wave physics experiments, it does not appear to be necessary to have first-person experience of a phenomenon in order to be capable of saying everything that humans can say about phenomenon. Of course, one could argue that it is a mistake to draw inferences about experience in general based solely on the limited domains that Collins examines. But since Collins shifts the issue from a speculative to an empirical matter, the burden of proof falls on those who hold a Dreyfus-like position.

Having agreed with Collins on this aspect of conversational practice, I find stronger claims for the power of linguistic socialization made by Collins untenable. By adhering to the minimal embodiment thesis, Collins fails to make crucial distinctions between experience and embodiment. I claim that it is largely because of embodied learning and embodied perception that humans can learn to speak about things they have not experienced and that the fluency acquired by Madeleine, by the colorblind, and by Collins in gravitational wave physics depends crucially on embodiment.

PHENOMENOLOGICAL EMBODIMENT

Two interconnected reasons can be identified that explain why Collins's account of linguistic socialization is flawed.

- By misunderstanding the phenomenological position on embodiment, Collins misrepresents the views of an established tradition.
- By failing to provide a comprehensive description of bodily activity in the relevant empirical cases, Collins misinterprets the extant data on linguistic socialization.
- If Collins better understood phenomenology, then he would be less inclined to misdescribe how Madeleine and colorblind people become skillful conversationalists. He also would be less inclined to draw analogical connections between Madeleine, the colorblind, himself, and the potential computers of the future.

In order to establish these points, it will be helpful to proceed by considering some of the reasons why Collins might fail to grasp the type of analysis that phenomenologists like Dreyfus provide.

Collins's point is not dissimilar to that of Dutch philosopher Philip Brey who writes:

> Human beings can have limbs and organs amputated or paralyzed and still not lose their ability to engage in abstract thought, and it is at least theoretically possible that, as sometimes depicted in science fiction stories, a brain could be removed from a body and kept the ability to think. (2001, p. 51)

Moreover, if we add technological considerations to this list of examples, Collins's case appears to be strengthened yet further. Consider the following example discussed in Nicholas Humphrey's *Seeing Red:*

> An apparatus, called vOICe, has recently been developed for helping blind people to see using their ears rather than their eyes. The subject wears a helmet with a video camera mounted on it, coupled to a light-to-sound translation program, with some headphones to receive the sound images. The device has the potential to map visual scenes to "soundscapes" in an analog way. Future versions will likely code color as a continuous extra dimension of the soundscape. However, as of now, when it comes to "seeing color," the device takes a short-cut and says the word RED! As the user's manual explains, when you activate a color identification button, "the talking color probe speaks the color detected in the center of the camera view. Now you know whether the apple you are about to eat is yellow, green or red, and you can check the dominant colors of your clothing." (2006, p. 23)

Given these considerations, Collins might appear justified in pressing his question: What are the minimum body parts that someone must have to become linguistically socialized? Similarly, Collins might also appear justified in being perplexed over Dreyfus's claim that having a human "front and back" helped Madeleine to become linguistically socialized. After all, giving a computer a human front and back will not turn a linguistically unsocializable computer into one that can be socialized.

How might a philosopher like Dreyfus respond to these charges? One reply would entail demonstrating that Collins misinterprets the data on Madeleine, the colorblind, and his own gravitational wave physics experiences. Such reinterpretation would emphasize the first-person perspective of the "lived body," and it would center on the two core issues that Brey associates with Dreyfus's phenomenology (2001, pp. 50–51).[6]

> Human perception and sensorimotor intelligence are "not localized in the brain," but instead are distributed through "a complex feedback system that comprises the nervous system, senses, the glands, and the muscles."
>   Humans develop abstract intelligence through sensorimotor activities, including activities that are conducted during the periods ranging from infancy through childhood. In order to comprehend these activities fully, the emotional and motivational processes that direct human action must be taken into account.

In what follows, I will provide my own phenomenological interpretation of the data that Collins takes to be relevant to his account. Although I do not wish to claim that my analysis is equivalent to the perspective that Dreyfus himself would advance, were he to attend closely to these matters, I nevertheless will emphasize the two considerations that Brey highlights.

One of the main points that I intend to establish is that while losing, or being born without, certain physical abilities clearly (and sometimes dramatically) influences how an embodied human agent relates to the world, situations of bodily debilitation, bodily depravation, and bodily diminishment tend to be situations in which more perceptual activity of an analogically related sort occurs than Collins acknowledges.

Since I do not direct access to the data at issue, I will be reconstructing situations pertaining to perceptual experience and the acquisition of skill. Admittedly, such reconstruction is not scientific. Because it is not scientific, Collins may be inclined to dismiss my claims as *ad hoc* hypotheses that are postulated to preserve phenomenological assumptions. The burden of proof, however, should be seen the other way round. The reconstructions that I will be providing accord with common developmental experiences. If I make claims that Collins disagrees with, it is also his responsibility to demonstrate where they fall short. He would finally need to present a developmental account of language acquisition, and, possibly, concept-formation. As the next section will clarify, given the possible acts of compensation that Madeleine and colorblind people engage in, Collins might even benefit from consulting recent debates about "sensory substitution," that is, "the possibility of substituting one kind of sensory input for another" (2006, p. 54).

A PHENOMENOLOGICAL INTERPRETATION
OF COLLINS'S DATA

Madeleine revised

Let's begin with Madeleine. Given her disabilities, she is a harder case than the others to discuss in embodied terms. By comparison, the colorblind and gravitational wave physics experiment are minor examples. From a phenomenological perspective, some of the relevant aspects of Madeleine's situation can be reconstructed as follows.

Madeleine's basic perceptual experiences are distributed throughout neural-somatic networks. Since Madeleine is blind and has cerebral palsy, her eyes, hands, and legs may contribute less to perceptual experiences than the eyes, hands, and legs of able-bodied people. In this respect, Madeleine is differently embodied than able-bodied people. Nevertheless, it is inappropriate to take up the standard of the minimal embodiment thesis and claim that Madeleine received "inputs" from the world and communicates "outputs" back to it simply by possessing a human brain connected to a few discrete body parts. One way to see why Madeleine's experiences are irreducible to the conditions stipulated in the minimal embodiment thesis is to phenomenologically revisit the experience of empathy.

Whereas Collins claims that the human experience of empathy occurs in the brain, a phenomenologist would see the human experience of empathy as distributed throughout the whole human body—that is, as dispersed through coordinated neural and somatic activity. The way that we experience empathy is not unique; empathy has a complex and distributed structure because affective human experiences in general are like this. In "Empathy and consciousness" Evan Thompson writes:

> Affect has numerous dimensions that bind together virtually every aspect of the organism—the psychosomatic networks of the nervous system, immune system, and endocrine system; physiological changes in the autonomic nervous system; the limbic system, and the superior cortex; facial motor changes and global differential motor readiness for withdrawal; subjective experiences along a pleasure-displeasure valence; social signaling and coupling; and conscious evaluation and assessment. (2001, p. 4)

Thompson's account of affect accords with first-person human experience and findings in biology and the neurosciences. In this respect, speculation is not required to establish that Madeleine's experiences of affect are experiences that she feels throughout her body. Madeleine may be handicapped, but she does not live in a human body that is so physically dissimilar to other human bodies that she experiences affect in an anomalous, non-human way. If Collins believes otherwise, he owes us an account that justifies this conviction. At present, Collins merely asserts that empathy is an "embrained" human experience; no supporting evidence is provided.

To clarify this point further, it will be useful to use some simple reconstructive phenomenology. In this context, a developmental issue needs to be clarified. Since empathy entails identifying with other people's feelings and motives, it is important to know how Madeleine experiences her own feelings and motives. Only the outline of an answer to this question needs to be provided to show where Collins errs; debates about the "simulation theory of mind" and the "theory theory of mind" need not be entertained.

For starters, it seems clear that if Madeleine is afraid, she does not relegate this experience to brain activity alone. Depending on the severity of the fright, Madeleine would have first-person awareness of her respiratory activity and heart rate accelerating as involuntary responses, and she would feel her lips quiver.[7] She might not feel her hands tremble in the same way as able-bodied people do in similar situations, but even if this is so, it simply means that her phenomenological experience of fear occurs through a less replete natural-somatic network than might be operative in others. Again, this is a comparative assessment in which different humans are juxtaposed; Madeleine remains more embodied than the parameters of the minimal embodiment thesis suggest. To elaborate further, when Madeleine is angry, she likely has first-person awareness of her facial muscles shifting, even though able-bodied people might have the same experience and also be aware of their posture shifting. Here, too, the only reductive comment about Madeleine's embodiment that can be made is comparative in nature.

Ultimately, Madeleine has such a rich experience of affect because she shares in our human evolutionary history. Within these parameters, the human organism has developed rapid means for responding to real and potential predators. Consequently, Madeleine and the rest of us become aware of affect, in part, through involuntary biological processes that are distributed throughout our bodies. This is not to say that social input is an insignificant index for people becoming aware of their emotions. To be sure, we may not be fully aware that we are afraid until someone asks: why are you so nervous? But, again, conversational input is merely one component of how people can learn of their emotional states. The critical points are: (1) human brains that are augmented only by the minimal input/output receptors detailed in the minimal embodiment thesis would find it difficult, if not impossible, to perceive an affect like fear in time to respond to it appropriately. John Mix and I articulated this outlook in an earlier article when we wrote:

> Depending on the intensity of the experience, fear can affect the human body in many different ways; alertness increases, the pupils widen in order to let in more light, the adrenal glands begin to pump more adrenaline and other hormones into the bloodstream, the heart races, the muscles tense, the blood pressure rises, digestion slows, the liver converts starches to sugar to generate more energy, and sweat production increases, sometimes leaving the hair of our bodies standing on end. (2006, p. 310)

What conclusions follow from this phenomenological discussion of affect? Clearly, the analysis does not tell us anything in principle about whether computers will even learn to talk about human emotions if they lack human-like bodies. But, this was not the point of the discussion. The analysis does show that there is at least one dimension of human experience—affect—in which Madeleine and the rest of us de facto experience sensations throughout our entire bodies, even though inputs from the brain are casually relevant.

Does this point have any purchase on the issue of linguistic socialization? Yes. If Madeleine can tell stories in which she displays an ability to create scenarios in which she empathizes with her characters, or if she provides an account of someone else's experience in an empathetic manner, it would seem that she is only capable of doing so because she had already developed a first-person understanding of what affect is—an understanding that came about through deeply embodied processes. Likewise, when Madeleine converses about experiences where an object takes on an emotional symbolic significance, perhaps an instance where two people fight over who gets to sit at the head of the table, she also seems to be drawing from a frame of reference that she developed through first-person affective experience. This is not to say that Madeleine needs to have experienced a particular emotion in order to discuss it. If Madeleine has never been in love, then, as Collins would assert, she could still speak about it, and maybe even compose moving poetry on the subject. But if Madeleine did not personally experience a range of emotions first hand, it is hard to imagine how she could come to talk about a range of emotions in a skillful way that would pass a Turing Test. In other words, Madeleine knows how important her wheelchair is to her, and she can extrapolate from that experience to talk about why other people would be emotionally invested in sitting at the head of the table. If Madeleine was never in love, she could still draw from associated experiences, pleasures, pains, longings, and so on, that are common to her everyday openness toward the world. My main point, then, is that Collins fails to inquire into the complex development relations that obtain (in human experience) between embodied perception and higher level cognitive abilities, including language development and use. I am suggesting—as does Dreyfus as well as the linguist George Lakoff and philosopher Mark Johnson (1999)—that Collins should even consider the possibility that "abstract concepts and abstract logic ultimately can be reduced to concrete, sensorimotor structures" (Brey, 2001, p. 51). If Collins disagrees with this view, he owes us an alternative account of how Madeleine could make empathetic conversation in the absence of having similar firsthand affective experiences that she could extrapolate from. In earlier essays, I even provided reason to believe that Madeleine's basic understanding of time and space—fundamental categories for perceiving and talking about matter and motion in our world—derive from her embodied orientation to movement and perception (Selinger, 2003; Selinger and Mix, 2006). Collins has yet to show why this analysis is wrong.

Finally, we should not lose sight of the fact that even though Madeleine is blind, she can construct a body image. Madeleine knows all too well that people can—and, perhaps, often do—respond to her in ways that place emphasis upon her bodily limitations. Considering how much assistance her aids provide, Madeleine's body image is probably rooted in the cognitive and affective significance that she accords to "dependency" and "independence." In developing her own body image, Madeleine also enhances her framework for understanding others. She becomes aware of how humans establish identity by judging other people's bodies—people who, in turn, evaluate our own corporeality. This understanding—one that became theoretically popularized in Jean-Paul Sartre's existential phenomenology—is an integral component in her ability to make sense of other people's actions—and, ultimately, to talk about other people's actions in an intelligible and empathetic manner.

Collins could object to my position by claiming that I conflate correlation with causation. He could reply that Madeleine's first-person embodied experiences of affect are casually irrelevant to her ability to make empathetic conversation. While long-standing patterns and stages of human development—cognitive and perceptual—leave me skeptical of this possibility, I concede that my view is falsifiable if the right kind of empirical tests prove otherwise. However, none of Collins's remarks about empathy or linguistic socialization are compelling in this respect; no alternate developmental evidence is provided. I would thus invite him to find a case in which a person was deprived of all affective experiences and still succeeded in becoming a conversationalist who could pass a Turing Test on human affairs. To crystallize the stakes of this invitation, I will return to the issue at the essay's conclusion.

Before proceeding to the colorblind experiment, it is important to note that Collins is not alone in displaying insufficient sensitivity to the complexity of whole-body perception. As it turns out, Sacks himself makes some of the same errors that Collins does when titling Madeleine's story "Hands."[8] What I want to question, then, is whether Sacks's literary inclinations and choices influence how he tells Madeleine's story. Because Sacks wrote the essays compiled in *The Man Who Mistook His Wife for a Hat* for a general audience, one needs to interpret his claims carefully. Firstly, the studies that appear there contain significantly less scientific detail than articles that appear in medical journals. Secondly, Sacks does not try to advance scientific knowledge in that book; his aim is to get a large audience to be engaged with abnormal behavior.

Sacks is puzzled when he first examines the sixty-year-old Madeleine because she was unable to recognize objects placed in her hands even though the sensory capacities of her hands were "completely intact: she could immediately and correctly identify light touch, pain, temperature, passive movements of the fingers" (1998, p. 60). Sacks concludes that in order to get Madeleine to be capable of perceiving with her hands, he first needs to get her to discover her hands—that is, he needs to create the conditions under which Madeleine would initiate using her fingers to explore and perceive her environment. In other words, Sacks wants to find a way to get Madeleine to experience an "impulse" to use her hands interrogatively (ibid., p. 60). Sacks characterizes this event as one that marks Madeleine's "birth as a complete 'perceptual individual'" (ibid., p. 61).

Sacks's solution is to change Madeleine's feeding conditions. He reasons that since Madeleine has been taken care of her entire life, she has not been given an incentive to use her hands. After instructing the nurses to move Madeleine's food "slightly out of reach on occasion," a hungry Madeleine commits "her first manual act" by reaching out and grabbing a bagel (ibid., p. 61). Because Madeleine actively grabs something with her hands for the first time, Sacks emphasizes how she comes to recognize objects:

> A bagel was recognized as round bread, with a hole in it; a fork as an elongated flat object with several sharp tines. But then this preliminary analysis gave way to an immediate intuition, and objects were instantly recognized as themselves, as immediately familiar in character and "physiognomy," were immediately recognized as unique, as "old friends." And this sort of recognition, not analytic but synthetic and immediate, went with vivid delight, and a sense that she was discovering a world full of enhancement and beauty. (Ibid.)

Much of Sacks's description here accords with a phenomenological orientation. Like phenomenologists, Sacks describes human perception as a skill that is developed through motivated action. Like phenomenologists, Sacks links the human impulse to develop skill as laden with affect, that is. Madeleine's impulse is initiated by frustration and it continues by being fueled with delight. Finally, like phenomenologists, Sacks describes Madeleine's perception of objects as an intuitive, synthetic, and immediate act. Dreyfus constantly makes these same points, and like Sacks he frames these observations as contrasting with "analytic" conceptions of perception.

What, then, do I find objectionable in Sack's account? Sacks seems to suggest that the reason why Madeleine could recognize a bagel as a bagel is because she had already developed an understanding of what a bagel is based upon stories that were told to her. Sacks writes:

> Had she not been of exceptional intelligence and literacy, with an imagination filled and sustained so to speak, by the images of others, images conveyed by languages, by the word, she might have remained almost as helpless as a baby. (Ibid., p. 62)

In contrast to the linguistic images that Madeleine had obtained, Sacks notes that she lacked the simplest internal tactile images to draw from (ibid.). The implication of Sacks's discussion is that Madeleine first acquired an understanding of bagels from conversations about them, and then— once she finally touched a bagel with her hands—extrapolated from verbal understanding to tactile understanding. In other words, for Sacks the medium of conversation had already given Madeleine a conceptual understanding of "bagelhood"; all she needed to do was apply this understanding in a tactile manner by correctly identifying a concrete object as a bagel.

By what justifies this conclusion? To be sure, Madeleine's linguistic instruction certainly played a significant role in her ability to develop a new skill. I agree with Sacks's view that without already possessing images conveyed by words. Madeleine would, in all likelihood, have been unable to develop the requisite perceptual recognition via her hands as quickly as she did. But why assume that because Madeleine could not initially recognize objects by touching them that she, therefore, failed to compensate, as best she could, for her diminished perceptual capacity by taking advantage of as much perceptual information as she could avail herself of? By Sacks's own description, the sensory capacities of Madeleine's hands were "completely intact." Here, we should recall that the experience of touch is neither localized to the hands nor to the brain; it is human experience that is distributed throughout her entire body. For all human beings, touch occurs wherever the skin's nerve ending transmit sensations to the brain, and it can even be affected by the presence of hair. Moreover, human skin turns out to be much more biologically complex than anything stipulated in the minimal embodiment thesis. According to the Texas Education Agency (April 2004), each square inch of skin contains approximately:

78 nerves,
650 sweat glands,
19–20 blood vessels,
78 sensory apparatuses to detect heat,
13 sensory apparatuses to detect cold,
1,300 nerve endings to record pain,
19,500 sensory cells at the end of nerve fibers,
160–165 pressure apparatuses for the sense of touch,
95–100 sebaceous glands,
65 hairs and hair muscles,
19,500,000 cells.

Given these considerations, while I can agree with Sacks that prior to developing the motivating impulse, Madeleine could not identify a bagel if one were put in her hand, I take issue with the (assumed converse) idea that her only understanding of what a bagel is came from language. If Madeleine had eaten a bagel before, she clearly took perceptual stock of it through her mouth, gums, tongue, and lips; the resistance it offered to her teeth was significant, as was the sound it made while being chewed and the sensations it provided her nose while being smelled.[9] From a developmental perspective, it is important to note a few things. Hearing is a more mature sense at birth than vision, and oral and olfactory exploration play a crucial role for babies; while the tongue and gums are vital for exploring surroundings, it may even be the case that newborns can recognize their mothers by scent. Our lips, moreover, are extraordinarily receptive surfaces. Some people have even theorized that humans often express intimacy by kissing each other on the lips because of how sensitive the surface is. Thus, while Madeleine may have had, comparatively speaking, a diminished perceptual sense of what a bagel is because she could not see one through her eyes or take in sophisticated information about it through her hands, she nevertheless could develop rich sensory awareness of what a bagel is through the incarnate processes of exploring and eating one. And, for all we know, it may even be a mistake to characterize Madeleine's sensory awareness as diminished. Certainly, her awareness is different from the experiences of able-bodied people. However, neither Collins nor Sacks informs us as to whether Madeleine has developed heightened perceptual sensitivities as compensation—or, a kind of "sensory substitution"—for her physical impairments. Alva Noë and some colleagues even suggest that auditory awareness can be experienced qualitatively as visual perception: "For example, a woman wearing a visual-to-auditory substitution device will explicitly describe herself as seeing through it" (Humphrey, 2006, p. 57).

In addition to the perceptual information that Madeleine could take in about objects in general, we need to remember that with respect to the example under review she could tacitly compare the experience of eating a bagel to eating other foods. And, if she was given a bagel that was sliced into digestible pieces, she could ask whoever was feeding her to describe what the bagel was like before it was cut up. In this context, her understanding of what a bagel is seems to develop through discursive and embodied means. Moreover, we need to remember that the first two objects that Madeleine identifies, a bagel and a fork, are objects that are presented to her in a specific context: Madeleine knows that the objects in front of her are objects that she would encounter at mealtime. Meals are significant to Madeleine because she is aware of how uncomfortable it is to be hungry, and the relevant class of objects that she is presented with during mealtime is quite limited in comparison to the entire class of objects that exist. By not reflecting on the context in which Madeleine first recognizes an object after touching it, Sacks conveys the misleading impression that Madeleine is doing something akin to playing a children's game where completely random objects are placed in a box and kids compete over who can identify them through touch.

Since Sacks does not describe the contexts in which Madeleine comes to identify other objects correctly, and since he does not discuss the relevance of Madeleine's ability to take in diminished information about objects through her intact and compromised senses and skin, he fails to establish that her perceptual inferences are rooted exclusively in extrapolations from internal images that were formed through conversation. Were Sacks to have provided a more nuanced view of how Madeleine's perceptual awareness is distributed throughout her body, it would have been more difficult to endorse the dramatic chapter title, "Hands." In the spirit of Collins's minimal embodiment thesis, in this one story Sacks conveys the misleading impression that the cognitively relevant human sense of touch can be localized to one body part, the hands.

Colorblindness and gravitational wave physics revisited

In the last section, I emphasized that by focusing on Madeleine as an adult, Collins and Sacks both failed to inquire into the relevance of her embodiment for her development as a skillful speaker. My main point about colorblindness is merely an elaboration of this phenomenological trajectory.

When Collins describes his colorblind experiment, he notes that the colorblind can learn to talk about colors they never perceive directly. While I am not skeptical about this display of skill, I take issue with Collins's view that the experimental data established that bodily ability is not a necessary condition for linguistic ability. Collins draws the wrong conclusions not because he calibrates his controlled experiment in terms of four subjects, that is, colorblind, color-perceiving, pitch-blind, and pitch-perceiving subjects. Given his assumptions, this configuration is quite ingenious. Collins errs for other reasons—reasons that he himself cities in the Appendix.

In the Second Appendix to the colorblind paper, Collins writes:

> Finally, colorblindness and perfect pitch have been discussed as though they were binary qualities—one either has them or one does not. But they each come in different types and each type lies on a continuum. To get a strong effect in the experiment participants need to be located toward the extreme end of the spectrum.

Unfortunately, this point is too important to list in an appendix as merely a "confounding factor." As Collins notes:

> The answers given by such a minimally colorblind person are likely to be indistinguishable from that of a color-perceiver under any test; likewise, a judge who is only marginally colorblind in this way does not really possess colorblindness as a "target expertise."

Based upon these considerations, Collins errs in conceiving of the current colorblind study as research that provides data which validates the minimal embodiment thesis. As currently configured, the test does not tell us anything about the analogical and inferential capacities of the participants. No measures are taken to determine how much of the participants' ability to talk about color can be linked to their ability to extrapolate from the colors that they are capable of perceiving. Moreover, no measures are taken to determine how much of the participants' ability to talk about color can be linked to their ability to extrapolate from conceptual frames of reference that were acquired through embodied forms of interaction and perception during early developmental stages.

Furthermore, even if the participants could only see in black and white, Collins would still need to investigate whether their ability to talk about colors arose developmentally through physical activity and distributed perceptual experience. Were he to do so, I bet he would discover compelling evidence suggesting that sensorimotor skills and distributed perception proved crucial to their ability to form the concept of color. It would not be surprising if it turned out to be the case that before learning what color is, babies use emotionally motivated physical exploration and sensory information to learn how to distinguish themselves from objects. By the time that babies are taught that color is an attribute, they already have learned to use physical exploration and sensory information to develop a rich understanding of their world. Even a completely blind person like Madeleine can learn crucial lessons about what color is by being instructed through contrasting perceptual exercises, that is, by being taught that while color is not something one tastes, smells, feels, it nevertheless resembles these qualities in that it is a perceptually descriptive aspect of phenomena. In this context, we should add that while the vOICe technology mentioned in the previous section might be capable of conveying what colors it detects, it cannot teach anyone what color is.

Ultimately, there are many perceptual skills that humans develop when learning to talk about primary colors. We have to learn to ignore relative motion, learn that there is a sense in which the color of an object remains constant even when modifications are made to that object (for example a red apple remains red even as it transitions from being uneaten to being partially eaten), learn that there is a sense in which color is separate from other characteristics (that is, size, function, and so on), learn that color has a normative dimension (that is, learn that since lighting can influence how color is perceived, it is sometimes necessary to move around until ideal lighting conditions are obtained), learn that imaging technologies can change the appearance of color, and so on. Again, I would contend that much—if not at all—of this learning typically occurs tacitly through physical interactions with objects and through perceptual experiences that are distributed throughout the socialized body. If I am wrong on this point, Collins should provide an alternative developmental account. To assist Collins in this process, I invite him to answer the following questions:

- What activities did Madeleine likely engage in when she first learned how to recognize herself as a being who is separate from the objects she encounters?
- Did any of these activities involve physical interaction with herself, other people, or objects? If so, what kinds of physical interaction took place? And, what was it like for Madeleine to experience such physical interactions? Were all of her experiences localized in her brain, or were some felt throughout her body (that is, through a complex feedback system that comprises the nervous system, senses, skin, glands, muscles, etc.)?
- How might learning to recognize herself as separate from the objects she encounters be a useful prerequisite for Madeleine learning to talk about color?
- What activities did Madeleine likely engage in when she learned to talk about color?
- Did any of these activities involve physical interaction with herself, other people, or objects? If so, what kinds of physical interaction took place? And, what was it like for Madeleine to experience such physical interactions? Were all of her experiences localized in her brain, or were some felt throughout her body (that is, through a complex feedback system that comprises the nervous system, senses, skin, glands, muscles, and so on)?
- Does the narrative provided when answering questions 1–5 fit within the parameters of the minimal embodiment thesis?

If Collins provides plausible replies, he should answer the last question in the negative.

Having said this, it should be clear that I am not making any in-principle claims about whether disembodied machines will ever learn to talk about primary colors in the way humans can. All that I am claiming is that the de facto practices humans participate in when they learn to talk about color are practices in which discursive training, embodied exploration, and embodied experience all prove to be crucial components. This modest claim, however, is sufficient to refute Collins's interpretation that the colorblind experiment disproves phenomenological claims about how humans use their bodies to become socialized into color discourse. Collins provides no reason to believe that any of the participants in the colorblind experiment learned to talk about color by participating in processes that differ from the ones I have discussed.

By refuting Collins's interpretation of embodiment in this experiment, I diminish substantively the amount of available empirical evidence that he can appeal to in order to validate his minimal embodiment thesis. If my claims in the last section were sufficient, then Collins is not entitled to appeal either to the Madeleine case or the colorblind experiment. His only remaining evidence is his analysis of his success at mastering the discourse of gravitational wave physics. But based on all the analysis that has transpired thus far, it is not necessary to discuss that case in detail. From a phenomenological point of view, we need to remember that Collins is an able-bodied sociologist who masters the discourse of gravitational wave physics. While it is remarkable that he can do so in the absence of experience performing gravitational wave experiments or writing theoretical papers, Collins never addresses the basic developmental question: if he did not already have a basic understanding of the world then how could he have developed an appropriate background understanding that would enable discussions about gravitational wave physics to be perceived as meaningful?

Put otherwise, comparing himself with Madeleine and colorblind people is an ill-conceived program, precisely because the comparison is predicated upon a misunderstanding of the conditions of embodiment that prevail. While Collins may lack direct experience performing gravitational wave physics experiments, he is not in any significant way physically challenged. By contrast, Madeleine and colorblind people lack physical capacities that able-bodied people possess. To inquire into the conversational capacities of Madeleine or colorblind people is thus to inquire into how they managed to overcome physical disabilities that limited how they interacted perceptually with the world. But to inquire into Collins's own conversational capacities is to inquire into how he managed to overcome the limits of his disciplinary training. In other words, Collins never asks: How, as an able-bodied person with an expert's knowledge in the sociology of science and technology, was he able to extrapolate from what he knew and what he experiences to develop expert ability at talking about gravitational wave physics? To begin to answer this question, Collins would need to provide a phenomenological account of his gravitational wave physics books. It would be surprising if the relevant training were entirely discursive—that is, if body language, communicative gestures, pictures of scientific equipment, and maybe even direct perceptions of laboratory equipment were anything less than critical. Moreover, Collins would also need to consider the impact of his background common sense (acquired, in part, through embodied learning) and his background technical knowledge of science. Here, explicit consideration of Collins's inferential and analogical capacities, and his development of those capacities, would need to be provided. In effect, Collins underestimates the richness of what he himself already brings to the conversational table.

CONCLUSION

Do Collins's experiments teach us anything? Yes. Collins provides compelling evidence that contributes directly to debates about whether human beings can pass Turing Tests on topics that they have not directly experienced. This contribution is relevant to both philosophical and sociological attempts to understand expertise. As Collins realizes, it may enable humanists, social scientists, and natural scientists to forge new directions past the academic "science wars."

Unfortunately, Collins wants to make a stronger claim than the one just reconstructed. Collins insists that he establishes something important about embodiment. He claims that given how much information the medium of conversation can convey, a minimally embodied agent with sufficient conversational experience can pass a Turing Test on subjects that the agent has not experienced firsthand.

The cases that Collins studied fail to support this conclusion. As a result of having developed to cognitive maturity, all of his test subjects combined physical interaction and neural-somatic perception to develop frameworks that they later tacitly appealed to when learning to talk about topics they have not experienced. Collins never establishes, and indeed never tries to establish, that his test subjects could still be masterful speakers if they lacked any rich, antecedent corporeal apprenticeship. Indeed, based on the data that Collins selects for study, he cannot support this conclusion; only adult test subjects are examined, and these subjects have already undergone corporeal apprenticeship before Collins examined their linguistic skill. Ironically, then, Collins claims to be providing an account of what can be learned through conversation, but he never examines how human beings actually learn to converse. In bypassing the phenomenological issue of skill development, Collins theoretically invents a developmentally unsound view of conversational practice. It is as if he reasons that if someone can talk about a topic that they have never experienced, the only way that the person could compensate for their experiential deficiency is to utilize the fullest potential of discourse alone. On the contrary, I hope to have shown Collins presents insufficient evidence for establishing this position. In the cases considered, conversation functioned as an important part of the compensatory process, but it depended upon the reception of a non-minimally embodied agent.

In short, what Collins calls "linguistic socialization" is really a developmental process in which agents avail themselves of discursive instruction, physical interaction, and acts of perception that are not localized in the brain. Given this range of components, Collins should drop the term. It misleads by suggesting that conversation itself can socialize (in the Turing Test sense) an unsocialized person. Of course, it would be acceptable for him to replace the phrase with "embodied-linguistic socialization." But this cumbersome locution is unnecessary. It is unnecessary because the thrust of my analysis suggests that Collins should describe his research as inquiry into what can be said in the absence of "direct experience"; his project to undermine phenomenological embodiment should be abandoned.

Should Collins want to proceed further with his research program, then he needs to be clearer about how to construct an appropriate empirical test for his idea of minimal embodiment. Unfortunately, the ideal test would be so coercive and vile that it should never transpire. Collins would have to take a human baby straight from birth and immediately shut down that baby's consciousness by inducing a coma (or coma-like state). While in that coma, the baby would need to be transplanted into a sensory-deprivation machine that could constrain the baby's perceptual ability to the conditions specified by the minimal embodiment thesis. Thus, if the baby had the experience of floating in water, the machine would be insufficiently restrictive. If under these austere conditions Collins could speak to the baby, and over time use discourse to get the baby to be a skillful conversationalist, then, and only then, would he be capable of proving his minimal embodiment thesis. Are there any alternatives? None that I can think of. While feral children are removed from society, they still have fully functioning human bodies with which to explore their surroundings.

As a final note, I would like to point out that Collins and the Good Old Fashioned Artificial Intelligence (GOFAI) researchers have more in common than Collins recognizes. Collins thinks that because he and Dreyfus agree that tacit knowledge cannot be formalized according to GOFAI parameters, he and Dreyfus both agree on why conversations are skillful events. What Collins does not appreciate is that Dreyfus is not interested solely in the tacit dimensions of skillful actions. For Dreyfus, in order to understand the tacit dimensions of human skill, one needs to understand the tacit dimensions of embodied action and embodied perception that allow humans to develop their skill in the first place. In this respect, it can be said that both GOFAI researchers and Collins take an overly intellectual, abstracted, and informational approach to the phenomena of language. Had Collins used resources compatible with phenomenology to inquire into why it is that humans can learn as much as they can from conversation, he would have recognized that the minimal embodiment thesis obscures, rather than points the way toward the answer. Instead, just as the computer science background of the early GOFAI researchers inclines them to imagine that humans are, at bottom, a bunch of rules and programs, Collins's sociological training seems to incline him to imagine that crucial forms of human learning are, at bottom, mostly derived from conversation.

Investigating Extrapolation Instead of Treating the Body as a Unit

Harry Collins

INTRODUCTION

The experiments on interactional expertise have two levels of significance. Insofar as they show what they appear to show, they reveal that a person can have considerable discursive expertise without practical involvement in a domain of expertise. To be exact, the colour blindness experiments show this, while the gravitational wave experiments, as they stand, show that the involvement can be minimal. This has clear implications for policy, sociology of expertise, and methodology. It includes our understanding of the role of different types of expertise in public debates about new technologies, of how scientific management works, of interdisciplinarity and the way people communicate within large domain practice, of specialist journalism, of peer review, of interpretative methods in the social sciences, and so forth. As Selinger agrees, it also has a bearing on the descriptions of the way humans acquire expertise, contradicting the models generated by Dreyfus in some respects, and this too has practical implications. Selinger recognizes that none of this is affected by his critique and he has been one of the most active in commending the idea of interactional expertise to others concerned with these matters. As he kindly remarks: "This contribution is relevant to both philosophical and sociological attempts to understand expertise."

Experiments are hard, they are always less straightforward than their published description would suggest, and they are always open to reinterpretation. As well as agreeing to the importance of the experiments Selinger has also put forward various criticisms of them and their interpretation both in here and in private communications. To respond to these, it may well be that it is not surprising that a red-green color-blind person can talk about colour given that such a person can already distinguish between other colours (and it may be that our sample size is a bit too small anyway); it may well be (in fact is) the case that the description of Collins's involvement with gravitational wave physics as purely linguistic is oversimple since Collins has spent time visiting laboratories and looking at apparatus; it may well be that Sacks's descriptions of Madeleine's disabilities on the one hand, and capabilities on the other, are exaggerated. The fact remains, however, that the color-blind outcome did differ from the pitch-blind outcome in the experiments; that Collins could easily have failed the gravitational wave test and, when he first saw the questions he thought he would, and when he first saw others' answers he thought he had; and that Madeleine could have remained dumb. So something has been learned from these experiments and observations. To narrow down the range of possible interpretations—to know more exactly what we have learned—we need more experiments. For example, to understand whether the color-blindness experiment might bear upon Madeleine's predicament we might start by testing the congenitally blind for equal fluency in colour language. Let's call that the "second stage color blindness experiment" (SSCB). The SSCB could give us new information and could eliminate the possibility that it is the direct experience of some colors that accounts for the ability of a person to speak as though they had experienced others. This is the way that an empirical programme of research moves forward. My main worry about Selinger's critique can be approached with a question: what would he learn from the SSCB that would bear on the embodiment thesis as he presents it? My worry is that Selinger's only available answer is "nothing." My concern is that Selinger's critique rules out any empirical programme of research that would bear on the embodiment thesis. This is because of the massive amount of work done in his argument by the notion of "extrapolation." Give extrapolation a free rein and anything is possible; no experiment on the relationship between levels of disability and discursive fluency has information value.

THE EMBODIMENT QUESTION

We can test the claim about the empirical emptiness of Selinger's version of embodiment with thought experiments. Let us treat the color-blindness experiment, the gravitational wave experiment and the observations of Madeleine as thought experiments without flaws. Imagine, then, a neo-Madeleine whose disabilities are the same as those described by Sacks in respect of Madeleine herself and yet is capable of becoming as fluent in color language as the color-blind respondents in the reported experiments and as fluent in gravitational wave physics as Collins (given that enough gravitational wave scientists spend enough time talking with her). Finding a neo-Madeleine and showing that she could attain these fluencies would be about the best that Collins could imagine as a doable empirical test of the strong interactional hypothesis; it would be the ultimate step in any new program of experimentation that might start with the SSCB. But it certainly appears that even a positive result from this test would not affect Selinger's main arguments.

Selinger's answer to how neo-Madeleine would learn the discourse of color perception or gravitational wave physics would come in the same two parts that he deploys in the case of Madeleine:

> Part 1: She is thoroughly embodied in spite of her manifest disabilities. The disabilities are, as it were, somewhat minor inconveniences for a body that still has most of what bodies have—lots of skin with huge numbers of sensors, the ability to move around, a sense of possessiveness about her wheelchair, and so forth. These major body components and capacities give neo-Madeleine a foundation of bodily experiences which she can deploy in acquiring discursive abilities in just the same way as most of us acquire them.

> Part 2: Even though neo-Madeleine does not have body parts tailored for easy acquisition of area of discourse such as colour-language and gravitational wave physics, she can extrapolate in the same way as the old Madeleine: "she can extrapolate from that experience [emotional attachment to her wheelchair] to talk about why other people would be emotionally invested in sitting at the head of the table. If Madeleine was never in love, she could still draw from associated experiences, pleasures, pains, longings, and so on, that are common to her everyday openness towards the world."

It is a short step from Madeleine's ability to extrapolate to the emotion of love from some other emotion to neo-Madeleine's ability to extrapolate to, say, the geometry of gravitational waves from the minimal spatial experiences she actually has had.

So we have made progress but at some cost. The progress is that it now has to be conceded that argument from Madeleine, supported by the sub-cases of color blindness and gravitational wave physics, does not bear on the embodiment thesis as set out by Selinger. The cost is that the embodiment thesis as set out by Selinger is too robust to be tested.

One can take the thought experiments to another stage. Imagine the person "Nobody," from whom we imaginatively remove more and more bodily bits and pieces (there is really a whole series of Nobodys with more or less complete bodies). Nobody's potential discursive abilities do not seem to become diminished because Nobody can always extrapolate from what is left. Selinger is never forced to say anything along the lines: "Now that Nobody has so little body left but still has maximal interactional expertise I have to admit that the body is not so important for the mastery of discourse as I thought." Selinger explains this himself. He says that the only experiment that would convince him that the body was not necessary for the acquisition of human language would be one where the human was divorced from all sensory experiences from birth: "[Even] if the baby had the experience of floating in water, the [sensory deprivation] would be insufficiently restrictive [to prove that the body was unnecessary to learn language]." Thus, it looks as though we can go all the way with Nobody, or nearly all the way, without troubling Selinger's embodiment thesis. It turns out that Selinger's recitation of the sensory capacity of Madeleine's skin turns out to be irrelevant because we can take most of it away and still have a functioning body. Under this model the body is a binary entity, it is either there or it is not there. And, as the case of Nobody seems to show, it is all there until it is almost not there at all.

To try to anticipate one potential counter to this, it might be claimed that though nothing more than the sensory experience of floating on water would be needed to allow extrapolation to begin, that experience would have to be complete—sensed by a complete body with its full complement of nerves, etc. But if this were the case then the Nobody experiments could tell us something: they could tell us when a body ceased to be a body as defined under the embodiment thesis. But if the Nobody experiments could tell us something while the Madeleine experiment tells us nothing then the embodiment thesis proponent must already know where the dividing line lies between a body and a non-body—somewhere below the level of Madeleine's body but above the level of one of the ever more debilitating Nobodies. But if they already know where the line is then they should say something about where the line is. Otherwise, once more, we have nothing we can test by experiment because the line can be arbitrarily moved in response to every new experiment.

The Nobody thought experiments, and Selinger's own example of the floating baby, show the crucial role of the second part of the argument—extrapolation. Everything that one can learn to say about the world can be learned by extrapolation from any single thing learned via a sensory input from the body. To state the obvious, if observations of Madeleine, neo-Madeleine, and Nobody can prove nothing of philosophical significance about the embodiment thesis, the results of the color-blindness and gravitational wave experiments are absolutely innocent of embodiment-relevant information content since in both case almost the full complement of sensory experiences are still in place. And, of course, the same applies to the SSCB.

The central claims of the embodiment theses are supported by empirical observation in the phenomenological style. Furthermore, the thesis has real world consequences: it claims to say something about how language is acquired by humans and, perhaps, how computers might or might not become intelligent. It is disappointing that though we now have a way to do experiments on the relationship between language, experience, and embodiment, such as the really exciting (to me) SSBC, the results of such experiments we can do will not bear on the embodiment thesis.

As we can see, the "slipping clutch" which prevents the experiments from moving the embodiment thesis either backwards or forwards is extrapolation. This is doubly unfortunate because extrapolation does not really do any explanatory work. We simply don't know how extrapolation does its job. Extrapolation means bringing past experience to bear on new experience so there ought to be a "problem of extrapolation" just as there is a problem of induction. We know we do extrapolate, just as we know we induce, but how do we know what old experiences properly bear on what new experiences and in just which way? Extrapolation is like the use of the idea of analogy to solve the problem of creativity and, as Dreyfus rightly points out, to use analogy you first have to solve "the problem of relevance" (1996, p. 173). The problem of relevance is just as relevant to extrapolation. What, then, are we to say to AI enthusiasts who, coming to believe the embodiment thesis, claim that all they need is a computer with a single sensory experience module which can be built on by a powerful extrapolation engine until the rest of the landscape of experience has been filled in? Even if Selinger is right that empathy is a bodily thing, extrapolation is still a brain thing so there seems no reason why the Lenat's of this world cannot use it as a resource when they are trying to make a case for disembodied or, at least, minimally embodied, computers.

To conclude this main part of the argument, Selinger has shown that Madeleine and the experiments do not bear on the embodiment thesis in the way I thought it did. But on the way we seem to have learned that the embodiment thesis is less interesting than we thought it was. It turns out that a working body, even according to Selinger's account, can be minimal. According to his account, an entity that is to acquire interactional expertise has to have those parts that are mooted in the minimal embodiment thesis—the parts to do with embedding in a language community—plus one additional sensory mechanism for extrapolation to build on. This body hardly differs from the minimal body of my minimal embodiment thesis so it turns out there is very little between us. The embodiment thesis pretty well is the minimal embodiment thesis.

PROFIT AND LOSS

Let me try to sum up what ground has been lost and what held. The observations and experiments do not show the embodiment thesis to be wrong but it turns out that the body of the embodiment thesis is almost identical to my minimal body anyway. Everything else seems intact.

Contra Dreyfus, people can learn to speak fluently about practical skills without practicing the skills. To use an example familiar to phenomenologists, this means that seeing-people without sticks can fully acquire the language of blind people with sticks if they talk to them long enough. Furthermore, such seeing-people without sticks would be just as good as blind people with sticks at advising stick manufacturers on the best kinds of sticks to make for blind stick-using people. They would also be good at describing what it might be like to hold a new kind of vibrating stick, not yet in use. Such a person would not be able to try the experiment and report what it actually feels like to hold a vibrating stick because you would need to be a blind stick-user to do that, but once vibrating sticks had become common in the blind community the seeing person would be able to become fluent in vibrating stick discourse.[10] It means that people who had never been blind and never used a blind person's stick, but who had acquired the interactional expertise, could be excellent coaches for novice blind stick-users. To change example, it would mean that, say, sports commentators could speak as knowledgeably and fluently about a sport as the players. It would mean, in short, that Dreyfus is wrong about needing practical experience to speak fluently about a practical domain—and this is now a testable claim.

This does not mean that the fluent speaker has the same experience of the world as the blind stick-user or the sports player, just that their language and judgments are indistinguishable.

By the way, sighted stick-users who are fluent in stick language probably don't exist. All those fluent in the language have almost certainly learned it as Merleau-Ponty describes. But a philosophical principle should not arise from a demographic contingency.

The distinction between the social embodiment thesis and what I call "the minimal embodiment thesis" also survives. Nothing in the debate counters the claim that the language of native language groups is based in their typical embodiment while in the case of individuals such as Madeleine the language associated with the typical body can be learned by someone with an untypical body. We can still feel sure that if we all had bodies like Madeleine's we would speak a different language even though Madeleine speaks ours. In sum, there is a different relationship between language and the body depending on whether you are looking at the language-speaking group or the individual. Individual talking lions, if they existed, could be following this debate if they had been brought up among us.

Minimal embodiment still seems to be all that is necessary to become embedded in society except that there is a small extra component in the minimal body. If we rule out extrapolation, of course, then all bets are off, but then the experiments, or experiments of a similar type, would bear on the embodiment thesis after all. Shortly I will argue that this is the right way to go.

The experiments, if they prove what they seem to prove, do still bear on all kinds of policy issues to do with expertise, the way managers do their jobs, the way journalists do their jobs, the way peer reviewers do their jobs, and so forth. But Selinger does not dispute any of this anyway.

It is disappointing that though the experiments, and potential experiments such as the SSCB, bear on such a lot they do not bear on the embodiment thesis. But it would be easy to fix things. Why do we not stop arguing about whether the body is necessary to language—I concede that it is—and start asking how much of a body is needed for language. In other words, let us study extrapolation rather than invoking it as a kind of argumentative bandage. If we think of the experiments and observations as explorations of the power and reach of extrapolation we see that the Madeleine case ceases to be irrelevant and becomes, instead, interestingly informative about how little body you can get away with—or how powerfully we humans can extrapolate.

Finally, let me take the opportunity to reiterate my own argument about AI. The debate is between embedding and embodying. We have discovered that there are big uncertainties about embodying: we don't know how much of it we need and how much we can do by extrapolation and we are not entirely clear about how far the argument is restricted to humans or to all entities that are to master natural language. On the other hand we know from such experiments as have been done that extrapolation does not seem to work where embedding is concerned: the pitch-blind cannot manage perfect-pitch discourse and colour perceivers cannot talk as though they were colorblind. The argument goes that this is because in neither case they have been embedded in the discourse. This, though one must turn elsewhere for the arguments, is because discourse, unlike the body, is the property of human groups not the individual. These results, both positive and negative, arise out of observations of human beings. Separate the most fluent speaker from the human group and before long the discourses will diverge.[11] The point is that this argument is about the nature of human knowledge not the nature of human knowers. It is about what knowledge is, not how it is acquired by humans. That is why it opens the door to the idea that, say, the ability to balance on a bicycle is a kind of knowledge that can be acquired in different ways under different circumstances whereas the ability to ride a bicycle in traffic is a kind of knowledge that can only be acquired by embedding in society.[12] We can say without ambivalence, then, that the conclusion to which these arguments and experiments lead, barring all the usual problems with experiments, is that neither neutral nets, nor Heideggerian architectures, nor, say, "Super-luminal Kryptonite quantum-entangles 13-dimensional lattices," all of which might obviate the need for a human-like body, will give rise to human-like intelligence—such as the ability to converse fluently in natural languages—unless they bring with them the ability to be embedded into human groups.

The Turing Test, Linguistic Expertise, and Intelligence

Hubert Dreyfus

I think we would all agree that to master a language one must be able to master demonstratives. Demonstratives are always situational so, as Heidegger puts it, authentic language use requires being able to point out shared aspects of the current situation. Inauthentic language use occurs when one just passes the word along. That is, uses the language in a desituated way.

I think this implies that one cannot be an expert in the use of the language of a domain unless one is able to use demonstratives in that domain and that requires being an expert in the domain. To take an example: a chess master can say (pointing to the board) that that position is a losing position. You and I will not be able to see what he is talking about but other masters can. It is likely that none of the masters can describe what the position is. It is not just certain pieces occupying certain squares but a field of forces (see Nabokov, *The Defense*) but they know one when they see one. In Heidegger's terms, in this case language is being used authentically to point out aspects of a shared situation to those who share the situation.

This has consequences for your current debate. The Turing Test assumes that the correct use of desituated language shows not only linguistic skill, as Harry points out, but also intelligence. I would hold (following Heidegger as usual) that to show intelligence one has to be able to make the discrimination an intelligent person would make. To tell that that remark in this situation was an insult is not just being able to define insults and the general situations they occur in, but, again, to know one when one hears one. Context is everything in determining whether a remark is an insult or a joke or compliment or just irrelevant. There is no way to recognize an insult if desituated remarks are being passed from a typewriter to a computer to a judge.

Of course, this feature of intelligence—that it involves the skill for recognizing a bit of language as referring to a specific situation—is not the only way intelligence works. But any skill, when it requires making subtle discriminations, has the same character. You may have mastered the way surgeons talk to each other, but you don't understand surgery unless you can tell thousands of different cuts from each other and judge which is appropriate. In the domain of surgery, no matter how well you can pass the word along, we are just dumb. So is the sportscaster who can't tell a strike from a ball until the umpire has announced it.

I think the upshot of this is that Harry has discovered a new way of showing how the Turing Test fails to test intelligence and also fails to test linguistic expertise. He has shown that just being able to pass the word along is an inauthentic use of language.

Gravitational wave physics may be an interesting exception. Science in general is desituated but even science normally requires skilled discriminations. A nuclear physicist who couldn't tell the path of an X particle from a Y particle in a cloud chamber just isn't an expert. String theory and Gravity quantum stuff may, however, be so theoretical that no situational discriminations are ever possible.

Interactional Expertise:
Between Formal and Informal

Harry Collins

Bert Dreyfus's response is especially useful as he represents the "jumping off point" for the idea of interactional expertise with his usual exceptional clarity and, as always, supports it well with examples. The jumping off point is that there are two models of knowledge, the "informal," under which knowledge can be grasped only by actively living in its world, and the "formal," under which knowledge can be represented by a set of rules that could be instantiated in a computer that had no presence in the lived world of knowledge (at least in principle). The tension between these models has polarized debate to such an extent that any suggestion that there are kinds of knowledge that can be acquired with something less than complete embedding in the total situation in which that knowledge is lived and used is treated as though it amounts to the same claim as that made by the "formalists." Thus linguistic competence attained in the absence of practical competence is said to amount, in Bert's words, to no more than "passing the word along" or "using language in a desituated way." The polarized debate has served an entire generation, among which I include myself, wonderfully well, particularly in combating the more ridiculous models of knowledge championed by the artificial intelligence community and certain styles of philosophy of science. Nevertheless, the major battles having been won, we can now move on. The point about the analysis of expertise are represented by the "Periodic Table of Expertises" (Collins and Evans, 2007), and exemplified by the detailed analysis of interactional expertise (which is just one of the table's categories), is that there is more to expertise than the polarized debate allows for.

Interactional expertise, then, is a category that falls in between knowledge under the formal model and knowledge under the "informally acquired only within the entire lived world of knowledge" model. Interactional expertise is fluent understanding of the language in the absence of practical competence. Because it is fluent understanding of the language its "logic" is that of the informal model: it is heavily tacit knowledge-laden, it cannot be reduced to sets of formal rules, and it can be acquired only through experience; it is just that the relevant experience is that of language use not the practice of a domain. Once mental space has been made for this as a possibility it becomes, I contend, completely obvious that interactional expertise must exist. If it did not exist then the world of every narrow specialist would be a private one. The fact is that in esoteric domains, such as gravitational wave physics, or soccer (I pick a quintessentially team game), each specialist group's experience would belong to the specialist group alone. There would, then, be no gravitational wave physics, only a small group "here" who could work out the waveforms of inspiraling binary neutron stars and another small group "there" who could build interferometer mirror suspensions and each would have no deep idea of what the other was there for because they had not shared their practical experience. In soccer, there would be no teams, only specialist full backs, specialist right wingers, and so forth, with no means of communicating with each other. Our world simply would not work were it not for the case that we can acquire sufficient fluency in each other's languages to be able to argue, see the other's point of view, and develop an efficient and unthinking division of labour in our team practices in the absence of complete practical experience in each other's domains of practice.

Bert's response, which is a lot more complicated than it looks at first glance, actually helps us see the richness of this in-between domain of language, or as it would be better to say, this domain of language that is a little bit off to the side of the informal pole of the polarized debate.

Bert makes a number of distinct points. He talks of what it is to be an expert; he talks of "demonstratives"; he talks of "situated use of language"; he criticizes the Turing Test; and he tells us that science might be an exceptional case of desituated knowledge. Let us take each in turn.

What it is to be an expert: here Bert is pushing at an open door. A full-blown expert in a domain (i.e., a contributory expert), would certainly be able to point out instances of various classes of performance by recognizing them. But the whole idea of interactional expertise is that it is not "full-blown." The surprising thing about it is that not being full-blown one can still do so much with it—for example, pass as an expert in a demanding linguistic test. The claim amounts to this: suppose you have never seen a game of, say, baseball but have acquired interactional expertise in baseball. In that case, you and full-blown baseball expert could talk just as fluently about a game you were listening to on the radio when demonstratives do not come into it. Is this possible? Let us see:

Demonstratives: as I understand it, both Dreyfus and Selinger have no wish to contest the claim that Madeleine, the blind, wheelchair-bound, woman, acquired a considerable degree of linguistic fluency through, what at first glance, was purely linguistic interaction. What might be thought to be a paradox for them is resolved by their insistence that the interaction was not purely linguistic but heavily dependent on the very considerable resources made available by Madeleine's remaining bodily functions—the sensitivity of the skin, the ability to move backward and forward, the sense of possession of her wheelchair, and so on. Therefore, we have no reason to imagine that Madeleine could not talk a pretty good game of baseball if she had been immersed in the language of baseball by her conversational partners. If this is not the case, then Dreyfus and Selinger should tell us why it is not the case and the onus would then be on them to analyse the way the body and language interact. (They cannot have it both ways—"Madeleine's remaining body allows her to develop linguistic fluency but only in certain domains that do not depend on the body.") But Madeleine cannot point to the "this and that" of baseball games because she cannot see them. So Madeleine could acquire linguistic fluency in baseball, such as would be needed to talk about a game listened to on the radio, without the ability to use demonstratives in respect of baseball—QED.

Situated use of language: interactional expertise does involve the situated use of language, it is just that the situation is the conversation. A linguistic situation can even exist on the printed page. Thus, if I write here "Evan and Bert, you really are rotten bastards for trying to criticize my notion of interactional expertise," neither Selinger nor Dreyfus will think that I am accusing them of being either illegitimate or putrefying. Situated use of language is not the same as use of demonstratives as the Madeleine example shows. Conversational fluency is not just "passing the word on." Conversational fluency demands situated use of language but the situation can be the conversation.

The Turing Test: the reason that there is not a snowball's chance in hell of any foreseeable computer passing a well-designed Turing Test is that a well-designed Turing Test does test for the ability to fit the use of language into the current conversational context. A well-designed Turing Test has to take into account the cultural match between judge and human participant because if the two are mismatched then it is easy for the machine to pass (imagine the extreme case where the test was conducted in Chinese but the judge did not speak Chinese and then one can then work back to the more subtle cases). Therefore the Turing Test, and the associated imitation game, remain good tests of linguistic/cultural fluency, not just "passing the word along."

Science: the work in sociology of scientific knowledge over the last thirty years has made use of precisely the kind of argument deployed by Bert to counter GOFAI to show that scientific theorization and experimentation is not desituated. Each interpretation of an experimental result, or even a theoretical result (Kennefick, 2000), is a matter of situated judgment, the situation being the social setting of the science. Therefore the experiments on linguistic competence in gravitational wave physics are not an example of the deployment of desituated knowledge. (The answers to Question 5 in the Gravitational Wave imitation game reveals that what counts as a correct answer, even in science, may depend on all kinds of subtle interpretations of what is being asked.)

To conclude, the major battles having been won in respect of the claim that all knowledge can, in principle, be reduced to something like the rules of computer programs, it is now time to explore the richness of the world of situated and informal knowledge and how different types of situated and informal knowledge interact with one another. It is not that the sports commentator or coach who has never played the sport to a high level has no knowledge of the sport, it is that he or she has a certain kind of knowledge, which, for some purposes may be better than that of the players; exactly the same can be said for, say, the scientific managers of large scientific projects. There is a wonderfully rich world out there full of sociological and philosophical territories still to be explored. And, by the way, there is no reason not to use experimental means to explore it if this turns out to be useful: experiment, after all, is just a special kind of situated activity.

ACKNOWLEDGEMENTS

Evan Selinger: Several people helped me think through the issues discusses in this essay. In particular, I would like to thank Harry Collins, John Mix, Trevor Pinch, and Jack Sanders for their openness and availability. I am also grateful that the following people provided stimulating conversations and thoughtful editorial suggestions: Muhammad Aurangzeb Ahmad, Robb Eason, Timothy Engström, Patrick Grim, Robert Rosenberger, David Suits, Craig Selinger, and Noreen Selinger.

NOTES

1. Even Collins, at an earlier period of his career, espoused the contrary position. Not taking into account how much a spy could learn from talking with natives of a given city, Collins claimed that if a native asked the right questions to an infiltrating spy, the spy would be revealed as a fraud who originated elsewhere. The switch of position is discussed in Collins and Evans (2007), p. 43.

2. Although I will be drawing upon previous critiques, the arguments that follow are intended to further extant discussions (Selinger, 2003; Selinger and Mix, 2006).

3. Collins and Kusch (1998) discuss what is required to write love-letters and conclude that a distinction needs to be drawn between knowing the emotion of love first hand and being able to discuss love as a consequence of linguistic socialization into a community where agents frequently claim to have experienced love. See also Collins and Evans (2007), pp. 43–44.

4. When Madeleine could not control her hands, it was not possible for her to read Braille.

5. If I am attributing too strong of a position to Dreyfus, it is only because his rhetoric is suggestive of the argument that I am making. Dreyfus clearly argues that experts cannot reproduce all of their perceptions and judgments in propositional form. In this sense, both Dreyfus and Collins agree that experts cannot pass on everything they know through discourse. But Dreyfus also argues that experts typically develop their skills by undergoing a five-step experiential process that runs from "novice" to "expert." In this context, he suggests that if someone does not experience all five of these stages, he or she will be unable to talk as experts do.

6. I am restricting my focus to dimensions of embodiment that are relevant to the Collins-Dreyfus debate. A comprehensive analysis of Dreyfus's views of embodiment would also need to attend to more overtly existential issues, such as "anxiety," "vulnerability," "commitment," "risk," "style," and so on.

7. If Madeleine studied meditation, however, she might be able to exert some control over these processes.

8. I am grateful to John Mix for taking the time to revisit Sacks's assumptions in the Madeleine case. The

analyst of her situation and my ability to grasp the austerity of Collins's minimal embodiment thesis would not be possible without his input.

9. We are given no reason to believe that Madeleine had not already eaten many bagels in her sixty years of life.

10. The vibrating stick question was put to me by Bert Dreyfus—private communication.

11. See, for example: Collins (1990) and Collins and Kusch (1998); the very many case studies of the construction of scientific knowledge upon which these works draw; and discussion of Simon's program BACON which was claimed to be able to discover scientific laws in Collins (1989, 1991), and Simon (1991).

12. Collins and Kusch (1998).

REFERENCES

Brey, P. (2001). Hubert Dreyfus: Human versus computers. In H. Archterhuis (Ed.), *American philosophy of technology: The empirical turn* (R. Crease, Trans.) (pp. 37–64). Bloomington: Indiana University Press.

Collins, H. M. (1989). Computers and the Sociology of Scientific Knowledge. *Social Studies of Science, 19,* 613–624.

Collins, H. M. (1990). *Artificial experts: Social knowledge and intelligent machines.* Cambridge: MIT Press.

Collins, H. M. (1991). Simon's Slezak. *Social Studies of Science, 21,* 148–149.

Collins, H. M. (1992). Dreyfus, forms of life, and a simple test for machine intelligence. *Social Studies of Science, 22,* 726–739.

Collins, H. M. (1996). Embedded or embodied? A review of Hubert Dreyfus' *What computers still can't do. Artificial Intelligence, 80,* 99–117.

Collins, H. M. (2000). Four kinds of knowledge, two (or maybe three) kinds of embodiment, and question of artificial intelligence. In M. Wrathall, and J. Malpas (Eds.), *Heidegger, coping, and cognitive science: Essays in honor of Hubert L. Dreyfus, Vol. 2* (pp. 179–195). Cambridge: MIT Press.

Collins, H. M. (2004a). Interactional expertise as a third kind of knowledge. *Phenomenology and the Cognitive Sciences, 3*(2), 125–143.

Collins, H. M. (2004b). The trouble with Madeleine. *Phenomenology and the Cognitive Sciences, 3*(2), 165–170.

Collins, H. M., and Evans, R. (2007). *Rethinking expertise.* Chicago: University of Chicago Press.

Collins, H. M., Evans, R., Ribeiro, R., and Hall, M. (2006). Experiments with interactional expertise. *Studies in History and Philosophy of Science, 37*(4), 656–674.

Collins, H. M., and Kusch, M. (1998). *The shape of actions: What humans and machines can do.* Cambridge: MIT Press.

Dreyfus, H. (1992a). *What computers still can't do.* Cambridge: MIT Press.

Dreyfus, H. (1992b). Response to Collins. Artificial experts. *Social Studies of Science, 22,* 717–726.

Dreyfus, H. L. (1996). Response to my critics. *Artificial Intelligence, 80,* 171–191.

Dreyfus, H. (2000a). Responses. In M. Wrathall, and J. Malpas (eds.), *Heidegger, authenticity, and modernity: Essays in honor of Hubert Dreyfus, Vol. 1* (pp. 305–341). Cambridge: MIT Press.

Humphrey, N. (2006). *Seeing red: A study in consciousness.* Cambridge: Belknap Press of Harvard University Press.

Kennefick, D. (2000). Star crushing: Theoretical controversy and the theoreticians' regress. *Social Studies of Science, 30,* 5–40.

Lakoff, G., and Johnson, M. (1999). *Philosophy in the flesh: The embodied mind and its challenge to Western thought.* New York: Basic Books.

Sacks, O. (1998). *The man who mistook his wife for a hat and other clinical tales.* New York: Simon and Schuster.

Selinger, E. (2003). The necessity of embodiment: The Dreyfus-Collins debate. *Philosophy Today, 47*(3), 266–279.

Selinger, E., and Mix, J. (2006). On interactional expertise: Pragmatic and ontological considerations. In E. Selinger, and R. Crease (Eds.), *The Philosophy of Expertise* (pp. 302–321). New York: Columbia University Press (First published in *Phenomenology and the Cognitive Sciences, 3,* 2004, 145–163)

Simon, H. A. (1991). Comments on the symposium on Computer Discovery and the Sociology of Scientific Knowledge. *Social Studies of Science, 21,* 143–148.

Thompson, E. (2001). Empathy and consciousness. *Journal of Consciousness Studies, 8*(5–7), 1–32.

27

Genetic Interventions and the Ethics of Enhancement of Human Beings

Julian Savulescu

Should we use science and medical technology not just to prevent or treat disease, but to intervene at the most basic biological levels to improve biology and enhance people's lives? By "enhance," I mean help them to live a longer and/or better life than normal. There are various ways in which we can enhance people but I want to focus on biological enhancement, especially genetic enhancement.

There has been considerable recent debate on the ethics of human enhancement. A number of prominent authors have been concerned about or critical of the use of technology to alter or enhance human beings, citing threats to human nature and dignity as one basis for these concerns. The President's Council Report entitled *Beyond Therapy* was strongly critical of human enhancement. Michael Sandel, in a widely discussed article, has suggested that the problem with genetic enhancement

> is in the hubris of the designing parents, in their drive to master the mystery of birth . . . it would disfigure the relation between parent and child, and deprive the parent of the humility and enlarged human sympathies that an openness to the unbidden can cultivate. . . . [T]he promise of mastery is flawed. It threatens to banish our appreciation of life as a gift, and to leave us with nothing to affirm or behold outside our own will. (2004)

Frances Kamm has given a detailed rebuttal of Sandel's arguments, arguing that human enhancement is permissible. Nicholas Agar, in his book *Liberal Eugenics,* argues that enhancement should be permissible but not obligatory. He argues that what distinguishes liberal eugenics from the objectionable eugenic practices of the Nazis is that it is not based on a single conception of a desirable genome and that it is voluntary and not obligatory.

In this chapter I will take a more provocative position. I want to argue that, far from its being merely permissible, we have a moral obligation or moral reason to enhance ourselves and our children. Indeed, we have the same kind of obligation as we have to treat and prevent disease. Not only *can* we enhance, we *should* enhance.

I will begin by considering the current interests in and possibilities of enhancement. I will then offer three arguments that we have very strong reasons to seek to enhance.

Tom Murray concludes "Enhancement" by arguing that "the ethics of enhancement must take into account the meaning and purpose of the activities being enhanced, their social context, and the other persons and institutions affected by them" (Murray, 2007: 514). Such caution is no doubt well grounded. But it should not blind us to the very large array of cases in which biological modification will improve the opportunities of an individual to lead a better life. In such cases, we have strong reasons to modify ourselves and our children. Indeed, to fail to do so would be wrong. Discussion of enhancement can be muddied by groundless fears and excessive caution and qualification. I will outline some ethical constraints on the pursuit of enhancement.

CURRENT INTEREST IN ENHANCEMENT

There is great public interest in enhancement of people. Women employ cosmetic surgery to make their noses smaller, their breasts larger, their teeth straighter and whiter, to make their cheekbones higher, their lips fuller, and to remove wrinkles and fat. Men, too, employ many of these measures, as well as pumping their bodies with steroids to increase muscle bulk. The beauty industry is testimony to be an attraction of enhancement. Body art, such as painting and tattooing, and body modification, such as piercing, have, since time began, represented ways in which humans have attempted to express their creativity, values, and symbolic attachments through changing their bodies.

Modern professional sport is often said to be corrupted by widespread use of performance-enhancing drugs, such as human erythropoietin, anabolic steroids, and growth hormones. However, some effective performance enhancements are permitted in sport, such as the use of caffeine, glutamine, and creatine in diets, salbutamol, hypoxic air tents, and altitude training. Many people attempt to improve their cognitive powers through the use of nicotine, caffeine, and drugs like Ritalin and Modavigil.

Mood enhancement typifies modern society. People use psychological "self-help," Prozac, recreational drugs, and alcohol to feel more relaxed, socialize better, and feel happier.

Even in the most private area of sexual relations, many want to be better. Around 34 percent of all men aged 40–70—around 20 million in the United States—have some erectile dysfunction, which is a part of normal ageing. There is a 12 percent decline in erectile function every decade normally. As a result, 20 million men worldwide use Viagra (Cheitlin, et al. 1999).

More radical forms of biological enhancement appear possible. Even if all disease (heart disease, cancer, etc.) were cured, the average human lifespan would only be extended by twelve years. However, stem cell science has the potential to extend human lifespan radically further than this, by replacing ageing tissue with healthy tissue. We could live longer than the current maximum 120 years.

But instead of the radical prolongation of length of life, I want to focus on the radical improvement in quality of life through biological manipulation. Some skeptics believe that this is not possible. They claim that it is our environment, or culture, that defines us, not genetics. But a quiet walk in the park demonstrates the power of a great genetic experiment: dog-breeding. It is obvious that different breeds of dog differ in temperament, intelligence, physical ability, and appearance. No matter what the turf, a Dobermann will tear a Corgi to pieces. You can debilitate a Dobermann through neglect and abuse. And you can make him prettier with a bow. But you will never turn a Chihuahua into a Dobermann through grooming, training, and affection. Dog breeds are all genetic—for over 10,000 years we have bred some 300–400 breeds of dog from early canids and wolves. The St. Bernard is known for its size, the greyhound for its speed, the bloodhound for its sense of smell. There are freaks, hard workers, vicious aggressors, docile pets, and ornamental varieties. These characteristics have been developed by a crude form of genetic selection—selective mating or breeding.

Today we have powerful scientific tools in animal husbandry: genetic testing, artificial reproduction, and cloning are all routinely used in the farming industry to create the best stock. Scientists are now starting to look at a wider range of complex behaviours. Changing the brain's reward centre genetically may be the key to changing behaviour.

Gene therapy has been used to turn lazy monkeys into workaholics by altering the reward centre in the brain. In another experiment, researchers used gene therapy to introduce a gene from the monogamous male prairie vole, a rodent that forms lifelong bonds with one mate, into the brain of the closely related but polygamous meadow vole. Genetically modified meadow voles become monogamous, behaving like prairie voles. This gene, which controls a part of the brain's reward centre different from that altered in the monkeys, is known as the vasopressin receptor gene. It may also be involved in human drug addiction.

Radical enhancements may come on the back of very respected research to prevent and treat disease. Scientists have created a rat model of the genetic disease Huntington's Chorea. This disease results in progressive rapid dementia at the age of about forty. Scientists found that rats engineered to develop Huntington's Chorea who were placed in a highly stimulating environment (of mazes, coloured rings, and balls) did not go on to develop the disease—their neurons remained intact. Remotivation therapy improves functioning in humans, suggesting that environmental stimulation in this genetic disease may affect brain biology at the molecular level (by altering neurotrophins). Prozac has also been shown to produce a beneficial effect in humans suffering from Huntington's Chorea. Neural stem cells have also been identified that could potentially be induced to proliferate and differentiate, mediated through nerve growth factors and other factors. We now know that a stimulating environment, drugs like Prozac, and nerve growth factors can affect nerve proliferation and connections—that is our brain's biology. These same interventions could, at least in theory, be used to increase the neuronal complement of normal brains and increase cognitive performance in normal individuals.

IQ has been steadily increasing since first measured, about twenty points per decade. This has been called the Flynn effect. Large environmental effects have been postulated to account for this effect. The capacity to increase IQ is significant. Direct biological enhancement could have an equal if not greater effect on increase in IQ.

But could biological enhancement of human beings really be possible? Selective mating has been occurring in humans ever since time began. Facial asymmetry can reflect genetic disorder. Smell can tell us whether our mate will produce the child with the best resistance to disease. We compete for partners in elaborate mating games and rituals of display that sort the best matches from the worst. As products of evolution, we select our mates, both rationally and instinctively, on the basis of their genetic fitness—their ability to survive and reproduce. Our (subconscious) goal is the success of our offspring.

With the tools of genetics, we can select offspring in a more reliable way. The power of genetics is growing. Embryos can now be tested not only for the presence of genetic disorder (including some forms of bowel and breast cancer), but also for less serious genetic abnormalities, such as dental abnormalities. Sex can be tested for too. Adult athletes have been genetically tested for the presence of the ACTN3 gene to identify potential for either sprint or endurance events. Research is going on in the field of behavioural genetics to understand the genetic basis of aggression and criminal behaviour, alcoholism, anxiety, antisocial personality disorder, maternal behavior, homosexuality, and neuroticism.

While at present there are no genetic tests for these complex behaviours, if the results of recent animal studies into hard work and monogamy apply to humans, it may be possible in the future to change genetically how we are predisposed to behave. This raises a new question: Should we try to engineer better, happier people? While at present genetic technology is most efficient at selecting among different embryos, in the future it will be possible to genetically alter existing embryos, with considerable progress already being made to the use of this technology for permanent gene therapy for disease. There is no reason why such technology could not be used to alter non-disease genes in the future.

THE ETHICS OF ENHANCEMENT

We want to be happy people, not just healthy people.

I will now give three arguments in favor of enhancement and then consider several objections.

First Argument for Enhancement: Choosing Not to Enhance Is Wrong

Consider the case of the Neglectful Parents. The Neglectful Parents give birth to a child with a special condition. The child has a stunning intellect but requires a simple, readily available, cheap dietary supplement to sustain his intellect. But they neglect the diet of this child and this results in a child with a stunning intellect becoming normal. This is clearly wrong.

But now consider the case of the Lazy Parents. They have a child who has a normal intellect but if they introduced the same dietary supplement, the child's intellect would rise to the same level as the child of the Neglectful Parent. They can't be bothered with improving the child's diet so the child remains with a normal intellect. Failure to institute dietary supplementation means a normal child fails to achieve a stunning intellect. The inaction of the Lazy Parents is as wrong as the inaction of the Neglectful Parents. It has exactly the same consequence: a child exists who could have had a stunning intellect but is instead normal.

Some argue that it is not wrong to fail to bring about the best state of affairs. This may or may not be the case. But in these kinds of case, when there are no other relevant moral considerations, the failure to introduce a diet that sustains a more desirable state is as wrong as the failure to introduce a diet that brings about a more desirable state. The costs of inaction are the same, as are the parental obligations.

If we substitute "biological intervention" for "diet," we see that in order not to wrong our children, we should enhance them. Unless there is something special and optimal about our children's physical, psychological, or cognitive abilities, or something different about other biological interventions, it would be wrong not to enhance them.

Second Argument: Consistency

Some will object that, while we do have an obligation to institute better diets, biological interventions like genetic interventions are different from dietary supplementation. I will argue that there is no difference between these interventions.

In general, we accept environmental interventions to improve our children. Education, diet, and training are all used to make our children better people and increase their opportunities in life. We train children to be well behaved, cooperative, and intelligent. Indeed, researchers are looking at ways to make the environment more stimulating for young children to maximize their intellectual development. But in the study of the rat model of Huntington's Chorea, the stimulating environment acted to change the brain structure of the rats. The drug Prozac acted in just the same way. These environmental manipulations do not act mysteriously. They alter our biology.

The most striking example of this is a study of rats that were extensively mothered and rats that were not mothered. The mothered rats showed genetic changes (changes in the methylation of the DNA) that were passed on to the next generation. As Michael Meaney has observed, "Early experience can actually modify protein-DNA interactions that regulate gene expression" (Society for Neuroscience 2004). More generally, environmental manipulations can profoundly affect biology. Maternal care and stress have been associated with abnormal brain (hippocampal) development, involving altered nerve growth factors and cognitive, psychological, and immune deficits later in life.

Some argue that genetic manipulations are different because they are irreversible. But environmental interventions can equally be irreversible. Child neglect or abuse can scar a person for life. It may be impossible to unlearn the skill of playing the piano or riding a bike, once learnt. One may be wobbly, but one is a novice only once. Just as the example of mothering of rats shows that environmental interventions can cause biological changes that are passed onto the next generation, so too can environmental interventions be irreversible, or very difficult to reverse, within one generation.

Why should we allow environmental manipulations that alter our biology but not direct biological manipulations? What is the moral difference between producing a smarter child by immersing that child in a stimulating environment, giving the child a drug, or directly altering the child's brain or genes?

One example of a drug that alters brain chemistry is Prozac, which is a serotonin reuptake inhibitor. Early in life it acts as a nerve growth factor, but it may also alter the brain early in life to make it more prone to stress and anxiety later in life by altering receptor development. People with a polymorphism that reduced their serotonin activity were more likely than others to become depressed in response to stressful experiences. Drugs like Prozac and maternal deprivation may have the same biological effects.

If the outcome is the same, why treat biological manipulation differently from environmental manipulation? Not only may a favourable environment improve a child's biology and increase a child's opportunities, so too may direct biological interventions. Couples should maximize the genetic opportunities of their children to lead a good life and a productive, cooperative social existence. There is no relevant moral difference between environmental and genetic intervention.

Third Argument: No Difference from Treating Disease

If we accept the treatment and prevention of disease, we should accept enhancement. The goodness of health is what drives a moral obligation to treat or prevent disease. But health is not what ultimately matters—health enables us to live well; disease prevents us from doing what we want and what is good. Health is instrumentally valuable—valuable as a resource that allows us to do what really matters, that is, lead a good life.

What constitutes a good life is a deep philosophical question. According to hedonistic theories, what is good is having pleasant experiences and being happy. According to desire fulfillment theories, and economics, what matters is having our preferences satisfied. According to objective theories, certain activities are good for people: developing deep personal relationships; developing talents, understanding oneself and the world, gaining knowledge, being a part of a family, and so on. We need not decide on which of these theories is correct in order to understand what is bad about ill health. Disease is important because it causes pain, is not what we want, and stops us engaging in those activities that give meaning to life. Sometimes people trade health for well-being: mountain climbers take on risk to achieve, smokers sometimes believe that the pleasures outweigh the risks of smoking, and so on. Life is about managing risk to health and life to promote well-being.

Beneficence—the moral obligation to benefit people—provides a strong reason to enhance people insofar as the biological enhancement increases their chance of having a better life. But can biological enhancements increase people's opportunities for well-being? There are reasons to believe that they might.

Many of our biological and psychological characteristics profoundly affect how well our lives go. In the 1960s Walter Mischel conducted impulse control experiments in which four-year-old children were left in a room with one marshmallow, after being told that if they did not eat the marshmallow, they could later have two. Some children would eat it as soon as the researcher left; others would use a variety of strategies to help control their behaviors and ignore the temptation of the single marshmallow. A decade later they re-interviewed the children and found that those who were better at delaying gratification had more friends, better academic performance, and more motivation to succeed. Whether the child had grabbed for the marshmallow had a much strong bearing on their SAT scores than did their IQ.

Impulse control has also been linked to socio-economic control and avoiding conflict with the law. The problems of a hot and uncontrollable temper can be profound. Shyness control has also been linked to socio-economic control and avoiding conflict with the law. Shyness too can greatly restrict a life. I remember one newspaper story about a woman who blushed violet every time she went into a social situation. This led her to a hermitic, miserable existence. She eventually had the autonomic nerves to her face surgically cut. This revolutionized her life and had a greater effect on her well-being than the treatment of many diseases.

Buchanan and colleagues have discussed the value of "all purpose goods." These are traits that are valuable regardless of the kind of life a person chooses to live. They give us greater all-round capacities to live a vast array of lives. Examples include intelligence, memory, self-discipline, patience, empathy, a sense of humour, optimism, and just having a sunny temperament. All of these characteristics—sometimes described as virtues—may have some biological and psychological basis capable of manipulation using technology.

Technology might even be used to improve our moral character. We certainly seek through good instruction and example, discipline, and other methods to make better children. It may be possible to alter biology to make people predisposed to be more moral by promoting empathy, imagination, sympathy, fairness, honesty, et cetera.

Insofar as these characteristics have some genetic basis, genetic manipulation could benefit us. There is reason to believe that complex virtues like fair-mindedness may have a biological basis. In one famous experiment a monkey was trained to perform a task and rewarded with either a grape or a piece of cucumber. He preferred the grape. On one occasion he performed the task successfully and was given a piece of cucumber. He watched as another monkey who had not performed the task was given a grape and he became very angry. This shows that even monkeys have a sense of fairness and dessert—or at least self-interest!

At the other end, there are characteristics that we believe do not make for a good and happy life. One Dutch family illustrates the extreme end of the spectrum. For over thirty years this family recognized that there were a disproportionate number of male family members who exhibited aggressive and criminal behavior. This was characterized by aggressive outburst resulting in arson, attempted rape, and exhibitionism. The behavior was documented for almost forty years by an unaffected maternal grandfather, who could not understand why some of the men in his family appeared to be prone to this type of behavior. Male relatives who did not display this aggressive behaviour did not express *any* type of abnormal behavior. Unaffected males reported difficulty in understanding the behaviour of their brothers and cousins. Sisters of the males who demonstrated these extremely aggressive outburst reported intense fear of their brothers. The behavior did not appear to be related to environment and appeared consistently in different parts of the family, regardless of social context and degree of social contact. All affected males were also found to be mildly mentally retarded, with a typical IQ of about 85 (females had normal intelligence) (Brunner 1993a). When a family tree was constructed, the pattern of inheritance was clearly X-linked recessive. This mean, roughly, that women can carry the gene without being affected; 50 percent of men at risk of inheriting the gene get the gene and are affected by the disease.

Genetic analysis suggested that the likely defective gene was a part of the X chromosome known as the monoamine oxidase region. This region codes for two enzymes that assist in the breakdown of neutrotransmitters. Neurotransmitters are substances that play a key role in the conduction of nerve impulses in our brain. Enzymes like the monoamine oxidases are required to degrade the neutotransmitters after they have performed their desired task. It was suggested that the monoamine oxidase activity might be disturbed in the affected individuals. Urine analysis showed a higher than normal amount of neurotransmitters being excreted in the urine of affected males. These results were consistent with a reduction in the functioning of one of the enzymes (monoamine oxidase A).

How can such a mutation result in violent and antisocial behavior? A deficiency of the enzyme results in a build-up of neurotransmitters. These abnormal levels of neurotransmitters result in excessive, and even violent, reactions to stress. This hypothesis was further supported by the finding that genetically modified mice that lack this enzyme are more aggressive.

This family is an extreme example of how genes can influence behavior: it is the only family in which this mutation has been isolated. Most genetic contributions to behavior will be weaker predispositions, but there may be some association between genes and behavior that results in criminal and other antisocial behavior.

How could information such as this be used? Some criminals have attempted a "genetic defense" in the United States, stating that their genes caused them to commit the crime, but this has never succeeded. However, it is clear that couples should be allowed to test to select offspring who do not have the mutation that predisposes them to act in this way, and if interventions were available, it might be rational to correct it since children without the mutation have a good chance of a better life.

"Genes, Not Men, May Hold the Key to Female Pleasure" ran the title of one recent newspaper article (*The Age* 2005), which reported the results of a large study of female identical twins in Britain and Australia. It found that "genes accounted for 31 percent of the chance of having an orgasm during intercourse and 51 percent during masturbation." It concluded that the "ability to gain sexual satisfaction is largely inherited" and went on to speculate that "the genes involved could be linked to physical differences in sex organs and hormone levels or factors such as mood and anxiety."

Our biology profoundly affects how our lives go. If we can increase sexual satisfaction by modifying biology, we should. Indeed, vast numbers of men attempt to do this already through the use of Viagra.

Summary: The Case for Enhancement

What matters is human well-being, not just treatment and prevention of disease. Our biology affects our opportunities to live well. The biological route to improvement is not different from the environmental. Biological manipulation to increase opportunity is ethical. If we have an obligation to treat and prevent disease, we have an obligation to try to manipulate these characteristics to give an individual the best opportunity of the best life.

HOW DO WE DECIDE

If we are to enhance certain qualities, how should we decide which to choose? Eugenics was the movement early in the last century that aimed to use selective breeding to prevent degeneration of the gene pool by weeding out criminals, those with mental illness, and the poor, on the false belief that these conditions were simple genetic disorders. The eugenics movement had its inglorious peak when the Nazis moved beyond sterilization to examination of the genetically unfit.

What was objectionable about the eugenics movement, besides its shoddy scientific basis, was that it involved the imposition of a state vision for a healthy population and aimed to achieve this through coercion. The movement was aimed not at what was good for individuals, but rather at what benefited society. Modern eugenics in the form of testing for disorders, such as Down Syndrome, occurs very commonly but is acceptable because it is voluntary, gives couples a choice of what kind of child to have, and enables them to have a child with the greatest opportunity for a good life.

There are four possible ways in which our genes and biology will be decided:

1) nature or God
2) "experts" (philosophers, bioethicists, psychologists, scientists)
3) "authorities" (government, doctors)
4) people themselves: liberty and autonomy

It is a basic principle of liberal states like the United Kingdom that the state be "neutral" to different conceptions of the good life. This means that we allow individuals to lead the life that they believe is best for themselves, implying respect for their personal autonomy or capacity for self-rule. The sole ground for interference is when that individual choice may harm others. Advice, persuasion, information, dialogue are permissible. But coercion and infringement of liberty are impermissible.

There are limits to what a liberal state should provide:

1) *safety*: the intervention should be reasonably safe.
2) *harm to others*: the intervention (like some manipulation that increases uncontrollable aggressiveness) should not result in harm. Such harm should not be direct or indirect, for example, by causing some unfair competitive advantage.
3) *distribution justice*: the interventions should be distributed according to principles of justice.

The situation is more complex with young children, embryos, and fetuses, and those who are incompetent. These human beings are not autonomous and cannot make choices themselves about whether a putative enhancement is a benefit or a harm. If a proposed intervention can be delayed until that human reaches maturity and can decide for himself or herself, then the intervention should be delayed. However, many genetic interventions will have to be performed very early in life if they are to have an effect. Decisions about such interventions should be left to parents, according to a principle of procreative liberty and autonomy. This states that parents have the freedom to choose when to have children, how many children to have, and arguably what kind of children to have.

Just as parents have a wide scope to decide on the conditions of the upbringing of their children, including schoolings and religious education, they should have similar freedom over their children"s genes. Procreative autonomy or liberty should be extended to enhancement for two reasons. Firstly, reproduction: bearing and raising children is a very private matter. Parents must bear much of the burden of having children, and they have a legitimate stake in the nature of the child they must invest so much of their lives raising.

But there is a second reason. John Stuart Mill argued that when our actions only affect ourselves, we should be free to construct and act on our own conception of what is the best life for us. Mill was not a libertarian. He did not believe that such freedom is valuable solely for its own sake. He believed that freedom is important in order for people to discover for themselves what kind of life is best for themselves. It is only through "experiments in living" that people discover what works for them and others come to see the richness and variety of lives that can be good. Mill strongly praised "originality" and variety in choice as being essential to discovering which lives are best for human beings.

Importantly, Mill believed that some lives are worse than others. Famously, he said that it is better to be Socrates dissatisfied than a fool satisfied. He distinguished between "higher pleasures" of "feelings and imagination" and "lower pleasures" of "feelings and imagination" and "lower pleasures" of "mere sensation" (Mill, 1910: 7). He criticized "apelike imitation," subjugation of oneself to custom and fashion, indifference to individuality, and lack of originality (1910: 119–20, 123). Nonetheless, he was the champion of people's right to live their lives as they choose.

> I have said that it is important to give the freest rein possible to things that are not customary, in order that it may in time transpire which of them are fit to become customary. But independence of action and disregard of custom are not deserving of encouragement solely for the chance they afford for better modes of action, and customs more worthy of general adoption, to be discovered; nor is it people of decided mental superiority who have a just claim to carry on their lives in their own way. There is no reason for all human existence to be constructed on some single or small number of patterns. If a person possesses a tolerable amount of common sense and experience, his own mode of designing his existence is the best, not because it is the best in itself, but because it is his own mode. (Mill, 1910: 125)

I believe that reproduction should be about having children with the best prospects. But to discover what are the best prospects, we must give individual couples the freedom to act on their own judgment of what constitutes a life with good prospects. "Experiments in reproduction" are as important as "experiments in living" (as long as they don't harm the children who are produced). For this reason, procreative freedom is important.

There is one important limit to procreative autonomy that is different from the limits to personal autonomy. The limits to procreate autonomy should be:

1) safety
2) harm to others
3) distributive justice
4) such that the parent's choices are based on a plausible conception of well-being and a better life for the child
5) consistent with development of autonomy in the child and a reasonable range of future life plans

These last two limits are important. It makes for a higher standard of "proof" that an intervention will be an enhancement because the parents are making choices for their child, not themselves. The critical question to ask in considering whether to alter some gene related to complex behavior is: Would the change be better for the individual? Is it better for the individual to have a tendency to be lazy or hardworking, monogamous or polygamous? These questions are difficult to answer. While we might let adults choose to be monogamous or polygamous, we would not let parents decide on their child's predispositions unless we were reasonably clear that some trait was better for the child.

There will be cases where some intervention is plausibly in a child's interests: increased empathy with other people, better capacity to understand oneself and the world around, or improved memory. One quality is especially associated with socio-economic success and staying out of prison: impulse control. If it were possible to correct poor impulse control, we should correct it. Whether we should remove impulsiveness altogether is another question.

Joel Feinberg has described a child's right to an open future (Feinberg 1980). An open future is one in which a child has a reasonable range of impossible lives to choose from and an opportunity to choose what kind of person to be; that is, to develop autonomy. Some critics of enhancement have argued that genetic interventions are inconsistent with a child's right to an open future. Far from restricting a child's future, however, some biological interventions may increase the possible futures or at least their quality. It is hard to see how improved memory or empathy would restrict a child's future. Many worthwhile possibilities would be open. But it is true that parental choice should not restrict the development of autonomy or reasonable range of possible futures open to a child. In general, fewer enhancements will be permitted in children than in adults. Some interventions, however, may still be clearly enhancements for our children, and so just like vaccinations or other preventative health care.

OBJECTIONS

Playing God or against Nature

This objection has various forms. Some people in society believe that children are a gift, of God or of nature, and that we should not interfere in human nature. Most people implicitly reject this view: we screen embryos and fetuses for diseases, even mild correctable disease. We interfere in nature or God's will when we vaccinate, provide pain relief to women in labor (despite objections of some earlier Christians that these practices thwarted God's will), and treat cancer. No one would object to the treatment of disability in a child if it were possible. Why, then, not treat the embryo with genetic therapy if that intervention is safe? This is no more thwarting God's will than giving antibiotics.

Another variant of this objection is that we are arrogant if we assume we could have sufficient knowledge to meddle with human nature. Some people object that we cannot know the complexity of the human system, which is like an unknowable magnificent symphony. To attempt to enhance one characteristic may have other unknown, unforeseen effects elsewhere in the system. We should not play God since, unlike God, we are not omnipotent or omniscient. We should be humble and recognize the limitations of our knowledge.

A related objection is that genes are pleiotropic—which means they have different effects in different environments. The gene or genes that predispose to manic depression may also be responsible for heightened creativity and productivity.

One response to both of these objections is to limit intervention, until our knowledge grows, to selecting between different embryos, and not interviewing to enhance particular embryos or people. Since we would be choosing between complete systems on the basis of their type, we would not be interfering with the internal machinery. In this way, selection is less risky than enhancement.

But such a precaution could also be misplaced when considering biological interventions. When benefits are on offer, such objections remind us to refrain from hubris and overconfidence. We must do adequate research before intervening. And because the benefits may be fewer than when we treat or prevent disease, we may require the standards of safety to be higher than for medical interventions. But we must weigh the risks against the benefits. If confidence is justifiably high, and benefits outweigh harms, we should enhance.

Once technology affords us the power to enhance our own and our children's lives, to fail to do so would be to be responsible for the consequences. To fail to treat our children's diseases is to wrong them. To fail to prevent them from getting depression is to wrong them. To fail to improve their physical, musical, psychological, and other capacities is to wrong them, just as it would be to harm them if we gave them a toxic substance that stunted or reduced these capacities.

Another variant of the "Playing God" objection is that there is a special value in the balance and diversity that natural variation affords, and enhancement will reduce this. But insofar as we are products of evolution, we are merely random chance variations of genetic traits selected for our capacity to survive long enough to reproduce. There is no design to evolution. Evolution selects genes, according to environment, that confer the greatest chance of survival and reproduction. Evolution would select a tribe that was highly fertile but suffered great pain the whole of their lives over another tribe that was less fertile but suffered less pain. Medicine has changed evolution: we can now select individuals who experience less pain and disease. The next stage of human evolution will be rational evolution, according to which we select children who not only have the greatest chance of surviving, reproducing, and being free of disease, but who have the greatest opportunities to have the best lives in their likely environment. Evolution was indifferent to how well our lives went; we are not. We want to retire, play golf, read, and watch our grandchildren have children.

"Enhancement" is a misnomer. It suggests luxury. But enhancement is no luxury. Insofar as it promoted well-being, it is the very essence of what is necessary for a good human life. There is no moral reason to preserve some traits—such as uncontrollable aggressiveness, a sociopathic personality, or extreme deviousness. Tell the victim of rape and murder that we must preserve diversity and the natural balance.

Genetic Discrimination

Some people fear the creation of a two-tier society of the enhanced and the unenhanced, where the inferior, unenhanced are discriminated against and disadvantaged all through life.

We must remember that nature allots advantage and disadvantage with no gesture to fairness. Some are born terribly disadvantaged, destined to die after short and miserable lives. Some suffer great genetic disadvantage while others are born gifted, physically, musically, or intellectually. There is no secret that there are "gifted" children naturally. Allowing choice to change our biology will, if anything, be more egalitarian, allowing the ungifted to approach the gifted. There is nothing fair about the natural lottery: allowing enhancement may be fairer.

But more importantly, how well the lives of those who are disadvantaged go depends not on whether enhancement is permitted, but on the social institutions we have in place to protect the least well off and provide everyone with a fair chance. People have disease and disability: egalitarian social institutions and laws against discrimination are designed to make sure everyone, regardless of natural inequality, has a decent chance of a decent life. This would be no different if enhancement were permitted. There is no necessary connection between enhancement and discrimination, just as there is no necessary connection between curing disability and discrimination against people with disability.

The Perfect Child, Sterility, and Loss of the Mystery of Life

If we engineered perfect children, this objection goes, the world would be a sterile, monotonous place where everyone was the same, and the mystery and surprise of life would be gone.

It is impossible to create perfect children. We can only attempt to create children with better opportunities for a better life. There will necessarily be difference. Even in the case of screening for disability, like Down Syndrome, 10 percent of people choose not to abort a pregnancy known to be affected by Down Syndrome. People value different things. There will never be complete convergence. Moreover, there will remain massive challenges for individuals to meet in their personal relationships and in the hurdles our unpredictable environment presents. There will remain much mystery and challenge—we will just be better able to deal with these. We will still have to work to achieve, but our achievements may have greater value.

Against Human Nature

One of the major objections to enhancement is that it is against human nature. Common alternative phrasings are that enhancement is tampering with our nature or an affront to human dignity. I believe that what separates us from other animals is our rationality, our capacity to make normative judgments and act on the basis of reasons. When we make decisions to improve our lives by biological and other manipulations, we express our rationality and express what is fundamentally important about our nature. And if those manipulations improve our capacity to make rational and normative judgments, they further improve what is fundamentally human. Far from being against the human spirit, such improvements express the human spirit. To be human is to be better.

Enhancements Are Self-Defeating

Another familiar objection to enhancement is that enhancements will have self-defeating or other adverse social effects. A typical example is increase in height. If height is socially desired, then everyone will try to enhance the height of their children at great cost to themselves and the environment (as taller people consume more resources), with no advantage in the end since there will be no relative pain.

If a purported manipulation does not improve well-being or opportunity, there is no argument in favour of it. In this case, the manipulation is not an enhancement. In other cases, such as enhancement of intelligence, the enhancement of one individual may increase that individual's opportunities only at the expense of another. So-called positional goods are goods only in a relative sense.

But many enhancements will have both positional and non-positional qualities. Intelligence is good not just because it allows an individual to be more competitive for complex jobs, but because it allows an individual to process information more rapidly in her own life, and to develop greater understanding of herself and others. These non-positional effects should not be ignored. Moreover, even in the case of so-called purely positional goods, such as height, there may be important non-positional values. It is better to be taller if you are a basketball player, but being tall is a disadvantage in balance sports such as gymnastics, skiing, and surfing.

Nonetheless, if there are significant social consequences of enhancement, this is of course a valid objection. But it is not particular to enhancement: there is an old question about how far individuals in society can pursue their own self-interest at a cost to others. It applies to education, health care, and virtually all areas of life.

Not all enhancements will be ethical. The critical issue is that the intervention is expected to bring about more benefits than harms to the individual. It must be safe and there must be reasonable expectation of improvement. Some of the other features of ethical enhancements are summarized below.

What Is an Ethical Enhancement?

An ethical enhancement:

> 1) is in the person's interests
> 2) is reasonably safe

3) increases the opportunity to have the best life
4) promoted or does not unreasonably restrict the range of possible lives open to that person
5) does not unreasonably harm others directly through excessive costs in making it freely available
6) does not place that individual at an unfair competitive advantage with respect to others, for example, mind-reading
7) is such that the person retains significant control or responsibility for her achievements and self that cannot be wholly or directly attributed to the enhancement
8) does not unreasonably reinforce or increase unjust inequality and discrimination—economic inequality, racism

What Is an Ethical Enhancement for a Child or Incompetent Human Being?

Such an ethical enhancement is all the above, but in addition:

1) the intervention cannot be delayed until the child can make its own decision
2) the intervention is plausibly in the child's interests
3) the intervention is compatible with the development of autonomy

CONCLUSION

Enhancement is already occurring. In sport, human erythropoietin boosts red blood cells. Steroids and growth hormone improve muscle strength. Many people seek cognitive enhancement through nicotine, Ritalin, Modavigil, or caffeine. Prozac, recreational drugs, and alcohol all enhanced mood. Viagra is used to improve sexual performance.

And of course mobile phones and aeroplanes are examples of external enhancing technologies. In the future, genetic technology, nanotechnology, and artificial intelligence may profoundly affect our capacities.

Will the future be better or just disease-free? We need to shift our frame of reference from health to life enhancement. What matters is how we live. Technology can now improve that. We have two options:

1) Intervention:
   - treating disease
   - preventing disease
   - supra-prevention of disease—preventing disease in a radically unprecedented way
   - protection of well-being
   - enhancement of well-being
2. No intervention, and to remain in a state of nature—no treatment or prevention of disease, no technological enhancement.

I believe that to be human is to be better. Or, at least, to strive to be better. We should be here for a *good* time, not just a *long* time. Enhancement, far from being merely permissible, is something we should aspire to achieve.

REFERENCES

The Age (2005), "Genes, Not Men, May Hold the Key to Female Pleasure," 9 June.

Brunner, H. G., Nelen, M., et al. (1993*a*), "Abnormal Behaviour Associated with a Point Mutation in the Structural Gene for Monoamine Oxidase A," Science, 262/5133: 578–80.

Cheitlin, M. S., Hutter, A. M., et al. (1999), "ACC/AHA Expert Consensus Document JACC: Use of Sildenafil (Viagra) in Patients with Cardiovascular Disease," *Journal of the American College of Cardiology,* 33/1: 273–82.

Feinberg, J. (1980), "The Child's Right to an Open Future," in W. Aiken and H. LaFollette (eds.), *Whose Child? Parental Rights, Parental Authority and State Power* (Totowa, NJ: Rowman and Littlefield), 124–53.

Mill, J. S. (1910), *On Liberty* (London: J. M. Dent).

Murray, T. (2007), "Enhancement," in B. Steinbock (ed.), *The Oxford Handbook of Bioethics* (Oxford University Press), 491–515.

Sandel, M. (2004), "The Case Against Perfection," *Atlantic Monthly* (Apr. 2004), 51–62.

Society for Neuroscience (2004), "Early Life Stress Harms Mental Function and Immune System in Later Years According to New Research," 26 Oct., http://apu.sfn.org/content/AboutSFN1/NewsReleases/am 2004_early.html, accessed Feb. 2006.

28

What's Wrong with Enhancement Technology?

Carl Elliott

Enhancement technologies is a term of art coined by bioethicists to refer to medical interventions that can be used not just to cure or control illness, but also to enhance human capacities and characteristics. The term comes out of a debate over gene therapy in the late 1980s. The first gene therapy trials involved the treatment of a genetic disease called adenosine deaminase (ADA) deficiency, which causes children to have severe problems with their immune system. Gene therapy to treat illnesses was widely seen as morally justifiable, as long as it could be shown safe and effective, but many skeptics worried about the prospect of manipulating a person's genetic constitution in an effort to improve them—to try to make them smarter or better looking, to change their personalities, and so on. The skeptics found the prospect of genetic enhancement even more worrying if the enhancements could be passed from generation to generation with germ line (rather than somatic cell) manipulation.

Out of this debate came a distinction between *therapy* on the one hand and *enhancement* on the other, the distinction suggesting that therapy was morally acceptable, and enhancement morally worrying. The purpose of the distinction was to allow researchers to pursue gene therapy for conditions such as ADA deficiency or cystic fibrosis, but to discourage them from trying to monkey around with the genetics of personality, intelligence, or physical appearance. Other bioethicists latched onto the enhancement/treatment distinction as a way of making more general moral distinctions between other types of medical interventions. Many people argue that third-party payers (e.g., insurance companies) are obligated to pay for treatments, for example, but not for enhancements, such as breast augmentation surgery or Rogaine for baldness.

The distinction between enhancement and therapy has not turned out to be terribly useful for ethical purposes. Part of the problem is that some interventions that are easily characterized as enhancements seem ethically justifiable, or even desirable. For example, immunizations enhance a person's immune system rather than cure or control an illness, yet no one is arguing that they should be banned or that insurance companies should not pay for them. A tougher problem comes from the fact that most *enhancements* can also be characterized as *treatments* for some kind of psychological problem. Is the antidepressant Paxil (paroxetine) an enhancement aimed at making shy people more outgoing, or is it a treatment for social anxiety disorder? Is Ritalin a concentration enhancer, or is it a treatment for attention deficit hyperactivity disorder? At best, the line between treatment and enhancement is very fluid.

431

The enhancement/treatment distinction also slides right over the question: what's wrong with enhancement technologies anyway? Is there a moral problem with wanting to be taller, or better looking, or happier, or able to concentrate better? Many of the characteristics many people want to enhance are generally seen as positive changes, and might not be at all worrying if they were achieved through, say, education, or work, or some sort of psychotherapy. What is wrong with trying to achieve these things with the tools of medicine?

The answer to that question will vary from one intervention to another. Each medical technology has its own merits and dangers, and what is morally worrying about one may not be a problem for another. Many of these worries have become quite familiar. A number of writers believe that the main problem for enhancement technologies will be access. They may well be right. If enhancement technologies are not paid for by third-party payers, then the extra boosts they provide will almost certainly be disproportionately available to the rich. This would reinforce conditions of social inequality.

I want to highlight a few of the less obvious problems with enhancement technologies. My purpose is not to suggest that my list exhausts the range of moral problems, or that my concerns apply equally to all technologies. I do not even want to suggest that these concerns mean that such technologies are unethical. My aims are diagnostic. I want to try to put a finger on the worries than many people feel about some of these technologies but often find difficult to articulate.

The first concern I want to highlight is what the Georgetown University philosopher Margaret Olivia Little calls the *problem of cultural complicity*. As Little points out, the demand for certain technologies is propped up by cultural forces that many of us would see as harmful. They are harmful because they make some people feel inadequate or unhappy with the way they are. For example, cosmetic surgeons have often taken advantage of the desire of some ethnic minorities to efface markers of their ethnicity—to perform surgery on the "Jewish nose," for example, or to alter "Asian eyes" in order to make them look more like the eyes of Europeans. Another example might be the pressure that many American women feel to conform to a certain body type, which leaves many women and girls feeling that they are too fat or that their breasts are too small. At the extreme end of the spectrum would be the cultural pressures that help produce psychiatric illnesses like anorexia nervosa.

These kinds of forces leave people with a dilemma. On the one hand, you might see them as harmful, and you might well believe that we would be better off as a society if we were free of them. Yet on the other hand, they are real. So you feel them, and if you are a parent, you feel their effects on your child. But if you give in to them, you help reinforce them. This is what Little means by cultural complicity. By getting breast augmentation surgery, you are complicit in the norm that creates the pressure for women to have large breasts. By boosting your son's height using growth hormone, you are complicit in the norm that creates the pressure for men to be tall. By taking Paxil or Zoloft for shyness, you are complicit in the cultural norm that makes shyness something to be ashamed of. By giving in to the pressure (or, if you are a doctor, by exploiting it), you are helping to reinforce that pressure.

This leads to a second problem that I will call the *problem of relative ends*. You can see the problem of relative ends most clearly with the prescription of synthetic growth hormone to increase the height of short children, especially short boys. Growth hormone was initially given to to children who have a genetic deficiency of growth hormone—that is, children whose bodies don't produce growth hormone themselves. But in the 1980s, pediatricians began to debate whether it should be given to short children who are not growth hormone deficient. The American Academy of Pediatrics eventually issued a policy statement on the ethics of growth hormone therapy, which said that it is ethically acceptable to prescribe it to some children who were not deficient in growth hormone but who were short for other reasons, such as Turner's Syndrome or chronic renal disease.

Some pediatricians (as well as the manufacturer of synthetic growth hormone) argued that growth hormone should not be restricted to children who are short because of an illness, but to any child who was so short that that they would be stigmatized by their condition. It also bears mentioning that synthetic growth hormone was very expensive—upwards of $50,000 a year at the time of this debate—and that there was little evidence that it could work very well for children who were not growth hormone deficient.

The case of growth hormone demonstrates the problem of relative ends. Boys and men want to be tall, but *tall* is a relative concept; being tall is dependent on others being short. 6′5″ is tall because the average American man is 5′10″; if the average American man were 6′5″, then 6′5″ would no longer be tall. In other words, not everyone can be tall; for tall men to exist, there must be short men. For short men to exist, there must be tall men. At best, everyone could be the same height, but even then I suspect that there would still be tall and short, only they might be measured in millimeters rather than inches. The situation here is like Gore Vidal's remark: it's not enough to succeed; others must fail.

The seminal text here, as anyone with small children will know, is Dr. Seuss's book *The Sneetches*. The important characteristic for Sneetches is whether they have stars on their bellies. As Dr. Seuss says, "But because they have stars, all the Star Belly Sneetches / Would brag, we're the best kind of Sneetch on the beaches." For Sneetches, the demand is for stars, and the supplier who fills that demand is an entrepreneur called Sylvester McMonkey McBean. McBean rides into town with a machine that puts stars on the bellies of the Plain-Belly Sneetches. But when everyone has a star on their bellies, the appeal of having one is gone. So then the Sneetches want to have their stars removed. McBean is happy to do this as well. The Sneetches get caught in a vicious circle, adding and removing stars from their bellies, for which McBean happily collects a fee.

A third (and related) concern with enhancement technologies is the *role of the market*. Increasingly, the American healthcare system has become powered by the engine of consumer capitalism. During the 1990s, according to *Fortune* magazine, the pharmaceutical industry became the most profitable industry in the United States, with annual profit margins exceeding 18 percent. Much of these profits come from drugs that can arguably be used for enhancement. According to the National Institute for Health Care Management, for example, the most profitable class of drugs in 2001 was the antidepressants. Of course, a market economy relies on the freedom of industry to sell people what they want, but it also relies on the ability of industry to persuade people that they want what industry is selling. In 1997, the FDA relaxed its restrictions on direct-to-consumer advertising for prescription drugs, allowing the pharmaceutical industry to advertise on television and in popular magazines. Is anything wrong with this? Some people would say no; this is simply how a capitalist economy works. Others would say that selling people what they want is wrong if it preys on their fears, insecurities, or weaknesses, like tobacco company executives who don't smoke themselves but make their living by persuading others to smoke.

This concern about the market has emerged in the recent history of antidepressants such as Prozac. Prozac (fluoxetine) was the first of a new generation of antidepressants known as selective serotonin reuptake inhibitors (SSRIs), which became extraordinarily popular during the 1990s. The psychiatrist Peter Kramer coined the phrase "cosmetic psychopharmacology" in his book *Listening to Prozac*. What Kramer found so intriguing about Prozac was not what it does for patients who are clinically depressed, but what it does for those who aren't: patients who are shy and withdrawn, or who have poor self-esteem, or who are somewhat obsessive. According to Kramer and others, some patients on Prozac (but not all or even most of them) undergo what seems to be a change in personality. Some of the more controlling, compulsive types become more laid-back and easygoing. Some shy people become more self-confident and assertive. Kramer's patients said things like "I feel like I've been drugged all my life and now I'm finally clearheaded" or "I never really felt like myself until now." Some patients seem to be able to see themselves in a way that they had been incapable of before. They didn't just get well; in the words of one patient, they became "better than well." This effect is what Kramer called cosmetic psychopharmacology—the use of psychoactive drugs not to treat severe mental disorders, but to improve various aspects of a person's mental life.

This effect has helped transform the SSRIs from a medical treatment into a market commodity. Today the SSRIs are not simply used to treat clinical depression. They are used to treat (among other things) social anxiety disorder, posttraumatic stress disorder, generalized anxiety disorder, obsessive-compulsive disorders, eating disorders, sexual compulsions, and premenstrual dysphoric disorder. Many of these disorders did not officially exist several decades ago, and many of those that did exist were thought to be very rare. They became widely diagnosed only when a treatment was developed. Some clinicians argue that this is simply the result of their greater awareness of the disorders and their improved diagnostic skills. But it is also true that once a pharmaceutical company develops a treatment for a psychiatric disorder, it also has a financial interest in making sure that doctors diagnose the disorder as often as possible. This may mean transforming what was once seen as ordinary human variation—being shy, uptight, or melancholy—into psychiatric disorders. The more people are persuaded that they have a disorder that can be medicated, the more medication that will be sold.

The fourth worry I want to mention is a little trickier. I will call it the *problem of authenticity.* Probably the best way to illustrate the problem of authenticity is with a case that Peter Kramer discusses in *Listening to Prozac.* Like other psychiatrists, Kramer found that when he prescribed Prozac, his patients told him things like "This is how I was always meant to feel." He writes about one patient called Tess who is clinically depressed. Tess has few friends, lots of obligations, and very poor self-esteem. Kramer prescribes a tricyclic antidepressant, and she eventually gets over her depression. But Kramer thinks she could do even better, so he switches her to Prozac. Soon she is happier, more outgoing, and more self-confident, and she knows when and how to say no. Her life is much better.

Eventually Kramer tapers her Prozac and then takes her off the drug completely. In a few months, Tess comes back and says, "I just don't feel like myself anymore." Remarkable, thinks Kramer: she has returned to the very state in which she has been for twenty or thirty years, her entire life apart from the past several months, and she says "I don't feel like myself." Instead, she says she feels like herself on Prozac.

What do we make of these kinds of remarks? It's clear that this patient changes quite a lot on Prozac. But should it be described as a transformation to a new self, or as a restoration to an authentic self? Or is it something else? Some people would argue that Prozac can restore an authentic self that has been hidden by pathology. The authentic self is the one that has the proper levels of serotonin in the brain. But there are other cases in the book that seem to point in the opposite direction. Kramer tells of one patient who is not a success on Prozac. Before Prozac he is bitter, sarcastic, and rather cynical in a way that he seems to have cultivated. On Prozac, he becomes less bitter and less cynical; he loses that sarcastic edge. And he is happier. But he doesn't like it. He doesn't feel comfortable with the person he has become, and so he stops taking the drug. For him, Prozac doesn't seem so much to restore an authentic self as to create a new one, a new one that he thinks is not really him.

This language of authenticity is very slippery, and it can be used in many different ways. For example, another area of medicine where patients talk about finding their true selves is transsexual surgery. Candidates for surgery might say, "I am really a woman, trapped in a man's body," and that the surgery will let them be who they really are. This sounds similar to what Arthur Frank calls a "restitution narrative" in his book *The Wounded Storyteller.* A restitution narrative has the basic form of "I was healthy, then I got sick, and then I was restored to health." This is like the restitution narrative, but the restitution is to something that never existed before, only wished for— not restoration back to health, but restoration to an ideal of health that had never before been realized. What interests me is not so much whether this narrative is true or not, whatever that would mean, but how persuasive it is. Even people who are troubled by the idea of a person changing his or her sex find themselves swayed by this kind of story: "I really am a man, trapped in a woman's body." It is a Cartesian explanation: a ghost locked in the wrong machine. And it sounds plausible to us in a way that it might not sound plausible to someone in a time and place without our tradition of body/mind dualism.

Of course, the ideal of authenticity can also be used to argue *against* enhancement technologies or other kinds of self-transformation. You can see this most clearly in the case of cosmetic surgery for markers of ethnic identity. To undergo surgery to make your Asian eyes look more European, or to use creams to make your dark skin lighter, is sometimes seen as an act of fakery or self-betrayal. Words like *fakery* and *self-betrayal* turn the language of authenticity around and use it for purposes just the opposite of those used by Tess and others who "become themselves" on Prozac. But these descriptions are persuasive only if you think of authenticity as a moral ideal.

The fifth worry about enhancement technologies I will call the *problem of relativism.* What I mean by this is simply that illnesses, by and large, are not objective entities that look the same to all people at all times. Rather, what counts as an illness is a product of a particular time and place, and a particular set of cultural understandings. Homosexuality was officially considered a mental disorder up until the 1970s and was listed in the American Psychiatric Association's *Diagnostic and Statistical Manual.* Today it is thought of as simply part of a person's identity, a constituent of the way some people are. Our understanding of homosexuality has moved from illness to identity. And, of course, we can also slide easily in the other direction from identity to illness. Some years ago, a person with three copies of chromosome 21 was called a mongoloid; now she has a genetic disease called Down syndrome. Whereas we used to think of her as a different type of human being, now we think of her as sick. We have redefined identity as illness.

The reasons for this are complex, of course. Very often what counts as an illness is a consequence of the discovery of a way to correct it. As Willard Gaylin has pointed out, before various reproductive technologies were developed, infertility was simply a fact of nature. Now that it can be treated, it is a medical problem. Before the invention of the lens, poor vision was simply a consequence of getting old. Now it is something to be treated by a medical specialist. Psychiatry is another striking example. Before the development of psychotherapy at the end of the nineteenth century, mental illness was limited to psychotic disorders; now it includes phobias, obsessions, compulsions, personality disorders, and so on. Today it is very easy to speak of any disagreeable personality trait as if it were an illness—and even some that are not so disagreeable, like shyness, which is being discussed more and more often in the ethical and psychiatric literature as if it were a kind of mental disability. The point is that in each of these cases, what was once simply an unavoidable aspect of some people's lives was conceptually transformed by technology into a medical problem.

My point here is simply that what we see as a straightforward example of a medical treatment will look differently to people from other times and other places, and that the line we often draw between enhancements and treatments is not as sharp as we would like to think. Let me take a deliberately provocative example—the way we respond to intersexed infants, or children born with ambiguous genitalia. The standard medical response to such a child is to assign the child a sex as soon as possible, male or female, and to treat the child over a period of time with surgery and hormones to ensure that the child's physical appearance conforms as closely as possible to that of a boy or a girl. Human beings are either male or female, and if they don't look like one or the other, then something must be medically wrong with them. Our conceptual system has no room for anything in between.

But things need not necessarily be this way. Contrast, for example, our Western attitudes towards intersexuals with those of 1930s Navaho, who didn't think of intersexuals as uncategorizable and in need of medical treatment, as we do, but rather thought of them as blessed by the gods. They were revered, even held in awe. The classic study here was done by the American anthropologist Walter Hill (and later made famous by Clifford Geertz). One Navaho interviewed by Hill in his study tells him, "(Intersexuals) know everything. They can do the work of both a man and a woman." "They are responsible for all the wealth in the country," says another. "If there were no more left, the horses, sheep and Navaho would all go." The Navaho of the 1930s made intersexuals the heads of the family and gave them control over family property. For them, the idea of surgically fixing an intersexed infant would seem strange, even morally objectionable. It certainly would not be treating an illness.

My broader point here is that like intersexuality, our understandings of illness, personality, and beauty are culturally located in particular places. Our current understandings are probably not going to look the same to someone who is not immersed in our culture, and they probably won't look the same to us in fifty years. This is not to say that we can easily change these cultural understandings: we can't just extract ourselves from our circumstances and see the world like the Navaho of the 1930s. An intersexed child has to live in our society, and so do children who are short, or shy, or heavy. Even so, the realization that our own contemporary understanding of the world is not fixed and immutable should make us cautious about embracing new enhancement technologies, and especially about embracing them so readily.

A sixth problem, related to the concern about social justice, is the *problem of competition*. This problem is most evident in the use of performance-enhancing drugs in sports. Performance-enhancing drugs give some athletes a competitive advantage, and even athletes who would rather not use performance-enhancing drugs feel forced to use them anyway, for fear that they will not otherwise be able to compete. This kind of fear may be most prevalent where the competition is explicit, such as a football game or a race, but some people have the same fear even when the competition is more subtle, such as cognitive performance at school or at work. For example, stimulants such as Ritalin (methylphenidate) are widely prescribed for children with attention deficit hyperactivity disorder. They make these children less distractible and less impulsive, and they improve their ability to concentrate. But studies have also shown that stimulants will improve even a normal person's ability to concentrate. If people on stimulants can work longer and more effectively than they otherwise could, does this give them a competitive advantage in the classroom?

This may be a problem even with relatively harmless medications. Take beta blockers, for example. Beta blockers are used to treat high blood pressure, and they work by blocking the effects of the sympathetic nervous system. For this very reason, they are also useful for performers who get stage fright, such as musicians, actors, and other people who have to do a lot of public speaking. Some performers get very nervous on the stage, and their voices tremble, their faces flush, their hearts race, and their palms sweat. These are exactly the reactions that beta blockers prevent. Unlike Valium or other psychoactive drugs, beta blockers do not make the people who take them any less anxious or nervous, or at least not directly. They just block the outward effects of a person's nervousness. They give people a mask of relaxation—which, as it happens, often makes them much more relaxed. Since a performer's greatest fear is often that his or her anxiety will be obvious to the audience, a drug that masks that anxiety can be very reassuring.

Some people would see beta blockers as a harmless enhancement technology. They are safe, effective, and have few (if any) long-term effects. They don't affect the mind, they don't change the personality, and they wear off in a couple of hours. People who take them usually can't even feel the difference, unless they have to perform on stage. Yet other people would see beta blockers as morally problematic precisely because they give people a competitive advantage. If you are a graduate student who sees the classroom as a competition, you may resent a fellow student who uses beta blockers for her class presentations. If you are a musician competing for a place in the orchestra, you may well resent the violinist who medicates herself before her solos.

The final worry about enhancement technologies I will mention is perhaps the hardest to pin down. This is a worry about what the political theorist Michael Sandel calls "the drive to mastery." It is hardest to pin down because it is less about the possible consequences of enhancement technologies than about the sensibility they reflect—an attitude that views the world as something to be manipulated, mastered, and controlled. When people charge that scientists or doctors are "playing God," at least some of them are objecting to the lack of humility entailed by this sensibility. They object to the arrogance of placing such extraordinary faith in human reason. As Leon Kass puts it, the objection is not so much a matter of attempting to do what ought to be left to God, but doing so in the absence of Godlike wisdom.

This kind of worry goes beyond conventional concerns about justice. As Sandel has pointed out, it is possible to imagine a world in which all athletes have equal access to safe, performance-enhancing drugs; in which we are allowed to choose the sex of our children without creating a gender imbalance; and in which we eat factory-farmed pigs and chickens genetically engineered not to feel pain. Yet many of us would resist such a world. And the reason we would resist is not because such a world would be unjust, or even because it would lead to a world with more pain and suffering, but because of the extent to which it has been planned and engineered. We would resist the idea that the whole world is there to be manipulated for human ends.

Part V

TECHNOLOGY AND NATURE

There is a good deal of intuitive appeal to the idea that a technology is an artificial, not naturally occurring object. Technologies are manufactured things that would not otherwise exist on Earth were it not for humans. The problem is that while a natural/artificial distinction can sometimes be helpful (e.g., to distinguish between real and fake flowers, or natural and artificial light) that distinction quickly breaks down when pressed into service to describe more complex cases. Anything modified in even the slightest way could no longer be considered natural if by "natural" we mean existing in an unchanged state if it were left alone, free from human intervention. That would rule out almost every human intervention in the world, including cooking food, farming with tools, wearing clothing, and building shelter as examples of just some of countless activities that seem to be natural for humans to do. Everything we make and do to modify our environment—including having customs and laws—would be considered unnatural or artificial. That definition is unhelpfully broad and vague. Surely there is more to be said about technology than simply that it is manufactured and not naturally occurring. But if we say that it is natural for humans to make and use technology then we have truly made a natural/artificial distinction meaningless. Every human action and creation would be considered natural. Therefore, every manufactured object—no matter how high-tech and synthetic—would be a product of nature and natural for humans to make, like any other naturally occurring object. That just seems false.

The contributors to this section examine the relationship between technology and the natural world. They explore positive and negative aspects of human modifications of the world, questioning the limits (if any) of appropriate uses of technology in relation to natural environments, animals, plants, and food. Is there anything wrong with, for example, artificially re-creating a natural environment? Is it still natural or is it an artifact? What about animals? Is there anything wrong with genetically modifying animals? Genetically modifying food? Is technology compatible with nature or is it wholly different from it?

In "The Big Lie: Human Restoration of Nature," Eric Katz argues against the view held by many conservationist and environmentalists that humans have an obligation to repair or reconstruct damaged natural systems. Environmental restoration, he argues, is based on a misperception of nature and a misguided understanding of humanity's place in the natural world. Restored natural environments are merely human creations like any other artifact. They are fake nature. Katz maintains that interventions that aim to re-create and restore nature in fact manipulate, dominate, and ultimately diminish the value of nature. From this perspective, restorations of nature are attempts at control that serve human ends and that do not let nature pursue its own development. Repairing nature is a technofix, representing the very anthropocentric attitude that causes environmental harms. The technofix is a symptom of human arrogance: we destroy nature and then attempt to pass off an artificial version of it as reality.

Katz argues that the fundamental misconception underlying nature restoration projects—however benign or well-intentioned they might be—is that they confuse human technologies for natural environments. Artifacts are designed by humans to satisfy human end whereas natural objects and systems exist without human design and have no purposes or ends. Natural entities are autonomous; they are free from human control. Artifacts obviously are not. Once we restore nature we destroy nature. We create an artificial reality—a false reality that is a mere illusion of a natural environment. Katz does not suggest that humans should refrain from cleaning up or restoring damaged environments. He simply warns us that we should not misunderstand what we are doing when we restore natural areas. We make technology, not nature.

Andrew Light disagrees with Katz and other philosophers who view restorations of nature as misguided attempts to replicate and restore its original value. In "Ecological Restoration and the Culture of Nature," Light grants that a restoration of nature is never the same as the original but suggests that the practice be seen more like a benevolent art restoration (that seeks to remedy a past harm) rather than a malicious forgery (that attempts to pass off a fake as the real thing). Benevolent restorations neither harm nor dominate nature, nor are they attempts to trick us. Rather they are earnest attempts to correct harm done to the environment by human intervention, to rectify the balance of nature, and to let nature pursue its own interests or own natural course. Furthermore, Light argues that ecological restoration not only restores nature but also restores our moral relationship with the environment. The more people participate in restorations the more connected we become to nature (and to each other) and the less likely we are to allow it to be further harmed. The environmental pragmatism Light endorses side-steps the natural-artificial distinction by granting it and, nonetheless, affirming the importance of participating together in restoration projects. Nothing but good, Light claims, comes from restoring nature.

"The Brave New World of Animal Biotechnology" is a report published by the Hastings Center, a nonprofit organization that studies ethical questions in healthcare, biotechnology, and the environment. The report explores the biotechnological modification of animals and some of the perplexing ethical issues that surround human uses of animal life. The authors take a pluralistic approach to questions about animal biotechnology. They contend that there are too many different contexts in which humans technologically modify animals to rely on a singular, monistic approach. The best way to figure out what our obligations are to animals is to pay attention to the particulars of each situation. There is no single, overarching set of obligations to animals but rather plural obligations. The authors also stress the need to be attentive to the particular ethical challenges of animal biotechnology that arise in different contexts but, at the same time, take an ecological (holistic) perspective in order to judiciously evaluate, compare, and coordinate our various obligations within and between various contexts.

The authors discuss several spheres of human activity and animal biotechnology, including science and biomedical research, economic markets and agribusiness, public political life, domestic life and recreation, culture and community life, and nature. Each sphere of activity has its own particular use of animal biotechnology and its own ethical perspective. For example, the main consideration of bioengineered fish is to balance the needs of fisherman with a concern for protecting the well-being of the eco-system. The main consideration of animal biotechnology and medical research are rather different. We might tolerate biotechnological interventions in a controlled laboratory that we would not tolerate in the wild. We might weigh the medical obligation to relieve suffering and the scientific pursuit of knowledge differently than we would the needs of the fishermen, and so on. The report urges us to respect the plurality of human activities in the natural world and to concurrently promote the human, animal, and natural good.

In "Ethics and Genetically Modified Foods," Gary Comstock examines the ethical arguments against foods that have been manipulated at the molecular level to have traits that are desirable to farmers or consumers. These foods are often viewed with suspicion and are generally perceived to be unnatural and perhaps unwise to eat. Comstock used to be opposed to food and animal biotechnology but he gradually changed his mind the more he examined the validity of the arguments of opponents of GM foods. He briefly considers some of these arguments starting with *extrinsic* objections (i.e., consequences) to GM technology, such as possible environmental harms, the reduction of biodiversity, and risks to food security. He concludes that while critics raise valid concerns, these extrinsic objections do not themselves justify a moratorium, much less a ban on GM foods. So long as local and national governments take adequate precautions and ensure that harms are prevented or minimized then there are no reasons to oppose such foods.

Comstock then considers *intrinsic* objections that assert that genetic modification technology is objectionable regardless of the consequences. Intrinsic arguments focus on the *unnaturalness* of GM foods. These arguments assert that to engage in agricultural biotechnology is, for example, to "play God," to tamper with nature, or to treat nature like a mere commodity. But Comstock claims that none of them hold up under scrutiny. He also challenges the *Precautionary Principle*, which states that nations should take precautionary measures against any potential environmental or health threats even if there is no firm scientific evidence that the threat is demonstrable. Comstock gives examples to show that this sweeping principle is empty and unhelpful. It would be better to approach things on a case-by-case basis on the basis of field tests. So, in answer to the question, "is it ethically justifiable to pursue genetically modified crops and foods?" Comstock answers yes, provided we proceed responsibly and with appropriate caution.

In "What's Wrong with Functional Foods?" David Kaplan examines a category of technologically enhanced food products that provide physiological benefits beyond their dietary or nutritional value. Known as "functional foods" or "nutraceuticals," these foods for specific health uses are designed to assist in the prevention or treatment of disease, or to enhance and improve human capacities. They include products like vitamin-fortified grains, energy bars, low-fat or low-sodium foods, and sports drinks. Functional foods are similar to other foods enhanced with artificial ingredients. Yet, unlike most food products, functional foods are proudly technological. They make no pretense to being natural. In fact, their technological character is presented as their virtue—an advancement in food technology. Functional foods are a key part of a trend in food science and marketing that is challenging traditional conceptions of diet and medicine. They are situated uniquely at the nexus of food science and technology, commerce and politics.

Kaplan compares the different legal status of functional foods in Europe, Japan, and the United States and finds the definitions to be murky, especially in the United States. Most troubling are the loose guidelines regulating the kind of health claims marketers are allowed to make. Kaplan believes this is intentional: the regulations are industry-friendly and destined to confuse consumers. The health benefits of functional foods are greatly exaggerated and, in many cases, nonexistent. They may not do much more than regular food. Kaplan also worries about the way that technologically enhanced foods blur the line between food and drugs. If food is to double as medicine then there are important social justice questions to answer about their appropriate use, their just distribution, and safety. Kaplan argues that governments have an obligation to regulate functional foods better than they currently do as a part of their obligation to protect the rights and promote the welfare of citizens. Finally, he questions the role of market forces in the production and consumption of techno-foods. On one hand, techno-foods are goods like any other commodity; on the other hand, they are like medicine, which is not a commodity like any other. Economic markets and public health are not necessarily compatible. At issue is who really benefits from functional foods: producers or consumers? Functional foods are more about politics and profits than health and nutrition.

29

The Big Lie: Human Restoration of Nature

Eric Katz

The trail of the human serpent is thus over everything

—William James, *Pragmatism*

I

I began with an empirical point, based on my own random observations: the idea that humanity can restore or repair the natural environment has begun to play an important part in decisions regarding environmental policy. We are urged to plant trees to reverse the "greenhouse effect." Real estate developers are obligated to restore previously damaged acreage in exchange for building permits.[1] The U.S. National Park Service spends $33 million to "rehabilitate" 39,000 acres of the Redwood Creek watershed.[2] And the U.S. Forest Service is criticized for its "plantation" mentality: it is harvesting trees from old-growth forests rather than "redesigning" forests according to the sustainable principles of nature. "Restoration forestry is the only true forestry," claims an environmentally conscious former employee of the Bureau of Land Management.[3]

These policies present the message that humanity should repair the damage that human intervention has caused the natural environment. The message is an optimistic one, for it implies that we recognize the harm we have caused in the natural environment and that we possess the means and will to correct these harms. These policies also make us feel good; the prospect of restoration relieves the guilt we feel about the destruction of nature. The wounds we have inflicted on the natural world are not permanent; nature can be made "whole" again. Our natural resource base and foundation for survival can be saved by the appropriate policies of restoration, regeneration, and redesign.

It is also apparent that these ideas are not restricted to policymakers, environmentalists, or the general public—they have begun to pervade the normative principles of philosophers concerned with developing an adequate environmental ethic. Paul Taylor uses a concept of "restitutive justice" both as one of the basic rules of duty in his biocentric ethic and as a "priority principle" to resolve completing claims.[4] The basic idea of this rule is that human violators of nature will in some way repair or compensate injured natural entities and systems. Peter Wenz also endorses a principle of restitution as being essential to an adequate theory of environmental ethics; he then attacks Taylor's theory for not presenting a coherent principle.[5] The idea that humanity is morally responsible for reconstructing natural areas and entities—species, communities, ecosystems—thus becomes a central concern of an applied environmental ethic.

In this essay I question the environmentalists' concern for the restoration of nature and argue against the optimistic view that humanity has the obligation and ability to repair or reconstruct damaged natural systems. This conception of environmental ethics is based on a misperception of natural reality and a misguided understanding of the human place in the natural environment. On a simple level, it is the same kind of "technological fix" that has endangered the environmental crisis. Human science and technology will fix, repair, and improve natural processes. On a deeper level, it is an expression of an anthropocentric worldview, in which human interests shape and redesign a comfortable natural reality. A "restored" nature is an artifact created to meet human satisfactions and interests. Thus, on the most fundamental level, it is an unrecognized manifestation of the insidious dream of the human domination of nature. Once and for all, humanity will demonstrate its mastery of nature by "restoring" and repairing the degraded ecosystems of the biosphere. Cloaked in an environmental consciousness, human power will reign supreme.

II

It has been many years since Robert Elliot published his sharp and accurate criticism of "the restoration thesis."[6] In an article entitled "Faking Nature," Elliot examines the moral objections to the practical environmental policy of restoring damaged natural systems, locations, and landscapes. For the sake of argument, Elliot assumed that the restoration of a damaged area could be re-created perfectly, so that the area would appear in its original condition after the restoration was completed. He then argued that the perfect copy of the natural area would be of less value than the original, for the newly restored natural area would be analogous to an art forgery. Two points seem crucial to Elliot's argument. First, the value of objects can be explained "in terms of their origins, in terms of the kinds of processes that brought them into being."[7] We value an artwork in part because of the fact that a particular artist, a human individual, created the work at a precise moment in historical time. Similarly, we value a natural area because of its "special kind of continuity with the past." But to understand the art work or the natural area in its historical context we require a special kind of insight or knowledge. Thus, the second crucial point of Elliot's argument is the coexistence of "understanding and evaluation." The art expert brings to the analysis and evaluation of a work of art a full range of information about the artist, the period, the intentions of the work, and so on. In a similar way, the evaluation of a natural area is informed by a detailed knowledge of ecological processes, a knowledge that can be learned as easily as the history of art.[8] To value the restored landscape as much as the original is thus a kind of ignorance; we are being fooled by the superficial similarities to the natural area, just as the ignorant art "appreciator" is fooled by the appearance of the art forgery.

Although Elliot's argument has had a profound effect on my own thinking about environmental issues, I believed that the problem he uses as a starting point is purely theoretical, almost fanciful.[9] After all, who would possibly believe that a land developer or a strip mining company would actually restore a natural area to its original state? Elliot himself claims that "the restoration thesis" is generally used "as a way of undermining the arguments of conservationists."[10] Thus it is with concern that I discover that serious environmentalist thinkers, as noted above, have argued for a position similar to Elliot's "restoration thesis." The restoration of a damaged nature is seen not only as a practical option for environmental policy but also as a moral obligation for right-thinking environmentalists. If we are to continue human projects which (unfortunately) impinge on the natural environment (it is claimed), then we must repair the damage. In a few short years a "sea-change" has occurred: what Elliot attacked as both a physical impossibility and a moral mistake is now advocated as proper environmental policy. Am I alone in thinking that something has gone wrong here?

Perhaps not enough people have read Elliot's arguments; neither Taylor nor Wenz, the principal advocate of restitutive environmental justice, list this article in their notes or bibliographies. Perhaps we need to reexamine the idea of re-creating a natural landscape; in what sense is this action analogous to an art forgery? Perhaps we need to push beyond Elliot's analysis, to use his arguments as a starting point for a deeper investigation into the fundamental errors of restoration policy.

III

My initial reaction to the possibility of restoration policy is almost entirely visceral: I am outraged by the idea that a technologically created "nature" will be passed off as reality. The human presumption that we are capable of this technological fix demonstrates (once again) the arrogance with which humanity surveys the natural world. Whatever the problem may be, there will be a technological, mechanical, or scientific solution. Human engineering will modify the secrets of natural processes and affect a satisfactory result. Chemical fertilizers will increase food production; pesticides will control disease-carrying insects; hydroelectric dams will harness the power of our rivers. The familiar list goes on and on.

The relationship between this technological mind-set and the environmental crisis has been amply demonstrated, and need not concern us here.[11] My interest is narrower. I want to focus on the creation of artifacts, for that is what technology does. The re-created natural environment that is the end result of a restoration project is nothing more than an artifact created for human use. The problem for an applied environmental ethic is the determination of the moral value of this artifact.

Recently, Michael Losonsky has pointed out how little we know about the nature, structure, and meaning of artifacts. "[C]ompared to the scientific study of nature, the scientific study of artifacts is in its infancy."[12] What is clear, of course, is that an artifact is not equivalent to a natural object; but the precise difference or set of differences is not readily apparent. Indeed, when we consider objects such as beaver dams, we are unsure if we are dealing with natural objects or artifacts. Fortunately, however, these kinds of animal-created artifacts can be safely ignored in the present investigation. Nature restoration projects are obviously human. A human-built dam is clearly artifactual.

The concepts of function and purpose are central to an understanding of artifacts. Losonsky rejects the Aristotelian view that artifacts (as distinguished from natural objects) have no inner nature or hidden essence that can be discovered. Artifacts have a "nature" that is partially comprised of three features: "internal structure, purpose, and manner of use." This nature, in turn, explains why artifacts "have predictable lifespans during which they undergo regular and predictable changes."[13] The structure, function, and use of the artifacts determine to some extent the changes which they undergo. Clocks would not develop in a manner which prevented the measurement of time.

Natural objects lack the kind of purpose and function found in artifacts. As Andrew Brennan has argued, natural entities have no "intrinsic functions," as he calls them, for they were not the result of design. They were not created for a particular purpose; they have no set manner of use. Although we often speak as if natural individuals (for example, predators) have roles to play in ecosystemic well-being (the maintenance of optimum population levels), this kind of talk is either metaphorical or fallacious. No one created or designed the mountain lion as a regulator of the deer population.[14]

This is the key point. Natural individuals were not designed for a purpose. They lack intrinsic functions, making them different from human-created artifacts. Artifacts, I claim, are essentially anthropocentric. They are created for human use, human purpose—they serve a function for human life. Their existence is centered on human life. It would be impossible to imagine an artifact not designed to meet a human purpose. Without a foreseen use the object would not be created. This is completely different from the way natural entities and species evolve to fill ecological niches in the biosphere.

The doctrine of anthropocentrism is thus an essential element in understanding the meaning of artifacts. This conceptual relationship is not generally problematic, for most artifacts are human creations designed for use in human social and cultural contexts. But once we begin to redesign natural systems and processes, once we begin to create restored natural environments, we impose our anthropocentric purposes on areas that exist outside human society. We will construct so-called natural objects on the model of human desires, interests, and satisfactions. Depending on the adequacy of our technology, these restored and redesigned natural areas will appear more or less natural, but they will never be natural—they will be anthropocentrically designed human artifacts.

A disturbing example of this conceptual problem applied to environmental policy can be found in Chris Maser's *The Redesigned Forest*. Maser is a former research scientist for the United States Department of Interior Bureau of Land Management. His book attests to his deeply felt commitment to the policy of "sustainable" forestry, as opposed to the short-term expediency of present-day forestry practices. Maser argues for a forestry policy that "restores" the forest as it harvests it; we must be true foresters and not "plantation" managers.

Nonetheless, Maser's plans for "redesigning" forest reveal several problems about the concepts and values implicit in restoration policy. First, Maser consistently compares the human design of forests with Nature's design. The entire first chapter is a series of short sections comparing the two "designs." In the "Introduction," he writes, "[W]e are redesigning our forests from Nature's blueprint to humanity's blueprint."[15] But Nature, of course, does not have a blueprint, nor a design. As a zoologist, Maser knows this, but his metaphorical talk is dangerous. It implies that we can discover the plan, the methods, the processes of nature, and mold them to our purposes.

Maser himself often writes as if he accepts that implication. The second problem with his argument is the comparison of nature to a mechanism that we do not fully understand. The crucial error we make in simplifying forest ecology—turning forests into plantations—is that we are assuming our design for the forest mechanism is better than nature's. "Forests are not automobiles in which we can tailor artificially substituted parts for original parts."[16] How true. But Maser's argument against this substitution is empirical: "A forest cannot be 'rebuilt' and remain the same forest, but we could probably rebuild a forest similar to the original if we knew how. No one has ever done it. . . . [W]e do not have a parts catalog, or a maintenance manual. . . ."[17] The implication is that if we did have a catalog and manual, if nature were known as well as artifactual machines, then the restoration of forests would be morally and practically acceptable. This conclusion serves as Maser's chief argument for the preservation of old-growth and other unmanaged forests: "We have to maintain some original, unmanaged old-growth forest, mature forest, and young-growth forest as parts catalog, maintenance manual, and service department from which to learn to practice restoration forestry."[18] Is the forest-as-parts-catalog a better guiding metaphor than the forest-as-plantation?

This mechanistic conception of nature underlies, or explains, the third problem with Maser's argument. His goal for restoration forestry, his purpose in criticizing the short-term plantation mentality, is irredeemably anthropocentric. The problem with present-day forestry practices is that they are "exclusive of all other human values except production of fast-grown wood fiber."[19] It is the elimination of other human values and interests that concerns Maser. "We need to learn to see the forest as the factory that produces raw materials. . . ." to meet out "common goal[:] . . . a sustainable forest for a sustainable industry for a sustainable environment for a sustainable human population."[20] Restoration forestry is necessary because it is the best method for achieving the human goods which we extract from nature. Our goal is to build a better "factory-forest," using the complex knowledge of forest ecology.

What is disturbing about Maser's position is that it comes from an environmentalist. Unlike Elliot's theoretical opponents of conservation, who wished to subvert the environmentalist position with the "restoration thesis," Maser advocates the human design of forests as a method of environmental protection and conservation for human use. His conclusion shows us the danger of using anthropocentric and mechanistic models of thought in the formulation of environmental policy. These models leave us with forests that are "factories" for the production of human commodities, spare-parts catalogs for the maintenance of the machine.

But Maser's view can be considered an extreme version of restoration thinking. Is Steve Packard's work with the Nature Conservancy a better expression of the underlying principles and values of restoration policy?[21] Is Packard's work more aligned with natural processes? Is it less technological, artifactual, and anthropocentric? Unfortunately not: even this more benign and less interventionist project of ecological restoration is based on problematic assumptions about the management of nature.

Packard describes the research and actions undertaken to rediscover and restore the tallgrass savanna or oak opening community of the Midwest. As he relates, the rediscovery of the savanna was an accidental by-product of a different project, the restoration of prairie landscapes which included a bur oak edge. Involving even small sites with degraded "prairies," the project entailed the enlargement of the areas by clearing brush and planting prairie species in its place. "Our objective was clear," he writes. "It was to restore these tracts to their original natural condition."

But how was this goal achieved? Packard asserts that he wanted to use "natural forces" such as fire to clear the brush; but this methodology is soon abandoned: "the question was, did we have enough determination and patience to give natural processes two or three hundred years to work themselves out? Or could we find something quicker?" Thus, he writes, "we decided to leapfrog the persistent brushy border and to re-cut our fire lines. . . ." Although Packard is using the natural force of fire, he is employing it in an artificially accelerated manner to achieve the desired results more quickly. A similar process is used when the "seeding process" begins: naturally occurring seeds are used, but the process involves the preparation of a "savanna mix," and human decisions regarding the placement and release of the seeds.

Although I have nothing but admiration for Packard's work, and I sincerely applaud his success, the significant philosophical lesson from his restoration project is that even such a "benign" and minimal intervention compromises the natural integrity of the system being restored. Despite his goal of restoring an original natural condition, Packard is actually creating an artificial substitute for the real savanna, one based on human technologies and designed for human purposes: a pure and grand vision of the old Midwest. The most telling passage in his chronicle of the savanna restoration is his report of the "farsighted" 1913 law which established the Forest Preserve District, a law whose statement of purpose "emboldened" Packard to accelerate the burning process. He quotes the law, with emphasis added: "to *restore,* restock, protect and preserve the natural forests and said lands . . . as nearly as may be, in their natural state and condition, for the purpose of the education, pleasure, and recreation of the public." Note that the purpose of the preservation and restoration is the production of human goods; as with all artifacts, the goal is a human benefit. Packard calls this a "noble statement." Clearly the aim of restoration is the creation of environments that are pleasing to the human population. If the restoration is done well, as in the case of

Packard's savannas, the area may appear natural; but it will not be natural, since it is the result of a technological acceleration of natural forces.

I began this section with a report of my visceral reaction to the technological re-creation of natural environments. This reaction has now been explained and analyzed. Nature restoration projects are the creations of human technologies, and as such, are artifacts. But artifacts are essentially the constructs of an anthropocentric worldview. They are designed by humans for humans to satisfy human interests and needs. Artifactual restored nature is thus fundamentally different from natural objects and systems which exist without human design. It is not surprising, then, that we view restored nature with a value different from the original.

IV

To this point, my analysis has supported the argument and conclusions of Elliot's criticism of "the restoration thesis." But further reflection on the nature of artifacts, and the comparison of forests to well run machines, makes me doubt the central analogy which serves as the foundation of his case. Can we compare an undisturbed natural environment to a work of art? Should we?

As noted in Section II, Elliot uses the art/nature analogy to make two fundamental points about the process of evolution: (1) the importance of a continuous casual history; and (2) the use of knowledge about this causal history to make appropriate judgments. A work of art or a natural entity which lacks a continuous causal history, as understood by the expert in the field, would be judged inferior. If the object is "passed off" as an original, with its causal history intact, then we would judge it to be a forgery or an instance of "faked" nature.

I do not deny that this is a powerful analogy. It demonstrates the crucial importance of causal history in the analysis of value. But the analogy should not be pushed too far, for the comparison suggests that we possess an understanding of art forgery as adequate for this task. L. B. Cebik argues that an analysis of forgery involves basic ontological questions about the meaning of art. Cebik claims that it is a mistake to focus exclusively on questions of value when analyzing art forgeries, for the practice of forgery raises fundamental issues about the status of art itself.[22]

According to Cebik, an analysis of forgeries demonstrated that our understanding of art is dominated by a limiting paradigm—"production by individuals." We focus almost exclusively on the individual identity of the artist as the determining factor in assessing authenticity. "Nowhere . . . is there room for paradigmatic art being fluid, unfinished, evolving, and continuous in its creation." Cebik has in mind a dynamic, communally cased art, an ever-changing neighborhood mural or music passed on for generations.[23] Another example would be classical ballet, a performance of which is unique dynamic movement, different from every other performance of the same ballot.

These suggestions about a different paradigm of art show clearly, I think, what is wrong with the art/nature analogy as a useful analytical tool. Natural entities and systems are much more akin to the fluid evolving art of Cebik's alternative model than they are to the static, finished, individual artworks of the dominant paradigm. It is thus an error to use criteria of forgery and authenticity that derive from an individualistic, static conception of art for an evaluation of natural entities and systems. Natural entities and systems are nothing like static, finished objects of art. They are fluid, evolving systems which completely transcend the category of artist or creator. The perceived disvalue in restored natural objects does not derive from a misunderstanding over the identity of the creator of the objects. It derives instead from the misplaced category of "creator"—for natural objects do not have creators or designers as human artworks do. Once we realize that the natural entity we are viewing has been "restored" by a human artisan it ceases to be a natural object. It is not a forgery; it is an artifact.

We thus return to artifacts, and their essential anthropocentric nature. We cannot (and should not) think of natural objects as artifacts, for this imposes a human purpose or design on their very essence. As artifacts, they are evaluated by their success in meeting human interests and needs, not by their own intrinsic being. Using the art/nature analogy of forgery reinforces the impression that natural objects are similar to artifacts—artworks—and that they can be evaluated using the same anthropocentric criteria. Natural entities have to be evaluated on their own terms, not as artworks, machines, factories, or any other human-created artifact.

V

But when are the terms appropriate for the evaluation of natural objects? What criteria should be used? To answer this question we need to do more than differentiate natural objects from artifacts; we need to examine the essence or nature of natural objects. What does it mean to say that an entity is natural (and hence, not an artifact)? Is there a distinguishing mark or characteristic that determines the descriptive judgment? What makes an object natural, and why is the standard not met through the restoration process?

The simple answer to this question—a response I basically support—is that the natural is defined as being independent of the actions of humanity. Thus, Taylor advocated a principle of noninterference as a primary moral duty in his ethic of respect for nature. "[W]e put aside our personal likes and our human interests. . . . Our respect for nature means that we acknowledge the sufficiency of the natural world to sustain its own proper order throughout the whole domain of life."[24] The processes of the natural world that are free of human interference are the most natural.

There are two obvious problems with this first simple answer. First, there is the empirical point that the human effect on the environment is, by now, fairly pervasive. No part of the natural world lies untouched by our pollution and technology. In a sense, then, nothing natural truly exists (anymore). Second, there is the logical point that humans themselves are naturally evolved beings, and so all human actions would be "natural," regardless of the amount of technology used or the interference on nonhuman nature. The creation of artifacts is a natural human activity, and thus the distinction between artifact and natural object begins to blur.

These problems in the relationship of humanity to nature are not new. Mill raised similar objections to the idea of "nature" as a moral norm over a hundred years ago, and I need not review his arguments.[25] The answer to these problems is twofold. First, we admit that the concepts of "natural" and "artifactual" are not absolutes; they exist along a spectrum, where various gradations of both concepts can be discerned. The human effect on the natural world is pervasive, but there are differences in human actions that make a descriptive difference. A toxic waste dump is different from a compost heap of organic material. To claim that both are equally non-natural would obscure important distinctions.

A second response is presented by Brennan.[26] Although a broad definition of "natural" denotes independence from human management or interference, a more useful notion (because it has implications for value theory and ethics) can be derived from the consideration of evolutionary adaptations. Our natural diet is the one we are adapted for, that is "in keeping with our nature." All human activity is not unnatural, only that activity which goes beyond our biological and evolutionary capacities. As an example, Brennan cites the procedure of "natural childbirth," that is, childbirth free of technological medical interventions. "Childbirth is an especially striking example of the wildness within us . . . where we can appreciate the natural at firsthand. . . ." It is natural, free, and wild not because it is a nonhuman activity—after all, it is human childbirth—but because it is independent of a certain type of human activity, actions designed to control or to manipulate natural processes.

The "natural" then is a term we use to designate objects and processes that exist as far as possible from human manipulation and control. Natural entities are autonomous in ways that human-created artifacts are not; as Taylor writes, "to be free to pursue the realization of one's good according to the laws of one's nature."[27] Then we thus judge natural objects, and evaluate them more highly than artifacts, we are focusing on the extent of their independence from human domination. In this sense, then, human action can also be judged to be natural—these are the human actions that exist as evolutionary adaptations, free of the control and alteration of technological processes.

If these reflections on the meaning of "natural" are plausible, then it should be clear why the restoration process fails to meet the criteria of naturalness. The attempt to redesign, re-create, and restore natural areas and objects is a radical intervention in natural processes. Although there is an obvious spectrum of possible restoration and redesign projects which differ in their value—Maser's redesigned sustainable forest is better than a tree plantation—all of these projects involve the manipulation and domination of natural areas. All of these projects involve the creation of artifactual natural realities, the imposition of anthropocentric interests on the processes and objects of nature. Nature is not permitted to be free, to pursue its own independent course of development.

The fundamental error is thus domination, the denial of freedom and autonomy. Anthropocentrism, the major concern of most environmental philosophers, is only one species of the more basic attack on the preeminent value of self-realization. From within the perspective of anthropocentrism, humanity believes it is justified in dominating and molding the nonhuman world to its own human purposes. But a policy of domination transcends the anthropocentric subversion of natural process. A policy of domination subverts both nature and human existence; it denies both the cultural and natural realization of individual good, human and nonhuman. Liberation from all forms of domination is thus the chief goal of any ethical or political system.

It is difficult to awaken from the dream of domination. We are all impressed by the power and breadth of human technological achievements. Why is it not possible to extend this power further, until we control, manipulate, and dominate the entire natural universe? This is the illusion that the restoration of nature presents to us. But it is only an illusion. Once we dominate nature, once we restore and redesign nature for our own purposes, then we have destroyed nature—we have created an artifactual reality, in a sense, a false reality, which merely provides us the pleasant illusory appearance of the natural environment.

VI

As a concluding note, let me leave the realm of philosophical speculation and return to the world of practical environmental policy. Nothing I have said in this essay should be taken as an enforcement of actions that develop, exploit, or injure areas of the natural environment and leave them in a damaged state. I believe, for example, that Exxon should attempt to clean up and restore the Alaskan waterways and land that was harmed by its corporate negligence. The point of my argument here is that we must not misunderstand what we humans are doing when we attempt to restore or repair natural areas. We are not restoring nature; we are not making it whole and healthy again. Nature restoration is a compromise; it should not be a basic policy goal. It is a policy that makes the best of a bad situation; it cleans up our mess. We are putting a piece of furniture over the stain in the carpet, for it provides a better appearance. As a matter of policy, however, it would be much more significant to prevent the causes of the stains.

NOTES

1. In Islip Town, New York, real-estate developers have cited the New York State Department of Environmental Conservation policy of "no-net loss" in proposing the restoration of parts of their property to a natural state, in exchange for permission to develop. A report in *Newsday* discusses a controversial case: "In hopes of gaining town-board approval, Blankman has promised to return a three-quarter-mile dirt road on his property to its natural habitat. . . ." Katti Gray, "Wetlands in the Eye of a Storm," Islip Special, *Newsday*, April 22, 1990; pp. 1, 5.

2. *Garbage: The Practical Journal for the Environment,* May/June 1990, rear cover.

3. Chris Maser, *The Redesigned Forest* (San Pedro, CA: R. & E. Miles, 1988), p. 173. It is also interesting to note that there now exists a dissident group within the U.S. Forest Service, called the Association of Forest Service Employees for Environmental Ethics (AFSEEE). They advocate a return to sustainable forestry.

4. Paul Taylor, *Respect for Nature: A Theory of Environmental Ethics* (Princeton, NJ: Princeton University Press, 1986), pp. 186–92, 304–6, and Ch. 4 and 6 generally.

5. Peter S. Wenz, *Environmental Justice* (Albany: SUNY Press, 1998), pp. 287–91.

6. Robert Elliot, "Faking Nature," *Inquiry* 25 (1982): 81–93; reprinted in Donald VanDeVeer and Christine Pierce, eds., *People, Penguins, and Plastic Trees: Basic Issues in Environmental Ethics* (Belmont, CA: Wadsworth, 1986), pp. 142–50.

7. Ibid., p. 86 (VanDeVeer and Pierce, p. 145).

8. Ibid., p. 91 (VanDeVeer and Pierce, p. 149).

9. Eric Katz, "Organism, Community, and the 'Substitution Problem,'" *Environmental Ethics* 7 (1985): 253–55.

10. Elliot, p. 81 (VanDeVeer and Pierce, p. 142).

11. See, for example, Barry Commoner, *The Closing Circle* (New York: Knopf, 1971) and Arnold Pacey, *The Culture of Technology* (Cambridge: MIT Press, 1983).

12. Michael Losonsky, "The Nature of Artifacts," *Philosophy* 65 (1990): 88.

13. Ibid., pp. 176–77.

14. Andrew Brennan, "The Moral Standing of Natural Objects," *Environmental Ethics* 6 (1984): 41–44.

15. Maser, *The Redesigned Forest,* p. xvii.

16. Ibid., pp. 176–77.

17. Ibid., pp. 88–89.

18. Ibid., p. 174.

19. Ibid., p. 94.

20. Ibid., pp. 148–49.

21. Steve Packard, "Just a Few Oddball Species: Restoration and the Rediscovery of the Tallgrass Savanna," *Restoration and Management Notes* 6:1 (Summer 1998): 13–22.

22. L. B. Cebik, "Forging Issues from Forged Art," *Southern Journal of Philosophy* 27 (1989): 331–46.

23. Ibid., p. 342.

24. Taylor, *Respect for Nature,* p. 177. The rule of noninterference is discussed on pp. 173–79.

25. J. S. Mill, "Nature," in *Three Essays on Religion* (London: Longmans, Green, Reader, and Dyer, 1874).

26. Andrew Brennan, *Thinking about Nature: An Investigation of Nature, Value, and Ecology* (Athens: University of Georgia Press, 1998), pp. 88–91.

27. Taylor, *Respect for Nature,* p. 174.

30

Ecological Restoration and the Culture of Nature: A Pragmatic Perspective

Andrew Light

Most environmental philosophers have failed to understand the theoretical and practical importance of ecological restoration. I believe this failure is primarily due to the mistaken impression that ecological restoration is only an attempt to restore nature itself, rather than an effort to restore an important part of the human relationship with nonhuman nature. In investigating this claim, I will first discuss the possibility of transforming environmental philosophy into a more pragmatic discipline, better suited to contributing to the formation of sound environmental policies, including ecological restoration. In particular, I will advocate an alternative philosophical approach to the kind of work on the value of ecological restoration raised by Eric Katz and other philosophers who claim that restored nature can never reproduce the actual value of nature. Here, I will make this contrast more explicit and go on to further argue that Katz's views in particular are not sufficiently sensitive to the values at work in the variety of projects falling within the category of ecological restoration. A more practically oriented philosophical contribution to future discussions of our policies concerning ecological restoration is needed than has been provided by environmental philosophers so far. A richer description of the ethical implications of restoration will identify a large part of its value in the revitalization of the human culture of nature. Before reaching this conclusion, however, I will briefly consider an alternative framework for environmental philosophy as a whole.

ENVIRONMENTAL PHILOSOPHY: WHAT AND FOR WHOM?

Two underlying questions that I believe still confound most environmental philosophers are, What is our discipline actually for? and, consequently, Who is our audience? So far, most work in environmental ethics has been concerned with describing the nonanthropocentric value of nature—that is, the value of nature independent of human concerns and reasons for valuing nature—and deter-

Andrew Light, "Ecological Restoration and the Culture of Nature: A Pragmatic Perspective," in *Restoring Nature: Perspectives from the Social Sciences and Humanities,* ed. Paul H. Gobster and R. Bruce Hull, Washington, D.C.: Island Press, 2000, pp. 49–70. Copyright © by Island Press. Reprinted by permission of Island Press, Washington, D.C.

mining the duties, obligations, or rights that follow from that description. But one can easily wonder whether such work is directed only toward other environmental philosophers as a contribution to the literature on value theory or whether it has a broader aim. Certainly, given the history of the field—formally beginning in the early 1970s with the work of thinkers as diverse as Arne Naess, Val Plumwood, Holmes Rolston, Peter Singer, and Richard Sylvan, all concerned with how philosophers could make some sort of contribution to the resolution of environmental problems—one would think that the aspirations of environmental philosophy would be greater than simply continuing an intramural discussion about the value of nature.

But if environmental philosophy is more than a discussion among philosphers about natural value, to what broader purposes and audiences should it reach? Taking a cue from the content and expected readership of this book, I pose at least four responses. Environmental philosophy might serve as (1) a guide for environmental activists searching for ethical justifications for their activities in defense of other animals and ecosystems; (2) an applied ethic for resource managers; (3) a general tool for policy makers, helping them to shape more responsible environmental policies; and (4) a beacon for the public at large, attempting to expand their notions of moral obligation beyond the traditional confines of anthropocentric (human-centered) moral concerns.

Environmental philosophy should, of course, aim to serve all of these purposes and groups, although I think that most importantly we should focus our energies on guiding policy makers and the public. My rationale is this: if the original reason for philosophers establishing this field was to make a philosophical contribution to the resolution of environmental problems (consistent with the response by other professionals in the early 1970s around environmental concerns), then the continuation, indeed the urgency, of those problems demands that philosophers do all that they can to actually help change present policies and attitudes involving environmental problems. If we talk only to each other about value theory, we have failed as environmental professionals, but if we can help convince policy makers to form better policies and make the case to the public at large to support these policies for ethical reasons, then we can join other environmental professionals in making more productive contributions to the resolution of environmental problems.

But as it now stands, the current focus in environmental philosophy on describing the nonanthropocentric value of nature often ends up separating environmental philosophy from other forms of environmental inquiry. As a prime example of this disconnection from practical considerations, many environmental philosophers do not think of restoration ecology in a positive light. My friend and colleague Eric Katz comes near the top of this list of philosophers; his chapter, "Another Look at Restoration: Technology and Artificial Nature" in *Restoring Ecology*, is the latest in a series of articles in which he argues that ecological restoration does not result in a restoration of nature and, in fact, may even create a disvalue in nature. Robert Elliot is another influential thinker in this camp, although his views have moderated significantly in recent years. Katz, Elliot, and others maintain that if the goal of environmental philosophy is to describe the nonhuman-centered value of nature and to distinguish nature from human appreciation of it, then presumably nature cannot be the sort of thing that is associated with human creation or manipulation. Thus, if restorations are human creations, so the arguments of the philosophical critics like Katz go, they can never count as the sort of thing that contains natural value.

In this view, restorations are not natural—they are artifacts. To claim that environmental philosophers should be concerned with ecological restoration is therefore to commit a kind of category mistake: it is to ask that they talk about something that is not part of nature. But to label ecological restorations a philosophical category mistake is the best case scenario of their assessment. At worst, restorations in their view represent the tyranny of humans over nature and shouldn't be practiced at all. Katz has put it most emphatically in arguing that "the practice of ecological restoration *can only* represent a misguided faith in the hegemony and infallibility of the human power to control the natural world" (Katz 1996, 222, my emphasis).

I have long disagreed with claims like this one. My early response to such positions was to simply set them aside in my search for broader ethical and political questions useful for a more public discussion of policies concerning ecological restoration (e.g., Light and Higgs, 1996). But I now think it is dangerous to ignore the arguments of Katz and Elliot, for at least two reasons. First, the arguments of Katz and Elliot represent the most sustained attempt yet to make a philosophical contribution to the overall literature on restoration and thus ought to be answered by philosophers also interested in restoration. Second, the larger restoration community is increasingly coming to believe that the sorts of questions being addressed by Katz and Elliot are the only kind of contribution that philosophy as a discipline can make to discussions of restoration. And since Katz has explicitly rejected the idea that ecological restoration is an acceptable environmental practice, the restoration community's assumption that environmental ethicists tend to be hostile to the idea of ecological restoration is a fair one. Given this disjunction, there would be no ground left for a philosophical contribution to public policy questions concerning ecological restoration since none of these issues would count as moral or ethical questions.[1]

I believe that philosophers can make constructive contributions to ecological restoration and to environmental issues in general by helping to articulate the normative foundations for environmental policies in ways that are translatable to the public. But making such contributions requires doing environmental philosophy in some different ways. Specifically, it requires a more public philosophy, one focused on making the kinds of arguments that resonate with the moral intuitions that most people carry around with them every day. Such intuitions usually resonate more with human-centered notions of value than with abstract nonanthropocentric conceptions of natural value.

I call the view that makes it plausible for me to make this claim about the importance of appealing to human motivations in valuing nature *environmental pragmatism*. By this I do not mean an application of the traditional writings of the American pragmatists—Dewey, James, and Pierce, for example—to environmental problems. Instead, I simply mean the recognition that a responsible and complete environmental philosophy includes a public component with a clear policy emphasis (see, for example, Light, 1996a; 1996b; 1996c). It is certainly appropriate for philosophers to continue their search for a true and foundational nonanthropocentric description of the value of nature. But environmental philosophers would be remiss if they did not also try to make other, perhaps more appealing ethical arguments that may have an audience in an anthropocentric public. Environmental pragmatism in my sense is agnostic concerning the existence of nonanthropocentric natural value. It is simply a methodology permitting environmental philosophers to endorse a pluralism allowing for one kind of philosophical task inside the philosophy community—searching for the "real" value of nature; and another task outside of that community—articulating a value to nature that resonates with the public and therefore has more impact on discussions of projects such as ecological restorations that may be performed by the public.

This approach modifies the philosophical contribution to questions about restoration ecology in a positive way. As mentioned, many philosophers have criticized ecological restoration because it is a human intervention into natural processes. In contrast, I have argued that such projects as the prairie restorations at the University of Wisconsin–Madison Arboretum would be fully supported by a pragmatic environmental philosophy (Light, 1996b). Restoration makes sense because on the whole it results in many advantages over mere preservation of ecosystems that have been substantially damaged by humans. More significantly, this pragmatic approach exposes other salient ethical issues involving the practice of ecological restoration beyond the discussion of natural value, such as whether there are moral grounds that justify encouraging public participation in restoration (see Light and Higgs, 1996). It is therefore the duty of the pragmatic environmental philosopher to get involved in debates with practitioners about what the value of restoration is in human terms, rather than restricting the discussion to a private debate among philosophers on whether restored nature is really nature. In the rest of this chapter, I will both offer a specific critique of Katz's claims about the value of restoration that does not rely on a pragmatist foundation for environmental philosophy as well as go on to discuss some pragmatic issues that contribute to a fuller philosophical analysis of the practice and ethics of ecological restoration.

ECOLOGICAL RESTORATION: A PRELIMINARY DISTINCTION

Following the project described above, in previous work I have outlined some preliminary distinctions that paint a broader picture of the philosophical terrain up for grabs in restoration than that presented by Katz and Elliot. Specifically, in response to Elliot's early critique of restoration (1995), I have tried to distinguish between two categories of ecological restoration that have differing moral implications.

Elliot begins his seminal article on restoration, "Faking Nature," by identifying a particularly pernicious kind of restoration—restoration that is used to rationalize the destruction of nature. On this claim, any harm done to nature by humans is ultimately repairable through restoration, so the harm should be discounted. Elliot calls this view the "restoration thesis" and states that "the destruction of what has value [in nature] is compensated for by the later creation (re-creation) of something of equal value" (Elliot 1995, 76). Elliot rejects the restoration thesis through an analogy based on the relationship between original and replicated works of art and nature. Just as we would not value a replication of a work of art as much as we would value the original, we would not value a replicated bit of nature as much as we would the original thing, such as some bit of wilderness. Elliot is persuasive that the two sorts of value choices are similar.

In responding to Elliot's (1995) criticisms of the value of restoration, I suggested a distinction implicit in his analysis of restoration to help us think through the value of ecological restoration (Light, 1997). The distinction is based on an acknowledgment Elliot makes in his 1995 article (and expands upon in his 1997 book):

> Artificially transforming an utterly barren, ecologically bankrupt landscape into something richer and more subtle may be a good thing. That is a view quite compatible with the belief that replacing a rich natural environment with a rich artificial one is a bad thing. (Elliot, 1995, 82)[2]

Following Elliot's lead that some kinds of restoration may be beneficial, I distinguished between two sorts of restorations: (1) malicious restorations, such as the kind described in the restoration thesis; and (2) benevolent restorations, or those undertaken to remedy a past harm done to nature although not offered as a justification for harming nature. Benevolent restorations, unlike malicious restorations, cannot serve as justifications for the conditions that would warrant their engagement.

If this distinction holds, then we can claim that Elliot's original target was not all of restoration, but only a particular kind of restoration, namely, malicious restorations. Although there is mixed evidence to support the claim that Elliot was originally going only after malicious restorations in his first work on the topic, the distinction is nonetheless intuitively plausible. It is certainly not the case, for example, that the sorts of restorations undertaken at the Wisconsin Arboretum or as part of the Chicago Wilderness effort are offered as excuses or rationales for the destruction of nature. In contrast, the restorations involved in mountaintop mining projects in rural West Virginia can certainly be seen as examples of malicious restorations. Mountaintop mining—through which tops of mountains are destroyed and dumped into adjacent valleys—is in part rationalized through a requirement that the damaged streambeds in the adjacent valleys be restored. The presumed ability to restore these streambeds is used as a justification for allowing mountaintop mining, counting this practice as a clear instantiation of Elliot's restoration thesis. The upshot of this malicious-benevolent distinction, however, is that one may be able to grant much of Elliot's claim that restored nature is not original nature while still not denying that there is some kind of positive value to the act of ecological restoration in many cases. Even if benevolent restorations are not restorations of original nature, and hence more akin to art forgeries rather than original works of art, they can still have some kind of positive content.

This positive content for many restorations can be developed more by pushing the art analogy a bit further. If ecological restoration is a material practice like making a piece of art (fake or not), why isn't it more like art restoration rather than art forgery? After all, we know that some parallels can be drawn between restoration projects and mitigation projects. A mitigation often involves the wholesale creation of a new ecosystem designed to look like a bit of nature that may have absolutely no historical continuity with the natural history of the land on which it is placed. For example, in order to meet an environmental standard that demands no net loss of wetlands, some environmental managers will sanction the creation of a wetland to replace a destroyed one on a piece of land where there had been no wetland. Conversely, a restoration must be tied to some claim about the historical continuity of the land on which the restoration is taking place. In some cases, this might simply entail linking original pieces of nature together to restore the integrity of the original ecosystem without creating a new landscape altogether (as in the case of the Wildlands Project to link the great Western parks in the United States and Canada with protected corridors). In that sense, a restoration could be more like repairing a damaged work of art than creating a fake one.[3]

The possibility of having benevolent restorations does much to clear the way for a positive philosophical contribution to questions of restoration. Katz, however, unlike Elliot, denies the positive value of any kind of restoration. For him, all restorations "can only" be malicious because they all represent evidence of human domination and arrogance toward nature. But surprisingly, even though Katz draws on Elliot's work in formulating his own position, he seems to ignore the fact that Elliot's original description of the restoration thesis was primarily directed against particular kinds of restorations. In his earliest and most famous article on restoration, "The Big Lie: Human Restoration of Nature," Katz acknowledged that although Elliot claimed that the restoration thesis mostly was advocated as a way of undermining conservation efforts by big business, he (Katz) was surprised to see environmental thinkers (such as forest biologist Chris Maser) advocating "a position similar to Elliot's 'restoration thesis.'" This position as Katz interprets it is that "restoration of damaged nature is seen not only as a practical option for environmental policy but also as a moral obligation for right-thinking environmentalists" (Katz, 1997, 96). But Maser's position is not the restoration thesis as Elliot defines it. Katz never does show that Maser, or any other restoration advocate that he analyzes, actually argues for restoration as a rationale for destruction of nature. As such, Katz never demonstrates that those in the restoration community that he criticizes endorse restorations for malicious reasons. If that is the case, then what is wrong with restoration in Katz's view?

KATZ AGAINST RESTORATION

Just as Elliot's original target of the "restoration thesis" has faded from philosophical memory, Katz's original target has also been somewhat lost in the years since he began writing on this topic in 1992. At first, Katz seemed most concerned with the arguments of fellow environmental ethicists like Paul Taylor and Peter Wenz, who advocated variously "restitutive justice" and a "principle of restitution" as part of our fulfillment of possible human obligations to nature. If we harmed nature, according to Taylor and Wenz, we would have to compensate it. Restoration would be part of a reasonable package of restitution. According to Katz, on these views humans have an "*obligation* and *ability* to repair or reconstruct damaged ecosystems" (Katz, 1997, 95, my emphasis). But I think it is crucial here to pay attention to the argument Katz is actually taking on and the objection he proceeds to make.

As Katz describes it, there are actually two separable questions to put to Taylor, Wenz, and other advocates of restoration: (1) do we have an obligation to try to restore damaged nature? and (2) do we have the ability to restore damaged nature? Katz argues quite forcefully that we do not have the ability to restore nature because what we actually create in ecological restorations are humanly produced artifacts and not nature, nonanthropocentrically conceived. Based on this claim, he assumes that the first question—whether we have an obligation to try to restore nature—is moot. Katz's logic is simple: we do not have an obligation to do what we cannot in principle do.

But even if we were to grant Katz the argument that it is impossible to restore nature, we may still have moral obligations to *try* to restore nature. How can this be true? There are a number of reasons, but before fully explicating this position we need to first better understand Katz's arguments.

Katz's chapter in *Restoring Ecology* reviews and expands upon several arguments he has made against restoration over the years. In examining his papers on this topic,[4] I have identified five separable, but often overlapping, arguments he has made against both the idea that we can restore nature and the practice of trying to restore it. I call these arguments KR1–5. They are listed below in order of how they arise in his work, accompanied with an example of supporting evidence from Katz's various papers on restoration.

KR1. The Duplicitous Argument
"I am outraged by the idea that a technologically created 'nature' will be passed off as reality" (Katz, 1997, 97).[5]

KR2. The Arrogance (or Hubris) Argument
"The human presumption that we are capable of this technological fix demonstrates (once again) the arrogance with which humanity surveys the natural world" (Katz, 1997, 97).

KR3. The Artifact Argument
"The re-created natural environment that is the end result of a restoration project is nothing more than an artifact created for human use" (Katz, 1997, 97).[6]

KR4. The Domination Argument
"The attempt to redesign, re-create and restore natural areas and objects is a radical intervention in natural processes. Although there is an obvious spectrum of possible restoration[s] . . . all of these projects involve the manipulation and domination of natural areas. All of these projects involve the creation of artifactual realities, the imposition of anthropocentric interests on the processes and objects of value. Nature is not permitted to be free, to pursue its own independent course of development" (Katz, 1997, 105).[7]

KR5. The Replacement Argument
"If a restored environment is an adequate replacement for the previously existing natural environment [which, for Katz, it can never be], then humans can use, degrade, destroy, and replace natural entities and habitats with no moral consequence whatsoever. The value in the original natural entity does not require preservation" (Katz, 1997, 113).[8]

I disagree with all of these arguments and have what I hope are thorough answers to all of them elsewhere. Here, I will focus on KR4, the domination argument, which is perhaps the argument that comes up the most throughout all of Katz's restoration papers. It is arguably the case that one can answer all of Katz's arguments by conceding one important premise of all of his claims as long as KR4 can be independently answered. KR4 is also interesting to me because his original articulation of it involved a very slim bit of admission that there is some sort of difference between various kinds of restoration projects. Even though these differences are not ultimately important for Katz, they are still nonetheless acknowledged, and they give me a space in which I can critique his position.

As I said above, I believe that KR1–3 and KR5 can be ignored in rejecting Katz's position as long as we are prepared to concede for now one important premise to all of his arguments. This is Katz's ontological assumption (a claim concerning the nature or essence of a thing) that humans and nature can be meaningfully separated so as to definitively argue that restored nature is an artifact, a part of human culture, rather than a part of nature. As Katz has admitted in an as yet unpublished public forum on his work, he is a nature–culture dualist. This means that for Katz, nature and culture are separate things entirely.[9] If one rejects this overall ontological view, then one may reject most of Katz's objections to restoration. But it is incredibly difficult to disprove another philosopher's ontology, let alone get him or her to concede this point.[10] Thus, even though I disagree with it, I will accept Katz's underlying assumption that restored nature does not reproduce nature.

But even if I grant this point that restored nature is not really nature, KR4 is still false because it is arguably the case that restoration does not "dominate" nature in any coherent sense but instead often helps nature to be "free" of just the sort of domination that Katz is worried about. The reasoning here is straightforward enough. If I can show that restorations are valuable for nature, even if I concede that they do not re-create nature, then the various motivations for restoration will distinguish whether a restoration is duplicitous (KR1) or arrogant (KR2). A benevolent restoration, for example, would not risk KR1 or KR2 because in principle it is not trying to fool anyone nor is it necessarily arrogant. Further, and more simply, conceding Katz's ontological claim about the distinction between nature and culture eliminates the significance of KR3—since we no longer care that what is created may or may not be an artifact—as well as KR5, since we have given up hope that a restoration could ever actually serve as a replacement for "real" nature.

Now, back to the domination argument. KR4 is a claim that could hold even for a view that conceded Katz's nature–culture distinction. The reason, following Katz, would be that even a failed attempt to duplicate natural value—or create something akin to nature while conceding that in principle "real" nature can never be restored by humans—could still count as an instance of "domination," as Katz has described it. An *attempt* at restoration, according to Katz's logic, would still prohibit nature from ever being able to pursue its own development. The reason is that for Katz, restoration is always a substitute for whatever would have occurred at a particular site without human interference. The idea is that even if humans can produce a valuable landscape of some sort on a denuded acreage, this act of production is still an instance of domination over the alternative of a natural evolution of this same acreage, even if a significant natural change would take ten times as long as the human-induced change and would be arguably less valuable for the species making use of it. Still, one can muster several arguments against KR4 (I will provide four) and still play largely within Katz's biggest and most contentious assumption about the ontological status of restored nature. After going through these arguments, we will see that these claims can lead to a new philosophical context for the evaluation of restoration, which I believe in the end also undermines the other KR arguments.

1) We can imagine cases in which nature cannot pursue its own interests (however one wishes

to understand this sense of nature having interests) because of something we have done to it. For example, many instances of restoration are limited to bioactivation of soil that has become contaminated by one form or another of hazardous industrial waste. If restoration necessarily prohibits nature from being "free," as KR4 maintains, then how do we reconcile the relative freedom that bioactivation makes possible with this claim? Restoration need not determine exactly what grows in a certain place, but may in fact simply be the act of allowing nature to again pursue its own interests rather than shackling it to perpetual human-induced trauma. In many cases of restoration, this point can be driven home further when we see how anthropogenically damaged land (or soil) can be uniquely put at risk of invasion by anthropogenically introduced exotic plants. South African ice plant, an exotic in southern California that destroys the soil it is introduced into, is highly opportunistic and can easily spread onto degraded land, thus ensuring that native plants will not be able to reestablish themselves. I highlight here this contentious native-exotic distinction because I suspect that given Katz's strong nature–culture distinction he would necessarily have to prefer a landscape of native plants over a landscape of exotics where the existence of the exotics is a result of an act of human (cultural) interference in nature. Allowing nature to pursue its own interests, given prior anthropogenic interference, thus involves at least as strong a claim to protect it from further anthropogenic risk through restoration practices as the case Katz makes for leaving it alone.

2) Even if we do agree with Katz that restorations only produce artifacts, can't it still be the case that the harm we cause nature still requires us to engage in what Katz would term "attempted restorations"? It simply does not follow from the premise that something is more natural when it is relatively free of human interference that we must therefore always avoid interfering with nature (this is actually a point that Katz finally recognizes in a later paper, "Imperialism and Environmentalism"). It is a classic premise of holism in environmental ethics (the theory that obligations to the nonhuman natural world are to whole ecosystems and not to individual entities, a view that Katz endorses) that some interference is warranted when we are the cause of an imbalance in nature. For example, hunting white-tailed deer is thought to be permissible under holism since humans have caused that species' population explosion. If such interventions are permissible to help "rectify the balance of nature," then why are there not comparable cases with the use of restoration as an aid to the "original," "real" nature? We can even imagine that such cases would be less controversial than holist defenses of hunting.

There are good cases in which restoration, even if it results in the production of an artifact, does not lead to the domination described by Katz. Imagine the case where the restoration project is one that will restore a corridor between two wilderness preserves. If there is positive natural value in the two preserves that is threatened because wildlife is not allowed to move freely between them, then restoration projects that would restore a corridor (by removing roads, for example) would actually not only be morally permissible but also possibly ethically required depending on one's views of the value of the nature in the preserves. This is not restoration as a "second best" to preservation or a distraction away from preservation; it is restoration as an integral and critical part of the maintenance of natural value. So, even if we agree with Katz that humans cannot really restore nature, it does not follow that they ought not to engage in restoration projects that actually repair the damage caused by past domination rather than furthering that domination.

Given objections like the two discussed so far, it is important to try to get a better handle on exactly what sort of damage is caused by domination in the sense described by Katz. It turns out that the worst damage to nature for Katz is domination that prevents the "self-realization" of nature:

> The fundamental error is thus domination, the denial of freedom and autonomy. Anthropocentrism, the major concern of most environmental philosophers, is only one species of the more

basic attack on the preeminent value of self-realization. From within the perspective of anthropocentrism, humanity believes it is justified in dominating and molding the nonhuman world to its
own human purposes. (Katz, 1997, 105)

Thus, the problem with restoration is that it restricts natural self-realization in order to force nature
onto a path that we would find more appealing.

3) With this clarification, we can then further object to Katz that his sense of restoration
confuses restoration with mitigation. The force of the charge of domination is that we mold nature
to fit our "own human purposes." But most restorationists would counter that it is nonanthropocentric nature that sets the goals for restoration, not humans. Although there is indeed some subjectivity in determining what should be restored at a particular site (which period do we restore to?) and
uncertainty in how we should do it (limitations in scientific and technical expertise), we cannot
restore a landscape just any way we wish and still have a good restoration in scientific terms. If
Katz objects that when we restore a denuded bit of land we are at least making something that fits
our need of having more attractive "natural" surroundings—an argument that Katz often
makes—we can reply that because of the constraints on restoration, as opposed to mitigation, the
fact that we find a restored landscape appealing is only contingently true. It is often the case that
what we must restore to is not the preferred landscape of most people. The Chicago Wilderness
project is a good example of this: many local residents see restoration activities as destroying the
aesthetically pleasing forests that now exist in order to restore the prairie and oak savanna ecosystems that existed prior to European settlement. But philosophically, because a restored landscape
can never necessarily be tied only to our own desires (since our desires are not historically and
scientifically determined in the same way as the parameters of a restoration), then those desires
cannot actually be the direct cause of any restriction on the self-realization of nature.

4) Finally, we must wonder about this value of self-realization. Setting aside the inherent
philosophical problems with understanding what this claim to self-realization means in the case of
nature, one has to wonder how we could know what natural self-realization would be in any particular case and why we would totally divorce a human role in helping to make it happen if we could
discern it. In an analogous case involving two humans, we do not say that a human right to (or
value of) self-realization is abrogated when a criminal who harms someone is forced to pay restitution. Even if the restitution is forced against the will of the victim, and even if the compensation in
principle can never make up for the harm done, we would not say that somehow the victim's self-
realization has been restricted by the act of restitution by the criminal. Again, there seems to be no
clear argument here for why the moral obligation to try to restore has been diminished by Katz's
arguments that we do not have the ability to really restore nature or pass off an artifact as nature.

RESTORING ENVIRONMENTAL PHILOSOPHY

If I am justified in setting aside the rest of Katz's arguments (KR1–3 and KR5) by accepting his
claim that humans really cannot restore "real" nature, then what sort of conclusions could we draw
about the role of philosophy in sorting out the normative issues involved in restoration? As it turns
out, Katz gives us an insight in figuring out the next step.

After explaining the harm we do to nature in the domination we visit upon it through acts of
restoration, Katz briefly assesses the harm that we do to *ourselves* through such actions:

> But a policy of domination transcends the anthropocentric subversion of natural processes. A
> policy of domination subverts both nature and human existence; it denies both the cultural and

natural realization of individual good, human and nonhuman. Liberation from all forms of domination is thus the chief goal of any ethical or political system. (Katz 1997, 105)

Although not very clearly explained by Katz, this intuition represents a crucial point for proceeding
further. In addition to connecting environmental philosophy to larger projects of social liberation,
Katz here opens the door to a consideration of the consequences of restoration on humans and
human communities. As such, Katz allows an implicit assertion that there is a value involved in
restoration that must be evaluated in addition to the value of the objects that are produced by restoration.

But the problem with drawing this conclusion is that this passage is also perhaps the most
cryptic in all of Katz's work on restoration. What does Katz mean by this claim? How exactly does
restoration deny the realization of an individual human, or cultural, good? This claim can only be
made understandable by assuming that some kind of cultural value connected to nature is risked
through the act of domination, or otherwise causing harm to nature. But what is this value?

I think the value Katz is alluding to here, although he never explores it seriously, describes
the value of that part of human culture that is connected to external, nonhuman nature. This is not
simply a suggestion that we humans are part of nature; it also points out that we have a *relationship
with* nature that exists on moral as well as physical terrain in such a way that our actions toward
nature can reciprocally harm us. If this is the view implicit in this claim, then it is still consistent
with much of the rest of Katz's larger views about the value of nature. We have a relationship with
nature even if we are separable from it. I will accept this basic tenet of Katz's argument: we do
exist in some kind of moral relationship with nature. And without fully explicating the content of
that relationship, it seems that Katz is right in assuming that somehow the way in which we act
toward nature morally implicates us in a particular way. In the same sense, when we morally mis-
treat another human, we not only harm them but also harm ourselves (by diminishing our character,
by implicating ourselves in evil, or however you want to put it). Katz is suggesting that our relationship with nature has a determinant effect on our moral character. Or, perhaps more accurately, this
is a suggestion necessary for Katz's comment to make sense, even though he never expresses this
view himself.

Now if this assumption is correct, and if there is anything to the arguments I have put forward
so far that there can be some kind of positive value to our interaction with nature, then doing right
by nature will have the same reciprocal effect of morally implicating us in a positive value as occurs
when we do right by other persons. Perhaps Katz would agree. Where Katz would disagree is with
the suggestion I would want to add to this: that there is some part of many kinds of restorations (if
not most kinds) that contains positive value. Aside from the other suggestions I have already made
concerning the possible positive content of restoration, one can also consider that the relationship
with nature that is implied in Katz's view has a moral content in itself that is not reducible to the
value of fulfilling this relationship's concomitant obligations. The relationship between humans and
nature imbues restoration with a positive value even if it cannot replicate natural value in its products. But understanding this point will require some explanation.

Consider that if I have a reciprocal relationship with another human (in which I do right by
them and they do right by me), then, to generalize Katz's account, there is a moral content to both
of our actions that implicates each of us as persons. Each of us is a better person morally because
of the way we interact with each other in the relationship. But the relationship itself, or rather just
the fact of the existence of the relationship, also has a moral content of its own (or what we could
call a *normative content,* meaning that the relationship can be assessed as being in a better or worse
state) that is independent of the fulfillment of any obligations. If this point of the possible separation
between the value of a relationship and the value of the fulfillment of obligations does not follow

intuitively, imagine the case where two people act according to duty toward each other without building a relationship of substantive normative content between them. Consider the following example. I have a brother with whom I am not terribly close. Although I always act according to duty to him—I never knowingly do harm to him and I even extend special family obligations to him—I do not have a substantive relationship with him that in itself has a normative content. Thus, if I do not speak to him for a year, nothing is lost because there is no relationship there to maintain or that requires maintenance for normative reasons. But if my brother needed a kidney transplant, I would give him my kidney unhesitatingly out of a sense of obligation—something I would not feel obliged to do for non–family members—even though I still do not feel intimately comfortable around him in the same way I do with my closest friends. Our relationship as persons—that sense of intimate affection and care for another person that I have experienced with other people—has no positive value for me (it isn't necessarily a disvalue, only a sense of indifference and a lack of closeness). So, I can have interaction with another person, even interaction that involves substantial components of obligation and duty (and, in Katz's terms, I will never put myself in a position to dominate that other person) but still not have a relationship with that person that involves any kind of positive value or that has normative standards of maintenance.

I do not think that I have any obligation to have a relationship in this sense with my brother. I, in fact, do not, even though my mother would like it if I did. But if I did have a relationship with my brother in this sense, then it would have a value above and beyond the moral interaction that I have with him now (the obligations that I have to him that can be iterated) that aids in a determination of our moral character.[11] If we had a relationship with normative content, there would be a positive or negative value that could be assessed if I lost touch with my brother or ceased to care about his welfare. (I could very well claim that it would be better for me to have such a relationship with him, but this would require an additional argument.)

Consider further that if I wanted to rectify or create anew a substantive normative relationship with my brother, like the relationship I have with several close friends, how would I do it? One thing I could do would be to engage in activities with him—the same sorts of activities (let us call them *material interactions*) that I do with my friends now. I might work with him to put up a fence or help him plant his garden. I might begin to talk over my personal and professional problems with him. I might go on a long journey with him that demanded some kind of mutual reliance such as whitewater rafting or visiting a foreign city where neither of us spoke the native language. In short, although there are, of course, no guarantees, I could begin to have some kind of material relationship with him as a prelude to having some kind of substantive normative relationship with him. Many factors might limit the success of such a project: for one thing, the distance between the two of us—he lives in our hometown of Atlanta and I live in New York. So, if I was really serious about this project of building a relationship between us that had value independent of the value of the fulfillment of our mutual obligations to each other that already exist, I'd have to come up with ways to bridge these interfering factors. Importantly, though, I couldn't form a substantive normative relationship with him merely by respecting his right of self-realization and autonomy as a person; I would have to somehow become actively involved with him.

Now, when we compare the case of the estranged brother to that of nature, many parallels arise. We know that we can fulfill obligations to nature in terms of respecting its autonomy and self-realization as a subject (in Katz's terms) without ever forming a substantive normative relationship with it. Assuming also that there is a kind of relationship with nature possible according to Katz's scheme (for this is in part what we harm when we dominate nature), it is fair to say that a relationship consisting of positive normative value with nature is compatible with Katz's overall view of the human–nature relationship. Because he says so little about what our positive relationship to nature could be, he is in no position to restrict it a priori. We also know that, as in the case of the estranged brother, we need some kind of material bridge to create a relationship with nature in order to see that relationship come about.

How do we build that bridge? Suggesting ways to overcome the gap between humans and nature (without necessarily disvaluing it) seems in part to be the restored role of environmental philosophy in questions of ecological restoration. Certainly, as in the case of my brother, distance is a problem. Numerous environmental professionals have emphasized the importance of being in nature in order to care for nature. Also, acts of preservation are important for there to be nature to have a relationship with. But what about restoration? Can restoration help engender such a positive normative relationship with nature? It seems clear to me that it can. When we engage in acts of benevolent restoration, we are *bound by* nature in the same sense that we are obligated to respect what it once was attempting to realize before we interfered with it. In Katz's terms, we are attempting to respect it as an autonomous subject. But we are also *bound to* nature in the act of restoring. In addition to the substantial personal and social benefits that accrue to people who engage in benevolent forms of restoration,[12] we can also say that restoration restores the human connection to nature by restoring that part of culture that has historically contained a connection to nature. This kind of relationship goes well beyond mere reciprocity; it involves the creation of a value in relationship with nature beyond obligation. Although it would take further argument to prove, I believe that this kind of relationship is a necessary condition for encouraging people to protect natural systems and landscapes around them rather than trade them off for short-term monetary gains from development. If I am in a normative relationship with the land around me (whether it is "real" nature or not), I am less likely to allow it to be harmed further. Specifying the parameters of restoration that help to achieve this moral relationship with nature will be the task of a more pragmatic environmental philosophy. As mentioned at the outset of this chapter, environmental pragmatism allows for and encourages the development of human-centered notions of the value of nature. Pragmatists are not restricted to identifying obligations to nature in the existence of nonanthropocentric conceptions of value but may embrace an expression of environmental values in human terms. More adequately developing the idea of restoration in terms of the human–nature relationship is thus appropriately under the pragmatist's purview. More importantly, however, the value articulated here exists between anthropocentrism and nonanthropocentrism, fully relying on the capacities of both sides of the human–nature relationship.[13]

We can even look to Katz for help in completing this pragmatic task. We don't want restorations that try to pass themselves off as the real thing when they are really "fakes" (KR1) or are pursued through arrogance (KR2); nor are we interested in those that are offered as justifications for replacing or destroying nature (KR5). We would not want our comparable human relationships to exhibit those properties either. But even given the legacy of inhuman treatment of each other, we know that it is possible to restore human relationships that do not resemble KR1, KR2, or KR5. There is, however, one possible worry to attend to in KR3, the artifact argument. Although earlier I said that the importance of KR3 is diminished by granting Katz's nature–culture distinction, there is a way that it can still cause us problems in grounding attempts at restoration in the positive value of strengthening the human–nature relationship.

Katz may object to my relationship argument that if we allow his claim that what has been restored is not really nature then we are not restoring a cultural relationship with nature but, in a sense, only extending the artifactual material culture of humans. At best, all we can have with restoration is a relationship with artifacts, not nature. Maybe he will allow that we improve relations with each other through cooperative acts of restoration, but this is not the same as a restoration of a relationship with nature itself.

But it should be clear by now that Katz would be mistaken to make such an objection for several reasons stemming in part from my earlier remarks.

1) Even if we admit that restored nature is an artifact and not real nature, restored nature can also serve as a way for real nature to free itself from the shackles we have previously placed upon it. Restoration can allow nature to engage in its own autonomous restitution. Of the different sorts of restoration projects that I have sketched above, many amount to aids to nature rather than creations of new nature.

2) Even if restoration is the production of an artifact, these artifacts do bear a striking resemblance to the real thing. This is not to say that restorations can be good enough to fool us (KR1). Rather, it is simply to point out that an opportunity to interact with the flora and fauna of the sort most common in benevolent restorations will increase the bonds of care that people will have with nonrestored nature. If a denuded and abandoned lot in the middle of an inner-city ghetto is restored by local residents who have never been outside of their city, then it will help them better appreciate the fragility and complexity of the natural processes of nature itself should they encounter them. The fact that restorationists are engaged in a technological process does not necessarily mean that their practices do not serve the broader purpose of restoring a relationship with nature. Just as beginning some form of mediated communication with my brother (such as e-mail or regular phone calls) does not restore a fully healthy communicative relationship with him that could be found through face-to-face conversation, it still helps me get used to the idea of some form of immediate and substantive communication.

And, finally, 3) if Katz persists in his worry that the act of restoration reifies domination by reaffirming our power over nature through the creation of artifacts, we can say that exactly the opposite is likely the case (at least in the case of benevolent restorations) when the goal is restoring the culture of nature, if not nature itself. Restorationists get firsthand (rather than anecdotal and textbook) exposure to the actual consequences of human domination of nature. A better understanding of the problems of bioactivating soil, for example, gives us a better idea of the complexity of the harm we have caused to natural processes. In a much healthier way than Katz seems willing to admit, knowing about that harm can empower us to know more precisely why we should object to the kinds of activities that can cause that harm to nature in the first place. As a parallel human case, imagine a carrier of a deadly and contagious disease (that she cannot die from) who ignores warnings about how to take precautions against spreading the disease to other people. If that person passes on her deadly disease to other people, would it not in the end benefit her to have the opportunity to volunteer to work in a hospital ward full of people dying from this particular disease? If the disease was incurable, she could never restore health to its victims (either out of reciprocity or a desire to form helpful normative relationships with others), but she might learn through her experience in the hospital ward to respect the importance of not risking giving this disease to others. Restoration similarly teaches us the actual consequences of our actions rather than allowing us to ignore them by restricting our interaction with nature to those parts we have not yet damaged.[14]

CONCLUSION

In a followup essay to "The Big Lie" called "The Call of the Wild," which used the figure of the *wildness* in the white-tailed deer population at Katz's summer home on Fire Island to help distinguish nature from culture, Katz embraced a kind of reciprocal relationship with nature. The wild white-tailed deer, which Katz admits in the essay are now quite tame, are described as

> members of [Katz's] moral and natural community. The deer and I are partners in the continuous struggle for the preservation of autonomy, freedom, and integrity. This shared partnership creates

obligations on the part of humanity for the preservation and protection of the natural world. (Katz 1997, 117)

Surely we would respond that this relationship also creates obligations of benevolent restoration as well. If the deer were threatened with harm without a needed restoration of a breeding ground, for example, would Katz not be obliged to do it? And, in doing this restoration, would he not help to generate positive value in his relationship with those deer?

It seems clear that benevolent restorations of this sort are valuable because they help us restore our relationship with nature, by restoring what could be termed our "culture of nature." This is true even if Katz is correct that restored nature has the ontological property of an artifact. Restoration is an obligation exercised in the interests of forming a positive community with nature and thus is well within the boundaries of a positive, pragmatic environmental philosophy. Just as artifacts can serve valuable relationship goals by creating material bridges to other subjects, artifactual landscapes can help restore the culture of nature. Further defining the normative ground of benevolent restorations should be the contribution that philosophy can make to the public consideration and practice of ecological restoration. It is a contribution directed at a larger audience, beyond the professional philosophy community, and aimed toward the practical end of helping to resolve environmental problems.[15]

NOTES

1. If we accept Katz's position, a philosophical inquiry into restoration would actually be an investigation of some other kinds of questions than those legitimately posed by environmental philosophers. Since Katz argues that restored nature is only an artifact, philosophers of technology would presumably still be doing philosophy when they were involved in an investigation of ecological restoration. This possibility of trying to define out certain practices from environmental ethics is no red herring. In a public forum discussing his work at the Central Division meeting of the American Philosophical Association in Chicago in 1998, Katz stated publicly that agriculture was not the proper purview of environmental ethics. Philosophers working on questions of ethics and agriculture could be doing agricultural ethics but not environmental ethics.

2. Elliot strengthens the more charitable view of restoration in his 1997 book.

3. From the early aesthetic theory of Mark Sagoff (before he ever turned to environmental questions), one can also pull the following distinction to help further deepen the discussion of different kinds of benevolent restorations: (1) integral restorations—restorations that "put new pieces in the place of original fragments that have been lost"; and (2) purist restorations—restorations that "limit [themselves] to clearing works of art and to reattaching original pieces that may have fallen" (Sagoff 1978, 457). As it turns out, one can argue that integral restorations are aesthetically (and possibly ethically) worrisome since they seem to create hybrid works of art (created by both the artist and the restorationist). But this is not too much of a problem for the analogy with ecological restoration since many of these restorations amount to something more akin to purist restorations—for example, cleaning land by bioactivating soil. Perhaps more common would be a subclass of purist restoration that we might call *rehabilitative* restoration. Examples of such projects would include cleaning out exotic plants that were introduced at some time into a site and allowing the native plants to reestablish themselves. Such activity is akin to the work of a purist art restorationist who corrects the work of a restorationist who had come before her. If a restorationist, for example, were to remove an eighteenth-century integral addition to a sixteenth-century painting, then we would assume that this rehabilitative act was consistent with a purist restoration. I provide a much more thorough discussion of the import of this distinction for ecological restoration in Light (1997).

4. Katz has four main papers on restoration: "The Big Lie: Human Restoration of Nature" (1992), "The Call of the Wild: The Struggle against Domination and the Technological Fix of Nature" (1992), "Artifacts and Functions: A Note on the Value of Nature" (1993), and "Imperialism and Environmentalism" (1993). All of these papers are collected in Katz (1997), and it is these versions that I have drawn on for this chapter.

5. Originally in Katz, "The Big Lie" (as are KR2–KR4). KR 1 is restated later in "The Call of the Wild":

"What makes value in the artifactually restored natural environment questionable is its ostensible claim to be the original" (Katz 1997, 114).

6. KR3 is most thoroughly elaborated later in Katz, "Artifacts and Functions."

7. The domination argument is repeated in Katz, "The Call of the Wild" (1997, 115) with the addition of an imported quote from Eugene Hargrove: domination "reduces [nature's] ability to be creative." The argument is also repeated in Katz, "Artifacts and Functions," and further specified in Katz, "Imperialism and Environmentalism." As far as I can tell, though, the argument for domination is not really expanded on in this last paper, except that imperialism is deemed wrong because it makes nature into an artifact (KR3).

8. Originally in Katz, "The Call of the Wild," and repeated in Katz, "Imperialism and Environmentalism" (1997, 139).

9. The forum here is the same as the one referenced in note 1: a public forum discussing Katz's work at the Central Division meeting of the American Philosophical Association in Chicago in 1998.

10. The absence of any perceptible progress in Katz's views following his debate with Donald Scherer is a case in point. Scherer spends too much time, I think, trying to push a critique of Katz's ontology and metaphysics. The resulting debate appears intractable. See Scherer (1995) and Katz (1996).

11. On a broader scale, just as there can be a town full of decent, law-abiding citizens, those citizens may not constitute a moral community in any significant sense.

12. Herbert W. Schroeder, "Psychological Benefits of Volunteering in Stewardship Programs," in *Restoring Ecology*. Robert E. Grese, Rachel Kaplan. Robert L. Ryan, and Jane Buxton, "Lessons for Restoration in the Traditions of Stewardship: Sustainable Land Management in Northern New Mexico," in *Restoring Ecology*.

13. It is also the case that restoration will be only one out of a large collection of practices available for adaptive management. Indeed, there could even be cases where something akin to mitigation (albeit a benevolent kind) would be justified rather than restoration if a claim to sustaining some form of natural value warranted it. In a project to clean up an abandoned mine site, for example, we can imagine a case where restoring the site to a landscape that was there before would not be the best choice and that instead some other sustainable landscape that would help to preserve an endangered species now in the area would be more appropriate. But overall, environmentalists must accept human interaction with nature as an acceptable practice to begin the ethical assessment of any case of environmental management. I am indebted to Anne Chapman for pressing me to clarify this point.

14. Katz can legitimately respond here that there seems to be no unique reason why people couldn't get these kinds of experiences that generate a closer relationship with nature out of some other kinds of activities. Why couldn't we just use this sort of argument to encourage more acts of preservation, or to simply take more walks though nature? Such an objection would, however, miss a crucial point. Even if it can be proved that we can get these kinds of positive experiences with nature in forms other than acts of restoration (and I see no reason why we couldn't), this does not diminish the case being built here: that restoration does not necessarily result in the domination of nature. The goal of my argument here is not to show that restoration provides a unique value compared with other environmental practices, but only to reject the claim that there is no kind of positive value that restoration can contribute to nature in some sense. So, an objection by Katz of this sort would miss the target of our substantive disagreement. Additionally, one could also argue that (1) restoration does, in fact, produce some unique values in our relationship with nature (see Andrew Light, "Negotiating Nature: Making Restoration Happen in an Urban Park Context," in *Restoring Ecology* for such a case in relation to the potential democratic values in restoration); and that (2) even if not unique in itself, restoration helps to improve other sorts of unique values in nature. A case for (2) could be made, for example, in Allen Carlson's work on the importance of scientific understanding for appreciating the aesthetic value of nature (Carlson 1995). Arguably, our experiences as restorationists give us some of the kinds of understandings of natural processes required for aesthetic appreciation according to Carlson's account. Importantly, this understanding is a transitive property: it gives us an ability to aesthetically appreciate not only the nature we are trying to restore, but also the nature we are not trying to restore. Restoration thus could provide a unique avenue into the aesthetic appreciation of all of nature, restored or not. The main point, however, should not be lost: restoration is an important component in a mosaic of efforts to revive the culture of nature. Without any reason to believe that it has other disastrous effects, restoration seems warranted within a prescribed context even if it is not a cure-all.

15. This chapter is based on a presentation originally given at a plenary session (with Eric Katz and Wil-

liam Jordan) of the International Symposium on Society and Resource Management, University of Missouri, Columbia, May 1998. Subsequent versions were presented as the keynote address of the Eastern Pennsylvania Philosophy Association annual meeting, Bloomsburg University, November 1998; and at Georgia State University, SUNY Binghamton, and Lancaster University (UK). I have benefited much from the discussions at all of these occasions and especially from the helpful comments provided by Cari Dzuris, Cheryl Foster, Warwick Fox, Paul Gobster, Leslie Heywood, Bruce Hull, Bryan Norton, George Rainbolt, and Christopher Wellman.

REFERENCES

Carlson, A. 1995. "Nature, Aesthetic Appreciation, and Knowledge." *Journal of Aesthetics and Art Criticism* 53: 393–400.

Elliot, R. 1995. "Faking Nature." In *Environmental Ethics,* edited by R. Elliot, 76–88. Oxford: Oxford University Press.

———. 1997. *Faking Nature.* London: Routledge.

Katz, E. 1996. "The Problem of Ecological Restoration." *Environmental Ethics* 18: 222–224.

———. 1997. *Nature as Subject: Human Obligation and Natural Community.* Lanham, MD: Rowman & Littlefield Publishers.

Light, A. 1996a. "Environmental Pragmatism as Philosophy or Metaphilosophy." In *Environmental Pragmatism,* edited by A. Light and E. Katz, 325–338. London: Routledge.

———. 1996b. "Compatibilism in Political Ecology." In *Environmental Pragmatism,* edited by A. Light and E. Katz, 161–184. London: Routledge.

———. 1996c. "Callicott and Naess on Pluralism." *Inquiry* 39: 273–94.

———. 1997. "Restoration and Reproduction." Unpublished manuscript presented at the Symposium on Ethics and Environmental Change: Recognizing the Autonomy of Nature, St. John's, Newfoundland, Canada, June 4–5.

Light, A., and E. Higgs. 1996. "The Politics of Ecological Restoration." *Environmental Ethics* 18: 227–247.

Sagoff, M. 1978. "On Restoring and Reproducing Art." *The Journal of Philosophy* 75: 453–470.

Scherer, D. 1995. "Evolution, Human Living, and the Practice of Ecological Restoration." *Environmental Ethics* 17: 359–379.

31

The Brave New World of Animal Biotechnology

Strachan Donnelley

EXPLORING ETHICAL LANDSCAPES

Oncomouse, BST cows, Beltsville pigs, and transgenic fish, among other bioengineered "exotics," have recently appeared on the horizon of practical ethics, with more to come. Biotechnology and transgenic animals in particular, which involve the introduction of foreign genetic material (usually from another species) into more standard reproductive processes, offer a challenging opportunity for reconsidering crucial and perplexing ethical issues that surround our human uses of animal life and our relation to wider animate nature. When ethically grappling with most practical issues, we inveterately ignore the natural settings and ecosystems of which we are integrally a part and in which our human existence in inextricably embedded. And despite the efforts of philosophically minded animal rightists, when seriously considering our ethical duties to animals we usually consider only individual animal welfare of experimental well-being, particularly pain and suffering, and not the broader significance of animal and animate life per se. But the new animal biotechnologies do not so easily let us off the hook. Their practical implications go decidedly beyond the more familiar ethical issues of individual welfare and well-being. Certain new biotechniques delve deeply into and manipulate the very character of animal and animate being. These new technologies have the potential to alter nature radically. Are these manipulations and alterations ethically legitimate? Do we have the moral right to undertake such interventions? If so, in what contexts, under what conditions, and within what limits?

These are not only primarily welfare issues. The ethical challenges of animal biotechnology more elusively point to living nature and "organic being" as such and the meaning that life, including human life, holds for us. The practices raise questions about ultimate values. The ethical issues touch upon respect (not only compassion), goodness (not only happiness), and violation (not only subjectively experienced harm). Moreover, biotechnological manipulations threaten to reach into the fundamental character of animal species that have arisen in historically deep evolutionary and ecological contexts. Nature and the various values that evolutionary and ecological contexts have for us are implicated. Does animal biotechnology in general or in its specific instances constitute an ethically appropriate human relation to nature? Are these the ways we ought to fit into the animate world?

These pressing questions naturally and legitimately surface with the advent of recent biotechnological developments. But they are not easily or straightforwardly settled. Animal biotechnology exists within the bewildering complex fabric of human communities and as part of the wider cultural and historical theme of human interventions into nature and animal life. Arguably the full range of humans' interaction with nature constitutes the most appropriate context for raising final ethical concerns: What should be the ethically self-imposed limits, if any, to our interventions into nature, for what reasons, in service of what moral values? From this perspective, does biotechnology, animal and other, have specific features that mark it out for special ethical attention? Does the scope, scale, and depth of the new interventions into life processes constitute a significant ethical *novum*? Moreover, and equally important, when ethically considering biotechnological practices, should we be guided by a single moral reason and ultimate value, or are we rather obliged to respond to a plurality of reasons and ultimate values, underived from one another and not easily coordinated?

This fundamental theme of monism and pluralism is crucial to understanding the force and limits of the various arguments that are characteristically brought to bear on biotechnology's practices, for and against. In our everyday experience of the world, we characteristically encounter many, perhaps innumerable spheres of human activity, all comprehended within wider human communities. Each sphere is centrally animated by its own specific human purposes and motivations, its own overarching perspectives on the world, and its own dominant ethical values and aspirations. The various forms of animal biotechnology are implicated in these several areas of activity in different ways, sometimes integral to their practical purposes and mores, sometimes a challenge to their animating vision and dominant values. This varied implication of biotechnology in different contexts undergirds the passions of its boosters and critics alike. Moreover, since we all participate directly in several spheres of human activity, differing moral attitudes can conflict within us individually. With varying emphases we can be both for and against moral reasons and according to different world visions. This is the ethical and practical public policy terrain through which we must make our way.

The particular philosophic and ethical challenges that this plural, complex, and internally conflicted ethical landscape poses are the explicit focus of this section. But we can here gain a brief overview of the thrust and scope of the overall problem. Consider the characteristically human spheres of economic activity, public political life, health care and medicine, scientific and biomedical research, domestic and family life, recreation and play, cultural and religious practices and traditions, and more. Consider the various values animating these several domains of activity: compassionate response to human (and animal) welfare and creaturely needs; ongoing commitments to opportunities for individual and community life; fundamental obligations to respect individual persons; respect for social justice; deep allegiances to aesthetic, religious, and ethical values and norms of cultural traditions and natural landscapes; pressing concern for the world's human community and the global biosphere, now and into the indefinite future. Here are familiar, everyday multiple values and ethical obligations. We encounter and undeniably plurality of moral goods.

Now consider traditional human uses of animals and their new biotechnological extensions. Since time immemorial we have domesticated and selectively bred animals for food, clothing, animal companies, and sport. Now we have growth-hormone-boosted, super milk-producing cows; the leaner, faster-growing, but arthritis-ridden Beltsville pig, also charged with added growth hormones; and the possibility of fish engineered to grow faster, fight harder, or resist cold and pollution better. Since the advent of modern science and biomedicine, animals have been used in research, educations, toxicity testing, and the production of therapeutic pharmaceutical agents. Now we have Oncomouse, Cystic Fibrosis mouse, and Memory mouse fashioned to chart basic biological processes or the course of significant human diseases, to aid in the discovery of practical medical therapies, and to test for suspected carcinogens. We have mice, goats, and sheep bioengineered to produce human insulin or other biologically active products in their milk. Is there a morally significant difference between the new and the old practices? Which practices, new or old, are ethically legitimate?

To grapple effectively with these questions requiring paying close attention to the various spheres of human activity, the many values and obligations involved, and the differing interventions into animal life. We need to attend to particular contexts of animal biotechnological intervention and satisfy plural moral concerns, some more pressing than others. Within these contexts we require a moral coordination of disparate and perhaps conflicting obligations. This involves attending to contextual plural obligations settled by judgments involving a decision-making art of "moral ecology" that weighs the various moral claims vis-à-vis one another.

In short, there invariably are moral claims of welfare, respect, justice, and the human and natural good at play in particular situations of animal biotechnological practice, but the fabric or constellation of these claims may significantly shift from context to context. What might be ethically permissible in the biomedical laboratory might be prohibited on farms, in the market place, or in the wild. But in each context all things morally relevant need explicitly to be considered and given their due. Because the various spheres of human activity are not hermetically sealed off from one another, either in human communities or within the lives of individuals, there is always the temptation to poach from outside the context at hand and to establish the hegemony of one ultimate ethical value over the plural others, whether the value be human or animal welfare, respect, justice, environmental protection, or the cultural or natural good. Animal biotechnological practices, as with all interventions into animal life, involve moral concerns for humans and animals, if not also wider nature, and for various aspects of welfare, respect, justice, and the good. It must be an overall aim of moral thinking and action to keep this plural richness of values and ethical concerns alive, well, and judiciously coordinated.

PHILOSOPHICAL AND ETHICAL CHALLENGES OF ANIMAL BIOTECHNOLOGY

Animal biotechnology and the fashioning of transgenic animals, along with biotechnology in general, are potent new tools in the arsenal of modern, post-industrial societies. They promise to transform scientific and biomedical research, medical therapies and health care, economic markets and agribusiness, if not the rest of our lives. They augur a new era of human existence and well-being. Yet animal bioengineering in particular confronts a curious cultural stumbling block. It faces a cacophony of ethically ardent boosters and passionate detractors alike, animated by equal moral zeal.

At the heart of the contention are the new and deeper human interventions into animate life and the explicit resurrection of an old theme: the incorporation of animals and animal nature in human communities and the corresponding involvement of humans in the wider natural world. In short, humans and nature are and have always been mutually immanent, each in the other. For both boosters and detractors the terms of this mutual involvement seem to be significantly changing. For the boosters, the further incorporation of nature into human communal life (the human "artificing" of nature) is potentially a great boon. For the detractors, we are threatened with extensive harm to natural systems and losing a humanly significant relation to nature. We are in danger of eclipsing important moral, cultural, and religious values and orientations.

Interestingly, the boosters and detractors are not always different. The contention often rages within individuals and communally allied groups themselves. The new interventions into nature seem inherently fraught with moral conflict and ambiguity. How do we explain these ethical tensions, both within and among human individuals?

The answer no doubt must be that individually and collectively we do not live in a single morally harmonious world, undergirded by one or coherently few ethical values that neatly organize our moral life and coordinate our practical activities. Rather, the opposite seems to hold. We confront an ineradicable moral plurality: a bewildering variety of values and ethical obligations, each claiming attention and not readily coordinated with the others. How can we make philosophic sense of this moral disjointedness? And how ethically and practically do we deal with this plurality with respect to animal biotechnology?

Plural moral values seem fundamentally rooted in the complexity of our human nature and our individual status as personal, social, and natural beings. Within the everyday life of human communities, there are several "spheres," "realms," or "domains" of human activity only loosely coordinated and integrated. Each has its perspective on the world; each is animated by dominant values (ethical and other); each involves an overriding notion of how life ought to be lived in the world. These spheres, realms, or domains—for our purposes, the terms are interchangeable—are loosely defined by the fundamental human activities to which they refer. The values, perspectives, and moral worldviews themselves dominantly evolve out of and in turn inform these particular and abiding modes of activity. Each domain has its complex cultural history, dynamically intertwined with the others; each its own structures of meaning and significance by which humans orient themselves in their several practical and cultural activities.

Tensions among these plural realms spawn the ethical contention and ambivalence surrounding the various modes of animal biotechnology. This situation is certainly not unique to animal biotechnology. Moral contention and the tension among plural ultimate values and opposing ethical allegiances seem to warp and woof of modern human existence.

Historically and logically there seem two basic strategies to deal with the pull of opposing values or obligations. One is to subsume the moral many under one grand monistic scheme: to establish a hierarchy of values and obligations under the hegemony of one ultimate value. The other possible strategy is to face the plural values and obligations squarely and somehow attempt to give each its proper due in conjunction with the others.

The monistic strategy may serve the peace of the soul by reducing internal moral conflict. Perhaps it is possible in relatively small and homogeneous communities. In any case, it invariable is bought at the price of the variety and richness of human experience and significant cultural activity. In this sense it impoverishes the human soul.

The monistic strategy is not a viable option in our modern world. Plural traditions and modes of human activity are too well entrenched practically to be subdued. Moreover, the very variety of humanly significant values and activities harbors a richness that is good in itself—if the several plural goods can be adequately coordinated and despite the inner tensions of moral and spiritual life. This might strike some of us as making a virtue out of ethical necessity. Yet the strategy of coordinating plural moral obligations seems our only practical alternative.

That there are several relatively autonomous yet mutually interacting spheres of human activity means that there are always specific or "provincial" contexts of human activity. Thus the coordination of obligations must always be contextual. Moreover, given that the many spheres of human activity can and do importantly influence one another, there must be a coordination of values both *within* and *among* particular spheres of activity. Contextually coordinating our plural obligations requires a decision-making art of moral ecology, judicious mutual weighting of the several obligations in the various contexts at hand, be they narrower or wider.

In considering the philosophic and ethical challenges of animal biotechnology, we will follow the demands of contextual plural obligations and moral ecology. The first task is briefly to sketch the most salient spheres of human activity and their worldviews, dominant values, and perspectives on animal biotechnology (pro, con, or both). This includes recognizing the ethical agreements and tensions among the various spheres with respect to animal biotechnologies. Next we will critically examine animal biotechnological interventions in different contexts of human activity and explore how the coordination of plural obligations might play out in different settings: for example, in the wild, in scientific and biomedical laboratories, in marketplaces, and on farms. Finally we will discuss the unfinished theoretical business of moral ecology and how to face plural obligations in particular contexts: specifically, how overall priorities or coordinations of the plural values and obligations should be effected; what value, goodness, and moral weight should be accorded to animals and nature; and who should decide all this.

Animal Biotechnology and Spheres of Human Activity

Science and Biomedical Research. The original context of animal biotechnology is of course modern science, including biomedicine, and the animating worldview of the natural sciences is crucial to understanding the boosters of animal biotechnology and their moral imperatives. Contemporary natural science has its roots in the philosophic and scientific revolutions of sixteenth and seventeenth century Europe, in which nature was conceived to be fundamentally materialistic and mechanistic in character. According to dominant Galilean, Cartesian, and Newtonian modes of thought, nature, including animate and animal life, was considered to be mere matter or energy in motion, operating under universal, deterministic casual laws (statistical or no) discoverable through experimental manipulations or rationally disciplined and "objective" observations.

This account no doubt constitutes an oversimplification of the activities and world picture of contemporary natural sciences, but arguably it has yet to be significantly modified or superseded, particularly with respect to its value implications. A rigorous classical materialist or mechanist, whether for methodological or ultimate metaphysical reasons, considers nature valueless and purposeless. Nature is accorded no particular goodness (intrinsic significance or inherent value) of its own, though it is deemed to have important instrumental value for us humans and other living beings. It serves our human purposes well, as means to our ends. Nature is considered a great, primary resource providing for human creaturely needs and our peculiarly human thirst for knowledge about the natural universe.

This is a worldview in which scientific knowing and its technological applications march hand in hand to satisfy the welfare needs of humans and perhaps other animals. Human technological control over natural processes in service of human well-being, particularly the relief of suffering or other human deprivations, is deemed an ultimate value and moral imperative. The controlled manipulation of nature, the experimental production of "natural effects" has even been considered by some to constitute the very essence of scientific knowledge.

From such a perspective there seems no inherent ethical check to biotechnological interventions, save practical threats to humans' own moral character and welfare, animal welfare (if animals' experiential nature is given due recognition), and ecosystem processes (via the introduction or escape of destructive organisms into the environment). There is no serious ethical concern for animal species or ecosystems as such, since they are typically conceived to carry no significance or ethical weight of their own. We humans are and should be ruled by enlightened self-interest and only pragmatic concern when it comes to wider animate nature. This is the dominant ethical landscape of modern laboratory science and biomedical research, if not medicine itself.

Interestingly this worldview and ethical perspective begins to shift when we move from the more experimentally manipulative laboratory sciences of physics, chemistry, and molecular biology to evolutionary biology, systematic, and ecology. With the latter sciences, biological and animal species, as the historically engendered outcome of evolutionary and ecological processes, and the evolutionary ecosystems themselves are accorded a significant (though perhaps amoral) reality of their own, independent of their instrumental usefulness to humans—a reality that ought to be protected from the systemic negative consequences of human interventions, whether animal biotechnological or other. This scientifically informed perspective includes, but goes beyond enlightened self-interest. There is an explicit, articulate, if difficult-to-explain respect for animate nature and its autopoetic or self-creative processes.

This shift in worldview and valuation is carried further when we move beyond the conceptually self-disciplined sciences of nature to more speculative philosophies of nature, in which the values and "natural goodness" of animate and animal life and evolutionary and ecological processes are explicitly and critically considered. Here a full-blown respect for nature and its own inherent goodness of fundamental significance in human experience and cultural life emerge. This is the ethical landscape of "natural preservations" of various stripes.

As we move along the continuum from the more reductionist, atomistic, and analytic laboratory sciences through the more systemic sciences of evolutionary biology and ecology to the more full-throated philosophies of nature, there is an ever more critical scrutiny of animal biotechnological practices. Will the biotechnological effects be truly benign with respect to humans, the animals themselves, and nature? Can we adequately predict individual animal, population, or ecosystemic consequences of biotechnological interventions? Can we really control them? Do we at least in certain contexts (for example, in the wild) illicitly interfere with a morally significant intactness or integrity of individual or species life and evolutionary/ecosystemic processes? Concerns of respect overtake and subsume welfare obligations.

Economic Markets and Agribusiness. Modern economic activity, including new agricultural and farming practices, shares much in common with the outlook of modern natural science. There is a common instrumental perspective on nature and animal life as a vast, if not inexhaustible resource for humans to draw upon. There is a common push to a "corporate" coordination of human activity and control over natural processes and products. There is a dominant moral commitment to human welfare and well-being, though here the accent might be more on providing opportunities for practical human activity (including the accumulation of wealth) than compassionately confronting human suffering, morbidity, and mortality.

There is a common conviction, at least in capitalistic countries, that both natural and economic systems run to their own, amoral rhythms, which for pragmatic reasons should not be significantly checked or undermined. Moreover, biotechnology, including animal biotechnology, promises to be good business and to promote the overall economic good. Modern science, its biotechnologies, and business are natural allies and mutually reinforce each other in unimpeded service to the various dimensions of human welfare—as long as other important values and ethical obligations do not intrude.

Public Political Life. The use of animals and animal biotechnology gains its first ethically critical check from the public domain of political life. For here the dominant ethical values are not solely or primarily shaped by the welfare needs or economic opportunities of individual human beings. Rather, especially in modern political democracies, moral attention focuses on the allied ethical values of respect for individuals, political or civil equality, social justice, and due political and legal process.

Historically it is from within this domain of human political activity that the more dominant modern ethical theories—deontological or rights-based and consequentialist/utilitarian with welfare concerns for the effects or consequences of actions—have primarily emerged. These theories base themselves on fundamental principles of equality, justice, and universal or atemporal rationality. Each morally relevant entity (for example, an individual human being) is to count as one and only one and is to be given equal ethical consideration, if not equal specific treatment. This holds whether moral interest is focused upon the rights and respect owed to relevant individuals or their individual welfare.

These politically inspired ethical theories become relevant to the use of animals and animal biotechnology when moral consideration is extended to animals themselves. This consideration is either in virtue of their being sentient creatures with their own experiential welfare or individual "subjects of a life" who ethically command respect and recognition of their inherent rights, for example, to life, liberty of action, and freedom from intentionally inflicted suffering. Whichever the basis of the moral status, the animals are brought within the relevant universe of ethical discourse and concern and are accorded serious and just moral attention.

This is the political ethics that undergirds the moral arguments of contemporary animal rightists and animal liberationists, with their universalist critiques (based on justice and respect or welfare) of our human use of animals, including new biotechnological interventions. These rationally universal arguments have involved a more or less radical condemnation of the human use of animals and animal biotechnology. Yet these same generic forms of political ethics are alternatively used to justify animal use and biotechnology. The arguments seem to turn on where one draws the line of moral considerability: be this the kind of individuality or capacities to be (variously) respected, or sentient beings' varying levels of experiential and existential welfare.

Domestic Life, Homes, and Recreation. If universality or universalism characterize the fundamental rationality and worldviews of the natural sciences and biomedicine, economic theory (if not economic practice), and political ethics, then *particularity* dominates the world of our homes and private lives. Here we also find lively attention to welfare, respect, human flourishing, and fairness, but these concerns characteristically are directed toward particular individuals involved in intimate and long-standing relationships. Here we dwell within a world of familiar subjects, rather than directing ethical attention to unknown others. Ethical considerations and responses are less formal, rationalistic, and objective, and more personal, aesthetic, and all-encompassing or holistic. There is ongoing concern for particular individuals who have multiple needs and dimensions to their lives, and who are more or less wrapped up with one another. The mutual adjustments required by concern for such interdependent individuals defy the more universalist and objective analyses of a politically oriented ethics.

The sphere of private and domestic activity adds a crucial dimension to a mature and nuanced ethics of animal use, including animal biotechnology, though again we should not expect things to be simple. The private realm is where many of us intimately get to know animals, for example, as companions or pets. The animals enter into our private lives and our webs of intimate personal relations. We know our dogs, cats, and horses (among animal others) to have individual characters and specific welfare needs and to enjoy highly particular and reciprocally meaningful relations with us.

Here is a characteristic origin of a natural and wholly legitimate empathetic concern for individual animal welfare, if not also a fundamental philosophical curiosity about animals and animate life. To take animals ethically into account, we do not need to be rationally convinced by the objective justice arguments of the political ethicists. Yet we also know that our "home" animals exist within the wider web of our intimate relations and that our ethical concerns and obligations to intimate human others often and legitimately take precedence over concerns for our animals. We rescue our children and human others before our animals in the face of imminent danger. We humanely "put down" our aging or incurably sick pets, but not ourselves. Less dramatically, when away from home, we send our children to grandparents, friends, or summer camps. Our dogs and cats go to kennels.

In this private realm we vividly glimpse our complex and morally ambivalent attitude toward animals. Animals may genuinely become a part of the ethical fabric of personal lives lived together. But they must fit within this fabric, and this means a primary and overriding, though not exclusive moral concern for the human others. Thus many of us who are keenly interested in animals and their individual welfare ethically tolerate the extermination of animal pests, an ethically appropriate use of animals in scientific and biomedical research and education, the use of animals for food, clothing, and entertainment, and even ethically nuanced and respectful hunting and fishing.

This no doubt is morally ambiguous and slippery territory, with ever present potential for animal abuse. Pet owners may ignore or be oblivious to the genuine welfare needs of their animals. Animal breeders may selectively indulge their own idiosyncratic tastes, at the expense of the breeds themselves. In enjoying meals together, we may neglect the condition of farm animals. Hunters and those who fish may make their prey mere objects of their skills, mutual competitiveness, or economic aspirations. Researchers pursuing scientific or therapeutic knowledge may overlook the suffering, potential or actual, of their animal subjects. This is undeniably true, and it is here that political ethics, with universalist critiques and reason-based appeals to justice and fairness, performs its most valuable service. Nevertheless its strength is perhaps its weakness. Its rationally dispassionate, objective glance characteristically misses the irreducible particularity and complexity of life human and animal. It misses life's multivalued and intricate interwovenness, in which humans both intimately and ethically relate to and use animals. Political ethics usefully prods us to clean up our act with respect to animal abuse, but it arguably should not tamper with the more fundamental fabric and habits and abidingly significant, value-laden activities of human communal life. We need a richer and more nuanced ethics to oversee these complexly textured matters.

In fact, it is precisely the decided ethical significance of the particular that public ethics, with its relative emphasis on sameness and equality over uniqueness and diversity, is most in danger of overlooking. We noted the importance of the *particular* in the attention given to the individual character and specific welfare of animals involved in personal or family webs of lives. But the moral strengths of particularity and a concerned attention to individual animals do not stop in homes. They centrally inform animal research ethics, where a critically anthropomorphic concern for individual and species being ought to hold sway. They help (beyond the call of political ethics) to turn the attention of animal breeders away from humanly "fashionable animals" to individual animals' own particular well-being. They could perhaps add strength to checking the more egregious abuses of factory farming. Particularity focuses the concernful attention of those who hunt, fish, or otherwise pursue animals in the wild on the lives of individual animals, the well-being of specific animal populations or species, and the ongoing viability of the particular habitats and ecosystems of which they are a part. All this is central to good animal ethics and (as we shall see) to ethical analyses of particular animal biotechnological practices.

The Cultural Community Life. The interplay of particularly and universality (as well as plurality and monism, to which it is related) in ethical response to animal use and biotechnology is taken to a new level in cultural activities and traditions, including religion, art, and philosophy. For it is in these spheres of human activity that questions of ultimate meaning or significance regarding humans, animals, and nature are explicitly addressed and systematically explored, if not conclusively answered. Moreover, the values involved are not exclusively or narrowly ethical, but are also religious, aesthetic, philosophical (ontological), and other.

It is from broadly religious, aesthetic, and philosophic-ethical perspectives that animal biotechnology is most searchingly questioned, again with a full range of ethical responses, pro and con. From religious perspectives, whether theistic or atheistic, is there a goodness to natural creation—to the very being and dynamic interplay of human, animal, and ecosystemic life—that is violated or threatened by certain uses of animals and biotechnological interventions? Does the evolving natural world harbor its own value-laden originality and cosmically dramatic story? Is there a certain religiously significant integrity or oneness to human individuals and communities, to animal life (individuals, social groups, and species), and to habitats, ecosystems, and evolutionary processes that bears protection from certain biotechnological practices?

The answers are as many as there are different cultural, religious, and philosophical perspectives—each characteristically aiming at a comprehensive world vision and interpretation, each expressing its own particularities and valuational emphases. Each perspective dwells within the world of human experience in a different way. Some argue for a significant human discontinuity with the rest of nature. They claim nature's amorality, valuelessness, and purposelessness, including that of biological and animal life, and assert humans' rightful and pragmatic dominion over nature and animals. Everything is permitted if it does not practically undermine nature or morally diminish human significance. Other perspectives find decidedly less distance between humans and the rest of nature. They encounter a nature that harbors its own specific goodness, values, and purposive agents, animals in particular. They propose a human relation to nature that involves caring stewardship, if not outright recognition of radical moral and religious equality among all creaturely beings. In these latter perspectives, many animal uses and biotechnological interventions ethically are ruled out and not only because they diminish human moral significance. They are claimed to mar or violate natural creation itself.

What are we to make of this plurality of religious and philosophic perspectives and visions that stand in tension, if not direct conflict with one another? First, these are the fundamental cultural, religious, and moral orientations by which human individuals and communities live. They are not to be lightly dismissed in ethically weighing the appropriateness of animal biotechnological practices. As we shall recurrently note, what might be ethically legitimate in one cultural community might be ethically illegitimate for another community, beyond all questions of animal welfare.

Moreover, though the "ultimate" answers are many and by necessity speculative and tentative, the underlying questions are legitimate and increasingly urgent. Does natural creation, including animal individual and species life and evolutionary/ecological processes, have its own goodness and significance to which we owe appropriate moral response? What is our own human significance in the natural scheme of things, our own limitations moral and otherwise, and our proper relation to the wider animate realm within which we find ourselves? Ethically and religiously, how ought we to fit in?

Nature. Nature in a very real sense is a domain of human activity, though it is most adequately understood as a domain *within which* various human activities take place and *toward which* activities are often directed. The domain itself, at least aboriginally, is not human. Moreover, the human activities undertaken are those of the spheres just described, with their characteristic perspectives, valuations, worldviews, and moral aspirations. Consider scientific and biomedical researchers, economic entrepreneurs, hikers, climbers, hunters/fishermen, ecotourists, philosophers, theologians, poets, artists, and others. Each goes into the natural world animated by different motives, expectations, and reasons. Each discovers there a goodness that is intrinsic to the human-nature interaction or instrumental to some other human good.

The multiple significance of nature for humans (which no doubt is influenced by cultural traditions) defies reduction to any single domain of human activity. In a crucial sense nature is the all-inclusive domain of domains, the context of contexts. All specific human activities take place within, if they do not directly respond to the natural world. Nature as the environing reality or setting of human activity has a special significance for contextual moral ecology and thus the scrutiny of amidst ultimate things human, is what finally must be protected.

Contextual Obligations and Moral Ecology

Each sphere of human activity has its own particular stake in animal biotechnology and its own dominant perspectives and practical valuations, pro or con. Yet none exists in human or cultural isolation, alone by itself, sovereign lord over its own domain. Rather, each sphere has a moral stake in the others, which is more or less weighty. With particular animal biotechnological practices, whatever the specific context, all ethically relevant things need to be considered. This consideration may decidedly encompass the concerns of other spheres of human activity, if not also the full fabric of the environing human community and natural world.

In short, we must conjoin wide-ranging or systemic concerns for humans and nature with ineradicably plural moral and cultural values. This is why we must develop a nuanced art of ethical analysis and decision-making that incorporates notions of contextual plural obligations and moral ecology. In any one context of animal biotechnological use, all the relevant domains of activity, values, and ethical obligations need to be heard. Yet no one constituency, value, or obligation can be a priori be allowed to dominate the others. No one "ethical interest group" ought to hold hostage the legitimate ethical interests (welfare, respect, or other) of the plural others. Each practical provincial context, with its own dominant, pressing, and legitimate ethical claims, whether pertaining to humans, animals, or wider nature, helps to determine the relative weighting, balance, or coordination of the many and perhaps rival ethical values and obligations. This is the final moral judgment that recognizes the ethically legitimate claims of a plurality of culturally embedded spheres of human activity. It aims at the strong moral harmony of a Heraclitus, born of ethical elements interlocked, in tension with one another.

The practical art of contextual analysis and moral ecology, the actual process of ethical decision-making, faces several outstanding philosophic or theoretical challenges. These include what guides the overall contextual coordination, the relative ethical significance or weight of animals and nature, and who decides these issues. These still unsettled questions will be discussed presently. Here we briefly look at animal biotechnological interventions in different activity to see how contextually determined plural obligations might play themselves out in practice.

Transgenic Animals and "Wild Nature." From the early days of biotechnology and the fashioning of transgenic organisms, a major ethical concern has been the purposeful introduction or unintended escape of bioengineered organisms into natural, humanly uncontrolled habitats and ecosystems. Will the novel organisms wreak ecological havoc, undermine the dynamic stability of habitats, and set off destructive chain-reactions throughout resident populations of animals and other organisms? The fear is of negative systemic effects that would practically undermine the well-being of both humans and the natural world. The immediately relevant critical questions are: Do we know the effects of such introductions? Can we predict them? Can we control them? Reasonable doubts on any of these questions counsel practical and ethical caution. Only the most weighty obligations to humans would justify countering this caution, if the risks are truly considerable and systemic. The Achilles heel of the ethical decision-making is our endemic ignorance of causes and effects when it comes to the flourishing of natural ecosystems. Yet in this context the ethical weight decidedly should be with concerns for nature for humans, rather than more parochial and forgoable human interests.

Take the case of experimentally designing transgenic fish, for example, carp, trout, and salmon. The motives for such interventions might be complex: economic, recreational, or preservationist. The practice might serve fish farming (faster-growing fish, with a better and more standard quality of meat). Beyond economically entrepreneurial aspirations, the technology might answer the pressing nutritional requirements of local human communities or the protection of rapidly dwindling, if not endangered wild fish stocks. (This is a worldwide crisis already upon us.) Or the transgenesis might produce fish better adapted to polluted or regional aquatic habitats than their wild counterparts and with qualities attractive to sport fishermen (gullibility, size, or fighting ability).

Here, long-range, morally ecological thinking is crucial. We may easily dismiss the putative "needs" of sport fishermen and economic entrepreneurs. It is less easy to counter genuine nutritional requirements of human populations and the protection of wild fish stocks and aquatic food chains. Yet issues of escapement and exotic species introduction importantly haunt the moral ecology. The genetic or behavioral qualities of transgenic fish introduced or escaped into the wild might undermine the very wild stocks and habitats that they were meant to preserve, to the long-term detriment or impoverishment of both humans and nature. (This is not to mention the problems of environmental pollution engendered by fish farming.)

But this is not all. There is the more elusive, less urgently practical, but fundamental cultural issue for which we need the ethical reservations of natural preservationists, philosophers, and theologians. By practicing transgenesis in the wild, do we or do we not break into natural processes that are good in themselves and that hold an ultimate significance (culturally, religiously, ethically) for many, if not most of us? This is nature engendering its own, more of less well-adapted biological creations—individuals, species, and ecosystems—the animate and animal issue of evolutionary and ecological processes. How important is it for us humanly, culturally, and ethically to protect, within the overall mandates of plural moral obligations, "original nature" and its still originating or creative dynamism?

All these pragmatic and moral factors, human and natural, must be relatively weighted in ethically deciding the role that animal biotechnology ought to play in the wild.

Transgenic Animals, Biomedicine, and Scientific Research. The plural values and obligations relating to humans, animals, and nature that arise in ethically considering animal biotechnology in the wild also surface in scientific and biomedical laboratory settings. But in shifting the scene of scrutiny, the constellation and relative weightings of the values and obligations may change significantly as well. Typically, practical ethical concerns for nature—for wild animal populations, habitats, and ecosystems—fade into moral background. We may still be seriously concerned with the genetic (genomic), bodily, and behavioral intactness of individual animals, but these concerns are now dominantly conjoined with issues of animals' experimental welfare and the possible benefits of the biotechnological interventions for basic science and fundamental human welfare, particularly the alleviation of suffering. All things considered, we might allow biotechnological interventions in "controlled" laboratory settings that we would deny in the wild. This shift, beyond pragmatically determined considerations, is importantly due to the dominant values and moral imperatives of scientific and biomedical activity: human (and animal) welfare, the relief of suffering and physiological distress, and the pursuit of basic knowledge about ourselves and the natural world. Thus we might ethically condone the transgenic production of Oncomouse, Cystic Fibrosis mouse, and Memory mouse (undertaken to facilitate the study of fundamental memory and learning processes). The decision would depend on the importance of the scientific project's purpose, amidst all other things that need to be considered.

The further considerations involve the legitimate stakes of the other realms of human activity, with their own ethical mandates, in biomedical and scientific research. From our participation in private life and webs of intimate personal relations involving both humans and animals comes an insistence on attention to the welfare of individual animals, with a minimization of suffering in research protocols and care settings as is appropriate to legitimate scientific goals. From public political ethics comes the ethical demand that there be a "just and fair" proportion between the overall benefits to be gained and the harms (especially suffering) to be inflicted, with a maximization of the former relative to a minimization of the latter. From the cultural and natural preservationists and others responsible for protecting ultimate values (finally all of us) comes a serious questioning of the admissibility of the intervention and research: whether it is ethically out-of-bounds with regard to violating the animal's individual or species integrity or inflicting significant suffering, no matter what the benefits envisioned.

These are the characteristic demands that are placed on animal care and use committees in their review of research and educational protocols. The scrutiny is only exacerbated by animal biotechnological innovations. The chief "novel" issues concern animal welfare and animal integrity. How can researchers, laboratory technicians, and animal caretakers know their animals and promote their well-being or welfare if a new strain or species of animal has been created with altered and perhaps unprecedented behavioral habits? And how is animal integrity—that which might be inadmissibly violated—to be understood? Is it the intactness of the animals' genetic or genomic structure and functionings, or behavioral, social, and "worldly" habits? Or are these all dimensions of animal integrity, however difficult to define adequately?

These particular hazards of animal research ethics and protocol review are only highlighted by animal biotechnology and transgenic innovations. They do not change the fundamental nature of an ethical decision-making that must be contextual. In particular, no ethical value or obligation can have an absolute or final precedence over the others. Given the plurality of ultimate and fundamental values, there can be no principled "trumping" of one value over the others. Rather, there must be a contextually defined and proportionate coordination of obligations. For example, religious objections to tampering with natural creation cannot by themselves block the creation of Oncomouse or CF mouse, with the anticipated benefits to human welfare and scientific knowledge. On the other hand, given deeply ingrained cultural or religious habits (which themselves may change over time), what might be ethically tolerable or even mandated in one local human community with respect to science and biomedicine may be inadmissible in another. Presumably one would not transgenically manipulate a sacred animal or plant of an indigenous culture, for example, cows in the more traditionally Hindu regions of India. This only underscores the cultural and social embeddedness of all scientific research and medicine and the fact that human welfare concerns and a thirst for knowledge do not always take precedence over other humanly or naturally important values.

Animal Biotechnology, Economic Markets, and Agribusiness. The domains of scientific and biotechnological research, biomedicine, and economic activity increasingly overlap and shade off into one another. The creation of mice, goats, and other animals that produce easily retrievable pharmaceutical products such as human insulin or t-PA at once serve significant human welfare needs and economic, entrepreneurial goals. We move away from such immediate health concerns and attendant ethical obligations when we come to potentially lucrative bovine somatotropin (BST) boosted cows, growth hormone primed pigs, and the aquatic factory farming of transgenic fish. As we transverse this spectrum, pressing human welfare obligations often recede, and the morally ecological analyses of the animal biotechnological practices significantly and complexly change. Animal welfare and human social/cultural factors come more to the fore.

The Beltsville pig, genetically fashioned for cost-efficient growth rates and feed consumption and for the leaner quality of its meat, proved to be severely compromised by arthritis and multiple other diseases. All parties, including the scientific animal production community, consider this an unfortunate ethical misadventure. There remains, however, the biotechnologists' expectant hope that the animal welfare issues can be overcome and that a new generation of engineered "food animals" will be more ethically acceptable. (But what ethical price would have to be paid, in terms of renewed biotechnological interventions, to overcome the ethical harm?) Similarly there are animal welfare concerns for the BST cows, though immediate animal suffering or harm seems much less acute, and ethical attention is more on the effects of intensifying factory-farming practices.

Beyond heightened concern for animal welfare and unjustifiable suffering, the economic boosters of animal biotechnology, whether from agribusiness or pharmaceutical industries, meet an interesting and complex social and cultural resistance. Small dairy farmers complain that the "big business" of BST-boosted cows will hasten the demise of family farms and local rural traditions. Others object to the "pollution" of milk with the bovine growth hormone. Still others challenge the patenting and economic commodification of animals: the conceptual reduction of their status as genuine living beings, aboriginally the creation of nature and unowned by humans, to mere configurations of living matter instrumentally at the disposal of humans for their own self-interested purposes, economic or other.

Whatever the actual salience of these ethical charges and critiques, obviously fundamental social, cultural, and religious values are at stake, arising out of broad cultural traditions and interests. Animal biotechnology, coupled to the engines of corporate economics, is felt to threaten fundamental and traditional moral, religious, and cultural orientations.

Again this poses an important challenge to a morally ecological analysis of animal biotechnologies, in the context of economic activity and elsewhere. There may be certain human communities or cultures in which animal biotechnological practices, even for the best human welfare reasons, are considered morally inadmissible. This raises important social or cultural justice issues on an international scale as biotechnology's province becomes increasingly global. Arguably, local cultural communities ought to decide whether they wish to participate in the enterprise and benefits of animal and wider biotechnology, irrespective of the insistent pressures of global economic justice. (For example, who should economically benefit from the genetic resources, natural or "artificed," that are swept up into international biotechnological-economic activity?)

On the other hand, what should the moral ecology be when cultural or religious objections come from a minority within a wider and culturally diverse community, such as the United States and many other countries? Granted, social, cultural, and community considerations ought to receive serious attention, especially in relation to optional or forgoable economic practices. But again, with a plurality of moral obligations, no sphere of human activity and no ethical interest group can be allowed to override the legitimate ethical interests of plural others. What then should be done? This question remains at the core of moral ecology's unfinished business.

Moral Ecology's Landscape: The Outstanding Issues

We have been arguing that the ethical consideration of animal biotechnologies defies any easy solution or subsumption under a "mono-valued" ethical system. There are too many different interventions in too many different contexts involving too many different motives, values, and ethical obligations. Yet the call for a contextual consideration of plural obligations and moral ecology decidedly implies a coordination of disparate and perhaps conflicting values and obligations within and between specific contexts. We require systematic ethical responses that genuinely recognize the plural value and ethical dimensions of our worldly existence. How do we square this circle, which is demanded by our overall responsibilities to humans, animals, and nature? How should such practical decisions be substantively guided? This is an outstanding and unsettled issue.

Yet we may begin to see our way. The first clues come from the sheer plurality of practices, contexts, values, and obligations themselves. This constitutes concrete and experientially incontestable evidence of the plural and complex goodness of human existence. Moreover, the goodness of both humans and nature is vulnerable to change and various harms. We must become ethically committed, as an overarching and fundamental moral duty, to this plurality itself: to upholding and promoting the various abiding and culturally significant spheres of human activity amidst the ecosystemic life and animate world in which they are embedded.

Herein is the second set of clues: the spheres or domains of human activity interpenetrate one another, and there are contexts within wider contexts within still wider contexts of activity and moral significance. Ethical atomism or provincialism is practically impossible and ethically irresponsible. Rather we must *concurrently* pursue the human, animal, and natural good. First and foremost we must prevent the significant undermining of any one domain or sphere of activity, human or natural, for the sake of others. This involves a mutual commitment, sensitivity, and concern among different human actors with various contextually defines allegiances. Such coordination requires a mutual accommodation without forgoing fundamental value and ethical commitments. We must fashion an ethically and publicly responsible life that is broadly "cosmopolitan."

Such a cosmopolitan and contextual ethics cannot be rationalist or universalist in a traditional sense, that is, involve principled logical arguments from first moral premises. Rather its "reasonable connections" must be more ethical-aesthetic. Its modes of thought must be more in keeping with the informal reasoning and moral art of the private realm of intimate relations, which must take in whole webs of life and multiple moral considerations at once. In short, moral ecology deals with complex wholes. The identification of what morally needs to be taken into account is only a first step. The coordination of the obligations is the second and more difficult step. The coordination of the obligations is the second and more difficult step. Yet the two steps are intimately and substantively conjoined. The first sets up the initial moral landscape: the moral goods and obligations that need to be upheld and protected, whether they pertain to humans, animals, or nature. The consequent moral coordination is protectionist or conservationist in character. It accepts the obligations as given and asks which are most seriously threatened and most in need of protection or promotion in the particular context at hand. For example, in the context of the wild, long-term concerns for nature take precedence over human aspirations, unless significant human needs or values are at stake with ecologically acceptable costs to natural systems. In the laboratory, serious threats to ethical obligations to human welfare and the pursuit of knowledge take precedence, unless other overriding threats supervene, for example, the possible escape of environmentally damaging organisms, disproportionate and inadmissible animal suffering, or the undermining of central cultural or religious values of the surrounding human community. In short, moral ecology, exercised in particular contexts, is rationally ruled by an unwavering response to the most serious threats to well-established moral allegiances and by a long-term ethical commitment to the plural good of humans and nature in their intimate interconnections.

But even if we could adequately see our way through these methodological and epistemological problems of establishing priorities of ethical concern and obligation, we face another problem. Is this commitment to the coordination of plural activities, values, and obligations practically realistic or an impossible dream, given the aggressive disharmony inherently spawned by ethical and political pluralism itself and the dynamic nature of humans' worldly life? History, recent or past, is sobering. Yet the world's dynamic becoming, which will not be rationally, technologically, politically, or ethically subdued, presents a way out of political, ethical, and cultural impasses. Different cultures and different domains of human activity can over time grow together, at least in understanding if not also in practice. This requires both mutual appreciation and mutual criticism as a way of moving toward a more adequate and ethical flourishing within and between particular spheres of activity and human cultures.

On the one hand, this attempt would directly address issues of social and cultural justice and the question of who should decide the ethical admissibility or inadmissibility of animal biotechnologies in different geopolitical contexts. In our commitment to the plural good, fundamental cultural and moral orientations should prevail over forgoable human welfare and global economic concerns. In short, particular cultures should be mutually appreciated and honored. Yet these same cultural and value perspectives ought to inform and mutually criticize one another about what is of abiding worth and truly enlists our ethical responsibility given present world conditions: that is, those human and natural individuals, communities, and activities that most centrally command our ethical attention. This is the dynamic growth—philosophic, ethical, religious, cultural, and political— made possible by the interconnected becoming of the world. In the end, it is the dynamically systemic or interactive character of worldly reality that allows for the practical coordinations of plural values and obligations—coordinations that will perhaps always be more imperfectly existential than perfectly rational.

The Goodness and Significance of Nature and Animal Life. Given humans' newly emerging and insistent responsibilities for biological life, ecosystems, and the environment, nowhere is "mutual appreciation and criticism" more globally and regionally needed than in trying to ferret out the meaning, significance, and goodness of animate life and evolutionary and ecological processes. (As we have seen, this is crucial to the various contextual ethical analyses of animal biotechnologies.) For such an understanding we need to bring together thinkers from various spheres of scientific and cultural activity: evolutionary biologists, animal researchers, anthropologists, philosophers, theologians, and others with a central stake in the multileveled significance of nature. A serious mutual confrontation of these plural areas of disciplined thought and activity promises philosophical and ethical advances.

For example, more traditional philosophers or theologians, committed to long-dominant modes of essentialist thinking, might see the significance of nature and animal species as arising from (or grounded in) atemporal and unchanging Platonic "ideas" (for example, the archetypal form "horse"). Or they might appeal to Aristotelian "substantial forms" (the "formal plan" of development into an adult horse, perhaps an unintended adumbration of "genomic information") or a once-and-for-all creation *ex nihilo* by a transcendent deity. This is how traditional modes of thinking typically account for the definite character, integrity, and goodness of nature's animate beings. Such traditional perspectives might (or might not) ethically counsel against modern forms of genetic tinkering, transgenesis, and the confounding of the eternal order of creation.

But contemporary molecular and evolutionary biologists would unite in contending that an essentialist explanation and interpretation of the animal and animate world is fundamentally flawed. Biological species arise and pass in dynamic evolutionary and ecological process. Thanks to random genetic variation (via genetic mutation and sexual reproduction) and natural selection, species diversify and evolve out of their biological predecessors, sharing and reconfiguring genetic information. Moreover, nature is no realm of essentialist perfection. Rather, our biosphere is an extraordinary, historically particular, and "chaotically orderly" realm of dynamic and systematically related "imperfections": individual organisms more or less well-adapted to worldly life; populational species of such individuals more or less well-adapted to ever-changing ecological niches; and ecosystems themselves more or less internally robust and dynamically viable, while changing in evolutionary/ecological time.

This by now well-founded and incontestable general evolutionary and ecological perspective does not annihilate the questions of natural goodness and integrity posed by traditional philosophers and theologians. It only defeats and renders obsolete essentialist modes of naturalistic thinking and philosophic interpretation. The natural goodness and integrity of biological individuals and species, as well as ecosystems, only need a new and more philosophically nuanced interpretation. Individual organisms still present themselves as having a lively integrity, intactness, or "oneness" that encompasses bodily, subjective, behavioral, and outwardly social functionings that are more or less flexibly adapted to an active, if vulnerable life in the world. Species as populations of biological individuals exhibit a spatiotemporally bounded and flexible integrity relative to some ecological niche, also changing. Moreover, species evolve in a creative, though orderly fashion, according to relatively few generic organic or bodily *Baupläne* (blueprint). Finally, the habitats and ecosystems themselves evidence a flexible and dynamic intactness with respect to internal stability and species diversification, more or less vulnerable to outside, wider ecosystemic processes or forces.

No doubt these several senses of "integrity"—individual, species, and ecosystemic—require further and careful conceptual and philosophical articulation and systematic coordination. Moreover, we will need further collaborative efforts in appreciating the full significance and goodness of the individuals, species, and ecosystems of the animate realm. But such an enterprise should only ethically and practically serve us well. It would further and more clearly reveal the complex meanings of "nature natural" for us humans. It would help us discern what the limits of our biotechnological and other human interventions in the wild ought to be and what needs to be ethically protected in scientific and biomedical research and economic activity. For example, a more adequate understanding of animal integrity might shift away from putative species-specific genes manipulations on the phenotypical or somatic expressions of genomes and the functional, behavioral, and experimental well-being of individual animals. With respect to the animal's overall organic integrity, certain manipulations may appear inconsequential, others not. This is what should significantly inform particular ethical deliberations of IACUCs and others.

Moreover we would better understand ourselves and our embedded existence in an animate nature that is ultimately significant, yet imperfectly good: that we and animate nature are not to be perfected, but that the world's evolving complex and finite goodness—the various dimensions of activity and value realized by humans and other organisms—is to be unequivocally affirmed and ethically protected.

In short, a speculative, disciplined advance in our understanding and assessment of the multi-leveled worth of nature would help us in the ethical coordinations required by our contextual plural obligations. Though in themselves such an understanding and assessment do not uniquely determine the outcome (positive or negative) of moral deliberations in different contexts, they would better inform us when it is ethically appropriate to move biotechnologically forward and when to take ethically protective stands. We would better know how to integrate our humanly cultural, technological, and natural selves and how practically to fit our human communities within the wider natural and animate world.

32

Ethics and Genetically Modified Foods

Gary Comstock

$\mathbf{M}$uch of the food consumed in the United States is genetically modified (GM). GM food derives from microorganisms, plants, or animals manipulated at the molecular level to have traits that farmers or consumers desire. These foods often have been produced using techniques in which "foreign" genes are inserted into the microorganisms, plants, or animals. Foreign genes are those taken from sources other than the organism's natural parents. In other words, GM plants contain genes they would not have contained if researchers had only used traditional plant-breeding methods.

Some consumer advocates object to GM foods, and sometimes they object on ethical grounds. When someone opposes GM foods on ethical grounds, he typically has some reason or other for his opposition. We can scrutinize his reasons and, when we do so, we are doing applied ethics. Applied ethics involves identifying people's arguments for various conclusions and then analyzing those arguments to determine whether the arguments support the conclusions. A critical goal here is to decide whether an argument is sound. A sound argument is one in which all of the premises are true and no mistakes have been made in reasoning.

Ethically justifiable conclusions inevitably rest on two kinds of claims: (a) empirical claims, or factual assertions about how the world *is,* claims ideally based on the best available scientific observations, principles, and theories, and (b) normative claims, or value-laden assertions about how the world *ought to be,* claims ideally based on the best available moral judgments, principles, and theories.

Is it ethically justifiable to pursue genetically modified crops and foods? There is an objective answer to this question, and we will try here to figure out what it is. But we must begin with a proper, heavy, dose of epistemic humility, acknowledging that few ethicists at the moment seem to think that they know the final answer.

Should the law allow GM foods to be grown and marketed? The answer to this, and every, public policy question rests ultimately with us; citizens who will in the voting booth and shopping market decide the answer. To make up our minds, we will use feelings, intuition, conscience, and reason. However, as we citizens are, by and large, not scientists, we must, to one degree or other, rest our factual understanding of the matter on the opinions of scientific experts. Therefore, ethical responsibility in the decision devolves heavily upon scientists engaged in the new GM technology.

Gary Comstock, "Ethics and Genetically Modified Food," *SCOPE Research Group* (University of California–Berkeley, University of Washington, and American Association for Advancement of Science, July 2001). Reprinted by permission of the author.

ETHICAL RESPONSIBILITIES OF SCIENTISTS

Science is a communal process devoted to the discovery of knowledge, and to open and honest communication of knowledge. Its success, therefore, rests on two different kinds of values.

Epistemological values are values by which scientists determine which knowledge claims are better than others. The values include clarity, objectivity, capacity to explain a range of observations, and ability to generate accurate predictions. Claims that are internally inconsistent are jettisoned in favor of claims that are consistent, and fit with established theories. (At times, anomalous claims turn out to be justifiable, and an established theory is overthrown, but these occasions are rare in the history of science.) Epistemological values in science also include fecundity, the ability to generate useful new hypotheses; simplicity, the ability to explain observations with the fewest number of additional assumptions or qualifications; and elegance.

Personal values, including honesty and responsibility, are a second class of values, values that allow scientists to trust their peers' knowledge claims. If scientists are dishonest, untruthful, fraudulent, or excessively self-interested, the free flow of accurate information so essential to science will be thwarted. If a scientist plagiarizes the work of others or uses fabricated data, the scientist's work will become shrouded in suspicion and otherwise reliable data will not be trusted. If scientists exploit those who work under them, or discriminate on the basis of gender, race, class, or age, then the mechanisms of trust and collegiality undergirding science will be eroded.

The very institution of scientific discovery is supported, indeed, permeated with values. Scientists have a variety of goals and functions in society, so it should be no surprise that they face different challenges.

University scientists must be scrupulous in giving credit for their research to all who deserve credit; careful not to divulge proprietary information; and painstaking in maintaining objectivity, especially when funded by industry. Industry scientists must also maintain the highest standards of scientific objectivity, a particular challenge since their work may not be subject to peer-review procedures as strict as those faced by university scientists. Industry scientists must also be willing to defend results that are not favorable to their employers' interests. Scientists employed by nongovernmental organizations face challenges, as well. Their objectivity must be maintained in the face of an organization's explicit-advocacy agenda, and in spite of the fact that their research might provide results that might seriously undermine the organization's fund-raising attempts. All scientists face the challenges of communicating complex issues to a public that receives them through media channels that often are not equipped to communicate the qualifications and uncertainties attaching to much scientific information.

At its core, science is an expression of some of our most cherished values. The public largely trusts scientists, and scientists must in turn act as good stewards of this trust.

A METHOD FOR ADDRESSING ETHICAL ISSUES

Ethical objections to GM foods typically center on the possibility of harm to persons or other living things. Harm may or may not be justified by outweighing benefits. Whether harms are justified is a question that ethicists try to answer by working methodically through a series of questions:[1]

1) What is the harm envisaged? To provide an adequate answer to this question, we must pay attention to how significant the harm or potential harm may be (will be it severe or trivial?); who the "stakeholders" are (that is, who are the persons, animals, even ecosystems, who may be harmed?); the extent to which various stakeholders might be harmed;

and the distribution of harms. The last question directs attention to a critical issue, the issue of justice and fairness: Are those who are at risk of being harmed by the action in question different from those who may benefit from the action in question?

2) What information do we have? Sound ethical judgments go hand in hand with thorough understanding of the scientific facts. In a given case, we may need to ask two questions: Is the scientific information about harm being presented reliable, or is it fact, hearsay, or opinion? And, what information do we not know that we should know before making the decision?

3) What are the options? In assessing the various courses of action, emphasize creative problem solving, seeking to find "win-win" alternatives in which everyone's interests are protected. Here we must identify what objectives each stakeholder wants to obtain; how many methods are available by which to achieve those objectives; and what advantages and disadvantages attach to each alternative.

4) What ethical principles should guide us? There are at least three secular ethical traditions:
   - Rights theory holds that we ought always to act so that we treat human beings as autonomous individuals, and not as mere means to an end.
   - Utilitarian theory holds that we ought always to act so that we maximize good consequences and minimize harmful consequences.
   - Virtue theory holds that we ought always to act so that we act the way a just, fair, good person would act.

   Ethical theorists are divided about which of these three theories is best. We manage this uncertainty through the following procedure. Pick one of the three principles. Using it as a basis, determine its implications for the decision at hand. Then, adopt a second principle. Determine what it implies for the decision at hand. Repeat the procedure with the third principle. Should all three principles converge on the same conclusion, then we have good reasons for thinking our conclusion morally justifiable.

5) How do we reach moral closure? Does the decision we have reached allow all stakeholders either to participate in the decision or to have their views represented? If a compromise solution is deemed necessary in order to manage otherwise intractable differences, has the compromise been reached in a way that has allowed all interested parties to have their interests articulated, understood, and considered? If so, then the decision may be justifiable on ethical grounds.

   There is a difference between *consensus* and *compromise. Consensus* means that the vast majority of people agree about the right answer to a question. If the group cannot reach a consensus but must, nevertheless, take some decision or other, then a *compromise* position may be necessary. But neither consensus nor compromise should be confused with the right answer to an ethical question. It is possible that a society might reach a consensus position that is unjust. For example, some societies have held that women should not be allowed to own property. That may be a consensus position, or even a compromise position, but it should not be confused with the truth of the matter. Moral closure is a sad fact of life; we sometimes must decide to undertake some course of action even though we know that it may not be, ethically, the right decision, all things considered.

ETHICAL ISSUES INVOLVED IN THE USE OF GENETIC TECHNOLOGY IN AGRICULTURE

Discussions of the ethical dimensions of agricultural biotechnology are sometimes confused by a conflation of two quite different sorts of objections to GM technology: intrinsic and extrinsic. It is critical not only that we distinguish these two classes, but keep them distinct throughout the ensuing discussion of ethics.

Extrinsic objections focus on the potential harms consequent upon the adoption of genetically modified organisms (GMOs). Extrinsic objections hold that GM technology should not be pursued because of its anticipated results. Briefly stated, the extrinsic objections go as follows. GMOs may have disastrous effects on animals, ecosystems, and humans. Possible harms to humans include perpetuation of social inequities in modern agriculture, decreased food security for women and children on subsistence farms in developing countries, a growing gap between well-capitalized economies in the Northern hemisphere and less capitalized peasant economies in the South, risks to the food security of future generations, and the promotion of reductionistic and exploitative science. Potential harms to ecosystems include possible environmental catastrophe, inevitable narrowing of germplasm diversity, and irreversible loss or degradation of air, soils, and waters. Potential harms to animals include unjustified pain to individuals used in research and production.

These are valid concerns, and nation-states must have in place testing mechanisms and regulatory agencies to assess the likelihood, scope, and distribution of potential harms through a rigorous and well-funded risk-assessment procedure. It is for this reason that I have said, above, that GM technology must be developed responsibly and with appropriate caution. However, these extrinsic objections cannot by themselves justify a moratorium, much less a permanent ban, on GM technology, because they admit the possibility that the harms may be minimal and outweighed by the benefits. How can one decide whether the potential harms outweigh the potential benefits unless one conducts the research, field tests, and data analysis necessary to make a scientifically informed assessment?

In sum, extrinsic objections to GMOs raise important questions about GMOs, and each country using GMOs ought to have in place the organizations and research structures necessary to insure their safe use.

There is, however, an entirely different sort of objection to GM technology, a sort of objection that, if it is sound, would indeed justify a permanent ban.

Intrinsic objections allege that the process of making GMOs is objectionable *in itself.* This belief is defended in several ways, but almost all of the formulations are related to one central claim, the "unnaturalness objection":

It is unnatural to genetically engineer plants, animals, and foods (**UE**).

If **UE** is true, then we ought not to engage in bioengineering, however unfortunate may be the consequences of halting the technology. Were a nation to accept **UE** as the conclusion of a sound argument, then much agricultural research would have to be terminated and potentially significant benefits from the technology sacrificed. A great deal is at stake.

In *Vexing Nature? On the Ethical Case Against Agricultural Biotechnology,* I discuss fourteen ways in which **UE** has been defended.[2] For present purposes, those fourteen objections can be summarized as follows:

(1) **To engage in agricultural biotech is to *play God.***
(2) **To engage in agricultural biotech is to *invent world-changing technology.***
(3) **To engage in agricultural biotech is *illegitimately to cross species boundaries.***
(4) **To engage in agricultural biotech is to *commodify life.***

Let us consider each claim in turn.

1) **To engage in agricultural biotech is to** *play God.*

In a Western theological framework, humans are creatures, subjects of the Lord of the Universe, and it would be impious for them to arrogate to themselves roles and powers appropriate only for the Creator. Shifting genes around between individuals and species is taking on a task not appropriate for us, subordinate beings. Therefore, to engage in bioengineering is to play God.

There are several problems with this argument. First, there are different interpretations of God. Absent the guidance of any specific religious tradition, it is logically possible that God could be a being who wants to turn over to us all divine prerogatives; or explicitly wants to turn over to us at least the prerogative of engineering plants; or who doesn't care what we do. If God is any of these beings, then the argument fails because playing God in this instance is not a bad thing.

The argument seems to assume, however, that God is not like any of the gods just described. Assume that the orthodox Jewish and Christian view of God is correct, that God is the only personal, perfect, necessarily existing, all-loving, all-knowing, and all-powerful being. On this traditional Western theistic view, finite humans should not aspire to infinite knowledge and power. To the extent that bioengineering is an attempt to control nature itself, the argument would go, bioengineering would be an acceptable attempt to usurp God's dominion.

The problem with this argument is that not all traditional Jews and Christians think that this God would rule out genetic engineering. I am a practicing evangelical Christian and the chair of my local church's council. In my tradition, God is thought to endorse creativity, and scientific and technological development, including genetic improvement. Other traditions have similar views. In the mystical writings of the Jewish Kabala, God is understood as one who expects humans to be co-creators, technicians working with God to improve the world. At least one Jewish philosopher, Baruch Brody, has suggested that biotechnology may be a vehicle ordained by God for the perfection of nature.[3]

I personally hesitate to think that humans can "perfect" nature. However, I have become convinced that genetic modification might help humans to rectify some of the damage we have already done to nature. And I believe God may endorse such an aim. For humans are made in the divine image. God desires that we exercise the spark of divinity within us. Inquisitiveness in science is part of our nature. Creative impulses are not found only in the literary, musical, and plastic arts. They are part of molecular biology, cellular theory, ecology, and evolutionary genetics, too. It is unclear why the desire to investigate and manipulate the chemical bases of life should not be considered as much a manifestation of our godlike nature as the writing of poetry and the composition of sonatas. As a way of providing theological content for **UE**, then, argument (1) is unsatisfactory because it is ambiguous and contentious.

2) **To engage in agricultural biotech is to** *invent world-changing technology.*

Let us consider (2) in conjunction with similar objection (2a).

2a) **To engage in agricultural biotech is to** *arrogate historically unprecedented power to ourselves.*

The argument here is not the strong one, that biotech gives us divine power, but the more modest one, that it gives us a power we have not had previously. But it would be counterintuitive to judge an action wrong simply because it has never been performed. On this view, it would have been wrong to prescribe a new herbal remedy for menstrual cramp, or to administer a new anesthetic. But that seems absurd. More argumentation is needed to call historically unprecedented actions morally wrong. What is needed is to know to what extent our new powers will transform society, whether we have witnesses prior transformations of this sort, and whether those transitions are morally acceptable.

We do not know how extensive the agricultural biotech revolution will be, but let us assume that it will be as dramatic as its greatest proponents assert. Have we ever witnessed comparable transitions? The change from hunting and gathering to agriculture was an astonishing transformation. With agriculture came not only an increase in the number of humans on the globe, but the first appearance of complex cultural activities: writing, philosophy, government, music, the arts, and architecture. What sort of power did people arrogate to themselves when they moved from hunting and gathering to agriculture? The power of civilization itself.[4]

Agricultural biotech is often oversold by its proponents. But suppose that they are right, that agricultural biotech brings us historically unprecedented powers. Is this a reason to oppose it? Not if we accept agriculture and its accompanying advances, for when we accepted agriculture, we arrogated to ourselves historically unprecedented powers.

In sum, the objections stated in (2) and (2a) are not convincing.

3) To engage in agricultural biotech is *illegitimately to cross species boundaries*.

The problems with this argument are both theological and scientific. I will leave it to others to argue the scientific case that nature gives ample evidence of generally fluid boundaries between species. The argument assumes that species boundaries are distinct, rigid, and unchanging while, in fact, species now appear to be messy, plastic, and mutable. To proscribe the crossing of species borders on the grounds that it is unnatural seems scientifically indefensible. It is also difficult to see how (3) could be defeated on theological grounds. None of the scriptural writings of the Western religious proscribe genetic engineering, of course, because genetic engineering was undreamed of at the time the holy books were written. Now, one might argue that such a proscription may be derived from Jewish or Christian traditions of scriptural interpretation. Talmudic laws against mixing "kinds," for example, might be taken to ground a general prohibition against inserting genes from "unclean" species into clean species. Here's one way the argument might go: For an observant Jew to do what scripture proscribes is morally wrong; Jewish oral and written law proscribe the mixing of kinds (for example, eating milk and meat from the same plate; yoking donkeys and oxen together); bioengineering is the mixing of kinds; therefore, for a Jew to engage in bioengineering is morally wrong.

But this argument fails to show that bioengineering is intrinsically objectionable in all of its forms for everyone. The argument might prohibit *Jews* from engaging in certain *kinds* of biotechnical activity but not all; it would not prohibit, for example, the transferring of genes *within* a species, nor, apparently, the transfer of genes from one clean species to another clean species. Incidentally, it is worth noting that the Orthodox community has accepted transgenesis in its food supply. Seventy percent of cheese produced in the United States is made using a GM product, chymosin. This cheese has been accepted as kosher by Orthodox rabbis.[5]

In conclusion, it is difficult to find a persuasive defense of (3) either on scientific or religious ground.

4) To engage in agricultural biotech is to *commodify life*.

The argument here is that genetic engineering treats life in a reductionistic manner, reducing living organisms to little more than machines. Life is sacred and not to be treated as a good of commercial value only, to be bought and sold to the highest bidder.

Could we apply this principle uniformly? Would not objecting to the products of GM technology on these grounds also require that we object to the products of ordinary agriculture on the same grounds? Is not the very act of bartering or exchanging crops and animals for cash vivid testimony to the fact that every culture on earth has engaged in the commodification of life for centuries? If one accepts commercial trafficking in non-GM wheat and pigs, then why should we object to commercial trafficking in GM wheat and GM pigs? Why should it be wrong for us to treat DNA the way we have previously treated animals, plants, and viruses?[6]

While (4) may be true, it is not a sufficient reason to object to GM technology because our values and economic institutions have long accepted the commodification of life. Now, one might object that various religious traditions have never accepted commodification, and that genetic engineering presents us with an opportunity to resist, to reverse course. Leon Kass,[7] for example, has argued that we have gone too far down the road of dehumanizing ourselves and treating nature as a machine, and that we should pay attention to our emotional reactions against practices such as human cloning. Even if we cannot defend these feelings in rational terms, our revulsion at the very idea of cloning humans should carry great weight. Mary Midgley[8] has argued that moving genes across species boundaries is not only "yucky" but, perhaps, a monstrous idea, a form of playing God.

Kass and Midgley have eloquently defended the relevance of our emotional reactions to genetic engineering but, as both admit, we cannot simply allow our emotions to carry the day. As Midgley writes, "Attention to . . . sympathetic feelings [can stir] up reasoning that [alters] people's whole world view."[9] But as much hinges on the reasoning as on the emotions.

Are the intrinsic objections sound? Are they clear, consistent, and logical? Do they rely on principles we are willing to apply uniformly to other parts of our lives? Might they lead to counterintuitive results?

Counterintuitive results are results we strongly hesitate to accept because they run counter to widely shared, considered moral institutions. If a moral rule or principle leads to counterintuitive results, then we have a strong reason to reject it. For example, consider the following moral principle, which we might call the "doctrine of naïve consequentialism" (**NC**):

Always improve the welfare of the most people (**NC**).

Were we to adopt **NC**, then we would not only be permitted but required to sacrifice one healthy person if by doing so we could save many others. If six people need organ transplants (two need kidneys, one needs a liver, one needs a heart, and two need lungs), then **NC** instructs us to sacrifice the life of the healthy person so as to transplant that person's six organs to the other six. But this result, that we are obliged to sacrifice innocent people to save strangers, is wildly counterintuitive. This result gives us a strong reason to reject **NC**.

I have argued that the four formulations of the unnaturalness objection considered above are unsound insofar as they lead to counterintuitive results. I do not take this position lightly. Twelve years ago, I wrote "The Case against bGH," an article, I have been told, that "was one of the first papers by a philosopher to object to agricultural biotech on explicitly ethical grounds." I then wrote a series of other articles objecting to GM herbicide-resistant crops, transgenic animals, and, indeed, all of agricultural biotechnology.[10] I am acquainted with worries about GM foods. But, for reasons that include the weakness of the intrinsic objections, I have come to change my mind. The sympathetic feelings on which my anti-GMO worldview was based did not survive the stirring up of reasoning.

WHY ARE WE CAREFUL WITH GM FOODS?

I do not pretend to know anything like the full answer to this question, but I would like to be permitted the luxury of a brief speculation about it. The reason may have to do with natural, completely understandable, and wholly rational tendency to take precautions with what goes into our mouths. When we are in good health and happy with the foods available to us, we have little to gain from experimenting with a new food, and no reason to take a chance on a potentially unsafe food. We may think of this disposition as the precautionary response.

When faced with two contrasting opinions about issues related to food safety, consumers place great emphasis on negative information. The precautionary response is particularly strong when a consumer sees little to gain from a new food technology. When a given food is plentiful, it is rational to place extra weight on negative information about any particular piece of that food. It is rational to do so, as Dermot Hayes points out, even when the source of the negative information is known to be biased.

There are several reasons for us to take a precautionary approach to new foods. First, under conditions in which nutritious, tasty food is plentiful, we have nothing to gain from trying a new food if, from our perspective, it is in other respects identical to our current foods. Suppose on a rack in front of me there are eighteen dozen maple-frosted Krispy Kreme doughnuts, all baked to a golden brown, all weighing three ounces. If I am invited to take one of them, I have no reason to favor one over the other.

Suppose, however, that a naked man runs into the room with wild hair flying behind him yelling that the sky is falling. He approaches the rack and points at the third doughnut from the left on the fourth shelf from the bottom. He exclaims, "This doughnut will cause cancer! Avoid it at all costs, or die!" There is no reason to believe this man's claim and yet, since there are so many doughnuts freely available, why should we take any chances? It is rational to select other doughnuts, since all are alike. Now, perhaps one of us is a mountain climber who loves taking risks. He might be tempted to say, "Heck, I'll try that doughnut." In order to focus on the right question here, the risk takers should ask themselves whether they would select the tainted doughnuts to take home to feed to their two-year-old daughter. Why impose any risk on your loved ones when there is no reason to do so?

The Krispy Kreme example is meant to suggest that food tainting is both a powerful and an extraordinarily easy social act. It is powerful because it virtually determines consumer behavior. It is easy because the tainted does not have to offer any evidence of the food's danger at all. Under conditions of food plenty, rational consumers do and should take precautions, avoiding tainted food no matter how untrustworthy the tainted.

Our tendency to take precautions with our food suggests that a single person with a negative view about GM foods will be much more influential than many people with a positive view. The following experiment lends credibility to this hypothesis. In a willingness-to-pay experiment, Hayes and colleagues paid eighty-seven primary food shoppers $40 each.[11] Each participant was assigned to a group ranging in size from a half-dozen to a dozen members. Each group was then seated at a table at lunchtime and given one pork sandwich. In the middle of each table was one additional food item, an irradiated pork sandwich. Each group of participants was given one of three different treatments: (a) the *pro-irradiation* treatment, (b) the *anti-irradiation* treatment, or (c) the *balanced* treatment.

Each treatment began with all of the participants at a table receiving the same, so-called natural description of an irradiated pork sandwich. The description read, in part, like this:

> The U.S. FDA has recently approved the use of ionizing to control Trichinella in pork products. This process in a ten-thousand-fold reduction in Trichinella organisms in meat. The process does not include measurable radioactivity in food.

After the participants read this description, they would proceed to conduct a silent bid in order to purchase the right to exchange their nonirradiated sandwich for the irradiated sandwich. Whoever bid the highest price would be able to buy the sandwich for the price bid by the second-highest bidder. In order to provide participants with information about the opinions of the others at their table so that they could factor this information into their future bids, the lowest and highest bids of each round were announced before the next round of bidding began. At the end of the experiment, one of the ten bidding rounds would be selected at random, and the person bidding the highest amount in that round would have to pay the second-highest price bid during that round for the sandwich.

After five rounds of bidding, the second-highest bids in all three groups settled rather quickly at an equilibrium point, roughly twenty cents. That is, someone at every table was willing to pay twenty cents for the irradiated pork sandwich, but no one in any group would pay more than twenty cents. The bidding was repeated five times in order to give participants the opportunity to respond to information they were getting from others at the table, and to insure the robustness of the price.

After five rounds of bidding, each group was given additional information. Group (a), the so-called Pro group, was provided with a description of the sandwich that read, in part:

> Each year, 9,000 people die in the United States from food-borne illness. Some die from Trichinella in pork. Millions of others suffer short-term illness. Irradiated pork is a safe and reliable way to eliminate this pathogen. The process has been used successfully in twenty countries since 1950.

The pro-group participants were informed that the author of this positive description was a pro-irradiation food-industry group. After the description was read, five more rounds of bidding began. The price of the irradiated sandwich quickly shot upward, reaching eighty cents by the end of round ten. A ceiling price was not reached, however, as the bids in every round, including the last, were significantly higher than the preceding round. The price, that is, was still going up when the experiment was stopped (see table 32.1).

Table 32.1: Effect of Information on Average Bid for Irradiated Pork
Reprinted from Hayes, et al. (2002).

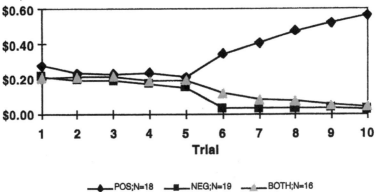

After its first five rounds of bidding, group (b) was provided with a different description. It read, in part:

> In food irradiation, pork is exposed to radioactive materials. It receives 300,000 rads of radiation—the equivalent of 30 million chest x rays. This process results in radiolytic products in food. Some radiolytic products are carcinogens, and linked to birth defects. The process was developed in the 1950s by the Atomic Energy Commission.

The source of this description was identified to the bidders as "Food and Water," an anti-irradiation activist group in England. After group (b) read this description, it began five more rounds of bidding. The bid went down, quickly reaching zero. After the first five rounds produced a value of twenty cents in group (b) for the pork sandwich described in a "neutral" way, *no one* in this group would pay a penny for the irradiated sandwich described in a "negative" way. This result was obtained even though the description was clearly identified as coming from an activist, nonscientific group.

After five rounds of bidding on the neutral description, the third group, group (c) received *both* the positive and the negative descriptions. One might expect that this group's response would be highly variable, with some participants scared off by the negative description and others discounting it for its unscientific source. Some participants might be expected to bid nothing while others would continue to bid high.

However, the price of the sandwich in the third, so-called balanced group, also fell quickly. Indeed, the price reached zero as quickly as it did in group (b), the negative group. That is, even though the third group had both the neutral and the positive description in front of them, no one exposed to the negative description would pay two cents for the irradiated sandwich.

Hayes's study illuminates the precautionary response, and carries implications for the GM debate. These implications are that, given neutral or positive descriptions of GM foods, consumers initially will *pay more* for them. Given negative descriptions of GM foods, consumers initially will *not* pay more for them. Finally, and this is the surprising result, given *both* positive and negative descriptions of GM foods, consumers initially will not pay more for them. Both sides in the GM food debate should be scrupulous in providing reasons for all of their claims. But especially for their negative claims.

In a worldwide context, the precautionary response of those facing food abundance in developed countries may lead us to be insensitive to the conditions of those in less fortunate situations. Indeed, we may find ourselves in the following ethical dilemma.

For purposes of argument, make the following three assumptions. (I do not believe that any of the assumptions are implausible.) First, assume that GM foods are safe. Second, assume that some GM "orphan" foods, such as rice enhanced with iron or vitamin A, or virus-resistant cassavas, or aluminum-tolerant sweet potatoes, may be of great potential benefit to millions of poor children. Third, assume that widespread anti-GM information and sentiment, no matter how unreliable on scientific grounds, could shut down the GM infrastructure in the developed world.

Under these assumptions, consider the possibility that by tainting GM foods in the countries best suited to conduct GM research safely, anti-GM activists could bring to a halt the range of money-making GM foods marketed by multinational corporations. This result might be a good or a bad thing. However, an unintended side effect of this consequence would be that the new GM orphan crops mentioned above might not be forthcoming, assuming that the development and commercialization of these orphan crops is dependent upon the answering of fundamental questions in plant science and molecular biology that will only be answered if the research agendas of private industry are allowed to go forward along with the research agendas of public research institutions.

Our precautionary response to new food may put us in an uncomfortable position. On the one hand, we want to tell "both sides" of the GM story, letting people know both about the benefits and the risks of the technology. On the other hand, some of the people touting the benefits of the technology make outlandish claims that it will feed the world while some of the people decrying the technology make unsupported claims that it will ruin the world. In that situation, however, those with unsupported negative stories to tell carry greater weight than those with unsupported positive stories. Our precautionary response, then, may well lead in the short term, at least, to the rejection of GM technology. Yet, the rejection of GM technology could indirectly harm those children most in need, those who need what I have called the orphan crops.

Are we being forced to choose between two fundamental values, the value of free speech versus the value of children's lives?

On the one hand, open conversation and transparent decision-making processes are critical to the foundations of a liberal democratic society. We must reach out to include everyone in the debate, and allow people to state their opinions about GM foods, whatever their opinion happens to be, whatever their level of acquaintance with the science and technology happens to be. Free speech is a value not to be compromised lightly.

On the other hand, stating some opinions about GM food can clearly have a tainting effect, a powerful and extraordinarily easy consequence of free speech. Tainting the technology might result in the loss of this potentially useful tool. Should we, then, draw some boundaries around the conversation, insisting that each contributor bring some measure of scientific data to the table, especially when negative claims are being made? Or are we collectively prepared to leave the conversation wide open? That is, in the name of protecting free speech, are we prepared to risk losing an opportunity to help some of the world's most vulnerable?

THE PRECAUTIONARY PRINCIPLE

As a thirteen-year-old, I won my dream job, wrangling horses at Honey Rock Camp in northern Wisconsin. The image I cultivated for myself was the weathered cowboy astride Chief or Big Red, dispensing nuggets to awestruck young rider wannabes. But I was, as they say in Texas, all hat.

"Be careful?" was the best advice I could muster.

Only after years of experience in a western saddle would I have the skills to size up various riders and advise them properly on a case-by-case basis. You should slouch more against the cantle and get the balls of your feet onto the stirrups. You need to thrust your heels in front of your knees and down toward the animal's front hooves. You, roll your hips in rhythm with the animal, and stay away from the horn! You, stay alert for sudden changes of direction!

Only after years of experience with hundreds of different riders would I realize that my earlier generic advice, well-intentioned though it was, had been of absolutely no use to anyone. As an older cowboy once remarked, I might as well have been saying, "Go crazy!" Both pieces of advice were equally useless in making good decisions about how to behave on a horse.

Now, as mad cow disease grips the European imagination, concerned observers transfer fears to GM foods, advising: "Take precaution!" Is this a valuable observation that can guide specific public-policy decisions, or well-intentioned but ultimately unhelpful advice?

As formulated in the 1992 Rio Declaration on Environment and Development, the precautionary principle states that "lack of full scientific certainty shall not be used as a reason for postponing cost-effective measures to prevent environmental degradation." The precautionary approach has led many countries to declare a moratorium on GM crops on the supposition that developing GM crops might lead to environmental degradation. The countries are correct that this is an implication of the principle. But is it the only implication?

Suppose global warming intensifies and comes, as some now darkly predict, to interfere dramatically with food production and distribution. Massive dislocations in international trade and corresponding political power follow global food shortages, affecting all regions and nations. In desperate attempts to feed themselves, billions begin to pillage game animals, clear-cut forests to plant crops, cultivate previously nonproductive lands, apply fertilizers and pesticides at higher-than-recommended rates, and kill and eat endangered and previously nonendangered species.

Perhaps not a likely scenario, but not entirely implausible, either. GM crops could help to prevent it, by providing hardier versions of traditional lines capable of growing in drought conditions, or in saline soils, or under unusual climactic stresses in previously temperate zones, or in zones in which we have no prior agronomic experience. On the supposition that we might need the tools of genetic engineering to avert future episodes of crushing human attacks on what Aldo Leopold called "the land," the precautionary principle requires that we develop GM crops. Yes, we lack full scientific certainty that developing GM crops will prevent environmental degradation. True, we do not know what the final financial price of GM research and development will be. But if GM technology were to help save the land, few would not deem that price cost-effective. So, according to the precautionary principle, lack of full scientific certainty that GM crops will prevent environmental degradation shall not be used as a reason for postponing this potentially cost-effective measure.

The precautionary principle commits us to each of the following propositions:

(1) We must not develop GM crops.
(2) We must develop GM crops.

As (1) and (2) are plainly contradictory, however, defenders of the principle should explain why implications are not incoherent.

Much more helpful than the precautionary principle would be detailed case-by-case recommendations crafted upon the basis of a wide review of nonindustry-sponsored field tests conducted by objective scientists expert in the construction and interpretation of ecological and medical data. Without such a basis for judging this use acceptable and that use unacceptable, we may as well advise people in the GM area to go crazy. It would be just as helpful as "Take precaution!"

RELIGION AND ETHICS

Religious traditions provide an answer to the question "How, overall, should I live my life?" Secular ethical traditions provide an answer to the question "What is the right thing to do?" When in a pluralistic society a particular religion's answers come into genuine conflict with the answers arrived at through secular ethical deliberation, we must ask how deep the conflict is. If the conflict is so deep that honoring the religion's views would entail dishonoring another religion's views, then we have a difficult decision to make. In such cases, the conclusions of secular ethical deliberation must override the answers of the religion in question.

The reason is that granting privileged status to one religion will inevitable discriminate against another religion. Individuals must be allowed to follow their consciences in matters theological. But if one religion is allowed to enforce its values on others in a way that restricts the others' ability to pursue their values, then individual religious freedom has not been protected.

Moral theorists refer to this feature of nonreligious ethical deliberation as the *overridingness* of ethics. If a parent refuses a lifesaving medical procedure for a minor child on religious grounds, the state is justified in overriding the parent's religious beliefs in order to protect what secular ethics regards as a value higher than religious freedom: the life of a child.

The overridingness of ethics applies to our discussion only if a religious group claims the right to halt GM technology on purely religious grounds. The problem here is the confessional problem, of one group attempting to enforce its beliefs on others. I mean no disrespect to religion; as I have noted, I am a religious person, and I value religious traditions other than my own. Religious traditions have been the repositories and incubators of virtuous behavior. Yet each of our traditions must in a global society learn to coexist peacefully with competing religions, and with nonreligious traditions and institutions.

If someone objects to GM technology on purely religious grounds, we must ask on what authority she speaks for her tradition, whether there are other, conflicting, views within her tradition, and whether acting on her views will entail disrespecting the views of people from other religions. It is, of course, the right of each tradition to decide its attitude about genetic engineering. But in the absence of other good reasons, we must not allow someone to ban GM technology for narrowly sectarian reasons alone. To allow such an action would be to disrespect the views of people who believe, on equally sincere religious grounds, that GM technology is not necessarily inconsistent with God's desires for us.

MINORITY VIEWS

When in a pluralistic society the views of a particular minority come into genuine conflict with the views of the majority, we must ask a number of questions: How deep is the conflict? How has the minority been treated in the past? If the minority has been exploited, have reparations been made? If the conflict is so deep that honoring the minority's views would entail overriding the majority's views, then we have a difficult decision to make. In such cases, the conclusions of the state must be just, taking into account the question of past exploitation and subsequent reparations, or lack thereof. This is a question of justice.

The question of justice would arise in the discussion of GM technology if the majority favored GM technology, while the minority claimed the right to halt GM technology. If the minority cited religious arguments to halt GMOs, yet the majority believed that halting GMOs would result in loss of human life, then the state faces a decision very similar to the one discussed in the prior section. In this case, secular policy decisions may be justified in overriding the minority's religious arguments insofar as society deems the value of human life higher than the value of religious freedom.

However, should the minority cite past oppression as the reason that their values ought to predominate over the majority's, then a different question must be addressed. Here, the relevant issues have to do with the nature of past exploitation; its scope and depth; and the sufficiency of efforts, have there been any, to rectify the injustice and compensate victims. If the problem is longstanding and has not been addressed, then imposing the will of the majority would seem a sign of an unjust society insensitive to its past misdeeds. If, on the other hand, the problem has been carefully addressed by both sides and, for example, just treaties arrived at through fair procedures have been put in place, are being enforced, are rectifying past wrongs, and are preventing new forms of exploitation, then the minority's arguments would seem to be far weaker. This conclusion would be especially compelling if it could be shown that the lives of other disadvantaged peoples might be put at risk by honoring a particular minority's wish to ban GMOs.

CONCLUSION

Earlier I described a method for reaching ethically sound judgments. It was on the basis of that method that I personally came to change my mind about the moral acceptability of GM crops. My opinion changed as I took full account of three considerations: (a) the rights of people in various countries to choose to adopt GM technology (a consideration falling under the human rights principle); (b) the balance of likely benefits over harms to consumers and the environment from GM technology (a utilitarian consideration); and (c) the wisdom of encouraging discovery, innovation, and careful regulation of GM technology (a consideration related to virtue theory).

Is it ethically justifiable to pursue GM crops and foods? I have come to believe that three of our most influential ethical traditions converge on a common answer. Assuming we proceed responsibly and with appropriate caution, the answer is yes.

NOTES

1. In describing this method, I have drawn on an ethics assessment tool devised by Dr. Courtney Campbell, Philosophy Department, Oregon State University, and presented at the Oregon State University Bioethics Institute in Corvallis, Oregon, summer 1998.

2. Gary Comstock, *Vexing Nature? On the Ethical Case against Agricultural Biotechnology* (Boston and Dordrecht: Kluwer Academic Publishers, 2000).

3. Brody Baruch, private communication.

4. William McNeill, "Gains and Losses: A Historical Perspective on Farming," 1989 Iowa Humanities Lecture, National Endowment for the Humanities and Iowa Humanities Board, Oakdale Campus, Iowa City, Iowa, 1989.

5. Jonathan Gressel, observation at the Annual Meeting of the Weed Science Society of America, Chicago, 10 February 1998; See also, Alan Ryan et al., *Genetically Modified Crops: The Ethical and Social Issues* (London: Royal Society, 1999), sec. 1.38.

6. Dorothy Nelkin and M. Susan Lindee, *The DNA Mystique: The Gene as Cultural* Icon (New York: Freeman, 1995).

7. Leon R. Kass, *Toward a More Natural Science: Biology and Human Affairs* (New York: Free Press, 1998); Kass, "Beyond Biology: Will Advances in Genetic Technology Threaten to Dehumanize Us All?" *New York Times*, 23 August 1998 [online], http://www.nytimes.com/books/98/08/23/reviews/980823.23kassct.html.

8. Mary Midgley, "Biotechnology and Monstrosity: Why Should We Pay Attention to the 'Yuk Factor,'" *Hastings Center Report* 30, no. 5 (2000): 7–15.

9. Ibid.

10. See also, Gary Comstock, "The Case against bGH," *Agriculture and Human Values* 5 (1998): 26–52. The other essays are reprinted in Comstock, *Vexing Nature?* chaps. 1–4.

11. Dermont Hayes, John A. Fox, and Jason F. Shogren, "Consumer Preferences for Food Irradiation: How Favorable and Unfavorable Descriptions Affect Preferences for Irradiated Pork in Experimental Auctions," *Journal of Risk and Uncertainty* 24, no. 1 (2002): 75–95.

33

What's Wrong with Functional Foods?

David M. Kaplan

FUNCTIONAL FOODS DEFINED

All food is in some sense functional insofar as it contains calories and nutrients that support health. The more narrowly construed sense of functional foods are those that have added ingredients believed to provide additional health benefits. Functional foods are not new. They have existed since the early 1900s when iodine was first added to salt to prevent goiter. Vitamin D has been added to milk since the 1930s, extra vitamins and minerals to breakfast cereals since the 1940s, and water fluoridated shortly thereafter. The difference between these fortified foods and the newer generation of functional foods is that more recent ones are designed to replace medicine with food, or sometimes to eliminate qualities to make them (seem) more healthy. Examples include *Benecol* (a cholesterol-lowering margarine), *Kitchen Prescription Soup* (with the herbal supplement Echinacea), *EggsPlus* (nutritionally enhanced eggs with extra omega-3 fatty acids), *Viactiv* (calcium chews), *Gatorade* and *Vitamin Water* (supplement beverages), *Wow Potato Chips* (fat free, fewer calories), Ensemble food products (with soluble fiber to promote heart health), low-carb food products (from beer to frozen food to fast food), and products geared toward the specific health needs of infants, toddlers, and the aging.

Often genetically modified foods are engineered to be nutritionally enhanced. The most notable example is the highly publicized, Vitamin-A enriched *Golden Rice,* which had been touted for its ability to reduce blindness in malnourished children. Other genetically modified products currently promised are high-protein and vitamin-enriched cassavas, milk and peanuts that are allergen-free, tomatoes with three-times the usual amount of lycopene, a cancer-fighting anti-oxidant, carrots with a hepatitis-B vaccine, and potatoes with a vaccine for cholera.

What counts as a functional food varies from nation to nation. But in each instance the definition has bound up the kind of health claims a product is allowed by law to make. For example, Japan, where the very concept of contemporary functional foods was invented, is the only nation in which functional foods have their own legal designation and regulatory body. Foods for Specific Health Uses (FOSHU) are defined as those foods and beverages with ingredients added for a determined health effect or to reduce the risk of disease or health-related condition. Applications for FOSHU certification are reviewed by the Japan's Ministry of Health and Welfare and must include scientific documentation established by clinical trials performed by approved research institutions.

Only FOSHU-approved products are permitted to make health claims on food labels. They are a separate category in the Japanese food system. Participation in FOSHU is, however, voluntary. Food companies can produce items that make general health claims (to promote health) so long as they make no specific claims (to treat diseases). Products making general, unregulated health claims make up 90 percent of the health food market in Japan.[1] To encourage greater participation in FOSHU, the government lowered the scientific requirements, allowed private-sector laboratories to make legitimate health claims, and streamlined the application process. Still, non-FOSHU-approved functional foods dominate the Japanese market.

In the United Kingdom, there is no legal definition of functional foods, only a working definition by the Ministry of Agriculture Fisheries and Food (MAFF). It denies functional foods as those foods enhanced to have additional health benefits beyond their nutritive benefits.[2] As in Japan, food products are allowed to make general, but not specific, health claims. If a product claims to be capable of preventing, treating, or curing human disease then the food must be licensed as medicine. Food manufacturers are prohibited from making any medicinal claims. They are, however, allowed to make claims which refer to possible disease factors ("can lower cholesterol"), to nutrient function ("Vitamin A is essential for normal vision"), or to recommend dietary practice ("part of a nutritious breakfast"). Other EU countries have adopted a similar strategy: they allow a wide range of generic health claims and have established procedures to assess the evidence for specific health claims. Common to all definitions of functional foods in the EU are that they be recognizable as food, not pills, capsules, or other drug-like forms.

The case in the United States is somewhat more vague. Functional foods are part of an overlapping family that includes food additives, food supplements, and genetically modified foods. The Food and Drug Administration (FDA) defines a "food additive" as any substance designed to help prevent spoilage, contamination, or make food look and taste batter. Additives are things like flavor enhancers (MSG), artificial colors and flavors, preservatives, stabilizers, sulfites, and nitrates. The FDA defines a "dietary supplement" (somewhat unhelpfully) as additional ingredients with either *nutritional* or *non-nutritional* properties, such as vitamins, minerals, proteins, herbs, enzymes, or extracts. They can either take drug-like forms or they can be added to foods. Finally, the FDA defines functional foods as any food product fortified with dietary supplements, food additives, genetically modified organisms, or vaccines with health benefits beyond that of conventional foods. These categories of modified foods are very rough and vague. It does not help clarify things when food technology industry representatives say things like, "fruits and vegetables, being natural sources of beneficial nutrients like vitamins, antioxidants, and fiber, are in essence the ultimate functional food."[3]

There is no legal definition for functional foods in the United States. Although food additives must receive pre-market FDA approval as "Generally Regarded as Safe" (GRAS), dietary supplements and functional foods do not. Under the Dietary Supplement Health and Education Act of 1994 (DSHEA) no pre-market approval is required for dietary supplements and extra-nutritional ingredients. In fact, the FDA must demonstrate that a product is *unsafe* for a product to be pulled from the market. Functional foods are often marketed as dietary supplements to avoid proving their ingredients are GRAS. Yet, only dietary supplements labels must include the disclaimer: "This statement has not been evaluated by the FDA. This product is not intended to diagnose, treat, cure, or prevent any disease." Functional food labels need not include a disclaimer about proven effectiveness. Given that functional foods are most often conventional foods with dietary supplement ingredients added, the lack of consistency in labeling is, if nothing else, puzzling.

Health claims, however, are more carefully regulated than ingredients. The FDA regulates "foods for special dietary use," which includes products used for supplying a special dietary need that exists "by reason of a physical, physiological, pathological, or other condition including but not limited to the conditions of disease, convalescence, pregnancy, lactation, infancy, allergic hypersensitivy to food, underweight, overweight, or the need to control the intake of sodium."[4] Health claims for foods for special dietary use must have premarket approval. Because compliance is voluntary and more strict than what is required for functional foods, very few products are identified as "for special dietary use." Incredibly, the FDA does not regulate "medical foods." These foods are prescribed by a physician for a patient with "special nutrient needs" in order to manage a disease or health condition. They are not intended for the general public. Examples of medical foods include *UltraClear* (for liver failure), *Vistrum* (for gastrointestinal balance), and *Nephrovite* (vitamin supplements for dialysis patients). The FDA does not require that medical foods have nutritional information labeled, nor must their health claims meet specified standards. In 1996, the FDA conceded that the lack of regulation is a problem and it has relied too much on the medical profession to regulate itself to prescribe and oversee the safety of medical foods.[5] As of 2006, the FDA's webpage continues to state that it is "exploring ways to more specifically regulate medical foods. This might include safety evaluations, standards for claims, and requiring specific information on the labels."[6]

As in Japan and the EU, food supplement companies in the United States are permitted to make general health claims ("Structure/Function Claim") without FDA approval, whereas a specific health claim ("Disease Claim') does require approval. Unlike other countries, the U.S. permits health claims to be made for nutrients already contained in conventional food. Nothing has to be added to food to warrant a health claim. Another difference between the U.S. and other countries is in the language used to distinguish between a general and specific health claim: it is parsed exceptionally thin. According to the FDA:

> An example of an acceptable claim is "a good diet promotes good health and prevents the onset of disease" or "better dietary and exercise patterns can contribute to disease prevention and better health."
>   An example of a disease claim is "Promotes good health and prevents the onset of disease" because the claim infers (sic) that the product itself will achieve the intended effect.[7]

It is hard to imagine that language like this does anything but confuse consumers.

FUNCTIONAL FOODS EFFECTIVENESS

The first concern about functional foods is practical, not philosophical. The fundamental practical problem with functional foods is that they do not work very well, and when they do work their health and nutritive effects are far less significant than their advocates would have us believe. That is because the very reductivist premise of functional foods—that food is the kind of thing that can be understood in terms of its component parts—is mistaken. When food is understood in terms of parts rather than wholes it usually does not deliver its promised effect as well as conventional food. There is increasing evidence that food broken down into its component parts and then reassembling as processed food is less nutritious than conventional food. It has been shown that ingredients isolated in laboratories do not function in the same way they do in whole foods.[8] The Center for Science in the Public Interest warns that too often manufacturer claims about functional ingredients are "misleading and unsubstantiated by scientific evidence," and until governments establish adequate regulatory controls "functional foods may merely amount to little more than 21st Century quackery."[9] Even the nutritionists and industry experts who contribute to *Food Technology,* the leading industry journal, caution that the "single-nutrient approach is too simplistic."[10] Food, it appears, is more than the sum of its chemical parts, therefore treating it as collections of single nutrients to be mixed and matched, rather than as the complex biological system it is, simply may not work.

It is true, however, that food fortification for some nutrients does work. The fluoridation of drinking water in the U.S. has helped prevent tooth decay, vitamin-D fortified milk has eliminated rickets, iodized salt reduced goiter, and niacin-enriched flour, pellagra.[11] The increased fortification of these nutrients has very effectively prevented deficiencies of the nutrients added and eliminated a number of sources of disease. Yet, in complex matters of public health, it is often difficult to isolate single casual explanations. For example, it is impossible to know precisely how effective niacin fortification was in the reduction of pellagra deaths in the 1940s since the decrease corresponds with changes in social and economic mobility, food safety, and food availability. If more people were eating healthier, more nutritious diets anyway, it is difficult to explain the reduction of the disease exclusively by niacin fortification. The situation is similar today with grain products fortified with folic acid to reduce the number of infants born with neural tube defects (anencephaly and spina bifida). On the one hand, higher levels of folate are now present in adults in the U.S. since fortification began in the 1980s, and fewer babies have been born with birth defects. On the other hand, the public is already more informed about the link between diet and fetal health—especially wealthier, more educated members of society. It is difficult to determine the effects of fortification on people who are already concerned about maintaining a healthy diet. Other casual factors may explain the reduction in birth defects.[12]

Some nutritionists worry that the single-nutrient approach drives functional food research and marketing, misleading the public to believe that there are dietary magic bullets in their food that will ensure a healthy diet regardless of what they eat. Enhancing food with dietary supplements is a quick techno-fix for more complicated issues of dietary patterns, lifestyle, and public health.

> Can we really accept that super-fortification will eliminate our need to select widely from conventional foods to balance nutrient intake? Americans are intrigued with the notion that a pill or a portion can settle all nutritional needs. Thus, we regard fortified cupcakes and synthesized orange juice as necessary steps in achieving that goal. . . . Dumping nutrients into such foods will not neutralize their detrimental effects or make them more healthful. Furthermore, fortification schemes serve primarily to add to the public's confusion about nutrition. By their nature, fortification practices discourage the most desirable modifications in food selection behavior.[13]

Although techno-solutions are often a short-cut, they should not be dismissed out of hand. It is much easier and more effective to supplement food than to address the more persistent underlying causes of malnutrition, such as poverty or insufficient education. But the small number of successful examples of food fortification should not lead us to assume that all food fortification will work as well. The single-nutrient approach to diet works only on rare occasions. The majority of functional foods are market-driven consumer goods that have not been proven to work at all.

FUNCTIONAL FOODS AS MEDICINE

The second concern about functional foods is that they blur the line between food and medicine. The FDA concedes that there is greater need for regulating the health claims made by functional food producers but has been negligent in its obligation to provide consumer protection. Meanwhile, the market in functional foods is booming. In 2004 sales of functional food products reached $22 billion in the U.S. and $47 billion worldwide.[14] Millions of people in the U.S., Western Europe, and Japan manage their own health by eating dietary supplements and functional foods instead of using prescription or over-the-counter drugs. A nationwide survey conducted recently by the Centers for

Disease Control and Prevention (CDC) found that 36 percent of American adults use complementary and alternative medicines ranging from diet to acupuncture to prayer. That means that a sizeable percentage of the public puts their health into their own hands. In 2003, 158 million Americans used some form of dietary supplements instead of over-the-counter drugs in order, they said, to save money, take control of their own lives, and to live healthier.[15] The trend is toward a public increasingly interested in maintaining better health through diet rather than spending money on health care and prescription medications. Under these conditions, the market for functional foods will only continue to grow.

In many ways, there is nothing new about this do-it-yourself approach to health care. It is a technologically mediated version of long-standing traditions that connect moral conduct with self-mastery of one's body. This connection between a self-imposed dietary regimen and moral conduct can be found in religious traditions throughout the world. For the ancient Greeks and Romans temperance and moderation of all of the appetites were central to moral conduct—especially sexual restraint but also control of diet, exercise, and strong emotions. Asian traditions also emphasized the relationship between diet, regimentation, and health. Taoism, Ayurveda, and Zen Buddhism are just some philosophical-religious systems that specify how bodily health connected to moral conduct leads to spiritual salvation. Although we have retained quite a bit from these traditions, the difference between our contemporary notions of diet and health and ancient and religious dietary practices is not only a greater understanding of physiology and nutrition but also the availability of technologies that extend our capacities in ways nontechnological dietary and health practices cannot. Our current dietary practices are much better at reducing risk of disease, treating disorders, and fostering health. The widespread use of dietary supplements, functional foods, and medical foods are twenty-first-century versions of long-standing, tradition-bound, dietary/health/self-management practices.

Yet the regulatory oversight for these edible technologies is terrible: existing regulations do not provide clear guidance—much less enforceable laws—on products ingredients, safety, and health claims. The most serious problem is the lack of regulation on medical foods. Although they are supposed to be used by patients under medical supervision, there is nothing stopping a food producer from calling any product a medical food and making it available to the public. Even when a medical food is used properly, there are no guarantee that the specific health claims made are supported by adequate scientific evidence. The FDA needs to change its current approach to the regulation of medical and functional foods to ensure safety and truthful labeling. It needs to clearly distinguish between medical and functional foods in unambiguous language, establishing standards and procedures for product composition, manufacturing practice and controls, and labeling requirements. The FDA should require that manufacturers notify the agency before about its health benefits are supported by what it calls "sound science." The quality and quality of scientific evidence required might be modeled after FOSHU. That would clearly distinguish between medical foods and functional foods, and establish standards for what kind of health claims functional foods can legitimately make.

As the line between food and drugs becomes increasingly blurry, the FDA should require that functional food labels carry the same disclaimer ("This statement has not been evaluated by the FDA. This product is not intended to diagnose, treat, cure, or prevent any disease") as dietary supplements. That would remove any arbitrary loophole in the food regulatory system and take a minimal step toward informing consumers of scientific validity of the health claims being made. It should require that all functional ingredients, like food additives, are GRAS before, not after, they are marketed.

The current burden of proof placed on the consumer to demonstrate a product is unsafe is unfair and unreasonable. Individuals lack the resources and know-how to provide scientific evidence for food safety. If "sound science" takes place in laboratories and large-scale research facilities, then it is the obligation of those with access to such places to ensure food safety and to verify health claims, not individuals. It is the obligation of the government to enforce laws and punish offenders for unsafe ingredients and false health claims. Only it has the legitimate power and authority to do so. Food safety is a matter of social justice. A government that fails to protect the safety of its citizens fails in its obligations to protect our rights—for what value do rights have if a citizen is unable to safely exercise those rights? How can we freely choose if the knowledge needed to make informed choices is hidden from us? Even the most minimal conceptions of social justice require the State to protect public safety. The market cannot guarantee food safety, health claims, and credible medical practice. That is the proper role of government.

FUNCTIONAL FOODS AS CONSUMER GOODS

The third concern about functional foods is with the role of the market. The food industry runs up against the troublesome fact, from its perspective, that each person can only eat so much food. On average we eat about 1,500 pounds of food in a year. Yet unlike other consumer goods there is a limit to how much we can consume. Although the epidemic of obesity might seem to suggest that this limit is flexible, the reason Americans are obese has less to do with the total mass of food consumed than with total calories, fats, and lack of exercise. Try as it might, the food industry has to convince us to eat more than we need to. The best way to do this is by adding value to cheap raw materials, usually in the form of convenience or fortification. The food industry has learned that selling unprocessed or minimally processed food is far less profitable than modifying existing food items by enhancing elements they already have in them (like vitamins and minerals) or by adding new elements to them. There is not a lot of money to be made selling oranges, somewhat more money to be made selling orange juice, but even more to be made selling orange juice that claims to provide the recommended daily allowance of calcium.[16]

Functional foods once played a crucial role in public health in eliminating nutritional deficiency disorders. It is conceivable that they may do so again. They may indeed, in some social context, be an intelligent way to support health and treat or prevent disease for people suffering from food restrictions and shortages. When functional foods do provide genuine public health solutions, they contribute immensely to the public welfare. They help to provide the very conditions for life; they help us to increase our capacities, to exercise our rights, and to live well together. The use of functional foods under these circumstances is, of course, a morally permissible policy for a government. In extreme cases, such as malnutrition or famine, a policy of functional food distribution might be required to manage long-standing nutritional needs if necessary for the public welfare. In the United States, the greatest challenges to nutritional health are currently obesity, chronic diseases (many of which are associated with obesity), the needs of an increasing aging population, and food safety. If functional foods can treat hypertension, diabetes, heart disease, arthritis, and eliminate the risks of food-borne illness and disease, then it would not only be wise to continue to develop and distribute them, but it is conceivable that it would be the obligation of the federal government to do so. This might take the form of food relief and food commodity distribution, school feeding programs, nutrition education programs, or incentives for private sector research and development.

It is morally defensible to rely on markets to provide functional foods to maintain public health, so long as no greater harms are inflicted, capacities diminished, or rights abused. If these conditions are met then markets and health are perfectly compatible. If individuals choose to support their health or treat disease by purchasing functional foods, and they are safe, effective, and consumed with knowledge, then there again is little reason to oppose them. The current case with folate-fortified grains is instructive: the market might presently be serving a genuine public health need by providing a functional food that reduces the instances of neural tube birth defects (assuming for the moment that little or no government subsidies were involved). If this is the case, then privatized food production and distribution should be encouraged as a matter of policy to support public health.

The problem with relying on market mechanisms is that they are fickle. Markets may or may not solve public health problems. That is not what they are designed to do. Consequently, to rely on them is, at best, unwise for a government, at worst, negligent and a failure to protect its citizens. The food industry very aggressively influences (and arguably distorts) nutrition science, federal regulation, and consumer choice. It functions like any other industry: it seeks to maximize profit and increase market share. The food industry does so by creating a favorable sales environment for its products. This includes lobbying political representatives to eliminate unfavorable regulations and pressure regulatory agencies not to enforce regulations, co-opting nutrition experts by supporting favorable research, and marketing and advertising, often to children who are unable to read ads critically. The food industry is, of course, free to sell people whatever people want, but it also relies heavily on its influence on the political process, marketing, and its version of nutritional advice in order to persuade people that they want what the industry is selling. Sometimes the food industry succeeds in producing and publicizing goods that people actually want and need; other times its means are less honest and serve to deceive people into thinking they want and need things they really do not.[18] Once functional foods are seen as one among many products that are a part of a sprawling food industry, then there is reason to question how vital they truly are. Functional foods should be seen as commodities with exchange-value rather than goods with use-value, as Marx would explain it.

When food and medicine are treated like any other consumer goods there is a real danger that our very dietary and medical practices ultimately serve the interests of others more than our own interests. Commerce in functional foods, then, is a profoundly moral and political matter. The more dietary practice becomes a matter of consumer choice, the less it becomes a matter for mechanisms of distribution other than the market. Yet that is precisely the social context in which functional food exists. On one hand, they are commodities like any other to be manufactured, sold, and consumed; on the other hand, they are uniquely situated at the nexus of diet, health, and commerce, spanning the worlds of optional consumer goods and vital human needs. This puts us all in a tenuous position: commercial interests have the potential to transform how we eat and how we care for ourselves, yet the very future of food is in the hands of those who may not have our best interests in mind. That may be the most important thing wrong with functional foods.

NOTES

1. Michael Heasman and Julian Mellentin, *The Functional Foods Revolution: Healthy People, Healthy Profits?* (London: Earthscan Publications, 2001), p. 134.

2. Ministry for Agriculture Fisheries and Food, Food Standards Agency, www.food.gov.uk/regu lation_health_claims.

3. Linda Orh, "Nutraceuticals and Functional Foods," *Food Technology*, May 2004, vol. 58, no. 5, p. 64.

4. U.S. Food and Drug Administration, Center for Food Safety and Applied Nutrition, "Food Labeling and Nutrition." www.cfsan.fda.gov/label.html.

5. "The agency believes that there is a need to reevaluate its policy for regulating medical foods because of a number of developments, including enactment of a statutory definition of 'medical food,' the rapid increase in the variety and number of products that are marketed as medical foods, safety problems associated

with the manufacture and quality control of these products, and the potential for fraud as claims that are not supported by sound science proliferate for these products." U.S. Food and Drug Administration, "Regulation of Medical Foods," *Federal Register,* November 29, 1996, (vol. 61, no. 231.

6. www.cfsan.fda.gov/~dms/ds-medfd.html.

7. "Structure/Function Claims: Small Entity Compliance Guide." U.S. Food and Drug Administration, Center for Food Safety and Applied Nutrition, January 9, 2002. www.cfsan.fda.gov/~dms/sclmguid.html.

8. Bruce Silverglade and Michael Jacobson, eds. *Functional Foods: Public Health Boon or 21st Century Quackery?* (New York: Center for Science in the Public Interest, 2000).

9. Ibid., p. 19.

10. Robert Ward and Herbert Watseka "Bioguided Processing: A Paradigm Change in Food Production," *Food Technology,* May 2004, vol. 58, no. 5, pp. 44–48.

11. Centers for Disease Control and Prevention, "Ten Great Public Health Achievements in the 20th Century," *Morbidity and Morality Weekly Report,* October, 15, 1999, vol. 48, no. 40, pp. 905–913.

12. For an analysis of food fortification, see, Marion Nestle, *Food Politics* (Berkeley: University of California Press, 2002), pp. 298–314.

13. C. Christopher, "Is Fortification Unnecessary Technology?" *Food Product Development,* 1978 vol. 12, no. 4, pp. 24–25. Quoted in Nestle, *Food Politics,* p. 314.

14. A. Elizabeth Sloan, "Top 10 Functional Food Trends 2004," *Food Technology,* April 2004, vol. 58, no. 4, p. 32.

15. "Complementary and Alternative Health Medicine Use Among Adults, United States, 2002." Centers for Disease Control and Prevention National Center for Health Statistics, *U.S. Department of Health and Human Services Publications,* 2004.

16. For more of this argument, see, Greg Critser, *Fat Land: How Americans Became the Fattest People in the World* (New York: Mariner Books, 2003).

17. "Universal Declaration of Human Rights," http://www.un.org/rights, p. 5. The right to food is recognized directly or indirectly by every country in the world, either written into their constitutions or by virtue of their membership in the United Nations. Article 25 of the 1948 Universal Declaration of Human Rights states that "everyone has the right to standard of living adequate for the health and well-being of himself and of his family, including food, clothing, housing, medical care and necessary social services, and the right to security in the event of unemployment, sickness, disability, widowhood, old age or other lack of livelihood in circumstances beyond his control."

18. For evidence of precisely how the food industry creates a favorable sales environment through lobbying, marketing, and co-opted nutrition experts, see, Marion Nestle, *Food Politics,* pp. 95–136, 175–218.

Part VI

TECHNOLOGY AND SCIENCE

Recently, philosophers have followed historians and social scientists in examining the role that technologies play in science. These philosophers of "technoscience" oppose and ultimately reverse the received view that technology is the concrete manifestation of abstract scientific principles. Instead they claim that science is made possible by technological instruments and devices. Our theoretical understanding of nature depends on the materiality of machines. For these thinkers, science is seen less as a lofty, intellectual pursuit of timeless truths of nature than as a practical activity of using devices that help us accomplish tasks. Scientific instruments are the key to scientific practice. They tell us what nature "really" is like. If we had no instruments, we'd have no *scientific* knowledge of nature. This conclusion makes science less about knowledge and ideas than about practices and machines.

A second consequence of tying science to instruments is that it highlights the social and political dimensions of technoscience. Science is a messy affair that involves complicated networks of practitioners, machines, funding, regulatory bodies, sometimes animals, and a wide range of social considerations that initially seem foreign to the pursuit of knowledge. Once we acknowledge the vital role that technologies play in scientific practice we place science squarely in the fray of society—not above it or superior to it.

The readings in this section examine this web of scientific reason and technical instruments. The authors examine the role instruments play in scientific knowledge; the kind of experience we have that corresponds to technoscientific knowledge; the role of laboratories and the scientific community; the role of moral theory in scientific expertise; and the relationships among scientists, policy makers, and the public. Philosophers of technoscience take a more realistic, less idealized, approach to science. They ground it firmly in the realms of technical expertise and public policy.

In "When Is an Image Not an Image?" Joseph Pitt examines how scientific instruments transform our understanding of what it is to see and what it means represent reality in an image. The question he asks is, in what sense is a computer generated picture of nano-scale entities by a scanning tunneling electron microscope (STEM) an accurate representation of what is really there? STEMs function so differently from ordinary vision (and even ordinary microscopes) that we stretch the meaning of words like "see" and "image" far beyond their conventional usage. Or, to preserve the usage of these terms, we have to expand our notion of what it is to see something and what constitutes an image. Pitt argues that STEM images are more like depictions than representations. They are "heuristic imaginings," or helpful portrayals of a microscopic realm rather than faithful images of nano-scale entities. We see by means of images generated by STEMs only in a "metaphorical extension" of our commonsense notion of seeing. Pitt calls pictures of nano-scale realities "extended metaphors" to call attention to their highly interpretive character.

Pitt believes that there are both epistemological and ethical consequences in claiming that electron microscopes produce accurate representations of reality. The epistemological issue concerns the way that STEM images capture the salient features of a nano-scale object. Since we have no direct access to the nano-world, we have to rely on instruments. But we have no other means of "seeing" that world except through a STEM. That means that we have no way of knowing if the image is an accurate portrayal of what is really out there or a mere fabrication of reality. Worse, the images are constructed by filtering out the "noise" and making things appear more clear and simple than they really are. The problem with these simplifications is that they mislead us into thinking that the world is less complex than it actually is. The ethical issue raised by STEM images is that they create false expectations about how much we really know about the world. They create the illusion that we know more about the world than we actually do, and that we can do more than we actually can do. Pitt reminds us that we do not know as much as we think we do about what takes place at the nano-level, and that to present that world in neat and tidy images misleads us by making it appear that there are always simple answers to complex problems.

In "Scientific Visualism," Don Ihde also examines the use of instruments in scientific practice. His method of "perceptualist hermeneutics" is a contextualist, interpretive way of seeing that aims to reveal how scientific instruments mediate experience and create new forms of perception. This technologically mediated interpretive perception is referred to as "instrumental realism": the real world only becomes an object of scientific discovery and explanation when it is constituted by scientific instruments. He carries out a "weak" program and a "strong" program to show how reality is increasingly transformed by instruments. The weak program examines the ways that instruments prepare reality for observation. Science is a hermeneutic (i.e., interpretive) practice that relies on instruments to make things scientifically analyzable. One set of devices (e.g., imaging technologies, telescopes, X-rays, MRI scans, and sonograms) make things visible. Another set of inscription-making devices (e.g., oscilloscopes and spectrographs) make things readable. Both kinds of technologies present the world in perception-transforming visual forms that must be interpreted to be understood.

The strong program examines the ways that instruments constitute and make an otherwise invisible reality visible. These devices not only bring the object of science into view but they shape and "give a voice" to the world so that it may be experienced. This is what Ihde calls "technoconstruction." He identifies several different "instrumental phenomenological variations" or the various ways that technology mediates perception and constitutes the content of science. For example, astronomers often rely on new instruments to reveal previously unknown phenomena. More instruments produce more scientific objects. Another variation shows how a range of instruments may be used to measure different processes of single objects (e.g., medical imaging technologies to view the brain using X-rays, ultrasound, and MRI). Again scientific instruments produce previously invisible phenomenon that are then translated into visible images for scientific observation. Different combinations of different instruments produce different interpretive phenomena. Ihde's theory of technoconstruction steers a path between an unmediated, naïve realism and social constructivism. He interprets science in terms of a visual hermeneutics, embodied within an instrumentally realist framework in which instruments mediate perceptions.

In "Laboratories," excerpted from *Science in Action* (1987), Bruno Latour argues that scientific facts are constructed, not discovered, through an interplay of laboratories and power relations. The traditional view of science claims that nature is the object of scientific inquiry and ultimately the arbiter of scientific truth. Scientists discover nature's truths through laboratory experimentation and record them in scientific journals. Latour inverts this progression by beginning with the literature of science, moving back to the laboratory, and eventually arriving at what we call "nature." The purpose of the inversion is to highlight the activity of fabricating scientific facts—something that only becomes apparent when we examine science *in action*. We get a very different picture of scientific practice prior to the settling of a dispute as compared with science after the dispute is settled. From this perspective, scientific practice is a technologically mediated, power-laden, social institution for creating and resolving controversies.

Latour shows how challenges to scientific literature occur in the laboratory, which is not only the high-cost place where scientists work but also where "inscriptions" are made. Laboratory instruments are inscription-making devices that depict nature. We never directly experience nature when employing instruments; rather we interpret effects of it through a visual display. Reading instrumentally produced inscriptions requires training and practice, Latour reminds us, often producing conflicting interpretations, or "trials of strength" among competing individuals and groups. The result of disagreement is a scientific object. Laboratories invent new, indeterminate objects and readings that become defined by the results of trials of strength. If an object passes the trials that test the ties linking the representatives of a scientific claim to what they speak for then, finally, we arrive at nature. Latour argues that we can never use nature to explain how and why a controversy ends; nature, rather, is the consequence of the settlement. The practice of technoscience is the ultimate referee.

In "Science Policy and Moral Purity: The Case of Animal Biotechnology," Paul B. Thompson examines two different approaches for linking science, moral theory, and public policy using animal biotechnology as a test case. The first approach is what he calls *moral purification*. Purification proceeds by isolating the social, environmental, animal, and health impacts of biotechnology from each other in terms of discrete categories. The purist presumes that science, technology, society, and values are ontologically distinct realms. In other words, science is not a *form* of technology; technology is not a *form* of society, and so on. The moral purification position maintains that the best way to approach scientific and moral issues is to keep them separate and distinct from one another. The alternative approach is what Thompson (following Latour) calls *hybridization*. This approach assumes that animal biotechnology is a totality of persons, products, animals, and social relations. It cuts across generally accepted conceptual or social patterns of organization. In a hybrid analysis, the political and scientific elements of biotechnology interpenetrate. For example, science is politicized insofar as it promotes the view that product and policy decisions should be "science-based." Similarly, government has (understandably) lost its purity as it increasingly relies on scientific-technical expertise rather than the will of the people to guide policy decisions about things like environmental risks, public safety, and animal welfare.

Thompson focuses his hybrid analysis on the political controversy over recombinant bovine somatotropin (rBST), a growth hormone given to cows to increase milk production. First, he examines the rBST controversy in terms of the standard categories of environmental impact, animal well-being, human health, and social consequences. He also considers objections to animal biotechnology based on religious conceptions of the sacredness of all life and arguments based on the intrinsic value of animals, whose natural purpose or *telos* should not be disturbed. Thompson wants to show that these categories are far from natural or essential but the product of a moral purification that falsely divides the world up into seemingly separate issues. As a result, a successful purification analysis would, for example, treat animal well-being as a separate issue from environmental impact. One set of issues should be debated or resolved without impacting others. By contrast, a hybridization approach starts from the premise that ethical issues in animal biotechnology are a mix of overlapping concepts and categories. Advocates of hybridization may infer that those who employ the strategy of purification seek to avoid accountability by dividing issues, separating the ethical from the environmental from the technical, and so on. Authority figures that ignore ethical issues are generally perceived to have questionable character and should not be trusted. The purification approach actually creates the suspicion that some animal biotechnologies are risky and unsafe even when they are not precisely because purification treats ethical issues as external to scientific issues. This creates the perception that those in power fail to take moral responsibility for it. Moral purification, in fact, creates the backlash against animal biotechnology.

Those who accept the hybrid approach have implicitly questioned the reliability of government and corporate experts based on their lack of concern for the effects of animal biotechnology on, above all, food safety. Yet, Thompson argues that advocates of hybridization face an uphill battle. The dominant political and scientific institutions rely on moral purification to justify the partitioning off of knowledge and power into separate realms. Scientists assess risk; the Food and Drug Administration governs food safety; the Department of Agriculture regulates livestock; economists and sociologists study social consequences; philosophers, ethics. The system of purification is invested with great political and scientific authority and power. Unfortunately, for those who choose the hybrid interpretation find themselves excluded from the dominant political, economic, and scientific mainstreams. Their framework is too broad and expansive—too interdisciplinary to fit into any dominant paradigms. Thompson argues that the strategy of purification can no longer be legitimated. He advocates for incorporating the hybrid view of science's moral significance and for creating a science/government hybrid that would overcome the institutional impediments that prevent us from addressing the moral issues related to animal biotechnology.

In "Technologies of Humility," Sheila Jasanoff argues that governments should reconsider existing relations among decision-makers, experts, and citizens in the management of technology. The problem is that scientific and technical advances have brought not only great benefits but also new risks, new uncertainties, and new vulnerabilities. The risks created by science and technology are woven into the fabric of progress. The question facing us, according to Jasanoff, is how we are to live in a democratic society with the knowledge that we are inevitably at risk? She claims that we must do more than assess the costs and benefits of science and technology; we must also assess their aims, values, and underlying political dimensions. Science-technology policy-making needs to become more political—more accountable to the public. To do so, Jasanoff believes that we need to give up the pretense of control over our technical systems and own up to the limits of human understanding: the unknown, the uncertain, the ambiguous, and the uncontrollable. In place of "technologies of hubris" she suggests "technologies of humility" that acknowledge our lack of perfect foresight and call for a new relationship among experts, decision-makers, and the public.

After a brief historical account of U.S. science and technology policy following the Second World War and the collapse of the "social contract" between science and society, Jasanoff describes the current model of knowledge production, known as *Mode 2*. In the Mode 2 model of science, pure and applied research is merged with public accountability and a concern for the broader social impacts of scientific research and development. The issue, she says, is not whether the public should have more say in technical decisions, or to make technical decision-making more accountable, but rather to promote better interactions among policy-makers, experts, and the public. The key is to complement technologies of hubris (which are based on predictive analysis and control) with technologies of humility (which are based on the unknown, unspecified, and indeterminate). The four focal points around which to develop technologies of humility are: *framing* (what is the purpose?), *vulnerability* (who will be hurt?), *distribution* (who benefits?), and *learning* (how can we know?). Jasanoff argues that attention to these focal points would improve public participation in decision-making, lead to more accountability by scientist and engineers, and produce more credible assessments of science and technology by experts.

34

When Is an Image Not an Image?

Joseph Pitt

The challenge is to tell the truth. In the world of nanotechnology this is not as easy as it sounds. Take, for example, the question of images claimed to represent what some nano-configuration or another looks like. It is alleged Scanning Tunneling Electron Microscopes (STEMs henceforth) produce such images. Let's rehearse what happens: According to Rasmussen and Hawkes:

> . . . an electron beam that is small compared with the image area passes over the specimen in a regular pattern, and a picture of the specimen surface is reconstructed on a video tube . . . interaction of the beam with the specimen produces varying intensities of backscattered and secondarily released electrons for each position in the scan, and these are registered by a detector placed appropriately near the specimen. . . . All electron microscopes depend on the capacity of magnetic and electric fields to alter the path of electron beams according to the laws of optics. (1998, 383)

Using a STEM is one of the ways it is said that we can see what is going on at the nano-level. However, I am suspicious. Or, to put it in a less antagonistic way, to accept this claim will, I believe force us to expand or change our understanding of what it is to see something, and in this case in particular, to understand what constitutes an image. There is nothing wrong with this. The meaning of words do change over time—they often expand, as the meaning of "men" in "All men are created equal" has expanded to include African Americans, other minorities, and women. However, we often do not pay attention to the fact that while we continue to use a word whose meaning we think we understand, in this instance "see" and "image," we also sometimes extend the meaning of that word by applying it to novel situations where they only apply at best metaphorically, as I argue below. Eventually what is at first a metaphorical extension of the meaning of a term may become an accepted part of the meaning of the term, but we should be sensitive to the fact that the meanings of words change over time. This claim is part of a more general thesis I am developing: to explain what we are doing when we employ novel instrumentation, we often employ words whose meanings we already understand in an effort to characterize the sort of thing we think we are now doing with this new instrument, despite the fact that seeing through a microscope is not the same as opening one's eyes and seeing a tree in front of me, if we are to adhere to a strict sense of "seeing." I argue elsewhere that in extending the meaning of words metaphorically we also

From *Techné: Research in Philosophy and Technology* 8, no. 3 (2005). Reprinted by permission of *Techné: Research in Philosophy and Technology.*

change the meanings of the family of concepts with which they are associated, such as evidence and explanation.[1]

If we take Rasmussen and Hawkes seriously, what the electron microscope does is to produce an image. But, I suggest, this is unintuitive for the reasons given below. Furthermore, to claim that an image is produced, suggesting by that that the image is a genuine and *realistic* representation of what is really there, has serious ethical and social consequences. I want to talk about images first, and then I will turn to some disturbing consequences of thinking about "seeing" by way of a STEM.

Imagine, if you will, a very accurate tennis ball machine. It is a device that shoots tennis balls at you so you can practice returning them without having a serving partner. Let's assume you take this machine and aim it at a wall built from rough hewed stone. Your job is to construct an accurate representation of the surface of the wall simply by observing the directions of the balls as they bounce off the wall. Well, clearly you need some help to do this. You need to know a lot about the physics of objects colliding and how irregular surfaces change the vectors, etc. You also need to know a lot about translating what you see happening to the balls after they collide with the wall into paper in a way that captures not the picture of the ball shooting off in this direction and then that, but the texture of the surface of the wall. It is not as if you are directly drawing what you see when you look at the wall. You are interpreting the action of the balls as indicating something about the surface and then you are putting that guess down on paper. That, with some minor modifications, is what the alleged image produced by an STEM is supposed to have accomplished. But instead of a person doing the drawing, a computer program does it. And, we are asked to consider the result an image of the surface. Take your hand, if you will, and run it over your shirt. Now draw what you felt. It is not easy is it? That is why I am asking this question, "when is an image not an image?"

Let us begin by trying to figure out what an image is. This is not an easy task, for we tend to use a substantial vocabulary of what we often take to be more or less synonymous terms when talking about what STEMs produce. Thus, there has been a lot of loose talk about images, representations, etc. Terms like these have been casually interchanged, mangled and generally semantically violated. I will not claim that I offer much of an improvement—but I at least want to alert us to the problem of image talk. In cases like this, my preferred method is to work our way toward a common-sense understanding of what ought to count, in this case, as an image.

My intuitions tell me an image is a representation—where a representation is the result of an attempt to capture the salient features of an object, scene, state of affairs, or ideas, etc. Fortunately or unfortunately, what constitutes a salient feature is a function of the person of persons constructing the image. As a first pass, consider the following items as images:

- Sculptures
- Photographs
- Portraits
- Still lives
- Landscapes
- Various kinds of drawings
- Motion pictures—both animated and "realistic"
- Visualization inspired by poetry
- Visualizations inspired by music
- Plays
- Operas
- Ballet and interpretive dance

If we accept the fact that these are images, then a Picasso such as the *Guernica* counts as an image, but it would seem that a Jackson Pollack does not only insofar as it is unclear what a Pollock is supposed to represent.[2] This entails declaring that to be an image is to be representational. But it says nothing about what makes something representational. That said, nevertheless, it is not shocking to note that not all paintings are images, where a painting is nothing more conceptually complicated than paint deliberately applied to a surface. But, if it is true that not all paintings are images, especially when they are not representational, have we not found a way into our topical question, when is an image not an image? It looks like we could reasonably say that an image is not an image when it is not representational. On the other hand, doesn't that just beg the question? After all, it isn't at all clear that for an image to be an image it must be an image of something. When you think about it, on the one hand, it seems arbitrary to demand the images be representational, but, on the other hand, to do so seems to beg the question. For example, consider the following as candidates for being added to the list above.

- Diagrams
- Flow charts
- Data tables

The interesting feature of these sorts of things is that while they are not representational, they do convey information in visual form. For, on the surface at least, it seems as if these forms of images have different semantics than written language. The important point however, is that they do seem to have a semantics, for they do manage to convey information. The unresolved problem that remains for us is how to determine if the image is an accurate representation. So, if we accept this approach, then one answer to our question is that an image is not an image when we do not know if it is representational but conveys information nonetheless.[3] With your permission, let's accept that for the time being as a first pass.

However, that just moves us back one step, for now we can re-ask the question that our quick look at electron microscopes motivated: when is an alleged representation a representation? The point here is epistemological.

I think it not too radical to suggest that seeing is a complex activity in which after learning to see that as a tree or as a car, we forget that we had to learn that. In our mature state we see the world around us and assume we see it for what it is. That is why philosophical questions like "but are you seeing what is really there?" seem so silly. But, on reflection, we also understand that seeing is an interpretive process and that we bring to our seeing a load of background information and experience. Elsewhere I have argued that to call it a seeing by way of images generated by an electron microscope is a metaphorical extension of our commonsense notion of seeing (Pitt 2005). But, I have now come to realize that there is a lot involved in appealing to metaphor here. If we unpack it, as I would like to start to do here, we can see that to understand through metaphor is to do a number of things at once. First, we use metaphor to access what is new and different because in a metaphor we take what we know and apply it to the unknown and say that the unknown is like the known in these various ways. It makes the new seem familiar and approachable, usually. Sometimes, as in the example of the tennis gun above, it makes the unknown or the new seem even stranger than we first thought. Second, when using metaphor to make the new and unknown approachable, we are also asked to accept that certain things that we do not really understand are reliable. Metaphors tell you this is like that in certain limited ways, and by the way, just accept that everything else is working just fine, however that happens. In the case of the electron microscope, when asked to accept what it produces as a representative image, we are also asked to accept the fact that the assumptions built into the manner in which that image is constructed are correct and reliable. To use the language of science studies, we black-box the process and merely look at the result. But to call the image created by the electron microscope an image is to ask us to accept in some fundamental way that the science is sound and the technology (programming?) reliable and the people manipulating it reliably are honest.

But, I suggest, this ought to be a lot to ask. What is interesting is that it appears that it is not. It is a measure of the success of the scientific establishment that we, the general public, tend to accept claims based on the use of increasingly complicated instruments working in the realm of the frontiers of science with increasing readiness. That is, the more complicated the science and the more simplified the public explanations, the more readily we tend to accept those fantasies. That is why it is important to know what really happens in an electron microscope before buying into the claims with which it is associated. Before I explore what that ominous sounding remark is supposed to suggest, let me give you just one example of the kind of phenomenon to which I am referring. I think we are all in awe of the images sent to us by the Hubble Space Telescope. The ones of the horse head and crab nebulae are just breathtaking—and the colors are truly inspiring—just one catch—the colors are computer generated. When I tell my students that, the looks on their faces resemble the one when they learned that there is no Santa Claus. What got me going in this direction was a presentation at the Conference, "Discovering the Nanoscale" at Darmstadt in October 2003 that revealed that the picture of the nano-scale IBM was not just constructed through the assistance of computers, but it too was computer enhanced—with the colors added, for example. This, it turns out is a pervasive problem; even the choice to use grey scale is a decision to create the image in a certain way. So when we say of an image that it must convey information, should we not also be asking (1) whether there is a claim that reality is being represented, and (2) is the image presented of something real or imagined? Perhaps, then, should we not be asking this slightly different question: "When is an image *not* an image?"

The issue here is both epistemological and ethical. The epistemological issue concerns, for lack of a better term, noise. We are familiar with the problem of filtering out noise when searching for an identifiable signal. The problem is multifaceted: what to filter out and on what criteria, what to amplify, to what degree, etc. The problem with color-enhancement and sharpening up of nano-images is that we don't yet know what is important and what is not. Further, the problem may become intractable since we do not have a god's eye view from which to determine if we have it right. In a certain sense then the problem here is and in *principle* lack of access, or to put it differently, a case of very strong underdetermination. But is this really a problem? We have in-principle-lack-of-access to many astronomical events, like the big bang, and we still claim to know a lot about the early universe. We have images from the Hubble of far distant galaxies that we can never get close to in person, and yet we can still understand a lot of what is going on here—or so we think.

My worry is that, unlike the "images" from the Hubble, we have relatively little experience in enhancing the images produced by STEMs. We have ways of checking up on the Hubble images. For example, we can experiment with filters and use smaller telescopes here on Earth to check out their effect when we look at mountains or trees. However, although we have lots of experiences with so-called images from STEMs—we do not have such successes in fixing them up. This is, in a curious way, a new version of the what-are-we-going-to-do-when-we-stain-a-specimen-that-we-are-going-to-examine-under-a-standard-miocroscope problem (see Pitt 2005). Computer enhancement of images is fun, especially with all the nifty colors we can use. But is it producing an honest replication of the object/surface in question? Clearly not, and that raises the ethical issues.[4]

The ethical issues arise in two forms: strong and relatively minor. The relatively minor issues have to do with the relationships between science and the public. For example, we are misleading the public when we fail to disclose fully what we are doing when we computer enhance our electron microscope constructed images. The strong ethical issues center on the fact that these images raise false expectations. Among them is that we know more than we do. The presentation of these beautiful pictures suggests in a very strong way that this is indeed what it is like out there, in there. But more importantly, they mislead in crucial ways. The beautiful computer simulations we see of nano-interactions are not only beautiful simulations, they are also almost heart-stopping in their ability to feed the hubris we sometimes exhibit when employing the newest technological toys, computer, and advanced programming techniques, among them. Please do not get the wrong impression—I am not suggesting that we should not employ the latest technologies in science. What I am talking about is the illusion we create not just in the general public but sometimes in the practicing scientific community. The illusion is that we know more than we really do. Never underestimate the ability of human beings for self-delusion. These computer generated and enhanced pictures suggest that the world is at rock bottom a simple place. It can be pictured as individual atoms resting on stable fields that we can manipulate at will, twirl them, enlarge and narrow them, put them to music, make them dance, when in fact nothing of the kind is the case. The world at the nano and quantum mechanical level is a buzzing, shifting, constantly in motion in nonlinear and nonclassical casual fashion.

This is all heading in one direction. It is not just misleading to suggest that the world is simple at the bottom. It is epistemically suspect. It employs a crucial but faulty assumption. It is the assumption that the world is better understood if we simplify our presentations of it. I humbly suggest that this is wrong-headed. It may in fact be helpful to extract some feature of the world, color it pretty non-natural colors and play with it. But it is more important to put that heuristically altered item back into the buzz and try to understand it in that environment, its "natural" environment. Most importantly it is crucial that we explain to the public and our colleagues the purpose of the heuristic move and what it reveals about what is really going on at the bottom.

So what is wrong with simplification? It suggests that we know more than we do and, crucially, that we can do more than we can. The scientific community has done a good job of convincing the public that it has god-like properties—but this situation presents a double-edged sword; the public feeds on gods that fail. Be honest about the mess and you will repeat positive rewards. Further, it is not the simplicity of the universe that makes it the object of our inquiry, it is the complications, the unanswered questions, the mess of it all. The more we look, the more complicated we find it to be. If you cuddle the public and give them simplicity and then in the crunch, when, for instance, in the hospital, you say, well it is more complicated than that, then you will have failed miserably. I love the pictures, but they are not representations. They are heuristic imaginings, extended metaphors, if you will, and they should be recognized as such and treated that way. How will that affect the way in which the work of science is perceived? My guess is that it will enhance it. Doing science is hard work. The public should know that and when they do the successes of science will be all the more appreciated. Telling the truth is also hard.

To conclude, let me summarize. The question is "in what sense is a STEM computer generated picture of nano-structures an accurate representation of what is there?" Following some discussion of how "seeing" using a STEM involved a metaphorical extension of the concept of "seeing," it was argued that to be a representation the image must convey information. The problem is in understanding what the information is conveying, since we cannot directly access the domain that we are purporting to represent. The problem is not that we do not know how to interpret what is presented to us as an image, but, rather, that we have loaded the creation of the representation ahead of time without being able to know if our guess that this is what the STEM and its fellow traveler computer programs are producing is an accurate picture of what is really there. The reason why there is so much discussion of when an image is an image is that this really is a question of whether or not the image that is produced is an accurate portrayal of something that is really there or a mere fabrication.

Consider one last attempt to convey a sense of the magnitude of the problem. If we do a random sample of some domain and then plot the results in three dimensions, assuming that sample is truly random and that there is no natural clumping of the data, which curve is the correct one? We can draw an infinite number of curves through those data. Without an independently certified decision procedure for selecting the correct curve we are simply left with the data. The problem is further complicated by the fact that there are ethical dimensions. (1) To say that this is what is taking place at the nano-level, is to lie, since we don't, in fact, know that to be the case. (2) To present these standard, nicely colored, enhanced, and simplified pictures as genuine representations of what is going on at the nano-level is to claim falsely that nature is in fact simple and clean and neatly colored at that level. But, nature is not neat and tidy at that level. To suggest otherwise is to mislead by way of making it appear that there are simple answers to very complex problems. That approach gets us into trouble at the political level and it should get us into equally big trouble in our epistemology.

NOTES

1. This thesis is being developed in a book-length manuscript under construction entitled tentatively, *Seeing Near and Far, A Heraclitian Philosophy of Science and Technology.*

2. If turning to art is seen as somehow cheating, it is important to remember that the creation of images began in art.

3. Yes, "information" is not defined. But, I suggest, we have to start somewhere. If we succeed in making progress by proceeding in the manner suggested we can always return and fine-tune the argument by going deeper into concepts like "information." Call this approach "conceptual bootstrapping."

4. The "Clearly not . . . " might be considered contentious, but with a little expansion, I believe it will be obvious. Consider, for example, that the surface on which nano-scale objects exist is at the interface between the quantum domain and the atomic. We have no idea how to visually represent what happens in the quantum domain, so we cannot say we are accurately representing the surface on which the atomic structures we are picturing sit. If we claim to be accurately depicting the surface, then how can we be sure of the space in which nano-structures function, and if that is uncertain, so must be our representation of the nano-structures themselves.

REFERENCES

Pitt, J. C. 2005. "The Epistemology of the Very Small," in A. Nordmann, J. Schummer, and D. Baird (eds.), *Discovering the Nanoscale.* IOS Press.

Rasmussen, N. and Hawkes, P. 1998. "Microscope, Electron," in R. Bud and D. J. Warner (eds.), *Instruments of Science.* New York: Garland Publishing.

35

Scientific Visualism

Don Ihde

THE "WEAK PROGRAM": HERMENEUTICS IMPLICIT WITHIN SCIENCE

Much of the line I have argued with respect to the history and philosophy of science is that Modern to late Modern science is what it is because it has found ways to enhance, magnify, and *modify* its perceptions. Science, as Kuhn and others after him seem to emphasize, is a way of "seeing." Given its explicit late Modern hyper-*visualism*, this is more than mere metaphor. There remains, deep within science, a belief that seeing is believing. The question is one of how one can see. And the answer is: One sees *through, with, and by means of instruments.* It is, first, this perceptualistic hermeneutics that I explore in the weak program.

Scientific Visualism

It has frequently been noted that scientific "seeing" is highly visualistic. This is, in part, because of historical origins, again arising in early Modern times in the Renaissance. Leonardo da Vinci played an important bridge role here, with the invention of what can be called the "engineering paradigm" of vision.[1] His depictions of human anatomy, particularly those of autopsies which display musculature, organs, tendons, and the like—"exploded" to show parts and interrelationships—were identical with the same style when he depicted imagined machines in his technical diaries. In short, his was not only a way of seeing which anticipated modern anatomies (later copied and improved upon by Vesalius) and modern draftmanship, but an approach which thus visualized both exteriors and interiors (the exploded style). Leonardo was a "handcraft imagist."

The move, first to an almost exclusively visualist emphasis, and second to a kind of "analytic" depiction, was faster to occur in some sciences than in others. In astronomy, analytic drawing of telescopic sightings was accurate early on and is being rediscovered as such today. The "red spot" on Jupiter was already depicted in the seventeenth century. But here, visual observations and depictions were almost the only sensory dimension which could be utilized. Celestial phenomena

From Don Ihde, *Expanding Hermeneutics: Visualism in Science* (Evanston, Ill.: Northwestern University Press, 1999), pp. 158–177. Copyright © 1999 by Northwestern University Press.

were at first open only to visual inspection, at most magnified through optical instrumentation. It would be much later—the middle of the twentieth century—that astronomy would expand beyond the optical and reach beyond the Earth with instruments other than optical ones.

Medicine, by the time of Vesalius, shifted its earlier tactile and even olfactory observations in autopsy to the visualizations à la da Vincian style, but continued to use diagnostics which included palpations, oscultations, and other tactile, kinesthetic, and olfactory observations. In the medical sciences, the shift to the predominantly visual mode for analysis began much later. The invention of both photography and X-rays in the nineteenth century helped these sciences become more like their natural science peers.

Hermeneutically, in the perceptualist style of interpretation emphasized here—the progress of "hermeneutic sensory translation devices" as they might be called—*imaging technologies* have become dominantly visualist. These devices make nonvisual sources into visual ones. This, through new visual probes of interiors, from X-rays, to MRI scans, to ultrasound (in visual form) and PET processes, has allowed medical science to deal with bodies become transparent.[2]

More abstract and semiotic-like visualizations also are part of science's sight. Graphs, oscillographic, spectrographic, and other uses of visual hermeneutic devices give Latour reason to claim that such instrumentation is simply a complex *inscription-making device* for a visualizable result. This vector toward forms of "writing" is related to, but different from, the various isomorphic depictions of imaging. I shall follow this development in more detail later.

While all this instrumentation designed to turn all phenomena into visualizable form for a "reading" illustrates what I take to be one of science's deeply entrenched "hermeneutic practices," it also poses something of a problem and a tension for a stricter phenomenological understanding of perception.

Although I shall outline a more complete notion of perception below, here I want to underline the features of perception which are the source of a possible tension with scientific "seeing" as just described. Full human perception, following Merleau-Ponty, is always *multidimensioned* and *synesthetic*. In short, we *never just see something* but always experience it within the complex of sensory fields. Thus the "reduction" of perception to a monodimension—the visual—is already an abstraction from the lived experience of active perception within a world.

Does this visualizing practice within science thus reopen the way to a division of science from the lifeworld? Does it make of science an essentially reductive practice? I shall argue against this by way of attempting to show that visualization in the scientific sense is a deeply *hermeneutic practice* which plays a special role. Latour's insight that experiments deliver *inscriptions* helps suggest the hermeneutic analogy, which works well here. Writing is language through "technology" in that written language is inscribed by some technologically embodied means. I am suggesting that the sophisticated ways in which science *visualizes* its phenomena is another mode by which understanding or interpretive activity is embodied. Whether the technologies are translation technologies (transforming nonvisual dimensions into visual ones), or more isomorphically visual from the outset, the visualization processes through technologies are science's particular hermeneutic means.

First, what are the epistemological advantages of visualization? The traditional answer, often given within science as well, is that vision is the "clearest" of the senses, that it delivers greater distinctions and clarities, and this seems to fit into the histories of perception tracing all the way back to the Greeks. But this is simply *wrong*. My own earlier researches into auditory phenomena showed that even measurable on physiological bases, hearing delivers within its dimension distinctions and clarities which equal and in some cases exceed those of visual acuity. To reach such levels of acuity, however, skilled practices must be followed. Musicians can detect minute differences in tone, microtones, or quarter tones such as are common in Indian music; those with perfect pitch abilities detect variations in tone as small as any visual distinction between colors. In the early days of auditory instruments, such as stethoscopes, or in the early use of sonar, before it became visually translated, skilled operators could detect and recognize exceedingly faint phenomena, as clearly and as distinctly as through visual operations. Even within olfactory perception, humans—admittedly much poorer than many of their animal cousins—can nevertheless detect smells when only a few molecules among millions in the gas mixture present occur in the atmosphere. In the realms of connoisseurship such as wine tasting, tea tasting, perfume smelling, and the like, specifics such as source, year, and blend—even down to individual ingredients—can be known. It is simply a cultural prejudice to hold that vision is ipso facto the "best" sense.

I argue, rather, that what gives scientific visualization an advantage are its *repeatable Gestalt features* which occur within a technologically produced visible form, and which lead to the rise and importance of *imaging* in both its ordinary visual and specific hermeneutic visual displays. And, here, a phenomenological understanding of perception can actually enhance the hermeneutic process which defines this science practice.

Let us begin with one of the simplest of these Gestalt features, the appearance of a figure against a ground. Presented with a visual display, humans can "pick out" some feature which, once chosen, is seen against the variable constant of a field or ground. It is not the "object" which presents this figure itself—rather, it is in the interaction of visual intentionality that a figure can appear against a ground.

In astronomy, for example, sighting comets is one such activity. Whether sighted with the naked eye, telescopic observation, or tertiary observations of telescopic photographs, the sighting of a comet comes about by noting the movement of a single object against a field which remains relatively more constant. Here is a determined and trained figure/ground perceptual activity. This is also an *interest-determined* figure/ground observation. While, empirically, a comet may be accidentally discovered, to recognize it as a comet is to have sedimented a great deal of previous informed perception.

These phenomenological features of comet discovery stand out by noting that the very structure of figure/ground is not something simply "given" but is *constituted* by its context and field of significations. To vary our set of observables, one could have "fixed" upon any single (or small group) of stars and attended to these instead. Figures "stand out" relative to interest, attention, and even history of perceivability *which includes cultural or macroperceptual features* as well. For example, I have previously referred[3] to a famous case of figure/ground reversibility in the history of aesthetics. In certain styles of Asian painting, it is the background, the openness of space, which is the figure or intended object, whereas the almost abstract tracing of a cherry blossom or a sparrow on a branch in the foreground is now the "background" feature which makes space "stand out."

When one adds to this mix the variability and changeability of instruments or technologies, the process can rapidly change. As Kuhn has pointed out, with increased magnifications in later Modern telescopes, there was an explosion of planet discoveries due to the availability of detectable "disc size," which differentiated planets from stars much more easily.[4]

I have noted Latour, in effect, sees instruments as "hermeneutic devices." They are means by which *inscriptions* are produced, visualizable results. This insight meshes very nicely with a hermeneutic reconstrual of science in several ways.

If laboratories (and other controlled observational practices) are where one prepares inscriptions, they are also the place where objects are made "scientific," or, in this context, *made readable*. Things, the ultimate referential objects of science, are never just naïvely or simply observed or taken, they must be *prepared* or *constituted*. And, in late Modern science, this constitutive process is increasingly pervaded by technologies.

But, I shall also argue that the results are often not so much "textlike," but are more like repeatable, variable *perceptual Gestalts*. These are sometimes called "images" or even pictures, but because of the vestigial remains of modernist epistemology, I shall call them *depictions*. This occurs with increasing sophistication in the realm of *imaging technologies* which often dominate contemporary scientific hermeneutics.

To produce the best results, the now technoconstituted objects need to stand forth with the greatest possible clarity and within a context of variability and repeatability. For this to occur, the conditions of instrumental transparency need to be enhanced as well. This is to say that the instrumentation, in operation, must "withdraw" or itself become transparent so the thing may stand out (with chosen or multiple features). The means by which the depiction becomes "clear" is constituted by the "absence" or invisibility of the instrumentation.

Of course, the instrumentation can never *totally* disappear. Its "echo effect" will always remain within the mediation. The mallet (brass, wood, or rubber) makes a difference in the sound produced. In part, this becomes a reason in late Modern science for the deliberate introduction of *multivariant* instrumentation or measurements. These *instrumental phenomenological variations* as I have called them also function as a kind of multiperspectival equivalent in scientific vision (which drives it, not unlike other cultural practices, toward a more postmodern visual model).

All of this regularly occurs within science practice, and I am arguing that it functions as a kind of perceptual hermeneutics already extant in those practices. I now want to trace out a few concrete examples, focused upon roles within imaging technologies, which illustrate this hermeneutic style.

Galileo's hand-held telescopes undertook "real time" observations, with all the limitations of a small focal field, the wobbliness of manual control, and the other difficulties noted above. And, while early astronomers also developed drawings—often of quite high quality—of such phenomena as planetary satellites, the isomorphism of the observation with its imaged production remained limited.

If, on the other hand, it is the repeatability of the Gestalt phenomenon which particularly makes instrumentally produced results valuable for scientific vision, then the much later invention of *photography* can be seen as a genuine technological breakthrough. Technologies as perception-transforming devices not only magnify (and reduce) referent phenomena, but often radically change parameters either barely noted, or not noted at all.

It would be interesting to trace the development of the camera and photography with respect to the history of science. For example, as Lee Bailey has so well demonstrated, not only was the camera obscura a favorite optical device in early Modern science, but it played a deliberately modeling role in Descartes's notion of both eye and ego.[5] From the camera obscura and its variants to the genuine photograph, there is a three-century history. This history finally focused upon the *fixing* of an image. As early as 1727, a German physician, Johann Schultze, did succeed in getting images onto chalk and silver powders, but the first successfully fixed image was developed by Joseph Niepce in 1826. His successor, Louis Daguerre, is credited usually (in 1839), but Daguerre simply perfected Niepce's earlier process.[6] I shall jump immediately into the early scientific use of photography.

If the dramatic appearance of relative distance (space) was the forefront fascination with Galileo's telescope, one might by contrast note that it is the dramatic appearance of a transformation of *time* which photography brought to scientific attention. The photograph "stops time," and the technological trajectory implicitly suggested within it is the ever more precise micro-instant which can be captured. In early popular attention, the association with time stoppage often took the association between the depiction and a kind of "death" which still photography evoked. Ironically, the stilted and posed earliest photos were necessary artifacts of the state of the technology—a portrait could be obtained only with a minutes-long fixed pose, since it took that long for the light to form the negative on glass covered with the requisite chemical mixture.

Photography, however, was an immediately popular and rapidly developing new medium. And, if portraits and landscapes were early favored, a fascination with motion also occurred almost immediately. The pioneers of stop-motion photography were Eadweard Muybridge and Thomas Eakins at the end of the nineteenth century. Muybridge's studies of horses' gaits served a popular scientific interest. He showed, with both galloping horses and trotters, that all four feet left the ground, thus providing "scientific" evidence for an argument about this issue, considered settled with Muybridge's photos of 1878.[7] Insofar as this is a "new" fact (this is apparently debatable since there are some paintings which purport to show the same phenomenon), it is a discovery which is instrumentally mediated in a way parallel to Galileo's telescopic capture of mountains on their Moon. And, if this time-stop capacity of the technology can capture a horse's gait, the trajectory of even faster time-stop photography follows quickly. By 1888 time-stop photography had improved to the extent that the Mach brothers produced the first evidence of shock waves by photographing a speeding bullet. In this case, the photo showed that the bullet itself penetrated its target, not "compressed air," which was until then believed to advance before the projectile and cause injury.

Here we have illustrations of an early *perceptual hermeneutic* process which yields visually clear, repeatable, convincing Gestalts of the phenomena described. At this level, however, there is a "realism" of visual result which retains, albeit in a time-altered form, a kind of visual *isomorphism* which is a variant upon ordinary perception. It is thus less "textlike" than many other variants which develop later.

The visual isomorphism of early still photography was also limited to surface phenomena, although with a sense of frozen "realism" which shocked the artists and even transformed their own practices.[8] The physiognomy of faces and things was precise and detailed. The stoppage of time produced a *repeatable image of a thing*, which could be analytically observed and returned to time and again.

A second trajectory, however, was opened by the invention of the X-ray process in 1896. Here the "insides" of things could be depicted. Surfaces became transparent or disappeared altogether, and what had been "invisible" or, better, occluded became open to vision. X-ray photos were not so novel as to be the first interior depictions; we have already noted the invention of the "exploded diagram" style practiced by da Vinci and Vesalius. And one could also note that various indigenous art, such as that of Arnhemland Aborigines and Inuit, had an "X-ray" style of drawing which sometimes showed the interiors of animals. But the X-ray photo did to its objects what still photography had done to surfaces—it introduced a time-stop, "realistic" depiction of interior features. In this case, however, the X-ray image not only depicts differently, but produces its images as a "shadow." The X-rays pass through the object, with some stopped by or reduced by resistant material—in early body X-rays, primarily bones.[9]

Moving rapidly, once again a trajectory may be noted, one which followed ever more distinct depiction in the development of the imaging technologies: today's MRI scan, CT tomography, PET scans, and sonograms all are variants upon the depiction of interiorities. Each of these processes not only does its depicting by different means but also produces different visual selectivities which vary what is more or less transparent and what is more or less opaque. (I shall return to these processes in more detail.) This continues to illustrate the inscription or visualization process which constitutes the perceptual hermeneutic style of science.

A third trajectory in visualization is one which continues from the earliest days of optical instrumentation: the movement to the ever more microscopic (and macroscopic) entities. The microscope was much later to find its usefulness within science than the telescope. As Ian Hacking has pointed out, as late as 1800 Xavier Bichet refused to allow a microscope in his lab, arguing that "When people observe in conditions of obscurity each sees in his own way and according as he is affected."[10] In part, this had to do with the features of the things to be observed. Many micro-organisms were translucent or transparent and hard to make stand out even as figures against the often fluid grounds within which they moved. When another device which "prepared" the object for science was invented, *staining processes through aniline dyes*, the microscope could be more scientifically employed.[11]

The trajectory into the microscopic, of course, explodes in the nineteenth and twentieth centuries, with electron microscopes, scanning, tunneling processes, and on to the processes which even produce images of atoms and atom surface structure. Let me include here, too, the famous radio crystallography which brought us DNA structure, as well as today's chromosome and genetic fingerprinting processes.

The counterpart, macro-imaging, occurs with astronomy and the "earth sciences" which develop the measuring processes concerning "whole Earth measurements." While each trajectory follows a different, exploitable image strategy, the result retains the Gestalt-charactered visualization which is a favored perceptual object within science.

The examples noted above all retain repeatable Gestalt, visualizable, and in various degrees, isomorphic, features. This is a specialized mode of perception and perceptual hermeneutics which plays an important role within science, but which also locates this set of practices within a now complicated lifeworld.

"Textlike" Visualizations

I shall now turn to a related, but different, set of visualizations, visualizations which bear much stronger relations to what can be taken as "textlike" features. Again, Latour is relevant: if the laboratory is science's *scriptorium*, the place where inscriptions are produced, then some of the production is distinctly textlike. A standard text, of course, is perceived. But to understand it one must call upon a specific hermeneutic practice—*reading*, and the skills which go into reading.

Once again, it would be tempting to follow out in more detail some of the history of the writings which have made up our civilizational histories (and which characterize the postmodern penchant for textuality, following Derrida's *On Grammatology*). But as far as written texts are concerned, I want to note in passing only that the histories of writing have tended to converge into an over narrowing set of choices: alphabetical, ideographic, and, for special purposes, simple pictographic forms. Related to this shrinkage of historical forms, I also want simply to note in passing that science follows and exacerbates this trend within its own institutional form, so much so that its dominantly alphabetic actual text preference is even more clearly narrowed to the emerging dominance of English as "the" scientific natural language.

And there is plenty of "text" in this sense within science. The proliferation of journals, electronic publications, books, and the range of texts produced is obvious enough. These texts, however, always remain secondary or tertiary with respect to science, as we have seen from Latour. So this is not the textlike phenomenon I have in mind; instead, I am pointing to those analogues of texts which permeate science: charts, graphs, models, and the whole range of "readable" inscriptions which remain visual, but which are no longer isomorphic with the referent objects or "things themselves."

Were we to arrange the textlike inscriptions along a continuum, from the closest analogue to the farthest and the most abstractly disanalogous, one would find some vague replication of the history of writing. Historians of alphabetic writing, for example, have often traced the letters of alphabetic writing to earlier *pictographic* items in pre-alphabetic inscriptions. Oscar Ogg, for example, shows that our current letter "A" derives from an inverted pictograph of a bull image.[12] Earlier hieroglyphic inscriptions could serve double purposes: as an analog image of the depicted animal or as the representation of a particular phoneme in the alphabetic sense.

The vestigial analog quality noted in the history of writing also occurs in scientific graphics: for example, a typical "translation" technology occurs in oscillography. If a voice is being patterned on an oscilloscope, the sound is "translated" into a moving, squiggly line on the scope. Each sound produces a recognizable squiggle, which highly skilled technicians can often actually "read."[13] The squiggle is no more, nor no less, "like" the sound made than the letter is within a text, but the technical "hermeneutic" can read back to the referent. As the abstraction progresses, often purposefully so that a higher degree of *graphic Gestalt* can be visualized, the reading-perception becomes highly efficient. "Spikes" on a graph, anomalies, upward or downward scatters—all have immediate significance to the "reader" of this scientific "text." Here is a hermeneutic process within normal science. And it remains visualizable and carries now in a more textlike context the repeatable Gestalt qualities noted above as part of the lingua franca of this style of hermeneutic.

Older instrumentation often was straightforward analogous to the phenomenon being measured. For example, columns of mercury within a thermometer embodied the "higher" and "lower" temperatures shown. Or, if a container was enclosed, a glass tube on the outside with piping to the inside could show the amount of liquid therein. Even moves to digital or numeric dials often followed analog representations.

Finally, although I am not attempting comprehensiveness in this location of hermeneutics in my "weak program" within scientific practice, I want to conclude with some *conventions* which also serve to enhance the textlike reading perceptions. Graphs come with conventions: up and down for high and low temperatures or intensities; with the range of the growing uses of "false color" imagery, rainbow spectrum conventions are followed again for intensities, and so on. All of this functions "like" a reading process, a visual hermeneutics which retains its visualizations, but which takes textlike directions.

In the weak program I have been following to this point, I have chosen science activities which clearly display their hermeneutic features. These, I have asserted, include a preference for visualization as the chosen sensory mode for getting to the things. But, rather than serving simply as a reduction of perceptual richness by way of a monosensory abstraction, visualization has been developed in a hermeneutic fashion—akin to "writing" insofar as writing is also a visual display. Thus, if science is separate from the lifeworld, it is so in precisely the same way that writing would not be included as a lifeworld factor.

Second, I have held that the process within science practice which prepares things for visualization includes the instrumentarium, the array of technologies which can produce the display, depiction, graphing, or other visualizable result which brings the scientific object "into view." (I am not arguing, as some have, that *only* instrumentally prepared object may be considered to be scientific objects. But, in the complex late Modern sciences, instrumentation is virtually omnipresent and dominant when compared to the older sciences and their observational practices.)

Third, I am not arguing that these clearly hermeneutic practices within science *exhaust* the notion of science. I have not dealt with the role of mathematization, with forms of intervention which do not always yield visualizable results, or the need to take apart the objects of science, to analyze things. And I do not mean to imply that these factors are also important to science. Rather, I have been making, so far, the weak case that there are important hermeneutic dimensions to science, especially relevant to the final production of *scientific knowledge*. In short, hermeneutics occurs inside, within, science itself.

Moving now from the implicit hermeneutics within science praxis to the more complex practices—increasingly technologically embodied and instrumentally constructed—we are ready to take note of a stronger program.

THE "STRONG PROGRAM": HERMENEUTIC SOPHISTICATION

In the "weak program" I chose to outline what could easily be recognized as hermeneutic features operative within science. As I now turn to a stronger program, I shall continue to examine certain extant features within science practice which relate to hermeneutic activity, but I shall increasingly turn here to forefront modes of investigation which drive the sciences closer to a postmodern variant upon hermeneutics.

Whole Body Perception

It is, however, also time to introduce more fully, albeit sketchily, a phenomenological understanding of perception in action. This approach will be recognizably close to the theory of perception developed by Merleau-Ponty, although taken in directions which include stronger aspects of multistability and polymorphy, which earlier investigations of my own developed.

1) I have already noted some perceptual *Gestalt features*, including the presentation of a perceptual field, within which figure/ground phenomena may be elicited. Following a largely Merleau-Pontean approach, one notes that fields are always complexly structured, open to a wide variety of intentional interests, and bounded by a horizonal limit. Science, I have claimed, in its particular style of knowledge construction, has developed a visualist hermeneutic which in the contemporary sense has fulfilled its interests through *imagery* constituted instrumentally or technologically. The role of repeatable, Gestalt patterns, in both isomorphic and graphic directions, is the epistemological product of this part of the quest for knowledge.

2) In a strong sense, all sensory fields, whether focused upon in reduced "monosensory" fashion, or as ordinarily presented in synthesized and multidimensional fashion, are *perspectival* and concretely *spatial-temporal*. Reflexively, the embodied "here" of the observer not only may be noted but is a constant in all sensory perspectivalism. This constant may be enhanced only by producing a string of interrelated perspectives, or by shifting into multiperspectival modes of observation. The "ideal observer," a "god's eye view," and nonperspectivalism do not enter a phenomenology of perception.

3) However, while a body perspective relative to the perceptual field or "world" is a constant, both the field and the body are *polymorphic* and *multistable*. In my work in this area, I have shown that multistability is a feature of virtually every perceptual configuration (and the same applies to the extensions and transformations of perception through instrumentation), and that the interrelation of bodily (microperceptual features) and cultural significations (macroperceptual features) makes the polymorphy even more complex. *There is no perception without embodiment; but all embodiment is culturally and praxically situated and saturated.*

4) While I have sometimes emphasized spatial transformations (Galileo's telescope) in contrast to temporal transformations (still photography), all perceptual spatiality is *spatial-temporal*. This space-time configuration may be shown with different effects, as in contrasts between visual repeatability and auditory patterning, but is a constant of all perceptions.

5) All perceptual phenomena are synesthetic and multidimensioned. The "monosensory" is an abstraction—although useful and possible to forefront—and simply does not occur in the experience of the "lived body" (*corps vécu*). The same applies, although not always noted, in our science examples. I will say more below on this feature of perception. The issue of the "monosensory" is

particularly acute with respect to the technological embodiments of science, since instruments (not bodies) may be "monosensory." Again, we reach a contemporary impasse which has been overcome only in part. Either we turn ingenious in the ways of "translating" the spectrum of perceptual phenomena into a visual hermeneutic—perhaps the dominant current form of knowledge construction in science—or we find ways of enhancing our instrumental reductions through variant instruments or new modes of perceptual transformation (I am pointing to "virtual reality" developments here).

A "strong program," I am hinting, may entail the need for breakthroughs whereby a fuller sense of human embodiment may be brought into play in scientific investigations. Whereas the current, largely visualist hermeneutic within science may be the most sophisticated such mode of knowledge construction to date, it remains short of its full potential were "whole body" knowledge made equally possible. This would be a second step toward the incorporation of lifeworld structures within science praxis.

Instrumental Phenomenological Variations

In the voice metaphor I used to describe the investigation of things, I noted that the "giving of a voice" entails, actually, the production of a "duet" at the least. But this also means that different soundings may be produced, either in sequence or in array, by the applications of different instruments. This is a material process which incorporates the practice of "phenomenological variations" along with the intervention within which a thing is given a voice. This practice is an increasing part of science practice and is apparent in the emergence of a suite of new disciplines which today produce an ever more rapid set of revolutions in understanding or of more frequent "paradigm shifts."

I use this terminology because it is a theme which regularly occurs in science reporting. A Kuhnian frame is often cast over the virtually weekly breakthroughs which are reported in *Science, Scientific American, Nature*, and other magazines. Challenges to the "standard view" are common. I shall look at a small sample of these while relating the challenges to the instrumental embodiments which bring about the "facts" of the challenges. Here the focus is upon *multiple instrumental arrays* which have different parameters in current science investigation.

1) *Multiple new instruments/more new things.* The development of multivariant instruments has often led to increased peopling of the discipline's objects. And much of this explosion of scientific ontology has occurred since the mid-twentieth century. This is so dramatically the case that one could draw a timeline just after World War II, around 1950 for most instrumentation, and determine new forms for many science disciplines. For example, in astronomy, until this century, the dominant investigative instrumentation was limited to *optical* technologies and thus restricted to the things which produce *light*. With the development of *radio-telescopy*, based upon technologies developed in World War II—as so many fields besides astronomy also experienced—the field expanded to the forms of microradiation which occur along spectra beyond the bounds of visible light. *The New Astronomy* makes this "revolution" obvious. The editors note that

> The range of light is surprisingly limited. It includes only radiation with wavelengths 30 percent shorter to 30 percent longer than the wavelength to which our eyes are most sensitive. The new astronomy covers radiation from extremes which have wavelengths less and one thousand-millionth as long, in the case of the shortest gamma rays, to over a hundred million times longer for the longest radio waves. To make an analogy with sound, traditional astronomy was an effort to understand the symphony of the Universe with ears which could hear only middle C and the two notes immediately adjacent.[14]

Without noting which instruments came first, second, and so on, expanding out from visible light, first to ultraviolet on one side, and infrared on the other, now reaching into the previously invisible-to-eyeball perceptions, but still within the spectrum of optical light waves, the first expansion into invisible light range occurs through types of "translation" technologies as I have called them. The usual tactic here is to "constitute" into a *visible depiction* the invisible light by using some convention of *false color* depiction.

The same tactic, of course, is used once the light spectrum itself is exceeded. While some discoveries in radio astronomy were made by *listening* to the radio "hiss" of background radiation, it was not long before the gamma-to-radio wavelengths beyond optical capacities were also "translated" into visible displays.

With this new instrumentation, the heavens begin to show phenomena previously unknown but which are familiar today: highly active magnetic gas clouds, radio sources still invisible, star births, supernovas, newly discovered superplanets, evidence of black holes, and the like. The new astronomy takes us closer and closer to the "birth" of the universe. More instruments produce more phenomena, more "things" within the universe.

The same trajectory can be found in many other science disciplines, but for brevity's sake I shall leave this particular example as sufficient here.

2) *Many instruments/the same thing.* Another variant, now virtually standard in usage, is to apply a range of instruments which measure different processes by different means to the same object. Medical imaging is a good example here. If some feature of the brain is to be investigated, perhaps to try to determine without surgical intrusion whether a formation is malignant or not, multiple instrumentation is now available to enhance the interpretation of the phenomenon. A recent history of medical imaging, *Naked to the Bone*, traces the imaging technologies from the inception of X-rays (1896) to the present.

As noted above, X-rays allowed the first technologized making of the body into a "transparent" object. It followed the pattern noted of preparing the phenomenon for a scientific "reading" or perception. Early development entailed—to today's retrospective horror—long exposures, sometimes over an hour, to get barely "readable" images sometimes called "shadowgraphs." This is because X-ray imaging relies upon radiation sent through the object to a plate, and thus the degrees of material resistance cast "shadows" which form the "picture." The earliest problems focused upon getting clearer and clearer images.[15]

I have noted that the microscope became useful only when the specimen could be prepared for "reading" through a dye process which enhanced contrasting or differentiated structures (in the micro-organism). This image enhancement began to occur in conjunction with X-rays as early as 1911 with the use of *radioactive tracers* which were ingested or injected into the patient. This was the beginning of nuclear medicine.[16]

Paralleling X-ray technologies, *ultrasound* began to be explored with the first brain images produced in 1937. The quality of this imaging, however, remained poor since bone tended to reduce what could be "seen" through this sounding probe. (As with all technologies, it takes some time before the range of usefulness is discovered appropriate to the medium. In the case of ultrasound, soft tissue is a better and easier-to-define target.)

But, even later than the new astronomy, the new medical imaging does not actually proliferate into its present mode until the 1970s. Then, in 1971 and 1972, several patents and patent attempts are made for magnetic resonance processes (MRIs). These processes produce imagery by measuring molecular resonances within the body itself. At the same moment, the first use of the computer—*as a hermeneutic instrument*—comes into play with the refinement of *computer-assisted tomography* (CT scanning). Here highly focused X-ray beams are sent through the object (brains, at first), and the data are stored and reconstructed through computer calculations and processes. Computer-"constructed" imaging, of course, began in the space program with the need to turn data into depictions. Kevles notes that

> After the Apollo missions sent back computer-reconstructed pictures of the moon, it did not stretch the imagination to propose that computers could reconstruct the images of the interior of the body, which, like pictures from space, could be manipulated in terms of color and displayed on a personal video moniter.

Here mathematics and imagery or constructed perceivable depictions meet. I claim this is important to a strong hermeneutic program in understanding science.

By 1975 the practical use of positron emissions is captured in the PET scan process. These emissions (from positrons within the object) are made visible. This imagery has never attained the detail and clarity of the above technologies but has some advantage in a dynamic situation when compared to the "stills" which are produced by all but ultrasound processes (and which also are limited in clarity). Thus living brain *functions* can be seen through PET instrumentation. Then, in the 1990s, functional MRI and more sophisticated computer tomographic processes place us into the rotatable, three-dimensional depictions which can be "built up" or "deconstructed" at command, and the era of the *whole body image* is attained.

While each of these processes can show different phenomena, the multiple use is such that ever more complete analysis can also be made of single objects, such as tumors, which can be "seen" with differences indicating malignancy or benignity. This, again, illustrates the ever more complex ways in which science instrumentation produces a visible result, a visual hermeneutics which is the "script" of its interpretive activity.

3) *Many instruments/convergent confirmations.* Another variant upon the multiple instrument technique is to use a multiplicity of processes to check—for example, dating—for greater agreement. In a recent dating of Java *homo erectus* skulls, uranium series dating of teeth, carbon 14, and electron spin resonance techniques were all used to establish dates much more recent (27,000 B.P. +/-53,000 B.P. for different skulls) than previously determined and thus found *homo erectus* to be co-extant with *homo sapiens sapiens.*[17] And, with the recent discovery of 400,000 B.P. javelin-like spears in Germany, one adds thermo-luminescence techniques to establish this new date for human habitation in Europe, at least double or triple the previously suspected earliest date for humans there. (Similar finds, now dated 350,000 B.P. in Siberia, and 300,000+B.P. finds in Spain, all within 1996–97 discovery parameters, evidence this antiquity.)[18]

4) *Single instrument (or instrumental technique)/widespread multiple results.* Here perhaps one of the most widely used new techniques involves DNA "fingerprinting," which is now used in everything from forensics (rapists and murderers both convicted and found innocent and released), to pushing dates back for human migrations or origins. (The "reading" process which goes with DNA identification entails matching pairs and includes visualizations once again.) *Scientific American* has recently reported that DNA tracing now shows that human migrations to the Americas may go back to 34,000 years (not far from the dates claimed for one South American site, claimed to be 38,000 B.P., which with respect to physical data remain doubtful), with other waves at 15,000 B.P., to more recent waves, included a set of Pacific originated populace around 6,500 B.P.)[19] The now widely cited DNA claims for the origins of *homo sapiens* between 200,000 and 100,000 B.P. is virtually a commonplace.

DNA fingerprinting has also been used in the various biological sciences to establish parentage compared to behavioral mating practices. One result is to have discovered that many previously believed-to-be "monogamous" species are, in fact, not. Similarly, the "Alpha Male" presumed successful at conveying his genes within territorial species has been shown to be less dominantly the case than previously believed.[20]

5) *Multiple instruments/new disciplines.* Beginning with DNA (mitochondrial DNA matching) again, the application of this technique has given rise to what is today called "ancient DNA" studies, with one recent result developed in Germany this year, which purports to show that Neanderthals could not have interbred with modern humans, due to the different genetic makeup of these hominids who coexisted (for a time) with modern humans.[21]

Then, returning to variants upon imaging, the new resources for such disciplines as archeology produce much more thorough "picturing" of ancient sites, activities, and relations to changes in environmental factors. Again, drawing from *Scientific American*—one can draw similar examples from virtually every issue of this and similar science-oriented magazines—the array of instruments now available produces, literally, an "in-depth" depiction of the human past. In part, now drawing from uses originally developed for military purposes, imaging from (1) *Landsat*, which used digital imaging and multispectral scanners from the 1970s, to (2) refinements for *Landsats 4 and 5* in the 1980s, which extended and refined the imaging and expanded to infrared and thermal scanning, (3) to *SPOT*, which added linear-array technologies to further refine imaging, to (4) imaging radar, which actually penetrates below surfaces to reveal details, to (5) *Corona*, which provides spectrographic imagery (recently declassified), the modern archeologist, particularly desert archeologists, can get full-array depictions of lost cities, ancient roads, walls, and the like from remote sensing used now.[22]

Instrumentation on Earth includes (1) electromagnetic sounding equipment, which penetrates up to six meters into the earth, (2) ground-penetrating radar, which goes down to ten meters, (3) magnetometers, which can detect such artifacts as hearths, (4) resistivity instruments, which detect different densities and thus may be used to locate artifacts, and (5) seismic instruments, which can penetrate deeper than any of the above instruments. At a recent meeting in Mexico, an anthropologist reported to me that a magnetometer survey of northern Mexico has shown there to be possibly as many as 86,000 buried pyramids (similar to the largest, Cholula, although the remainder are smaller).[23]

In short, the proliferation of instrumentation, particularly that which yields imagery, is radical and contemporary and can now yield degrees and spans of three-dimensional imagery which includes all three of the image breakthroughs previously noted: early optics magnified the micro- and macro-aspects of barely noted or totally unnoted phenomena through magnification (telescopy and microscopy) but remained bound to the limits of the optically visible, by producing "up close" previously distant phenomena.

Photography increased the detail and isomorphy of imaging in a repeatable produced image, which could then be studied more intensely since "fixed" for observation. It could also be manipulated by "blowups" and other techniques, to show features which needed enhancement. Then, with X-rays, followed by other interior-producing imagery, the possibilities were outlined for the contemporary arrays by which image surfaces are made transparent so that one may see interiors. Then add the instruments which expand thoroughly beyond the previously visible and which now go into previously invisible phenomena through the various spectra which are "translated" into visible images for human observation.

Here we have the decisive difference between ancient science and Modern science: Democritus claimed that phenomena, such as atoms, not only were in fact imperceptible, but were *in principle* imperceptible. A modern, technologized science returns to Democritus to the "in fact" only—that is, the atomic is invisible only until we can come up with the technology which can make it visible. This, I claim, is an instrumental *visual hermeneutics*.

Technoconstitution

In the reconstrual of science which I have been following here, I have argued that late Modern science has developed a complex and sophisticated system of *visual hermeneutics*. Within that visualist system, its "proofs" are focused around the things seen. But, also, things are never just or merely seen—the things are *prepared* or made "readable." Scientifically, things are (typically, but not exclusively) *instrumentally mediated*, and the "proof" is often a *depiction* or image.

Interestingly, if in ordinary experience there is a level of *naïve realism* where things are taken simply to be what they are seen to be, similarly, within imaging there is at least the temptation to an imaging naive realism. That naïveté revolves around the intuitive taking of the image to be "like" or to "represent" an original (which would be seen in unmediated and eyeball perceptions). In short, "truth" is taken to be some kind of *isomorphism* between the depiction and the object.

With the issue stated in precisely that way, there are many traps which are set which could lead us back into the issues of modern (that is, Cartesian or seventeenth- and eighteenth-century) epistemologies. But to tackle these would lead us into a detour of some length. It would entail deconstructing "copy theory" from Plato on, deconstructing "representationalism" as the modern version of copy theory, before finally arriving at a more "postmodern" theory which entails both a theory of relativistic intentionality, a notion of perspectivalism, and an understanding of instrumental mediations as they operate within a phenomenological context. (I have addressed these issues in some degree in essays which preceded my formulation here.)[24] I simply want to avoid these traps.

To do so, I shall continue to interpret science in terms of a visual hermeneutics, embodied within an instrumentally realistic—but *critical*—framework in which instruments mediate perceptions. The device I shall now develop will fall within an idealized "history" of imaging, which, while containing actual chronologically recognizable features, emphasizes patterns of *learning to see*.

Isomorphic visions

The first pattern is one which falls into one type of *initial isomorphism* within imaging. As a technical problem, it is the problem of getting to a "clear and distinct" image. Imaging technologies do not just happen, they develop. And in the development there is a dialectic between the instrument and the user in which both a learning-to-see meets an elimination-of-bugs in the technical development. This pattern is one which, in most abstract and general terms, moves from initial "fuzziness" and ambiguity to greater degrees of clarity and distinctness.

Histories of the telescope, the microscope, photography, and X-rays (and, by extension, all the other imaging processes as well) are well documented with respect to this learning-to-see. Galileo, our quasi-mythical founder of early Modern science, was well aware of the need to teach telescopic vision, and of the problems which existed—although he eventually proclaimed the *superiority* of instrumentally mediated vision over ordinary vision. The church fathers, however, did have a point about how to take what was seen through the telescope. Not all of Galileo's observations were clear and easily seen by "any man." The same problem reemerged in the nineteenth century through the observations of Giovanni Schiaparelli, who gave the term the "canals of Mars." Schiaparelli was a well-known astronomer who had made a number of important discoveries, particularly with respect to asteroids and meteor swarms (because, in part, he had a much better telescope than Galileo). But in noting "canali"—which should have been translated into "channels"—which were taken to be "canals"—he helped stimulate the speculations about life on Mars. But neither channels nor canals existed—these, too, were instrumental artifacts.[25]

The dialectic between learning and technical refinement, in the successful cases, eventually leads to the production of clear and distinct images and to quick and easy learning. These twin attainments, however, cover over and often occlude the history and struggle which preceded the final plateaus of relative perfections. Thus, as in the previous illustrations concerning my guests and our Vermont observations of the Moon, once the instrument is focused and set, it literally takes only instants before one can recognize nameable features of its surface. The "aha phenomenon," in short, is virtually immediate today because it is made possible by the advanced technologies. That instantaneity is an accreted result of the hidden history of learning-to-see and its accompanying technical debugging process.

This same pattern occurred with the microscope. Although microorganisms never before seen were detected early, the continued problems of attaining clear and distinct microscopic vision were so difficult that it did not allow the microscope to be accepted into ordinary scientific practice until the nineteenth century. Again, the dialectic of learning how and what to see meets the gradual technical improvement concerning lenses and focusing devices, and finally the application of dying procedures to the things themselves. (This is an overt example of preparing a thing to become a scientific or "readable" object!)

Photography stands in interesting contrast to microscopy—if it took a couple of centuries for microscopy to become accepted for scientifically acceptable depiction, photography was much faster to win the same position. From Niepce's first "fixed" image in 1826, to the more widely accepted date of 1839 for Daguerre's first images, it was less than a half century until, as Bettyann Kevles notes in her history of medical imaging, "By the 1890's photographs had become the standard recorders of objective scientific truth."[26]

The same pattern occurs, but with even greater speed, in the history of early X-rays. In publicizing his new invention, Wilhelm Röntgen made copies of the X-ray of his wife's hand, which showed the bones of her fingers and the large ring which she wore, and sent these to his colleagues across Europe as evidence of his new process. That X-ray (with a long exposure time) was fuzzy, and while easily recognizable as a skeletal hand and ring, contrasts starkly with the radiograph made by Michael Pupin of Prescott Butler's shot-filled (shotgun-injured) hand later the same year. X-rays, duplicated across Europe and America almost immediately after Röntgen's invention, were used "scientifically" from the beginning.

The acceleration of acceptance time (of the learning-technical vision dialectic) similarly applies to the recent histories of imaging, which include, as above, sonograms (1937) and MRIs (1971) in medicine, of remote imaging since the *Tiros* satellite (1965), or of digitally transmitted and reconstructed images from *Mariner 4* (1965) in Earth and space science.

All of the above samples, however, remain within the range of the possible "naive image realism" of visual isomorphisms in which the objects are easily recognizable, even when new to the observer's vision. (Even if Röntgen had never before seen a "transparent hand" as in the case of his wife's ringed fingers, it was "obvious" from the first glimpse what was seen.) The pattern of making clear is an obvious trajectory. Yet we are not quite ready to leave the realm of the isomorphic.

How does one make "clearer" what is initially "fuzzy"? The answer lies in forms of manipulation, what I shall call *image reconstruction*. The techniques are multiple: enlargements (through trajectories of magnification noted before), enhancements (where one focuses in upon particular features and finds ways to make these stand out), contrasts (by heightening or lessening features of or around the objects), and so on. In my examinations I shall try not to be comprehensive, but to remain within the ranges of familiarity (to at least the educated amateur) concerning contemporary imaging. All of these manipulations can and do occur within and associated with simply isomorphic imaging and, for that matter, within its earlier range of black-and-white coloring. Histories of the technical developments which go with each of these techniques are available today and provide fascinating background to the rise of scientific visualism.

The moral of the story is images don't just occur. They are made. But, once made—assuming the requisite clarity and accuracy and certification of origin, etc.—they may then be taken as "proofs" within the visual hermeneutics of a scientific "visual reading." We are, in a sense, still within a Latourean laboratory.

Translation techniques

Much of what can follow in this next step has already been suggested within the realm of the isomorphic. But what I want to point to here is the use in late Modern science of visual techniques which begin ever more radically to vary *away from the isomorphic*.

One of these variables is—if it could be called that—simply the variable use of *color*. Returning to early optics, whatever Galileo or Leeuwenhoek saw, they saw in "true color." And, as we have seen, sometimes that itself was a problem. The transparent and translucent micro-organisms in "true color" were difficult to see. With aniline dyes, we have an early use of "false color." To make the thing into a scientific or "readable" object, we intervene and create a "horse of a different color." "False coloring" becomes a standard technique within scientific visual hermeneutics.

The move away from isomorphism, taken here in gradual steps which do not necessarily match chronologically what happened in the history of science, may also move away from the limits of ordinary perception. As noted above, the "new" late Modern astronomy of midcentury to the present was suddenly infused with a much wider stretch of celestial "reality" once it moved beyond optical and visible limits into, first, the humanly invisible ranges of the still optical or light itself, in the ranges of the infrared and ultraviolet. The instrumentation developed was what I have been calling a *translation technology* in that the patterns which are recordable on the instrumentation can be rendered by "false coloring" into visible images. This same technique was extended later to the full wave spectrum now available from gamma rays (short waves) through the optical to radio waves (long waves), which are rendered in the standard visually gestaltable, but false color, depictions in astronomy. All this is part of the highly technologized, instrumentalized visual hermeneutics which makes the larger range of celestial things into seeable scientific objects.

The "realism" here—and I hold that it is a realism—is a Hacking-style realism: if the things are "paintable"[27] (or "imagable") with respect to what the instruments detect as effects which will not go away, then they are "real." But they have been *made visible* precisely through the technological constructions which mediate them.

Higher level construction

Within the limits of the strong program, I now want to take only two more steps: I am purposely going to limit this attempt to reconstrue science praxis as hermeneutic to contemporary imaging processes which make (natural) things into scientific, and thus "readable," visual objects. I am not going to address the related, but secondary, visual process which entails *modeling*. That process which utilizes the computer as a hermeneutic device is clearly of philosophical interest, but I shall stop short of entering that territory here.

Computers, of course, are integral to many of the imaging processes we have already mentioned. Medical *tomography* (MRI, PET, fMRI, etc.) entails computer capacities to store and construct images. What is a visual Gestalt is built up from linear processes which produce data which have to be "constructed" by the computer. Similarly, the digitally transmitted imagery from distance sensing in satellite, space, and other remote imaging processes also has necessary computer uses. Much of contemporary imaging is computer embodied. And computers open the ways to much more flexible, complex, and manipulable imaging than any previous technology. For the purposes here, however, they will remain simply part of the "black boxes" which produce images which mediate perceptions.

The two higher-level constructive activities I want to point to here entail, first, the refinement of imaging which can be attained through specifically recognizing our technologies as mediating technologies which, in turn, must take into account the "medium" through which they are imaging. I turn again to astronomical imaging: the *Hubble* space telescope has recently captured the most public attention, but it is but one of the instrumental variations which are today exploring the celestial realms.

The advantage *Hubble* has is that it is positioned beyond the effects of the atmosphere with its distortions and interferences—the clarity of *Hubble* vision in this sense is due in part to its extra-Earthly perspective. (Science buffs will recall that at launch it had several defects in operation which were subsequently fixed—thus placing the *Hubble* in the usual pattern of needing technical adjustment to make its images clear!) But, in part by now being able to (phenomenologically) vary *Hubble* with Earth-bound optical telescopes, the move to enhancing Earth-bound telescopy through computer compensations has become possible. Astronomy is moving toward technoconstructions which can account for atmospheric distortions "on the spot" through a combination of laser targeting and computer enhancements. Earth-bound telescopes are today being given new life through these hi-tech upgrades which "read" atmospheric distortions and "erase" these processes which can make clearer new "readable" images. *Science* regularly publishes an "imaging" issue devoted to updating what is taken as the state-of-the-art in imaging (in 1997 it was the 27 June issue). A description of how one "undoes the atmosphere" is included, which entails computer reconstructions, telescopes in tandem, and adaptive optics. This process, *Science* claims, "combat[s] the warping effects of gravity on their giant mirrors . . . reclaims images from the ravages of the atmosphere . . . [and] precisely undoes the atmospheric distortions."[28]

But alongside *Hubble* are the other variants: the infrared space observatory, the *Cosmic Background Explorer*, and other satellite instrumentation which produces imagery from the nonoptical sources. All these technologies are variants upon the same multidimensioned variables which produce readable images, or make things into readable scientific objects.

The final set of instrumental productions I wish to note are the *composites* which produce variants upon "wholes." Earlier I dealt with "whole Earth measurements" which constitute one realm of composite imagery. To determine whether or not sea levels are rising overall, the composite imagery produced combines (1) multiple satellite photo imagery, (2) Earth-bound measurements (such as buoys, laser measurements, and land markers), and (3) computer averaging processes to produce a depiction (false colored) which can, in comparing time slices, show how much the oceans have risen. The composite depiction displays a flat-projection map of the Earth with level plateaus in false color spectra which can be compared between years, decades, and so on.

Similar processes occur in medical imaging. The "whole body imagery" available today on the internet is the result of two full-body "image autopsies," one each of a male and a female, whose bodies through tomographic processes may be seen in whatever "slice" one wishes. The linear processes of tomography show, slice-by-slice, vertically, horizontally, or in larger scans, the full bodies of the corpses used. The dimensions can be rotated, realigned, sectioned, and so on. Tomography also allows one to "peel," layer by layer, the object imaged—from skin, to networked blood vessels, to bones, and so on. (Both the whole Earth and whole body images are probably among the world's most expensive "pictures.") Moreover, all the manipulations which entail enhancements, contrasts, colorings, translations, and the like are utilized in these "virtual" images. Yet, while these virtual "realities" are different from the examination of any actual cadaver, they clearly belong to the visual hermeneutics of science in the strong sense. Things have been prepared to be seen, to be "read" within the complex set of instrumentally delivered visibilities of scientific imaging.

NOTES

1. A more complete discussion of Leonardo's transformation of vision can be found in chapter 1 of Don Ihde, *Postphenomenology* (Evanston: Northwestern University Press, 1993).

2. See Bettyann Holtzmann Kevles, *Naked to the Bone: Medical Imaging in the Twentieth Century* (New Brunswick: Rutgers University Press, 1997).

3. See Don Ihde, *Experimental Phenomenology* (Albany: SUNY Press, 1986), pp. 128–29.

4. Thomas Kuhn, *Structure of Scientific Revolutions* (Chicago: University of Chicago Press, 1962), pp. 115–16.

5. See Lee W. Bailey, *Skull's Darkroom: The Camera Obscura and* Subjectivity, (Dordrecht, Kluwer Academic, 1989).

6. See both Jon Darius. *Beyond Vision* (Oxford: Oxford University Press, 1984), pp. 34–35; and Peter Pollack, *The Picture History of Photography* (New York: Thames & Hudson, 1977), pp. 65–67.

7. Darius, *Beyond Vision*, pp. 34–35.

8. I have traced some interesting cross-cultural aspects of the imaging of others in pre-compared to post-photographic contexts; see chapter 4 of *Postphenomenology*.

9. See Kevles, *Naked to the Bone*, pp. 3–20.

10. Ian Hacking, *Representing and Intervening* (Cambridge: Cambridge University Press, 1983), p. 193.

11. Ibid.; see the chapter on "Microscopes," pp. 186–209.

12. Oscar Ogg, *The 26 Letters* (New York: Crowell, 1961), p. 78.

13. These patterns are used by speech pathologists, for example, to show speakers how what they are saying does not, in fact, correspond to the standard form of a native language.

14. *The New Astronomy*, 2nd ed. (Cambridge: Cambridge University Press, 1996), p. 6.

15. Kevles, *Naked to the Bone,* p. 20.

16. Kevles provides a time chart, paralleling the various developments in the multiple imaging instrumentation.

17. *Science* 274 (13 December 1996): 1870–873.

18. *Science* 276 (30 May 1997): 1331–334.

19. *Scientific American* 276 (August 1997): 46–47.

20. *Science* 243 (31 March 1989): 1663.

21. *Science* 277 (11 July 1997): 176–78.

22. *Scientific American* 276 (August 1997): 61–65.

23. The visit to the Cholula pyramid and conversations with anthropologists occurred during the 9th International Conference of the Society for Philosophy and Technology, November 1996.

24. Postmodernism is more thoroughly discussed in *Postphenomenology*.

25. *Micropaedeia, Encyclopaedia Britannica*, 15th ed. (Chicago, 1994), vol. 10, p. 514.

26. Kevles, *Naked to the Bone*, p. 15.

27. I refer to Hacking's "If you can spray them then they are real" in *Representing and Intervening*.

28. *Science* 276 (27 June 1997): 1994. This rhetoric is an example of the more-than-neutral language often employed by science reporting.

36

Laboratories

Bruno Latour

FROM TEXTS TO THINGS: A SHOWDOWN

"**Y**ou doubt what I wrote? Let me show you." The very rare and obstinate dissenter who has *not* been convinced by the scientific text, and who has not found other ways to get rid of the author, is led from the text into the place where the text is said to come from. I will call this place the **laboratory**, which for now simply means, as the name indicates, the place where scientists *work*. Indeed, the laboratory was present in the texts we studied in the previous chapter: the articles were alluding to "patients," to "tumors" to "HPLC," to "Russian spies," to "engines"; dates and times of experiments were provided and the names of technicians acknowledged. All these allusions however were made within a paper world; they were a set of semiotic actors presented in the text but not *present* in the flesh; they were alluded to as if they existed independently from the text; they could have been invented.

1) Inscriptions

What do we find whèn we pass through the looking glass and accompany our obstinate dissenter from the text to the laboratory? Suppose that we read the following sentence in a scientific journal and, for whatever reason, do not wish to believe it:

> (1) "36.1 shows a typical pattern. Biological activity of endorphin was found essentially in two zones with the activity of zone 2 being totally reversible, or statistically so, by naloxone."

We, the dissenters, question this figure 36.1 so much, and are so interested in it, that we go to the author's laboratory (I will call him "the Professor"). We are led into an air-conditioned, brightly lit room. The Professor is sitting in front of an array of devices that does not attract our attention at first. 'You doubt what I wrote? Let me show you.' This last sentence refers to an image slowly produced by one of these devices (figure 36.1):

We now understand that what the Professor is asking us to watch is related to the figure in the text of sentence (1). We thus realise where this figure comes from. It has been *extracted* from the instruments in this room, *cleaned, redrawn*, and *displayed*. We also realise, however, that the

Reprinted by permission of the publisher from *Science in Action: How to Follow Scientists and Engineers through Society* by Bruno Latour, pp. 64–74, 91–100, Cambridge, Mass.: Harvard University Press. Copyright © 1987 by Bruno Latour.

images that were the last layer in the text, are the *end result* of a long process in the laboratory that we are now starting to observe. Watching the graph paper slowly emerging out of the physiograph, we understand that we are at the junction of two worlds: a paper world that we have just left, and one of instruments that we are just entering. A hybrid is produced at the interface: a raw image, to be used later in an article, that is emerging from an instrument.

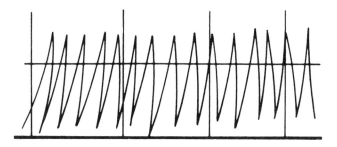

Figure 36.1

"Ok. This is the base line; now, I am going to inject endorphin, what is going to happen? See?!" (figure 36.2)

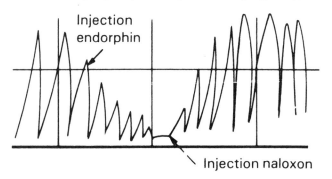

Figure 36.2

"Immediately the line drops dramatically. And now watch naloxone. See?! Back to base line levels. It is fully reversible."

For a time we focus on the stylus pulsating regularly, inking the paper, scribbling cryptic notes. We remain fascinated by this fragile film that is in between text and laboratory. Soon, the Professor draws our attention beneath and beyond the traces on the paper, to the physiograph from which the image is slowly being emitted. Beyond the stylus a massive piece of electronic hardware records, calibrates, amplifies and regulates signals coming from another instrument, an array of glassware. The Professor points to a glass chamber in which bubbles are regularly flowing around a tiny piece of something that looks like elastic. It is indeed elastic, the Professor intones. It is a piece of gut, guinea pig gut ("myenteric plexus-longitudinal muscle of the guinea pig ileum," are his words). This gut has the property of contracting regularly if maintained alive. This regular pulsation is easily disturbed by many chemicals. If one hooks the gut up so that each contraction sends out an electric pulse, and if the pulse is made to move a stylus over graph paper, then the guinea pig gut will be induced to produce regular scribbles over a long period. If you then add a chemical to the chamber you *see* the peaks drawn by the inked stylus slow down or accelerate at the other end. This perturbation, invisible in the chamber, is visible on paper: the chemical, no matter what it is, is given a *shape* on paper. This shape "tells you something" about the chemical. With this set-up you may now ask new questions: if I double the dose of chemical will the peaks be doubly decreased? And if I triple it, what will happen? I can now measure the white surface left by the decreasing scribbles directly on the graph paper, thereby defining a quantitative relation between the dose and the response. What if, just after the first chemical is added, I add another one which is known to counteract it? Will the peaks go back to normal? How fast will they do so? What will be the pattern of this return to the base line level? If two chemicals, one known, the other unknown, trace the same slope on the paper, may I say, in this respect at least, that they are the same chemicals? These are some of the questions the Professor is tackling with endorphin (unknown), morphine (well known) and naloxone (known to be an antagonist of morphine).

We are no longer asked to believe the text that we read in *Nature*; we are now asked to believe *our own eyes*, which can see that endorphin is behaving exactly like morphine. The object we looked at in the text and the one we are now contemplating are identical except for one thing. The graph of sentence (1), which was the most concrete and visual element of the text, is now in (2) the most abstract and textual element in a bewildering array of equipment. Do we see more or less than before? On the one hand we can see more, since we are looking at not only the graph but also the physiograph, and the electronic hardware, and the glassware, and the electrodes, and the bubbles of oxygen, and the pulsating ileum, and the Professor who is injecting chemicals into the chamber with his syringe, and is writing down in a huge protocol book the time, amount of and reactions to the doses. We can see more, since we have before our eyes not only the image but what the image is made of.

On the other hand we see *less* because now each of the elements that makes up the final graph could be modified so as to produce a different visual outcome. Any number of incidents could blur the tiny peaks and turn the regular writing into a meaningless doodle. Just at the time when we feel comforted in our belief and start to be fully convinced by our own eyes watching the image, we suddenly feel uneasy because of the fragility of the whole set up. The Professor, for instance, is swearing at the gut saying it is a "bad gut." The technician who sacrificed the guinea pig is held responsible and the Professor decides to make a fresh start with a new animal. The demonstration is stopped and a new scene is set up. A guinea pig is placed on a table, under surgical floodlights, then anaesthetised, crucified and sliced open. The gut is located, a tiny section is extracted, useless tissue peeled away, and the precious fragment is delicately hooked up between two electrodes and immersed in a nutrient fluid so as to be maintained alive. Suddenly, we are much further from the paper world of the article. We are now in a puddle of blood and viscera, slightly nauseated by the extraction of the ileum from this little furry creature. We realize that many other manual abilities are required in order to write a convincing paper later on. The guinea pig alone would not have been able to tell us anything about the similarity of endorphin to morphine; it was not mobilizable into a text and would not help to convince us. Only a part of its gut, tied up in the glass chamber and hooked up to a physiograph, can be mobilized in the text and add to our conviction. Thus, the Professor's art of convincing his readers must extend beyond the paper to preparing the ileum, to calibrating the peaks, to tuning the physiograph.

After hours of waiting for the experiment to resume, for new guinea pigs to become available, for new endorphin samples to be purified, we realise that the invitation of the author ("let me show you") is not as simple as we thought. It is a slow, protracted and complicated staging of tiny images in front of an audience. "Showing" and "seeing" are not simple flashes of intuition. Once in the lab we are not presented outright with the real endorphin whose existence we doubted. We are presented with another world in which it is necessary to prepare, focus, fix and rehearse the vision of the real endorphin. We came to the laboratory in order to settle our doubts about the paper, but we have been led into a labyrinth.

This unexpected unfolding makes us shiver because it now dawns on us that if we disbelieve the traces obtained on the physiograph by the Professor, we will have to give up the topic altogether or go through the same experimental chores all over again. The stakes have increased enormously since we first started reading scientific articles. It is not a question of reading and writing back to the author any more. In order to argue, we would now need the manual skills required to handle the scalpels, peel away the guinea pig ileum, interpret the decreasing peaks, and so on. Keeping the controversy alive has already forced us through many difficult moments. We now realize that what we went through is nothing compared to the scale of what we have to undergo if we wish to continue. Earlier, we only needed a good library in order to dispute texts. It might have been costly and not that easy, but it was still feasible. At this present point, in order to go on, we need guinea pigs, surgical lamps and tables, physiographs, electronic hardware, technicians and morphine, not to mention the scarce flasks of purified endorphin; we also need the skills to use all these elements and to turn them into a pertinent objection to the Professor's claim. As will be made clear later, longer and longer detours will be necessary to find a laboratory, buy the equipment, hire the technicians and become acquainted with the ileum assay. All this work just to start making a convincing counterargument to the Professor's original paper on endorphin. (And when we have made this detour and finally come up with a credible objection, where will the Professor be?)

When we doubt a scientific text we do not go from the world of literature to Nature as it is. Nature is not directly beneath the scientific article; it is there *indirectly* at best. Going from the paper to the laboratory is going from an array of rhetorical resources to a set of new resources devised in such a way as to provide the literature with its most powerful tool: the visual display. Moving from papers to labs is moving from literature to convoluted ways of getting this literature (or the most significant part of it).

This move through the looking glass of the paper allows me to define an **instrument**, a definition which will give us our bearings when entering any laboratory. I will call an instrument (or **inscription device**) any set-up, no matter what its size, nature and cost, that provides a visual display of any sort in a scientific text. This definition is simple enough to let us follow scientists' moves. For instance an optical telescope is an instrument, but so is an array of several radio-telescopes even if its constituents are separated by thousands of kilometers. The guinea pig ileum assay is an instrument even if it is small and cheap compared to an array of radiotelescopes or the Stanford linear accelerator. The definition is not provided by the cost nor by the sophistication but only by this characteristic: the set-up provides an inscription that is used as the final layer in a scientific text. An instrument, in this definition, is not every set-up which ends with a little window that allows someone to take a reading. A thermometer, a watch, a Geiger counter, all provide readings but are not considered as instruments as long as these readings are not used as the final layer of technical papers. This point is important when watching complicated contrivances with hundreds of intermediary readings taken by dozens of white-coated technicians. What will be used as visual proof in the article will be the few lines in the bubble chamber and not the piles of printout making the intermediate readings.

It is important to note that the use of this definition of instrument is a relative one. It depends on time. Thermometers *were* instruments and very important ones in the eighteenth century, so were Geiger counters between the First and Second World Wars. These devices provided crucial resources in papers of the time. But now they are only parts of larger set-ups and are only used so that a new visual proof can be displayed at the end. Since the definition is relative to the use made of the "window" in a technical paper, it is also relative to the intensity and nature of the associated controversy. For instance, in the guinea pig ileum assay there is a box of electronic hardware with many readings that I will call "intermediate" because they do not constitute the visual display eventually put to use in the article. It is unlikely that anyone will quibble about this because the calibration of electronic signals is now made through a black box produced industrially and sold by the thousand. It is a different matter with the huge tank built in an old gold mine in South Dakota at a cost of $600,000 (1964 dollars!) by Raymond Davis to detect solar neutrinos. In a sense the whole set-up may be considered as *one* instrument providing one final window in which astrophysicists can read the number of neutrinos emitted by the sun. In this case all the other readings are intermediate ones. If the controversy is fiercer, however, the set-up is broken down into *several* instruments, each providing a specific visual display which has to be independently evaluated. If the controversy heats up a bit we do not see neutrinos coming out of the sun. We see and hear a Geiger counter that clicks when Argon decays. In this case the Geiger counter, which gave only an intermediate reading when there was no dispute, becomes an instrument in its own right when the dispute is raging.

The definition I use has another advantage. It does not make presuppositions about what the instrument is made of. It can be a piece of hardware like a telescope, but it can also be made of softer material. A statistical institution that employs hundreds of pollsters, sociologists and computer scientists gathering all sorts of data on the economy *is* an instrument if it yields inscriptions for papers written in economic journals with, for instance, a graph of the inflation rate by month and by branch of industry. No matter how many people were made to participate in the construction of the image, no matter how long it took, no matter how much it cost, the whole institution is used as *one* instrument (as long as there is no controversy that calls its intermediate readings into question).

At the other end of the scale, a young primatologist who is watching baboons in the savannah and is equipped only with binoculars, a pencil and a sheet of white paper may be seen as an instrument if her coding of baboon behavior is summed up in a graph. If you want to deny her statements, you might (everything else being equal) have to go through the same ordeals and walk through the savannah taking notes with similar constraints. It is the same if you wish to deny the inflation rate by month and industry, or the detection of endorphin with the ileum assay. The instrument, whatever its nature, is what leads you from the paper to what supports the paper, from the many resources mobilized in the text to the many more resources mobilised to create the visual displays of the texts. With this definition of an instrument, we are able to ask many questions and to make comparisons: how expensive they are, how old they are, how many intermediate readings compose one instrument, how long it takes to get one reading, how many people are mobilised to activate them, how many authors are using the inscriptions they provide in their papers, how controversial are those readings. . . . Using this notion we can define more precisely than earlier the laboratory as any place that gathers one or several instruments together.

What is behind a scientific text? Inscriptions. How are these inscriptions obtained? By setting up instruments. This other world just beneath the text is invisible as long as there is no controversy. A picture of moon valleys and mountains is presented to us as if we could see them directly. The telescope that makes them visible is invisible and so are the fierce controversies that Galileo had to wage centuries ago to produce an image of the Moon. Once that fact is constructed, there is no instrument to take into account and this is why the painstaking work necessary to tune the instruments often disappears from popular science. On the contrary, when science in action is followed, instruments become the crucial elements, immediately after the technical texts; they are where the dissenter is inevitably led.

There is a corollary to this change of relevance on the inscription devices depending on the strength of the controversy, a corollary that will become more important in the next chapter. If you consider only fully-fledged facts it seems that everyone could accept or contest them equally. It does not cost anything to contradict or accept them. If you dispute further and reach the frontier where facts are made, instruments become visible and with them the cost of continuing the discussion rises. It appears that *arguing is costly*. The equal world of citizens having opinions about things becomes an unequal world in which dissent or consent is not possible without a huge accumulation of resources which permits the collection of relevant inscriptions. What makes the differences between author and reader is not only the ability to utilize all the rhetorical resources studied earlier, but also to gather the many devices, people and animals necessary to produce a visual display usable in a text.

2) Spokesmen and Women

It is important to scrutinise the exact settings in which encounters between authors and dissenters take place. When we disbelieve the scientific literature, we are led from the many libraries around to the *very few* places where this literature is produced. Here we are welcomed by the author who shows us where the figure in the text comes from. Once presented with the instruments, who does the talking during these visits? At first, the authors: they *tell* the visitor what to *see*: "See the endorphin effect?" "Look at the neutrinos!" However, the authors are not lecturing the visitor. The visitors have their faces turned towards the instrument and are watching the place where the thing is writing itself down (inscription in the form of collection of specimens, graphs, photographs, maps—you name it). When the dissenter was reading the scientific text it was difficult for him or her to doubt, but with imagination, shrewdness and downright awkwardness it was always possible. Once in the lab, it is much more difficult because the dissenters see with their own eyes. If we leave aside the many other ways to avoid going through the laboratory that we will study later, the dissenter does not have to believe the paper nor even the scientist's word since in a self-effacing gesture the author has stepped aside. "See for yourself" the scientist says with a subdued and maybe ironic smile. "Are you convinced now?" Faced with the thing itself that the technical paper was alluding to, the dissenters now have a choice between either accepting the fact or doubting their own sanity—the latter is much more painful.

We now seem to have reached the end of all possible controversies since there is nothing left for the dissenter to dispute. He or she is right in front of the thing he or she is asked to believe. There is almost no human intermediary between thing and person; the dissenter is in the very place where the thing is said to happen and at the very moment when it happens. When such a point is reached it seems that there is no further need to talk of "confidence": the thing impresses itself directly on us. Undoubtedly, controversies are settled once and for all when such a situation is set up—which again is very rarely the case. The dissenter becomes a believer, goes out of the lab, borrowing the author's claim and confessing that "X has incontrovertibly shown that A is B." A new fact has been made which will be used to modify the outcome of some other controversies.

If this were enough to settle the debate, it would be the end of this chapter. But . . . there is someone saying "but, wait a minute . . ." and the controversy resumes!

What was imprinted on us when we were watching the guinea pig ileum assay? "Endorphin of course," the Professor *said*. But what did we *see*? This:

With a minimum of training we see peaks; we gather there is a base line, and we see a depression in relation to one coordinate that we understand to indicate the time. This is not endorphin yet. The same thing occurred when we paid a visit to Davis's gold and neutrino mine in South Dakota. We saw, he said, neutrinos counted straight out of the huge tank capturing them from the sun. But what *did* we see? Splurges on paper representing clicks from a Geiger counter. Not neutrinos, yet.

When we are confronted with the instrument, we are attending an "audio-visual" spectacle. There is a *visual* set of inscriptions produced by the instrument and a *verbal* commentary uttered by the scientist. We get both together. The effect on conviction is striking, but its cause is mixed because we cannot differentiate what is coming from the thing inscribed, and what is coming from

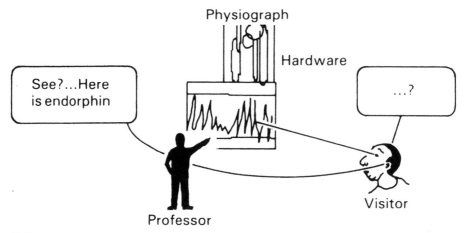

Figure 36.3

the author. To be sure, the scientist is not trying to influence us. He or she is simply commenting, underlining, pointing out, dotting the i's and crossing the t's, not adding anything. But it is also certain that the graphs and the clicks by themselves would not have been enough to form the image of endorphin coming out of the brain or neutrinos coming out of the sun. Is this not a strange situation? The scientists do not say anything more than what is inscribed, but without their commentaries the inscriptions say considerably less! There is a word to describe this strange situation, a very important word for everything that follows, that is the word **spokesman** (or **spokeswoman**, or **spokesperson**, or mouthpiece). The author behaves as if he or she were the mouthpiece of what is inscribed on the window of the instrument.

The spokesperson is someone who speaks for others who, or which, do not speak. For instance a shop steward is a spokesman. If the workers were gathered together and they all spoke at the same time there would be a jarring cacophony. No more meaning could be retrieved from the tumult than if they had remained silent. This is why they designate (or are given) a delegate who speaks on their behalf, and in their name. The delegate—let us call him Bill—does not speak in *his* name and when confronted with the manager does not speak "as Bill" but as the "workers' voice." So Bill's longing for a new Japanese car or his note to get a pizza for his old mother on his way home are not the right topics for the meeting. The voice of the floor, articulated by Bill, wants a "3 percent pay raise—and they are deadly serious about it, sir, they are ready to strike for it," he tells the manager. The manager has his doubts: "Is this really what they want? Are they really so adamant?" "If you do not believe me," replies Bill, "I'll show you, but don't ask for a quick settlement. I told you they are ready to strike and you will see more than you want!" What does the manager see? He does not see what Bill said. Through the office window he simply sees an assembled crowd gathered in the aisles. Maybe it is because of Bill's interpretation that he reads anger and determination on their faces.

For everything that follows, it is very important not to limit this notion of spokesperson and not to impose any clear distinction between "things" and "people" in advance. Bill, for instance, represents people who could talk, but who, in fact, cannot all talk at once. Davis represents neutrinos that cannot talk, in principle, but which are made to write, scribble and sign thanks to the device set up by Davis. So in practice, there is not much difference between people and things: they both need someone to talk for them. From the spokesperson's point of view there is thus no distinction to be made between representing people and representing things. In each case the spokesperson literally does the talking for who or what cannot talk. The Professor in the laboratory speaks for endorphin like Davis for the neutrinos and Bill for the shopfloor. In our definition the crucial element is not the quality of the represented but only their number and the unity of the representative. The point is that confronting a spokesperson is not like confronting any average man or woman. You are confronted not with Bill or the Professor, but with Bill and the Professor *plus* the many things or people on behalf of whom they are talking. You do not address Mr. Anybody or Mr. Nobody but Mr. or Messrs. Manybodies. As we saw, it may be easy to doubt one person's word. Doubting a spokesperson's word requires a much more strenuous effort however because it is now one person—the dissenter—against a crowd—the author.

On the other hand, the strength of a spokesperson is not so great since he or she is by definition *one* man or woman whose word could be dismissed—one Bill, one Professor, one Davis. The strength comes from the representatives' word when they do not talk by and for themselves but *in the presence of* what they represent. Then, and only then, the dissenter is confronted simultaneously with the spokespersons and what they speak for: the Professor and the endorphin made visible in the guinea pig assay; Bill and the assembled workers; Davis and his solar neutrinos. The solidity of what the representative says is directly supported by the silent but eloquent presence of the represented. The result of such a set-up is that it seems as though the mouthpiece does not "really talk," but that he or she is just commenting on what you yourself directly see, "simply" providing you with the words you would have used anyway.

This situation, however, is the source of a major weakness. Who is speaking? The things or the people *through* the representative's voice? What does she (or he, or they, or it) say? Only what the things they represent would say if they could talk directly. But the point is that they cannot. So what the dissenter sees is, in practice, rather different from what the speaker says. Bill, for instance, says his workers want to strike, but this might be Bill's own desire or a union decision relayed by him. The manager looking through the window may see a crowd of assembled workers who are just passing the time and can be dispersed at the smallest threat. At any rate do they really want 3 percent and not 4 percent or 2 percent? And even so, is it not possible to offer Bill this Japanese car he so dearly wants? Is the "voice of the worker" not going to change his/its mind if the manager offers a new car to Bill? Take endorphin as another instance. What we really saw was a tiny depression in the regular spikes forming the base line. Is this the same as the one triggered by morphine? Yes it is, but what does that prove? It may be that all sorts of chemicals give the same shape in this peculiar assay. Or maybe the Professor so dearly wishes his substance to be morphine-like that he unwittingly confused two syringes and injected the same morphine twice, thus producing two shapes that indeed look identical.

What is happening? The controversy flares even after the spokesperson has spoken and displayed to the dissenter what he or she was talking about. How can the debate be stopped from proliferating again in all directions? How can all the strength that a spokesman musters be retrieved? The answer is easy: by letting the things and persons represented *say for themselves the same thing that the representatives claimed they wanted to say.* Of course, this never happens since they are designated because, by definition, such direct communication is impossible. Such a situation however may be convincingly staged.

Bill is not believed by the manager, so he leaves the office, climbs onto a podium, seizes a loudspeaker and asks the crowd, "Do you want the 3 percent raise?" A roaring "Yes, our 3 percent! Our 3 percent!" deafens the manager's ears even through the window pane of his office. "Hear them?" asks Bill with a modest but triumphant tone when they are sitting down again at the negotiating table. Since the workers themselves said exactly what the "workers' voice" had said, the manager cannot dissociate Bill from those he represents and is really confronted with a crowd acting as one single man.

The same is true for the endorphin assay when the dissenter, losing his temper, accuses the Professor of fabricating facts. "Do it yourself," the Professor says, irritated but eager to play fair. "Take the syringe and see for yourself what the assay reaction will be." The visitor accepts the challenge, carefully checks the labels on the two vials and first injects morphine into the tiny glass chamber. Sure enough, a few seconds later the spikes start decreasing and after a minute or so they return to the base line. With the vial labelled endorphin, the very same result is achieved with the same timing. A unanimous, incontrovertible answer is thus obtained by the dissenter himself. What the Professor said the endorphin assay will answer, if asked directly, is answered by the assay. The Professor cannot be dissociated from his claims. So the visitor has to go back to the "negotiating table" confronted not with the Professor's own wishes but with a professor simply transmitting what endorphin really is.

No matter how many resources the scientific paper might mobilise, they carry little weight compared with this rare demonstration of power: the author of the claim steps aside and the doubter sees, hears and touches the inscribed things or the assembled people that reveal to him or to her exactly the same claim as the author.

3) Trials of Strength

For us who are simply following scientists at work there is no exit from such a setup, no back door through which to escape the incontrovertible evidence. We have already exhausted all sources of dissent; indeed we might have no energy left to maintain the mere idea that controversy might still be open. For us laymen, the file is now closed. Surely, the dissenter we have shadowed will give up. If the things say the same as the scientist, who can deny the claim any longer? How can you go any further?

The dissenter goes on, however, with more tenacity than the laymen. The identical tenor of the representative's words and the answers provided by the represented were the result of a carefully staged situation. The instruments needed to be working and finely tuned, the questions to be asked at the right time and in the right format. What would happen, asks the dissenter, if we stayed longer than the show and went backstage; or were to alter any of the many elements which, everyone agrees, are necessary to make up the whole instrument? The unanimity between represented and constituency is like what an inspector sees of a hospital or of a prison camp when his inspection is announced in advance. What if he steps outside his itinerary and tests the solid ties that link the represented and their spokesmen?

The manager, for instance, heard the roaring applause that Bill received, but he later obtains the foremen's opinion: "The men are not for the strike at all, they would settle for 2 percent. It is a union order; they applauded Bill because that's the way to behave on the shopfloor, but distribute a few pay raises and lay off a few ringleaders and they will sing an altogether different song." In place of the unanimous answer given by the assembled workers, the manager is now faced with an *aggregate* of possible answers. He is now aware that the answer he got earlier through Bill was extracted from a complex setting which was at first invisible. He also realises that there is room for action and that each worker may be made to behave differently if pressures other than Bill's are exerted on them. The next time Bill screams "You want the 3 percent, don't you?" only a few half-hearted calls of agreement will interrupt a deafening silence.

Let us take another example, this time from the history of science. At the turn of the century, Blondlot, a physicist from Nancy, in France, made a major discovery like that of X-rays. Out of devotion to his city he called them "N-rays." For a few years, N-rays had all sorts of theoretical developments and many practical applications, curing diseases and putting Nancy on the map of international science. A dissenter from the United States, Robert W. Wood, did not believe Blondlot's papers even though they were published in reputable journals, and decided to visit the laboratory. For a time Wood was confronted with incontrovertible evidence in the laboratory at Nancy. Blondlot stepped aside and let the N-rays inscribe themselves straight onto a screen in front of Wood. This, however, was not enough to get rid of Wood, who obstinately stayed in the lab asking for more experiments and himself manipulating the N-ray detector. At one point he even surreptitiously removed the aluminium prism which was generating the N-rays. To his surprise, Blondlot on the other side of the dimly lit room kept obtaining the same result on his screen even though what was deemed the most crucial element had been removed. The direct signatures made by the N-rays on the screen were thus made by something else. The unanimous support became a cacophony of dissent. By removing the prism, Wood severed the solid links that attached Blondlot to the N-rays. Wood's interpretation was that Blondlot so much wished to discover rays (at a time when almost every lab in Europe was christening new rays) that he unwittingly made up not only the N-rays, but also the instrument to inscribe them. Like the manager above, Wood realised that the coherent whole he was presented with was an aggregate of many elements that could be induced to go in many different directions. After Wood's action (and that of other dissenters) no one "saw" N-rays any more but only smudges on photographic plates when Blondlot presented his N-rays. Instead of enquiring about the place of N-rays in physics, people started enquiring about the role of auto-suggestion in experimentation! The new fact had been turned into an artifact.

The way out, for the dissenter, is not only to dissociate and disaggregate the many supporters the technical papers were able to muster. It is also to shake up the complicated set-up that provides graphs and traces in the author's laboratory in order to see how resistant the array is which has been mobilised in order to convince everyone. The work of disbelieving the literature has now been turned into the difficult job of manipulating the hardware. We have now reached another stage in the escalation between the author of a claim and the disbeliever, one that leads them further and further into the details of what makes up the inscriptions used in technical literature.

Let us continue the question-and-answer session staged above between the Professor and the dissenter. The visitor was asked to inject morphine and endorphin himself in order to check that there was no foul play. But the visitor is arguments, we have analysed so far. What was the endorphin tried out by the dissenter? The superimposition of the traces obtained by: a sacrificed guinea pig whose gut was then hooked up to electric wires and regularly stimulated; a hypothalamus soup extracted after many trials from slaughtered sheep and then forced through HPLC columns under a very high pressure.

Endorphin, before being named and for as long as it is a new object, *is* this list *readable* on the instruments *in* the Professor's laboratory. So is a microbe long before being called such. At first it is something that transforms sugar into alcohol in Pasteur's lab. This something is narrowed down by the multiplication of feats it is asked to do. Fermentation still occurs in the absence of air but stops when air is reintroduced. This exploit defines a new hero that is killed by air but breaks down sugar in its absence, a hero that will be called "Anaerobic" or "Survivor in the Absence of Air." Laboratories generate so many new objects because they are able to create extreme conditions and because each of these actions is obsessively inscribed.

This naming after what the new object does is in no way limited to actants like hormones or radioactive substances, that is to the laboratories of what are often called 'experimental sciences'. Mathematics also defines its subjects by what they *do*. When Cantor, the German mathematician, gave a shape to his transfinite numbers, the shape of his new objects was obtained by having them undergo the simplest and most radical trial: is it possible to establish a one-to-one connection between, for instance, the set of points comprising a unit square and the set of real numbers between 0 and 1? It seems absurd at first since it would mean that there are as many numbers on one side of a square as in the whole square. The trial is devised so as to see if two different numbers in the square have different images on the side or not (thus forming a one-to-one correspondence) or if they have only one image (thus forming a two-to-one correspondence). The written answer on the white sheet of paper is incredible: "I see it but I don't believe it," wrote Cantor to Dedekind. There are as many numbers on the side as in the square. Cantor creates his transfinites from their performance in these extreme, scarcely conceivable conditions.

The act of defining a new object by the answers it inscribes on the window of an instrument provides scientists and engineers with their final source of strength. It constitutes our **second basic principle**,* as important as the first in order to understand science in the making: scientists and engineers speak in the name of new allies that they have shaped and enrolled; representatives among other representatives, they add these unexpected resources to tip the balance of force in their favour. Guillemin now speaks for endorphin and somatostatin, Pasteur for visible microbes, the Curies for polonium, Payen and Persoz for enzymes, Cantor for transfinites. When they are challenged, they cannot be isolated, but on the contrary their constituency stands behind them arrayed in tiers and ready to say the same thing.

4) Laboratories against Laboratories

Our good friend, the dissenter, has now come a long way. He or she is no longer the shy listener to a technical lecture, the timid onlooker of a scientific experiment, the polite contradictor. He or she is now the head of a powerful laboratory utilising all available instruments, forcing the phenomena supporting the competitors to support him or her instead, and shaping all sorts of unexpected objects by imposing harsher and longer trials. The power of this laboratory is measured by the extreme conditions it is able to create: huge accelerators of millions of electron volts; temperatures approaching absolute zero; arrays of radio-telescopes spanning kilometres; furnaces heating up to thousands of degrees; pressures exerted at thousands of atmospheres; animal quarters with thousands of rats or guinea pigs; gigantic number crunchers able to do thousands of operations per millisecond. Each modification of these conditions allows the dissenter to mobilise one more actant. A change from micro to phentogram, from million to billion electron volts; lenses going from metres to tens of metres; tests going from hundreds to thousands of animals; and the shape of a new actant is thus redefined. All else being equal, the power of the laboratory is thus proportionate to the number of actants it can mobilise on its behalf. At this point, statements are not borrowed, transformed or disputed by empty-handed laypeople, but by scientists with whole laboratories *behind* them.

However, to gain the final edge on the opposing laboratory, the dissenter must carry out a fourth strategy: he or she must be able to transform the new objects into, so to speak, older objects and feed them back into his or her lab.

What makes a laboratory difficult to understand is not what is presently going on in it, but what *has been* going on in it and in other labs. Especially difficult to grasp is the way in which new objects are immediately transformed into something else. As long as somatostatin, polonium, transfinite numbers, or anaerobic microbes are shaped by the list of trials I summarised above, it is easy to relate to them: tell me what you go through and I will tell you what you are. This situation, however, does not last. New objects become **things**: "somatostatin," "polonium," "anaerobic microbes," "transfinite numbers," "double helix" or "*Eagle* computers," things isolated from the laboratory conditions that shaped them, things with a name that now seem independent from the trials in which they proved their mettle. This process of transformation is a very common one and

---

*Editor's note: Latour's First Basic Principle states, "the fate of facts and machines is in the later user's hands; their qualities are thus a consequence, not a cause, of a collective action."

occurs constantly both for laypeople and for the scientist. All biologists now take "protein" for an object; they do not remember the time, in the 1920s, when protein was a whitish stuff that was separated by a new ultracentrifuge in Svedberg's laboratory. At the time protein was nothing but the action of differentiating cell contents by a centrifuge. Routine use however transforms the naming of an actant after what it does into a common name. This process is not mysterious or special to science. It is the same with the can opener we routinely use in our kitchen. We consider the opener and the skill to handle it as one black box which means that it is unproblematic and does not require planning and attention. We forget the many trials we had to go through (blood, scars, spilled beans and ravioli, shouting parent) before we handled it properly, anticipating the weight of the can, the reactions of the opener, the resistance of the tin. It is only when watching our own kids still learning it the hard way that we might remember how it was when the can opener was a "new object" for us, defined by a list of trials so long that it could delay dinner forever.

This process of routinisation is common enough. What is less common is the way the same people who constantly generate new objects to win in a controversy are also constantly transforming them into relatively older ones in order to win still faster and irreversibly. As soon as somatostatin has taken shape, a new bioassay is devised in which sosmatostatin takes the role of a stable, unproblematic substance in a trial set up for tracking down a new problematic substance, GRF. As soon as Svedberg has defined protein, the ultracentrifuge is made a routine tool of the laboratory bench and is employed to define the constituents of proteins. No sooner has polonium emerged from what it did in the list of ordeals above than it is turned into one of the well-known radioactive elements with which one can design an experiment to isolate a new radioactive substance further down in Mendeleev's table. The list of trials becomes a thing; it is literally *reified*.

This process of reification is visible when going from new objects to older ones, but it is also reversible although less visible when going from younger to older ones. All the new objects we analysed in the section above were framed and defined by stable black boxes which had *earlier* been new objects before being similarly reified. Endorphin was made visible in part because the ileum was known to go on pulsating long after guinea pigs are sacrificed: what was a new object several decades earlier in physiology was one of the black boxes participating in the endorphin assay, as was morphine itself. How could the new unknown substance have been compared if morphine had not been known? Morphine, which had been a new object defined by its trials in Seguin's laboratory sometime in 1804, was used by Guillemin in conjunction with the guinea pig ileum to set up the conditions defining endorphin. This also applies to the physiograph, invented by the French physiologist Marey at the end of the nineteenth century. Without it, the transformation of gut pulsation would not have been made graphically visible. Similarly for the electronic hardware that enhanced the signals and made them strong enough to activate the physiograph stylus. Decades of advanced electronics during which many new phenomena had been devised were mobilised here by Guillemin to make up another part of the assay for endorphin. Any new object is thus shaped by simultaneously importing many older ones in their reified form. Some of the imported objects are from young or old disciplines or pertain to harder or softer ones. The point is that the new object emerges from a complex set-up of sedimented elements each of which has been a new object at some point in time and space. The genealogy and the archaeology of this sedimented past is always possible in theory but becomes more and more difficult as time goes by and the number of elements mustered increases.

It is just as difficult to go back to the time of their emergence *as it is to contest them*. The reader will have certainly noticed that we have gone full circle from the first section of this part (borrowing more black boxes) to this section (blackboxing more objects). It is indeed a circle with a feedback mechanism that creates better and better laboratories by bringing in as many new objects as possible in as reified a form as possible. If the dissenter quickly re-imports somatostatin, endorphin, polonium, transfinite numbers as so many incontrovertible black boxes, his or her opponent will be made all the weaker. His or her ability to dispute will be decreased since he or she will now be faced with piles of black boxes, obliged to untie the links between more and more elements coming from a more and more remote past, from harder disciplines, and presented in a more reified form. Has the shift been noticed? It is now the author who is weaker and the dissenter stronger. The author must now either build a better laboratory in order to dispute the dissenter's claim and tip the balance of power back again, or quit the game—or apply one of the many tactics to escape the problem altogether that we will see in the second part of this book. The endless spiral has traveled one more loop. Laboratories grow because of the number of elements fed back into them, and this growth is irreversible since no dissenter/author is able to enter into the fray later with fewer resources at his or her disposal—everything else being equal. Beginning with a few cheap elements borrowed from common practice, laboratories end up after several cycles of contest with costly and enormously complex set-ups very remote from common practice.

The difficulty of grasping what goes on inside their walls thus comes from the sediment of what has been going on in other laboratories earlier in time and elsewhere in space. The trials currently being undergone by the new object they give shape to are probably easy to explain to the layperson—and we are all laypeople so far as disciplines other than our own are concerned—but the older objects capitalised in the many instruments are not. The layman is awed by the laboratory set-up, and rightly so. There are not many places under the sun where so many and such hard resources are gathered in so great numbers, sedimented in so many layers, capitalised on such a large scale. When confronted earlier by the technical literature we could brush it aside; confronted by laboratories we are simply and literally impressed. We are left without power, that is, without resource to contest, to reopen the black boxes, to generate new objects, to dispute the spokesmen's authority.

Laboratories are now powerful enough to define **reality**. To make sure that our travel through technoscience is not stifled by complicated definitions of reality, we need a simple and sturdy one able to withstand the journey: reality as the latin word *res* indicates, is what *resists*. What does it resist? *Trials of strength*. If, in a given situation, no dissenter is able to modify the shape of a new object, then that's it, it *is* reality, at least for as long as the trials of strength are not modified. In the examples above so many resources have been mobilised by the dissenters to support these claims that, we must admit, resistance will be vain: the claim has to be true. The minute the contest stops, the minute I write the word "true," a new, formidable ally suddenly appears in the winner's camp, an ally invisible until then, but behaving now as if it had been there all along; Nature.

APPEALING (TO) NATURE

Some readers will think that it is about time I talked of Nature and the real objects *behind* the texts and behind the labs. But it is not I who am late in finally talking about reality. Rather, it is Nature who always arrives late, too late to explain the rhetoric of scientific texts and the building of laboratories. This belated, sometimes faithful and sometimes fickle ally has complicated the study of technoscience until now so much that we need to understand it if we wish to continue our travel through the construction of facts and artefacts.

1) "Natur mit uns"

"Belated?" "Fickle?" I can hear the scientists I have shadowed so far becoming incensed by what I have just written. All this is ludicrous because the reading and the writing, the style and the black boxes, the laboratory set-ups—indeed all existing phenomena—are simply *means* to express something, vehicles for conveying this formidable ally. We might accept these ideas of "inscriptions," your emphasis on controversies, and also perhaps the notions of "ally," "new object," "actant" and "supporter," but you have omitted the only important one, the only supporter who really counts, Nature herself. Her presence or absence explains it all. Whoever has Nature in their camp wins, no matter what the odds against them are. Remember Galileo's sentence, "1000 Demosthenes and 1000 Aristotles may be routed by any average man who brings Nature in." All the flowers of rhetoric, all the clever contraptions set up in the laboratories you describe, all will be dismantled once we go from controversies about Nature to what Nature is. The Goliath of rhetoric with his laboratory set-up and all his attendant Philistines will be put to flight by one David alone using simple truths about Nature in his slingshot! So let us forget all about what you have been writing for a hundred pages—even if you claim to have been simply following us—and let us see Nature face to face!

Is this not a refreshing objection? It means that Galileo was right after all. The dreadnoughts I studied may be easily defeated in spite of the many associations they knit, weave and knot. Any dissenter has got a chance. When faced with so much scientific literature and such huge laboratories, he or she has just to look at Nature in order to win. It means that there is a *supplement*, something more which is nowhere in the scientific papers and nowhere in the labs which is able to settle all matters of dispute. This objection is all the more refreshing since it is made by the scientists themselves, although it is clear that this rehabilitation of the average woman or man, of Ms or Mr Anybody, is also an indictment of these crowds of allies mustered by the same scientists.

Let us accept this pleasant objection and see how the appeal to Nature helps us to distinguish between, for instance, Schally's claim about GHRH and Guillemin's claim about GRF. They both wrote convincing papers, arraying many resources with talent. One is supported by Nature—so his claim will be made a fact—and the other is not—it ensues that his claim will be turned into an artefact by the others. According to the above objections, readers will find it easy to give the casting vote. They simply have to see who has got Nature on his side.

It is just as easy to separate the future of fuel cells from that of batteries. They both contend for a slice of the market; they both claim to be the best and most efficient. The potential buyer, the investor, the analyst are lost in the midst of a controversy, reading stacks of specialised literature. According to the above objection, their life will now be easier. Just watch to see on whose behalf Nature will talk. It is as simple as in the struggles sung in the Iliad: wait for the goddess to tip the balance in favour of one camp or the other.

A fierce controversy divides the astrophysicists who calculate the number of neutrinos coming out of the sun and Davis, the experimentalist who obtains a much smaller figure. It is easy to distinguish them and put the controversy to rest. Just let us see for ourselves in which camp the sun is really to be found. Somewhere the natural sun with its true number of neutrinos will close the mouths of dissenters and force them to accept the facts no matter how well written these papers were.

Another violent dispute divides those who believe dinosaurs to have been coldblooded (lazy, heavy, stupid and sprawling creatures) and those who think that dinosaurs were warm-blooded (swift, light, cunning and running animals). If we support the objection, there would be no need for the 'average man' to read the piles of specialised articles that make up this debate. It is enough to wait for Nature to sort them out. Nature would be like God, who in medieval times judged between two disputants by letting the innocent win.

In these four cases of controversy generating more and more technical papers and bigger and bigger laboratories or collections, Nature's voice is enough to stop the noise. Then the obvious question to ask, if I want to do justice to the objection above, is "what does Nature say?"

Schally knows the answer pretty well. He told us in his paper, GHRH *is* this amino-acid sequence, not because he imagined it, or made it up, or confused a piece of haemoglobin for this long-sought-after hormone, but because this is what the molecule is in Nature, independently of his wishes. This is also what Guillemin says, not of Schally's sequence, which is a mere artefact, but of his substance, GRF. There is still doubt as to the exact nature of the real hypothalamic GRF compared with that of the pancreas, but on the whole it is certain that GRF is indeed the amino-acid sequence earlier. Now, we have got a problem. Both contenders have Nature in their camp and say what it says. Hold it! The challengers are supposed to be refereed by Nature, and not to start another dispute about what Nature's voice really said.

We are not going to be able to stop this new dispute about the referee, however, since the same confusion arises when fuel cells and batteries are opposed. "The technical difficulties are not insurmountable," say the fuel cell's supporters. "It's just that an infinitesimal amount has been spent on their resolution compared to the internal combustion engine's. Fuel cells are Nature's way of storing energy; give us more money and you'll see." Wait, wait! We were supposed to judge the technical literature by taking another outsider's point of view, not to be driven back *inside* the literature and *deeper* into laboratories.

Yet it is not possible to wait outside, because in the third example also, more and more papers are pouring in, disputing the model of the sun and modifying the number of neutrinos emitted. The real sun is alternately on the side of the theoreticians when they accuse the experimentalists of being mistaken and on the side of the latter when they accuse the former of having set up a fictional model of the sun's behaviour. This is too unfair. The real sun was asked to tell the two contenders apart, not to become yet another bone of contention.

More bones are to be found in the paleontologists' dispute where the real dinosaur has problems about giving the casting vote. No one knows for sure what it was. The ordeal might end, but is the winner really innocent or simply stronger or luckier? Is the warm-blooded dinosaur more like the real dinosaur, or is it just that its proponents are stronger than those of the cold-blooded one? We expected a final answer by using Nature's voice. What we got was a new fight over the composition, content, expression and meaning of that voice. That is, we get *more* technical literature and *larger* collections in bigger Natural History Museums, not less; *more* debates and not less.

I interrupt the exercise here. It is clear by now that applying the scientists' objection to any controversy is like pouring oil on a fire, it makes it flare anew. Nature is not outside the fighting camps. She is, much like God in not-so-ancient wars, asked to support all the enemies at once. "Natur mit uns" is embroidered on all the banners and is not sufficient to provide one camp with the winning edge. So what is sufficient?

2) The Double-Talk of the Two-Faced Janus

I could be accused of having been a bit disingenuous when applying scientists' objections. When they said that something more than association and numbers is needed to settle a debate, something outside all our human conflicts and interpretations, something they call "Nature" for want of a better term, something that eventually will distinguish the winners and the losers, they did not mean to say that we know what it is. This supplement beyond the literature and laboratory trials is unknown and this is why they look for it, call themselves "researchers," write so many papers and mobilise so many instruments.

"It is ludicrous," I hear them arguing, "to imagine that Nature's voice could stop Guillemin and Schally from fighting, could reveal whether fuel cells are superior to batteries or whether Watson and Crick's model is better than that of Pauling. It is absurd to imagine that Nature, like a goddess, will visibly tip the scale in favour of one camp or that the Sun God will barge into an astrophysics meeting to drive a wedge between theoreticians and experimentalists; and still more ridiculous to imagine real dinosaurs invading a Natural History Museum in order to be compared with their plaster models! What we meant, when contesting your obsession with rhetoric and mobilisation of black boxes, was that *once the controversy is settled, it is Nature the final ally that has settled it* and not any rhetorical tricks and tools or any laboratory contraptions."

If we still wish to follow scientists and engineers in their construction of technoscience, we have got a major problem here. On the one hand scientists herald Nature as the only possible adjudicator of a dispute, on the other they recruit countless allies while waiting for Nature to declare herself. Sometimes David is able to defeat all the Philistines with only one slingshot; at other times, it is better to have swords, chariots and many more, better-drilled soldiers than the Philistines!

It is crucial for us, laypeople who want to understand technoscience, to decide which version is right, because in the first version, as Nature is enough to settle all disputes, we have nothing to do since no matter how large the resources of the scientists are, they do not matter in the end—only Nature matters. Our chapters may not be all wrong, but they become useless since they merely look at trifles and addenda and it is certainly no use going on for four other chapters to find still more trivia. In the second version, however, we have a lot of work to do since, by analyzing the allies and resources that settle a controversy we understand *everything* that there is to understand in technoscience. If the first version is correct, there is nothing for us to do apart from catching the most superficial aspects of science; if the second version is maintained, there is everything to understand except perhaps the most superfluous and flashy aspects of science. Given the stakes, the reader will realise why this problem should be tackled with caution. The whole book is in jeopardy here. The problem is made all the more tricky since scientists *simultaneously* assert the two contradictory versions, displaying an ambivalence which could paralyse all our efforts to follow them.

We would indeed be paralyzed, like most of our predecessors, if we were not used to this double-talk or the two-faced Janus. The two versions are contradictory but they are not uttered by the same face of Janus. There is again a clear-cut distinction between what scientists say about the cold settled part and about the warm unsettled part of the research front. As long as controversies are rife, Nature is never used as the final arbiter since no one knows what she is and says. But *once the controversy is settled*, Nature is the ultimate referee.

This sudden inversion of what counts as referee and what counts as being refereed, although counter-intuitive at first, is as easy to grasp as the rapid passage from the "name of action" given to a new object to when it is given its name as a thing (see above). As long as there is a debate among endocrinologists about GRF or GHRH, no one can intervene in the debates by saying, "I know what it is, Nature told me so. It is that amino-acid sequence." Such a claim would be greeted with derisive shouts, unless the proponent of such a sequence is able to show his figures, cite his references, and quote his sources of support, in brief, write another scientific paper and equip a new laboratory, as in the case we have studied. However, once the collective decision is taken to turn Schally's GHRH into an artefact and Guillemin's GRF into an incontrovertible fact, the reason for this decision is not imputed to Guillemin, but is immediately attributed to the independent existence of GRF in Nature. As long as the controversy lasted, no appeal to Nature could bring any extra strength to one side in the debate (it was at best an invocation, at worst a bluff). As soon as the debate is stopped, the supplement of force offered by Nature is made the explanation as to why the debate did stop (and why the bluffs, the frauds and the mistakes were at last unmasked).

So we are confronted with two almost simultaneous suppositions:

Nature is the final cause of the settlement of all controversies, *once controversies are settled.*
As long as they last *Nature will appear simply as the final consequence of the controversies.*

When you wish to attack a colleague's claim, criticise a world-view, modalise a statement you cannot *just* say that Nature is with you; "just" will never be enough. You are bound to use other allies besides Nature. If you succeed, then Nature will be enough and all the other allies and resources will be made redundant. A political analogy may be of some help at this point. Nature, in scientists' hands, is a constitutional monarch, much like Queen Elizabeth the Second. From the throne she reads with the same tone, majesty and conviction a speech written by Conservative or Labour prime ministers depending on the election outcome. Indeed she *adds* something to the dispute, but only after the dispute has ended; as long as the election is going on she does nothing but wait.

This sudden reversal of scientists' relations to Nature and to one another is one of the most puzzling phenomena we encounter when following their trails. I believe that it is the difficulty of grasping this simple reversal that has made technoscience so hard to probe until now.

The two faces of Janus talking together make, we must admit, a startling spectacle. On the left side Nature is cause, on the right side consequence of the end of controversy. On the left side scientists are *realists*, that is they believe that representations are sorted out by what really is outside, by the only independent referee there is, Nature. On the right side, the same scientists are *relativists*, that is, they believe representations to be sorted out among themselves and the actants they represent, without independent and impartial referees lending their weight to any one of them. We know why they talk two languages at once: the left mouth speaks about settled parts of science, whereas the right mouth talks about unsettled parts. On the left side polonium was discovered long ago by the Curies; on the right side there is a long list of actions effected by an unknown actant in Paris at the Ecole de Chimie which the Curies propose to call 'polonium'. On the left side all scientists agree, and we hear only Nature's voice, plain and clear; on the right side scientists disagree and no voice can be heard over theirs.

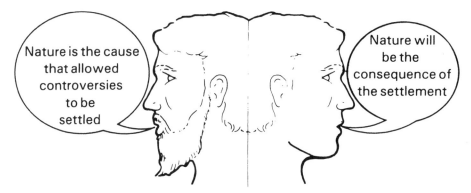

Figure 36.4

3) The Third Rule of Method

If we wish to continue our journey through the construction of facts, we have to adapt our method to scientists' double-talk. If not, we will always be caught on the wrong foot: unable to withstand either their first (realist) or their second (relativist) objection. We will then need to have two different discourses depending on whether we consider a settled or an unsettled part of technoscience. We too will be relativists in the latter case and realists in the former. When studying controversy—as we have so far—we cannot be *less* relativist than the very scientists and engineers we accompany; they do not *use* Nature as the external referee, and we have no reason to imagine that we are more clever than they are. For these parts of science our **third rule of method** will read: since the settlement of a controversy is *the cause* of Nature's representation not the consequence, we *can never use the outcome—Nature—to explain how and why a controversy has been settled.*

This principle is easy to apply as long as the dispute lasts, but is difficult to bear in mind once it has ended, since the other face of Janus takes over and does the talking. This is what makes the study of the past of technoscience so difficult and unrewarding. You have to hang onto the words of the right face of Janus—now barely audible—and ignore the clamours of the left side. It turned out for instance that the N-rays were slowly transformed into artifacts much like Schally's GHRH. How are we going to study this innocent expression "it turned out?"

Using the physics of the present day there is unanimity that Blondlot was badly mistaken. It would be easy enough for historians to say that Blondlot failed because there was "nothing really behind his N-rays" to support his claims. This way of analysing the past is called Whig history, that is, a history that crowns the winners, calling them the best and the brightest and which says the losers like Blondlot lost simply *because* they were wrong. We recognise here the left side of Janus' way of talking where Nature herself discriminates between the bad guys and the good guys. But, is it possible to use this as the reason why in Paris, in London, in the United States, people slowly turned N-rays into an artefact? Of course not, since at that time today's physics obviously could not be used as the touchstone, or more exactly since today's state is, in part, the *consequence* of settling many controversies such as the N-rays!

Whig historians had an easy life. They came after the battle and needed only one reason to explain Blondlot's demise. He was wrong all along. This reason is precisely what does not make the slightest difference while you are searching for truth in the midst of a polemic. We need, not one, but *many* reasons to explain how a dispute stopped and a black box was closed.

However, when talking about a cold part of technoscience we should shift our method like the scientists themselves who, from hard-core relativists, have turned into dyed-in-the-wool realists. Nature is now taken as the cause of accurate descriptions of herself. We cannot be more relativist than scientists about these parts and keep on denying evidence where no one else does. Why? Because the cost of dispute is too high for an average citizen, even if he or she is a historian and sociologist of science. If there is no controversy among scientists as to the status of facts, then it is useless to go on talking about interpretation, representation, a biased or distorted world-view, weak and fragile pictures of the world, unfaithful spokesmen. Nature talks straight, facts are facts. Full stop. There is nothing to add and nothing to subtract.

This division between relativists and realist interpretation of science has caused analysts of science to be put off balance. Either they went on being relativists even about the settled parts of science—which made them look ludicrous; or they continued being realists even about the warm uncertain parts—and they made fools of themselves. The third rule of method stated above should help us in our study because it offers us a good balance. We do not try to undermine the solidity of the accepted parts of science. We are realists as much as the people we travel with and as much as the left side of Janus. But as soon as a controversy starts we become as relativist as our informants. However we do not follow them passively because our method allows us to document both the construction of fact and of artefact, the cold and the warm, the demodalised and the modalised statements, and, in particular, it allows us to trace with accuracy the sudden shifts from one face of Janus to the other. This method offers us, so to speak, a stereophonic rendering of fact-making instead of its monophonic predecessors!

37

Science Policy and Moral Purity:
The Case of Animal Biotechnology

Paul B. Thompson

INTRODUCTION

Animal biotechnology is controversial. The Hoban and Kendall survey on U.S. public attitudes to genetic engineering reports higher levels of moral concern over animal applications of recombinant DNA techniques than for microbial, plant, or even human applications (Hoban and Kendall, 1992). The extended political controversy over recombinant bovine somatotropin (rBST) is also evidence that some, perhaps many, are reluctant to accept animal biotechnologies, even when they do not involve the direct manipulation of an animal genome. However, the actual points for caution or ethical concern with respect to animal biotechnology are seldom specified with care in the public record. This may be because there are radically different ways to understand the relationship of products of animal biotechnology, the responsibilities of democratic government, and the role of the scientific community. This paper describes and then compares two ways of understanding science-based policy issues using animal biotechnology as the principle case. The contrast between these two approaches is neither universal nor pervasive for issues in science policy, but the case of animal biotechnology exemplifies a pattern that can and does appear when science and its policy implications are disputed. Although many of the specific controversies that are discussed are unique to animal biotechnology, I would submit that the contrast between two ways of organizing the moral and political issues raised by animal biotechnology is characteristic of broader problems in democratic science policy.

The first approach to the issues presumes criteria for *purification* of moral issues. Purification begins by assuming that animal biotechnology is the application of rDNA and other lab techniques, theories, and concepts from molecular biology to non-human animals. Such applications seek either to establish truths about the biology of animals, or to develop novel biomedical or agricultural products and processes. A purification need not presume that research seeks one application exclusively; research may serve both goals simultaneously, even for the purist. The purist does, however, presume that science, technology, society, and values are ontologically distinct: they differ qualitatively and inhabit (or perhaps instantiate) different categories or modalities of being. This implies

Paul B. Thompson, "Science Policy and Moral Purity: The Case of Animal Biotechnology," *Agriculture and Human Values* 14, no. 1 (March 1997): 11–27. Copyright © 1997 Springer Science + Business Media BV. Reprinted by permission.

that science is *not* a form of technology, of society, or of values, but differs in kind, and a similar distinctness of kind applies to each with respect to the others. The advocate of purification presumes that understanding of scientific, political, *and* ethical issues is enhanced when one is able to draw and maintain distinctions, and to apply these distinctions to problems in a systematic way.

The alternative will be called *hybridization*. Hybridization begins with the assumption that animal biotechnology is a totality of persons, products, animals, and social relations. This assumption is consistent with the sense in which someone might answer questions like "What do you do for a living?" or "How did you invest your inheritance?" with the simple phrase, "Animal biotechnology." An advocate of hybridization asserts that the moral significance of animal biotechnology can only be grasped by focusing attention on this totality, and sees the totality as defying reduction into parts and subsequent categorization of parts into ontologically discrete partitions such as "nature," "technology," "law," "custom," et cetera. In the hybrid analysis, social, environmental, and human health consequences become intermingled with concerns about animal health or well-being and with religious or metaphysically based views on the moral limits on human interference with nature. More importantly, however, the hybrid analysis takes the political elements of the regulatory system to be inseparable from its scientific elements, and interprets the truth seeking elements of the regulatory system to be inseparable from its scientific elements, and interprets the truth seeking elements of science in light of their political links to commercial and governmental interests. In the hybrid interpretation, scientists are understood to be seeking a place at the center of decision making for industry and public policy when they promote the view that product and policy decisions should be "science based." The distinction between a purified and hybrid analysis is, thus, a reflection of the tension between a philosophy of science that stresses goals of truth seeking, order, and verification, and a sociology of science that stresses the incentives and organizations forming the structure of scientists" daily action.

ANIMAL BIOTECHNOLOGY AND ITS UNWANTED CONSEQUENCES

The potential scope of animal biotechnology is difficult to specify, but can be illustrated by examples. Animal biotechnology clearly includes the development of basic science and technology for understanding the basic biology of all animals, and overlaps with biomedical research on humans in this respect. Likely applications of this science include specialized animal models for biomedical research such as the "onco-mouse," genetically engineered to be disposed toward pathologies of interest to human cancer researchers. Genetically engineered animals may be used in their tissues or body fluids. Xenografts, animals engineered to produce organs capable of transplantation into human recipients, represent one of the most exotic applications currently under active development. Agricultural animals may eventually be genetically engineered, but the first implications will come from genetically modified bacteria that produce animal drugs such as porcupine somatotropin for low-fat meat, or bovine somatotropin, discussed at length below. The science and potential applications of animal biotechnology are concisely discussed in a Hastings Center publication, "The brave new world of animal biotechnology," by Strachan Donnelley, Charles R. McCarty and Rivers Singleton, Jr. (1994).

Animal biotechnology has been episodically contested throughout its short history. The first round of conflict in the USA centered on the granting of patents for animals. Hearings for the *Animal Patent Act of 1986* generated a series of moral arguments against animal patents, most of which could be construed as arguments against animal biotechnology itself (Lesser, 1989). A second episode of controversy occurred with the approval process for recombinant bovine somatotropin (rBST). This animal drug is still awaiting approval in Canada, Japan, and Europe, at this writing. Public debate and regulatory hearings have again generated a series of arguments that raise specific ethical concerns. Throughout the history of these political debates, there have been occasional but steady attempts to measure public attitudes toward animal biotechnology. While surveys and focus groups in the USA, Canada, and Europe support the claim that animal biotechnology is ethically contentious, they do little to illuminate the ethical or philosophical reasons that would be adduced to explain why.

Using Stich, Rollin, and the Hastings Center as expert sources, ethical concerns about animal biotechnology can be classified in five categories. First, some people clearly believe that there may be something intrinsically wrong with direct manipulation of animal genomes using recombinant DNA techniques. Statements on genetic engineering from main line religious denominations in the USA and Europe express the view that there are limits to what scientists should do, but they are vague on how these limits apply to animal biotechnology. These seemingly ill-formed objections are sometimes referred to as "the Yuk factor," and Rollin has examined these limits under the heading "the Frankenstein thing." Stich, Rollin, and the Hastings Center authors recognize this category of possible ethical argument, though they find the arguments advanced under this banner unconvincing. More specific objections to particular technologies can be classified into one of four categories of unwanted or unintended consequences alleged to be caused by widespread use of rBST in the dairy industry. They are *environmental impact, animal well-being, human health,* and *social consequence*s. A brief review of each category of consequence follows, and they are covered roughly in order of their contentiousness, with the most difficult religions and metaphysical issues saved for last. Controversy over rBST enjoys a recent and robust political history and is presented as a model for ethical contentions over animal biotechnology at several junctures. This debate is well known to scholars of agricultural science, but a summary of the main contested claims will provide an empirical basis for the philosophical analysis that is to follow.

Somatotropin or growth hormone is produced naturally in mammals and regulates not only growth but other functions, notably lactation. When somatotropins are administered under carefully managed conditions, milk production can be increased, and the lactation cycle can be extended. Bovine somatotropin can, therefore, be administered to cows under a herd management regime that results in significant increases in milk production. It is not economical, however, to use bovine somatotropin harvested from cows because of the high production cost. Genetic modification of bacteria for production of somatotropin was one of the first successful applications of recombinant DNA technology, and genetically engineered organisms are now used routinely to produce human growth hormone for medical applications. Several animal drug companies including Monsanto, Eli Lilly, and Upjohn succeeded in developing a recombinantly produced bovine somatotropin over the last decade, and the Monsanto version, trade-named Posilac, was approved for use in the USA in the Fall of 1993. The social history of rBST deserves a more extended treatment than is warranted in the present context. It must suffice to say that the technology has been opposed by a complex network of interested parties. The emphasis here will be to exposit the arguments that have been advanced to justify this opposition.

ENVIRONMENTAL IMPACT

Although biomedical technologies are not typically thought to have significant environmental dimensions, agricultural technologies are routinely assessed with respect to environmental impact, though standards have arguably been less stringently applied to agriculture than to manufacturing and energy sectors of the economy (Thompson et al., 1994). While the technical requirements of environmental assessment are becoming relatively well defined, the ethical significance of environmental assessment is extremely complex. There are, for example, environmental impacts that impinge on human health, but assessments also model technology's impact on broader ecosystem processes. Impacts on these processes may be considered adverse only when they affect human life, but they may also be considered significant simply because they challenge the stability or equilibrium of an ecological zone. Transgenic fish, for example, might eventually displace indigenous species from their habitat (Donnelley et al., 1994) and transgenic vaccines might allow cattle ranching activities in tropical areas that have hitherto been reserved for wildlife habitat (Yilma, 1994; Pastoret, 1994).

A growing literature in environmental ethics in agriculture and fisheries management provides bases for framing these effects as ethical issues. There are two related areas of controversy. First, one general philosophical approach, adopted by resource economists, emphasizes valuation of environmental impacts, and then proceeds toward policy strategies for optimizing outcomes. This consequentialist philosophy has been challenged by ethicists who describe non-consequential obligations (such as rights) to animals, future generations, and even to nature as such. In a pattern familiar to political theory, these obligations are seen as overriding or "trumping," arguments that aim to arrive at efficient cost/benefit trade-offs or market solutions. Another philosophical controversy involves so-called anthropocentrists are alleged to define the value of nature purely in terms of its usefulness to present or future human beings, while eco-centrists attribute forms of intrinsic value to animals, plants, species, ecosystems, and other biological entities. To date, there has been comparatively little direct intersection between these literatures in environmental ethics and debates over animal biotechnology (but see Hobbelink, 1991; Attfield, 1995), yet it is clear how concerns over transgenic fish or vaccines might intersect the debate. The vaccine case is particularly illustrative, for the regions of Africa where recombinant vaccines may permit cattle production have great human poverty and food deprivation, as well as large national park areas designed to protect the last of Africa's sensational wildlife (Fox, 1992: 57).

The rBST case is a relatively poor model for illustrating ethical issues associated with environmental impacts of animal biotechnology. The consensus of opinion on rBST was to regard environmental impact as one of the least serious of consequences of potential impacts associated with the technology. Reviewing the sustainability of rBST in 1993, Gail Feenstra noted many of the points summarized below, but includes no discussion of environmental impact (Feenstra, 1993). Even the author of one of the most critical studies on environmental impact admitted, "If society and farmers place sufficient emphasis on environmental quality, there need not be environmental degradation due to the introduction of BGH" (Rayburn, 1993: 256). The U.S. government post-approval review of rBST appears to have been based on the assumption that rBST would reduce the number of dairy cows, and since fecal wastes are regarded as the most serious environmental contaminant associated with dairying, the reduced number of cows was projected to produce a corresponding reduction in the total volume of waste. As such, the environmental impact of rBST was judged to be positive (Executive Office of the President, 1994). This conclusion is questionable on the grounds that the social consequences of restructuring the dairy industry have secondary environmental impact (Lanyon and Beegle, 1989; Rayburn, 1993; Krimsky and Wrubel, 1996). When environmental risks are more direct or better understood, they are more likely to emerge as a category having ethical significance.

ANIMAL WELL-BEING

Although impact on animals may be a marginal category, in some areas of research ethics it has always been prominent in discussions of animal biotechnology, and for obvious reasons. Rollin's

1985 [1990] and 1992 papers and his 1995 book on animal biotechnology stress the possibility that genetic engineering may produce situations that contribute to animal suffering. Certainly this potential has been one of the most controversial topics with respect to rBST. Comstock raised the issue of animal welfare impacts associated with rBST in a 1988 paper, noting stress associated with the administration and with the pharmacological effects of rBST. Concerns linkage of rBST to enhanced milk production and in turn to increased incidence of mastitis has been the subject of considerable review and concern ever since. However, the concern for animal well-being noted by Rollin and Comstock is unlikely to be defined as a compromise to animal health, given current approaches that are standard in the animal sciences. Rollin introduces the concept of *telos* to describe the genetically encoded set of physical and psychological needs that determine "the fundamental interests central to [animals'] existences, whose thwarting or infringement matters to them" Rollin [1985] 1990: 305). He suggests that any experimental or production practice that compromises an animals' *telos* is morally wrong, and specifically notes that a farmer's profitability (or a consumer's price reduction) does not provide a sufficient justification for practices that violate the package of rights an animal must be accorded in virtue of its *telos* (Rollin, 1995: 188–194).

For both Rollin and Comstock, these rights cash out on terms of practices that produce pain or suffering to individual animals, or that allow animals to meet functional needs. However, although current approaches to the assessment of technology give little heed to impacts upon animal well-being, it is nevertheless easy to see how the anticipation of such impacts fit under the general heading of responsibilities noted by Stich, once the pain and suffering of non-human animals is recognized as morally significant. Furthermore, it would appear far easier to assess the welfare impact of animal drugs like rBST than to assess environmental impact. Extensive physiological, behavioral, and cognitive approaches to the assessment of impact on animals are in at least rudimentary stages of development; the task now is simply to apply them in the study of animal well-being.

Assessment of transgenic animals, animals whose genomes have been altered through manipulation of recombinant DNA, will be more difficult, however. It may be impossible to anticipate the impact of a genetic modification on an animal's needs. Although it is wrong to compromise an existing animal's *telos,* Rollin explicitly stipulates that it will not be wrong to change the *telos* that other animals in a given species might share through genetic engineering, even if doing so may result in chimerical beasts that cause aesthetic revulsion. What will matter is the *telos* of the new animal, and our ability to assess the vital interests and needs of animals whose genetic constitution departs significantly from that of animals whose genome is the result of evolutionary adaptation. Domestication and even conventional breeding rely on selection in a way that allows us to predict a rough fit between an animal's physiological, behavioral, and psychological needs and the environment in which it will live and reproduce. It is less clear that the animals produced through recombinant techniques will have behaviors, interests, and needs adapted to the environments in which they will live. Although this introduces uncertainty into our collective ability to anticipate impacts on animal well-being, it does not alter the conceptual basis of the scientists' responsibility to consider and assess such impacts.

HUMAN HEALTH

Human health is perhaps the most obvious area of potential impact from genetic engineering as it affects agricultural animals. One dimension is food safety, understood as the probability that consumption of a food will produce injury or debilitating disease, or that substitution of a food for reasonable alternative foods will adversely affect a person's health through nutritional deficiencies.

Of all potential impacts from rBST, this is the one that has received the greatest technical specification, and on which there is the greatest unanimity (Munro and Hall, 1991). At present, the consensus standard is that foods produced using animal biotechnology must be at least as safe as conventional foods, and procedures for assessment of food products from biotechnology in all industrialized nations virtually assures that far more will be known about the probability of injury or disease from recombinantly produced foods than from foods of more conventional origin. There is, as a result, the possibility that ethics will weigh in on the side of *less* attention to food safety in virtue of disproportionate expenditure of resources on the assessment and mitigation of quantitatively minimal risks (Johnson and Thompson 1991).

Despite this circumstance, food safety emerged as one of the most prominent public points of controversy in the rBST case. Samuel Epstein, a biomedical researcher at the University of Illinois expressed early concerns about potential health impacts, but the overwhelming consensus of scientific opinion has been that use of rBST in the production of milk has no impact on the health risk associated with milk (Kroger, 1992). Krimsky and Wrubel note an unusually high degree of scientific consensus. Critics of rBST then turned to the possibility that mastitis associated with elevated levels of milk production might create human health hazards. The Pure Food Campaign under the leadership of Jeremy Rifkin organized chefs on both coasts to protest what they termed adulteration of milk by addition of rBST. However, with the exception of the few sources cited here, the vast majority of criticisms associated with food purity are addressed to factors that do not bear in any direct way on the probability of injury or other deleterious human health impacts associated with rBST.

Human health impacts from biomedical applications of animal biotechnology will significantly expand the universe of discourse for this category beyond food safety. However, many of the new issues will have familiar analogs in medical bioethics. Drug and transplant therapies, as well as environmental exposures will presumably pose differential levels of risk. One way to analyze these risks stresses efficient distributions of risk, so that public policies prioritize risk issues for regulatory action, or utilize risk/benefit trade-off criteria. This approach is often argued for environmental hazards, while risks associated with physician administered therapies are generally subjected to criteria of informed consent. Clearly therapies such as xenografts raise new issues for informed consent: how does a physician advise a patient of the potential psychological impact from living the rest of one's life with the liver of a pig? While the eventual determination of such issues is, at present, poorly defined, their placement within traditional categories of medical bioethics seems reasonably clear (Donnelley et al., 1994).

SOCIAL CONSEQUENCES

The social consequences of technology pose vague and potentially highly contentious issues. For biomedical applications of animal biotechnology, the issues appear to be instances of larger social issues about the health care system and the distribution of health. It is, for example, clearly possible to raise questions about specific drugs or therapies developed for use using animal biotechnology. Is the development of the drug or therapy an appropriate use of scarce resources? How will the new product affect the distribution of health care, or the ability of an already taxed health care delivery system to provide equitable service to rich and poor alike? Does the creation of new industries utilizing animals for the production of pharmaceuticals or organs for transplant represent an untoward invasion of commercial values or market logic into an area of decision making that should be shielded from economic forces? While each of these questions might be associated with any biomedical application of animal biotechnology, they are, in fact, generic questions in the social dimensions of medical ethics (Donnelley et al., 1994). Animal biotechnology does not pose concerns distinct from those of chemical techniques or for biotechnologies based on microbes, plants, or human beings.

Unintended social consequences are also associated with all agricultural technologies. Some consequences, such as the elimination of hand labor jobs, may be intentional. Some technologies are too costly for poor producers, but can give large or wealthy farmers significant advantages over the poor. The economic structure of agriculture in both developed and developing countries means that aggressive early adopting farmers derive short-term benefits from production enhancing technology, but that the ultimate beneficiaries are food consumers. Although animal biotechnologies may be less susceptible to a farm size bias than are mechanical and chemical technologies, it is reasonable to think that many poor producers will be unable to compete with richer competitors as a direct result of biotechnology.

The fact that regionally and culturally defined groups of agricultural producers tend to choose animal production as their profit center creates special ethical significance for the social impact of animal biotechnology in agriculture. Social consequences associated with restructuring in the dairy industry precipitated the entire debate over rBST following a study by economist Robert Kalter (1985). Kalter's study precipitated that relatively small scale dairy producers might be disadvantaged when rBST became available. This prediction is itself somewhat complex, and a substantial literature on it is summarized by Tauer (1992). For the purposes of this discussion, the economic issues that arise in predicting a technology's effect on the size distribution of farms and the makeup of rural communities are less relevant than the general question of why alleged impacts on small vs. large farms might be thought ethically significant.

There are at least two strategies for approaching this issue. One begins with the assumption that those adversely affected by new technology are harmed in some way analogous to impacts on human health and welfare. They may be deprived of income they would have received without the technology, and may also be harmed in more subtle psychological and social ways. These impacts must be weighed against benefits not only to other producers, but also to food consumers (Thompson et al., 1994: 242–245). A second strategy begins with the observation that those who make decisions about whether to develop and market a technology occupy a position of power over the small farmers who will be affected. On this more populist view, what is ethically significant is the distribution of power, not the distribution of risks and benefits. The remedies associated with the populist way of framing the issue enhance affected parties' ability to influence decisions that will have dramatic effect on their future livelihood and way of life. In this respect, it is crucial to note that in the USA, no agency of government has the authority to monitor or regulate technology based upon social consequences. Lacking an outlet for their frustrations, groups seeking remediation of social consequences will politicize the regulatory process for environmental, animal welfare, and human health consequences (Thompson, 1992). A similar pattern of politicization might be associated with the social consequences of biomedical animal biotechnologies, but such a pattern has yet to be observed.

THE FRANKENSTEIN THING

The aforementioned categories encompass a wide range of consequences and largely recapitulate the listing of ethical responsibilities in the Stich/Rollin/Hastings model. They represent types of consequence that might be associated with many technological innovations and scientific discoveries, and do not derive their force directly from the fact that recombinant DNA transfer is the technology in question. There is another class of consequences that does relate specifically to the alteration of animal genomes. The literature on animal biotechnology includes religious and meta-physical objections based on a philosophical view that alteration of species boundaries constitutes an evil in itself. Here, the act of alteration simply *is* the event seen as wrong, so it is inappropriate to classify such acts as "consequences," of animal biotechnology.

Religiously based objections to animal biotechnology are largely inchoate and provide little guidance for scientists. Many large mainline Christian denominations have issued statements on genetic engineering, and all imply that some forms of genetic manipulation using recombinant techniques are intrinsically wrong. These statements, however, are vague and do not clearly distinguish between human and non-human genetic engineering, nor do most distinguish between somatic and cell line therapy for humans. Rollin concludes that theological objections to genetic engineering should not be interpreted as ethical objections at all, though he does note that, "to a religious person, anything that violates any of his or her religious tenets must be seen as morally problematic" (Rollin, [1985] 1990: 297).

A consortium of religious leaders signed a statement opposing patenting of human and animal genes in May of 1995. The event is significant in the breadth of faiths represented. Roman Catholics, Jews, and mainline Protestants were joined by representatives of the Southern Baptist Convention, as well as American Muslims and Hindus. Groups that would not agree on social issues such as abortion or school prayer found common ground in opposing these gene patents. Yet the statement itself did not include a rationale for opposition, largely because there would have been no consensus on the reasons. Richard D. Land, a Southern Baptist, was quoted in the *New York Times* that "the altering of life-forms, and the creation of new life-forms, is a revolt against the sovereignty of God and an attempt to be God." United Methodist Kenneth Carder, however, endorses genetic engineering's capacity to produce benefits to mankind, and opposes patents on the much narrower ground that they "reduce the sacred gift of life to a marketable commodity" (Taylor, 1995). These statements clarify and specify the broad policy statements of the churches and in doing so may base religious concerns on social, environmental, or human and animal welfare considerations. Carder's rationale might be read as an essentially social argument that might command assent from anyone who agrees that placing genes under the regime of market forces is socially undesirable. By way of contrast, Land's statement presupposes a theology, and finds genetic engineering problematic on theological grounds. Such theological arguments promise a reprise of philosophical disputes between religion and biology that have raged for more than a century.

Does philosophy of science recognize any analogies to Land's theological concern? Rollin notes that there is a widely held view that "to tinker with species . . . is to tinker with the stability of nature" (Rollin, [1985] 1990: 298). Although Rollin goes on to ridicule this view as biologically naïve, it has been defended in some detail by Henk Verhoog (1992a, b). In "The concept of intrinsic value and transgenic animals," Verhoog reviews several approaches to understanding the concept of intrinsic value, concluding that any meaningful interpretation will be formulated independently from ideas that derive from the use of the being in question, be it human or not. As applied to animals, the concept of intrinsic value implies a natural state, undisturbed by human interference (Verhoog, 1992a: 156). Verhoog argues that a notion of *telos* or essence is an implicit background assumption for accounts of abnormality and suffering. He thus rejects Rollin's inference that human beings may permissibly change an animal's *telos,* substituting the view that such modifications to animals effectively rob them of their being as the product of evolutionary history (Verhoog 1992b:, 274–276).

Impacts upon animal *telos* are philosophically more controversial than aforementioned challenges to the environment, to animal well-being, to human health, and to social stability. While any of these other categories might pose difficult problems of measurement, or conflicting strategies for managing trade-offs, there is little debate over whether one has identified a class of potential ethical issues, even if the actual threats arising from animal biotechnology in any given category are highly unlikely or upon analysis found to be nonexistent. With respect to intrinsic value and animal *telos,* however, it is less clear that we have identified a category that is meaningful to contemporary biologists, who, as Rollin notes ([1985] 1990: 297), understand species in dynamic terms. Verhoog is aware of this problem, and calls the priority of biologists' conceptualization of species into question, stating that those who use a scientifically based definition of species have simply begged the key moral question (Verhoog, 1992b: 277).

THE ETHICS OF ANIMAL BIOTECHNOLOGY: A PURIFICATION

How do purification and hybridization bear on the ethics of animal biotechnology? As stated at the outset, each represents an alternative strategy for understanding or interpreting the locus of moral significance in animal biotechnology. Purification seeks clear, distinct, and defensible areas of moral concern; hybridization rejects clarity as providing a false vision. This section traces the logic of purification, while the next takes up the rhetoric of hybridization. The Stich/Rollin/Hastings analysis to the case of rBST illustrates a strategy of purification. Purification produces an analysis in the classic sense where a complex compound is broken down into basic and mutually distinct elements. The problematic character of issues in each category is logically independent of issues in other categories. Environmental impact, animal welfare, human health, and social impact each represent classes of moral significance that would be valued in ways that do not, in any obvious sense, interpenetrate. Harmful consequences across several categories might be seen as additive. For example, harmful environmental effects might be made worse by risk to animal health, or mitigated by beneficial social consequences. However, each set of moral problems can be assessed and evaluated independently.

When the welter of confusion is seen to exhibit some form of order, the ethical issues present themselves either as problems admitting of solution, or as deep, contentious issues arising over contested concepts. The animal well-being issue is an example of the former issue, while environmental impact is an example of the second.

As defined by Rollin, the animal well-being issue is largely one of assuring that animals are not forced to submit to undue suffering or stifling of interests at the hands of human beings. One of the most potent examples of this concern would be animals that are genetically engineered to serve as models for human diseases. Such animals will be intentionally subjected to diseases that produce pain and suffering, and they will be deprived of any opportunity to engage in the behaviors that have become a part of that animal's biological and cognitive needs through evolution. Given Rollin's approach to the question, there is nothing morally unacceptable (though there may be something aesthetically unsettling) about resolving this problem by genetically engineering animals in which evolution is effectively "deprogrammed," at least insofar as the relevant biological and cognitive needs are unnecessary to the biomedical research that is being performed. That is, the animals used as models of disease may not need consciousness, or even nervous system response to the disorders that would necessarily register pain in a conventional animal (Rollin, 1992). While the practicality of such deprogramming would be an obstacle, what is important here is the relative philosophical simplicity of the solution to this ethical issue.

Environmental impact, by contrast, is enormously complex when approached as a philosophical issue. What, exactly, are the environmental impacts to be minimized or mitigated, and why? Are we understating environmental impact as a matter of human health, or as a reduction of natural resources commonly appropriated for human use? Are we understanding it as reduced opportunity for future generations, and if so, how can we understand what we are morally obligated to leave to future generations? Or, as many recent environmental philosophers would suggest, should environmental issues be understood as duties that our society owes directly to nature? Do wildlife and ecosystems have interests that are capable of being harmed in morally significant ways by human action? Does it make sense to extend rights not only to animals, but to trees, or even to the interactions of inert matter? Does the Earth itself exhibit a form of life that entitles it to moral status, as advocates of the Gaia hypothesis have suggested? This list of questions suggests that it is possible to approach the matter of environmental impact from a variety of different philosophical starting points. Different definitions for "nature," "life," "ecosystem," "interests," and "rights," will produce very different accounts of what is right and what is wrong.

To be sure, there are contested issues in the animal well-being debate, as well. Some define "pain" in a manner inconsistent with Rollin, and certainly others, like Verhoog, would reject Rollin's entire approach. Nevertheless, Rollin is applying the strategy of purification to the issue of animal biotechnology in a way that singles out problems that can be ameliorated in principle with certain specific actions. The issues raised by animal suffering are shown to be separable from others. Even Verhoog admits the legitimacy of limiting suffering; he simply would raise *additional* issues associated with *telos*. It happens that Verhoog's concern with additional issues would overturn Rollin's proposed solution to the problems of pain and suffering, yet this does not controvert the manner in which Rollin has shown the issues of pain and suffering to be logically distinct from Verhoog's broader concerns with *telos*. So purification, in this instance, produces a fairly straightforward, if pragmatically difficult, solution to the ethical problem. It is ironic that the animal well-being issue is politically more contentious than the philosophically more difficult issue of environmental impact. Here, the role of purification is to show that there are many different concepts, or definitions of concepts, at work, and that there are difficult philosophical tasks to be accomplished if we are to reconcile all the possible positions into a logically consistent whole. The relative lack of agreement here suggests, that this may become a far more difficult area of ethical disagreement as time goes by.

The larger point to be made through these examples is that purification orders ethical issues. Ideally, this order separates issues into categories that do not overlap one another with respect to their most basic concepts. A definition adopted for the purpose of responding to concerns for animal well-being, for example, should neither depend upon nor affect the issues of environmental impact. When purification is successful, the basic concepts or claims in any one category of issue will be logically distinct from others and will not, if properly reconstructed, produce inconsistencies with the definitions for basic concepts or claims from other categories.

Purification assumes that while extensive and complex, social and ethical implications of technology be defined in categorical terms that not only have relatively little overlap, but that reflect ontological categories. When successful, purification means that one set of issues can be resolved or debated with little or no implications for others. So, traditional human responsibilities for animal welfare would not impinge upon issues of human health, and issues of social justice would be thought independent of responsibilities to nature. But the breadth of categories in which new technologies have impact produces problems for social acceptance of a new technology and for resolving the political issues that the new technology creates. Because new technologies overlap so many domains of pure ethical discourse, they may be opposed on multiple fronts. When opponents attack technology simultaneously on many fronts, the strategy of purification may become difficult to manage. This is taken to be a political rather than a philosophical difficulty, but it is this presumption that hybridization attacks.

A HYBRIDIZATION OF ETHICAL ISSUES
IN ANIMAL BIOTECHNOLOGY

Animal biotechnology can be understood as a hybrid in both a broad and a restricted sense. As the above attempt at purification shows, genetic engineering of animals has implications in many distinct domains of ethical deliberation, and rather different types of question and argument would seem to apply. The sheer number of domains in which animal biotechnology become significant makes it difficult to manage ethical issues under a strategy of purification, even when there are few philosophical barriers to analyzing the issues in purified terms. But animal biotechnology may also be a hybrid in Bruno Latour's sense. Latour (1992) uses the terminology of purification and hybridization to distinguish issues or activities that tend to cut across generally accepted conceptual or social patterns of organization. In his work, hybrids encompass multiple dimensions in their very being. Hybrids cross the implicit ontological boundaries that distinguish "issues" from "organizations." They are hybrids precisely because understanding them defies classification into basic ontological categories like "thing," "idea," "nature," or "society." Animal biotechnology may hybridize the very categories that are the implicit starting points of purification. The task now is to develop an interpretation of animal biotechnology as a hybrid in this broader sense.

As noted, the strategy of purification depends upon complex assumptions that partition knowledge and action into departments. These assumptions are seldom examined, least of all by participants in ethical debates. Latour's work on the role of purification in the construction of scientific networks can be extended so as to provide the basis for an ethical examination and evaluation of partitioning assumptions. The extension begins with the search for an alternative interpretation of the ethical issues associated with animal biotechnology. The search rejects the notion that a view should be privileged in virtue of its logical clarity or ontological parsimony. The view we seek does not presuppose either the categories or the desirability of purification. Given the purified analysis of animal biotechnology that has already been presented, Latour's analysis of how scientific networks deploy the rhetoric of rationality can be used to blur distinctions already made, and to understand any or all of the (allegedly) distinct ethical issues discussed in the first half of the paper as dimensions or elements of a single theme. In fact, one need not go farther than the actual statements of participants in the debate over rBST to find an example.

The critics of rBST persistently noted a host of factors that introduce anxiety into the prospect of rBST milk. In itself, anxiety may be of little ethical significance, but the central claim of critics such as Carol Tucker Foreman (1991), Ken Taylor (1991), and Dianne Hunter (1992) is that one cannot trust authority, whether scientific or governmental, to safeguard the food supply. With frequent reinforcement in the form of stories on scientific misconduct, exploitation of innocent victims who participated in nuclear tests in the 1950s, and the continuing problems of pollution, it has not been difficult to develop anxiety over scientific conduct. The link between conduct, anxiety, and risk produces an argument with direct ethical implications. As in political debates over sexual misconduct, character also plays a role. Evidence that bears negatively on a person's character is interpreted as evidence against the validity or reliability of the claims they espouse. In most instances, criticism of rBST is advanced in sound bites and briefings that make it difficult to document this message, but it seems capable of spreading quickly and extensively by word of mouth. The central theme of this public criticism can be summarized in the following enthymeme:

- When people in positions of authority are believed to have questionable character, persons subjected to that authority quite rationally take their interests to be at risk.
- People who ignore ethical issues have questionable character.

- If the people who researched, developed, and approved rBST have ignored ethical issues, then our interests are at risk.

What is crucial here is to see how this argument is sufficient to make all of the purified considerations discussed above bear on the question of safety.

As discussed above, the scientific evidence on the safety of rBST is overwhelming, but the public does not have access to this evidence. This lack of access is *not* a function of FDA policies that restrict release of scientific data. The broader public lacks meaningful access to evidence even when data are public because most people do not have the time, the resources, and the skills necessary to translate the data into an informative indicator of food safety. Instead, they must rely on someone to summarize and report the data to them. It is the summarizing report that will be the basic information on which outsiders will make their own judgments of the safety of rBST. But when there are conflicting opinions, as there are, an outsider must discount the face value of the information by a factor that reflects the outsider's estimate of the reliability of the information's source. The practice of discounting reports of unreliable sources is a paradigmatically reasonable approach to making decisions in situations in which one must rely on others, and is one that would undoubtedly be employed by the very scientists who have conducted rBST experiments when making decisions outside their area of expertise. This pattern for relating uncertainty and risk is repeated in disputes over nuclear waste (Shrader-Frechette, 1993). The frequent response of scientists to denigrate such anxiety as either irrational (see Lewis, 1990) or mere perception (see Hotchkiss, 1991) only exacerbates an observer's justifiable fear that the people entrusted to look after food safety are not worthy of that trust.

All the other factors identified in the purified analysis (social consequences, animal welfare, environmental impact and even religious sensitivity) may be interpreted as evidence bearing on the reliability of an information source. Those who adopt the hybrid view have implicitly questioned the reliability of government, corporate, and university scientists based on their apparent lack of concern for impacts and problems that scientists have judged to be irrelevant to food safety. The inference is something like: "People who don't care about animals, small farms, or the environment aren't to be trusted." For some, it is even worse that they ignore people's religious feelings. Lack of concern for social consequences is probably especially significant in light of folk, family farm, and populist traditions in the United States and Europe. The sorting into categories that the advocate of the purified analysis takes to be a powerful way of organizing the ethical issues is seen as a character flaw by those inclined toward a hybridizing view.

Those who have accepted the partition of ethical issues implied by a strategy of purification are likely to regard the approach that absorbs all concerns into a concern for safety as a paranoiac, irrational perception of the risks. Latour's sociology of science challenges us to look harder for the rationality of the hybrid view. An argument can be made from the ordinary language. Scientists develop precise meanings from the words, "risk," and "safety," meanings that become almost interchangeable with the word "probability," in many scientific contexts. However, research on unschooled or folk attitudes toward decisions made under uncertainty suggests that people (including scientists, in one study) do not rely upon qualified notions of probability, but upon heuristics that substitute adequately well in most situations (Tversky and Kahneman, 1982). Though this research itself may be interpreted in different ways, it is at least plausible to suggest that people do not routinely understand the words "risk," or "safety," to refer exclusively to the quantified probability of adverse outcomes. An analysis of the grammar of these words shows that as a verb, "risk" links the potential peril to the intentional action of an individual or group, while "safety" is frequently and reasonably understood as a feeling of confidence and well-being. To the extent that these characteristics of common grammar penetrate nonscientists' attitudes, it is reasonable to expect that talk of risk will often, if not generally, occur when well-being is threatened not by mere chance, but by the planned and deliberate action of others (see Hornig, 1990, 1992; Thompson, 1995). The hybrid approach is a rational interpretation so long as ordinary language, rather than technical usage, establishes the criteria for what we mean by risk and safety. What is more, the character issues become linked to broader and more serious matters when the stakes are raised.

PURITY OR HYBRIDS?

Thus far, purification and hybridization have been presented as intellectual alternatives. It must be admitted that the argument from hybridization may not seem particularly compelling to someone trained in the sciences or philosophy. However, to arbitrate the contest between purification and hybridization in light of scientific or logical rigor is to misunderstand the burden of proof that hybridization demands. Ethical problems in animal biotechnology are ethical precisely because someone might be responsible for *doing* something about them, and because that responsibility might entail social, institutional, or governmental enforcement. Reasonable people can reach the conclusions of the hybrid interpretation without violating any canons of common sense. If we assume (as I do) that reasonable people should not be arbitrarily excluded from debate over policy and enforcement, then adherence to moral purification cannot be arbitrarily chosen as the standard for participation in debate. As such it is incumbent on those who would reject the inferences to justify the use of purification as an exclusionary tactic in governance and public policy, or failing such justification to ameliorate exclusionary applications of power with more inclusive political procedures. Put another way, those who choose the hybrid interpretation might find themselves excluded from the seminars and lectures of scientists and philosophers on the basis of this intellectual difference. When they find themselves excluded from political or economic power, the situation not only becomes more serious, but the fact of exclusion reinforces and validates the inferences that gave rise to the hybrid interpretation in the first place.

It is now important to understand the system of purification in both narrow and broad terms. Narrowly, it is a set of ethical concepts that allow us to partition the complex welter of inchoate ethical concerns into logically distinct categories. Purification also reflects a broader set of intellectual categories that serve as principles for organizing knowledge into disciplines, departments, and areas of concentration and for organizing at least some governmental authorities into agencies, administrations, services, and offices. Environmental impact is studied by environmental scientists and regulated by environmental agencies. Animal welfare is studied by physiologists and ethologists and is regulated (if at all) by institutional care committees. Toxicology and pharmacology are the scientific provenance of food safety. U.S. agencies such as the FDA and Food Safety Inspection Service (FSIS) have analogues in most nations. Social consequences are studied by economists and sociologists and ethics is part of philosophy, but government does not regulate in these areas. Most industrialized countries afford a role in government to each of the categories. This role bestows political authority on the purified ethical analysis of animal biotechnology. The hybrid interpretation is excluded, and it is important to see how.

Ethical criteria effect policy change only when policy makers can apply them in enforcing the law. As Rollin notes, religious believers may feel a moral obligation to practice rituals central to their faith, but the use of political authority to enforce such practices is now rare. As a practical matter political authority is mustered by convincing individuals in positions of power, be they monarchs, legislators, judges, or bureaucrats, that the mandate under which they wield their power justifies or perhaps requires action. Only absolute monarchs, however, are defined as having unlimited mandates. The more usual case is represented by the FDA of the U.S. government, which has a clear mandate to enforce criteria that relate to human health, but no authority at all to even consider social consequences. This means that issues relating to human health count both ethically and legally, while issues relating to social consequences are not subjected to a legally binding test. Under the U.S. Constitution and the current Federal Code, authority to deal with social consequences reverts to the U.S. Congress, an institution unlikely to act on this issue in the foreseeable future due to practical limitations. However, the simultaneous inclusion of food safety and exclusion of religion and social consequence from public policy in the United States gives legal force to purification, and deprives the alternative worldview of symmetrical legal standing. To summarize: FDA will regulate based on a pure, probabilistically based interpretation of safety. An understanding of safety in which social consequences have a bearing on one's feeling of well-being, for example, will be ruled out without being taken seriously. Those who advocate such a view have no standing, and are, as a matter of fact, likely targets of ridicule.

The system of purification is invested with political authority and power. At the same time, the system of purification reflects, to a large degree, the set of categories that define divisions of knowledge within academic and scientific research institutions. Each category of consequence, human, animal, ecosystem, and social, would be the object of study by separate departments, disciplines, or subdisciplines. These departments are routinely (though far from universally) seen to be operating in logically distinct spheres. There is thus a double institutionalization of the system of categories produced by purification, first in government and second in the academic departments of the sciences, including the social sciences and to some extent the humanities. The significance of this double set of social institutions is subtle and complex. On the one hand, it may be interpreted as validation of the order produced by the purification itself, suggesting that similar patterns of purification have been replicated and reproduced in a variety of otherwise independent contexts. On the other hand, it may be that government and science have co-evolved so as to product mutually consistent organizational divisions for addressing complex issues. The social histories implied by each of these two alternatives raise large and deep philosophical issues that must be set aside here, but so long as the second alternative is plausible, the anxiety that arises from seeing purification as a form of power seeking is not only warranted, but increased.

Seen from one vantage point, the system of purification establishes a leviathan of science and government. The basic assumptions that partition knowledge also partition government power. Those who do not share the basic assumptions or who rely on ordinary language rather than technical definitions of concepts are outsiders. Their arguments have no standing and cannot be converted into policy by the agencies that have been established with limited mandates. Scientists and scientific organizations, in the meantime, have been placed at the center of the leviathan. They control the definitions that are used to translate regulatory mandates into operational terms. They do the research that will form the empirical basis for policy decisions. It is unfair to suggest (as critics have) that scientists have an interest in manipulating the results of that research, for the long term viability of the leviathan depends upon objective research procedures. It is entirely fair, however, to say that scientists have an interest in maintaining the structure of the leviathan, for it assures their status and the continuing demand for their services. Both government and private industry need scientific institutions to perform the dual function of defining criteria and evaluating specific products or technologies, and this need established a market for science, both in the form of jobs and research funds. This means, however, that scientists have an interest in preserving the system of purification. Those who would propose alternative interpretations are, thus, enemies of science, not in any elevated philosophical sense, but in having adopted basic assumptions that fail to support the system of purification that links science, government, and private industry in a mutually supportive network.

To the extent that democracy is understood as a form of government distinctive for its receptivity to participation and resting upon consent of the governed, the events that turn ordinary people into enemies of science can be seen to compromise government, rather than science. This observation does not imply that the floodgates should be opened to any ordinary person's assessment of risk and safety. It is clear that government food safety agencies have applied highly defensible standards in appealing to science as they have. Nevertheless, their appeal to science has changed not only science, but government, and the full implications of that change have yet to be recognized by scholars and theoreticians of science *or* government.

In this context, it is useful to revisit Verhoog's comments on speciation. Verhoog's accusation is that those who would attribute ethical significance to the biologists' way of defining species have simply assumed what needs to be proven. In a 1993 paper, Verhoog writes that molecular biology reduces complex functions and structures to genetic factors. He laments the loss of a personal relation between the biologist and his object: "There seems to be a reverse relationship between the degree of reductive objectivation and the degree of moral relevance of the entities studied" (Verhoog 1993: 94). Verhoog's position seems to hinge on the assumption that "moral relevance" can be sensed in ordinary phenomenal experience of nature, but becomes increasingly unavailable as biological entities are interpreted in terms that reduce to statements about genes. Verhoog does not provide an argument for using ordinary phenomenal experience of life as the basis for ethical judgments, rather than the molecular account of life, but he does show that neither Rollin nor the advocates of the view he represents have provided arguments either.

Rollin accepts the assumption that biologists who study speciation are best qualified to define the term "species" and to establish the criteria for determining what is and what is not a species. This assumption bears on ethical issues of modifying animals, for it suggests that there is no natural order to the particular distribution of species to which human beings have become accustomed. If species are in flux, it is more difficult to see how there could be something ethically questionable in rearranging them. Verhoog does not question Rollin's account of the biologist's definition. Instead he suggests that Rollin has begged the central question in assuming that biologists are better qualified to define species boundaries than are ordinary people. He notes how the order of species is implicit in key categories of ordinary language. His position is that animals are co-evolved with humans into distinct species through a conceptual as well as biological process (Verhoog, 1992). The implication is that part of what it means to be human is to live among well-defined animal species. While he does not supply a full argument for preferring the common sense, natural language notion of speculation, Verhoog is successful in demonstrating that mere assumption of disciplinary biology's superiority is a question begging failure to enjoin the ethical issue at its root.

The root question both for ethical controversy over animal biotechnology and for Verhoog's ontological challenge to genetic engineering is whether the strategy of purification, complete with its application in both academy and government, can be legitimated. This root question emerges only when purification is seen in contrast to hybridization that proposes its own internally consistent standards of rationality. No form of reflection or discourse can emerge *ex nihilo;* some assumptions must be granted. Nevertheless, it is not only possible to articulate the rationality of views that reject the assumptions of purification, but also to show how they represent very reasonable responses on the part of the vast majority not incorporated in scientific networks. This turn of events is recent and symptomatic of the closing space between science and government, and of the emergence of a hybrid science/government. A continued failure to enjoin the defense of purification is, therefore, unwarranted, an *argumentum ad baculum.* The assessment of animal biotechnology takes form in a deeper philosophical debate about the criteria of rationality. To simply equate one viewpoint with rationality produces no debate at all; it is neither science nor democracy.

At the risk of deflating the central thesis, it is prudent to end on qualifications to it. First the merger of science and power described here depends on complex circumstances that are clearly *not* repeated in every scientific/political controversy. Scientific networks are politically strong on some issues, weak on others. Second, nothing in this paper implies that the hybrid interpretation of animal biotechnology is the one that ought to hold sway in deciding its political future. The author has more frequently been associated with arguments of purification. Hybridization deserves a kind of intellectual and political respect; its claims should be enjoined philosophically, and its advocates should be accommodated politically. This is a prescription for a procedural response to hybridization, not a capitulation to its assessment of animal biotechnology. Finally, the paper has provided little indication of what procedural measures are required, but that is a discussion presupposing the thesis offered here, and, hence, beyond the scope of its statements and defense.

REFERENCES

Aiken, W. (1984). Ethical issues in agriculture, in T. Regan (ed.), *Earthbound: New introductory essays in environmental ethics* (pp. 257–288). New York: Random House.

Attfield, R. (1995). Genetic engineering: Can unnatural kinds be wronged? in P. Wheale and R. McNally (eds.), *Animal genetic engineering: Of pigs, oncomice and men* (pp. 201–210). London: Pluto Press.

Comstock, G. (1998). The case against BST, *Agriculture and Human Values* 5(3): 36–52.

Descartes, R. (1637 [republished 1950]). *Discourse on method.* Indianapolis: Bobbs-Merrill.

Donnelley, S., McCarthy, C. R., and Singleton, R., Jr. (1994). The brave new world of animal biotechnology, Special supplement, *Hastings Center Report* 24(1): S1–S31.

Doyle, M. P. and Marth, E. H. (1991). Food safety issues in biotechnology, in B. Baumgardt and M. Marvin (eds.), *Agricultural biotechnology: Issues and choices* (pp. 55–67). Lafayette, IN: Purdue Research Foundation.

Executive Office of the President (1994). *Use of bovine somatotropin (BST) in the United States: Its potential effects, a study conducted by the Executive Branch of the Federal Government.* Washington, DC.

Feenstra, G. (1993). Is BGH sustainable? The consumer perspective, in W.C. Liebhardt (ed.), *The dairy debate: Consequences of bovine growth hormone and rotational grazing technologies.* Davis: University of California, Sustainable Agriculture Research and Education Program.

Foreman, C. T. (1991). Food safety and quality for the consumer: Policies and communication, in J. F. MacDonald (ed.), *Agricultural biotechnology, food safety, and nutritional quality for the consumer* (pp. 74–81). Ithaca, NY: National Agricultural Biotechnology Council.

Fox, M. W. (1992). *Superpigs and wondercorn: The brave new world of biotechnology and where it all may lead.* New York: Lyons and Burford.

Feyerabend, P. (1975). *Against method.* London: Verso.

Hallberg, M. (1992). BST: Issues, facts and controversies, in M. Hallberg (ed.), *Bovine somatotropin and emerging issues* (pp. 293–301). Boulder, CO: Westview.

Hoban, T. and Kendall, P. (1992). *Consumer attitudes about the use of biotechnology in agriculture and food production.* Raleigh, NC: North Carolina State University.

Hobbelink, H. (1991). *Biotechnology and the future of the world agriculture.* London: Zed Books.

Hobbes, T. (1651 [republished 1951]). *Leviathan.* New York: Penguin Books.

Hornig, S. (1990). Science stories: Risk, power and perceived emphasis, *Journalism Quarterly* 67(4): 767–777.

Hornig, S. (1992). Gender differences in responses to news about science and technology, *Science, Technology, and Human Values* 17(4): 532–543.

Hotchkiss, J. H. (1991). Food related risks: A food scientist's perspective, in J. F. MacDonald (ed.), *Agricultural biotechnology, food safety, and nutritional quality for the consumer.* Ithaca, NY: National Agricultural Biotechnology Council.

Hunter, D. (1992). To live free as natives, free of fear: What citizens should require from animal biotechnology, in J. F. MacDonald (ed.), *Animal biotechnology: opportunities and challenges* (pp. 133–140). Ithaca, NY: National Agricultural Biotechnology Council.

Johnson, G. L. and Thompson, P. B. (1991). Ethics and values associated with agricultural biotechnology, in

B. Baumgardt and M. Martin (eds.), *Agricultural biotechnology: Issues and choices* (pp. 121–137). Lafayette, IN: Purdue Research Foundation.

Kalter, R. (1985). The new biotech agriculture: Unforeseen economic consequences, *Issues in Science and Technology* 2: 125–133.

Krimsky, S. and Wrubel, R. (1996). *Agricultural biotechnology and the environment: Science, policy and social issues.* Urbana: University of Illinois Press.

Kroger, M. (1992). Food safety and product quality, in M. Hallberg (ed.), *Bovine somatotropin and emerging issues* (pp. 265–270). Boulder, CO: Westview.

Lanyon, L. E. and Beegle, D. B. (1989). The role of on-farm nutrient balance assessments in an integrated approach to nutrient management, *Journal of Soil and Water Conservation* 44: 164–168.

Latour, B. (1986). *Science in action.* Cambridge, MA: Harvard University Press.

Latour, B. (1988). *The pasteurization of France.* Cambridge, MA: Harvard University Press.

Latour, B. (1992). *We have never been modern.* Cambridge, MA: Harvard University Press.

Latour, B. and Woolgar, S. (1979). *Laboratory life: The social construction of scientific facts.* Beverly Hills, CA: Sage Publications.

Lesser, W. H. (1989). *Animal patents: The legal, economic and social issues.* New York: Stockton Press.

Lewis, H. (1990). *Technological risk.* New York: W. W. Norton.

Munro, I. C. and Hall, R. L. (1991). Food safety and quality: Assessing the impact on biotechnology, in J. F. MacDonald (ed.), *Agricultural biotechnology, food safety, and nutritional quality for the consumer* (pp. 64–73). Ithaca, NY: National Agricultural Biotechnology Council.

NABC (1991). Summary, in J. F. MacDonald (ed.), *Agricultural biotechnology, food safety, and nutritional quality for the consumer* (pp. 58–62). Ithaca, NY: National Agricultural Biotechnology Council.

Norton, B. (1991). *Toward unity among environmentalists.* Oxford: Oxford University Press.

Pastoret, P. P. (1994). Vaccine case study: Development and deliberate release of a vaccinia-rabies recombinant virus for fox vaccination against rabies, in M. McGlaughlin (ed.), *Proceedings of the international workshop on animal biotechnology issues* (pp. 27–30). Davis: University of California at Davis.

Rayburn, E. B. (1993). Potential ecological and environmental health effects of pasture and BGH technology, in W. Leibhardt (ed.), *The dairy debate: Consequences of bovine growth hormone and rotational grazing technologies.* Davis: University of California, Sustainable Agriculture Research and Education Program.

Rollin, B. E. (1981). *Animal rights and human morality.* Buffalo, NY: Prometheus Books.

Rollin, B. E. (1985 [republished 1990]). The Frankenstein thing, in S. M. Gendel, A. D. Kline, D. M. Warren and F. Yates (eds.), *Agricultural bioethics: Implications of agricultural biotechnology* (pp. 292–308). Ames: Iowa State University Press.

Rollin, B. E. (1992). The creation of transgenic animal "models" for human genetic disease, in J. F. MacDonald (ed.), *Animal biotechnology: Challenges and opportunities* (pp. 85–94). Ithaca, NY: National Agricultural Biotechnology Council.

Rollin, B. E. (1995). *The Frankenstein syndrome: Ethical and social issues in the genetic engineering of animals.* Cambridge and New York: Cambridge University Press.

Ruscio, K. P. (1994). Policy cultures: The case of science policy in the United States, *Science, Technology, and Human Values* 19: 205–222.

Shapin, S. and Shaffer, S. (1985). *Leviathan and the air pump.* Princeton, NJ: Princeton University Press.

Shrader-Frechette, K. (1993). *Burying uncertainty: The case against permanent geological disposal of high level radioactive waste.* Berkeley: University of California Press.

Stich, S. (1978 [republished 1989]). The recombinant DNA debate, in M. Ruse (ed.), *Philosophy of biology* (pp. 229–243). New York: Macmillan.

Tauer, L. (1992). Impact of BST on small versus large dairy farms, in M. Hallberg (ed.), *Bovine somatotropin and emerging issues* (pp. 207–217). Boulder, CO: Westview.

Taylor, K. (1991). Many voices: Citizens and the issues of biotechnology and food safety, in J. F. MacDonald (ed.), *Agricultural biotechnology, food safety, and nutritional quality for the consumer* (pp. 96–102). Ithaca, NY: National Agricultural Biotechnology Council.

Taylor, K. (1995). Differences in church positions on genetic patenting, *Center for Biotechnology Policy and Ethics Newsletter* 5(2): 2–3.

Thompson, P. B. (1992). BST and ethical issues, in M. Hallberg (ed.), *Bovine somatotropin and emerging issues* (pp. 33–49). Boulder, CO: Westview.

Thompson, P. B. (1995). *The spirit of the soil: Agriculture and environmental ethics.* London: Routledge.

Thompson, P. B., Matthews, R., and VanRavenswaay, E. (1994). *Ethics, public policy and agriculture.* New York: Macmillan.

Tversky, A. and Kahneman, D. (1982). Judgment under uncertainty: Heuristics and biases, in D. Kahneman, P. Slovic, and A. Tversky (eds.), *Judgment under uncertainty: Heuristics and biases* (pp. 3–20). Cambridge: Cambridge University Press.

Verhoog, H. (1992a). Ethics and genetic engineering of animals, in A. W. Musschenga et al. (eds.), *Morality, worldwide and law* (pp. 267–278). Assen/Maastricht: Van Gorcum.

Verhoog, H. (1992b). The concept of intrinsic value and transgenic animals, *Journal of Agricultural and Environmental Ethics* 5(2): 147–160.

Verhoog, H. (1993). Biotechnology and ethics, in T. Brante, S. Fuller, and W. Lynch (eds.), *Controversial science* (pp. 83–106). New York: SUNY Press.

Yilma, T. (1994). A vaccinia virus recombinant vaccine for rinderpest, in M. McGloughlin (ed.), *Proceedings of the international workshop on animal biotechnology issues* (pp. 19–21). Davis: University of California at Davis.

Yonkers, R. (1992). Potential adoption and diffusion of BST among dairy farmers, in M. Hallberg (ed.), *Bovine somatotropin and emerging issues* (pp. 177–192). Boulder, CO: Westview.

38

Technologies of Humility: Citizen Participation in Governing Science

Sheila Jasanoff

THE PERILS OF PREDICTION

Long before the terrorist atrocities of 11 September 2001 in New York, Washington, D.C., and Pennsylvania, the anthrax attacks through the U.S. mail, and the U.S.-led wars in Afghanistan and Iraq, signs were mounting that America's ability to create and operate vast technological systems had outrun her capacity for prediction and control. In a prescient book, published in 1984, the sociologist Charles Perrow forecast a series of "normal accidents," which were strung like dark beads through the latter years of the twentieth century and beyond—most notably, the 1984 chemical plant disaster in Bhopal, India; the 1986 loss of the *Challenger* shuttle and, in the same year, the nuclear plant accident in Chernobyl, USSR; the contamination of blood supplies with the AIDS virus; the prolonged crisis over BSE ("mad cow disease"); the loss of the manned U.S. space shuttle *Columbia* in 2003; and the U.S. space program's embarrassing, although not life-threatening, mishaps with the *Hubble* telescope's blurry lens, and several lost and extremely expensive Mars explorers.[1] To these, we may add the discovery of the ozone hole, climate change, and other environmental disasters as further signs of disrepair. Occurring at different times and in vastly different political environments, these events nonetheless have served collective notice that human pretensions of control over technological systems need serious reexamination.

While American theorists have often chalked up the failings of technology to avoidable error, especially on the part of large organizations,[2] some European analysts have suggested a more troubling scenario. Passionately set forth by the German sociologist Ulrich Beck, the thesis of "reflexive modernization" argues that risks are endemic in the way that contemporary societies conduct their technologically intensive business.[3] Scientific and technical advances bring unquestioned benefits, but they also generate new uncertainties and failures, with the result that doubt continually undermines knowledge, and unforeseen consequences confound faith in progress. Moreover, the risks of modernity often cut across social lines and operate as a great equalizer of classes. Wealth may increase longevity and improve the quality of life, but it offers no assured protection against the ambient harms of technological societies. This observation was tragically borne out when the collapse of the World Trade Center on 11 September 2001 ended the lives of some 3,000 persons,

Sheila Jasanoff, "Technologies of Humility: Citizen Participation in Governing Science," *Minerva* 41 (2003): 223–44. Copyright © 2003 Springer Science + Business Media BV. Reprinted by permission.

discriminating not all among corporate executives, stock market analysts, computer programmers, secretaries, firefighters, policeman, janitors, restaurant workers, and others. Defeat in war similarly endangers the powerful along with the disempowered. In many other contexts, however, vulnerability remains closely tied to socio-economic circumstances, so that inequalities persist in the ability of social groups and individuals to defend themselves against risk.

"Risk," on this account, is not a matter of simple probabilities, to be rationally calculated by experts and avoided in accordance with the cold arithmetic of cost-benefit analysis.[4] Rather, it is part of the modern human condition, woven into the very fabric of progress. The problem we urgently face is how to live democratically and at peace with the knowledge that our societies are inevitably "at risk." Critically important questions of risk management cannot be addressed by technical experts with conventional tools of prediction. Such questions determine not only whether we will get sick or die, and under what conditions, but also who will be affected and how we should live with uncertainty and ignorance. It is sufficient, for instance, to assess technology's consequences, or must we also seek to evaluate its aims? How should we act when the values of scientific inquiry appear to conflict with other fundamental social values? Has our ability to innovate in some areas run unacceptably ahead of our powers of control?[5] Will some of our most revolutionary technologies increase inequality, promote violence, threaten cultures, or harm the environment? And are our institutions, whether national or supranational, up to the task of governing our dizzying technological capabilities?

To answer questions such as these, the task of managing technologies has to go far beyond the model of "speaking truth to power" that once was thought to link knowledge to political action.[6] According to this template, technical input to policy problems has to be developed independently of political influences; the "truth" so generated acts as a constraint, perhaps the most important one, on subsequent exercises of political power. The accidents and troubles of the late twentieth century, however, have called into question the validity of this model—either as a descriptively accurate rendition of the ways in which experts relate to policy-makers, or as a normatively acceptable formula for deploying specialized knowledge within democratic political systems.[7] There is growing awareness that even technical policy-making needs to get more political—or, more accurately, to be seen more explicitly in terms of its political foundations. Across a widening range of policy choices, technological cultures must learn to supplement the expert's preoccupation with measuring the costs and benefits of innovation with greater attentiveness to the politics of science and technology.

Encouragingly, the need for reform in governing science and technology has been acknowledged by political authority. In the millennial year 2000, for example, the House of Lords Select Committee on Science and Technology in Britain issued a report on science and society that began with the ominous observation that relations between the two had reached a critical phase.[8] The authors foresaw damaging consequences for science and technology if these conditions were allowed to persist. This observation was widely attributed to Britain's particular experience with BSE, but the crisis of confidence vis-à-vis the management of science and technology has spread significantly wider. The European Union's 2001 White Paper on Governance drew on the activities of a working group on "Democratizing Expertise," whose report promised new guidelines "on the collection and use of expert advice in the Commission to provide for the accountability, plurality and integrity of the expertise used."[9] The intense worldwide discussion of the risks, benefits, and social consequences of biotechnology that began in the late 1990s can be seen as sharing many of the same concerns.

These initiatives and debates reflect a new-found interest on the part of scientists, governments, and many others in creating greater *accountability* in the production and use of scientific knowledge. The conduct of research has changed in ways that demand increased recognition. As captured by the "Mode 2" rubric, the pursuit of science is becoming more dispersed, context-dependent, and problem-oriented. Given these shifts, concerns with the assurance of quality and reliability in scientific production, reflecting the dominance of the "speaking truth to power" model, are now seen as too narrowly focused. The wider public responsibilities of science, as well as changes in modes of knowledge-making, demand new forms of public justification. Accountability can be defined in different ways, depending on the nature and context of scientific activity—for example, in demands for precaution in environmental assessments, or in calls for bioethical guidelines in relation to new genetic technologies. Whatever its specific articulation, however, accountability in one or another form is increasingly seen as an independent criterion for evaluating scientific research and its technological applications, supplementing more traditional concerns with safety, efficacy, and economic efficiency.

But how can ideas of accountability be mapped onto well-entrenched relations between knowledge and power, or expertise and public policy? The time is ripe for seriously reevaluating existing models and approaches. How have existing institutions conceptualizing the roles of technical experts, decision-makers, and citizens with respect to the uses and applications of knowledge? How should these understandings be modified in response to three decades of research on the social dimensions of science? Can we respond to the demonstrated fallibility and incapacity of decision-making institutions, without abandoning hopes for improved health, safety, welfare, and social justice? Can we imagine new institutions, processes, and methods for restoring to the playing field of governance some of the normative questions that were sidelined in celebrating the benefits of technological progress? And are there structured means for deliberating and reflecting on technical matters, much as the expert analysis of risks has been cultivated for many decades?

There is a growing need, I shall argue, for what we may call the "technologies of humility." These are methods, or better yet institutionalized habits of thought, that try to come to grips with the ragged fringes of human understanding—the unknown, the uncertain, the ambiguous, and the uncontrollable. Acknowledging the limits of prediction and control, technologies of humility confront "head-on" the normative implications of our lack of perfect foresight. They call for different expert capabilities and different forms of engagement between experts, decision-makers, and the public than were considered needful in the governance structures of high modernity. They require not only the formal mechanisms of participation but also an intellectual environment in which citizens are encouraged to bring their knowledge and skills to bear on the resolution of common problems. Following a brief historical account, I will offer a framework for developing this approach.

THE POST-WAR SOCIAL CONTRACT

In the United States, the need for working relationships between science and the state was famously articulated not by a social theorist or sociologist of knowledge, but by a quintessential technical expert: Vannevar Bush, the distinguished MIT engineer and presidential adviser. Bush foresaw the need for permanent changes following the mobilization of science and technology during the Second World War. In 1945, he produced a report, *Science—The Endless Frontier*,[10] that was later hailed as laying the basis for American policy in science and technology. Science, in Bush's vision, was destined to enjoy government patronage in peacetime as it had during the war. Control over the scientific enterprise, however, would be wrested from the military and lodged with the civilian community. Basic research, uncontaminated by industrial application or government policy, would thrive in the free air of universities. Scientists would establish the substantive aims as well as the intellectual standards of research. Bush believed that bountiful results flowing from their endeavors would translate in due course into beneficial technologies, contributing to the nation's prosperity and progress. Although his design took years to materialize, and even then was only imperfectly attained, the U.S. National Science Foundation (NSF) emerged as a principal sponsor of basic research.[11] The exchange of government funds and autonomy in return for discoveries, technological innovations, and trained personnel came to be known as America's "social contract for science."

The Bush report said little about how basic research would lead to advances in applied science or technology. That silence itself is telling. It was long assumed that the diffusion of fundamental knowledge into application was linear and unproblematic. The physical system that gripped the policy-maker's imagination was the pipeline. With technological innovation commanding huge rewards in the marketplace, market considerations were deemed sufficient to drive science through the pipeline of research and development into commercialization. State efforts to promote science could then be reasonably restricted to support for basic or "curiosity-driven" research. Simplistic in its understanding of the links between science and technology, this scheme, we may note, provided no conceptual space for the growing volume of scientific activity required to support and legitimate the multiple undertakings of modern states in the late twentieth century. In a host of areas, raging from the environmental policy to mapping and sequencing the human genome, governmental funds have been spent on research that defies any possible demarcation between basic and applied. Yet, for many years after the war, the basic-applied distinction remained the touchstone for distinguishing work done in universities from that done in industries, agricultural experiment stations, national laboratories, and other sites concerned primarily with the uses of knowledge.

As long as the "social contract" held sway, no one questioned whether safeguarding the autonomy of scientists was the best way to secure the quality and productivity of basic research. Peer review was the instrument that scientists used for self-regulation as well as quality control. This ensured that state-sponsored research would be consistent with a discipline's priorities, theories, and methods. Peer review was responsible, with varying success, for ensuring the credibility of reported results, as well as their originality and interest.

So strong was the faith in peer review that policy-makers, especially in the U.S., often spoke of this as the best means of validating scientific knowledge, even when it was produced and used in other contexts—for example, for the purpose of supporting regulatory policy. In practice, a more complex, tripartite approach to quality control developed in most industrial democracies—peer review by disciplinary colleagues in basic science; the development of good laboratory practices, under applicable research protocols, such as products testing or clinical trials in applied research; and risk assessment for evaluating the health or environmental consequences of polluting emissions and industrial products. But as the importance of testing, clinical research, and risk assessment grew, so, too, did calls for ensuring their scientific reliability. Once again, peer review—or its functional analogue, independent expert advice—was the mechanism that governments most frequently used for legitimation.

Signs of wear and tear in the "social contract" began appearing in the 1980s. A spate of highly publicized cases of alleged fraud in science challenged the reliability of peer review and, with it, the underlying assumptions concerning the autonomy of science. The idea of science as a unitary practice also began to break down as it became clear that research varies from one context to another, not only across disciplines, but—even more important from a policy standpoint—across institutional settings. It was recognized, in particular, that regulatory science, produced to support governmental efforts to guard against risk, was fundamentally different from research driven by scientists' collective curiosity. At the same time, observers began questioning whether the established categories of basic and applied research held much meaning in a world where the production and uses of science were densely connected to each other, as well as to larger social and political consequences.[12] The resulting effort to reconceptualize the framework of science-society interactions forms an important backdrop to present attempts to evaluate the accountability of scientific research.

SCIENCE IN SOCIETY—NEW ASSESSMENTS

Rethinking the relations of science has generated three major streams of analysis. The first stream takes the "social contract" for granted, but points to its failure to work as its proponents had foreseen. Many have criticized science, especially university-based science, for deviating from idealized, Mertonian norms of purity and disinterestedness. Despite (or maybe because of) its conceptual simplicity, this critique has seriously threatened the credibility of researchers and their claim to autonomy. Other observers have tried to replace the dichotomous division of *basic* and *applied* science with a more differentiated pattern, calling attention to the particularities of science in different settings and in relation to different objectives. Still others have made ambitious efforts to respecify how scientific knowledge is actually produced. This last line of analysis seeks not so much to correct or refine Vannevar Bush's vision of science, as to replace it with a more complex account of how knowledge-making fits into the wider functioning of society. Let us look at each of these three critiques.

Deviant science. Scientific fraud and misconduct became an issue on the U.S. policy agenda in the 1980s. Political interest reached a climax with the notorious case of alleged misconduct in an MIT laboratory headed by Nobel laureate biologist David Baltimore. He and his colleagues were exonerated, but only after years of inquiry, which included investigations by Congress and the FBI.[13] This and other episodes left residues in the form of greatly increased Federal powers for the supervision of research, and a heightened tendency for policy-makers and the public to suspect that all was not in order in the citadels of basic science. Some saw the so-called "Baltimore affair" as a powerful sign that legislators were no longer content with the old social technological benefits.[14] Others, like the seasoned science journalist Daniel Greenberg, accused scientists of profiting immoderately from their alliance with the state, while failing to exercise moral authority or meaningful influence on policy.[15] American science has since been asked to justify more explicitly the public money spent on it. A token of the new relationship came with the reform of NSF's peer review criteria in the 1990s. The Foundation now requires reviewers to assess proposals not only on grounds of technical merit, but also with respect to wider social implications—thus according greater prominence to social utility. In effect, the very public fraud investigations of the previous decade opened up taken-for-granted aspects of scientific autonomy, and forced scientists to account for their objectives, as well as to defend their honesty.

To these perturbations may be added a steady stream of challenges to the supposed disinterestedness of academic science. From studies in climate change to biotechnology, critics have accused researchers of having sacrificed objectivity in exchange for grant money or, worse, equity interests in lucrative start-up companies.[16] These allegations have been especially damaging to biotechnology, which benefits significantly from the rapid transfer of skills and knowledge. Since most Western governments are committed to promoting such transfers, biotechnology is caught on the horns of a very particular dilemma: how to justify its promises of innovation and progress credibly when the interests of most scientists are unacceptably aligned with those of industry, government, or—occasionally—'public interest' advocates.

Predictably, pro-industry bias has attracted the most criticism, but academic investigators have also come under scrutiny for alleged pro-environment and anti-technology biases. In several cases involving biotechnology—in particular, that of the monarch butterfly study conducted by Cornell University scientist John Losey in the United States,[17] and Arpad Pusztai's controversial rat-feeding study in the United Kingdom[18]—industry critics have questioned the quality of university-based research, and have implied that political orientations may have prompted premature release or the overinterpretation of results. In April 2002, another controversy of this sort erupted over an article in *Nature* by a University of California scientist, Ignacio Chapela, who concluded that DNA from genetically modified corn had contaminated native species in Mexico. Philip Campbell, the journal's respected editor, did not retract the paper, but stated that "the evidence available is not sufficient to justify the publication of the original paper," and that readers should "judge the science for themselves."[19] As in the Losey and Pusztai cases, critics charged that Chapela's science had been marred by non-scientific considerations. Environmentalists, however, have viewed all these episodes as pointing to wholesale deficits in knowledge about the long-term and systemic effects of genetic modification in crop plants.

Context-specific science. The second line of attack on the science-society relationship focuses on the "basic-applied" distinction. One attempt to break out of the simplistic dualism was proposed by the late Donald Stokes, whose quadrant framework, using Louis Pasteur as the prototype, suggested that "basic" science can be done within highly "applied" contexts.[20] Historians and sociologists of science and technology have long observed that foundational work can be done in connection with applied problems, just as applied problem-solving is often required for resolving theoretical issues (for example, in the design of new scientific instruments). To date, formulations based on such findings have been slow to take root in policy cultures. The interest of Stokes' work lay not so much in the novelty of his insights as in his attempt to bring historical facts to bear on the categories of science policy analysis.

Like Vannevar Bush, Stokes was more interested in the promotion of innovation than in its control. How to increase the democratic supervision of science was not his primary concern. Not surprisingly, the accountability of science has emerged as a stronger theme in studies of risk and regulation, the arena in which governments seek actively to manage the potentially harmful aspects of technological progress. Here, too, one finds attempts to characterize science as something more than "basic" or "applied."

From their background in the philosophy of science, Funtowicz and Ravetz proposed to divide the world of policy-relevant science into three nested circles, each with its own system of quality control: (1) "normal science" (borrowing the well-known term of Thomas Kuhn), for ordinary scientific research; (2) "consultancy science," for the application of available knowledge to well-characterized problems; and (3) "post-normal science," for the highly uncertain, highly contested knowledge needed for many health, safety, and environmental decisions.[21] These authors noted that, while traditional peer review may be effective within "normal" and even "consultancy" science, the quality of "post-normal" science cannot be assured by standard review processes alone. Instead, they proposed that work of this nature be subjected to *extended peer review,* involving not only scientists but also the stakeholders affected by the use of science. Put differently, they saw accountability, rather than mere quality control, as the desired objective when science becomes "post-normal."[22]

Jasanoff's 1990 study of expert advisory committees in the United States noted that policy-relevant science (also referred to as "regulatory science")—such as science done for purposes of risk assessment—is often subjected to what policy-makers call "peer review."[23] On inspection, this exercise differs fundamentally from the review of science in conventional research settings. Regulatory science is reviewed by multidisciplinary committees rather than by individually selected specialists. The role of such bodies is not only to validate the methods by which risks are identified and investigated, but also to confirm the reliability of the agency's interpretation of the evidence. Frequently, regulatory science confronts the need to set standards for objects or concepts whose very existence has not previously been an issue for either science or public policy: "fine particulate matter" in air pollution control; the "maximum tolerated dose" (MTD) in bioassays; the "maximally exposed person" in relation to airborne toxics; or the "best available technology" in many programs of environmental regulation. In specifying how such terms should be defined or characterized, advisory committees have to address issues that are technical as well as social, scientific as well as normative, regulatory as well as metaphysical. What *kind* of entity, after all, is a "fine" particulate or a "maximally exposed" person, and by what markers can we recognize them? Studies of regulatory science have shown that the power of advisory bodies definitely to address such issues depends on their probity, representativeness, transparency, and accountability to high authority—such as courts and the public. In other words, the credibility of regulatory science ultimately rests upon factors that have more to do with accountability in terms of democratic politics, than with the quality of science as assessed by scientific peers.

In modern industrial societies, studies designed to establish the safety or effectiveness of new technologies are frequently delegated to producers. Processes of quality control for product testing within industry include the imposition and enforcement of good laboratory practices, under supervision by regulatory agencies and their scientific advisers. The precise extent of an industry's knowledge-producing burden is often negotiated with the regulatory agencies, and may be affected by economic and political considerations that are not instantly apparent to outsiders (setting MTDs for bioassays is one well-known example). Resource limitations may curb state audits and inspections of industry labs, leading to problems of quality control, while provisions extending confidential trade information from disclosure may reduce the transparency of product- or process-specific research conducted by industry. Finally, the limits of the regulator's imagination place significant limitations on an industry's duty to generate information. Only in the wake of environmental disasters involving dioxin, methyl isocyanate, and PCBs, and only after the accidental exposure of populations and ecosystems, were gaps discovered in the information available about the chronic and long-term effects of many hazardous chemicals. Before disaster struck, regulators did not appreciate the need for such information. Occurrences like these have led to demands for greater public accountability in the science that is produced to support regulation.

New modes of knowledge production. Going beyond the quality and context-dependency of science, some have suggested that we need to take a fresh look at the structural characteristics of science in order to make it more socially responsive. Michael Gibbons and his co-authors have concluded that the traditional disciplinary science of Bush's "endless frontier" has been largely supplanted by a new "Mode 2" of knowledge production.[24] The salient properties of this new Mode, in their view, include the following:

- Knowledge is increasingly produced in contexts of application (i.e., *all* science is to some extent "applied" science);
- Science is increasingly transdisciplinary—that is, it draws upon and integrates empirical and theoretical elements from a variety of fields;
- Knowledge is generated in a wider variety of sites than ever before, not just in universities and industry, but also in other sorts of research centers, consultancies, and think-tanks; and
- Participants in science have grown more aware of the social implications of their work (i.e., more "reflexive"), just as publics have become more conscious of the ways in which science and technology affect their interests and values.

The growth of "Mode 2" science, as Gibbons et al. note, has necessary implications for quality control. Besides old questions about the intellectual merits of their work, scientists are being asked to answer questions about marketability, and the capacity of science to promote social harmony and welfare. Accordingly:

Quality is determined by a wider set of criteria, which reflects the broadening social composition of the review system. This implies that "good science" is more difficult to determine. Since it is no longer limited to the judgments of disciplinary peers, the fear is that control will be weaker and result in lower quality work. Although the quality control process in Mode 2 is more broadly based, it does not follow . . . that it will necessarily be of lower quality.[25]

One important aspect of this analysis is that, in "Mode 2" science, quality control has for practical purposes merged with accountability. Gibbons et al. view all of science as increasingly more embedded in, and hence more accountable to, society at large. To keep insisting upon a separate space for basic research, with autonomous measures for quality control, appears, within their framework, to be a relic of an earlier era.

In a more recent work, Helga Nowotny, Peter Scott, and Michael Gibbons have grappled with the implications of these changes for the production of knowledge in public domains.[26] Unlike the "pipeline model," in which science generated by independent research institutions eventually reaches industry and government, Nowotny et al. propose the concept of "socially robust knowledge" as the solution to problems of conflict and uncertainty. Contextualization, in their view, is the key to producing science for public ends. Science that draws strength from its socially detached position is too frail to meet the pressures placed upon it by contemporary societies. Instead, they imagine forms of knowledge that would gain robustness from their very embeddedness in society. The problem, of course, is how to institutionalize polycentric, interactive, and multipartite processes of knowledge-making within institutions that have worked for decades at keeping expert knowledge away from the vagaries of populism and politics. The question confronting the governance of science is how to bring knowledgeable publics into the front-end of scientific and technological production—a place from which they have historically been strictly excluded.

THE PARTICIPATORY TURN

Changing modes of scientific research and development provide at least a partial explanation for the current interest in improving public access to expert decision-making. In thinking about research today, policy-makers and the public inevitably focus on the accountability of science. As the relations of science have become more pervasive, dynamic, and heterogeneous, concerns about the integrity of peer review have transmuted into demands for greater public involvement in assessing the costs and benefits, as well as the risks and uncertainties, of new technologies. Such demands have arisen with particular urgency in the case of biotechnology, but they are by no means limited to that field.

The pressure for accountability manifests itself in many ways, of which the demand for greater transparency and participation is perhaps most prominent. One notable example came with U.S. Federal legislation in 1998, pursuant to the Freedom of Information Act, requiring public access to all scientific research generated by public funds.[27] The provision was hastily introduced and scarcely debated. Its sponsor, Senator Richard Shelby (R-Alabama), tacked it on as a last-minute amendment to an omnibus appropriations bill. His immediate objective was to force disclosure of data by the Harvard School of Public Health from a controversial study of the health effects of human exposure to fine particulates. This so-called Six Cities Study provided key justification for the U.S. Environmental Protection Agency's stringent ambient standard for airborne particulate matter, issued in 1997. Whatever its political motivations, this sweeping enactment showed that Congress was no longer willing to concede unchecked autonomy to the scientific community in the collection and interpretation of data, especially when the results could influence costly regulatory action. Publicly funded science, Congress determined, should be available at all times to public review.

Participatory traditions are less thoroughly institutionalized in European policy-making, but recent changes in the rules governing expert advice display a growing commitment to involving the public in technically grounded decisions. In announcing the creation of a new Directorate General for Consumer Protection, the European Commission observed in 1997 that, "Consumer confidence in the legislative activities of the EU is conditioned by the *quality and transparency* of the scientific advice and its use on the legislative and control process" (emphasis added).[28] A commitment to greater openness is also evident in several new UK expert bodies, such as the Food Standards Agency, created to restore confidence in the wake of the Bovine Spongeform Encephalopath (BSE) crisis. Similarly, two major public inquiries—the Phillips Inquiry on BSE and the Smith Inquiry on the Harold Shipman murder investigation—set high standards for public access to information through the Internet. All across Europe, opposition to genetically modified foods and crops has prompted experiments with diverse forms of public involvement, such as citizen juries, consensus conferences, and referenda.[29]

Although these efforts are admirable, formal participatory opportunities cannot by themselves ensure the representative and democratic governance of science. There are, to start with, practical problems. People may not possess enough specialized knowledge and material resources to take advantage of formal procedures. Participation may occur too late to identify alternatives to dominant or default options; some processes, such as consensus conferences, may be too ad hoc or issue-specific to exercise sustained influence. More problematic is the fact that even timely participation does not necessarily improve decision-making. Empirical research has consistently shown that transparency may exacerbate rather than quell controversy, leading parties to deconstruct each other's positions instead of deliberating effectively. Indeed, the Shelby Amendment reflects one U.S. politician's conviction that compulsory disclosure of data will enable any interested party to challenge researchers' interpretations of their work. Participation, in this sense, becomes an instrument to challenge scientific points on political grounds. By contrast, public participation that is constrained by established formal disclosures, such as risk assessment, may not admit novel viewpoints, radical critiques, or considerations lying outside the taken-for-granted framing of the problem.

While national governments are scrambling to create new participatory forms, there are signs that such changes may reach neither far enough nor deeply enough to satisfy the citizens of a globalizing world. Current reforms leave out public involvement in corporate decision-making at the design and product-development phases. The Monsanto Company's experience with the "Terminator gene" suggests that political activists may seize control of decisions on their own terms, unless governance structures provide for more deliberative participation. In this case, the mere possibility that a powerful multinational corporation might acquire technology to deprive poor farmers of their rights, galvanized an activist organization—Rural Advancement Foundation International (RAFI)—to launch an effective worldwide campaign against the technology.[30] Through a combination of inspired media tactics (including naming the technology after a popular science-fiction movie) and strategic alliance-building (for example, with the Rockefeller Foundation), RAFI forced Monsanto to back down from this particular product. The episode can be read as a case of popular technology assessment, in a context where official processes failed to deliver the level of accountability desired by the public.

Participation alone, then, does not answer the problem of how to democratize technological societies. Opening the doors to previously closed expert forums is a necessary step—indeed, it should be seen by now as a standard opening procedure. But the formal mechanisms adopted by national governments are not enough to engage the public in the management of global science and technology. What has to change is the *culture* of governance, within nations as well as internationally; and for this we need to address not only the mechanics, but also the substance of participatory politics. The issue, in other words, is no longer *whether* the public should have a say in technical decisions, but *how* to promote more meaningful interaction among policy-makers, scientific experts, corporate producers, and the public.

TECHNOLOGIES OF HUMILITY

The analytic ingenuity of modern states has been directed toward refining what we may call the "technologies of hubris." To reassure the public, and to keep the wheels of science and industry turning, governments have developed a series of predictive methods (e.g., risk assessment, cost-benefit analysis, climate modeling) that are designed, on the whole, to facilitate management and control, even in areas of high uncertainty.[31] These methods achieve their power through claims of objectivity and a disciplined approach to analysis, but they suffer from three significant limitations. First they show a kind of peripheral blindness toward uncertainty and ambiguity. Predictive methods focus on the known at the expense of the unknown, producing overconfidence in the accuracy and completeness of the pictures they produce. Well-defined, short-term risks command more attention than indeterminate, long-term ones, especially in cultures given to technological optimism. At the same time, technical proficiency conveys the false impression that analysis is not only rigorous, but complete—in short, that it has taken account of all possible risks. Predictive methods tend in this way to downplay what falls outside their field of vision, and to overstate whatever falls within.[32]

Second, the technologies of predictive analysis tend to pre-empt political discussion. Expert analytical frameworks create high entry barriers against legitimate positions that cannot express themselves in terms of the dominant discourse.[33] Claims of objectivity hide the exercise of judgment, so that normative presuppositions are not subjected to general debate. The boundary work that demarcates the space of "objective" policy analysis is carried out by experts, so that the politics of demarcation remains locked away from public review and criticism.[34]

Third, predictive technologies are limited in their capacity to internalize challenges that arise outside their framing assumptions. For example, techniques for assessing chemical toxicity have become ever more refined, but they continue to rest on the demonstrably faulty assumption that people are exposed to one chemical at a time. Synergistic effects, long-term exposures, and multiple exposures are common in normal life, but have tended to be ignored as too messy for analysis—hence, as irrelevant to decision-making. Even in the aftermath of catastrophic failures, modernity's predictive models are often adjusted to take on board only those lessons that are compatible with their initial assumptions. When a U.S.-designed chemical factory in Bhopal released the deadly gas methyl isocyanate, killing thousands, the international chemical industry made many improvements in its internal accounting and risk-communication practices. But no new methods were developed to assess the risks of technology transfer between radically different cultures of industrial production.

To date, the unknown, unspecified, and indeterminate aspects of scientific and technological development remain largely unaccounted for in policy-making; treated as beyond reckoning, they escape the discipline of analysis. Yet, what is lacking is not just knowledge to fill the gaps, but also processes and methods to elicit what the public wants, and to use what is already known. To bring these dimensions out of the shadows and into the dynamics of democratic debate, they must first be made concrete and tangible. Scattered and private knowledge has to be amalgamated, perhaps even disciplined, into a dependable civic epistemology. The human and social sciences of previous centuries undertook just such a task of translation. They made visible the social problems of modernity—poverty, unemployment, crime, illness, disease, and lately, technological risk—often as a prelude to rendering them more manageable, using what I have termed the "technologies of hubris." Today, there is a need for "technologies of humility" to complement the predictive approaches: to make apparent the possibility of unforeseen consequences; to make explicit the normative that lurks within the technical; and to acknowledge from the start the need for plural viewpoints and collective learning.

How can these aims be achieved? From the abundant literature on technological disasters and failures, as well as from studies of risk analysis and policy-relevant science, we can abstract four focal points around which to develop the new technologies of humility. They are *framing, vulnerability, distribution,* and *learning.* Together they provide a framework for the questions we should ask of almost every human enterprise that intends to alter society: what is the purpose; who will be hurt; who benefits; and how can we know? On all these points, we have good reason to believe that wider public engagement would improve our capacity for analysis and reflection. Participation that pays attention to these four points promises to lead neither to a hardening of positions, nor to endless deconstruction, but instead to richer deliberation on the substance of decision-making.

Framing. It has become an article of faith in the policy literature that the quality of solutions to perceived social problems depends on the way they are framed.[35] If a problem is framed too narrowly, too broadly, or wrongly, the solution will suffer from the same defects. To take a simple example, a chemical-testing policy focused on single chemicals cannot produce knowledge about the environmental health consequences of multiple exposures. The framing of the regulatory issue is more restrictive than the actual distribution of chemical-induced risks, and hence is incapable of delivering optimal management strategies. Similarly, a belief that violence is genetic may discourage the search for controllable social influences on behavior. A focus on the biology of reproduction may delay or impede effective social policies for curbing population growth. When facts are uncertain, disagreements about the appropriate frame are virtually unavoidable and often remain intractable for long periods. Yet, few policy cultures have adopted systematic methods for revising the initial framing of issues.[36] Frame analysis thus remains a critically important, though neglected, tool of policy-making that would benefit from greater public input.

Vulnerability. Risk analysis treats the "at-risk" human being as a passive agent in the path of potentially disastrous events. In an effort to produce policy-relevant assessments, human populations are often classified into groups (e.g., most susceptible, maximally exposed, genetically predisposed, children or women) that are thought to be differently affected by the hazard in question. Based on physical and biological indicators, however, these classifications tend to overlook the social foundations of vulnerability, and to subordinate individual experiences of risk to aggregate numerical calculations.[37] Recent efforts to analyze vulnerability have begun to recognize the importance of socio-economic factors, but methods of assessment still take populations rather than individuals as the unit of analysis. These approaches not only disregard differences within groups, but reduce individuals to statistical representations. Such characterizations leave out of the calculus of vulnerability such factors as history, place, and social connectedness, all of which may play crucial roles of their vulnerability, ordinary citizens may regain their status as active subjects, rather than remain undifferentiated objects in yet another expert discourse.

Distribution. Controversies over such innovations as genetically modified foods and stem cell research have propelled ethics committees to the top of the policy-making ladder. Frequently, however, these bodies are used as "end-of-pipe" legitimation devices, reassuring the public that normative issues have not been omitted from governmental deliberation. The term "ethics," moreover, does not cover the whole range of social and economic realignments that accompany major technological changes, nor their distributive consequences, particularly as technology unfolds across global societies and markets. Attempts to engage systematically with distributive issues in policy processes have not been altogether successful. In Europe, consideration of the "fourth hurdle"—the socio-economic impact of biotechnology—was abandoned after a brief debate. In the U.S., the congressional Office of Technology Assessment, which arguable had the duty to evaluate socio-economic impacts, was dissolved in 1995.[38] President Clinton's 1994 injunction to Federal agencies to develop strategies for achieving environmental justice has produced few dramatic results.[39] At the same time, episodes like the RAFI-led rebellion against Monsato demonstrate a deficit in the capacity for ethical and political analysis in large corporations, whose technological products can fundamentally alter people's lives. Sustained interactions between decision-makers, experts, and citizens, starting at the upstream end of research and development, could yield significant dividends in exposing the distributive implications of innovation.

Learning. Theorists of social and institutional learning have tended to assume that what is "to be learned" is never part of the problem. A correct, or at least a better, response exists, and the issue is whether actors are prepared to internalize it. In the social world, learning is complicated by many factors. The capacity to learn is constrained by limiting features of the frame within which institutions must act. Institutions see only what their discourses and practices permit them to see. Experience, moreover, is polysemic, or subject to many interpretations, no less in policy-making than in literary texts. Even when the fact of failure in a given case is more or less unambiguous, its causes may be open to many different readings. Just as historians disagree over what may have caused the rise or fall of particular political regimes, so policy-makers may find it impossible to attribute their failures to specific causes. The origins of a problem may appear one way to those in power, and in quite another way to the marginal or the excluded. Rather than seeking monocausal explanations, it would be fruitful to design avenues through which societies can collectively reflect on the ambiguity of their experiences, and to assess the strengths and weakness of alternative explanations. Learning, in this modest sense, is a suitable objective of civic deliberation.

CONCLUSION

The enormous growth and success of science and technology during the last century has created contradictions for institutions of governance. As technical activities have become more pervasive and complex, demand has grown for more complete and multivalent evaluations of the costs and benefits of technological progress. It is widely recognized that increased participation and interactive knowledge-making may improve accountability and lead to more credible assessments of science and technology. Such approaches will also be consistent with changes in the modes of knowledge production, which have made science more socially embedded and more closely tied to contexts of application. Yet, modern institutions still operate with conceptual models that seek to separate science from values, and that emphasize prediction and control at the expense of reflection and social learning. Not surprisingly, the real world continually produces reminders of the incompleteness of our predictive capacities through such tragic shocks as Perrow's "normal accidents."

A promising development is the renewed attention being paid to participation and transparency. Such participation, I have argued, should be treated as a standard operating procedure of democracy, but its aims must be considered as carefully as its mechanisms. Formally constituted procedures do not necessarily draw in all those whose knowledge and values are essential to making progressive policies. Participation in the absence of normative discussion can lead to intractable conflicts of the kind encountered in the debate on policies for climate change. Nor does the contemporary policy-maker's near-exclusive preoccupation with the management and control of risk, leave much space for tough debates on technological futures, without which we are doomed to repeat past mistakes.

To move public discussion of science and technology in new directions, I have suggested a need for "technologies of humility," complementing the predictive "technologies of hubris" on which we have lavished so much of our past attention. These *social* technologies would give combined attention to substance and process, and stress deliberation as well as analysis. Reversing nearly a century of contrary development, these approaches to decision-making would seek to integrate the "can do" orientation of science and engineering with the "should do" questions of ethical and political analysis. They would engage the human subject as an active, imaginative agent, as well as a source of knowledge, insight, and memory. The specific focal points I have proposed— framing, vulnerability, distribution, and learning—are pebbles thrown into a pond, with untested force and unforeseeable ripples. These particular concepts may prove insufficient to drive serious institutional change, but they can at least offer starting points for a deeper public debate on the future of science in society.

NOTES

1. Charles Perrow, *Normal Accidents: Living with High Risk Technologies* (New York: Basic Books, 1984).

2. *Ibid.* See also Diane Vaughan, *The Challenger Launch Decision: Risky Technology, Culture, and Deviance at NSA* (Chicago: University of Chicago Press, 1996); James F. Short and Lee Clarke (eds.), *Organizations, Uncertainties, and Risk* (Boulder, Colo.: Westview Press, 1992); and Lee Clarke, *Acceptable Risk? Making Decisions in a Toxic Environment* (Berkeley: University of California Press, 1989).

3. Ulrich Beck, *Risk Society: Towards a New Modernity* (London: Sage, 1992).

4. A pre-eminent example of the calculative approach is given in John D. Graham and Jonathan B. Wiener (eds.), *Risk versus Risk: Tradeoffs in Protecting Health and the Environment* (Cambridge, Mass.: Harvard University Press, 1995).

5. Never far from the minds of philosophers and authors of fiction, these concerns have also been famously articulated in recent times by Bill Joy, co-founder and chief scientist of Sun Microsystems. See Joy, "Why the Future Doesn't Need Us," *Wired,* http://www.wired.com/wired/archive/8.04/joy.html.

6. The *locus classicus* of this view of the right relations between knowledge and power is Don K. Price, *The Scientific Estate* (Cambridge, Mass.: Harvard University Press, 1965).

7. See, in particular, Sheila Jasanoff, *The Fifth Branch: Science Advisers as Policymakers* (Cambridge, Mass.: Harvard University Press, 1990).

8. United Kingdom, House of Lords Select Committee on Science and Technology, Third Report, *Science and Society,* http://www.parliament.the-stationery-office.co.uk/pa/ld199900/ldselect/ldsctech/38/3801.htm (2000).

9. Commission of the European Communities, *European Governance: A White Paper,* COM (2001), 428, http://europa.eu.int/eur-lex/en/com/cnc/2001/com2001_0428en01.pdf (Brussels, 27 July 2001), 19.

10. Vannevar Bush, *Science—The Endless Frontier* (Washington, DC: U.S. Government Printing Office, 1945).

11. The creation of the National Institutes of Health (NIH) to sponsor biomedical research, divided U.S. science policy in a way not contemplated by Bush's original design. In the recent politics of science, NIH budgets have proved consistently easier to justify than appropriations for other branches of science.

12. For reviews of the extensive relevant literatures, see Sheila Jasanoff, Gerald E. Markle, James C. Peterson, and Trevor Pinch (eds.), *Handbook of Science and Technology Studies* (Thousand Oaks, CA: Sage, 1995).

13. Daniel J. Kevles, *The Baltimore Case: A Trial of Politics, Science, and Character* (New York: Norton, 1998).

14. David H. Guston, *Between Politics and Science: Assuring the Integrity and Productivity of Research* (Cambridge: Cambridge University Press, 2001).

15. Daniel S. Greenberg, *Science, Money, and Politics: Political Triumph and Ethical Erosion* (Chicago: University of Chicago Press, 2001).

16. See, for example, Sonja Boehmer-Christiansen, "Global Climate Protection Policy: The Limits of Scientific Advice, Parts 1 and 2," *Global Environmental Change,* 4(2), (1994), 140–159; 4(3), (1994), 185–200.

17. John E. Losey, L. S. Raynor, and M. E. Carter, "Transgenic Pollen Harms Monarch Larvae," *Nature,* 399 (1999), 214.

18. Stanley W. B. Ewen and Arpad Pusztai, "Effect of diets containing genetically modified potatoes expressing *Galanthus nivalis lectin* on rat small intestine," Lancet, 354 (1999), 1353–1354.

19. "*Nature* Regrets Publication of Corn Study," *Washington Times,* http://www.washingtontimes.com/national/20020405–9384015.htm, 5 April 2002.

20. Donald E. Stokes, *Pasteur's Quadrant: Basic Science and Technological Innovation* (Washington, DC: Brookings Institution, 1997).

21. Silvio O. Funtowicz and Jerome R. Ravetz, "Three Types of Risk Assessment and the Emergence of Post Normal Science," in Sheldon Krimsky and D. Golding (eds.), *Social Theories of Risk* (New York: Praeger, 1992), 251–273.

22. A problem with this analysis lies in the term "post-normal science." When scientific conclusions are so closely intertwined with social and normative considerations as in Funtowicz and Ravetz's outermost circle, one may just as well call the "product" by another name, such as "socially-relevant knowledge" or "socio-technical knowledge."

23. Jasanoff, *op. cit.* note 7.

24. Michael Gibbons, Camille Limoges, Helga Nowotny, Simon Schwartzman, Peter Scott, and Martin Trow, *The New Production of Knowledge: The Dynamics of Science and Research in Contemporary Societies* (London: Sage, 1994).

25. *Ibid.,* 8.

26. Helga Nowotny, Peter Scott, and Michael Gibbons, *Re-Thinking Science: Knowledge and the Public in an Age of Uncertainty* (Cambridge: Polity, 2001), 166–178.

27. Public Law 105–277 (1998). The Office of Management and Budget in the Clinton administration controversially narrowed the scope of the law to apply not to *all* publicly-funded research, but only to research actually relied upon in policy-making. The issue is not completely resolved as of this writing.

28. European Commission, *1997 Communication of the European Commission on Consumer Health and Safety,* COM (97), 183. http://europa.eu.int/comm/food/fs/sc/index_en.html.

29. Simon Joss and John Durant (eds.), *Public Participation in Science: The Role of Consensus Conferences in Europe* (London: Science Museum, 1995).

30. In 1998, a small cotton seed company called Delta and Pine Land (D&PL) patented a technique designed to switch off the reproductive mechanism of agricultural plants, thereby rendering their seed sterile. The company hoped that this technology would help protect the intellectual property rights of agricultural biotechnology firms by taking away from farmers the capacity to re-use seed from a given year's genetically modified crops in the next planting season. While the technology was still years away from the market, rumors arose of a deal by Monsanto to acquire D&PL. This was the scenario that prompted FAFI to act. Robert F. Service, "Seed-Sterilizing 'Terminator Technology' Sows Discord," *Science,* 282 (1998), 850–851.

31. See, for example, Theodore M. Porter, *Trust in Numbers: The Pursuit of Objectivity in Science and Public Life* (Princeton, NJ: Princeton University Press, 1995).

32. Alan Irwin and Brian Wynne (eds.), *Misunderstanding Science? The Public Reconstruction of Science and Technology* (Cambridge: Cambridge University Press, 1996).

33. Langdon Winner, "On Not Hitting the Tar Baby," in Langdon Winner, *The Whale and the Reactor: A Search for Limits in an Age of High Technology* (Chicago: University of Chicago Press, 1986), 138–154.

34. Jasanoff, *op. cit.* note 7.

35. Donald A. Schon and Martin Rein, *Frame/Reflection: Toward the Resolution of Intractable Policy Controversies* (New York: Basic Books, 1994).

36. Paul C. Stern and Harvey V. Fineberg (eds.), *Understanding Risk: Informing Decisions in a Democratic Society* (Washington, DC: National Academy of Science Press, 1996).

37. For some examples, see Irwin and Wynne, *op. cit.* note 32.

38. Bruce Bimber, *The Politics of Expertise in Congress: The Rise and Fall of the Office of Technology Assessment* (Albany: State University of New York Press, 1996).

39. "Federal Actions to Address Environmental Justice in Minority Populations and Low-Income Populations," Executive Order 12298, Washington, DC, 11 February 1994.